Geology of the Norwegian Oil and Gas Fields

Organizing and Editing Committee

Elisabeth Holter, Conference Organizer (Norwegian Petroleum Society)
Anthony M. Spencer, Chairman (Statoil)
Colin J. Campbell (Fina Exploration Norway)
Sigmund H. Hanslien (Esso Norge)
Philip H. H. Nelson (Norske Shell)
Eigill Nysæther (Norsk Hydro)
Else G. Ormaasen (Norwegian Petroleum Directorate)

Geology of the Norwegian Oil and Gas Fields

An atlas of hydrocarbon discoveries, containing full descriptions of 37 of Norway's major oil and gas fields and finds. Prepared in conjunction with the conference proceedings—*Habitat of Hydrocarbons on the Norwegian Continental Shelf*—organized by the Nowegian Petroleum Society in Stavanger, Norway, 1–3 October 1985

Edited by

A. M. Spencer *et al.*

Published by
Graham & Trotman
(a member of the Kluwer Academic Publishers Group)
for the Norwegian Petroleum Society

First published in 1987 by

Graham & Trotman Limited
Sterling House
66 Wilton Road
London SW1V 1DE
UK

Graham & Trotman
101 Philip Drive
Assinippi Park
Norwell, MA 02061
USA

British Library Cataloguing in Publication Data

Geology of the Norwegian oil and gas field: an atlas of hydrocarbon discoveries, containing full descriptions of 37 of Norway's major oil and gas fields and finds.
1. Gas, Natural, in submerged lands—Norway
2. Petroleum in submerged lands—Norway
3. Continental shelf—Norway
I. Spencer, A.M. II. Norsk Petroleumsforening
553.2′8′09481 TN882.N8

ISBN 086010 908 9

LCCCN 87-27198

Typeset and Printed in Great Britain

Contents

Part III—Egersund Sub-Basin

Part IV—South Viking Graben

Part V—North Viking Graben—West

Part VI—North Viking Graben—East

Part VII—Halten Terrace

Part VIII—Hammerfest Basin

Part IX—Bibliography

Part X—Compilation Tables

Part XI—Wells drilled

Foreword

This volume contains articles on 37 of the oil and gas fields offshore Norway. The articles have been written in a standardized layout with the aim of providing an easy to use databook for the petroleum geologist. The idea for the compilation was conceived by the Geology and Geophysics Committee of the Norwegian Petroleum Society, namely F. Aamodt, D. W. Ellenor, K. Kolbjørnsen, E. Nysæther, P. Songstad, A. M. Spencer and J. Stangenes. The work of organizing and editing this compilation has been undertaken by a specially appointed subcommitee, consisting of E. Holter, C. J. Campbell, S. Hanslien, P. H. H. Nelson, E. Nysæther, E. Ormaasen and A. M. Spencer from 1984 to 1987, in parallel to the preparation of the conference proceedings from the *Habitat of Hydrocarbons Symposium* held in Stavanger from 1 to 3 October 1985.

The main aim of the present volume is to provide geologists and geophysicists working on petroleum geology in Norway with a full and convenient account of the main fields and finds. The idea of publishing such a compilation—the first of its kind in Europe—was suggested by the databooks on areas in the USA (e.g. *Selected Oil and Gas Fields in West Texas* (1982) West Texas Geological Society). In Norway the technical climate is one of considerable openness and this gave the hope that full geological information would be made available for a databook. The standardized layout adopted aimed to cover all aspects of the petroleum geology in both words and drawings. The subcommittee prepared two example articles and these, with the adopted format, were sent to operating companies in March 1985 requesting up to a total of 53 articles. The 37 articles included here arrived from November 1985 to February 1987. On receipt, they were refereed and edited by the subcommittee and several outside referees. In addition, an introductory article, setting the discoveries in a historical perspective, a bibliography and a list of exploration and appraisal wells are included. A/S Norske Shell are thanked for providing the index maps which appear as end-papers in this volume.

It is hoped that the aim of the volume—to help the petroleum geologist in his daily work—will be fulfilled. All who have helped to bring this project to fruition are thanked, especially the authors and the twelve companies whom they represent: Amerada Hess, Amoco, BP, Elf, Esso, Fina, Mobil, Norsk Hydro, Phillips, Saga, Shell and Statoil.

Organizing and Editing Committee

Principal abbreviations

mD	millidarcy
mkb	metres, kelly bushing
mss	metres, sub-sea
Sm^3	standard cubic metres
Sm^3d	standard cubic metres per day
$Sm^3/d/bar$	standard cubic metres per day per bar

Conversion factors

1 Sm^3 = 35.315 standard cubic feet
1 Sm^3 = 6.29 barrels
1 bar = 14.504 pounds per square inch

$$\text{API oil gravity} = \frac{141.5}{\text{specific gravity of oil}} - 131.5$$

Part I—Introduction

1

The discovery of oil and gas in Norway: an historical synopsis

C. J. Campbell* and E. Ormaasen†

**Fina Exploration Norway, Postboks 4055, Tasta, 4001, Stavanger, Norway*
† Norwegian Petroleum Directorate, Postboks 600, 4001 Stavanger, Norway

The exploration of the North Sea had its origins in the Netherlands where the discovery of the Gröningen Field in Permian sandstones prompted successful exploration offshore in the southern North Sea. This in turn drew attention to the possibilities of the southern Norwegian shelf, where in 1968 a well was drilled for a Permian objective but unexpectedly found hydrocarbons in the Late Cretaceous chalk and opened up the prolific Greater Ekofisk trend, with its most exceptional reservoirs.

Exploration was also proceeding northwards in United Kingdom and in Norwegian waters, and led to the discovery of hydrocarbons in Jurassic and Palaeogene sandstones: the former in rotated fault blocks; the latter in turbidite fan deposits. This phase culminated in Norway in the Fourth Round in 1979 that led to the discovery of several major oil and gas fields, including Statfjord, Troll, Oseberg, Gullfaks and Snorre. The knowledge gained thereby permitted an understanding of the generation of hydrocarbons in Upper Jurassic organic shales, which was then applied to the northern shelf when that was opened in the Fifth Round in 1980. It led to the discovery of two hydrocarbon provinces with Jurassic reservoirs: Haltenbanken with several light oil and gas accumulations; and the Hammerfest Basin with predominantly gas finds.

Looking to the future, attention is drawn to the importance of the Barents Sea Platform, a huge, gently deformed platform with mainly Palaeozoic and Triassic prospects.

Exploration on the Norwegian Shelf, an area of about one million square kilometres, has led to the discovery of 58 oil fields and 47 gas fields. It has been accomplished by a colossal seismic effort and the drilling of as many as 520 exploration and appraisal wells. Reserves of almost 4×10^9 tonnes of oil equivalent have been proven, and a high proportion of Europe's undiscovered hydrocarbon resources are thought to lie in Norway.

INTRODUCTION

During the early phases of exploration in a new province, discoveries are often made more by accident than by design. That is because there is simply insufficient geological information for a thorough evaluation. This has perhaps led to the often-quoted dictum that 'Geologists have to be optimists'. A geologist may indeed be an optimist, but it is not to be considered an essential professional qualification. Instead, he should take a dispassionate view of the known facts and apply relentless logic in extrapolating from the known to the unknown. It is a descriptive science, for he cannot change the nature of the rocks beneath the sea-bed. It is also an imaginative science for he must develop elegant conceptual hypotheses covering the complex range of circumstances responsible for the accumulation of oil and gas, and then invite the investor and the engineer to

Geology of the Norwegian Oil and Gas Fields © A. M. Spencer *et al.*, eds (Graham & Trotman, 1987) pp. 1–37.

test them by drilling. He is entitled to come forward with drilling proposals to test his ideas only to the extent that he has thoroughly investigated all available information. Another requirement is to record and document his studies, because each successive hypothesis is built on the results of its predecessor. Lastly, there is an obligation to publish results, as the industry benefits from this form of co-operation between its members, notwithstanding their competitive relationships.

This volume is a compilation of the results of both the successful geological hypotheses and the more fortuitous circumstances that have led to the discovery of oil and gas in Norway. It provides a database from which to understand the factors that have determined the known accumulations of hydrocarbons in Norway. It forms thus a most valuable basis from which to extrapolate and evaluate future prospects. It is also a ready reference of general information on the subject.

The authors of the articles on the individual fields were asked to provide strictly factual information to a standard format. In many cases they have also covered the events leading up to the discovery and have discussed the regional setting of the field. This inevitably leads to a certain repetition from article to article, for the regional setting of all fields in a given province is of course the same. Furthermore, location maps and stratigraphic classifications do not strictly require separate consideration for each field.

This introductory chapter aims, therefore, to provide the overall regional setting and the general historical background for the groups of fields in each province. It also furnishes statistical information on a number of topics, including licencing, wildcat drilling, reserve estimates, seismic coverage and such subjects, which document the progress of exploration in Norway over the past 20 years. The Appendix contains a listing of exploration wells.

The subject concerns the mapping of resources and the determination of reserves: these are terms that require definition, as there is a tendency to loose usage. The Norwegian Petroleum Directorate recognizes discovered and undiscovered resources of varying degrees of certainty, restricting the term *reserves* to those resources that are commercially producible (Fig. 1). That means that the figures are influenced by political and commercial decisions. The classification is used in official documents and is applied in this paper. The reserve and resource figures quoted in the

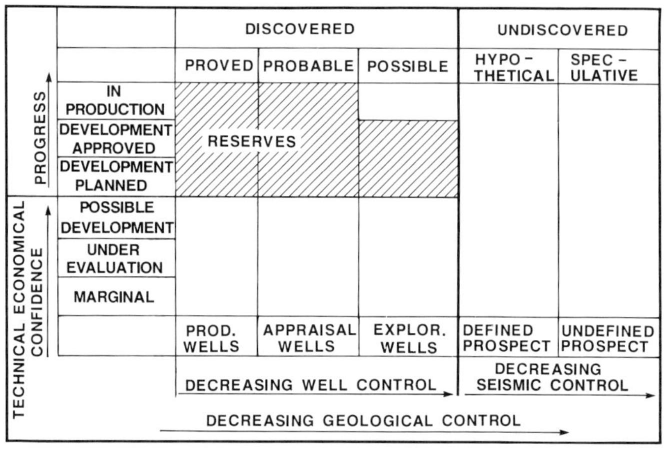

Fig. 1. Classification of petroleum reserves and resources (as adopted by the Norwegian Petroleum Directorate). Resources are shown white; reserves are marked by diagonal line ornament.

individual field description are, however, those quoted by the operator and may or may not follow the official classification.

THE EARLY DAYS

Most of the World's ultimate petroleum resources lie in a few well-known prolific provinces, of which the Middle East is, of course, the most important. Many of them had been already discovered more than half a century ago. Few people would have imagined at the end of the Second World War that Europe would contain another prolific basin beneath the stormy waters of the North Sea. Today, it produces three and a half million barrels of oil per day, which is about one-third of the production of the Middle East. There were in fact already, at that time, indications of promise: the start of the North Sea exploration campaign is sometimes attributed to the unexpected discovery of hydrocarbons in a well drilled by Shell as an exhibit in the World Petroleum Congress in Holland in 1936.

A second important milestone was the discovery in 1947, again in Holland, of gas in Permian sandstones of the Rotliegende Formation. These continental red-beds were not perceived to be a likely reservoir for hydrocarbons, and it was not until later that the source was identified in the underlying deeply buried Carboniferous coal measures. This in turn led to the discovery of the Gröningen Field in 1959, which by 1965 was realized to be one of the World's largest gas fields. From that discovery the industry turned its attention offshore into the southern part of the

North Sea, where several major Permian finds were made in both Dutch and British waters during the 1960s.

THE FIRST FINDS IN NORWAY

It was against this background that the industry began to take an interest in Norway's continental shelf. It was hoped that this part of the North Sea might also contain Carboniferous coal measures, sufficiently buried to yield gas, and that it would provide reservoirs in the form of Early Permian aeolian sandstones beneath a cover of Late Permian salt to act as a seal.

It was evidently a mistaken hypothesis, for no such accumulations have been found, despite an extensive subsequent search. Nevertheless, licences to explore and produce hydrocarbons in Norway were first awarded by the Government in 1965 to several companies and groups, including: Phillips–Fina–Agip Group, the Amoco Group, the Petronord Group, Shell and Esso (Table 1). Most of the blocks awarded lie in the southern limit of the shelf, in an area where the fields of the Ekofisk Complex were later discovered, but some of them lay farther north, mainly adjoining the Norway–United Kingdom median line.

Chalk Fields in the Central Graben

In 1968, the Amoco Group drilled Well 2/8-1 on the flanks of what appeared to be a large salt-induced anticline, with the intention of investigating possible reservoirs in the Permian. The well failed in its main objective but it made an unexpected discovery of hydrocarbons in the Upper Cretaceous chalk. This formation occurs over much of north-west Europe. It was regarded as a singularly unpromising reservoir for hydrocarbons, being deficient in porosity and permeability, although an earlier small discovery had been made in Denmark. This unexpected discovery paved the way for drilling in the following year, when the Phillips Group found a thick zone of highly porous and oil-impregnated chalk in the adjoining Block 2/4, thereby discovering the Ekofisk Field, the first giant, multibillion-barrel field in the North Sea.

Continued exploration led to the discovery of a number of other accumulations, together making up the Greater Ekofisk Complex which comprises Ekofisk, West Ekofisk, Edda, Eldfisk, Albuskjell, Tor, Tommeliten, Valhall, Hod and some satellites (Fig. 2).

The accumulation of oil and gas in the chalk fields of the Southern Norwegian Shelf results from a unique combination of circumstances, and involves processes of a special nature that even today are not fully understood. The details are discussed in the field descriptions, later in the volume, but some of the salient features are summarized below.

Hydrocarbon Generation and Migration

The Upper Jurassic section in the Greater Ekofisk area contains a thick development of organic-rich shales that have been buried sufficiently to yield oil and gas in prolific quantities. Maturation studies suggest that oil generation commenced in Early Tertiary times in the deeper catchment areas to the west of the producing trends, but occurred progressively later towards the east. Peak generation occurs at the present day in the vicinity of Valhall.

The Chalk reservoirs are separated from the Upper Jurassic source-rocks by several hundred metres of Lower Cretaceous shale that in other circumstances would be interpreted as an effective seal. It appears that the fluids in the source-rocks expand on being subjected to increasing temperature with burial. The resulting rise in pressure is sufficient to fracture the overlying shales creating a vertical migration path into the reservoirs above. Faults no doubt facilitate the process by providing lines of weakness.

The chalk reservoirs are capped by a seal of Paleocene shales, sufficient to hold most of the hydrocarbons in the reservoirs. However, these shales are imperfect seals, and several of the chalk fields are overlain by gas-charged layers in the Tertiary succession, which give rise to the so-called 'gas chimneys' or 'gas clouds' visible on the seismic records. Such shallow gas-charged sequences distort the seismic ray-paths and give rise to velocity effects, which in some cases impose severe difficulties to geophysical interpretation.

Structure

The Chalk fields lie in the axial part of the Central Graben, which itself resulted from Jurassic rifting. The area has been affected by halokinesis due to movement of Late Permian salt. The structures responsible for the Chalk fields have characteristically a domal configuration, commonly possessing minor crestal grabens. Some are due to deep-seated salt movement, although salt itself has been encountered in only a few of the wells. Others

Table 1. Interests in the First Licence Round, 1965

Block & Licence Numbers		Esso	Others	Shell	Gulf	Phillips	Texaco	Chevron	Fina	Total	Amerada	Amoco	Texas Eastern	Agip	Elf	Norsk Hydro	Noco
		Company Interest in Block (%)															
PL 001	16/1	100															
	16/2	100															
	16/7	100															
	25/11	100															
PL 002	16/9	100															
	17/8	100															
	17/9	100															
	18/7	100															
PL 003	8/2	100															
	8/3	100															
	9/7	100															
	9/8	100															
PL 004	6/3										28.3	28.3	28.3				15.1
	7/1										28.3	28.3	28.3				15.1
	7/4										28.3	28.3	28.3				15.1
PL 005	7/3										28.3	28.3	28.3				15.1
	7/6										28.3	28.3	28.3				15.1
	8/4										28.3	28.3	28.3				15.1
PL 006	2/2										28.3	28.3	28.3				15.1
	2/5										28.3	28.3	28.3				15.1
	2/8										28.3	28.3	28.3				15.1
	3/4										28.3	28.3	28.3				15.1
PL 007	16/3		47.6							26.6					13.3	12.5	
	16/5		47.6							26.6					13.3	12.5	
	16/6		47.6							26.6					13.3	12.5	
	17/4		47.6							26.6					13.3	12.5	
PL 008	2/6		47.6							26.6					13.3	12.5	
	7/2		47.6							26.6					13.3	12.5	
	18/10		47.6							26.6					13.3	12.5	
	18/11		47.6							26.6					13.3	12.5	
PL 009	9/5		47.6							26.6					13.3	12.5	
	10/8		47.6							26.6					13.3	12.5	
	10/11		47.6							26.6					13.3	12.5	
	11/9		47.6							26.6					13.3	12.5	
PL 010	16/12			100													
	17/10			100													
	17/11			100													
	25/12			100													
PL 011	1/3			100													
	1/6			100													
	8/7			100													
PL 012	9/12			100													
	10/7			100													
	10/9			100													

Table 1 *(continued)*

Company Interest In Block (%) — Block & Licence Numbers

Company	9/10	3/5	3/1	PL 022 2/3	11/10	PL 021 11/7	16/8	8/5	PL 020 7/9	PL 019B 7/12	11/8	7/12	PL 019 2/1
Esso													
Others	100	100	100	100	100	100							
Shell													
Gulf							100	100	100	100	100	100	100
Phillips													
Texaco													
Chevron													
Fina													
Total													
Amerada													
Amoco													
Texas Eastern													
Agip													
Elf													
Norsk Hydro													
Noco													

Company	7/11	2/7	2/4	PL 018 1/5	8/11	8/10	PL 017 8/8	17/12	16/11	8/1	PL 016 7/8
Esso											
Others											
Shell											
Gulf											
Phillips	50	50	50	50	50	50	50	50	50	50	50
Texaco											
Chevron											
Fina	30	30	30	30	30	30	30	30	30	30	30
Total											
Amerada											
Amoco											
Texas Eastern											
Agip	20	20	20	20	20	20	20	20	20	20	20
Elf											
Norsk Hydro											
Noco											

Company	10/12	10/10	10/5	PI 015 9/9	9/11	8/12	3/3	PL 014 3/2	9/4	8/9	PL 013 8/6	Total % interest
Esso												15.2
Others												14.8
Shell												12.7
Gulf												8.9
Phillips												7.0
Texaco	50	50	50	50	50	50	50	50	50	50	50	6.9
Chevron	50	50	50	50	50	50	50	50	50	50	50	6.9
Fina												4.2
Total												4.0
Amerada												3.6
Amoco												3.6
Texas Eastern												3.6
Agip												2.8
Elf												2.0
Norsk Hydro												1.9
Noco												1.9
												100.0

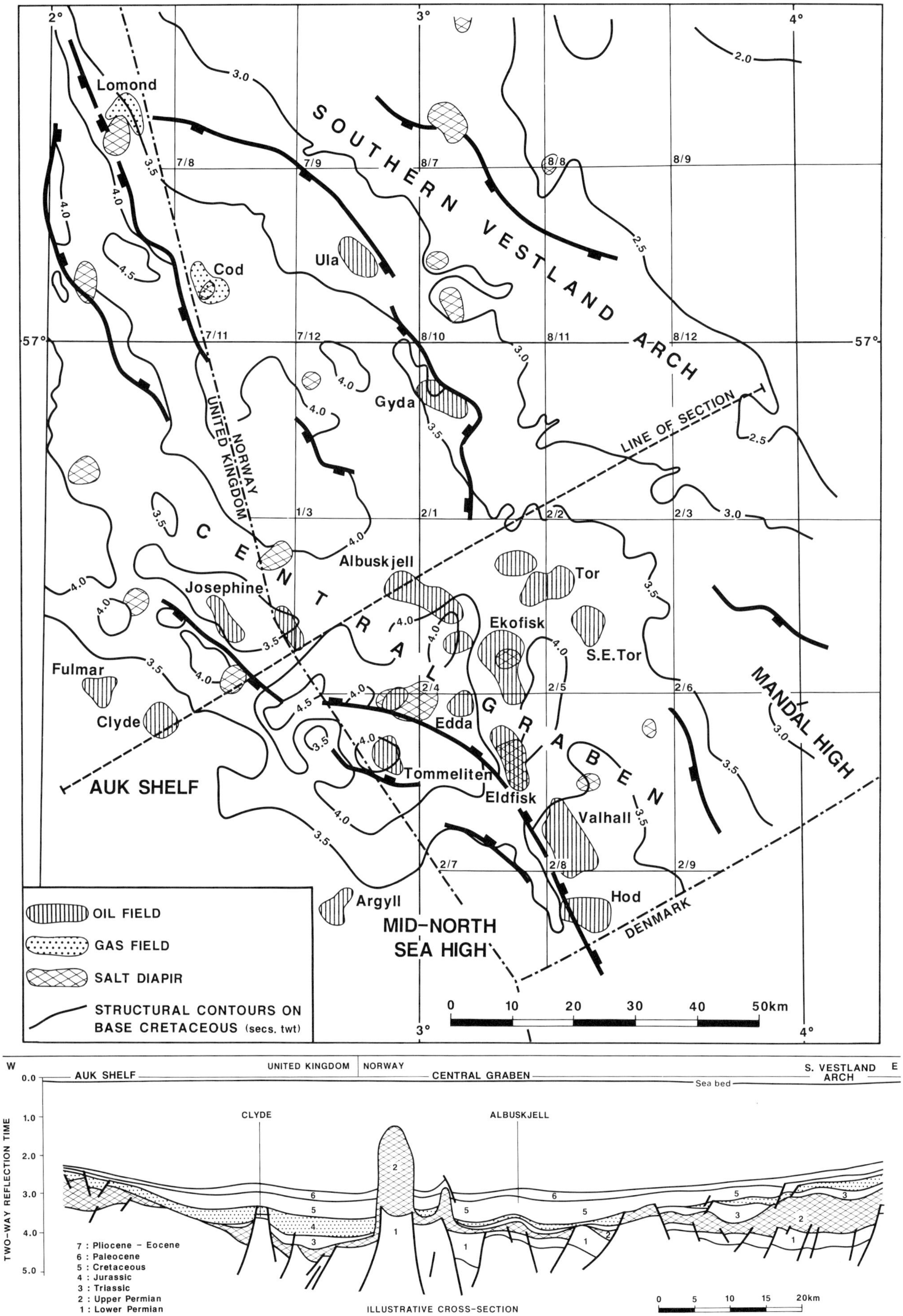

Lomond
Cod
Ula
Gyda
SOUTHERN VESTLAND ARCH
CENTRAL GRABEN
MANDAL HIGH
AUK SHELF
MID-NORTH SEA HIGH
LINE OF SECTION
NORWAY
UNITED KINGDOM
DENMARK
Albuskjell
Tor
Josephine
Ekofisk
S.E.Tor
Fulmar
Clyde
Edda
Tommeliten
Eldfisk
Valhall
Hod
Argyll
OIL FIELD
GAS FIELD
SALT DIAPIR
STRUCTURAL CONTOURS ON BASE CRETACEOUS (secs. twt)
0 10 20 30 40 50km
W
E
AUK SHELF
UNITED KINGDOM
NORWAY
CENTRAL GRABEN
S. VESTLAND ARCH
Sea bed
CLYDE
ALBUSKJELL
TWO-WAY REFLECTION TIME
7 : Pliocene - Eocene
6 : Paleocene
5 : Cretaceous
4 : Jurassic
3 : Triassic
2 : Upper Permian
1 : Lower Permian
ILLUSTRATIVE CROSS-SECTION
0 5 10 15 20km

Fig. 2 (a).

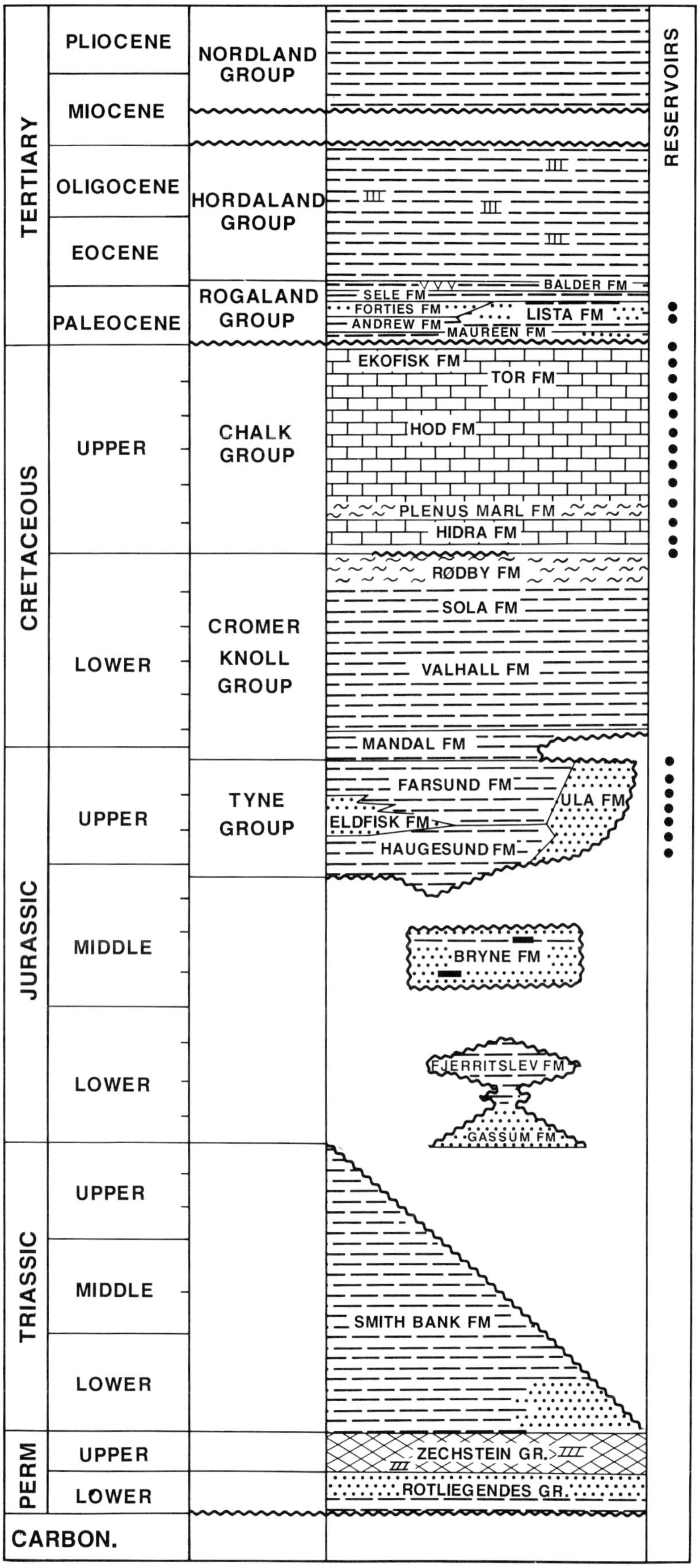

Fig. 2 (b).

Fig. 2. Regional setting, Central Graben.

are transpressional features associated with transcurrent faulting on the Lindesnes Ridge, which is an inverted structural uplift within the graben that has had an important influence on the location of the fields.

Reservoir

Chalk consists of calcilutite made up primarily of coccoliths, the tests of algae: the particles are no more than a few micrometres in size. Sedimentological studies have shown that the Chalk in the Ekofisk area occurs in two main facies: pelagic chalk and redeposited chalk. The latter occurs as various slump and turbidite types that were deposited as a consequence of instability on the margins of the graben. In the Ekofisk Field several such flow-deposits are stacked one upon another. The distribution of the flow-channels may be mapped. In addition there are a number of dense zones that were deposited as hard grounds.

Redeposited chalk has higher initial porosity than has pelagic chalk, and forms the better reservoir rock. The initial porosity of chalk can be extremely high, but porosity is normally lost by compaction with burial. High porosities, in the 35–50% range have, however, been preserved in the crests of the Chalk fields, mainly due to the development of overpressure. The early entry of hydrocarbons, where that has occurred, has also inhibited diagenesis and the consequent loss of porosity. Fracturing gives rise to secondary permeability, which is an important element in the reservoirs.

A characteristic of the Chalk fields is that the oil–water contacts are gradational: porosity and oil saturation decline gradually down-flank and are affected by subtle facies variations.

The extremely high-porosity chalks found in the crestal parts of the fields are vulnerable to formation collapse when the reservoir pressure drops following production. Seafloor subsidence, as has occurred at the Ekofisk Field, is a related phenomenon.

The cumulative production and reserves of the Chalk fields are given in Table 2.

The principal references on the Chalk Fields are: Brewster and Dangerfield (1984), D'Heur (1980), Hardman and Kennedy (1980), Skovbro (1983), and Van den Bark and Thomas (1981).

OTHER EARLY DISCOVERIES

Although the most important discoveries in the First-Round licences lay in the chalk province, some finds were made farther north along the eastern margin of the Central Graben. We may, however, treat them together with the Second- and Third-Round blocks which were awarded between 1969 and 1978 (see Table 3).

Two different exploration campaigns developed from the award of these licences, involving Palaeogene and Jurassic prospects.

Palaeogene Fields in the South Viking and Central Grabens

The first Palaeogene discovery in Norway and, indeed, the first oil discovery offshore Norway, was made by Esso in Well 25/11-1 in 1967. The accumulation, which was confirmed in 1974, is known as the Balder Field. It is characterized by a most irregular reservoir, and has not yet been developed. Another early discovery was the Cod Field, which was found

Table 2. Cumulative production and reserves of the Chalk Fields

	Cumulative production 31.12.85			*Reserves remaining*			*Total*		
Field	*Oil (10^6Sm^3)*	*Gas (10^9Sm^3)*	*NGL (10^6 tonne)*	*Oil (10^6Sm^3)*	*Gas (10^9Sm^3)*	*NGL (10^6 ton)*	*Oil (10^6Sm^3)*	*Gas (10^9Sm^3)*	*NGL (10^6 ton)*
Albuskjell	5.8	10.4	0.7	2.0	6.5	0.4	7.8	16.9	1.1
Edda	3.1	1.6	0.1	1.3	0.3	0.1	4.4	1.9	0.2
Ekofisk	126.7	49.1	3.6	110.3	76.9	9.9	237.0	126.0	13.5
Eldfisk	26.5	8.8	0.9	24.3	22.8	2.1	50.8	31.6	3.0
Tor	15.3	8.0	0.8	4.5	5.7	0.7	19.8	13.7	1.5
Valhall	6.3	1.0	0.3	13.1	15.2	1.0	19.4	16.2	1.3
West Ekofisk	10.2	18.6	0.8	1.7	7.6	0.5	11.9	26.2	1.3
Tommeliten	—	—	—	6.0	23.0	—	6.0	23.0	—
Total	193.9	97.5	7.2	163.2	158.0	14.7	357.1	255.5	21.9

Table 3. Interests in the Second and Third Licence Rounds (incl. PL. 037), 1969–78

Block & Licence Numbers — Company interest in Block (%)

Company	2nd round PL 023 3/7	PL 024 25/1	PL 025 15/3	PL 026 25/2	PL 027 25/8	PL 028 25/10	PL 029 15/16	PL 030 30/10	PL 031 2/10	PL 032 2/9	PL 033 2/11	PL 034 30/5	PL 035 30/11	PL 036 25/4
Statoil		5												
Esso					100	100	100	100						
Elf	43.6	41.4	53.2	43.6										
Norsk Hydro	34.6	32.9	10	34.6										
Shell												100	100	
Others														100
Total	21.8	20.7	21.8	21.8										
BP														
Phillips									51.7					
Amoco										25	25			
Texas Eastern										25	25			
Amerada										25	25			
Agip									18.3					
Noco										25	25			
Conoco														
Fina									30					
Deminex														
Mobil														
Saga														
Lasmo			15											

Company	Special award PL 037 33/9	PL 037 33/12	3rd round 15/12	15/11	PL 038 6/3	PL 039 24/9	30/7	PL 040 29/9	PL 041 35/3	PL 042 36/1	30/4	PL 043 29/6
Statoil	50		50	50	50	50	50	50	50	50	50	50
Esso		10	50	50	50							
Elf							28.8	28.8				
Norsk Hydro						10	6.8	6.8				
Shell		10										
Others									10.5			
Total							14.4	14.4				
BP									24.5		50	50
Phillips												
Amoco		1.1								10		
Texas Eastern		1.1								10		
Amerada		1.1								10		
Agip												
Noco												
Conoco		10				40						
Fina												
Deminex												
Mobil		15								15		
Saga		1.7							15			
Lasmo												

Company	PL 044 1/9	24/12	PL 045 24/11	15/9	PL 046 15/8	33/5	PL 047 33/2	15/5	PL 048 15/2	PL 049 33/6	Total% interest
Statoil	50	50	50	50	50	50	50	50	50	50	30.8
Esso				40	40						18.3
Elf						17.5	17.5	21.8	21.8		9.1
Norsk Hydro		10	10	10	10	13.8	13.8	17.3	17.3		6.8
Shell											6.0
Others		35	35								5.2
Total						8.7	8.7	10.9	10.9		4.4
BP											3.5
Phillips	25.9										2.2
Amoco											1.8
Texas Eastern											1.8
Amerada											1.8
Agip	9.1									30	1.6
Noco											1.4
Conoco											1.4
Fina	15										1.3
Deminex						10	10			20	1.1
Mobil											0.4
Saga		5	5								0.7
Lasmo											0.4
											100.0

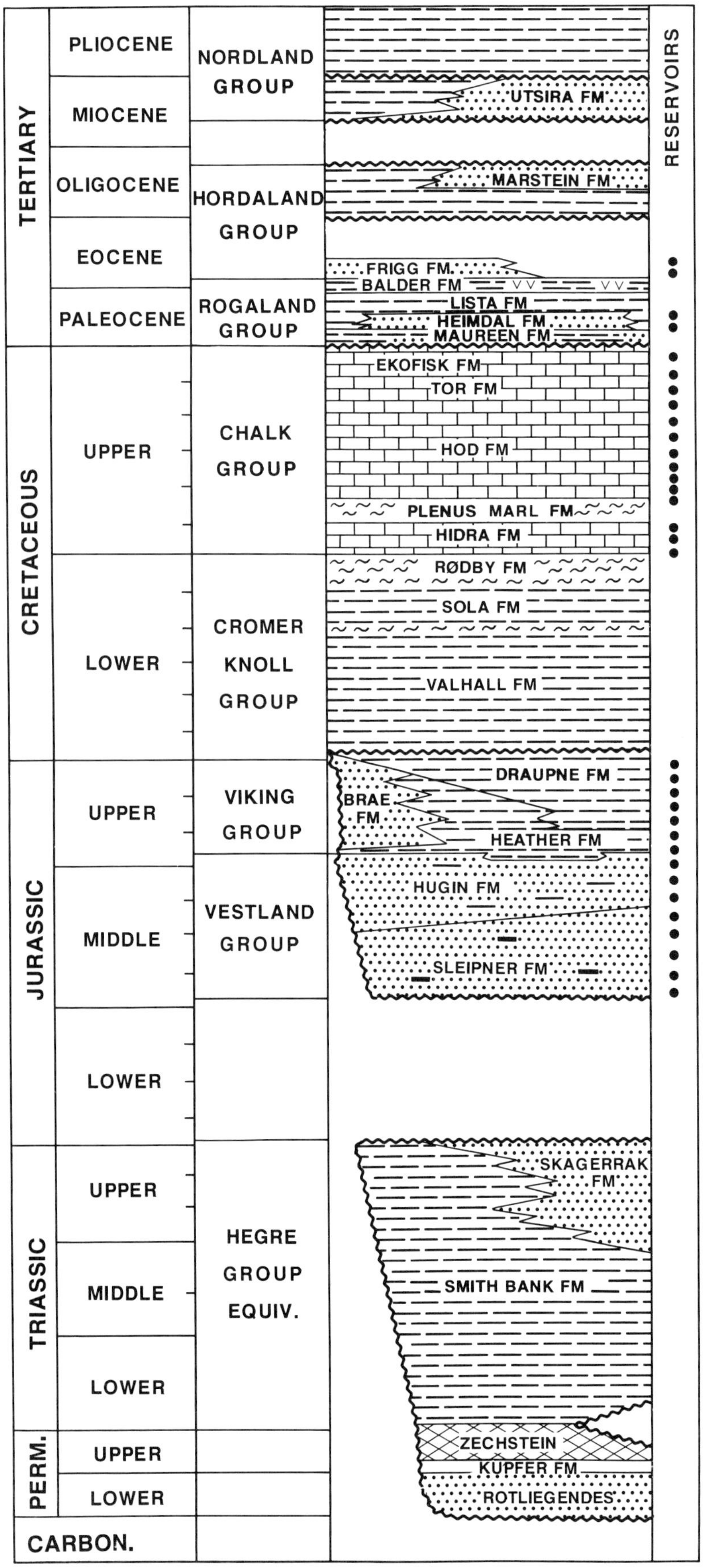

Fig. 3 (a).

Fig. 3. Regional setting, Sleipner Area.

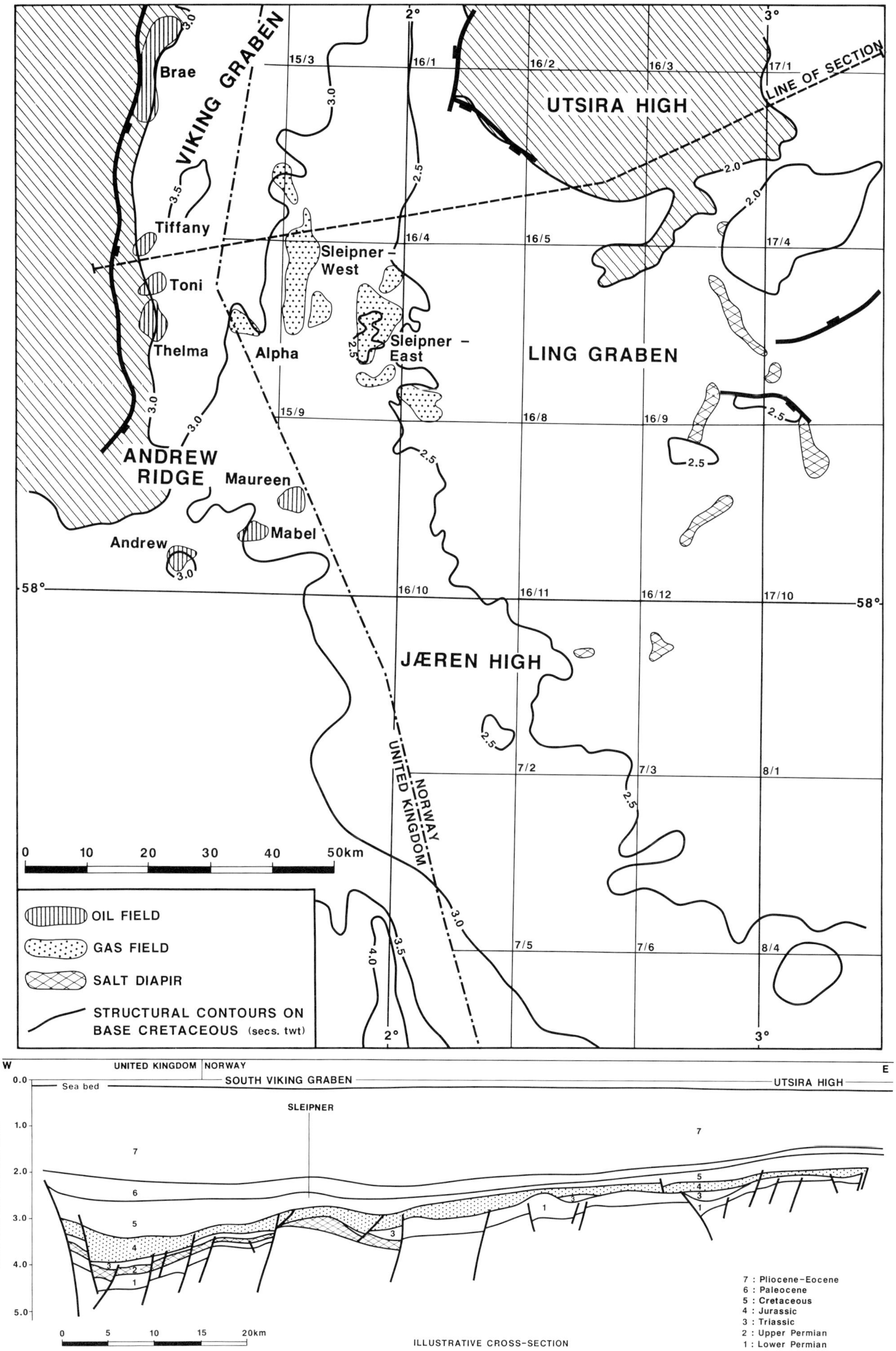
VIKING GRABEN
UTSIRA HIGH
LINE OF SECTION
LING GRABEN
ANDREW RIDGE
JÆREN HIGH
Brae
Tiffany
Toni
Thelma
Alpha
Sleipner - West
Sleipner - East
Maureen
Mabel
Andrew
15/3
16/1
16/2
16/3
17/1
16/4
16/5
17/4
15/9
16/8
16/9
16/10
16/11
16/12
17/10
7/2
7/3
8/1
7/5
7/6
8/4
2°
3°
58°
NORWAY
UNITED KINGDOM
0 10 20 30 40 50km
OIL FIELD
GAS FIELD
SALT DIAPIR
STRUCTURAL CONTOURS ON BASE CRETACEOUS (secs. twt)
W
E
UNITED KINGDOM
NORWAY
SOUTH VIKING GRABEN
UTSIRA HIGH
Sea bed
SLEIPNER
TWO-WAY REFLECTION TIME
0 5 10 15 20km
ILLUSTRATIVE CROSS-SECTION
7 : Pliocene-Eocene
6 : Paleocene
5 : Cretaceous
4 : Jurassic
3 : Triassic
2 : Upper Permian
1 : Lower Permian

Fig. 3 (b & c)

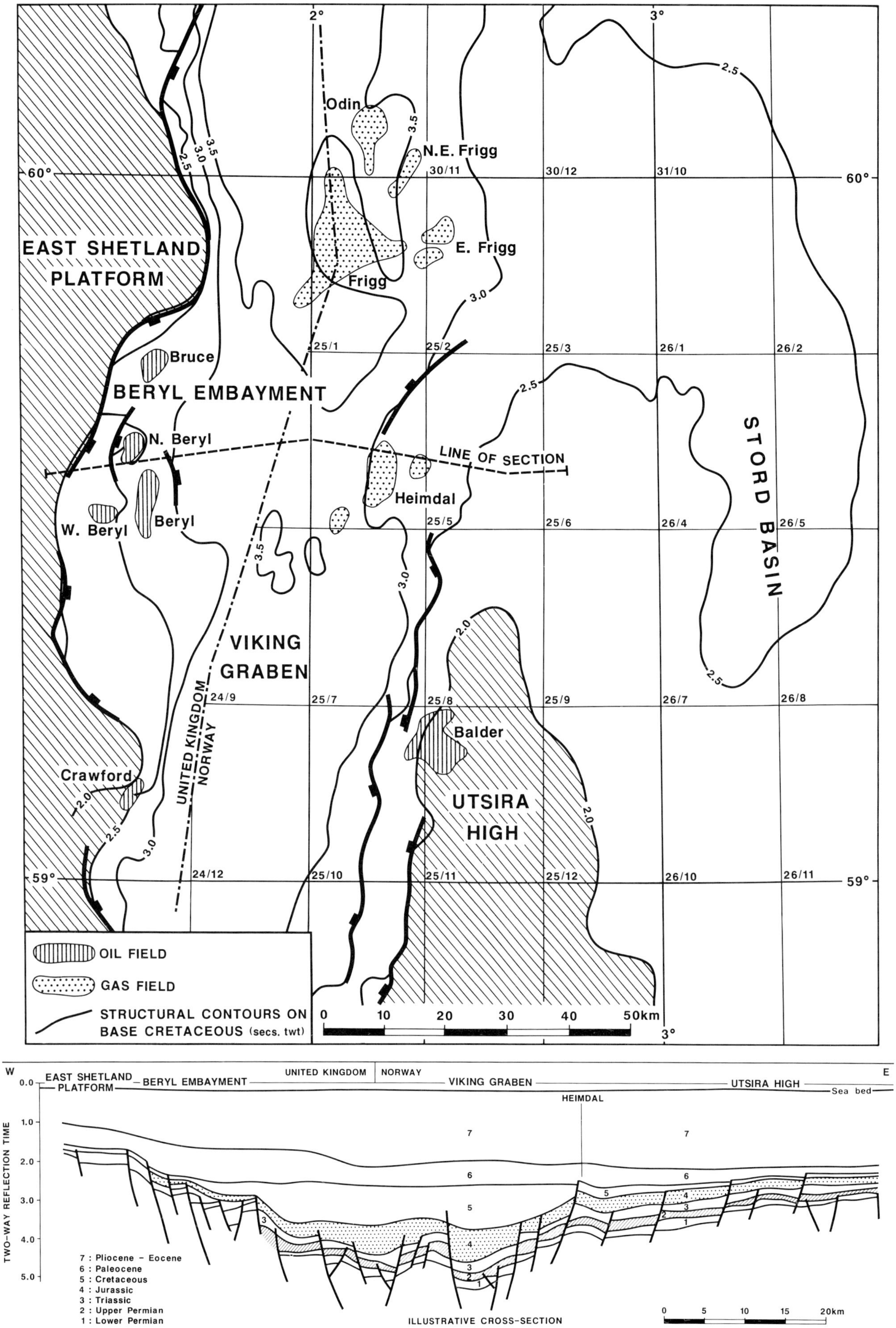

EAST SHETLAND PLATFORM
BERYL EMBAYMENT
VIKING GRABEN
UTSIRA HIGH
STORD BASIN
Odin
N.E. Frigg
E. Frigg
Frigg
Bruce
N. Beryl
Beryl
W. Beryl
Heimdal
Balder
Crawford
LINE OF SECTION
UNITED KINGDOM
NORWAY
60°
59°
2°
3°
30/11
30/12
31/10
25/1
25/2
25/3
26/1
26/2
25/5
25/6
26/4
26/5
24/9
25/7
25/8
25/9
26/7
26/8
24/12
25/10
25/11
25/12
26/10
26/11
2.0
2.5
3.0
3.5
OIL FIELD
GAS FIELD
STRUCTURAL CONTOURS ON BASE CRETACEOUS (secs. twt)
0 10 20 30 40 50km
W
E
EAST SHETLAND PLATFORM
BERYL EMBAYMENT
UNITED KINGDOM
NORWAY
VIKING GRABEN
UTSIRA HIGH
Sea bed
HEIMDAL
TWO-WAY REFLECTION TIME
0.0
1.0
2.0
3.0
4.0
5.0
7 : Pliocene - Eocene
6 : Paleocene
5 : Cretaceous
4 : Jurassic
3 : Triassic
2 : Upper Permian
1 : Lower Permian
ILLUSTRATIVE CROSS-SECTION
0 5 10 15 20km

Fig. 4 (a).

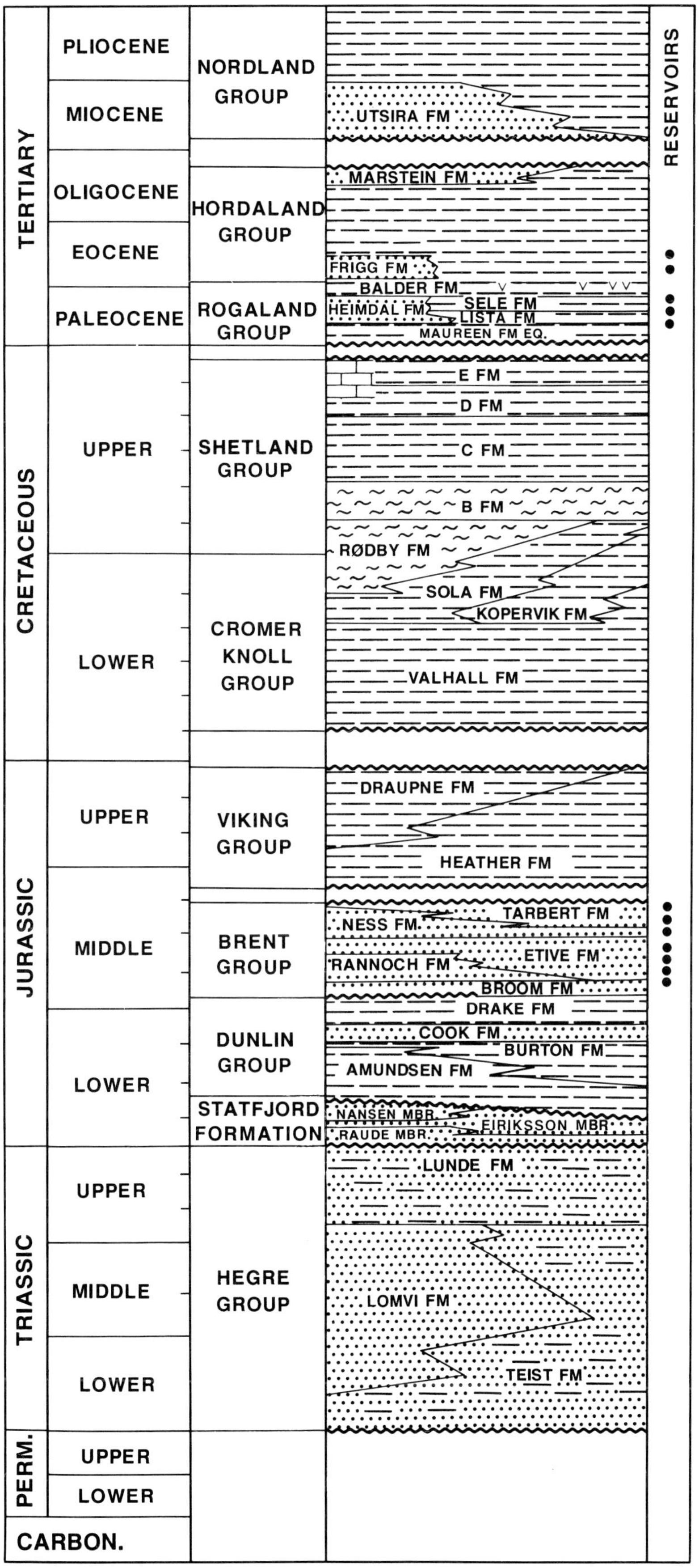

Fig. 4 (b).

Fig. 4. Regional setting, southern Viking Graben.

Table 4. Cumulative production and reserves of the Palaeogene fields in the southern Viking Graben

Field	*Cumulative production 31.12.85*			*Reserves remaining*			*Total*		
	Oil (10^6Sm^3)	*Gas (10^9Sm^3)*	*NGL (tonne)*	*Oil (10^6Sm^3)*	*Gas (10^9Sm^3)*	*NGL (tonne)*	*Oil (10^6Sm^3)*	*Gas (10^9Sm^3)*	*NGL (tonne)*
Frigg	0.3	72.2	—	1.0	127.0	—	1.3	199.2	—
Heimdal	—	—	—	3.0	34.0	—	3.0	34.0	—
North-east Frigg	—	3.4	—	0.1	8.0	—	0.1	11.4	—
Odin	—	5.8	—	0.1	22.0	—	0.1	27.8	—
East Frigg	—	—	—	—	13.0	—	—	13.0	—
Total	0.3	81.4	—	4.2	204.0	—	4.5	285.4	—

Table 5. Interests in the Fourth Licence Round, 1979

Company Interest In Block (%)	Block & Licence Numbers: PL 051 30/2	PL 052 30/3	PL 053 30/6	PL 054 31/2	PL 055 31/4	PL 056 34/2	PL 057 34/4	PL 058 35/8	Total % Interest
Statoil	50	50	56.4	50	50	50	50	50	50.8
Unocal	25	30							6.9
Norsk Hydro		10	12.2	5	15				5.3
Shell				35					4.4
Amoco						25	10		4.4
Gulf								30	3.7
Tenneco	25								3.1
Saga			7.3				15		2.8
Mobil			8			12.5			2.6
Deminex		5					15		2.5
Esso					20				2.5
Norske Getty								20	2.5
Elf			10.7						1.3
Arco					10				1.3
Others		5		5					1.3
Total			5.4						0.7
Phillips						5.8			0.7
BP					5				0.6
Conoco				5					0.6
Amerada							5		0.6
Texas Eastern							5		0.6
Agip						3.35			0.4
Fina						3.35			0.4
									100.0

by Phillips and partners in 1968 by Well 7/11-1. It produces from Paleocene sandstones.

The most important discoveries were the Frigg Field, made by the Petronord Group in 1971, and the Sleipner East Field, which was discovered by Statoil in 1982. The Odin, North-east Frigg, and Heimdal fields are other Palaeogene finds of lesser importance. The location of these fields is shown in Figs 3 and 4. These discoveries paralleled exploration in adjoining United Kingdom waters during this time when a number of important Palaeogene fields were found, including Forties (1970), Montrose (1969) and Maureen (1973).

Hydrocarbon Generation and Migration

The hydrocarbons in these fields are also believed to have been sourced from the Upper Jurassic shales. Generation commenced in Late Cretaceous times. Oil and gas later migrated upwards along faults and fractures into the Palaeogene reservoirs. These hydrocarbons subsequently migrated eastwards to accumulate in, for example, the Balder Field, where the oil is partly biodegraded. Continued subsidence during Tertiary times depressed the source-rocks to great depths in the axis of the graben, and increased temperatures led to the cracking of the liquid hydrocarbons into gas.

The upward migration of gas through hundreds of metres of shale, which generally have what would otherwise be considered as satisfactory sealing conditions, suggests an autofracturing process, brought about by the excessive pressure conditions.

Structure

In regional terms, the primary mechanism for trapping the hydrocarbons in these fields is probably stratigraphic, related to a huge fan complex that spread into the graben during Paleocene and Eocene times, although the fields themselves also lie in structural closures, calling for no particular comment.

The Frigg Field straddles the boundary with the United Kingdom, with 60.82% of the original reserves currently estimated to lie on the Norwegian side. The gas is transported to Scotland in two pipelines.

Reservoirs

The reservoirs in these fields consist of sands deposited by turbidity currents that flowed into the Viking Graben from the western borderlands, the Moray Firth Basin and the East Shetland Platform. The sands were deposited in a series of lobes under a wide variety of depositional conditions. The sands are accordingly lenticular, not readily identifiable by seismic means, and of variable quality.

There are two reservoir intervals that are variously productive in the different fields: one is Eocene, the other is Paleocene. The details are given in the field descriptions.

The principal references on the Palaeogene Fields are as follows: Conort (1986) and Héritier *et al.* (1979). The reserves of the fields in question are given in Table 4.

Jurassic Finds in the Central Graben

The Second and Third rounds of licencing, which followed Jurassic discoveries in the United Kingdom, opened the way for exploration along the eastern margin of the Central Graben. A number of prospects with Jurassic reservoirs down-thrown to the boundary faults were drilled. The initial results were disappointing but two important discoveries were eventually made: the Ula Field, discovered in 1976 by Well 7/12-2; and the Gyda Field, discovered in 1980 when BP drilled Well 2/1-3 (see Fig. 2). Ula has now been placed on production and Gyda has been declared commercial. Another discovery made by Well 7/11-5 is under appraisal (see p. 00). The prospects tend to be elusive, with a stratigraphic component, and are not normally very large, but further discoveries can be expected.

Structural Setting

The eastern margin of the Central Graben is cut by an important system of faults, downthrowing to the west, that were active during Late Jurassic times. The faults are normal in character, but may be partly transcurrent in nature. Anticlinal folds, due to local compression associated with the transcurrent faults, and/or halokinetic effects, are associated with the faults and give rise to hydrocarbon traps.

Hydrocarbon Generation and Migration

The Upper Jurassic organic-rich shales in the Central Graben provide the source of the hydrocarbons in these fields. Generation commenced in Early Tertiary times in the deeper parts, becoming progressively later towards the margins. The reservoirs are in direct communication with the overlying source-rocks. The Upper Jurassic and Lower Cretaceous shales provide a satisfactory top seal to the reservoirs.

Reservoir

Marine sands, eroded from the graben margins during Late Jurassic times, were deposited in structural depressions on the down-thrown side of the boundary faults. They form essentially local centres of deposition of variable character and thickness. Sealing conditions against the down-thrown faults are in some cases unsatisfactory.

The principal references are as follows: Bailey *et al.* (1981), and Spencer *et al.* (1986).

Hydrocarbons have also been found in Jurassic reservoirs in the Sleipner and Heimdal fields in the southern part of the Viking Graben under conditions similar to those discussed later in relation to the Fourth Round of licencing.

Of considerably more importance however were the results of Well 211/29-1 that was drilled in United Kingdom waters in 1971 by Shell and Esso at a location close to the

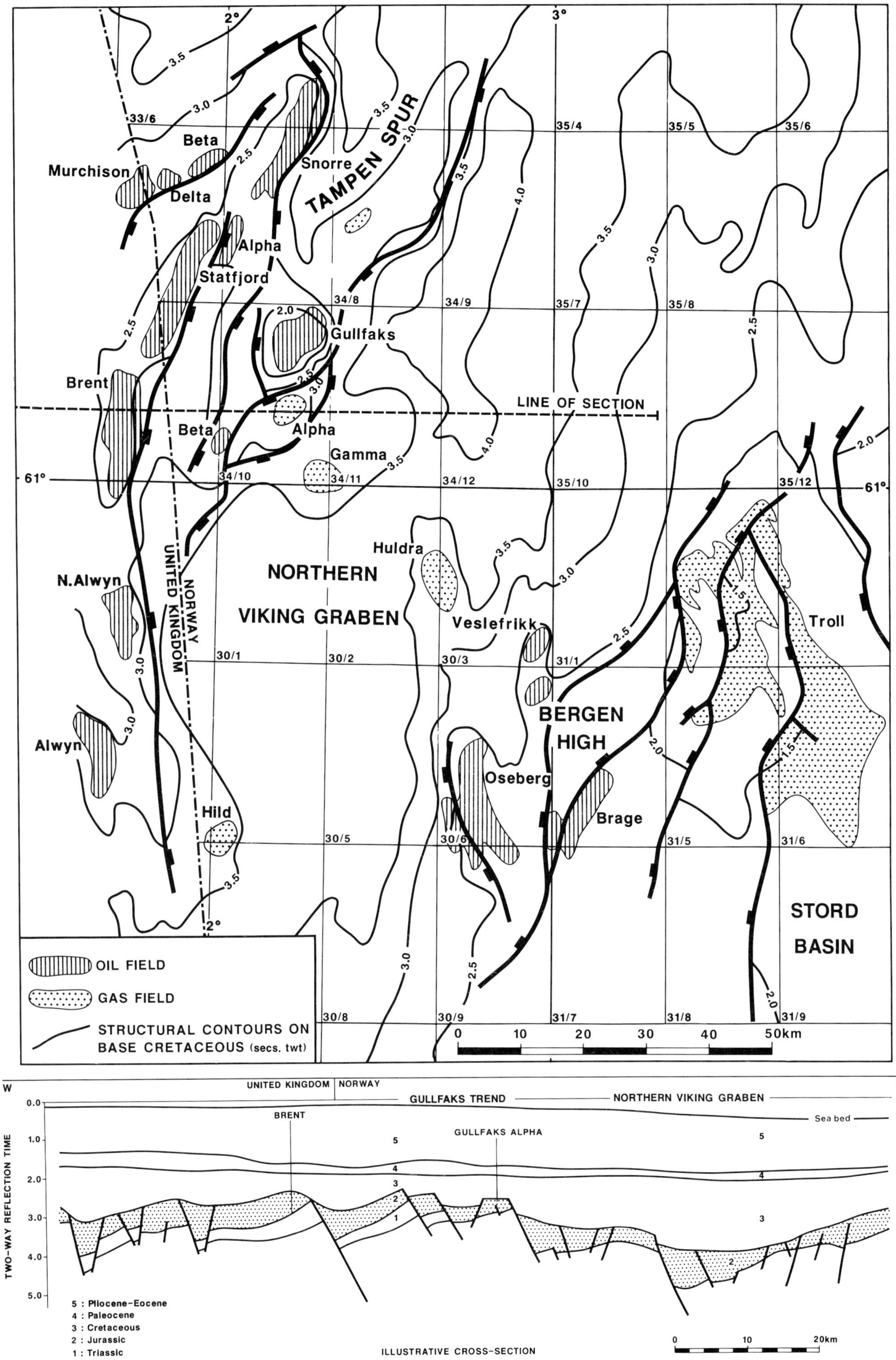

2°
3°
3.5
3.0
2.5
2.0
4.0
1.5
33/6
35/4
35/5
35/6
34/8
34/9
35/7
35/8
34/10
34/11
34/12
35/10
35/12
30/1
30/2
30/3
31/1
30/5
30/6
31/5
31/6
30/8
30/9
31/7
31/8
31/9
61°
Murchison
Beta
Delta
Snorre
TAMPEN SPUR
Alpha
Statfjord
Gullfaks
Brent
LINE OF SECTION
Beta
Alpha
Gamma
Huldra
NORTHERN
VIKING GRABEN
N.Alwyn
NORWAY
UNITED KINGDOM
Veslefrikk
Troll
BERGEN
HIGH
Alwyn
Oseberg
Brage
Hild
STORD
BASIN
OIL FIELD
GAS FIELD
STRUCTURAL CONTOURS ON
BASE CRETACEOUS (secs. twt)
0 10 20 30 40 50km
W
E
UNITED KINGDOM
NORWAY
GULLFAKS TREND
NORTHERN VIKING GRABEN
Sea bed
BRENT
GULLFAKS ALPHA
TWO-WAY REFLECTION TIME
0.0
1.0
2.0
3.0
4.0
5.0
5 : Pliocene–Eocene
4 : Paleocene
3 : Cretaceous
2 : Jurassic
1 : Triassic
ILLUSTRATIVE CROSS-SECTION
0 10 20km

Fig. 5 (a).

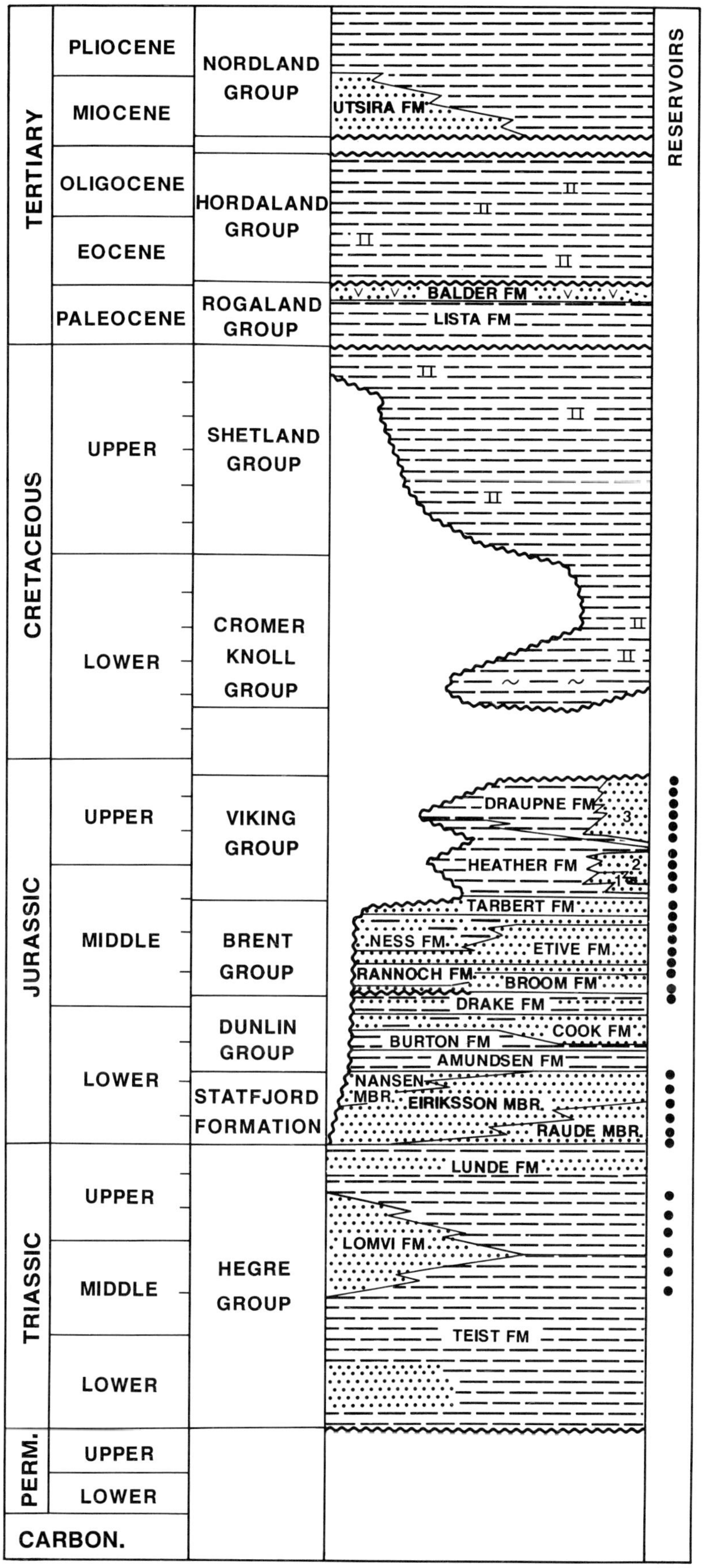

1 KROSSFJORD FM.

2 FENSFJORD FM.

3 SOGNEFJORD FM.

Fig. 5 (b).

Fig. 5. Regional setting, northern Viking Graben.

Norwegian median line, making a discovery in Jurassic reservoirs. Further drilling in the following years confirmed a major accumulation, which was named the Brent Field. It also led exploration into an adjoining structure in Blocks 33/9 and 33/12 in the Norwegian sector drilled by Mobil and its partners in 1974, discovering the Statfjord Field which remains one of Norway's largest fields. The details are described later in the volume. It heralded the Fourth Round of licencing.

THE FOURTH ROUND: NEW TERMS

The value of crude oil increased radically in 1973, and again in 1979, in circumstances of perceived World energy-shortage. In addition, as mentioned above, major discoveries were made during this epoch in adjoining United Kingdom waters on the western margin of the Viking Graben. The Brent and Statfjord fields lie on a major structural trend, the Tampen Spur, that extends into Norwegian waters, plunging northwards in increasing depths of water. There were hopes of other major discoveries.

In these circumstances the Norwegian authorities felt justified in reconsidering their licence terms. In the ensuing Fourth Round, which was held in 1979, Statoil, the state oil agency, which had been created in 1972, further strengthened its position. It was granted an interest of 50% in each of the prime licences, its costs during exploration and appraisal being carried by its partners. (In later years the other national companies were excused from this burden.) Statoil was also able to have a disproportionate influence on the conduct of the venture by voting its licence interest, despite being carried. It was also granted an option to increase its participation in the event of a commercial discovery on a sliding scale based on levels of production. The authorities encouraged separate applications from each of the interested companies, and most of the groups that had existed previously were disbanded.

After the initial round, the Fourth Round of licencing has been by far the most important round to be held in Norway. Ten blocks were awarded: some covered highly prospective tracts on trend with the United Kingdom discoveries; others opened up exploration in a new province on the eastern flank of the Viking Graben. Important discoveries have been made subsequently on almost all of these blocks. The list of licencees is given in Table 5.

The three Norwegian companies, Statoil, Norsk Hydro and Saga Petroleum, together with Norske Shell, became operators of the more promising blocks.

Jurassic Fields in the Northern Viking Graben

In geological terms, most of the discoveries made in the Fourth-Round blocks are characterized by having Jurassic reservoirs in traps formed by rotational fault-blocks. The regional structure and the location of the fields are shown in Fig. 5.

Hydrocarbon Generation and Migration

The Upper Jurassic organic-rich shales of the Draupne Formation form the principal hydrocarbon source-rock in the area. The organic content, however, changes across the basin, reflecting the different sedimentary conditions: in the west, sapropel capable of generating oil predominates; whereas to the east, increasing quantities of terrestrially derived organic matter make the rock progressively more gas prone.

Subsidence of the axial part of the Viking Graben and the deep Møre Basin to the north was such as to bury these source-rocks to the level of peak generation by Cretaceous times. Further subsidence during Tertiary times has depressed them further into the metagenic stage, giving rise to a major charge of gas–condensate.

Ideal conditions for oil generation occur on the western margin of the basin where the source-rocks reached peak generation in Late Tertiary times over a relatively wide area. Less satisfactory conditions occur to the east, both because the source-rocks themselves become more gas prone and because the belt of effective Late Tertiary generation is narrower.

A secondary hydrocarbon source-rock sequence occurs in the Lower Jurassic succession where partly carbonaceous shales and coals are thought to have generated gas and, to a lesser extent, perhaps liquid hydrocarbons. In volumetric terms, however, generation from these Lower Jurassic strata is of minor importance compared with that from the Upper Jurassic.

The pattern of migration is complex, as the main hydrocarbon source-rock in the Upper Jurassic overlies the several reservoirs in stratigraphic terms. In addition, the source-rocks are generally absent from the crests of the blocks due to subsequent erosion and/or

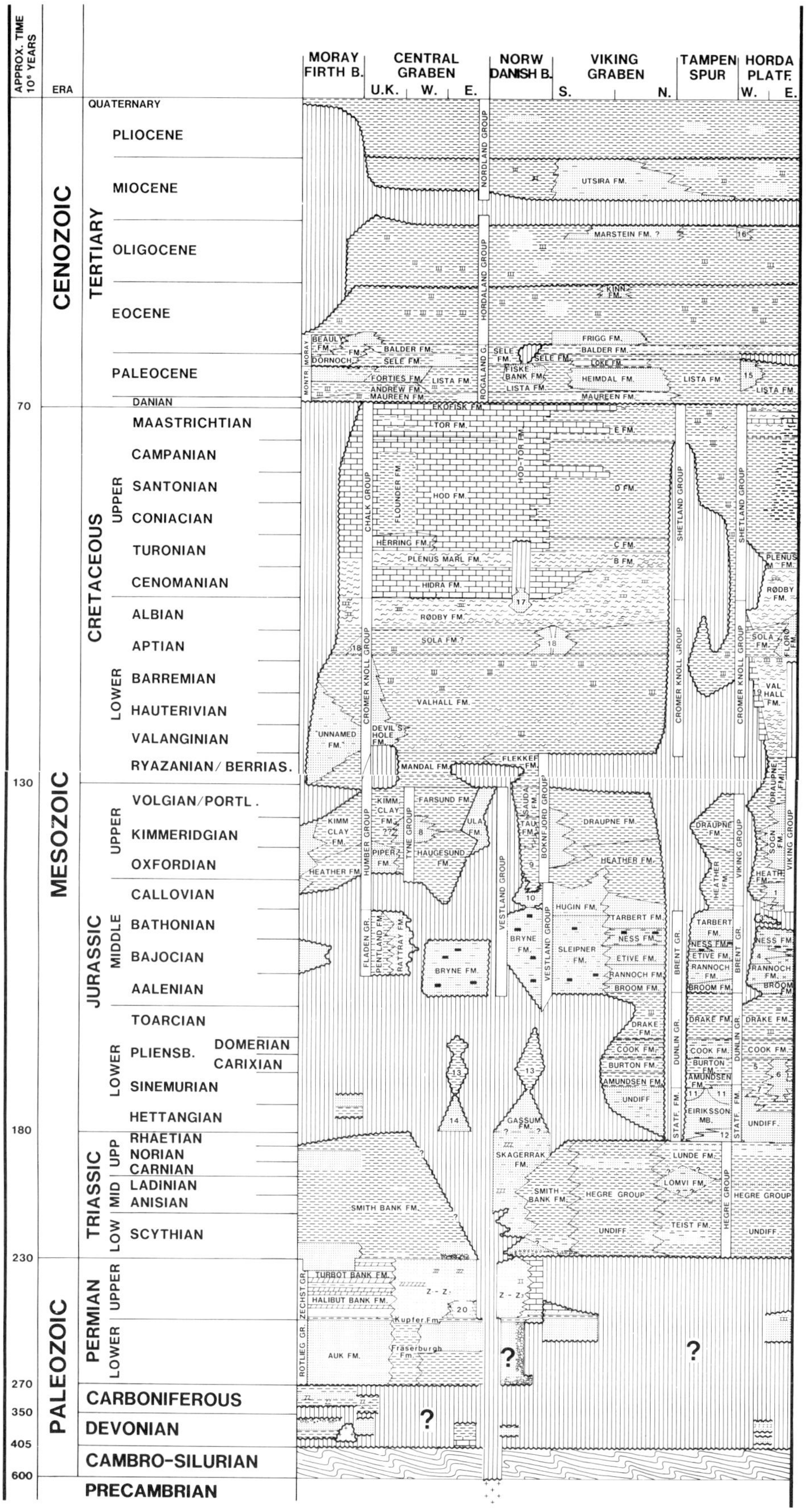

Fig. 6. Stratigraphic diagram of the northern North Sea. 1. Fensfjord Fm 2. Krossfjord Fm 3. Tarbert Fm 4. Etive Fm 5. Burton Fm 6. Johansen Fm 7. Amundsen Fm 8. Eldfisk Fm 9. Egersund Fm 10. Sandnes Fm 11. Nansen Mbr 12. Raude Mbr 13. Fjerritslev Fm 14. Gassum Fm 15. Ninian Fm 16. Marstein Fm 17. Klepp Fm 18. Kopervik Fm 19. Utvik Fm 20. Argyll Fm.

Table 6. Interests in licences north of 62°, to 1986 (incl. Round 10B)

Round	5/1 (1980)			5/2 (1981)			5/3 (1982)						7 (1982)					8 (1984)										9 (1985)							10 B (1986)								
Petroleum Licences & Blocks	PL 059 6507/12	PL 060 7119/12	PL 061 7120/12	PL 062 6507/11	PL 063 7117/9	PL 064 7120/8	PL 073 6407/1	PL 074 6407/2	PL 075 6507/10	PL 076 7119/7	PL 077 7120/7	PL 078 7120/9	PL 080 6609/5	PL 081 6609/7	PL 082 6609/10	PL 083 6609/11	PL 084 6610/7	PL 091 6406/3	PL 092 6407/6	PL 093 6407/9	PL 094 6506/12	PL 095 6507/7	PL 096 7119/9	PL 097 7120/6	PL 098 7120/10	PL 099 7121/4	PL 100 7121/7	PL 105 6406/6	PL 106 6407/4	PL 107 6407/7	PL 108 7120/1	PL 109 7120/2 +7120/3	PL 110 7120/5 +7121/5	PL 111 7121/1	PL 121 6407/5	PL 122 6507/2	PL 123 6507/6	PL 124 6507/8	PL 125 6508/5	PL 126 6607/5	PL 127 6607/12	PL 128 6608/10 +6608/11	Total % Interest
Company Interest in Blocks (%)																																											
Statoil	50	50	50	50	50	50	50	50	50	50	50	50	50	50	50	50	50	50	50	50	50	50	50	50	50	50	50	50	50	50	50	50	50	50	50	50	50	50	50	50	50	50	50
Norsk Hydro	5	10	15	5	20	15	10			25	15	25	5		10	25					5		5	20		10				20	5	5		15	20	20		10				15	8.4
Esso		25				25							15	20	30									10	50							10		35		10				35			6.3
Elf	20					5				15		15					25						35				35		20		5		20				10				35		5.7
Shell				25						10		10								30										20	40								35				4.0
Mobil																		45	40		15											15			20	10				15			3.8
Conoco			25				20															30				10			20				10					15					3.1
BP					20				30											20								40									15						3.0
Saga	10	5		10	5			10			5				10			5					10						10								15					15	2.6
Arco				10				15	10							20					10	10																10					2.0
Phillips						5					10		30	30																													1.8
Agip								15									25				10									10												10	1.7
Total					5						10															30											10						1.3
Tenneco																					10	10										10						10					1.0
Amerada																								10				10					5			10							0.8
Deminex	5	5						10																10			4																0.8
Amoco			10				20																																				0.7
Britoil																			10																10								0.5
Fina																																	5								15		0.5
Svenska																5											10																0.4
Occidental																																							15				0.4
Volvo	10																																										0.3
Others		5																									1											5					0.3
Unocal									10																																		0.2
Texaco											10																																0.2
Texas Eastern																																										10	0.2
																																											100.0

non-deposition. Hydrocarbons may in some cases migrate across the leading fault-scarps into the upthrown reservoirs. In other circumstances, migration occurs across the contact between the source and the underlying reservoir on the dipping flanks of the fault-blocks and along strike, although in some cases where there are seals between multiple reservoirs, migration into the deeper reservoirs may be impeded.

Structure

Rotational fault-blocks, due primarily to Late Jurassic rifting, form traps for hydrocarbons in the province under consideration. The fault-blocks dip away from the axis of the Viking Graben: westwards to the west and eastwards to the east. The fault-blocks themselves are affected by oblique transcurrent faulting that tends to offset the individual rift zones.

Considered in greater detail, the effectiveness of the trap depends on a fine balance between the strength of uplift on the individual fault-block and the degree of consequent erosion. Fault-blocks with relatively minor uplift tend to preserve more of the Jurassic succession beneath the late Kimmerian Unconformity so that the Brent Group, with its good characteristics, is available as a reservoir. Structural closure on such fault-blocks tends, however, to be reduced. By contrast, blocks that have been subject to strong uplift may have larger closure, yet tend to have been subject to intense erosion so that only the Lower Jurassic and Triassic, with their generally inferior characteristics, are preserved as possible reservoirs.

Reservoirs

The principal reservoir in the basin is the Middle Jurassic Brent Group, which was deposited in a delta prograding northwards and passing into marine shales at the northern extremity of the basin. The conditions are described in detail in connection with the Statfjord Field (see p. 319).

The lower Jurassic Statfjord Formation consists of fluvio-deltaic sandstones which form a secondary reservoir in most of the fields on the western flank of the basin.

Triassic reservoirs are locally important, where structural conditions allow migration into them, as for example is the case in the Snorre Field. The Triassic succession was deposited under fluvial conditions, and the sandstones are very variable in thickness, lateral extent and character. Secondary porosity due to leaching is a phenomenon recognized where the formation is in juxtaposition with the Base Cretaceous (Late Kimmerian) unconformity.

Somewhat different conditions occur on the eastern margin of the basin, where the Jurassic consists predominantly of coarse clastic material, and a thick sequence of Middle to Upper Jurassic sandstones form the reservoir in, for example, the Troll Field, described later in the volume.

General articles describing the Viking Graben fields include: Badley *et al.* (1984), Hellem *et al.* (1986), Karlsson (1986), Kirk (1980).

In a certain sense, the Fourth Round represented the climax of exploration on the Norwegian Shelf, for none of the blocks allocated subsequently have contained such prolific finds. It opened up the northern end of the North Sea basin, and permitted a definitive classification of the stratigraphy (see Fig. 6).

EYES TURN NORTH

Although the Fourth Round commanded the prime attention of the industry, it did not preclude consideration of new areas, north of the 62nd Parallel. The Norwegian Petroleum Directorate had already acquired some regional seismic coverage and had begun the process of evaluation. There remained, however, certain political difficulties to be overcome before serious exploration and drilling could commence in the waters of the far north, for it was feared that the entry of the oil industry might have some ill-defined adverse effect on the fisheries that dominated the economy of these remote areas. The approach to exploration in the north had to be a gradual one to allow the local interests to gain confidence that they would benefit rather than suffer from the entry of the oil industry.

The regional seismic surveys by the Petroleum Directorate drew attention to two areas of promise: Haltenbanken, off Mid-Norway; and the Troms I Area off the northern tip of Norway. Both areas appeared to contain a relatively thick Mesozoic succession with rotational fault-blocks capable of forming hydrocarbon traps. It was decided that the national companies should be entrusted to run more detailed surveys over these areas on behalf of the industry. It was perhaps surprising that the industry was willing to accept the onerous licence conditions introduced in the Fourth Round, considering the greatly increased geological risks and environmental hazards of the north. But the companies did accept the conditions, not wishing to preju-

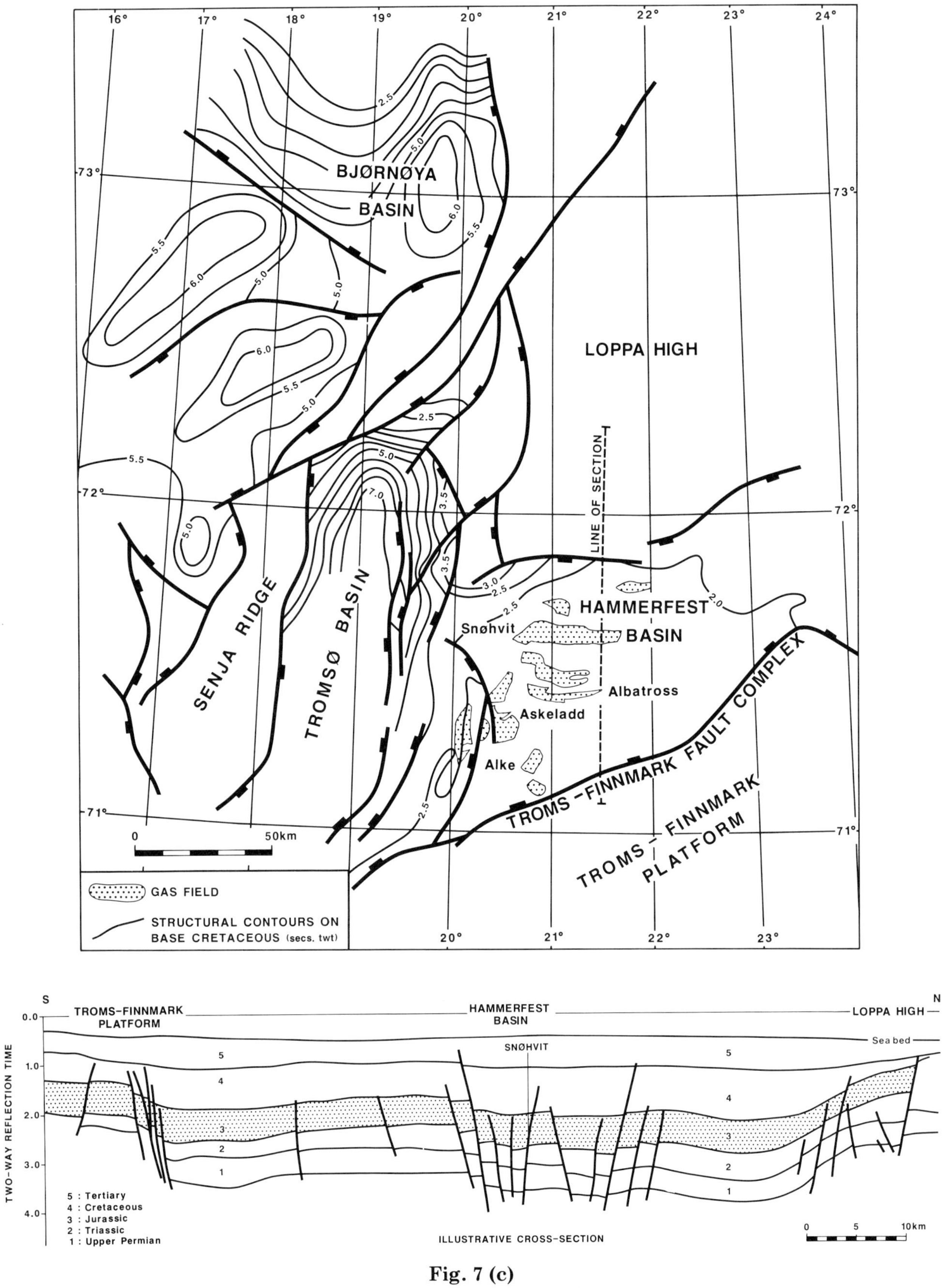
BJØRNØYA
BASIN
LOPPA HIGH
LINE OF SECTION
SENJA RIDGE
TROMSØ BASIN
HAMMERFEST
BASIN
Snøhvit
Albatross
Askeladd
Alke
TROMS-FINNMARK FAULT COMPLEX
TROMS-FINNMARK
PLATFORM
0
50km
GAS FIELD
STRUCTURAL CONTOURS ON
BASE CRETACEOUS (secs. twt)
S
N
TROMS-FINNMARK
PLATFORM
HAMMERFEST
BASIN
LOPPA HIGH
Sea bed
SNØHVIT
TWO-WAY REFLECTION TIME
5 : Tertiary
4 : Cretaceous
3 : Jurassic
2 : Triassic
1 : Upper Permian
ILLUSTRATIVE CROSS-SECTION
10km

Fig. 7 (c)

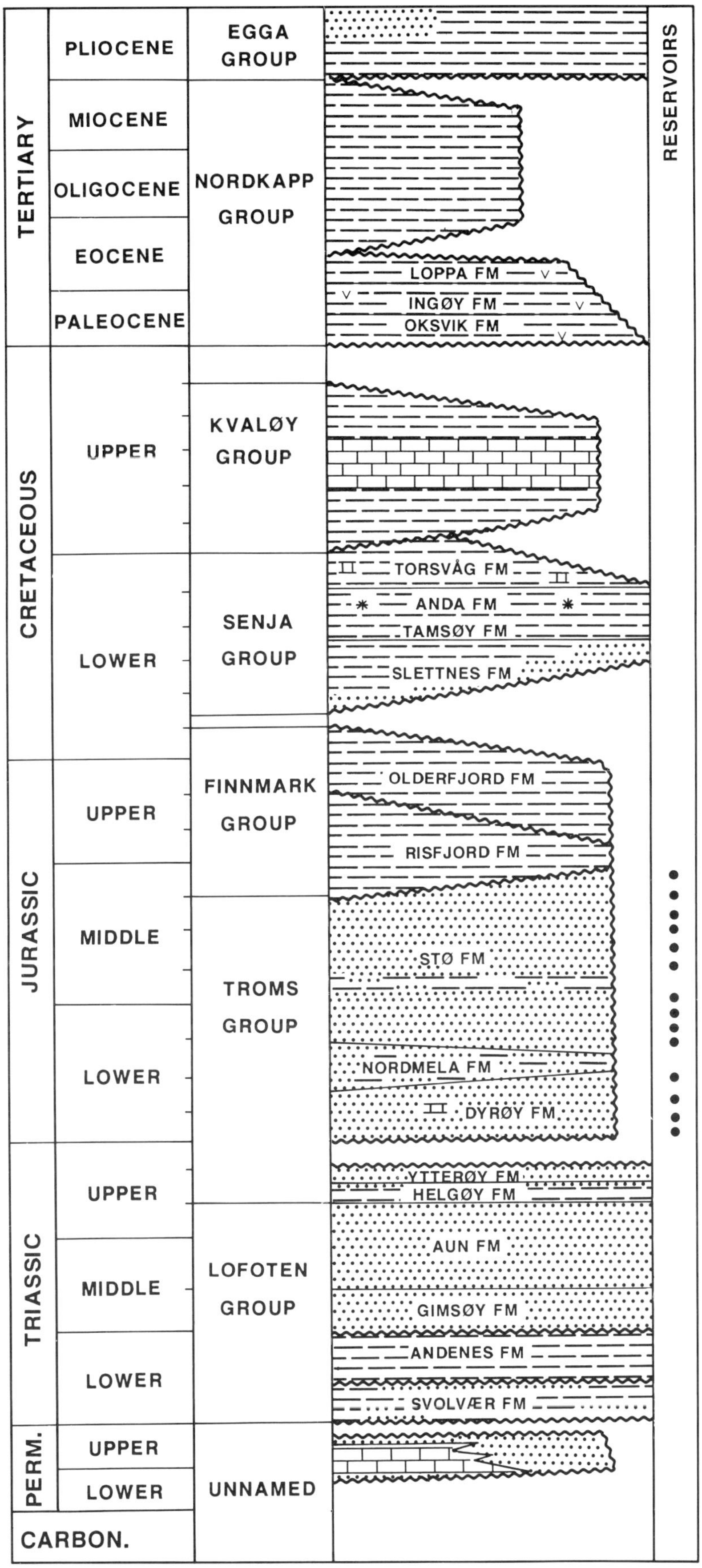

Fig. 7 (a)

Fig. 7. Regional setting of the Hammerfest Basin.

dice their chances of participating in what they hoped would develop into a major new province, and crude oil was at the time selling at very high prices.

The first licences north of the 62nd Parallel were awarded in the Fifth Round, which was itself divided into three parts. Three blocks were awarded in the Troms I and Haltenbanken areas in 1980 in Part 1; three more blocks were granted in both areas in 1981 in Part 2; and six further blocks were allocated in 1982 in Part 3. Later awards were made in the Seventh, Eighth, Ninth and Tenth Rounds, which are considered later. Table 6 lists the blocks and interests awarded in the area north of 62°.

DISCOVERIES IN THE NORTH

This exploration has led to the discovery of two productive provinces: the Hammerfest Basin in the Troms I Area, and Haltenbanken. The geology of these provinces may be summarized as follows:

Jurassic Finds in the Hammerfest Basin

The Hammerfest Basin is an ENE-trending structural depression in the eastern part of the Troms I designated licencing area. It is bounded to the south and north by stable platforms forming the Troms–Finnmark Platform and the Barents Sea Platform (Loppa High), respectively, where comparatively thin Mesozoic sequences overlie Palaeozoic successions with carbonates. The Hammerfest Basin passes westwards into the Tromsø Basin, where a very thick Mesozoic fill is invaded by salt piercements of considerable size. To the east, an arch separates it from another salt-invaded structural depression forming the Nordkapp Basin (see Fig. 7).

Hydrocarbon Generation and Migration

The hydrocarbons found so far in the Hammerfest Basin consist mainly of gas, although a thin column of liquids occurs in the Snøhvit Field.

The source of the hydrocarbons in the Hammerfest Basin is uncertain. Upper Jurassic source-rocks are present in the area, but the extent of maturity in the vicinity of the fields is not easily estimated because the degree of Tertiary inversion is not known exactly. It is possible that the hydrocarbons are derived from Upper Jurassic source-rocks in the Tromsø Basin to the west, where they are mature and overmature, implying a long-range migration. Alternatively, they may come from the Upper Jurassic in the vicinity of the fields, if it is more mature than its present depth of burial would suggest. A third possibility is that they are mainly derived from the gas-prone Lower Jurassic and Triassic succession in the Hammerfest Basin itself. More than one source sequence may have contributed.

Structure

A pronounced anticlinical arch runs along the axis of the Hammerfest Basin plunging towards the west-south-west. It is cut by a series of normal cross-faults, mainly downthrowing to the west, which are particularly pronounced at the western end of the arch adjoining the Tromsø Basin where they form a cuspate system, termed the Transition Zone. The arch is also cut by longitudinal faults, which may be partly transcurrent, giving rise to a number of horsts. The hydrocarbon accumulations lie on this arch, being controlled in detail by the individual faults. Several phases of faulting are recognized: movements on some of them occurred at least as late as in Mid-Tertiary times.

Major transcurrent faults bound the Hammerfest Basin, separating it from the Loppa High to the north and the Troms–Finnmark Platform to the south respectively.

Reservoirs

The principal reservoir in the Hammerfest Basin is a succession of Lower–Middle Jurassic deltaic and shallow-marine sandstones. The details of a representative reservoir succession are given in the Askeladden Field description in this volume.

The discovered gas resources of the Hammerfest Basin are shown in Table 7. The principal references on the Hammerfest Basin are as follows: Berglund *et al.* (1986), Rønnevik and Jacobsen (1984).

Jurassic finds in Haltenbanken

The Haltenbanken area comprises several geological provinces off Mid-Norway. In the east lies the shallow, mildly deformed, Mesozoic Trøndelag Platform, which gently downwarps into the Helgeland Basin to its north and leads into Traenabanken. Between the Vøring Basin in the west, where a thick Cretaceous and Tertiary fill may partly over-

Table 7. Discovered gas resources of the Hammerfest Basin (recoverable)

Field	*Gas ($10^9 Sm^3$)*
Albatross	34.3
Askeladd	52.0
Snøhvit	74.4
Snøhvit North	3.3
Other	42.7
Total	206.7

lie oceanic crust, and the Trøndelag Platform lies the highly faulted Halten Terrace. The latter passes northwards into the Nordland Ridge and is adjoined to the south by the Frøya High, made up of two Mesozoic/Palaeozoic uplifts. Some interpretations attribute the regional structure to oblique-slip tectonics, with zones of uplift and subsidence developing as a consequence of transpression or transtension, associated with curvature on a major transcurrent fault, the Kristiansund–Bodø fault complex.

A number of discoveries of oil and gas have been made in Haltenbanken in a variety of structural circumstances. The regional setting is summarized in Fig. 8.

Hydrocarbon Generation and Migration

The Upper Jurassic contains organic rich shales in Haltenbanken, as it does over much of the rest of the shelf, and is the primary source-rock. Information on it is, however, limited because it occurs in reduced thickness, having been subject to erosion on the structural uplifts where most of the wells have been drilled. It is believed to contain both oil- and gas-prone intervals, and may attain thicknesses in excess of 300 m in the undrilled structural depressions. It is immature on the eastern platform, but peak generation was reached during Late Tertiary times over much of the Halten Terrace. It is overmature to the west. The influx of metagenic gas from the west may explain the preponderance of gas, condensate or light oil, in all but the Draugen Field.

Organic shales associated with coal measures in the Lower Jurassic provide an additional source of mainly gaseous hydrocarbons, which may in some cases have migrated upwards to charge the reservoirs. The latter source-rock sequence may be important, particularly in the eastern areas.

Generation occurred during Mid- to Late Tertiary times over most of the generating zones. Most of the accumulations appear to have been charged from catchment areas in their immediate vicinity, although in some cases they were subject to gas-charge from the west as mentioned. Draugen relies on longer-range migration from the west, made possible by the fact that in this area the intervening fault-blocks dip westwards facilitating migration.

Pressure considerations may be of particular importance in Haltenbanken. The seals on the deep structures in the west may lack sufficient integrity to contain the accumulation of hydrocarbons.

Structure

Most of the discoveries lie on rotational fault-blocks or horsts at depths of up to 4500 m. The regional structure may involve oblique-slip tectonics, with zones of compression and extension being associated with curvature on a major transcurrent fault system. Salt tectonics are also important in some areas. The regional plunge is mainly to the south or south-west. In some cases cross-faulting provides an important structural control. All the fields are in structures exhibiting relief at the level of the Base Cretaceous (Late Kimmerian) unconformity, although the deeper structure, which is more complex, controls the accumulations. The Draugen Field, which lies on the eastern platform, is an exception insofar as it lies on a gentle anticlinal feature, in which stratigraphic factors may contribute to closure.

Reservoirs

The principal reservoirs in most of the fields occur in the Lower–Middle Jurassic and consist of deltaic and shallow-marine sandstones. Sandstones in the upper part of the succession generally have the better reservoir characteristics, although there is not a constant relationship between depth and porosity. The sandstones are relatively uniform across the basin, although the upper sandstones may be locally missing from structural uplifts due to Late Jurassic or later erosion. Some of the accumulations, especially in the west, contain multiple reservoirs, containing different types of hydrocarbons.

The Draugen Field is exceptional in that Upper Jurassic sandstones form the reservoir.

The discovered resources of the Haltenbanken area are given in Table 8. The principal references are: Aasheim and Larsen (1984), Aasheim *et al.* (1986), Ellenor and Mozetic (1986), Koenig (1986), Caselli (1986), Jackson and Hastings (1986).

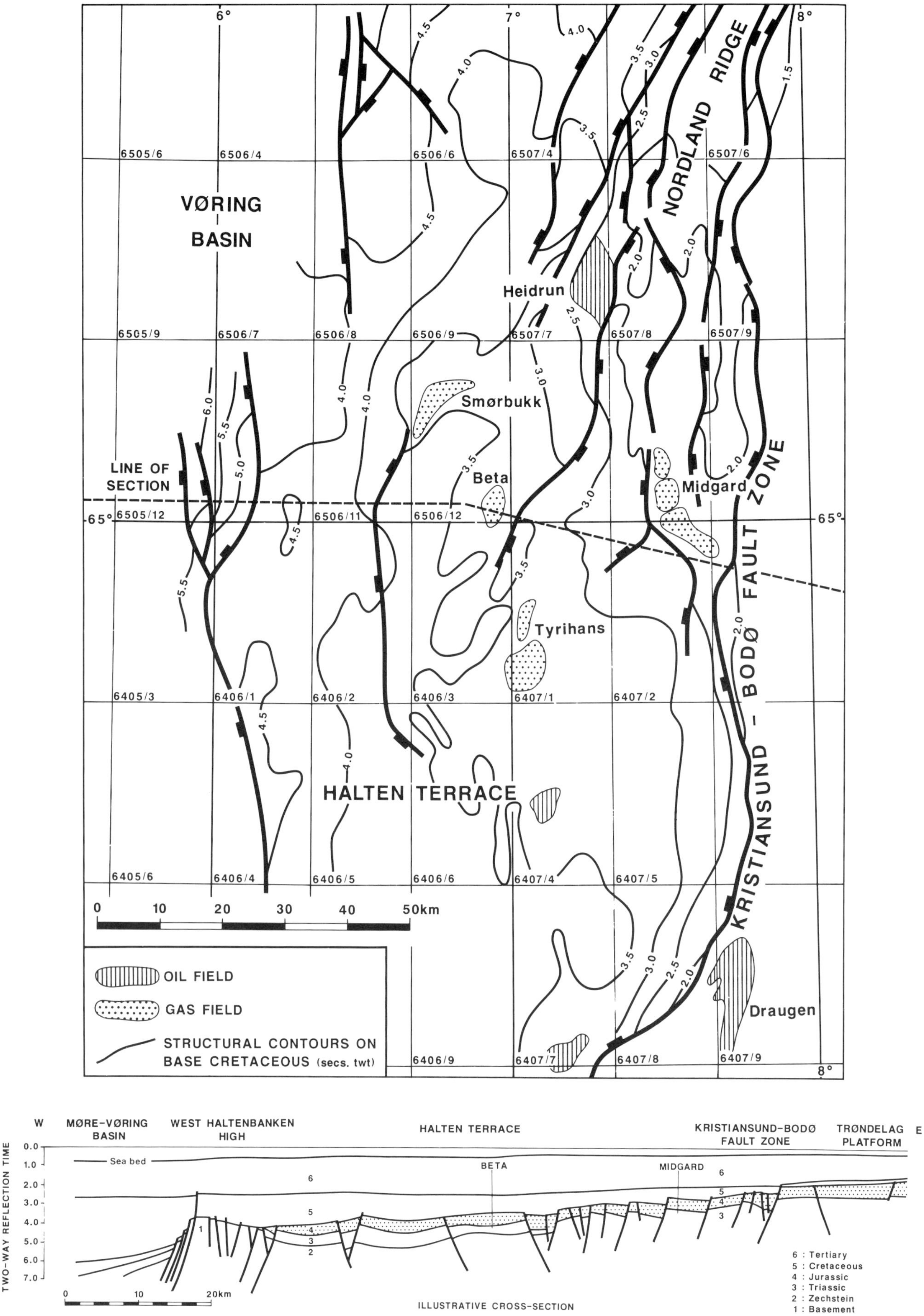
6°
7°
8°
65°
VØRING
BASIN
NORDLAND RIDGE
Heidrun
Smørbukk
Beta
Midgard
Tyrihans
HALTEN TERRACE
Draugen
KRISTIANSUND - BODØ FAULT ZONE
LINE OF
SECTION
6505/6
6506/4
6506/6
6507/4
6507/6
6505/9
6506/7
6506/8
6506/9
6507/7
6507/8
6507/9
6505/12
6506/11
6506/12
6405/3
6406/1
6406/2
6406/3
6407/1
6407/2
6405/6
6406/4
6406/5
6406/6
6407/4
6407/5
6406/9
6407/7
6407/8
6407/9
0 10 20 30 40 50km
OIL FIELD
GAS FIELD
STRUCTURAL CONTOURS ON
BASE CRETACEOUS (secs. twt)
W
E
MØRE-VØRING
BASIN
WEST HALTENBANKEN
HIGH
HALTEN TERRACE
KRISTIANSUND-BODØ
FAULT ZONE
TRØNDELAG
PLATFORM
TWO-WAY REFLECTION TIME
Sea bed
BETA
MIDGARD
0 10 20km
ILLUSTRATIVE CROSS-SECTION
6 : Tertiary
5 : Cretaceous
4 : Jurassic
3 : Triassic
2 : Zechstein
1 : Basement

Fig. 8 (c)

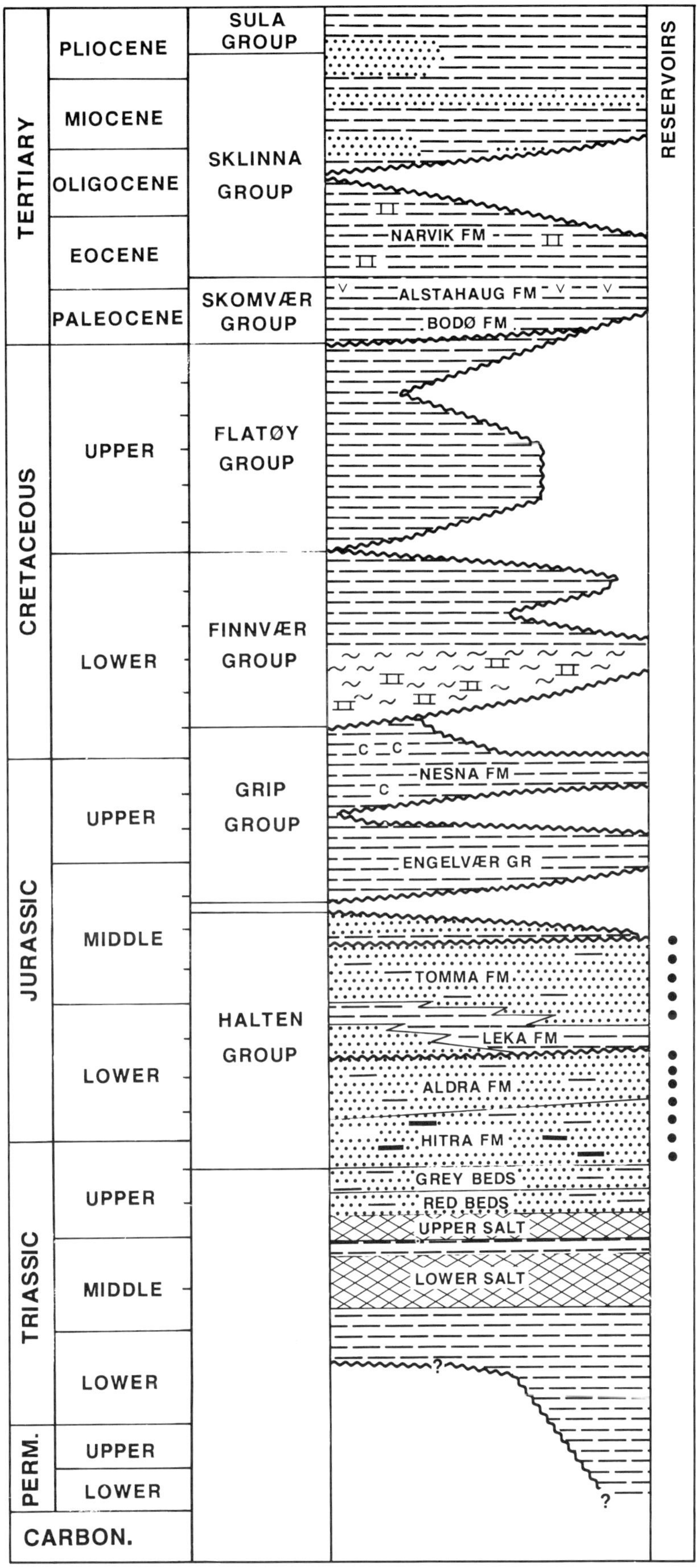

Fig. 8 (a).

Fig. 8. Regional setting of the Haltenbanken Area.

Table 8. Discovered oil and gas resources of Haltenbanken (recoverable)

Field	Oil (10^6 Sm^3)	Gas (10^9 Sm^3)
Draugen	39.0	—
Heidrun	87.0	31.0
Midgard	22.0	103.0
Smørbukk	60.0	83.0
Tyrihans	12.0	51.0
6506/12 Beta	76.0	54.0
Total	296.0	322.0

DISAPPOINTMENTS IN THE NORTH

Reference has also to be made to other provinces, which have so far failed to yield significant discoveries: disappointments that were not, however, wholly unforeseeable.

First, Traenabanken, to the north-east of Haltenbanken, which was licensed in the Seventh Round when five blocks were allocated (see Table 6). The failure of exploration so far may be attributed to the fact that the Upper Jurassic source-rocks in communication with most of the structures of interest, especially to the east, are immature.

Secondly, three licences, covering Blocks 7120/1, 7120/2, 7120/3 and 7121/1 were awarded in the Ninth Round in 1985 on the Loppa High, which forms the margin of an extensive Palaeozoic platform extending over the Barents Sea (see Fig. 9). It is reported that one of the wells, 7120/2-1, drilled in these blocks encountered residual hydrocarbons in Permo-Carboniferous carbonates.

The third province is the Nordland Ridge, between the Haltenbanken and Traenabanken off Mid-Norway, where nine blocks were licenced in Round 10B in 1985 (Table 6). The results of the initial drilling suggest that it resembles the barren Traenabanken to the north-east, rather than Haltenbanken.

Lastly, two wells on the Senja Ridge to the west of the Tromsø Basin showed it to be an inverted, predominantly argillaceous, Mesozoic basin.

OTHER DEVELOPMENTS IN THE NORTH SEA

In parallel with the developments north of 62°, described above, exploration continued in the North Sea. It is necessary to return to that area and pick up the story of the events following the Fourth Round. In 1981, nine previously relinquished blocks were reallocated in the Sixth Round, as listed in Table 9. The main discoveries to have been made are those represented by Well 1/3-4 (Elf) and Well 7/11-5 (Norsk Hydro).

In the Eighth Round in 1984, it was decided to remove the artificial barrier of the 62nd Parallel, and to treat the entire shelf as a single entity. In addition to the blocks north of 62°, listed in Table 6, seven blocks were allocated in the North Sea (see Table 10). Of these, particular importance attaches to Block 34/7, which covers an extension of the Snorre Field that had already been found by a well drilled in Block 34/4, close to the boundary. It is remarkable in that the Triassic forms the principal reservoir, although the overlying Statfjord Formation is also productive. The field contains substantial amounts of oil, but there are development difficulties due to the fact that it lies in 300–370 m of water. It is described in detail later in the volume. There are a number of satellites between Snorre and Statfjord awaiting evaluation and possible development.

Other blocks were allocated in Rounds 9 and 10A in the North Sea (see Table 10). None have so far yielded a major discovery. The greatest disappointment was Block 34/8, a huge structure down-thrown to the Gullfaks structure where closure was dependent on what is evidently a non-sealing fault. Also of significance were Blocks 9/2, 9/3, 26/4 and 25/6, which were designed to test structures outside the main productive trends of the Central and Viking grabens. It was hoped that any encouragement from these awards might herald future exploration in the Stord and Egersund basins, where it is necessary to rely on the discovery of source-rocks other than those in the Upper Jurassic which have been responsible for most of the oil found in north-west Europe to date. Some such encouragement has been forthcoming (Well 9/2-1), and some significance may attach to the hydrocarbons tested in Well 25/6-1.

SVALBARD

The foregoing description relates to the main part of the Norwegian shelf, adjoining the mainland, but there is another territory to consider, namely the Svalbard Archipelago. Svalbard was effectively a 'no-man's land' until its status was regularized by the Svalbard Treaty of 1920. This treaty granted sovereignty to Norway, provided that the 40

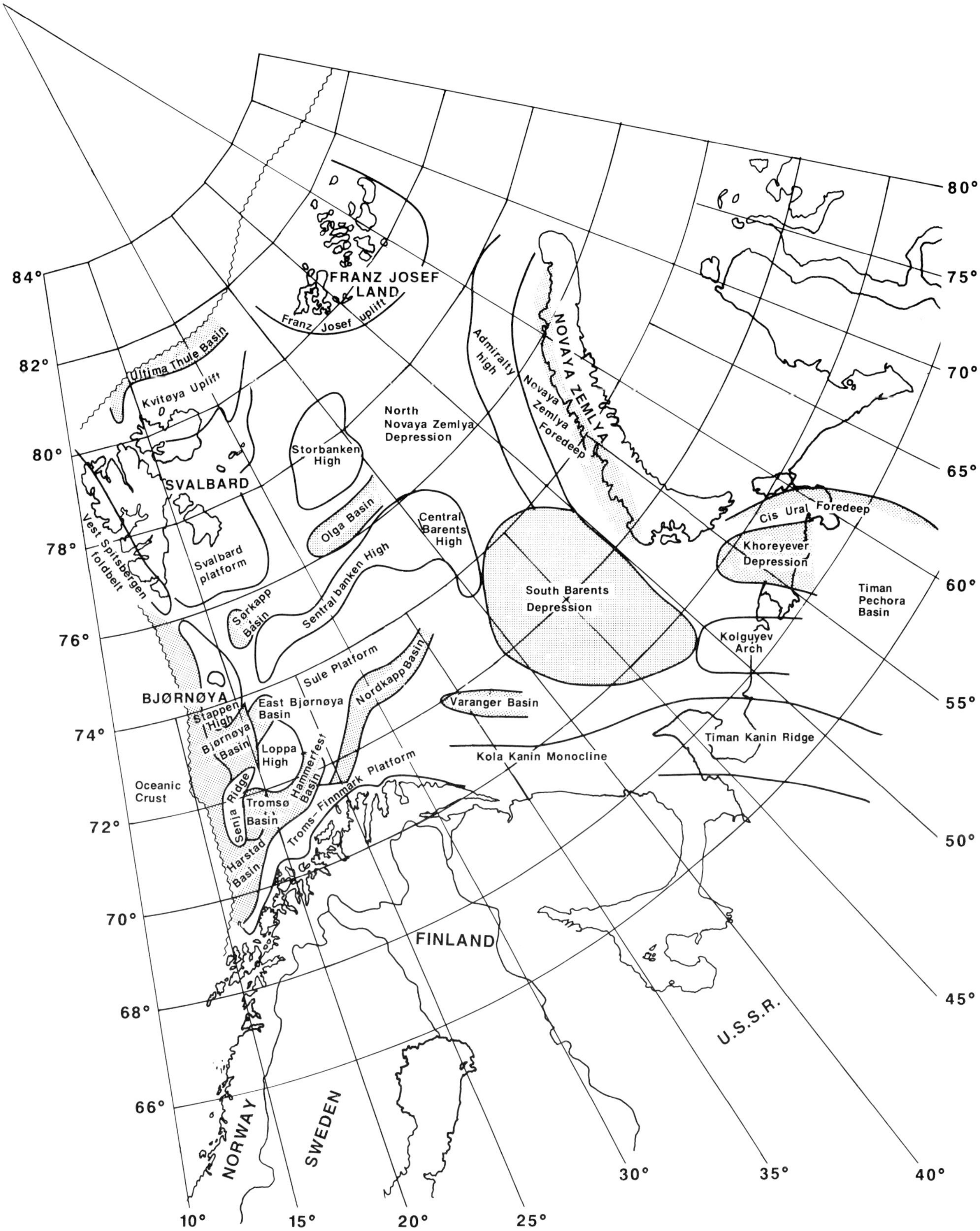

Fig. 9. Regional setting of the Barents Sea Area.

signatory-countries were granted equal opportunities to develop the territory's economic possibilities, including its minerals. Coal mining has been the principal activity in that regard: there are Soviet coal mines at Barentsburg and Pyramiden in addition to Norwegian mines at Longyearbyen and Svea. There has also been a certain interest in oil exploration. Caltex, Fina, Total and Norsk Hydro, together with a number of other inde-

Table 9. Interests in the Sixth Licence Round, 1981

Company Interest in Block (%)	Block & Licence Numbers: PL 065 1/3	PL 066 2/2	PL 067 2/5	PL 068 2/8	PL 068 2/11	PL 069 7/8	PL 070 7/11	PL 071 8/3	PL 072 16/7	Total % Interest
Statoil	50	50	50	50		50	50	50	50	50
Norsk Hydro		10			25	15	25		10	10.6
Shell	15		30					35		10
Mobil		25			7.5		7.5			5
Conoco					7.5	25	7.5			5
Esso									40	5
Saga		10			10		10			3.8
Elf	17									2.1
Texaco	10									1.3
Agip			10							1.3
Phillips			10							1.3
Total	8									1
Svenska								7.5		0.9
Volvo								7.5		0.9
BP						5				0.6
Deminex						5				0.6
Arco		5								0.6
										100.0

pendent companies, have taken up oil rights under the mining code, which is not a very satisfactory framework for the purpose. A total of 13 wells have been drilled over the past 20 years or so without any significant discovery. The failure is commonly attributed to the effects of inversion impairing reservoir and hydrocarbon source-rock conditions, but there has nevertheless been a resumption of interest by several companies recently. Svalbard comprises a very large territory, covering a land area of some 63 000 km^2, so there is perhaps room to find local areas with better prospects than regional considerations may suggest. A Devonian to Oligocene succession rests on Precambrian basement (Worsley and Aga, 1986), and the presence of exposed Triassic hydrocarbon source-rocks is noteworthy. The contract terms for petroleum activity differ from those on the rest of the Norwegian Shelf and are very much less onerous.

CONCLUSIONS ON THE RESULTS TO DATE

The foregoing account describes the principal chapters in the history of the exploration of the Norwegian Shelf, and it summarizes the essential findings. It is an impressive record, which is due primarily to the presence of prolific hydrocarbon source-rocks in the Upper Jurassic that have been buried sufficiently to generate oil along a belt that can be traced from Ekofisk in the south almost to Svalbard in the north. In many areas this zone dips abruptly beneath a thick Cretaceous and Tertiary cover to reach the depth of metagenic generation, causing a massive gas charge. In other areas, such as on the Tampen Spur, west of the Viking Graben, and on Haltenbanken, structural terraces preserve the source-rocks at a depth of peak oil generation over relatively wide areas, providing an adequate oil charge shielded from the massive metagenic gas influx from the basin deeps.

A GLIMPSE OF THE FUTURE

It remains for future exploration to show if there are other source-rock sequences capable of providing results as prolific as the Upper Jurassic. The distribution of this productive trend is now well defined, and it is unlikely that many more major oil fields remain to be found in it. On the other hand, there is considerable opportunity for finding additional step-outs and deeper reservoirs in the established producing provinces. The anticipated future production of the established fields is shown on Fig. 10.

The main hope for finding a major new oil trend lies in the Barents Sea where Triassic

Table 10. Interests in the North Sea blocks, allocated in the 8th, 9th and 10th Rounds (1980–1986)

Company Interest in Block (%)	PL 086 6/3 + 7/1	PL 087 16/4	PL 088 24/6 + 25/4	PL 089 34/7	PL 090 35/11	PL 101 16/10	PL 102 25/5	PL 103 25/7	PL 104 30/9	Special Award PL 112 25/2	PL 113 2/12	PL 114 9/2	PL 115 9/3	PL 116 15/12	PL 117 25/6	PL 118 26/4	PL 119 29/3	PL 120 34/8	Total % Interests
Statoil	50	50	50	50	50	50	50	50	50	50	50	50	50	50	50	50	50	**50**	**50**
Norsk Hydro	10	20		12	10				30	17.3	25			10				18	8.5
Conoco	30							30						30			15	13	6.6
Total			30							10.9						25	25		5.1
Shell		20					20					15	30						4.7
Elf				8			30			21.8								13	4.0
Amerada	10							10			25			10					3.1
Saga				10					5			15			15			6	2.8
Mobil					40				5										2.5
Agip						25									20				2.5
Fina						20									15				1.9
Tenneco		10							5				15						1.7
BP																25			1.4
Unocal			20																1.1
Deminex				4		5						10							1.0
Others				1					5				5				5		0.9
Esso				15															0.8
Britoil								10									5		0.8
Occidental												10							0.6
																			100.0

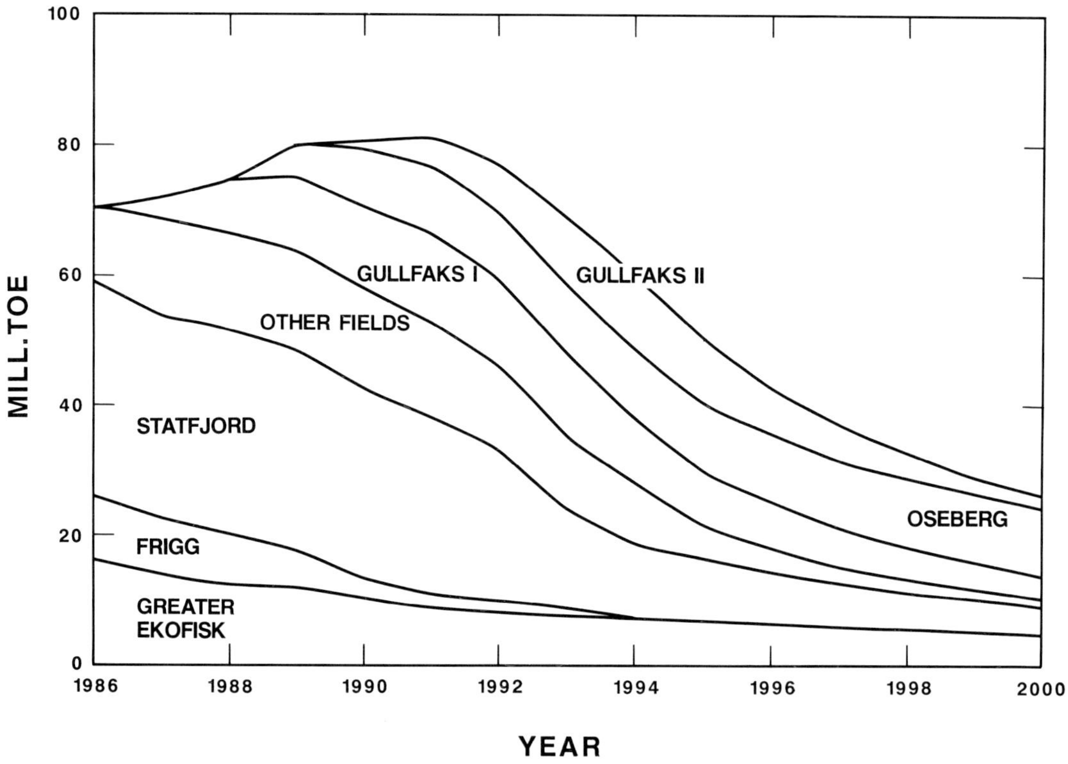

Fig. 10. Production forecast.

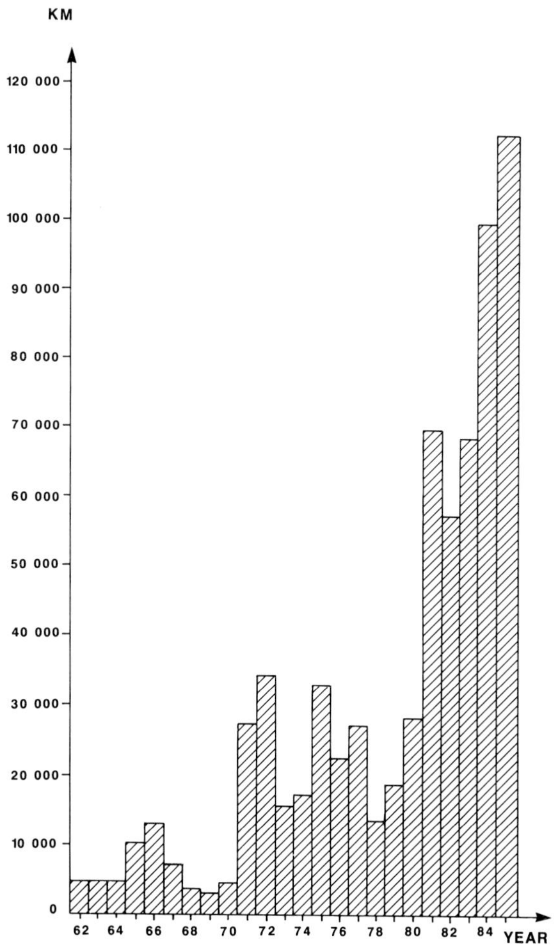

Fig. 11. Seismic surveys, 1962–85.

and Palaeozoic sources may be adequately developed to provide substantial oil accumulations. It is tempting to make an analogy with the Persian Gulf province where internal sinks in a Mesozoic carbonate platform generated prolific quantities of oil that accumulated in huge, gentle, salt-induced structures. It is also relevant to note that the easterly extensions of the Barents Sea province are productive in the Soviet Union. At the same time it has to be realized that most of the World's oil lies in strata younger than the Palaeozoic carbonates that form the principal reservoir in the Barents Sea. There is, furthermore, the difficulty of the regional inversion of the province during the Tertiary, meaning that the strata have previously been more deeply buried than is so today. The reservoirs may be heavily indurated. The results of drilling to date on the margins of the platform and on Svalbard have not been particularly promising.

The World's oil resources are decidedly finite: current estimates suggest an ultimate recovery of about 1700 billion barrels, of which no more than about 500 billion remain to be found. Norway's shelf has sustained a highly successful exploration effort over the past 20 years, but it is unrealistic to expect the pace to continue much longer unless a major new province is found, which in effect means

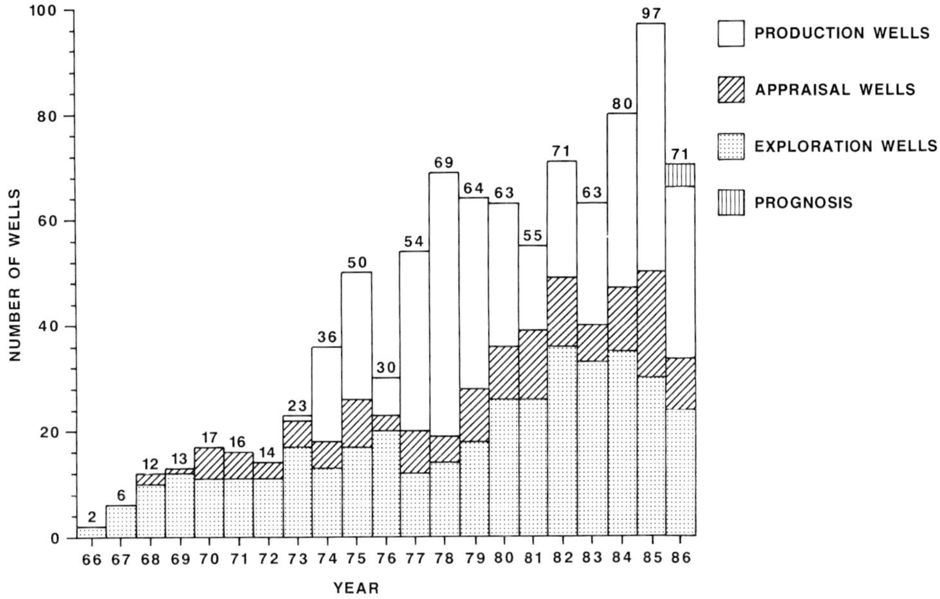

Fig. 12. Growth of drilling, 1966–86.

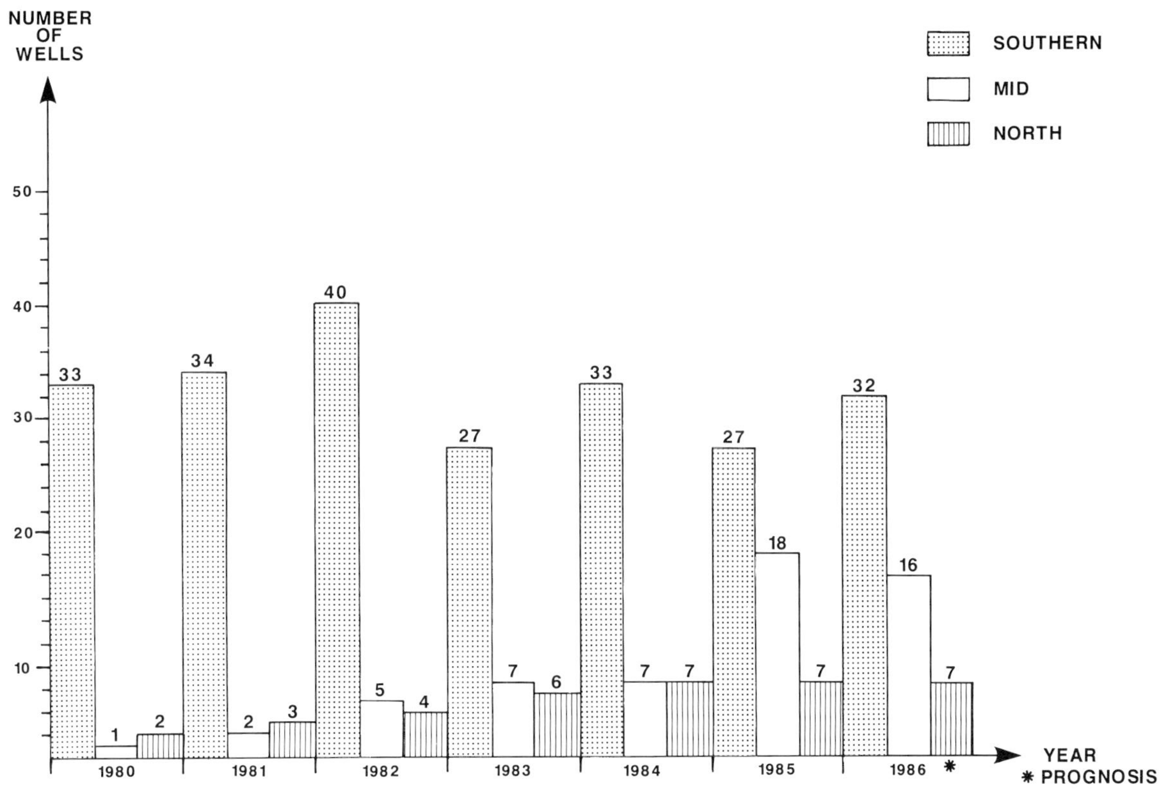

Fig. 13. Growth of drilling by area.

the Barents Sea. The problems of deep water and the harsh arctic environment are also factors that cannot be ignored. The amelioration of licence terms which has been accounced at the time of writing provides valuable new incentives to maintain the pace of activity, despite the adverse consequences of the radical fall in oil prices late in 1985.

The possibilities for gas, on the other hand, are very great indeed. Not only does the Upper Jurassic become gas prone towards the east, but there are several other intervals in the succession that are capable of generating gas. It seems reasonable to expect that Norway may well develop into one of Europe's principal sources of gas in the next century.

THE RECORD OF EXPLORATION: SOME STATISTICS

The exploration described above has been made possible by seismic surveys on a colossal scale. Figure 11 shows the amount of reflection seismic survey undertaken on the shelf.

The geological and geophysical work has led to the definition of a large number of prospects, which have been investigated by the drilling of 520 exploration and appraisal wells, listed in the Appendix, on pages 00–00. Figure 12 shows the growth of drilling to a peak in 1985, when 50 wells were drilled, which has been followed by a decline, consequent upon the collapse in oil prices.

Figure 13 shows the growth of drilling by area, illustrating the increasing importance of northern areas. The drilling has resulted in the discovery of 58 oilfields and 47 gas fields. Figure 14 shows the resources that have been discovered as a consequence of this activity. The rate of discovery is, however, declining, as shown in Fig. 15, following the normal pattern. The discovery per well drilled is also declining.

As discussed above, the undiscovered potential is less easily determined. It may be helpful to distinguish the potential of the well-defined trend of oil and gas generated in Upper Jurassic source-rocks, which is responsible for most of known oil and gas in northwest Europe (with the notable exception of the Southern North Sea Province). It is here thought that perhaps more than three-quarters of the ultimate recoverable oil from this trend in Norway has already been found, whereas perhaps less than half of the ultimate recoverable gas has been discovered. In addition, the huge unexplored Barents Sea province may contain new prolific oil and/or gas trends derived from other than Upper Jurassic source-rocks, which may be in the multi-billion-barrel category, but more drilling is required before the amounts can be quantified in even the most general terms. There may also be source-intervals in pre-Jurassic strata in the interior basins, such as the Stord, Egersund or Helgeland Basins, but that, for the moment, is pure speculation.

Lastly, it is well to relate the foregoing discussion to the geography of Norway, and to draw attention to the very considerable dimensions of its continental shelf. The southern boundaries of the shelf are clearly defined by the median lines with the United Kingdom and Denmark. The position is less clear in the north, where there is a disputed boundary with the Soviet Union and where lies the Svalbard Archipelago, a trust territory administered by Norway. The western boundaries with Greenland and Iceland are of little significance in terms of presently perceived exploration activities. In geographical terms,

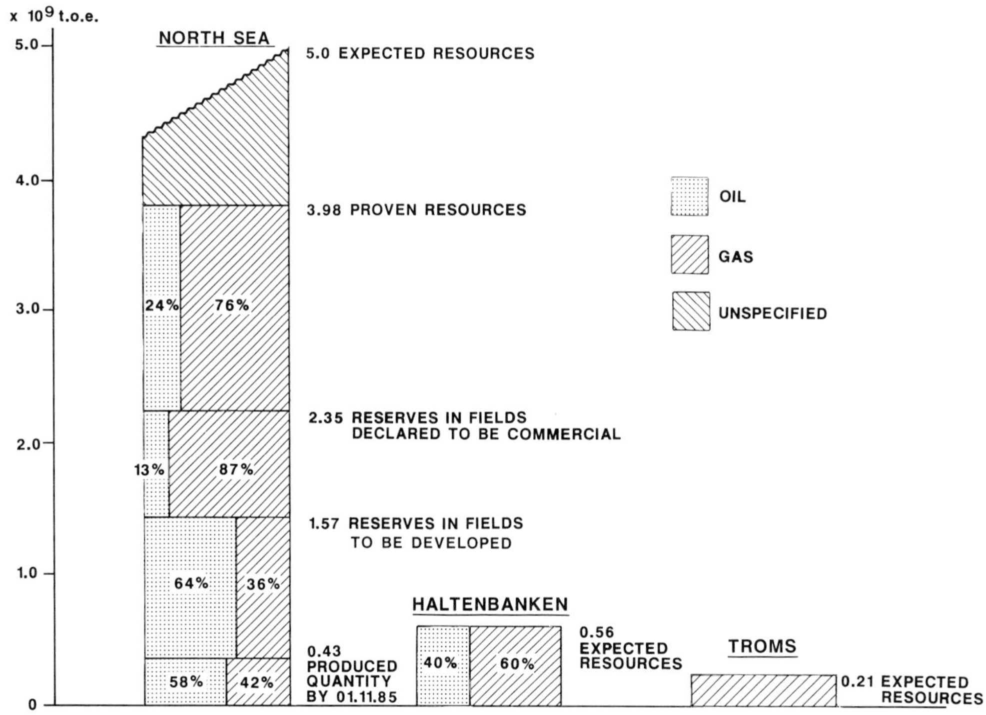

Fig. 14. The discovered petroleum resources of Norway.

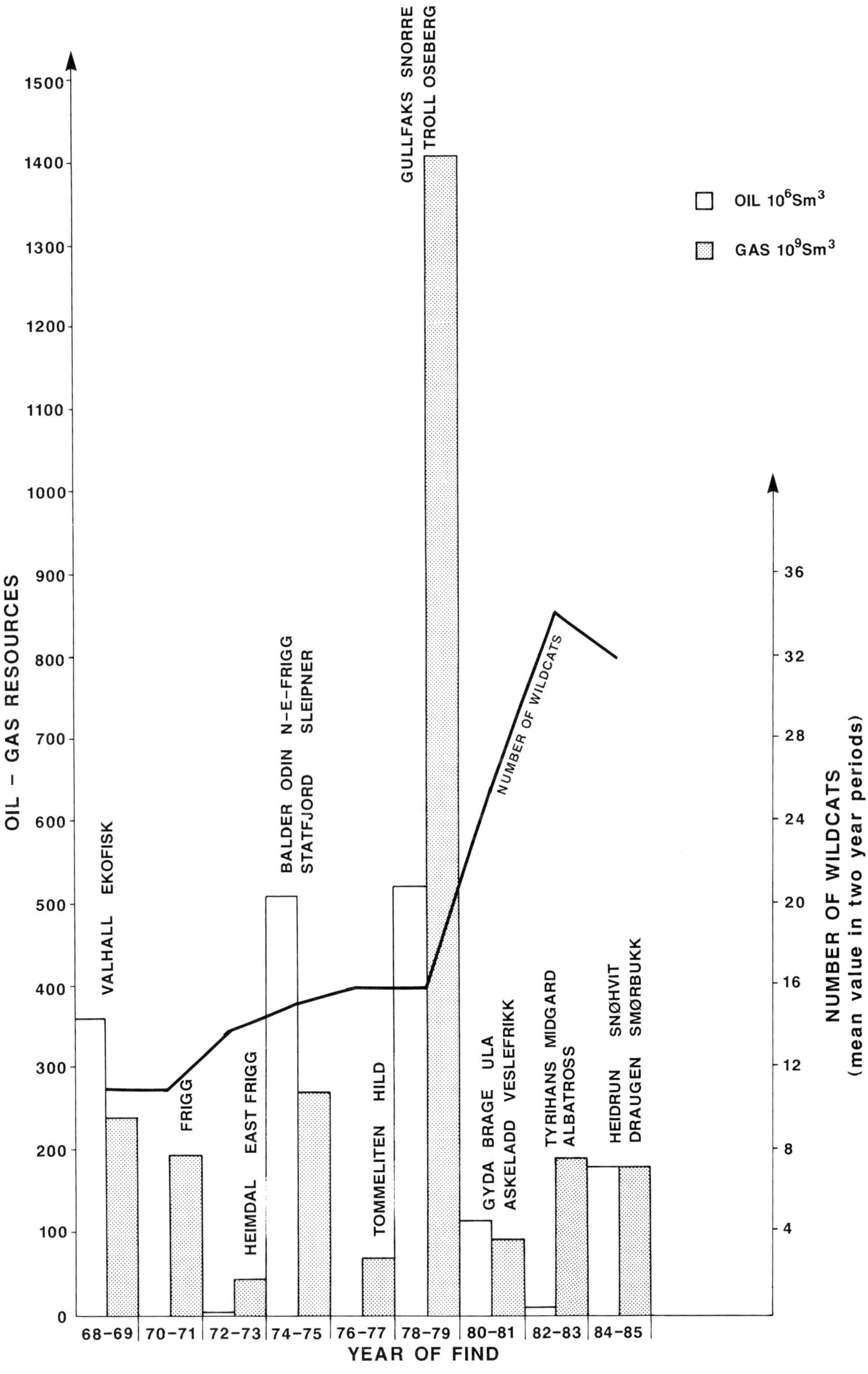

Fig. 15. Discovery by the year.

the shelf within the 500-m isobath, which covers most, if not all, of the area with petroleum potential, extends for a distance of almost 3000 km from the southern limit to the northern extremity of the Barents Sea. It ranges in width greatly: in the north lies the huge Barents Sea platform, some 1400 km in width; whereas it is relatively narrow along much of the northern coastline, being in places less than 75 km in width. It covers an area of approximately one million square kilometres.

Despite its high latitude, the Norwegian Shelf enjoys a relatively mild climate due to the effects of the Gulf Stream, although parts of the Barents Sea north of latitude 74° fall within the zone of permanent sea-ice cover. Although most of the shelf described above is covered by sedimentary rocks, which could in the most general sense be considered prospective for oil and gas, in practice only relatively restricted areas possess, or are likely to possess, the characteristics needed for commercial accumulations.

The oil and gas resources of Norway have great importance in a European context. Norway has been responsible for about one-fifth of the cumulative production of oil in north-west Europe. It was also responsible for about one-third of Europe's production in 1985. But in terms of reserves it accounts for about 40%. Its share of the undiscovered resources of the area is greater still, and could be very much greater if the huge, virtually untested, Barents Sea should prove to be a prolific oil province. Norway was responsible for about 15% of the production of natural gas in western Europe in 1985. It has almost half the proven reserves and probably a much greater proportion of the area's undiscovered gas.

The exploration environment in Norway is exceptionally open, and there is a wide dissemination of information on drilling and exploration results, which greatly facilitates the process of exploration.

It is to the furtherance of this positive policy that this volume is dedicated.

ACKNOWLEDGEMENTS

The authors express their appreciation for the helpful comments from A. M. Spencer, Eigil Nysæther and S. Hanslien. The contributions of J. Neish and S. J. D. Campbell are also acknowledged.

The authors would like to thank respectively the Directors of Fina Exploration Norway and the Director of the Norwegian Petroleum Directorate (NPD) for permission to publish this article. The views expressed are, however, those of the authors alone, and are subject to the limitations of their knowledge and understanding.

REFERENCES

Aasheim, S.M. and Larsen, V. 1984. The Tyrihans discovery—preliminary results From Well 6407/1-2, In Spencer, A.M. *et al.* (Ed.) *Petroleum Geology of the North European Margin*, Graham and Trotman, London, pp. 285–292.

Aasheim, S.M., Dalland, A., Netland, A. and Thon, A. 1986. The Smørbukk Gas/Condensate Discovery, Haltenbanken. In Spencer, A.M. *et al.* (Eds) *Habitat of Hydrocarbons on the Norwegian Continental Shelf*, Graham and Trotman, London, pp. 299–305.

Badley, M.E., Egeberg, T. and Nipen, O. 1984. Development of rift basins illustrated by the structural evolution of the Oseburg feature, block 30/6, offshore Norway, *Journal Geological Society London* **141** (4), 639–649. Discussion by D. Soutter and J. S. Ritchie **142** (5), 933–934.

Bailey, C.C., Price, I. and Spencer, A.M. 1981. The Ula Oil Field, block 7/12, Norway, *In Norwegian Symposium on Exploration* (Norsk Petroleumsforening), Article 18.

Berglund, L.T., Augustson, J., Færseth, R., Gjelberg, J. and Ramberg-Moe, H. 1986. Evolution of the Hammerfest Basin, In Spencer, A.M. *et al.* (Eds) *Habitat of Hydrocarbons on the Norwegian Continental Shelf*, Graham and Trotman, London, pp. 319–338.

Brewster, J. and Dangerfield, J.A. 1984. Chalk fields along the Lindesnes Ridge, Eldfisk, *Marine Petroleum Geology* **1** (3), 239–278.

Caselli, F. In press. Oblique-slip tectonics Mid-Norway shelf, In *3rd Conference on Petroleum Geology of North-west Europe*, 26–29 October 1986.

Conort, F. 1986. Habitat of Tertiary Hydrocarbons, South Viking Graben, In Spencer, A.M. *et al.* (Eds) *Habitat of Hydrocarbons on the Norwegian Continental Shelf*, Graham and Trotman, London, pp. 159–170.

D'Heur, M. 1980. Chalk reservoir of the West Ekofisk Field, In *The Sedimentation of the North Sea Reservoir Rocks* (Norsk Petroleumsforening), Article X.

Ellenor, D.W. and Mozetic, A. 1986. Draugen Oil Discovery, In Spencer, A.M. *et al.* (Eds), *Habitat of Hydrocarbons on the Norwegian Continental Shelf*, Graham and Trotman, London, pp. 313–316.

Hardman, R.F.P. and Kennedy, W.J. 1980. Chalk Reservoirs of the Hod Field, Norway, In *The Sedimentation of the North Sea Reservoir Rocks* (Norsk Petroleumsforening), Article XI.

Hellem, T., Kjemperud, A. and Øverbø, O.K. 1986.

The Troll Field: a Geological/Geophysical Model Established by the PL 085 Group. In Spencer, A.M. *et al.* (Eds), *Habitat of Hydrocarbons on the Norwegian Continental Shelf*, Graham and Trotman, London, pp. 217–238.

Héritier, F.E., Lossel, P. and Wathne, E. 1980. The Frigg gas field, In Illing, L.V. and Hobson, G.D. (Eds), *Petroleum Geology of the Continental Shelf of North-west Europe*, Institute of Petroleum, Heyden, London, pp. 380–391.

Jackson, J.S. and Hastings, D.S. 1986. The role of salt movement in the tectonic history of Haltenbanken, and its relationship to structural style. In Spencer A.M. *et al.* (Eds), *Habitat of Hydrocarbons on the Norwegian Continental Shelf*, Graham and Trotman, London, pp. 241–257.

Karlsson, W. 1986. The Snorre, Statfjord and Gullfaks oilfields and the habitat of hydrocarbons on the Tampen Spur, offshore Norway, In Spencer, A.M. *et al.* (Eds), *Habitat of Hydrocarbons on the Norwegian Continental Shelf*, Graham and Trotman, London, pp. 181–197.

Kirk, R.H. 1980. Statfjord Field—a North Sea Giant, *Memoirs American Association Petroleum Geologists*, **30,** 95–116.

Koenig, R.H. 1986. Oil Discovery in 6507/7, an initial look at the Heidrun Field, In Spencer, A.M. *et al.* (Eds), *Habitat of Hydrocarbons on the Norwegian Continental Shelf*, Graham and Trotman, London, pp. 307–311.

Rønnevik, H.C. and Jacobsen, H.P. 1984. Structures and basins in the Western Barents Sea, In Spencer, A.M. *et al.* (Eds), *Petroleum Geology of the North European Margin*, Graham and Trotman, London, pp. 19–32.

Skovbro, B. 1983. Depositional conditions during Chalk sedimentation in the Ekofisk area Norwegian North Sea, In Kaasschieter, J.P.H. and Reijers, T.J.A. (Eds), *Petroleum Geology of the South-eastern North Sea and the Adjacent Onshore Areas*, Geol. Mijnbouw **62** (1), 169–175.

Spencer, A.M., Home, P.C. and Wiik, V. 1986. Habitat of hydrocarbons in the Jurassic Ula Trend, Central Graben, Norway, In Spencer, A.M. *et al.* (Eds), *Habitat of Hydrocarbons on the Norwegian Continental Shelf*, Graham and Trotman, London, pp. 111–127.

Van Den Bark, E. and Thomas, O.D. 1980. Ekofisk: first of the giant oilfields in western Europe. *Memoirs American Association Petroleum Geologists* **30,** 195–224; reprinted in *Bulletin American Association Petroleum Geologists* **65** (11), 2341–2363.

Worsley, D. and Aga, O.J. 1986. *Evolution of an Arctic Archipelago—The Geological History of Svalbard*, Den norske stats oljeselskap a.s., Stavenger.

Part II—Central Graben

2

Albuskjell

Michel D'Heur

Petrofina, c/o Fina Exploration Norway, Skogstøstraen 37, Postboks 4055, Tasta, N-4001, Stavanger, Norway

The Albuskjell Field is a rich gas–condensate accumulation straddling Blocks 1/6 and 2/4 in the south part of the Norwegian sector of the North Sea. It is part of the 'Greater Ekofisk' area where the reservoirs are chalks of Danian and Maastrichtian ages with porosities of up to 40% or more. In Albuskjell the Danian reservoir has low permeabilities and the Maastrichtian reservoir has contributed nearly all the production. The source-rock is the Late Jurassic shale of the Mandal Formation. The reservoir is overlain by more than 3000 m of Tertiary clay and claystone. The discovery well was drilled in 1972, and the field came on-stream in 1979 from two drilling–production platforms linked to the Ekofisk Centre, some 15 km to the south-east. Albuskjell is expected to produce some 18×10^9 Sm^3 of gas and 8×10^6 Sm^3 of condensate by 1997.

INTRODUCTION

The Albuskjell gas–condensate field lies in 70 m of water, some 315 km south-west of Stavanger and 15 km north-west of the Ekofisk Field (Fig. 1). It straddles the boundary between Blocks 1/6 and 2/4. Albuskjell is one of the eight currently producing hydrocarbon accumulations belonging to the 'Greater Ekofisk' area, where chalk forms the reservoir. Initial recoverable reserves are estimated at approximately 18×10^9 Sm^3 of gas and 8×10^6 Sm^3 of condensate.

The field was named after a Norwegian sea-shell, by Norske Shell.

HISTORY

Blocks 1/6 and 2/4 were allocated in 1965 in the first round of licencing on the Norwegian continental shelf. Block 1/6 is part of production licence PL011 granted to the Norske Shell (100%), and Block 2/4 belongs to licence PL018 allocated to Phillips, Petrofina and Agip, later joined by the Petronord Group. Present participating interest in Block 2/4 is:

Phillips Petroleum Company of Norway (operator) 36.96%
Norske Fina A/S 30%
Norsk Agip A/S 13.04%
Elf Aquitaine Norge A/S 8.1%
Norsk Hydro A/S 6.7%
Total Marine Norsk A/S 4.05%
Eurafrep Norge A/S 0.45%
Coparex Norge A/S 0.4%
Cofranord A/S 0.3%

The Albuskjell structure was identified during the early stages of exploration in the mid-sixties as an elongated NW–SE-trending anticline with a culmination in Block 1/6 (see D'Heur *et al.*, 1985, Fig. 2).

The discovery of Ekofisk in late 1969 greatly stimulated exploration activities in the area, and the Phillips Group was already very active with the appraisal of the Ekofisk, West Ekofisk, Eldfisk, and Tor structures when Shell decided to drill the first exploration well, 1/6-1 in Block 1/6. Shell and the

Geology of the Norwegian Oil and Gas Fields © A. M. Spencer *et al.*, eds (Graham & Trotman, 1987) pp. 39–50.

Albuskjell Field data summary

Trap	
Type	Combined structural and stratigraphic
Depth to crest	3070 mss
Lowest closing contour	3260 mss (top chalk)
Gas–water contact	3400 mss or deeper
Pay zone	
Formation	Tor Fm (mainly), Ekofisk Fm
Age	Maastrichtian, Danian
Lithology	Chalk
Porosity (average)	21% (Tor Fm)
Hydrocarbon saturation (average)	65% (Tor Fm)
Permeability—matrix	0.01–1 mD
—effective	up to × 10 matrix permeability, or even more due to fractures
Hydrocarbons	
Condensate gravity	0.79 g/cm^3
Gas gravity (relative to air)	0.70
Gas/oil ratio (initial)	890 Sm^3/Sm^3 (2/4 area) 1600 Sm^3/Sm^3 (1/6 area)
Formation water	
—	
Reservoir conditions	
Temperature	138 °C (3200 mss)
Pressure (initial)	500 bar (3200 mss)
Field size	
Net area	*c.* 25 km^2
Recoverable condensate	8×10^6 Sm^3
Recoverable gas	18×10^9 Sm^3
Production	
Start-up	1979
Development	2 production platforms (18 wells) linked to Ekofisk Centre
Production rate (end 1986)	4×10^6 Sm^3/day
Cumulative production (end 1986)	13.5×10^9 Sm^3

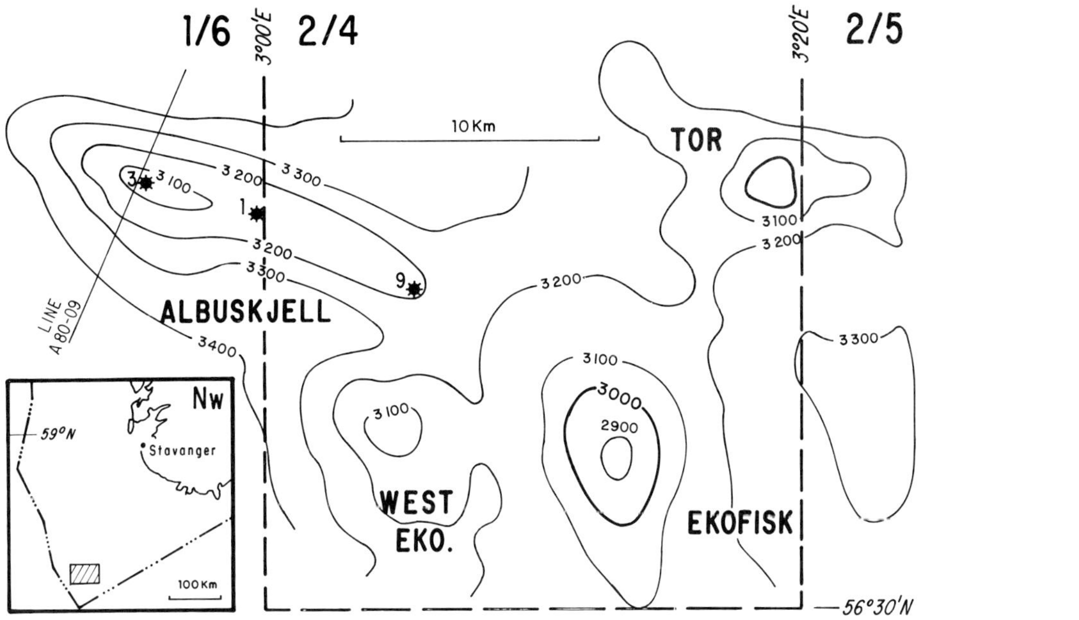

Fig. 1. Location map and depth to the top of the Chalk (Top Ekofisk Fm). Contour interval = 100 m.

Phillips Group shared equally the cost of this well, which was located close to the boundary between the two blocks. The well was spudded in July 1972, and encountered the top of the chalk at 3115 mss. The main target was the Danian chalk (Fig. 2) which in West Ekofisk, only 5 km to the south, was found to be 150 m thick, with an average porosity of 31%. Well 1/6-1, however, found less than 20 m of porous Danian which, when tested over an interval of 6 m, produced only 24 Sm^3/day of condensate and 50 000 Sm^3/day of gas (Interval A, Fig. 3).

The second objective, the Maastrichtian chalk, which had just been identified as an excellent reservoir in the area, comprised less than 6 m of 'net pay' in Well 1/6-1. This interval (B, Fig. 3) flowed at a rate of 350 Sm^3/day of condensate and 460 000 Sm^3/day of gas. The well was terminated in Zechstein salt at a depth of 4820 mss. Based on the results of Well 1/6-1, the field, which was named Albuskjell by Shell, appeared to be a modest find.

The discovery well was located in a slight saddle in the middle of an anticline, and it was hoped that the crestal locations both to the north-west and the south-east would contain better reservoirs. Well 1/6-2 was drilled by Shell in late 1972 on another prospect. Well 2/4-9, drilled by Phillips in the summer of 1973, was designed to appraise the south-eastern nose of the structure. The well gave unexpected results: the Danian chalk, although 90 m thick, contained no produceable reservoir, whilst the Maastrichtian, despite its low structural elevation, contained gas-bearing reservoirs with porosities of up to 30%, between 3260 and 3330 mss. The well was terminated in the Campanian at 3720 mss. The Maastrichtian reservoir was tested as a whole and flowed at a rate of 460 Sm^3/day of condensate and 450 000 Sm^3/day of gas (interval C, Fig. 3).

A third well, 1/6-3, drilled in 1974 to appraise the north-western apex also gave unexpected results: the Danian chalk, which was 120 m thick, contained an interval 90 m thick with porosities of as much as 40%, but the Maastrichtian was thinner and less porous than in Well 2/4-9.

The Danian interval was not tested successfully, and the Maastrichtian reservoir flowed at a rate of 410 Sm^3/day of condensate and 510 000 Sm^3/day of gas (interval D, Fig. 3).

The field was unitized in December 1975. Interests in the unitized area (Fig. 4) are:

Norske Shell A/S	50%
Phillips Petroleum Company Norway (operator)	18.48%
Fina Exploration Norway	15%
Norsk Agip A/S	6.52%
Norsk Hydro A/S	3.35%
Elf Norge A/S	4.05%
Total Marine Norsk A/S	2.024%
Erafrep Norge A/S	0.228%
Coparex Norge A/S	0.199%
Cofranord Norge A/S	0.152%

Two 20-slot drilling-production platforms, namely 1/6-Alpha and 2/4-Foxtrot, were installed. Development drilling commenced in early 1979. The development plan called for 15 wells to be drilled from each platform (Fig.

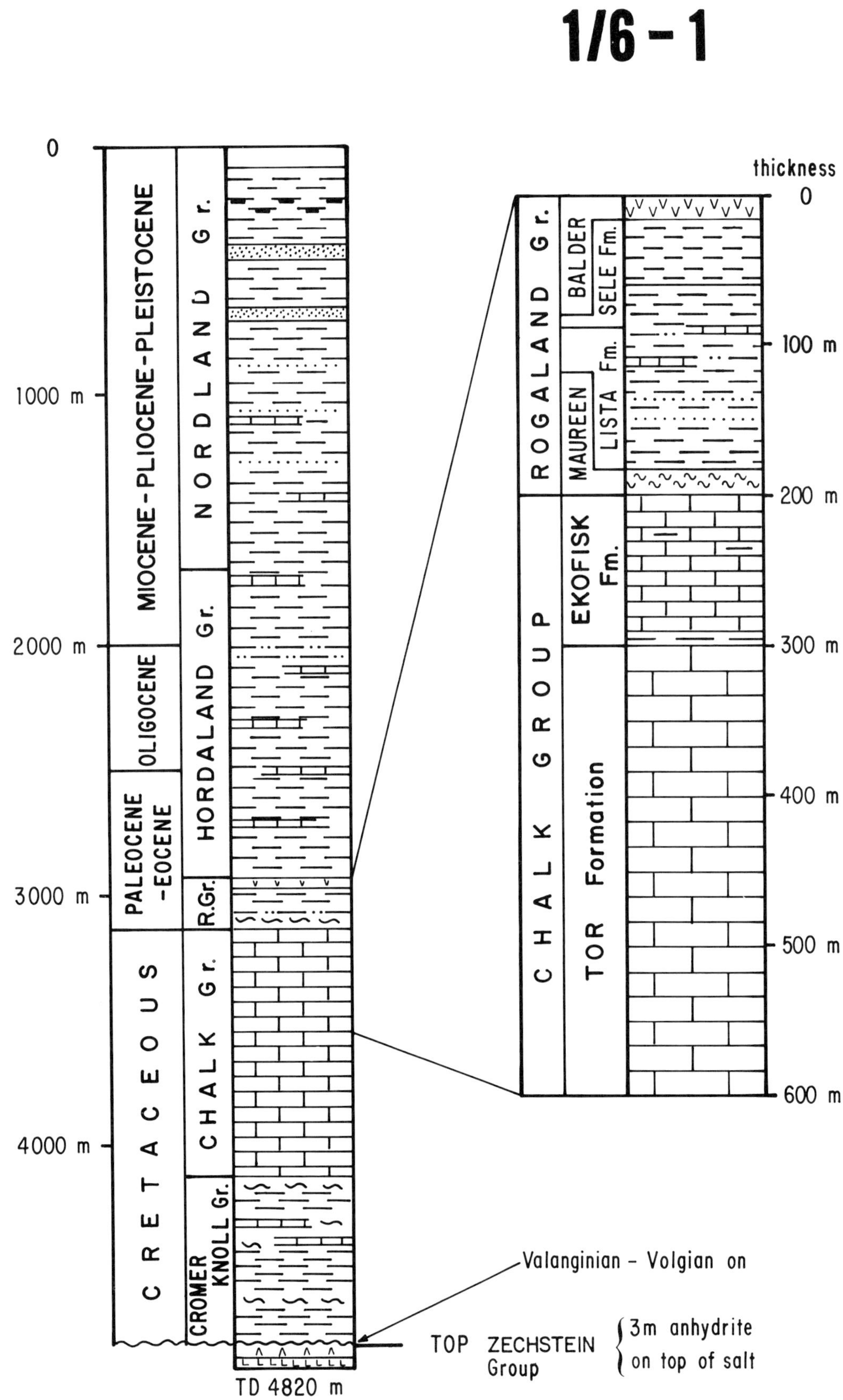

Fig. 2. Lithostratigraphy of the discovery well.

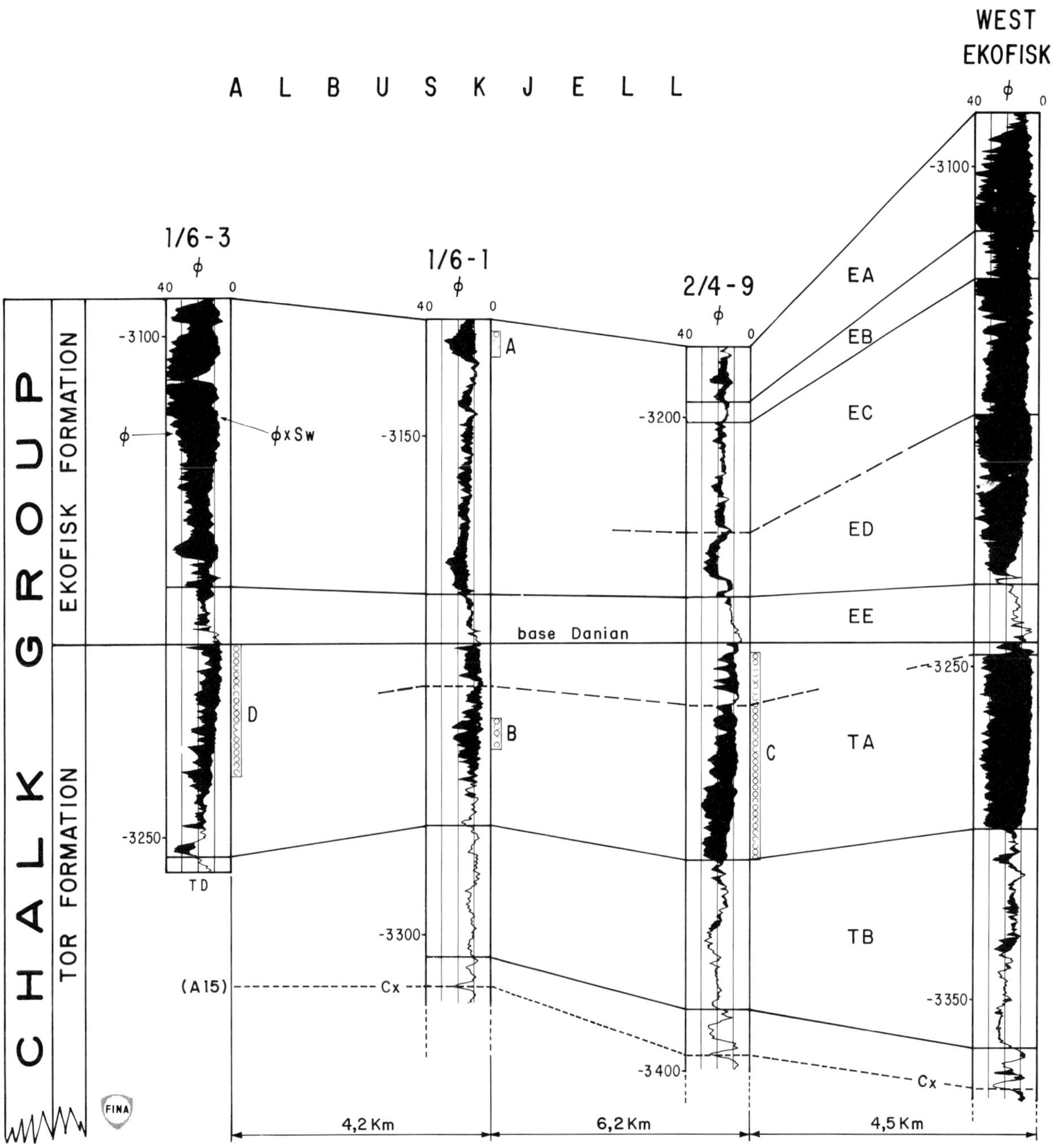

Fig. 3. Porosity and hydrocarbon saturation logs of the three exploration–appraisal wells and comparison with the West Ekofisk crestal well. (Datum = Base Danian).

4), but unexpected results from several wells led to a reduction in the number of wells.

The vertical holes from each platform, namely 1/6-A10 and 2/4-F10, were drilled to about 4100 mss to investigate the entire chalk section. Four thin porous intervals of the Hod and Hidra formations were tested. The best zone, 15 m thick (in the Lower Hod Member), flowed at a rate of 60 000 Sm^3/day gas plus condensate and water.

Full development was reached in 1981 with 18 producing wells (Fig. 5). Well 2/4-F1 was drilled in 1982 to test a possible porosity extension to the south-east but encountered no 'pay' zones.

Since then, it has become obvious that the production platforms have different economic lifespans. It has therefore been decided to accelerate the 1/6-Alpha production by well restimulation and additional drilling in the north-west crestal area of the field. Well 1/6-A7 was drilled in 1986 for such a purpose, and other locations are being evaluated. So far as production is concerned, a peak of 9×10^6 Sm^3/

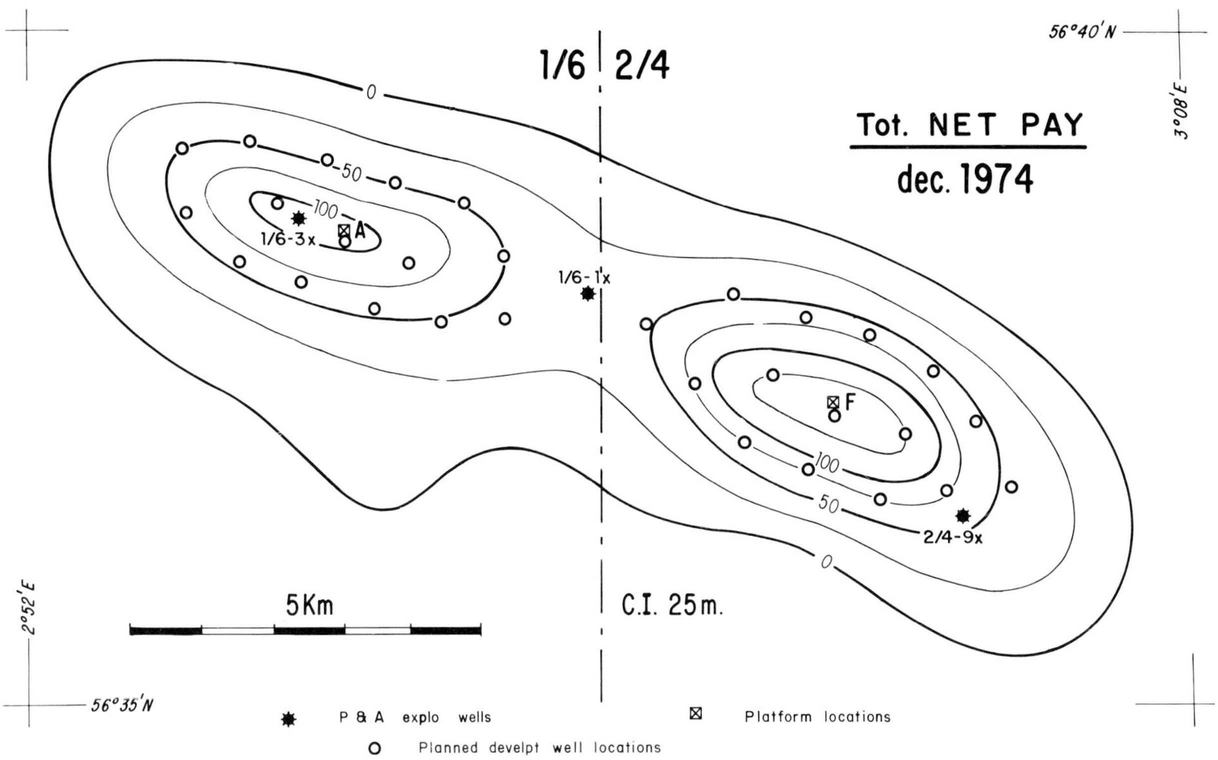

Fig. 4. A 1974 optimistic hypothesis of the 'net pay' thickness and the originally planned locations of the 30 development wells.

day of gas and 4000 Sm^3/day of condensate was reached in 1981. At the end of 1986, the daily rate was around 4×10^6 Sm^3 of gas.

GEOLOGICAL SETTING

Albuskjell is located in the Central Graben of the North Sea as are the other chalk fields of the Norwegian and Danish sectors. The Central Graben was initiated in Late Palaeozoic times during a period of rifting. The tectonic history of the graben, which is a complex NW–SE trough, has recently been described by Gowers and Sæbøe (1985). Near Albuskjell, Permian Zechstein salt lies below some 3000 m of terrigenous strata of Triassic, Jurassic and Early Cretaceous ages, overlain by up to 1000 m of chalk and 3000 m of Tertiary clay and claystones. The section, however, is not complete below the field (Fig. 2).

Four geologic episodes played a major role in the formation of large hydrocarbon accumulations in the Upper Cretaceous and Lower Paleocene chalks. These are:

1. Thick deposits of organic-rich clay of the Mandal Formation accumulated in the trough during Late Jurassic times.
2. Tectonic pulses in the graben during Maastrichtian and Danian times, especially along its faulted margins and on the nearby platform, caused remobilization of unstable carbonate deposits. The reworked material was then redistributed into the trough by gravity currents and deposited as allochthonous chalks which today make the best reservoirs. In contrast, the pelagic chalks (autochthonous) are generally not productive.
3. Halokinesis of the Zechstein salt has created the structural traps.
4. Continued subsidence and the accumulation of thick Tertiary clays provided adequate seal.

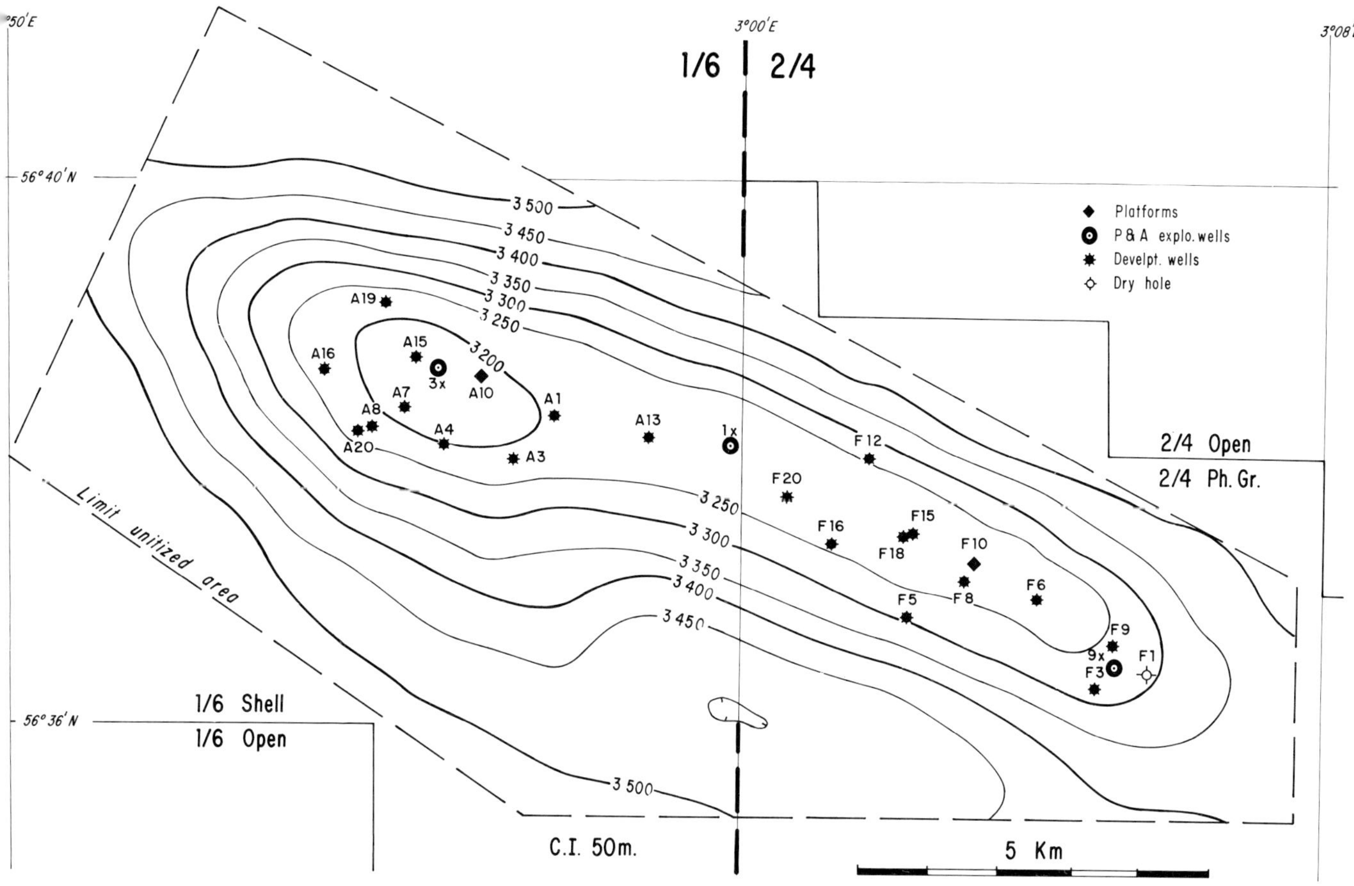

Fig. 5. Depth map of the top of the Tor Formation.

GEOPHYSICS

The Albuskjell structure was first identified in the mid-sixties on the basis of a 2 × 2 km seismic grid of trending N–S and E–W. The structure appeared to be an elongated anticline with a culmination in Block 1/6. Several seismic surveys have been shot since then. During the exploration drilling phase, one structural interpretation suggested the presence of two culminations, a north-west apex in Block 1/6 and a south-east apex in Block 2/4, with an intermediate saddle close to the block boundary. This interpretation led to an over-optimistic view of total net pay thickness (Fig. 4). The most recent survey, in 1979–80, consists of 37 transverse lines (NE–SW) with a 500-m spacing, 10 radial lines mainly on the western crest and three longitudinal lines totalling about 250 km.

The quality of the data is good except over the crest of the structure in Block 1/6 where the presence of gas in the overlying Tertiary claystone creates a velocity anomaly and distortion of the seismic signal over an area of some 8 km^2 (Fig. 6). The first prominent reflector below 3 s marks the top of the Paleocene; the next, about 200 ms below, is the top of the Ekofisk Formation. The structure is expressed in the overlying Tertiary succession as high as Middle Miocene level. Colour synlogs have been used to detect porosity patches between wells. Two seismically distinct porous zones can be mapped in the Tor Formation, 0–40 ms and 40 and 60 ms below the top of the Maastrichtian; the upper zone is the more extensive.

TRAP

The Albuskjell structure is an elongated, NW-trending, doubly plunging anticline, about 18 km long and 5 km wide (Fig. 5). At top Chalk level, its areal closure is 50 km^2, and the vertical relief reaches 260 m in Block 1/6. The field itself is, however, restricted to some 25 km^2. The area has been subjected to numerous tectonic pulses from Mesozoic to Late Tertiary times. The present-day structural shape is the result of movements in the underlying salt from Early Eocene to Middle Miocene. The

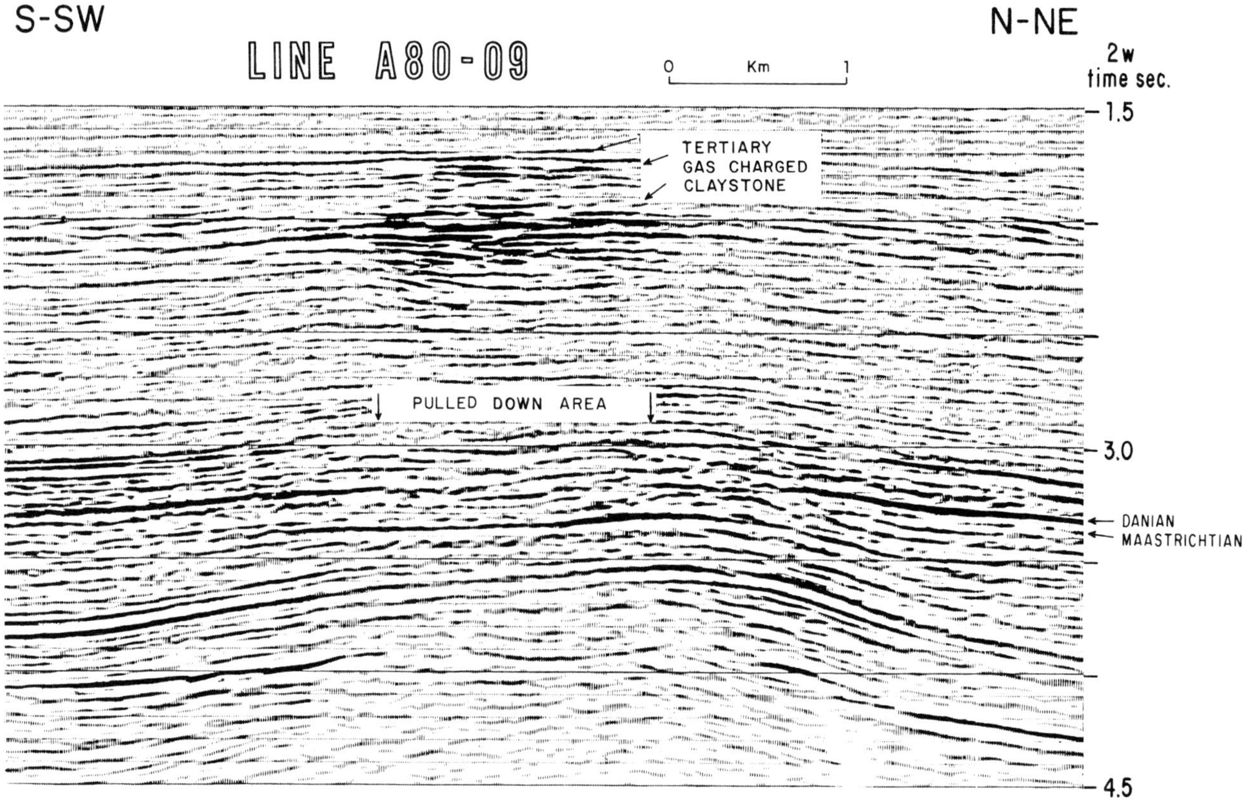

Fig. 6. A SW–NE seismic section through the crestal area in Block 1/6. See Fig. 1 for location.

trapping mechanism is neither purely structural nor purely stratigraphic, and is due in part to late calcite recrystallization in the aquifer.

Structural reconstructions by Dronyck, Connary and Knight (1980–85, company reports) have shown that the geometric configuration has changed little in Block 1/6 after the hydrocarbons invaded the trap (Oligocene to Early Miocene), whereas the palaeocrest moved hundreds of metres northwards in Block 2/4. This structural change is responsible for the asymmetrical distribution of hydrocarbons in the eastern part of the field. The present-day structural axis has a sigmoidal shape that is thought to be the result of several minor, left-lateral, strike-slip faults (Dronyck, 1980, company report).

The basal Danian layer, called the 'Ekofisk Tight Zone', consists of argillaceous, laminated chalk which forms the cap-rock over the main reservoir. Thick Tertiary shales and claystones form the seal of the overall chalk mass.

RESERVOIRS

The reservoirs of Albuskjell consist of reworked chalks of the Ekofisk and Tor formations of Danian (Early Paleocene) and Maastrichtian (Late Cretaceous) ages respectively. The petrology and sedimentology of the Central Graben chalks are now well documented (see, for example, Hancock, 1981 and Kennedy, 1980). Debris-flow type sediments with high initial porosity provide the best reservoirs where porosity has been preserved from the effect of 'chemical' compaction (i.e. pressure dissolution and reprecipitation of calcite in adjacent pore-space). Otherwise, pelagic chalks buried to over 3000 m would normally have their porosities reduced to $12\% \pm 6\%$. In the aquifer or palaeo-aquifer, even the favourable debris-flow facies has lost its high porosity as a result of intense 'chemical compaction'. In the hydrocarbon zone, this phenomenon is inhibited. The allochthonous chalk was transported in slumps and slides over short distances and by mass flows over greater distances, measured in tens of kilometres.

As is the case for most chalk reservoirs of the Greater Ekofisk Area, the Ekofisk and Tor formations at Albuskjell are of layer-cake type as a result of the alternation of chalk beds with distinct sedimentary facies. This gives rise to a 'porosity stratification' which is best preserved in the hydrocarbon-bearing zone (see West Ekofisk Field, this volume).

Based on this porosity stratification, the Ekofisk and Tor formations have been subdivided into eight units (Fig. 2). These units are described in the Ekofisk chapter in this volume.

The Ekofisk Formation is composed of alternating sequence of massive homogeneous chalk and laminated argillaceous chalk. The Tor Formation whose upper member is the main reservoir in Albuskjell is composed of very clean, massive chalk that is in marked contrast with the overlying Ekofisk Formation. These formations contain numerous stylolites. Typical sections have been described by Watts *et al.* (1980) for Well 1/6-A10, and by Kennedy (1985) for Well 2/4-F10.

The non-carbonate fraction of the chalk is 1–5% in the Tor Formation and 5–45% in the Ekofisk Formation, except for the classical 'Ekofisk Tight Zone' where it reaches 60%. The latter interval comprises numerous shale laminae and creates a vertical permeability-barrier above the main reservoir.

The Ekofisk and Tor formations differ from those of the neighbouring fields in several aspects related to the sedimentation and structural–diagenetic history:

(a) The 'reworked Maastrichtian' (zone ED) of the lower part of the Ekofisk Formation is poorly developed and has fair to low porosities, in contrast to the Ekofisk and West Ekofisk fields where it is very thick and possesses good porosities (Fig. 3). This is related to the distance from the source-areas of the massive debris-flows and to the sea-bed topography.

(b) There are numerous examples of in-situ slumping, due to synsedimentary halokinetic pulses of the Zechstein salt or regional seismicity.

(c) A slight palaeorelief probably influenced the distribution of the massive debris-flow deposits, as suggested by thickness variations from well to well. This remains true even if the effect of differential porosity preservation on present-day thickness is taken into account ('solid thickness', see West Ekofisk, this volume, p. 00). It is, however, acknowledged that variation in 'solid thickness' may result from erosion, slumping and input of cement through faults.

(d) In Block 1/6, the porosity trends are related to the structural shape, with the highest porosity in the crestal area, as is the case for most chalk reservoirs (Figs 7 and 8). In Block 2/4, however, the high-porosity zone is located on the southern flank, near Well 2/4-F5 (Figs 7 and 8). This probably means that this area was a structural high during most of the diagenetic history of the field, or at least at the time when the hydrocarbons moved into the reservoir. This interpretation of the variation of structural shape through time for the eastern part of the field has been confirmed by structural reconstructions by Dronyck, Connary and Knight (company reports, 1980–85).

(e) The relationship between porosity and permeability is unusual. Generally, in the central area of the 'Greater Ekofisk' chalk fields, net reservoir intervals are defined by a porosity cut-off of 15–20% for the Ekofisk Formation, and 10–15% for the Tor Formation. Higher-porosity cut-offs apply in the flank areas.

In Albuskjell Field, however, chalks of the Ekofisk Formation, with porosity up to at least 25% seem unable to produce hydrocarbons at commercial rates even in the crestal part of the structure. For the Tor Formation, however, the porosity cut-off is similar to those of the other fields.

(f) The fracture system, which is an important feature in the productivity of chalk reservoirs, is less developed at Albuskjell than in the Ekofisk Field. Detailed fracture analyses on Albuskjell cores have been made by Farell (company report) and Watts (1983). Stylolite-associated fractures are the most important for reservoir permeability enhancement. They are interpreted as extension fractures of predominantly radial trend, associated with diapirism. They are near vertical, rarely completely sealed, and form permeability zones adjacent to the stylolites. They are more abundant in the Tor than in the Ekofisk Formation, and they appear to be lithology dependent.

Tectonic fractures, predominantly vertical and presumably associated with tectonic activity in the Central Graben, are also present in both formations. Since they are generally mineralized, they contribute much less to production than is the case for the stylolite-associated fractures.

SOURCE-ROCK

The hydrocarbons contained in the Greater Ekofisk fields have been sourced from the prolific Upper Jurassic Mandal Formation. Almost nothing is known about the Jurassic section below and around Albuskjell as it has not been penetrated by any well drilled on the structure.

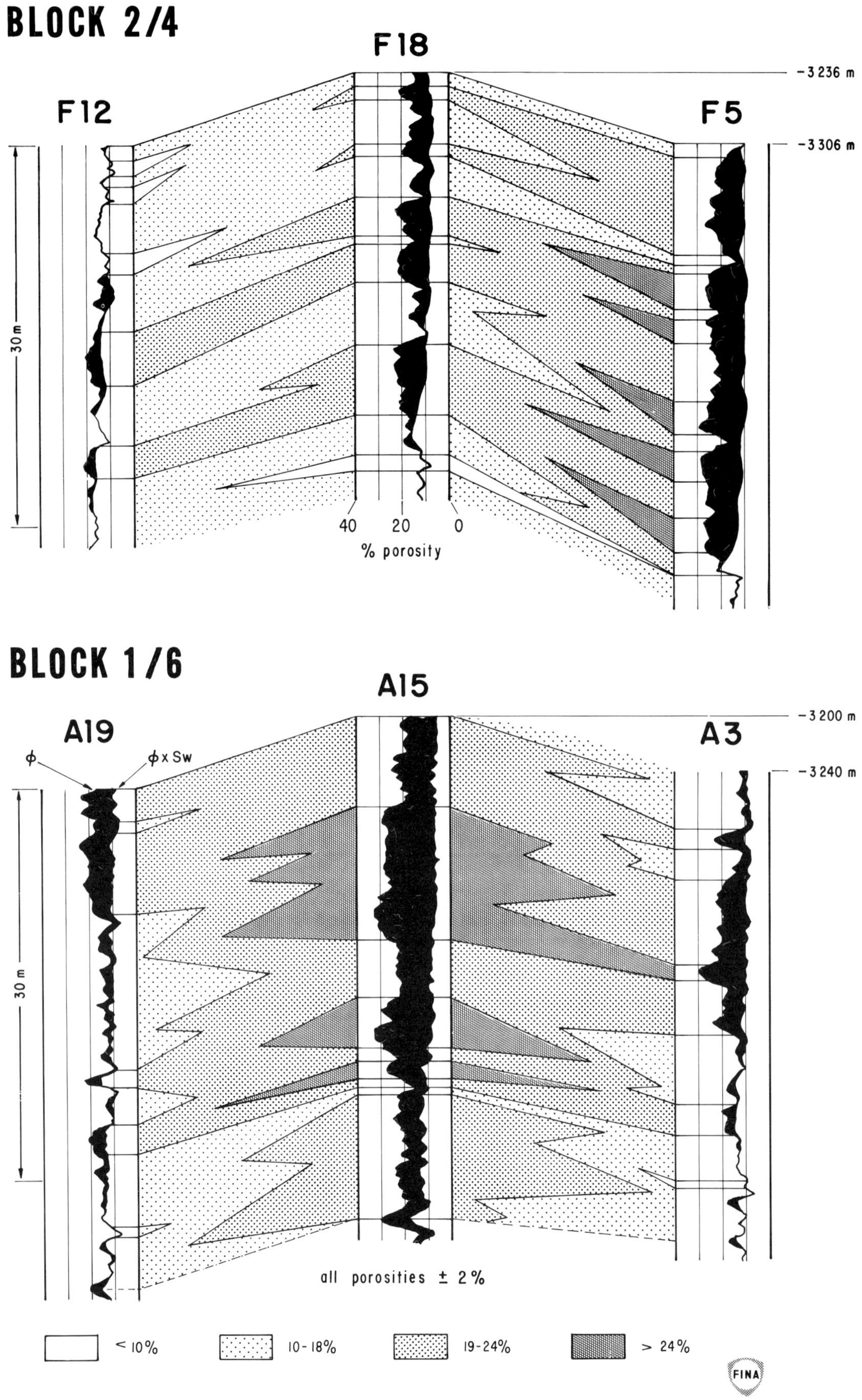

Fig. 7. SE–NW porosity cross-sections through the main reservoir in the eastern and western parts of the field. The top of the interval shown is 10–15 m below the top of the Tor Formation. Note that depth scale and thickness scale are not similar. Location of sections shown on Fig. 8, precise structural relationship between wells given in Fig. 5.

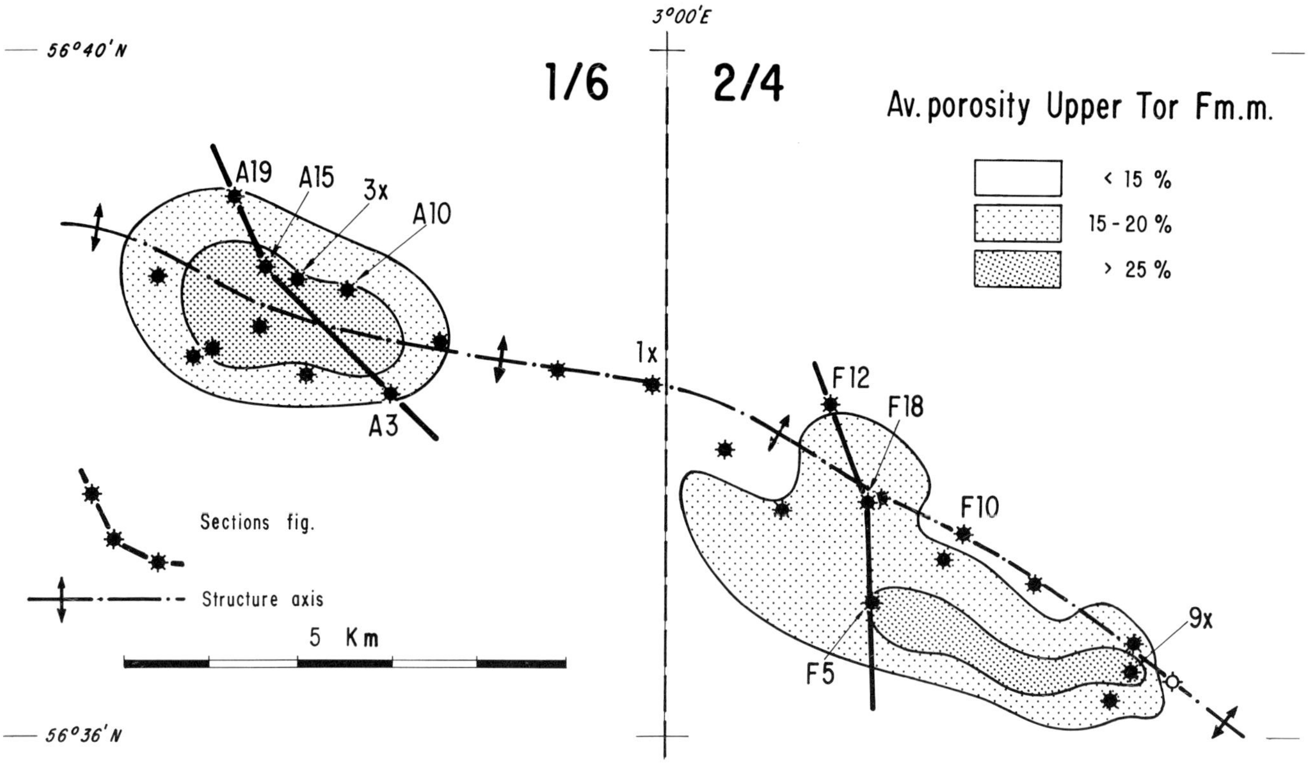

Fig. 8. Average porosity map of the Upper Member of the Tor Formation (TA).

Seismic indications suggest that the Jurassic is absent below the crestal area but that the stratigraphic section may be fairly complete in the flanks, with the top of the Mandal Formation at about 5000 mss. It is considered that hydrocarbons probably migrated into the reservoir during Oligocene to Early Miocene times.

HYDROCARBONS

Albuskjell is a rich gas–condensate field. After the start of production, fluid variation along the length of the structure was confirmed. The gas–oil ratios at the production platforms are different, with 1/6-Alpha initially producing about 1600 Sm^3/Sm^3 and 2/4-Foxtrot 890 Sm^3/Sm^3. The average gravity of the condensate is around 0.79 g/cc.

The distribution of hydrocarbons in the reservoir is complicated by late stage differential structural movements. Prediction of downflank hydrocarbon saturation therefore remains difficult. The 100% S_w line is located at different depths for the western crestal wells, the eastern crestal wells and the south-eastern flank wells.

RESERVES

The estimates of original reserves have been subjected to many revisions. For gas, recovery values as high as 75×10^9 Sm^3 were proposed following the three exploration–appraisal wells. The poor results of several development wells, however, led to a drastic reduction of the early estimates.

The complex field geology and the resulting variable distribution of hydrocarbons are responsible for the large variations in the reserve estimates, despite the 25 wells that have been drilled on the structure. One of the major remaining uncertainties is the recovery factor applicable to the Ekofisk Forma-

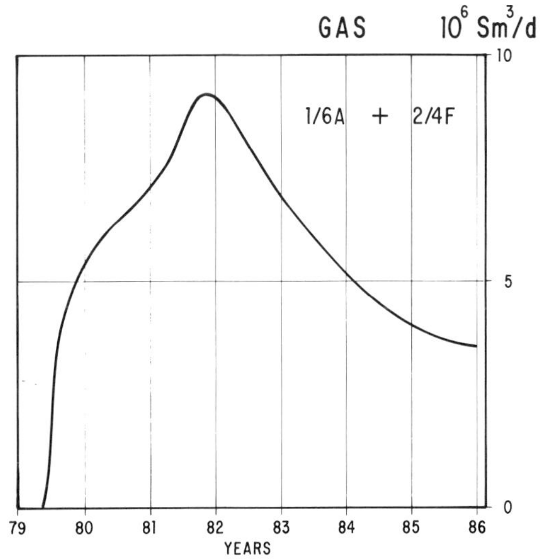

Fig. 9. Production history for gas (simplified).

tion reservoir. As yet, this interval has contributed very little to the production.

Recoverable gas from the Tor Formation reservoir is now estimated at about 18×10^9 Sm^3. By the end of 1986, gas production from the 1/6-Alpha and 2/4-Foxtrot platforms had totalled 6.6×10^9 Sm^3 and 6.9×10^9 Sm^3, respectively (Fig. 9). The production is transported from the 1/6 Alpha to the 2/4-Foxtrot platform (7.5 km) and then to the Ekofisk Centre (12.9 km) by a 18-inch oil line and a 24-inch gas line.

ACKNOWLEDGEMENTS

The author thanks Norske Shell, Phillips Petroleum Company of Norway, Petrofina and Fina Exploration Norway, Norsk Agip, Elf Aquitaine Norge, Norsk Hydro, Total Marine Norsk and the other co-venturers for their permission to publish this paper. The opinions expressed in this article are those of the author and do not necessarily reflect the opinions of the co-venturers.

REFERENCES

Gowers, M.B. and Sæbøe, A. 1985. On the structural evolution of the Central Trough in the Norwegian and Danish sectors of the North Sea, *Marine and Petroleum Geology* **2,** (4), 298–318.

Hancock, J.M. 1981. Cretaceous, In *Introduction to the Petroleum Geology of the North Sea*, chapter H, Joint Association for Petroleum Exploration Courses, London.

Kennedy, W.J. 1980. Aspect of chalk sedimentation in the southern Norwegian offshore, In *The Sedimentation of the North Sea Reservoir Rocks*, Norsk Petroleumsforening, Article IX.

Kennedy, W.J. 1985. Sedimentology of the Late Cretaceous and Early Paleocene Chalk Group, North Sea Central Graben, *Chalk Research Program*, proprietary report.

Watts, L. 1983. Microfractures in Chalks of the Albuskjell Field, Norwegian Sector, North Sea: possible origin and distribution, *American Association Petroleum Geologists Bulletin* **67,** 201–234.

Watts, L., Lapre, J.F., Schijndel-Goester, F.S. and Ford, A. 1980. Upper Cretaceous and Lower Tertiary Chalks of the Albuskjell Area, North Sea: deposition in a slope and a base-of-slope environment, *Geology* **8,** 217–221.

3

Cod

Michel D'Heur

Petrofina, c/o Fina Exploration Norway A/S, Skogstøstraen 37, N-4001 Stavanger, Norway

The gas–condensate Cod Field lies in Block 7/11, 285 km south-west of Stavanger. It was discovered in 1968 and came into production in 1977. The reservoirs are sandstones of Late Paleocene age culminating at a depth of 2870 mss. The sandstones were deposited in several episodes of submarine channel and fan–lobe progradation. The maximum net pay thickness exceeds 70 m. The hydrocarbon distribution in the field is complicated by reservoir discontinuity. The source rock for hydrocarbons is the prolific Upper Jurassic shales. The trap is of combined structural/stratigraphic type and Tertiary claystones provide the seal. Total recovery should be approximately 7.5×10^9 Sm^3 gas and 2.7×10^6 Sm^3 condensate.

INTRODUCTION

The gas–condensate Cod Field lies in the southern area of the Norwegian Sector, in Block 7/11, 1 km east of the median line between the United Kingdom and Norway. The field is also located 285 km south-west of Stavanger and 75 km north-north-west of Ekofisk (Fig. 1).

The water depth at Cod is 78 m. All prospects identified by the Phillips Group in Norway received a fish name starting with the letter that designated a specific area. Block 7/11 belonged to 'Area C' and so the name Cod was selected. The reservoir is sandstone of Late Paleocene age at a depth of 2870 mss. It is expected to produce about 7.5×10^9 Sm^3 of gas and 2.7×10^6 Sm^3 of condensate.

HISTORY

Blocks 7/11 and 7/8, designated as the 'C area', were granted as Production Licence PL018 to Phillips, Fina and Agip in 1965. Phillips and Agip crossed interest with the Petronord Group in January 1967 and the present participation interests in PL018 are:

Phillips Petroleum Company of Norway Operator,	36.96%
Norske Fina A/S	30%
Norsk Agip A/S	13.04%
Elf Aquitaine Norge A/S	8.1%
Norsk Hydro A/S	6.7%
Total Marine Norsk A/S	4.05%
Eurafrep Norge A/S	0.45%
Coparex Norge A/S	0.4%
Cofranord A/S	0.3%

Three 'en-échelon' structures were identified in Blocks 7/8 and 7/11 in the early stages of seismic reconnaissance (Fig. 2). In 1967, the southern feature was divided into two distinct prospects, namely Cod and North-east Cod. Well 7/11-1 was drilled on Cod in 1968. It was the second well drilled by the Group in the Central North Sea: only very few wells had been drilled by then to the north of the Mid-North Sea High. It is therefore not surprising that the objectives of the well were not very

Geology of the Norwegian Oil and Gas Fields © A. M. Spencer *et al.*, eds (Graham & Trotman, 1987) pp. 51–62.

Cod Field data summary

Trap	
Type	combined structural and stratigraphic
Depth to crest	2870 mss
Lowest closing contour	3070 mss
Gas–water contact	locally 3070 mss or deeper
Gas column	locally more than 170 m
Pay zone	
Formation	Forties Fm (or equivalent)
Age	Upper Paleocene
Gross thickness	up to 100 m
Net thickness	up to 70 m
Porosity (average; range)	17%; 15–24%
Hydrocarbon saturation (average)	57%
Permeability	15–100 mD
Hydrocarbons	
Condensate gravity	0.76 g/cm^3
Gas gravity (relative to air)	0.69
Gas/oil ratio (initial)	2353 Sm^3/Sm^3
Formation water	
—	

Reservoir conditions	
Temperature	132 °C
Pressure (initial)	366 bar
Field size	
Net area	13 km^2
Hydrocarbon pore volume	46×10^6 Sm^3
Recoverable condensate	2×10^6 Sm^3
Recoverable gas	8.3×10^9 Sm^3
Production	
Start-up	December 1977
Development scheme	1 production platform 9 slots 6 producing wells
Production rate (end 1986)	1.5×10^6 Sm^3/day
Cumulative production (12/1986)	6×10^9 Sm^3 gas 2.2×10^6 Sm^3 condensate

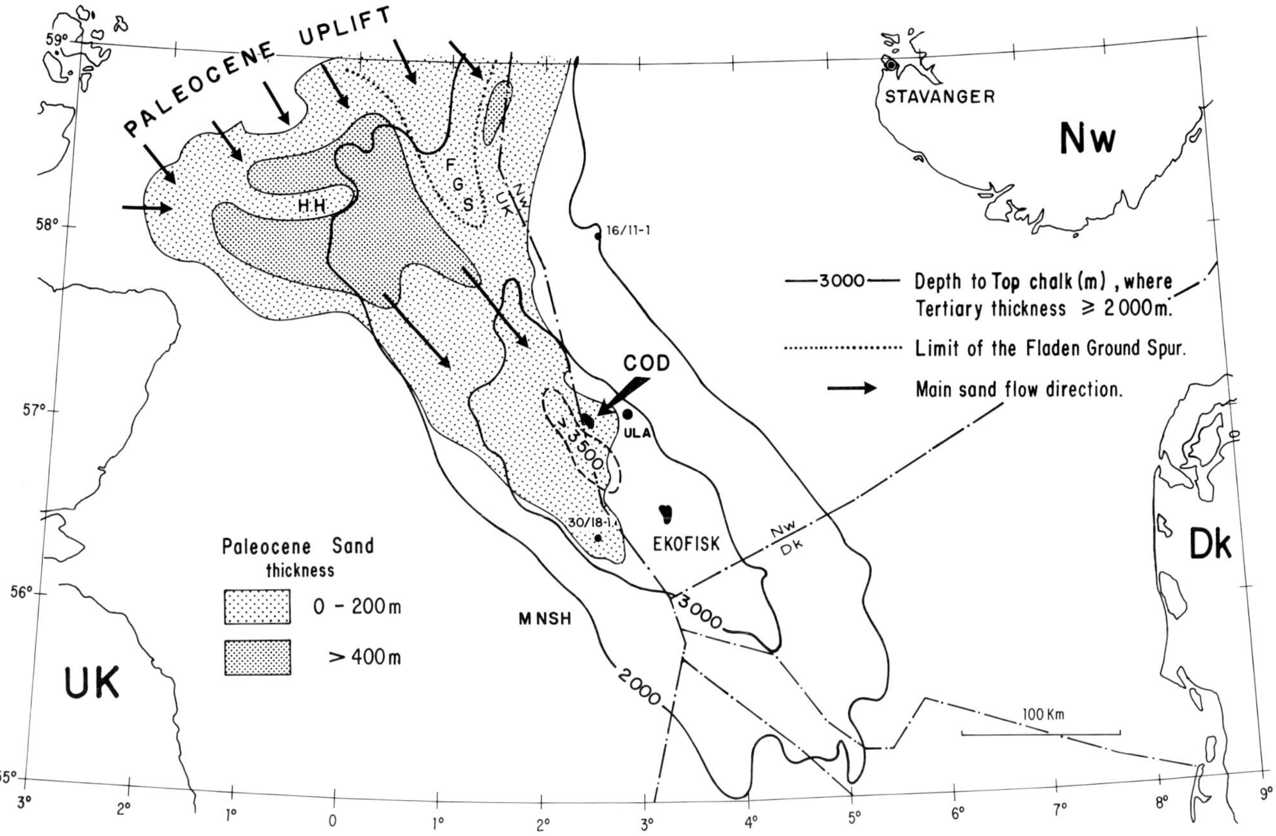

Fig. 1. Location map showing distribution of the sandstone of the Montrose Group and depth to the chalk in the Central Graben area. MNSH, Mid-North Sea High; HH, Halibut Horst; FGS, Fladen Ground Spur.

precise. The two nearest wells were UK-30/18-1 80 km to the south, where oil shows had been reported from thin Paleocene sandstones, and N-16/11-1 110 km to the north, where the Phillips Group had identified porosity zones in Cretaceous limestones and in Jurassic sandstones (Fig. 1). The source for the sand found in Well 30/18-1 was thought to be the Mid-North Sea High, and the sand was believed to pinch out toward the north between Wells UK-30/18-1 and N-16/11-1. Additional objectives of Well 7/11-1 were sandstones of Lower Cretaceous and Triassic 'Bunter' ages, known to the south.

Well 7/11-1 was located near the crest of the Cod prospect. It encountered a gas-bearing sequence of interbedded sandstone and shale of Paleocene age between 2845 mkb and 3025 mkb. The upper half of this interval, called the 'Main Cod Sand', contained good-quality reservoirs totalling 52 m of net pay. The best zone was perforated over 23 m and flowed 1.3×10^6 Sm^3/day gas and some condensate. Below the Paleocene sandstone, the well penetrated 650 m of chalk of Lower Paleocene and Upper Cretaceous age, a few metres of black Jurassic shale and white Permian anhydrite and 250 m of Zechstein salt. The well bottomed at 3885 mss without reaching the Rotliegend.

The first appraisal well, 7/11-2, was drilled in mid-1968 1.8 km east of the discovery well to appraise the eastern flank of the structure. This well reached the Paleocene 120 m deeper than Well 7/11-1. The reservoir was gas bearing down to its base, had a similar net pay thickness and gave flow rates similar to the discovery well. The well bottomed in the chalk at 3400 mss. Well 7/11-3 was drilled in late 1968 at a low structural position 3 km South-west of Well 7/11-1. It found the reservoir at 3060 mss. Most of the sandstone was of poor reservoir quality and an oil–water contact was encountered 3 m below the top of the reservoir interval. When tested, the uppermost part of the 'Main Cod Sand' flowed at rates of 3400 Sm^3/day gas, 14 Sm^3/day oil and 53 Sm^3/day water. This test suggested the presence of an oil leg below the gas accumulation.

Well 7/11-4 was drilled in 1969 on the North-east Cod structure (C, Fig. 2); the Paleocene comprised only some thin sandstone beds totalling less than 9 m and containing no hydrocarbons. Since no well in the area had proven hydrocarbons in the Mesozoic, Well

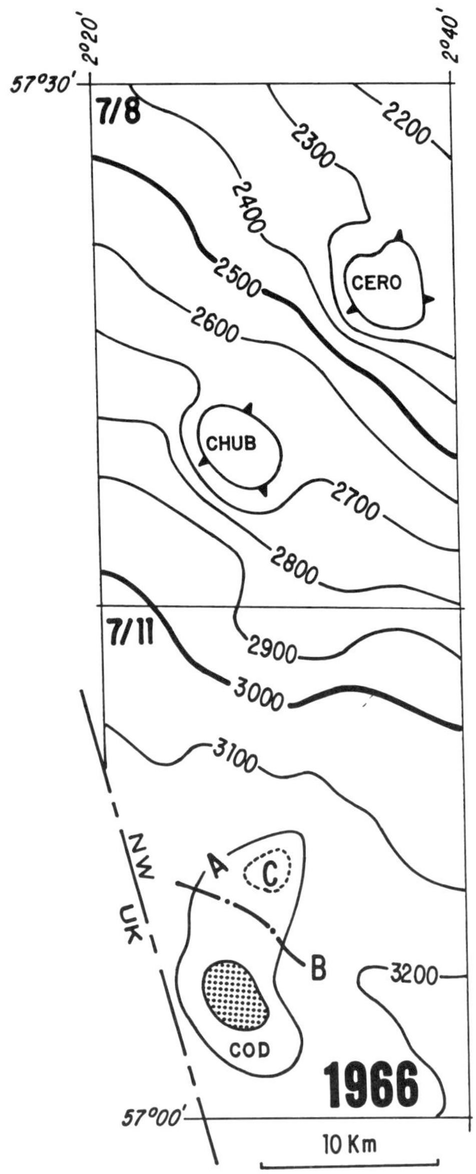

Fig. 2. Depth map of the base of the Tertiary before drilling. Three en-echelon structures were identified in Blocks 7/8 and 7/11, together called 'C Area'. The southern structure, delimited by contour A was later subdivided into two distinct features, namely Cod and North-east Cod (C). Contours in metres.

7/11-4 was stopped in the chalk. Based on the three exploration/appraisal wells, the amount of gas-in-place was estimated at about 28×10^9 Sm^3. A nine-slot drilling/production platform was installed in 1975 on the basis of the structural interpretation shown in Fig. 3 (1973).

Development drilling commenced in the summer of 1976 with Well A7. Despite being located only 1 km north-north-east of the discovery well, A7 encountered less than 15 m of reservoir in the 'Main Cod Sand' member and no hydrocarbons were present. This negative result was thought to result from a drastic shaling-out of the reservoir interval northwards. However, Well A2 drilled midway between 7/11-1 and 7/11-3 found thick water-bearing sandstone and demonstrated that Well 7/11-3 did not belong to the Cod accumulation. Well A5, located very close to the discovery well, found a thinner reservoir interval. To honour the results of these three development wells, a N–S sealing fault was arbitrarily placed as shown in Fig. 3 (1977). This interpretation led to a drastic reduction in the reserve estimate.

Five other wells drilled in 1976 and 1977 gave good results. A ninth well, A6 (also referred to as 7/11-7) was drilled in 1983 to evaluate the Mesozoic section on the eastern flank of the structure, following discoveries in Upper Jurassic sandstones at Ula and North-east Cod. Well A6 gave disappointing results.

The Cod Field was linked to the Ekofisk Centre and production began in December 1977 with a peak of 3.3×10^6 Sm^3/day of gas reached in 1980. The production rate early in 1986 was about 1.7×10^6 Sm^3/day gas and 270 Sm^3/day condensate. The cumulative production to 1986 was approximately 6.0×10^9 Sm^3 gas and 2.2×10^6 Sm^3 condensate.

GEOLOGICAL SETTING

The Cod Field lies near the centre of the Tertiary Central Trough of the North Sea. South-west of Cod, the thickness of the Tertiary exceeds 3500 m (Fig. 1). Permian Zechstein salt is the oldest rock penetrated below the Cod Field (Fig. 4). During the Zechstein, the area belonged to the 'Northern Permian Basin' which continued to subside during Triassic time. Three hundred metres of varicoloured shallow-marine claystone, shale, fluvial siltstone and sandstone have been encountered in Well 7/11-7 (A6), but much thicker deposits of the Smith Bank Formation probably exist downflank. Part of the Triassic and all Lower and Middle Jurassic strata have been eroded during several phases of Kimmerian movements. Marine sandstones and shales 40 m thick and of Kimmeridgian age rest on the Triassic rocks. They are considered to be equivalent to the upper unit of the Ula Formation (Home, this volume; Vollset and Doré, 1984).

This sandy interval is overlain by 120 m of dark to black shale of the Mandal Formation, which is the source-rock for the hydrocarbons of the Cod Field. During Late Jurassic and

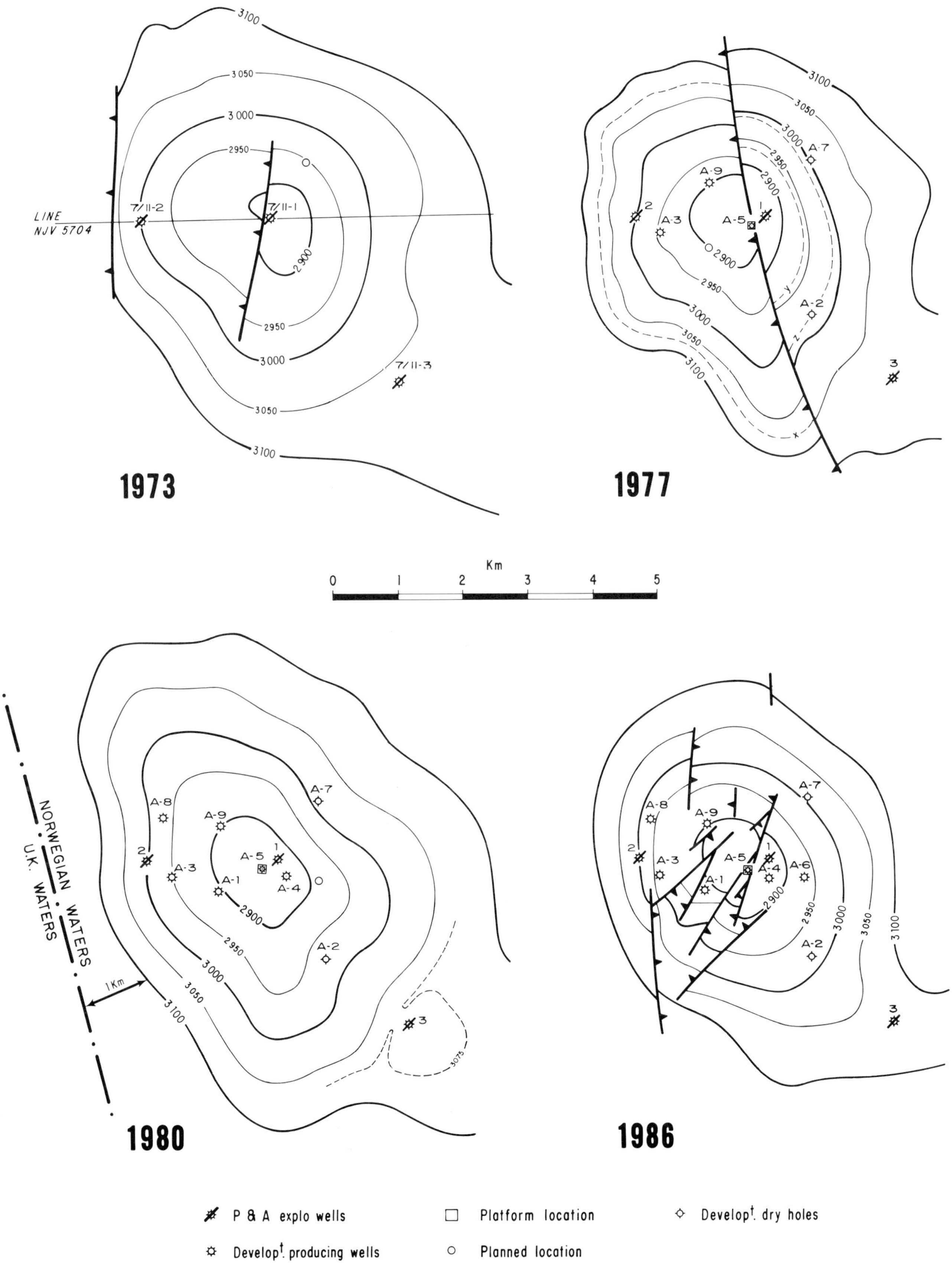

Fig. 3. Successive structural interpretations of Cod after discovery and appraisal. In 1977, (x) is a 'gas-down-to' in the western half of the structure, (y) is a 'gas-down-to' and (z) is a 'water-up-to' in the eastern part of the dome. Contours in metres subsea.

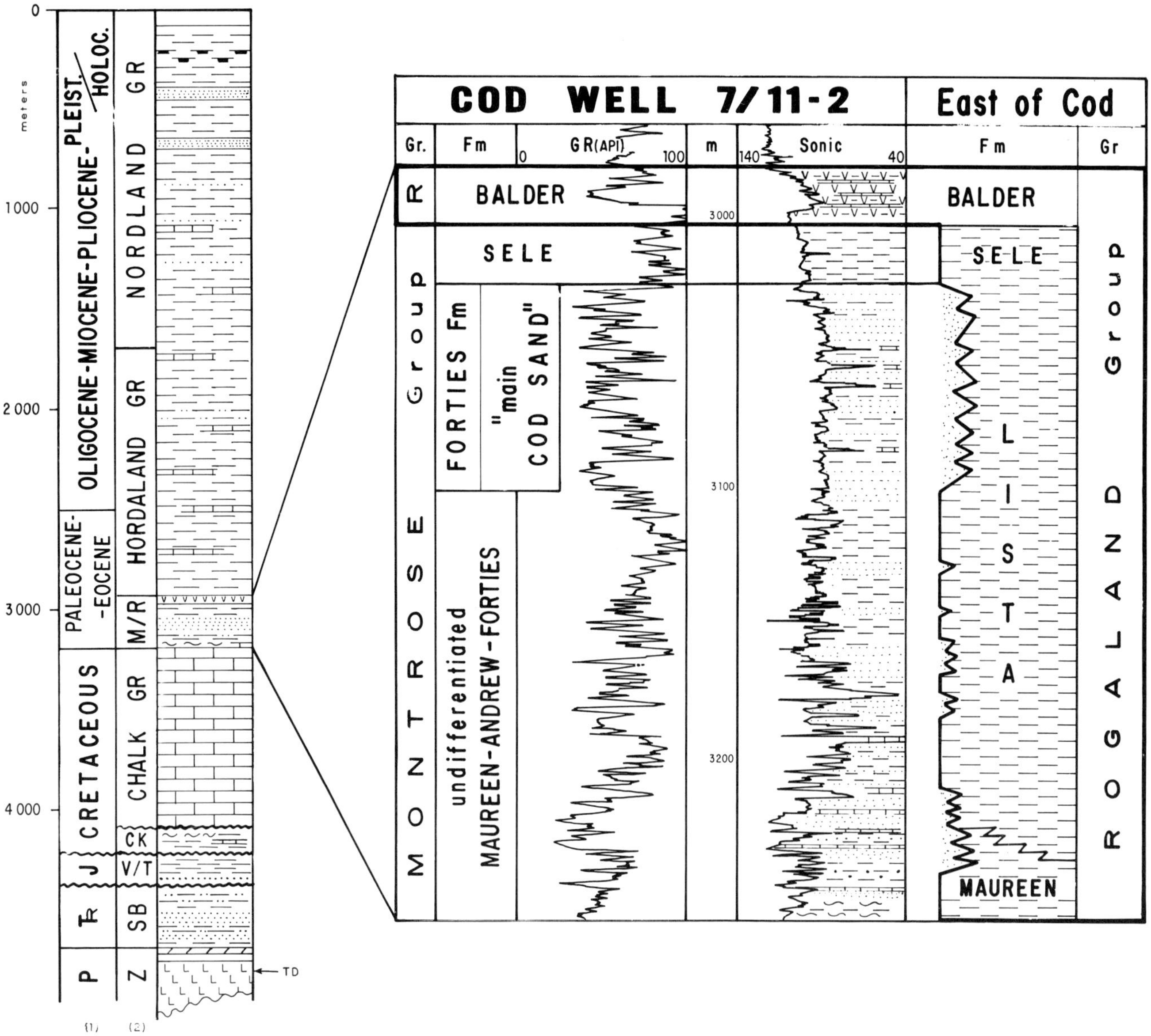

Fig. 4. Stratigraphy of Cod Field and Area. P, Permian; Tr, Triassic; J, Jurassic; Z, Zechstein; SB, Smith Bank Fm; V/T, transition Vestland to Tyne Gr; CK, Cromer Knoll Gr; M/R, transition Montrose to Rogaland Gr.

Early Cretaceous times, Cod was located between the Vestland Arch to the north-east and the Central Graben that had developed to the west. In Well 7/11-7 (A6), the Lower Cretaceous consists of 85 m of calcareous shale and limestone of the Cromer Knoll Group. These, as well as the Chalk Group (920 m thick) and the Tertiary sequence, conform to the general stratigraphy described by Deegan and Scull (1977). The Montrose Group of Middle to Late Paleocene age is of greater interest since it contains the reservoirs of the Cod Field. During Paleocene times, uplift, tilting and erosion of the East Shetland Platform occurred and enormous quantities of sand were shed into the Central Trough and Viking Graben (Fig. 1). Cod being located some 200 km south-south-east of the source area received only a reduced amount of Paleocene sand. The Cod Field reservoir is overlain by some 3000 m of younger Tertiary, consisting mainly of clay, claystone and shale.

GEOPHYSICS

The seismic lines shot in the mid-1960s allowed identification of three domal features in Area C (Fig. 2). In order to select the best location for the first well in Block 7/11, an additional SW–NE line was shot across the southern feature which appeared to be the largest. This line proved that the large NE–SW-trending anticline in fact consisted of two distinct features separated by a saddle (line B, Fig. 2). Seismic lines totalling 52 km were shot in 1969. Until development drilling commenced in 1976, Cod was covered by a 1 km × 1 km seismic grid orientated N–S and E–W.

The problematic results of the first development wells led to the further acquisition of 14 seismic lines (146 km) orientated SW–NE and NW–SE, at a spacing of 700 m. These new data did little to improve the structural definition, but were utilized in an attempt to determine sand-body continuity. They were also used to build the 'unfaulted' structural interpretation of 1980 (Fig. 3).

In 1983, 320 km of seismic were shot jointly with Norsk Hydro following their oil discovery in Upper Jurassic sandstones at North-east Cod. A seismic line through the Cod Field is shown on Fig. 5.

TRAP

Cod is a domal structure located over a salt plug. It is nearly circular in shape, with a diameter of some 6 km at Paleocene level. The

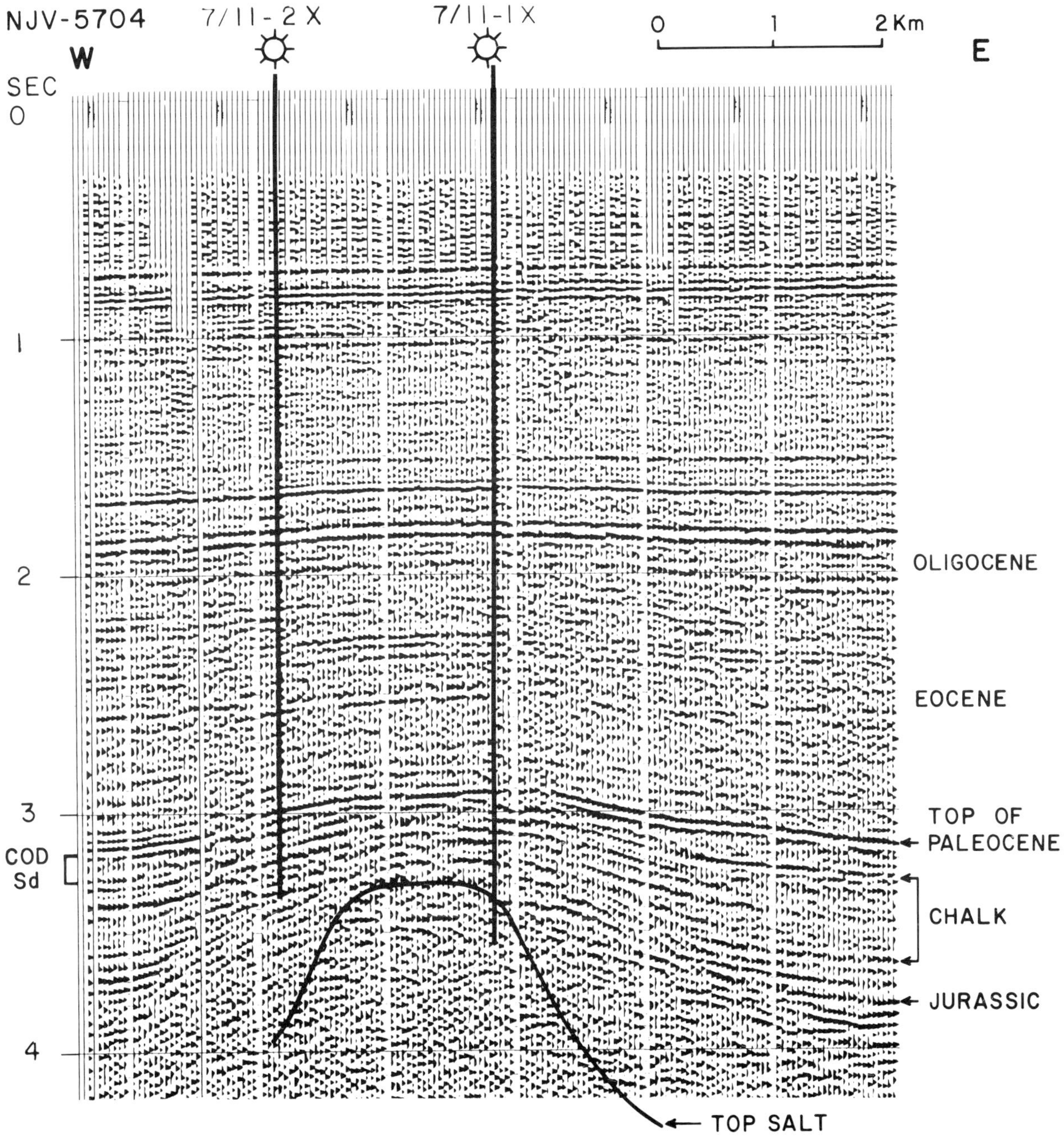

Fig. 5. Seismic line shot W–E (location on Fig. 3).

reservoir culminates at 2870 mss. The vertical closure exceeds 200 m. The eastern half of the structure contains only small quantities of hydrocarbons and the productive area is limited to approximately 13 km^2. The trap is a combined structural and stratigraphic closure of channel sandstone bodies cut by several faults.

A slight topographic high probably existed during the deposition of the reservoir sand, as suggested by thickness variations of the Paleocene interval. Halokinesis of the Zech-

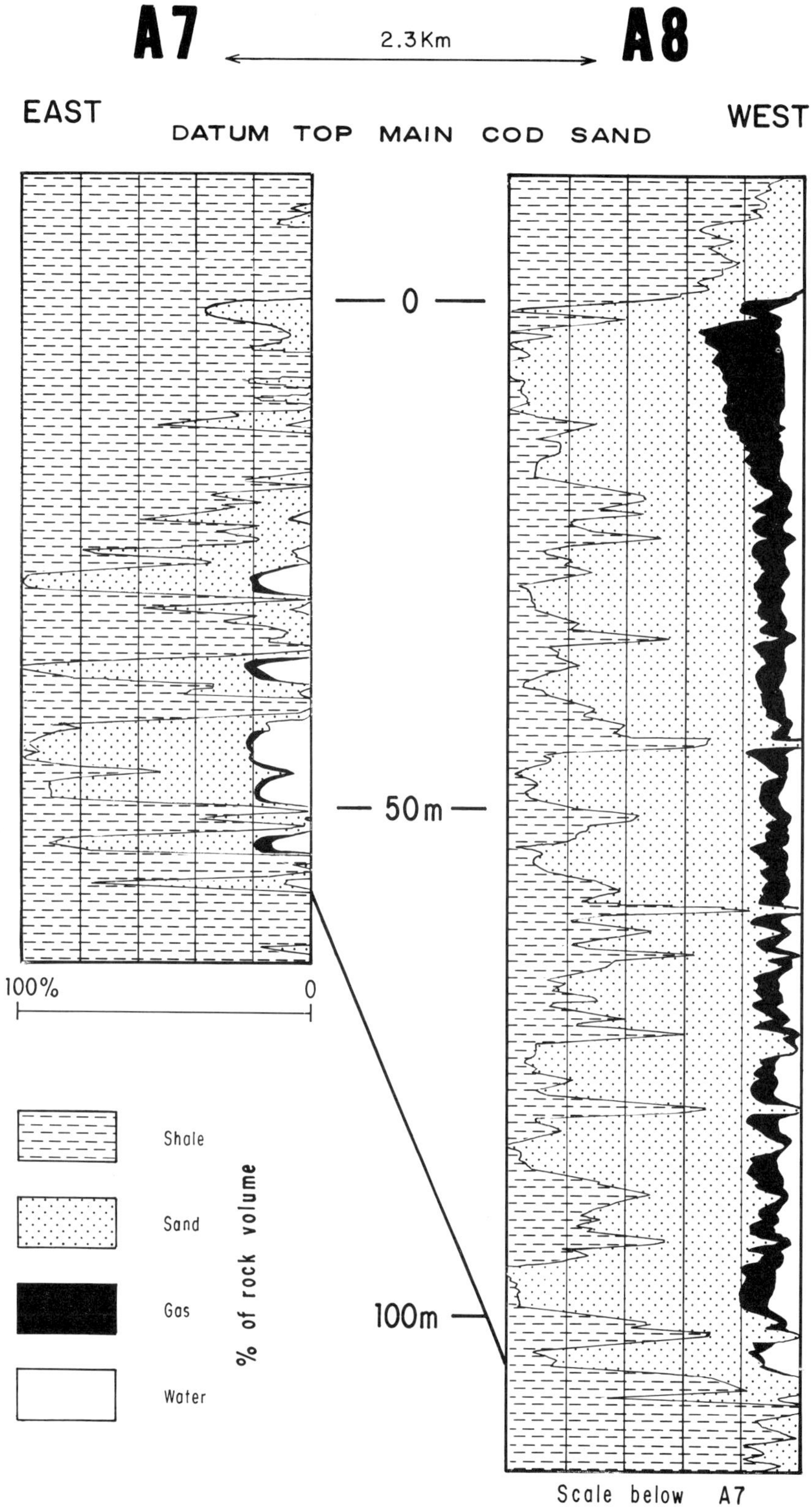

Fig. 6. Interpreted logs of development wells A7 and A8 showing thickness variation of the 'Main Cod Sand'.

stein salt persisted until mid-Miocene time and is responsible for the shape of the trap. The seal consists of Tertiary shale and claystone. Well 7/11-3 is probably separated from the main structure by a slight saddle, as shown in Fig. 3 (1980).

RESERVOIR

The reservoir of the Cod Field consists of sandstones of Late Paleocene age, belonging to the Montrose Group (Fig. 6). However, since Cod is located near the transition area between the coarse clastic Montrose Group and its shaly equivalent (Rogaland Group, Lista Formation), a precise equivalence of the Cod sandstone with previously defined formations is not yet certain. According to Deegan and Scull (1977, Fig. 33), the Andrew Formation extends more to the south than the Forties Formation. For that reason, it is proposed that the Cod reservoir (the 'Main Cod Sand') belongs to the Forties Formation, whereas the lower sand–shale sequence is classified as undifferentiated Maureen–Andrew–Forties Formation. Within the field itself correlation of the 'Main Cod Sand' is uncertain; problems of the reservoir continuity were identified during the early phase of development drilling. Lenticular-shaped sand bodies, which thin and shale out towards the north-east, as well as an increase in calcite cement south-eastwards and the presence of a sealing fault, are all features identified early in the appraisal of the field.

Detailed sedimentological studies were undertaken on core material and log response. Kessler *et al.* (1980) suggested that the sandstone interval was deposited in at least six major episodes of submarine channel and fan–lobe progradation. 'The resulting depositional sequences, each consists of a discrete package of thickening upward sandstone beds. The sequences range in thickness from 15 to 40 m. Individual sandstone beds within a sequence range in thickness from a few centimetres to 7 m. Thickening upward progradational sequences often immediately overlie each other, suggesting catastrophic abandonment of a channel and fan lobe system and almost immediate re-occupation and progradation by another similar system just above it. Other sequences are separated by an interval of thinning upward sandstone beds, indicative of gradual channel and fan lobe system abandonment' (Kessler *et al.*, 1980).

Kessler *et al.* (1980) described thin massive beds, graded beds with some associated planar bedding and ripples, and occasionally zones 0.5–2.0 m thick of disc structures (concave-upwards, water-release features). These sedimentary structures are consistent with various types of sediment gravity-flows, especially grain-flows.

As an example, the facies identified by Kessler *et al.* (1980) for Well 7/11-3X (Fig. 7) are (from top to base):

1. gradual fan–lobe abandonment (A);
2. fan-lobe progradation (P);
3. fan-lobe or channel abandonment (A);
4. fan-lobe or channel progradation (P);

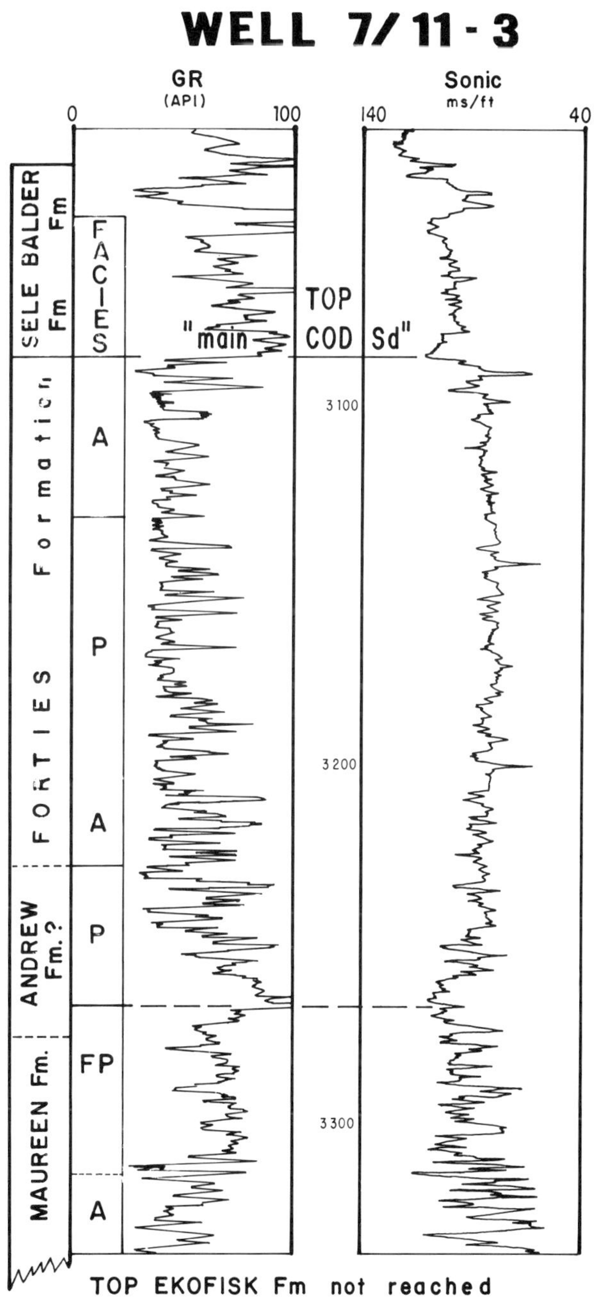

Fig. 7. Sandstone facies after Kessler *et al.* (1980), P, progradation; A, abandonment; FP, outer fan plain.

5. outer fan–plain, thin-bedded sandstone (FP);
6. fan–lobe or channel abandonment (A).

This sequence, observed in the south-easternmost well of the field, suggests a more distal depositional setting than those observed in wells to the north-west. Based on such a description, on seismic observation and on reservoir pressure data, Kessler *et al.* (1980) proposed a complex sand–lobe distribution in which no faults are needed to justify reservoir discontinuity (Fig. 8).

A recent Paleocene re-mapping based on new seismic has, however, shown evidence of several faults (Fig. 3, 1986). In addition, new pressure data suggest communication of sand-bodies which were considered not to be connected in the previous model. The porosity of the 'Main Cod Sand' varies generally between 15 and 24% and the permeability between 15 and 100 mD. A porosity/saturation log of the reservoir interval in Wells A7 and A8 is given in Fig. 6.

SOURCE-ROCK

Upper Jurassic shale of the Mandal Formation, equivalent to the Kimmeridge Clay Formation, is the source-rock for the hydrocarbons of the Cod Field. Brownish–black and grey–black shales belonging to the Mandal Formation have been penetrated between 4400 and 4520 m in Well 7/11-7 (A6). These source-rocks have a high potential for generation of wet gas, condensate and light oil. The upper 25 m have an average total organic carbon (TOC) of 6.5%. The kerogen content is mainly of Type III and has passed the peak of oil generation. This same interval is at a depth greater than 5000 m to the west of Cod.

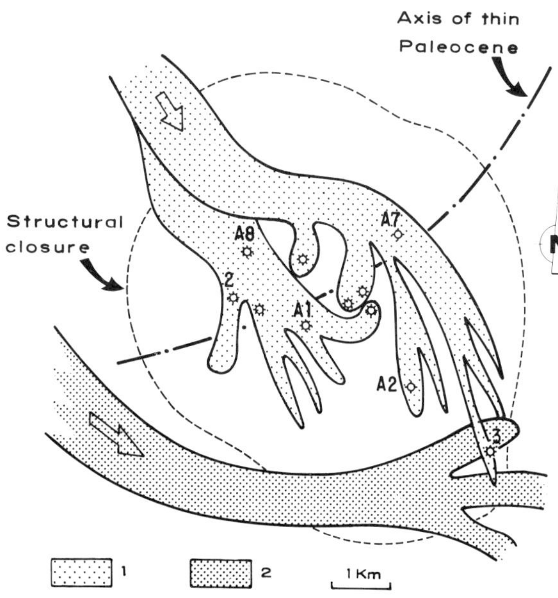

Fig. 8. Hypothetical lobe distribution, modified after Kessler *et al.* (1980). 1, main Cod system; 2, earlier abandoned system.

HYDROCARBONS

The gas–water contacts in the various Cod wells occur over a depth range of 170 m. As early as 1969, it was suggested that the turbidite nature of the sandstone was responsible for the five or more distinct accumulations identified by reservoir pressure data. In several wells, the 'Main Cod Sand' flowed at rates of 1.1–1.3 × 10^6 Sm^3/day gas and 240–310 Sm^3/day condensate (0.76 g/c^3). The presence of oil with a gravity of 0.80 g/cc was established in 1968 in Well 7/11-3. In Well A5, oil with a gravity of 0.86 g/cc flowed from a ten-metre interval at the base of the 'Main Cod Sand'. This oil zone extends down to 2920 mss, whereas the gas zone in Well 7/11-2 extends up to 3080 mss.

When the reservoir was found water bearing at 2975 mss in Well A2, the difference in gas–water elevation was accounted for by a major normal fault (Fig. 3, 1977). It also became obvious that Well 7/11-3 belonged to another small accumulation. Hydrodynamic studies in 1978 suggested that at least nine distinct accumulations exist at Cod. When the new seismic data led to an 'unfaulted version' of the reservoir (Fig. 3, 1980), Kessler *et al.* (1980) proposed the complex sand–lobe distribution to explain reservoir discontinuity (Fig. 8). It is probable that their work now needs to be updated, since recent pressure analyses suggest reservoir communication that was not recognized in 1978–80. In addition, new seismic data show evidence of faults (Fig. 3, 1986). A suite of 'net pay' maps produced between 1975 and 1979 (Fig. 9) shows the evolution of ideas on hydrocarbon distribution.

RESERVES

Original recoverable reserves for the Cod Field are now estimated at approximately 7.5 × 10^9 Sm^3 of gas and 2.7 × 10^6 Sm^3 of condensate. Production up to late 1986 was 6.1 × 10^9 Sm^3 gas and 2.2 × 10^6 Sm^3 condensate, which is 81% of the expected total recovery.

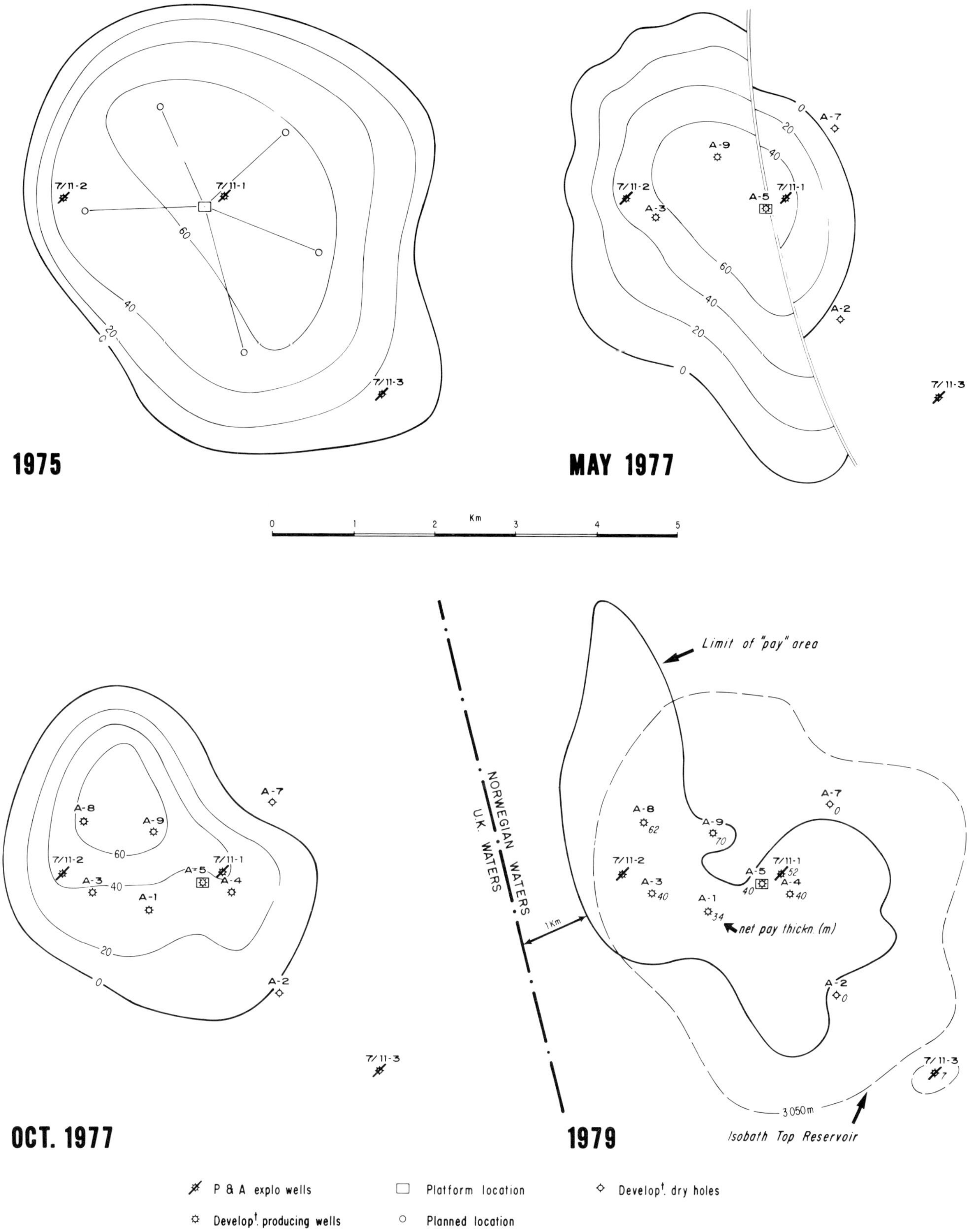

Fig. 9. Successive interpretations of net pay thickness between 1975 and 1979.

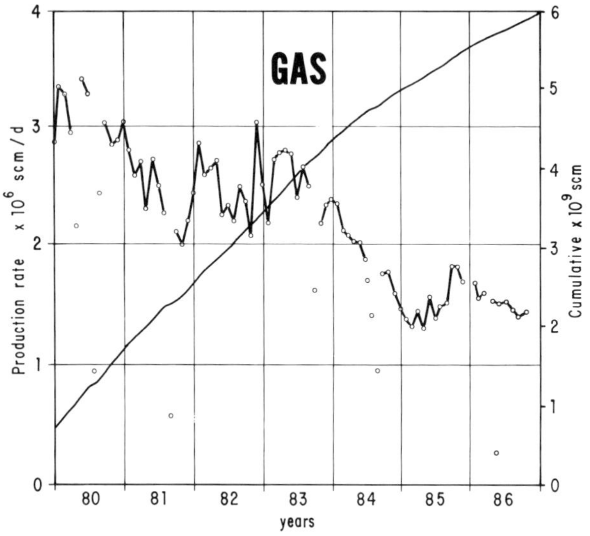

Fig. 10. Production history and cumulative production of gas since 1980.

The maximum flow-rate was obtained by Well A9 in 1978 at 1.2×10^6 Sm^3/day of gas, and a peak of production for the field was reached in 1980 with 3.3×10^6 Sm^3/day of gas.

Cod is linked to the Ekofisk Centre by a 16-inch oil and gas line, 75 km long. The production history of the gas is shown in Fig. 10.

ACKNOWLEDGMENTS

The author thanks Petrofina, Fina Exploration Norway, Phillips, Agip, Elf, Norsk Hydro and Total for their permission to publish this article. Although it contains some interpretations of geologists and geophysicists of the Phillips Stavanger office, mainly Keya Collins and Stephen Connary, the opinions expressed in this article do not necessarily represent those of any of the above-mentioned Companies. Thanks also to Dominique Moens, Joao de Lucena and Jean-Marie Questiaux for their help.

REFERENCES

Deegan, C.E. and Scull, B.J. 1977. A standard lithostratigraphic nomenclature for the Central and Northern North Sea, *Report Institute Geological Sciences UK* 77/25.

Kessler, L.G., Zang, R.D., Englehorn, J.A. and Eger, J.D. 1980. Stratigraphy and sedimentology of a Palaeocene submarine fan complex, Cod Field, Norwegian North Sea, In *Sedimentation of the North Sea Reservoir Rocks*, Norsk Petroleumsforening, Article VIII.

Vollset, J. and Doré, A.G. 1984. A Revised Triassic and Jurassic lithostratigraphic nomenclature for the Norwegian North Sea, *Norwegian Petroleum Directorate Bulletin 3.*

4

Edda

Michel D'Heur* and Frédérique Michaud†

* *Petrofina, c/o Fina Exploration Norway, Skogstøstraen 37, N-4001, Stavanger, Norway*
† *Phillips Petroleum Company Norway, Present address, Elf Italia, 8 Largo L. Mossa, I-00165 Roma, Italy*

The Edda Field, which was discovered in 1972, is located in the Central Graben of the North Sea in the southernmost part of the Norwegian sector. The reservoirs occur in chalk of the Ekofisk and Tor formations which are respectively of Danian (Early Paleocene) and Maastrichtian (Late Cretaceous) ages. The reservoir rock comprises reworked chalk whose main source was a coccolith ooze previously deposited on the graben margins. The best reservoirs occur when these allochthonous chalks were deposited as debris flows. Low-porosity pelagic chalks are also found mixed or interbedded within these reworked units.

The field is located on the Lindesnes Ridge, a half-horst type palaeorelief which strongly influenced the distribution of the various chalk facies. The existence of this tectonic feature, and its influence on the reservoir thickness and porosity in the Edda area, was not known prior to the field's development. This led to the drilling of several dry holes and to drastic reductions in the reserve estimates. Seven producing wells were drilled in 1979, and a peak production of 5200 Sm^3 oil/day was achieved in 1980. The field is expected to produce a total of 4×10^6 Sm^3 of oil and 2.2×10^9 Sm^3 of gas.

INTRODUCTION

The Edda Field lies in the south of the Norwegian sector, in Block 2/7, about 320 km from shore and 10 km south-west of the giant Ekofisk Field (Fig. 1). The water depth is about 75 m. The field is named after a fish. Blocks 2/4 and 2/7 comprised the Phillips Group 'E' area and all prospects were given names starting with 'E'. Edda is the smallest oil field yet developed in the North Sea. However, when it was decided to develop the field, the reserves were thought to be about twice the present-day estimate of 4×10^6 Sm^3 oil (original recoverable).

The present participating interests in the Field are:

Phillips Petroleum Company Norway	36.96%
Petrofina (Fina Exploration Norway)	30%
Norsk Agip	13.04%
Elf Aquitaine Norge	8.1%
Norsk Hydro	6.7%
Total Marine Norsk	4%
Others	1.2%

HISTORY

Production Licence PL018 which includes Block 2/7 was granted to Phillips, Petrofina and Agip in 1965. These three companies were joined by the so-called Petronord Group, comprising the other companies listed above. At that time, the area was covered by a wide seismic grid and the only prospect clearly identified in Block 2/7 was Eldfisk.

Edda, which has only a slight structural

Geology of the Norwegian Oil and Gas Fields © A. M. Spencer *et al.*, eds (Graham & Trotman, 1987) pp. 63–72.

Edda Field data summary

Trap	
Type	dome structure over salt swell
Depth to crest	3063 mss at top Ekofisk Formation
Lowest closing contour	3231 mss
Oil–water contact	variable, dependent upon reservoir characteristics and structural position
Oil column (average)	128 m
Pay zone	
Formation	Ekofisk and Tor Formation chalks
Age	Danian and Late Cretaceous
Gross thickness (range)	170 to 253 m
Net/gross ratio	
Porosity (average; range)	25%; 10–27%
Hydrocarbon saturation (average)	64%
Permeability	With effective permeability up to 24 mD (fractures)
Hydrocarbons	
Oil gravity (initial)	0.83–0.84 g/cm^3
Oil type	undersaturated volatile oil
Bubble point	348 bar
Gas/oil ratio (initial)	307 m^3/m^3
Formation volume factor	1.89
Formation water	
Salinity (average)	52000 ppm NaCl equivalent
Resistivity (average)	0.042 Ohm m at 118 °C
Reservoir conditions	
Temperature	132 °C
Pressure	490 bar
Pressure gradient in reservoir	0.150–0.160 bar/m
Field size	
Net area	10 km^2
Hydrocarbon pore volume	34.4×10^6 reservoir m^3
Recovery factor	22%
Recoverable oil	4×10^6 Sm^3
Recoverable gas	2.2×10^9 Sm^3
Production	
Start-up	December 1979
Development scheme	One production platform with 7 producing wells
Production rate (January 1986)	460 Sm^3/day oil; 230×10^3 Sm^3/day gas
Cumulative production (January 1986)	3×10^6 Sm^3 oil; 1.7×10^9 Sm^3 gas

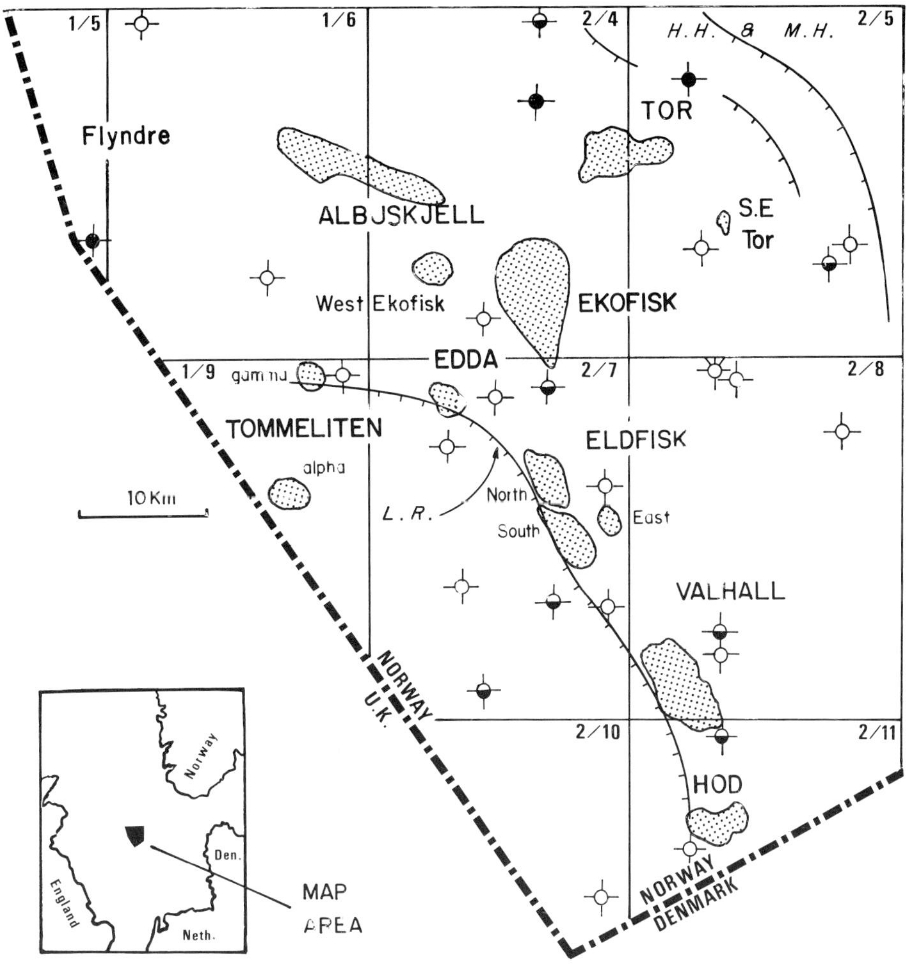

Fig. 1. Location map (L.R., Lindesnes Ridge; H.H. and M.H., Hidra and Mandal High faults).

relief, lies exactly within a cell of this seismic grid and was not detected. The structure was found after the shooting of additional seismic lines in the early 1970s (Fig. 2). The first well drilled on Block 2/7 discovered the Eldfisk Field while the second, 2/7-2X, tested a small prospect called Eel in the southern-most area of the block. The third, 2/7-3X, was designed to appraise the southern part of the large Eldfisk structure.

The fourth well on the block was located on the Edda prospect near a slightly pronounced apex at the top of the chalk level. This well, 2/7-4X, was spudded in June 1972. Since Ekofisk and three other chalk fields had already been discovered, the chalk of the Ekofisk Formation (Lower Paleocene) was regarded as the main target in Edda (Fig. 3). The well reached the top of the chalk at 3067 mss. The chalk of the Ekofisk Formation was found to be 93 m thick, with moderate to poor porosities and oil saturation exceeding 50% in only 16 m of the formation. The chalks of the Tor Formation were 167 m thick and comprised two porous intervals. The upper one (T1) was 36 m thick, had an average porosity of 26% and was oil bearing. It was tested at a rate of 1050 Sm^3/day oil and 275×10^3 Sm^3/day gas. The lower one (T2) contained no hydrocarbons.

In Well 2/7-4X, designated as the discovery well of Edda, oil was found down to 3207 m (base of T1) and water up to 3238 m (top of T2) (Fig. 4). At that time, the limit of the productive area was thought to correspond to contour A on Fig. 2.

Well 2/7-5X, was drilled late in 1972, 3 km north-east of the discovery well. This first appraisal well encountered the chalk less than 15 m deeper than was the case in Well 2/7-4X, which meant that most of the Ekofisk and Tor formations were above the base of the oil zone found in the discovery well. However, water saturation in Well 2/7-5X was everywhere higher than 70%, and the porosity of the chalk was only moderate to poor. Based on the negative results of this well, a permeability barrier was thought to exist in the minor saddle area between the wells (line B on Fig. 2).

A third well, 2/7-10X, was drilled in 1973 in a

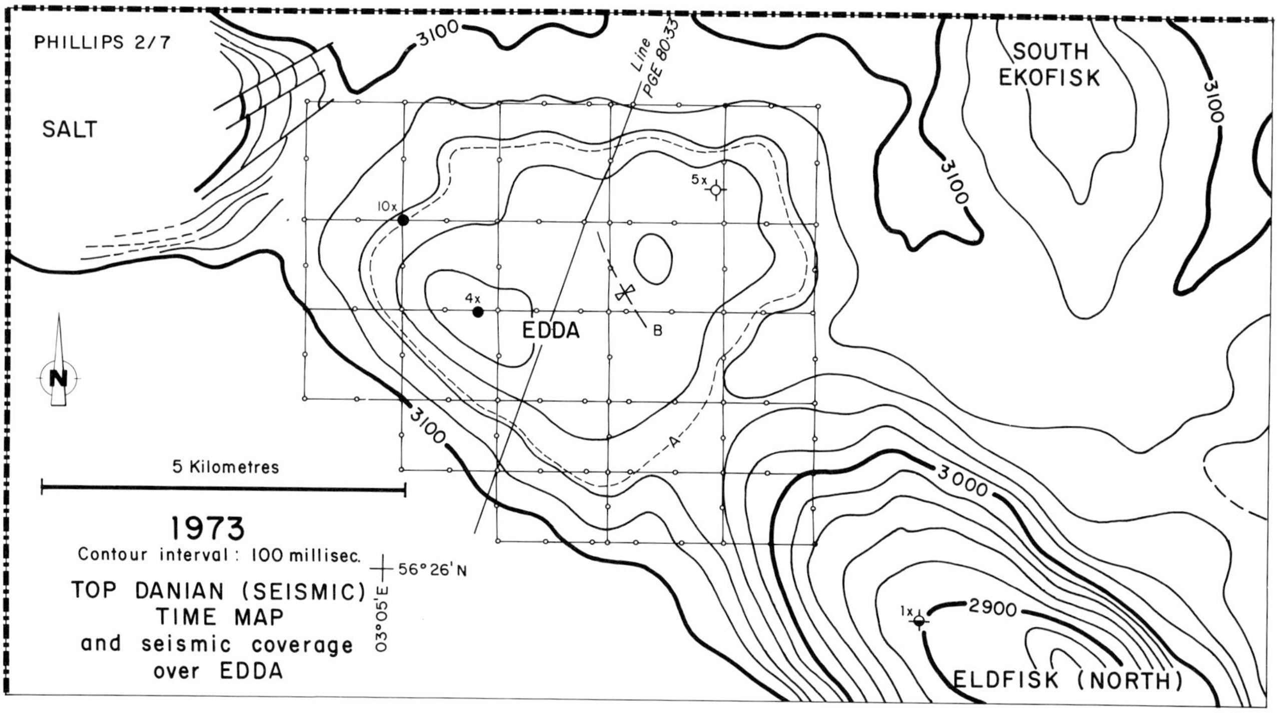

Fig. 2. 1972–73 Top Danian Time map for the north-western area of Block 2/7. The grid shows the seismic coverage over Edda. Contour A shows the field limit as interpreted after the 2/7-4X discovery well was drilled. Line B (saddle?) is the eastern limit of the reservoir based on the result of the first appraisal Well 2/7-5X and seismic line 0308 (after D'Heur *et al.*, 1985, by permission of the publishers, Butterworth ©).

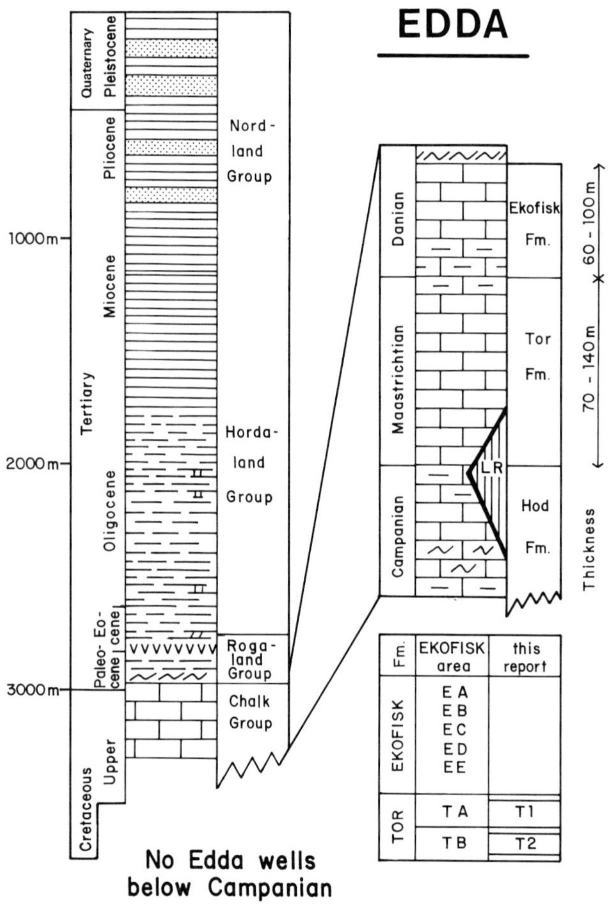

Fig. 3. Lithostratigraphy and nomenclature of the Edda area.

low structural position 1.3 km north-west of the discovery well. At that location, the Ekofisk Formation lacked porosity but the Tor Formation included a porous and oil-bearing chalk unit, 34 m thick, most of which appears to be in the oil–water transition zone. Since this reservoir contains oil deeper than the T2 water-bearing reservoir of the 2/7-4X Well, Edda is considered as a multilayer reservoir with distinct oil–water contacts.

Based on the three exploration/appraisal wells, the reserve estimate of the field was computed at about 8×10^6 Sm^3 of oil plus associated gas, and on this basis it was decided to develop the field in association with its giant neighbour.

A 15-slot drilling–production platform, 2/7-C, was installed near Well 2/7-4X, and development drilling started in late 1978.

Since it was well established at that time that the chalk porosity rapidly decreases from the crest towards the flank of a structure, a careful, step-by-step drilling sequence was planned for Edda in order to avoid drilling poor producing wells. The four wells of the first step gave much less pay section than expected but were nevertheless completed as producers. Well C8 gave unexpectedly poor results despite its rather high structural position and found no significant reservoir development in either the Ekofisk or the Tor

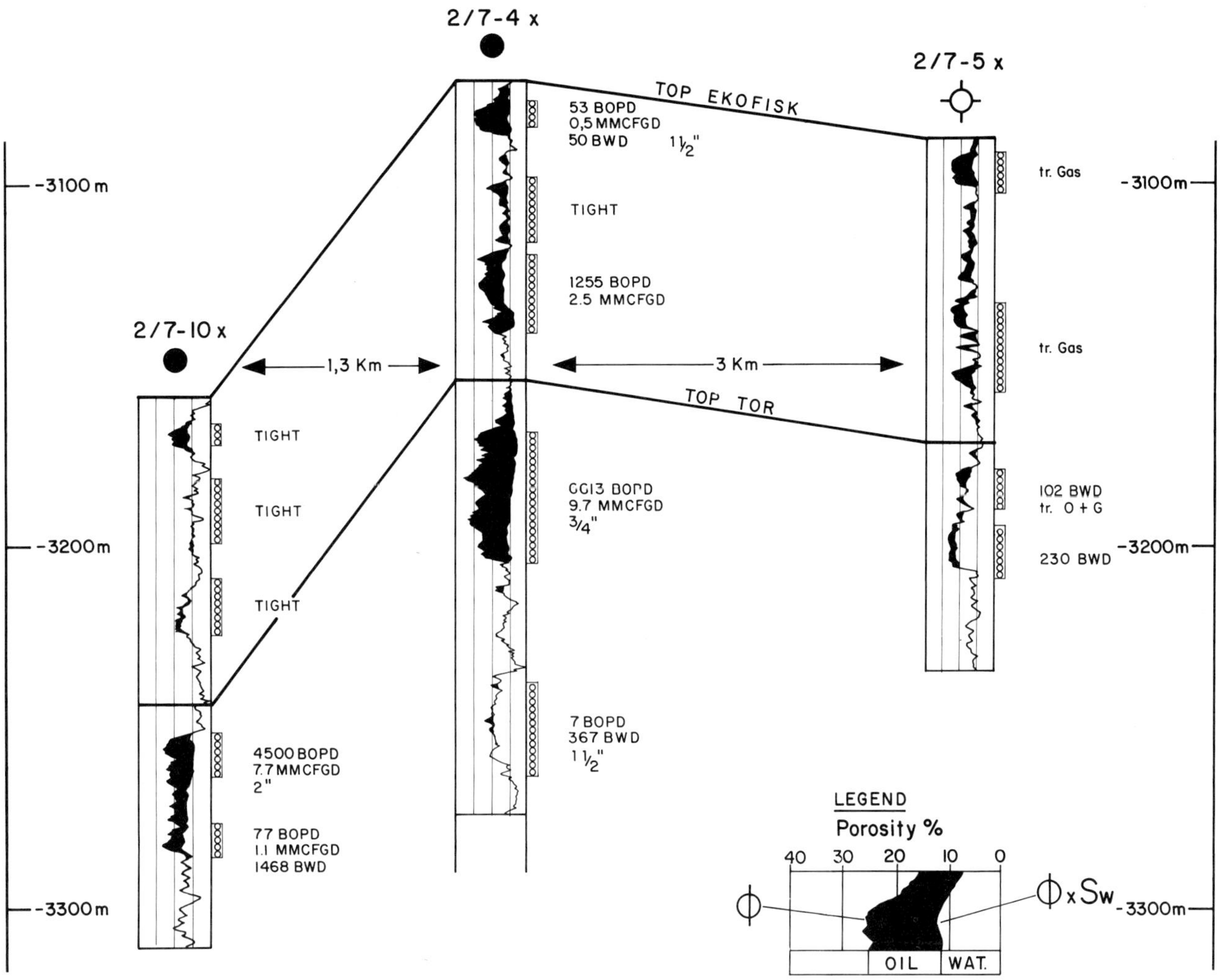

Fig. 4. Porosity and hydrocarbon distribution in the three exploration–appraisal wells on Edda (modified from D'Heur *et al.*, 1985).

formations. In addition, these were abnormally thin and their porosity was nowhere in excess of 20%. Two other dry holes were drilled before it was realized that a palaeohigh, later termed the Lindesnes Ridge, was located below the field, and was responsible for the thickness and porosity reduction in the reservoir intervals.

The three remaining wells came in satisfactorily, and by 1980 the peak oil production of 5200 Sm^3/day was reached with seven producing wells.

GEOLOGICAL SETTING

The Edda Field is located at the centre of the North Sea Tertiary Basin which itself overlies the Central Graben that formed during Mesozoic times. The tectonic history of the area has recently been described by Gowers and Sæbøe (1985). As far as chalk fields are concerned, it is interesting to emphasize only these episodes:

- Halokinetic movements within the thick Permian salt deposits caused the dome shape of most, if not all, the Greater Ekofisk area structures.
- After the deposition of the Triassic continental series and following the Mid-Jurassic uplift, a substantial low developed in the Central Graben which received very thick marine black shales (Mandal Formation), which form the prolific hydrocarbon source-rock in the area.
- Subsidence continued throughout Cretaceous and Tertiary times, and as much as 1300 m of Cretaceous (mainly chalk) and 3000 m of Tertiary (mainly shales) sediments were deposited in the area.
- Tectonic pulses, during Cretaceous and Early Paleocene times caused sea-floor instability, mainly on the dipping edges of the

graben margins; this resulted in considerable sliding, slumping and resuspension of previously unconsolidated chalk deposits. This material was carried into the deep areas of the Graben and redeposited as turbidites, debris-flows and mud-cloud sediments. These allochthonous chalk facies, which have been described by Kennedy (1980, 1985), make the best reservoirs.

Within this geological context, Edda has the particularity of being located over an inversion feature called 'the Lindesnes Ridge' and the reservoir units were deposited over a half-horst type palaeorelief. Since none of the Edda wells has been drilled below the Hod Formation, the deeper conditions are unknown. However, it is assumed that the Campanian half-horst model established for the Eldfisk Field (Brewster and Dangerfield, 1984; D'Heur, 1984) extends into the Edda area (see Eldfisk, this volume). The structural setting of Edda may then be as shown on Fig. 5. The Lindesnes Ridge had a marked influence on the deposition of the Ekofisk and Tor formations, greatly reducing their reservoir potential (see below). Unfortunately, the existence of this adverse factor was not known early enough to prevent the drilling of dry holes on Edda.

GEOPHYSICS

The seismic coverage available at the early exploration phase was too widely spaced to allow the identification of the Edda prospect. This low-relief structure was completely inside a 5 × 5 km cell when the first exploration well was drilled in Block 2/7 on the Eldfisk prospect. Some months later a 1.5 × 1.5 km grid was shot as a basis for further exploration and appraisal drilling. Additional seismic surveys were shot in 1980 on Edda and in 1982 on the whole Block.

The 1973 interpretation of the 'Top Chalk marker' is presented on Fig. 2 and a 1980 seismic line is shown on Fig. 6. The first (upper) prominent reflector around 3.0 s coincides with the Lower Eocene–Upper Paleocene boundary and corresponds to the volcanic ash marker. A second reflector, about 100 ms deeper, corresponds with the shale–chalk interface at the top of the Ekofisk Formation.

As in the Ekofisk and West Ekofisk Fields, a seismic porosity anomaly had been identified and mapped at Edda within the Upper Cretaceous when the Edda discovery well was drilled. While this well was located at the centre of the seismic porosity anomaly, the

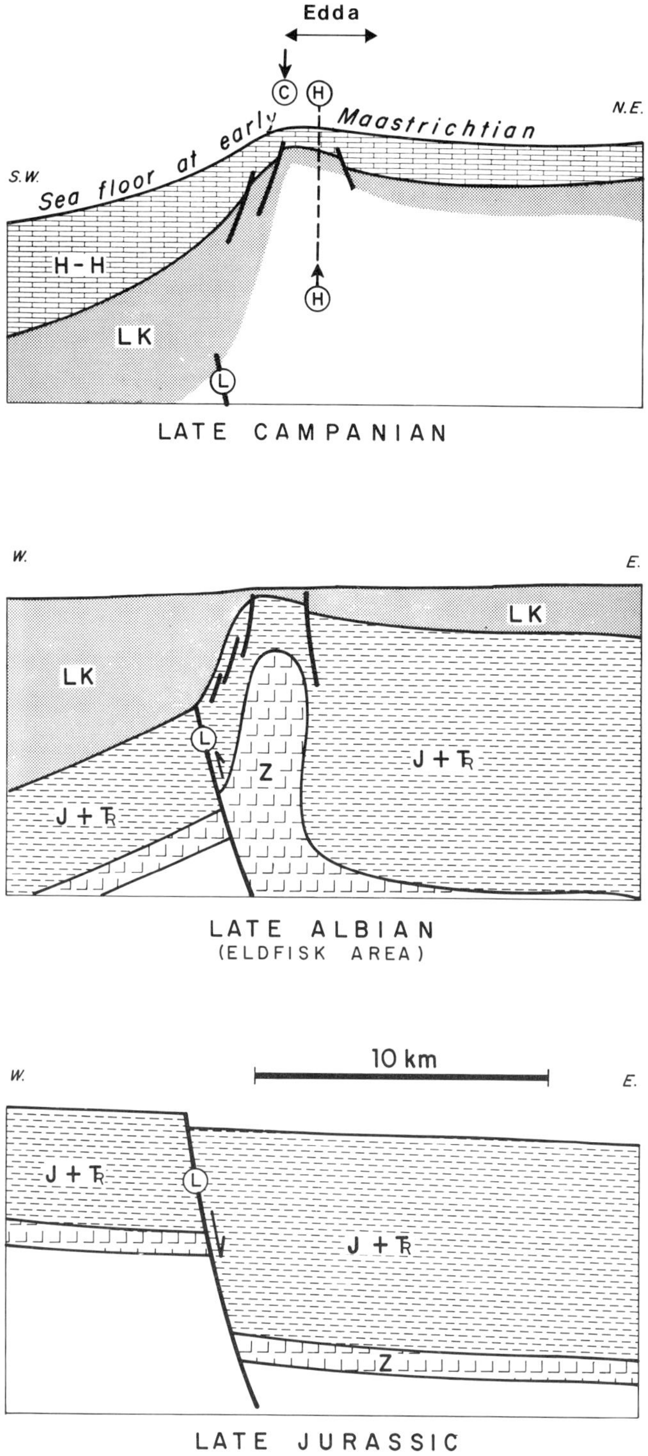

Fig. 5. Structural setting of the Edda Field and palaeoreconstruction at Late Jurassic and Late Albian in the Eldfisk area. Z, Zechstein salt; TR, Triassic; J, Jurassic; LK, Lower Cretaceous; H–H, Hidra to Hod formations; L, Lindesnes Fault; C, Palaeocrest during Maastrichtian and Danian; H, present-day structural axis (after D'Heur *et al.*, 1985, by permission of Butterworth ©).

first appraisal well, 2/7-5X was located outside it and, despite its high structural elevation, did not find producible hydrocarbons. In fact, the seismic porosity limit as mapped in 1972

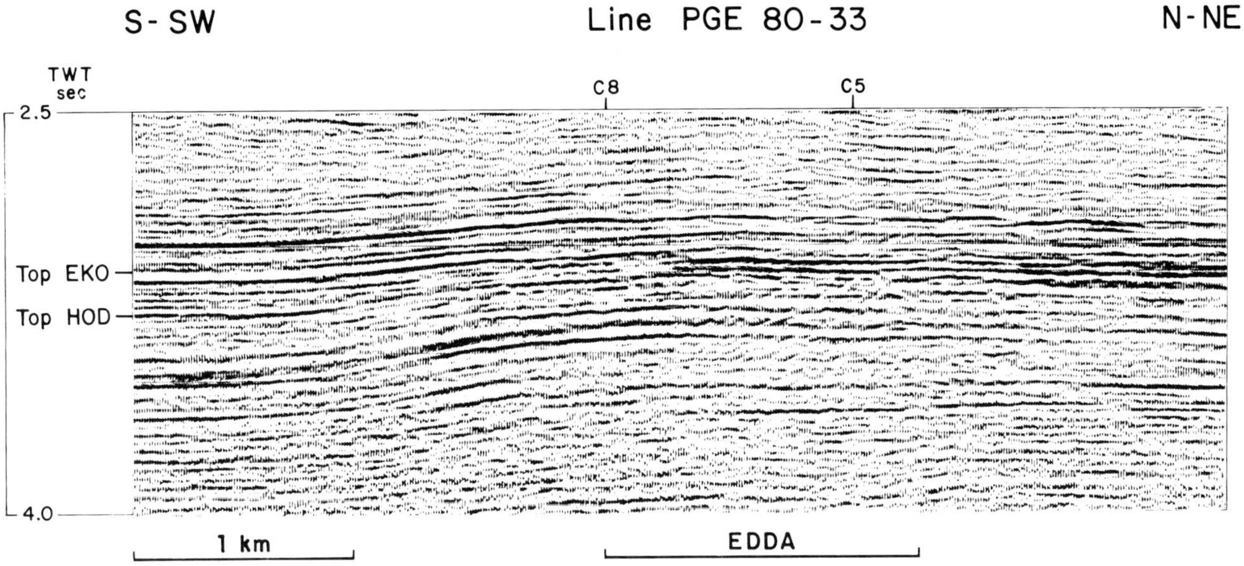

Fig. 6. Seismic line orientated SSW–NNE across the Edda Field (see Fig. 2 for location).

corresponds well with the limit of the field as established by appraisal and development drilling, except over the palaeocrest of the Lindesnes Ridge. It is worth noting that a 'gas cloud effect', such as exists in the Tertiary claystones above several chalk fields in the area, is not present above the Edda Field.

TRAP

Since some Edda wells are non-productive despite their high structural elevation, the trap is clearly a combined structural and stratigraphic type. The Lindesnes Ridge, which influenced the lateral facies variation of the chalk, is responsible for the stratigraphic trapping mechanism of the accumulation. The final shape of the reservoir is the result of low-amplitude halokinetic movements of the Permian salt during early to Mid-Tertiary times. The field covers an area of 10 km^2 (Fig. 7). It is bounded to the south-west by tight chalk above the palaeocrest of the Lindesnes Ridge. The eastern and south-eastern limits of the field also correspond with zones of reduced porosity, the cause of which is less obvious. The northern and north-eastern limit is a structural closure towards Ekofisk and West Ekofisk. The reservoir may be oil bearing down to a palaeospill-point between Edda and the salt plug located to the north-west. The seal is provided by thick Tertiary shales. The field contains at least two distinct oil accumulations separated by the Tight Zone (layer EE).

RESERVOIRS

The reservoirs are chalks of the Ekofisk and Tor formations. The petrology, sedimentology and reservoir characteristics of the North Sea chalk are now well documented (Hardman, 1982; Hancock, 1976; D'Heur, 1984, 1986; Kennedy, 1980). The general reservoir characteristics are summarized in the West Ekofisk chapter in this volume (p. 165). The detailed geological model for the Edda Field (D'Heur *et al.*, 1985) is summarized below. The Edda chalks exhibit porosities of up to 35%, whereas normal diagenesis of a coccolith ooze buried to 3000 m should have led to its porosity being reduced to about 10% ($\pm$5%). The combined effects of favourable sedimentary facies, overpressure and the early presence of oil in the pores, are the main causes for the abnormally high degree of porosity preservation (Scholle, 1977 and various researchers). The most porous sedimentary facies are the redeposited chalks, the best of which are the massive debris-flow deposits. Locally, they represent up to 50% of the sediment thickness. At the other end of the scale, the autochthonous chalks (normal pelagic coccolith ooze) exhibit poor to non-existent reservoir characteristics.

In Edda, the Tor Formation, which is 70–140 m thick, contains several debris-flows and other allochthonous chalk units with high original porosity. The source of allochthonous material for the two main reservoir units (T1 and T2), was probably located to the east, and these units progressively thin westwards

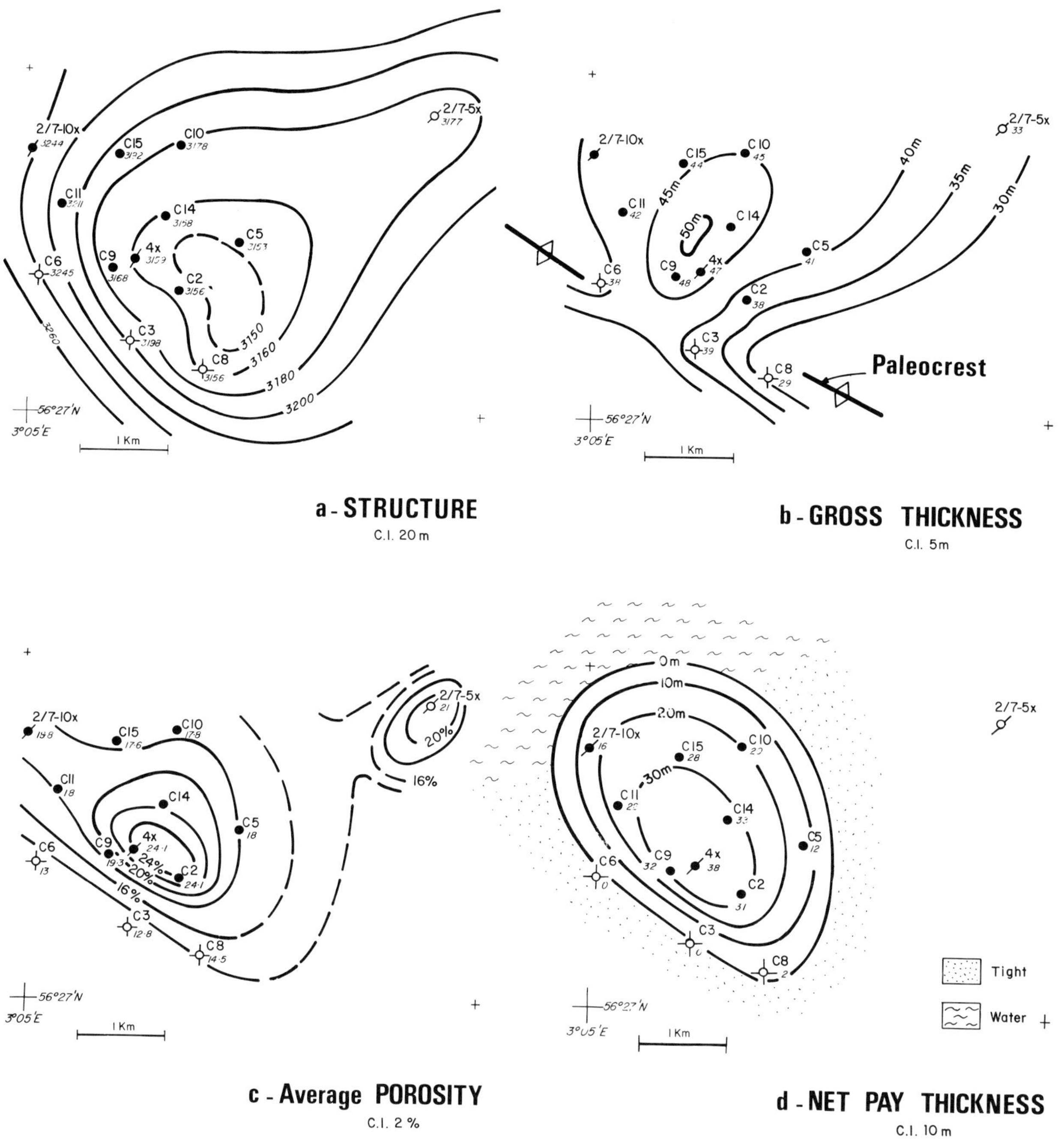

Fig. 7. Selected maps of the Edda Field for the main reservoir interval (upper Tor Formation).

over the Lindesnes Ridge. In area A on Fig. 8, only the Lower Tor Member is influenced by the half-horst slope and thins towards the west. In area B, the thickness of all the members is reduced from north-east to south-west. Over the palaeocrest (area C), the thin debris-flow beds of the Lower Tor member pinch-out, and the reservoir units T1 and T2 are half as thick as in area A. Off the crest, the Tor Formation rapidly thickens again towards the south-west. Thickness variations of the Danian layers are characterized by similar although less pronounced trends which implies that the ridge was not totally buried by the end of Maastrichtian times.

As far as porosity is concerned, three factors played a major role. First, the ratio of allochthonous (favourable) to autochthonous (tight) chalk decreases from east to west over the gently dipping eastern flank of the Lindesnes Ridge, because of the existence of the palaeoslope and of an easterly source for the favourable chalk facies. Secondly, the rock underwent a stronger mechanical compaction over the palaeocrest. Thirdly, differential porosity preservation by overpressure

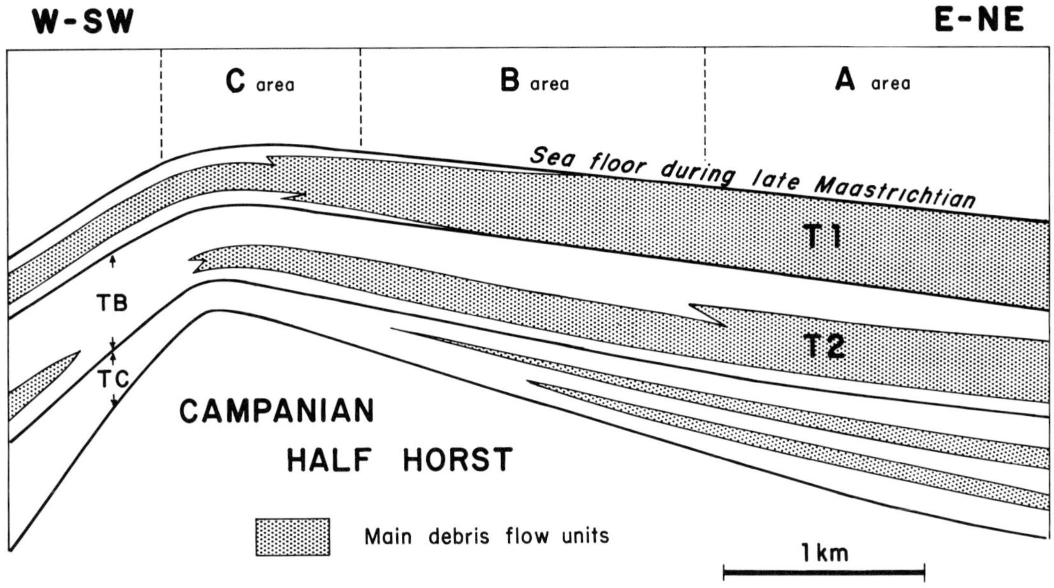

Fig. 8. Influence of the Lindesnes Ridge on the distribution of the reservoir formations.

and the presence of oil in the pores is a function of structural elevation in the trap. The present-day porosity distribution for the upper reservoir of the Tor Formation (Fig. 7) reflects: (1) the original average porosity decrease from east to west; (2) a drastic porosity reduction over the palaeocrest; and, (3) a porosity decrease on the flanks, northwards and eastwards. In the crestal area, the average porosity of the Upper Tor reservoir (T1) is 27% with a maximum of 35%. For this interval, a porosity value of 29% corresponds to a matrix permeability of 1 mD.

The present-day 'gross thickness' variation of any reservoir unit is influenced by depositional thickness variation and by differential porosity preservation (Fig. 7 and see West Ekofisk, this volume, p. 165). Compared to the Ekofisk and West Ekofisk Fields, the Tor Formation at Edda is thinner and significantly less porous. This is mainly due to the distal location of Edda relative to the source of allochthonous chalk, the existence of the palaeohigh and the low structural relief of the trap.

In Edda, the Upper Member of the Tor Formation contains approximately 80% of the initial reserves, the remaining 20% being distributed between three layers of the Ekofisk Formation.

The upper layer (EA) of the Ekofisk Formation consists of 10–15 m of clean porous chalk of very low permeability. This rock is non-productive when its porosity is lower than 22%. Two other Ekofisk Formation layers, EC and ED, contain small oil accumulations in allochthonous chalks of Danian and Maastrichtian ages which are interbedded with pelagic chalk. The basal member, EE, of the Ekofisk Formation is the typical Ekofisk Tight Zone, made up of alternating dark (clay-rich) and light (clay-poor) chalks interbedded with rare, thin, distal debris-flows and turbidites. This layer creates a continuous permeability barrier above the main reservoir.

Porosity and thickness variations of all the members of the Ekofisk and Tor formations are illustrated on the porosity cross-section (Plate 1).

SOURCE

The Edda structure is close to the axis of the Central Graben where the Upper Jurassic source-rock is often deeper than 5000 m. Its upper part, the Mandal Formation, comprises the 'Hot Shales' consisting of black marine sediments with very high organic contents. In the Edda area, peak oil generation is thought to have taken place during late Miocene times, while to the north-west significant amounts of oil were generated from the 'Hot Shales' as early as Paleocene times.

HYDROCARBONS

Edda contains undersaturated volatile oil. The initial GOR was 307 Sm^3/Sm^3 and the oil gravity was 0.83 g/cm^3. Over most of the field area, the hydrocarbon distribution in the

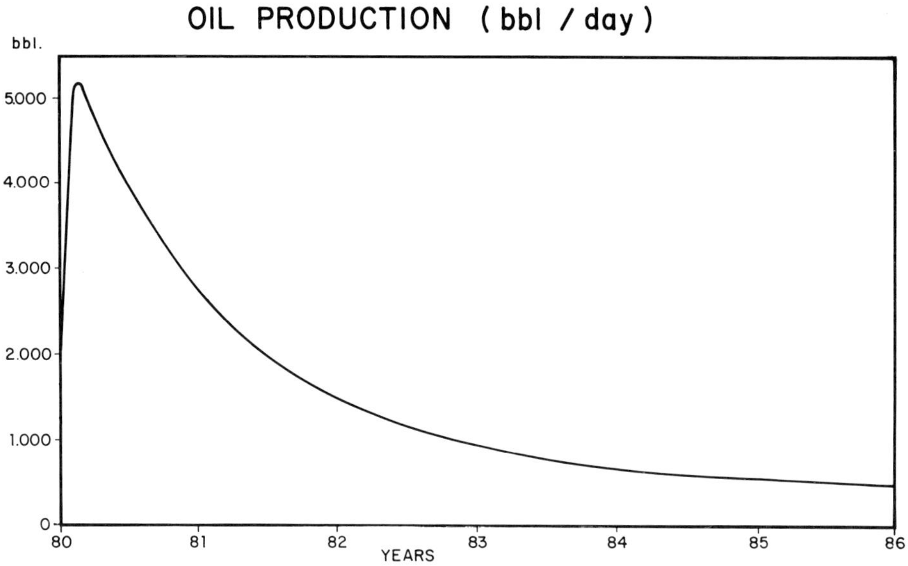

Fig. 9. Oil production history.

reservoir is mainly related to its porosity which, in turn, is influenced by the Campanian palaeoridge and by structural elevation. The two distinct oil accumulations in Edda have not shown significant variation in their oil properties. A limited active aquifer may exist on the northern flank of the structure.

RESERVES

The Edda reservoir has a very low matrix permeability and its productivity is low since it is less fractured than most of the other chalk fields. Improvements in well productivity have been obtained by hydraulic acid fracturing. The peak oil production of 5200 Sm^3/day was obtained with seven producing wells in 1980 (Fig. 9). By early 1986, the daily production rate was only 460 Sm^3 of oil and 230×10^3 Sm^3 of gas. The production is transported to the Ekofisk Centre by a 10-inch oil line and a 12-inch gas line. By the end of 1993, the field is expected to have produced a total of about 4×10^6 Sm^3 of oil and 2.2×10^9 Sm^3 of gas.

ACKNOWLEDGMENTS

The authors wish to thank Phillips, Petrofina, Agip, Elf Aquitaine, Norsk Hydro and Total for their permission to publish this paper.

REFERENCES

Brewster, J. and Dangerfield, J.A. 1984. Chalk fields along the Lindesnes Ridge, Eldfisk, *Marine and Petroleum Geology* **1,** (3), 239–278.

D'Heur, M. 1984. Porosity and Hydrocarbon Distribution in the North Sea Chalk Reservoirs, *Marine and Petroleum Geology* **1,** (3), 211–238.

D'Heur, M., de Walque, L. and Michaud, F. 1985. Geology of the Edda Field Reservoir, *Marine and Petroleum Geology* **2,** (4), 327–340.

D'Heur, M. 1986. The Norwegian chalk fields, In: Spencer, A.M. *et al.* (Eds), *Habitat of Hydrocarbons on the Norwegian Continental Shelf*, Graham and Trotman, London.

Gowers, M.B. and Sæbøe, A. 1985. On the structural evolution of the Central Trough in the Norwegian and Danish Sectors of the North Sea, *Marine and Petroleum Geology* **2** (4), 298–318.

Hancock, J.M. 1976. The petrology of the Chalk, *Proceedings Geologists Association* **86** (for 1975), 499–535.

Hardman, R.F.P. 1982. Chalk reservoirs of the North Sea, *Bulletin Geological Society Denmark* **30,** 119–137.

Kennedy, W.J. 1980. Aspect of chalk sedimentation in the Southern Norwegian Offshore, In *Sedimentation of the North Sea Reservoir Rocks*, Norsk Petroleumsforening, Article IX.

Kennedy, W.J. 1985. Sedimentology of the Late Cretaceous and Early Paleocene Chalk Group, North Sea Central Graben, *Chalk Research Program*, Symposium, Stavanger, May 21–22 (Proprietary Report).

Scholle, P.A. 1977. Chalk diagenesis and its relation to petroleum exploration: oil from chalks, a modern miracle? *American Association Petroleum Geologists Bulletin* **61,** 982–1009.

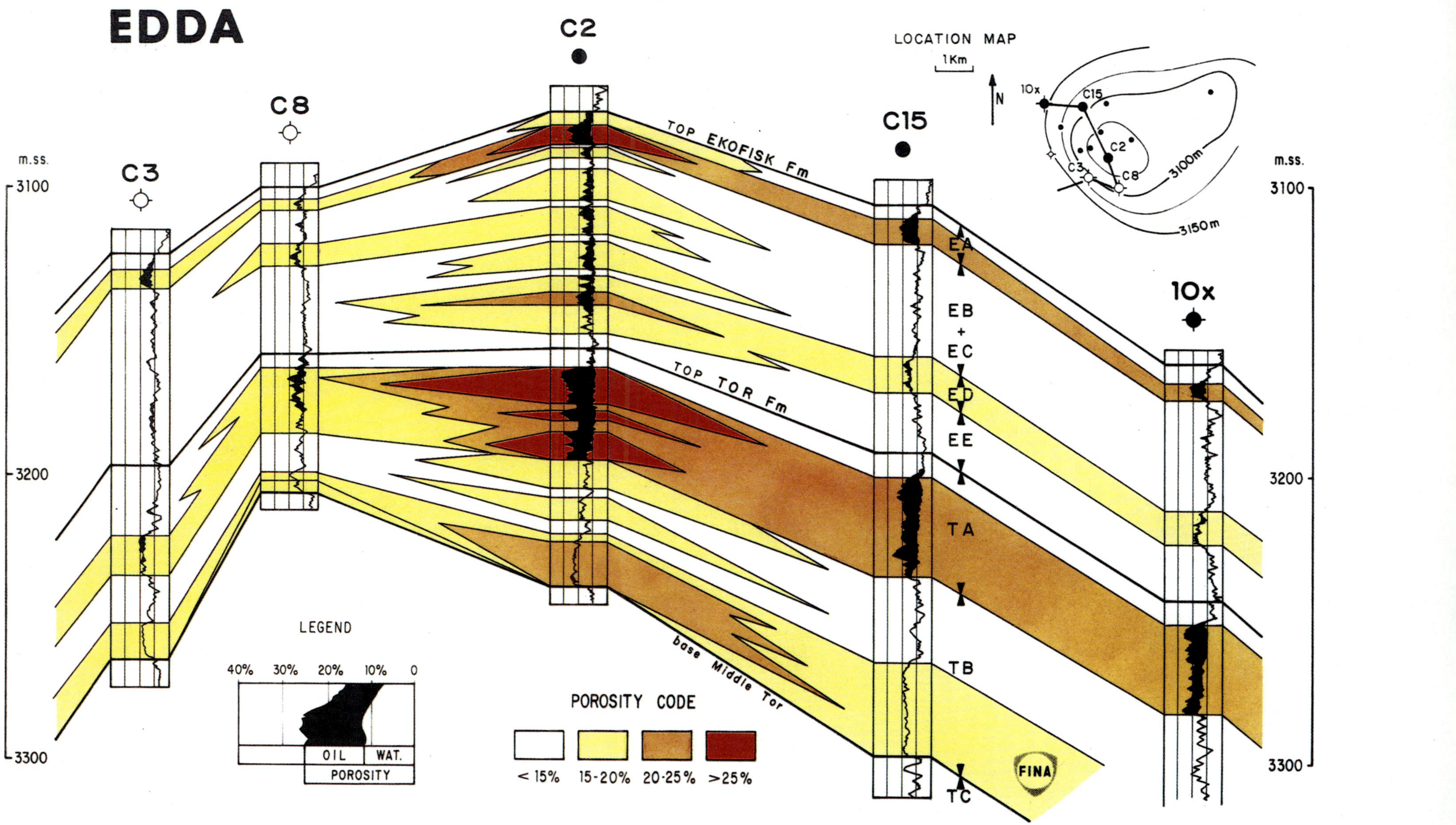

Plate 1. Porosity cross-section (see Fig. 3 for definition of intervals).

5

Ekofisk

Lawrence J. Pekot and Gregory A. Gersib

Phillips Petroleum Company Norway, PO Box 220, N-4056 Tananger, Norway

Ekofisk Field, located in Block 2/4 of the Norwegian sector, is the largest of the North Sea chalk fields, with an initial oil-in-place of over one billion standard cubic metres. Discovered in 1969 by the Phillips Norway Group, it was the first commercial discovery in Norway and has been on production since 1971. Ekofisk Centre serves as the central processing and transportation facility for all fields in the southern portion of the Norwegian North Sea. The hydrocarbons are present in Danian and Cretaceous age Ekofisk and Tor Formations. These chalks are described as a multilayered system. Reservoir layer characteristics are determined by sedimentology, depositional history, diagenetic history and hydrocarbon emplacement. The structure is a salt-induced, domal feature covering 49 km^2 with 300 m of vertical closure. The chalks within the reservoir possess porosities as high as 45%. Matrix permeability is low, less than 10 mD, but flow is enhanced through an extensive natural fracture system. Production was initially by solution gas drive with gas injection beginning in 1975 and water injection scheduled to begin in 1987. Current production is maintained through 45 production wells and five gas injectors. Twenty water injection wells are planned. Additional production techniques, problems and secondary recovery methods are discussed.

INTRODUCTION

Ekofisk Field is located in Block 2/4 of the Norwegian Sector of the North Sea, 310 km south-west of Stavanger, Norway (Fig. 1). Water depth in the area is approximately 70 m. It is the largest of the Central Graben chalk fields with initial in-place reserves of over 10^9 Sm^3 of oil and 280×10^9 Sm^3 of gas.

The Ekofisk Field received its name through a combination of two practices used at that time by Phillips for identifying prospects. All prospects in a licence area began with a common letter. The Phillips Norway Group identified the licence including Blocks 2/4 and 2/7 as the 'E' area. Field and prospect names in this area include Ekofisk, West Ekofisk, Eldfisk, Edda, Eel, Elver and Ebba. Secondly, many prospects in Norway were being named after fish (or *fisk* in Norwegian). Since no more fish names starting with 'E' were available when Ekofisk was mapped, the prospect was named after the seismic reflection 'echo' that is recorded. Echo translates into *Eko* in Norwegian. See also D'Heur *et al.* (1985).

HISTORY

Blocks 2/4 and 2/7 were awarded in 1965 as part of Production Licence 018. Participating interests in the licence at present are:

Phillips Petroleum Company Norway (operator)	36.960%
Norske Fina A/S	30.000%
Norsk Agip A/S	13.040%
Elf Aquitaine Norge A/S	8.094%
Norsk Hydro Production AS	6.700%
Total Marine Norsk A/S	4.047%
Eurafrep Norge A/S	0.456%
Coparex Norge A/S	0.399%
Cofranord A/S	0.304%

Geology of the Norwegian Oil and Gas Fields © A. M. Spencer *et al.*, eds (Graham & Trotman, 1987) pp. 73–87.

Ekofisk Field data summary

Trap type	dome structure over a salt swell
Depth to crest	2888 m at top Ekofisk Formation
Lowest closing contour	3155 m at top Ekofisk Formation
Hydrocarbon contact	3288 m but variable, dependent upon layer and structural position
Oil column	400 m
Pay zone	
Formation	Ekofisk and Tor Formation chalks
Age	Danian and Maastrichtian
Gross thickness	250 m average
Net/gross ratio (average; range)	0.70; 1.0–0.3
Cut-off for N/G	ø = 15%, shale = 30%, $S_w = 50\%$
Porosity (average; range)	32%; 48–15% Ekofisk Formation 30%; 41–15% Tor Formation
Hydrocarbon saturation	90% net interval
Permeability (average; range)	1.0 mD; 0.06–8.0 mD Ekofisk Formation matrix 1.0 mD; 0.1–8.0 mD Tor Formation matrix; effective permeability up to 150 mD, dependent upon fracture intensity
Hydrocarbons	
Oil gravity	0.844 g/cm^3
Gas gravity (relative to air)	0.68
Bubble point	383 bar
Gas/oil ratio	276 Sm^3/Sm^3
Formation volume factor	1.72
Formation water	
Salinity	61 000 ppm NaCl equivalents average, variable
Resistivity	0.038 ohm m at 131 °C
Reservoir conditions	
Temperature	131 °C
Pressure	485 bar at 3170 m
Pressure gradient in reservoir	0.155 bar/m
Field size	
Net area	49 km^2
Net hydrocarbon pore volume	1828×10^6 reservoir m^3
Recovery factor	21%
Drive mechanism	solution gas drive, gas injection, waterflood
Recoverable hydrocarbons	(NPD estimates)
—oil	237×10^6 Sm^3
—NGL/condensate	13.5×10^6 Sm^3
—gas	126×10^9 Sm^3
Production	
Start-up	July 1971
Development scheme	3 steel production platforms with a total of 45 production wells and 5 gas injectors; 20 water injectors planned
Production rate	as of February 1986, 13 000 Sm^3/day oil, 13×10^6 Sm^3/day gas
Cumulative production	as of February 1986, 120×10^6 Sm^3 oil, 63×10^9 Sm^3 gas

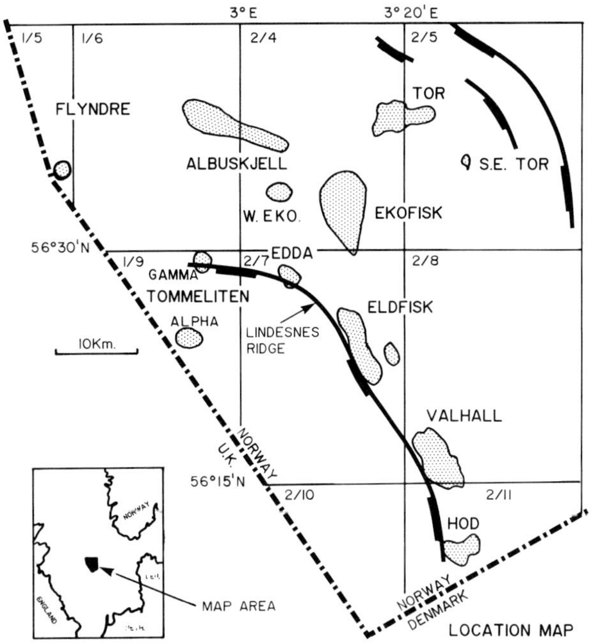

Fig. 1. Location map—Norwegian chalk fields and finds.

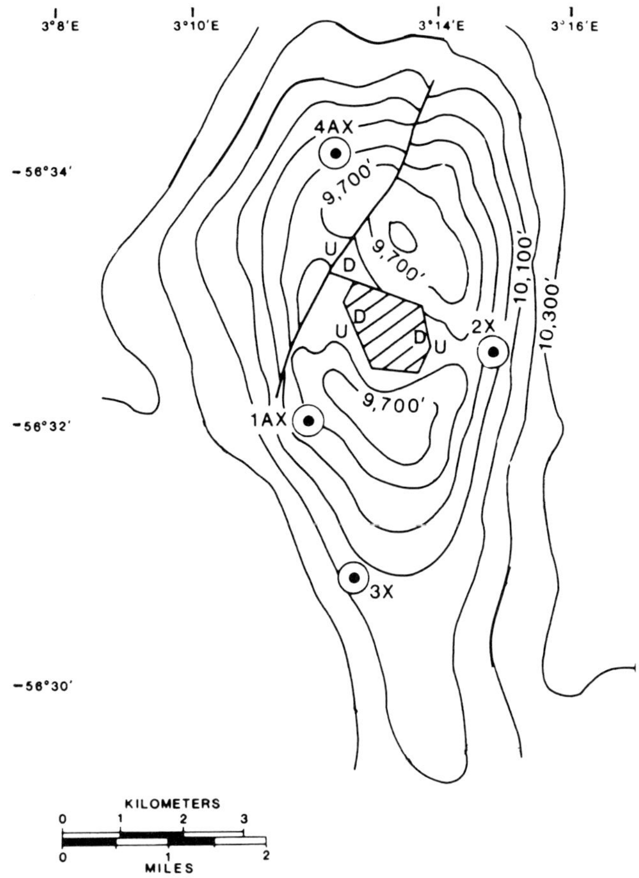

Fig. 2. Structure map—Top Ekofisk Formation, 1971 interpretation showing the central collapse area. Contours are in feet subsea.

The discovery, appraisal and initial development of Ekofisk have been previously documented (Heidbreder, 1978; Van den Bark and Thomas, 1981). Ekofisk was the first field on production in Norwegian waters and the important events of its history are summarized here.

In 1969, a seismic grid of approximately two kilometres existed in the area. At the top Ekofisk Formation, these data indicated a N–S-orientated elongate dome with its central area collapsed or down-faulted (Fig. 2). The location of the first well, 2/4-1X, was selected to penetrate the structure as high as possible without entering the apparent collapsed area. Mechanical problems forced abandonment of this well in Miocene age rocks at 1676 m. The well was respudded one kilometre south of the original location as the 2/4-1AX. Beginning at 3050 m subsea, the well penetrated 183 m of high-porosity pay in the Ekofisk and Tor Formation chalks. A drill-stem test flowed oil at 170 Sm^3/day, but was limited by the surface testing equipment. In 1970, Wells 2/4-2X, -3X and -4X were drilled on the east, south and north flanks of the structure, respectively. All these wells produced oil in commercial quantities and all had avoided the crestal area of the structure. It was not until after the seismic data were reprocessed using slower velocities that it was known that the apparent crestal graben feature did not exist. It was a seismic anomaly caused by the presence of hydrocarbons in the shallower Tertiary section above the reservoir.

Ekofisk was declared commercial and the jack-up drilling rig *Gulftide* was converted into a temporary production facility. The four existing wells were re-entered and completed using subsea wellhead equipment. Production was loaded onto tankers through two single-point mooring buoys. Ekofisk was onstream in July 1971, only 18 months after discovery. The early production system was in operation over two years before permanent facilities were available. During this period, flow rates averaged 4500 Sm^3/day with maximum rates of over 7000 Sm^3/day recorded.

In April, 1974, permanent facilities became operational. Three drilling platforms, production, quarters, and pumping and flare platforms, along with the concrete underwater storage tank, were commissioned. Re-injection of produced gas was begun in February 1975, when the oil pipeline to Teeside, England became operational. Operation of the gas sales pipeline to Emden, West Germany began in September 1977.

In April, 1981, a pilot water injection project was begun and in the summer of 1983 the decision was made to water-flood the Tor Formation reservoir. A new platform is under construction to carry water-injection facilities capable of delivering 60 000 Sm^3 of water per day. Predrilling of the 20 water-injection wells is in progress and the project is scheduled for a 1987 start-up.

The development of Ekofisk Field has been ongoing for 15 years and will continue into the future. Ekofisk Centre serves as the central processing and transport facility for all the fields in the southern part of the Norwegian sector, as well as the link to the continent for the Statpipe gas-gathering system (Fig. 3). In addition to the original platforms constructed for Ekofisk Field development, the north flare, riser, hotel, Amoco riser and Statpipe riser platforms have been added. The water-injection platform is under construction.

GEOLOGICAL SETTING

The tectonic history and geologic setting of the Central Graben have been adequately described in numerous papers (D'Heur, 1984; Gowers and Sæbøe, 1985; Van den Bark and Thomas, 1981; Ziegler, P.A., 1982; Ziegler, W.H., 1975) and will be briefly summarized here. The North Sea has undergone four major geological phases; geosynclinal phase (Cambrian–Carboniferous), intracratonic phase (Permian–Triassic), rifting phase (Triassic–Cretaceous), and post-rifting phase (Tertiary–Recent) (Ziegler, 1975). The Central Graben was initiated in Late Triassic times by a major period of rifting. This crustal extension and subsidence of the graben continued through Jurassic times creating a complex NW–SE fault-bounded trough. Associated movement along these faults influenced the deposition and erosion of sediments during this period. A thick accumulation of an organic-rich shale, the Mandal Formation, occurred during Late Jurassic times and it serves as the primary source of hydrocarbons in the North Sea. The increased sediment thickness and subsequent major tectonic phase, the Cimmerian, probably initiated halokinesis in the graben. Continued extensional movements and widespread basinal subsidence during the Cretaceous Period contributed to the change in deposition from a predominantly clastic environment to a predominantly carbonate environment of deposition. Pulses of halokinesis continued throughout Cretaceous and into Tertiary times and played a major role in the redistribution of sediment.

A generalized stratigraphic section for the area is shown in Fig. 4. The Cretaceous and Tertiary stratigraphy of the Ekofisk Field conforms to the nomenclature proposed by Deegan and Scull (1977). Of the lithostratigraphic intervals identified, the Ekofisk and Tor Formations contain the productive intervals in the Ekofisk Field. The Chalk Group is over 800 m thick at Ekofisk Field.

Chalk sedimentation ended during Late Danian times with the establishment of clastic sedimentation during a basinal regression associated with the Laramide movements. This relatively rapid deposition of clastic sediments was a factor in the formation of abnormally high pressures in the Chalk and Lower Tertiary shale section, and is partly responsible for the high porosity preservation in the chalk reservoirs.

GEOPHYSICS

The Ekofisk structure was originally defined by seven seismic lines shot on a grid approximately two kilometres square. The quality of this seismic was acceptable except where low-velocity, gas-charged Tertiary age shales caused the distortion or destruction of seismic information (Fig. 5). The poor-quality data over the crest led to an initial interpretation of a down-faulted structural crest, as shown in Fig. 2. The mapping of the structure has evolved over the years as data from development wells, and new surface seismic were integrated into the interpretation. More recently, Phillips has undertaken an extensive borehole seismic profile (VSP) programme in conjunction with new surface seismic to define further the faulting in the Ekofisk and Tor Formations. Borehole profile techniques have generally improved the resolution of the data in the Tor and Ekofisk Formations but these results are still slightly distorted by the low-velocity effects of the gas-charged shales. In 1984, 50 km of surface seismic, using waterguns, was shot in conjunction with three walk-away borehole profiles in the 2/4-B12. The borehole profile data were used to improve the processing technique for the surface seismic, and have resulted in a much improved structure map (Fig. 9(a)).

The borehole profiles show that rapid variation of rock characteristics occurs in the most productive intervals of the Tor and Ekofisk

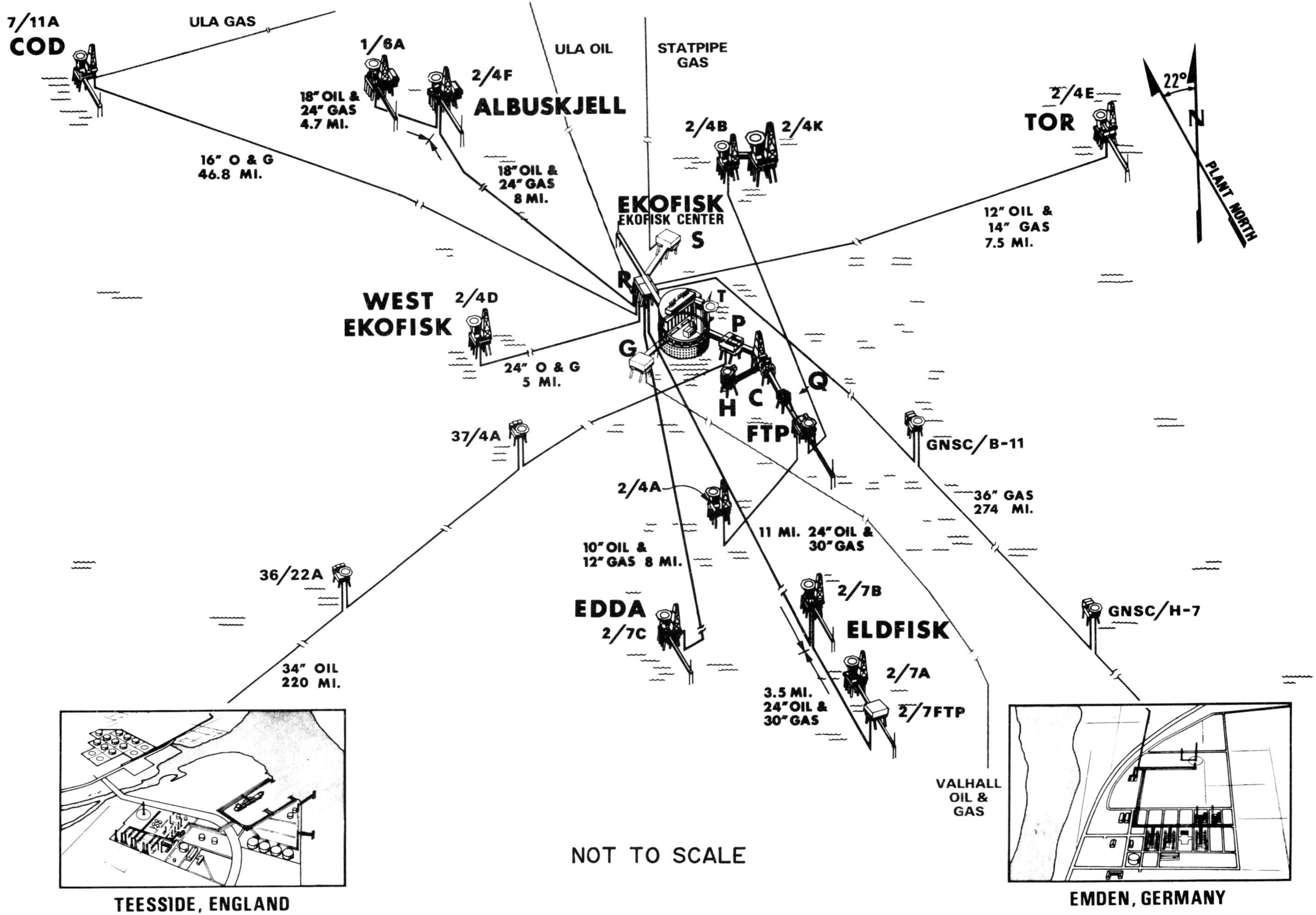

Fig. 3. Ekofisk Centre and the Greater Ekofisk development.

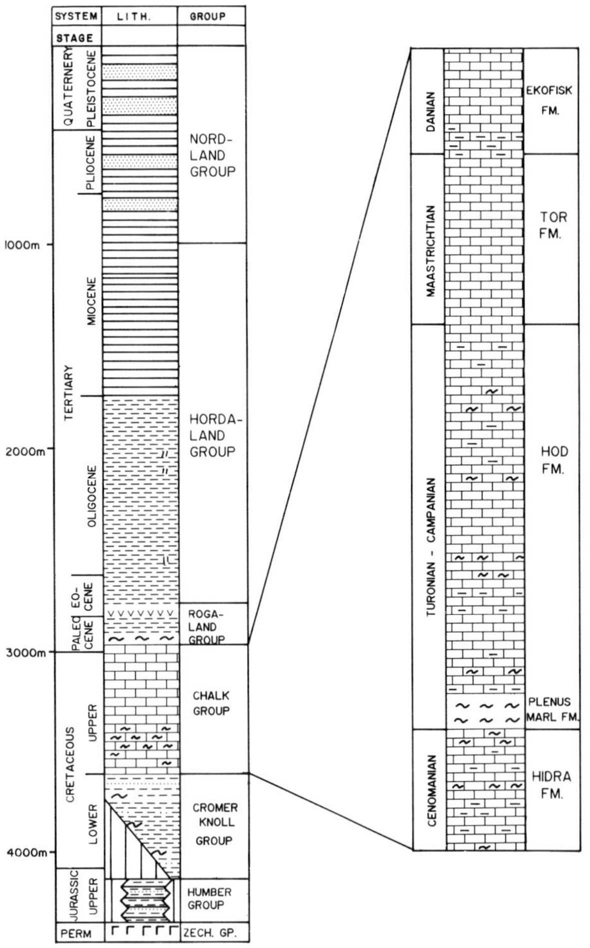

Fig. 4. Generalized stratigraphic column for the Greater Ekofisk Area.

Formations. Geophysical modelling is currently being completed with the intention of estimating the porosity variation and of locating zones of permeability enhancement or reduction that could enhance production or affect water-flood management.

TRAP

The Ekofisk structure was initiated during Late Jurassic time by halokinesis triggered by faulting associated with the Central Graben extension system. Structural reconstructions using both seismic and well logs, indicate that numerous episodes of movement have affected the Ekofisk structure. Watts (1983) concluded that the Albuskjell Field also had experienced repeated halokinetic movements. Reconstructions also show that the axial crest of the structure was not stationary through time. A similar phenomenon can be observed on the Albuskjell and Tor fields. Differential salt movement across the structure during Late Cretaceous and Early Paleocene times may help explain localized variations of facies during chalk deposition. Halokinesis continued until early Miocene times and, together with differential compaction, formed the present-day structure.

The seal is formed by the thick sequence of overpressured Tertiary shales and claystones. Gas in parts of the Tertiary section over the faulted crest of the structure can be identified on surface seismic sections as disturbed zones and is also noted when drilling. This gas is interpreted as leakage from the reservoir.

RESERVOIR

Several significant papers have been published on the sedimentology of the chalk of the North Sea (Hancock, 1976; Hancock and Scholle, 1975; Hardman, 1982; Kennedy, 1980, 1985; Nygaard *et al.*, 1983; Van den Bark and Thomas, 1981; Watts *et al.*, 1980).

The majority of the chalk is composed of poor- to well-cemented biomicrites consisting mainly of coccolithophorid skeletal debris. Mineralogically the skeletal debris is a physicochemically stable form of low magnesium calcite.

Detailed mineralogical analyses on cores in the Ekofisk Field indicate that silica (chert, chalcedony, cryptocrystalline and microcrystalline quartz) is the major acid-insoluble fraction in the chalk. Percentages range from 1–5% in the pay section of the Upper Tor Formation to 15–70% in the Ekofisk Formation. Minor amounts of glauconite, phosphates, barite, dolomite, pyrite, marcasite, detrital quartz, feldspar, clays and biogenic grains are present, but generally constitute less than 1–2% of the bulk volume. Clays are present in larger volumes (up to 15%) in the lower portion of the Ekofisk Formation (layer EE).

It is generally accepted that chalks of the Central Graben were deposited in water depths of 300–700 m (Hancock, 1976; Hancock and Scholle, 1975) and were initially deposited as faecal pellets produced by organisms feeding in the upper levels of the water column. Kennedy (1985) provided a detailed review of the sedimentological processes involved in chalk deposition in the Central Graben from which the following summary is largely taken.

Two general sedimentary facies are recognized in the Central Graben chalks—autochthonous pelagic chalks and allochthonous

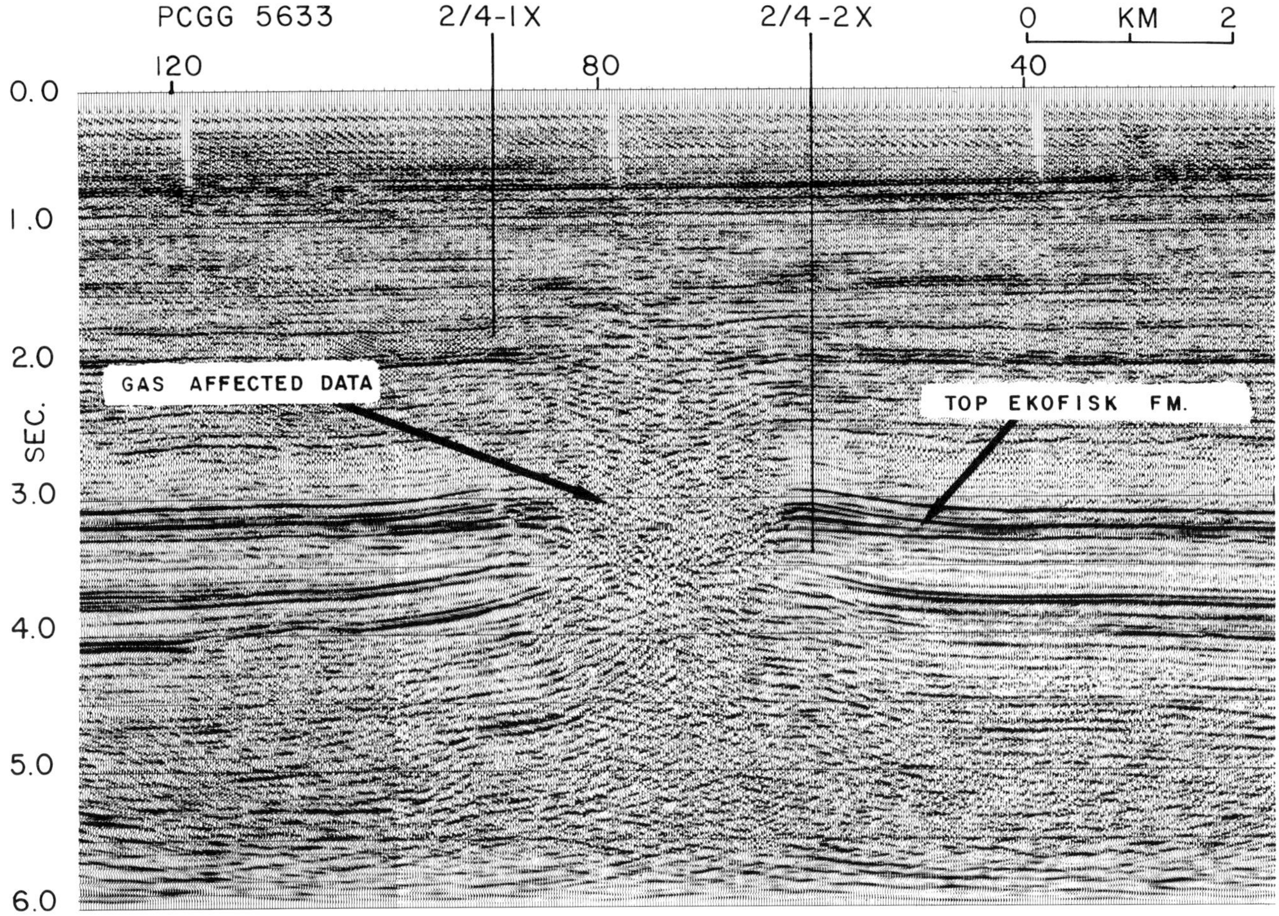

Fig. 5. East–West seismic cross-section, line PCGG5633, shot in 1969. Note the area of disturbed data which obscures seismic interpretation of the crestal area.

chalks. The two can be difficult to discern in cores because extensive bioturbation has destroyed much of the original depositional fabric.

The autochthonous chalks are composed mainly of intensely bioturbated beds that are alternately pale and dark. The cyclicity of the pale and dark units may reflect variations of the following: carbonate production in surface waters, carbonate dissolution, and relative abundance of terrigenous material.

The allochthonous chalk facies is composed of sediments that were deposited by mass movement processes such as slides, slumps, grain-flows, debris-flows, mud-flows, and turbidity currents. Bioturbation has often destroyed much of the primary depositional fabric and identification of the elements of this facies is difficult in cores.

Sediment movement was initiated by seismicity or by the creation of unstable slopes on the graben margin or on topographic highs in the graben formed by syndepositional halokinesis or faulting. The main sediment movement direction in Late Cretaceous and Early Paleocene times was north-east to south-west with minor input from the south. Local structural features in the graben, such as the Lindesnes Ridge, played an important role in localized reworking of sediment.

Phillips Petroleum Norway has developed an informal layering and correlation scheme based on biostratigraphy, log signature, mineralogy and core descriptions (Fig. 6). Regional correlations using the system have been developed from a well database of over 250 wells in the Ekofisk area and can be convincingly demonstrated for these major layers over an area covering hundreds of square kilometres (Fig. 7).

Well 2/4C-8 is near the crest of the structure and has one of the thickest reservoir sections in the field (Fig. 8). A short description of each of the layers is given below.

The Tor Formation is divided into three layers—TA, TB and TC, all of which are Maastrichtian in age.

Layer TC, the deepest Tor Formation layer, is characterized by its variable log signature. Several beds of increased-porosity rock can be correlated in wells throughout the Greater Ekofisk Area. These beds are generally fea-

STAGE	GROUP	FORMATION	PHILLIPS GROUP ZONATION	
PALEOCENE	MONTROSE	MAUREEN EQ.	25MA	MAUREEN FORMATION EQUIVALENT
	CHALK	EKOFISK	25EA	UPPER POROUS ZONE
			25EB	TOMMELITEN TIGHT ZONE
			25EC	REWORKED DANIAN ZONE
			25ED	REWORKED MAASTRICHTIAN ZONE
			25EE	EKOFISK TIGHT ZONE
MAASTRICHTIAN TO TURONIAN		TOR	30TA	UPPER
			30TB	MIDDLE
			30TC	LOWER
		HOD	31HA–32HD	UPPER
			33HE	MIDDLE
			34HF–34HH	LOWER
		PLENUS MARL	34PM	
CENOMANIAN		HIDRA	35HA	

Fig. 6. Stratigraphic correlation scheme used for the Chalk Group in the Greater Ekofisk Area.

tureless and may have burrowed tops. They are interpreted as individual or stacked debris-flows in a pelagic-dominated sequence. There is no production from this layer.

Layer TB shows an increased porosity (averages 20–30% on the crest) and a more continuous log signature compared to the layer below. From the available core, this layer consists mainly of a homogeneous biomicrite with poor laminations and some burrowed horizons. These are interpreted as debris-flows and mud-flow deposits. There is some evidence that slumping has taken place. A few wells in crestal positions have produced from this layer but most flank wells have high water saturations.

Layer TA consists of high-porosity (35–40%), homogeneous chalk with some laminations in units up to 10 m thick. A burrowed horizon is sometimes present at the top of these units. Dips in this layer are highly variable, and distorted internal features are common. Stylolitic surfaces are very common. This layer is interpreted as a complex sequence of slumped debris-flows and mud-flows with minor turbidites. Some flank wells have a lower-porosity section in the upper part of this layer. Mineralogical studies indicate that the average composition is: 95–98% calcite; 2–4% quartz and trace; 1% clays. Approximately 50% of the total oil production of the Ekofisk Field flows from this layer.

The Ekofisk Formation is divided into five layers: Upper Porous Layer (EA), Tommeliten Tight Zone (EB), Reworked Danian Layer (EC), Reworked Maastrichtian Layer (ED), Ekofisk Tight Zone (EE).

The EE layer is characterized by a rapid gamma-ray increase, a decreased porosity and an increased insoluble-residue fraction. The insoluble fraction is composed of quartz (10–70% of total rock), clay (2–12%), dolomite (1–3%) and feldspar (trace). The lithology is described as a laminated chalk, with shale partings, interbedded with homogeneous biomicrite chalks and thin calcarenites. All lithologies show extensive bioturbation and some diagenetic alteration. This layer is interpreted as being deposited pelagically, by debris-flows and by small turbidites. The contact of layers TA and EE represents the Cretaceous–Tertiary boundary.

The ED layer is characterized by an increase in average porosity (35–45%), a decrease in quartz content (2%) and the presence of predominantly Maastrichtian taxa. The lithology is predominantly homogeneous biomicrite chalks, with faint lamination and wackestones. Burrowed horizons cap some units and most show internal deformation. These are interpreted as stacked debris-flows that have been locally slumped. This layer is over 30 m thick in the northern part of the Ekofisk Field and thickens to over 75 m in the south-west part of the field. The source of the Maastricthian-age sediment was probably the Lindesnes Ridge with minor input from the graben margin to the north-east.

Layer EC is characterized by lower average porosities (30–35%) and an erratic log signature. It consists mainly of homogeneous and laminated chalks with individual units being interpreted as debris-flows. Erratic dips suggest that extensive slumping has taken place. The taxa in this layer are predominantly Danian in age. This layer thins to the south of the field and suggests a localized high in that area at the time of deposition. The mineralogy of this layer shows a marked increase in quartz (12–60%) from the layer below. This quartz is in the form of grain coatings, void fillings, fracture sealings and bioclasts.

At the Ekofisk Field, the boundary between layers EB and EC is difficult to pick on some logs. Layer EB is again variable but generally has a slight increase in average porosity and an increase in quartz content from the layer below. Available cores show a mixed litho-

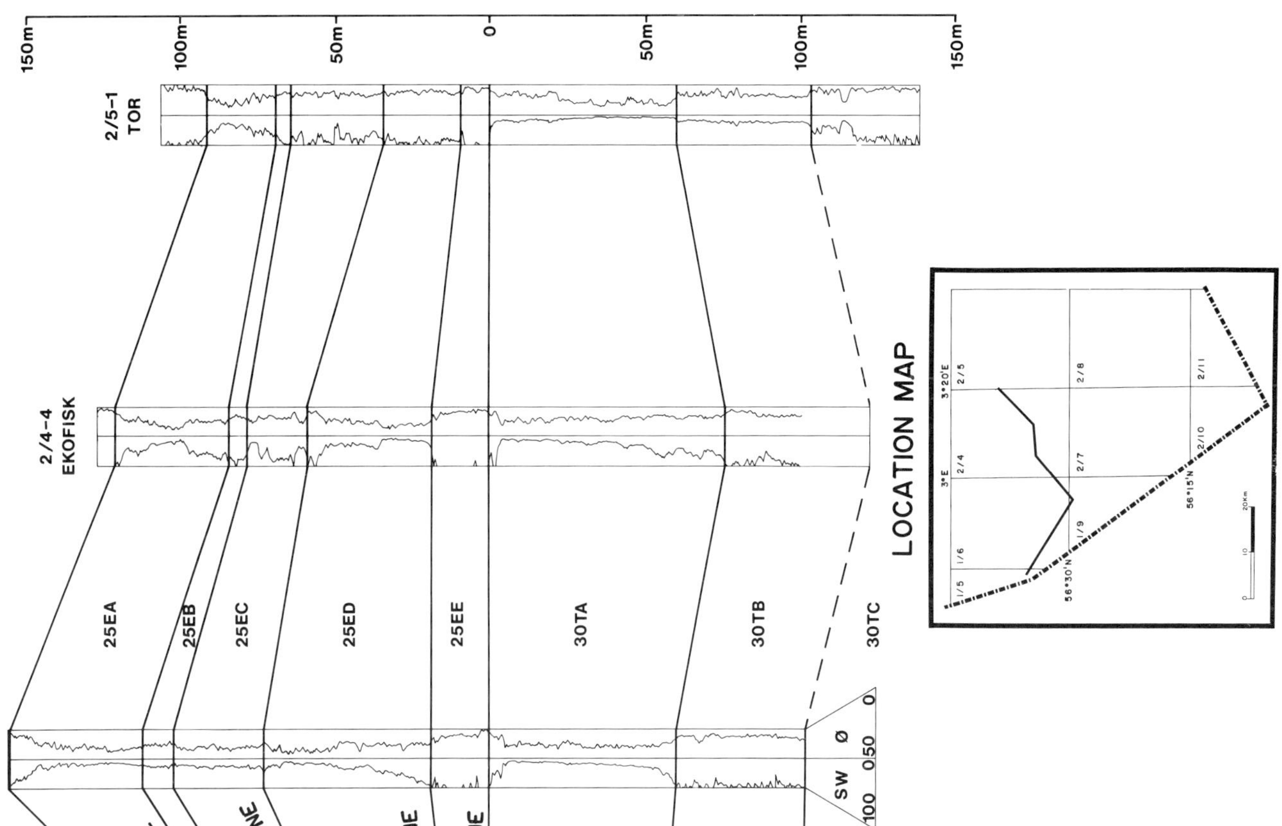

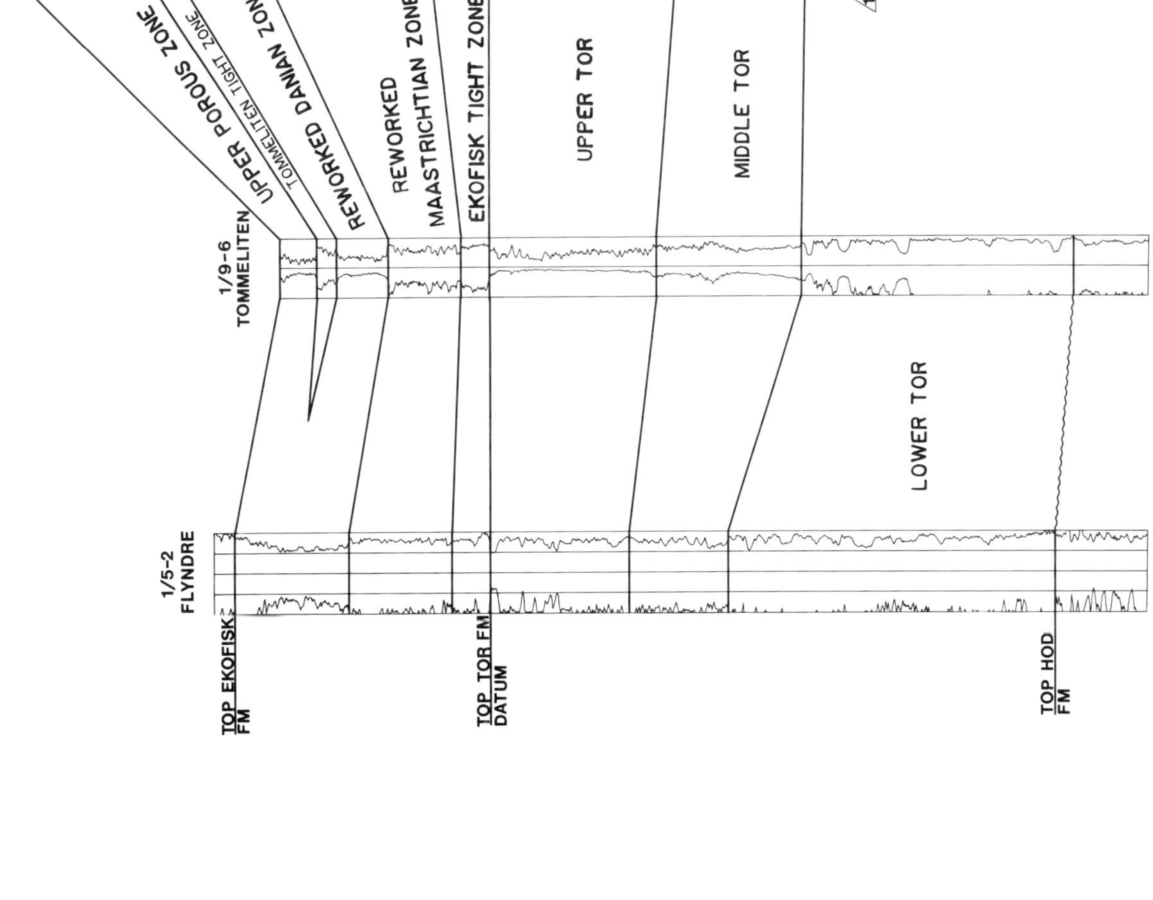

Fig. 7. Application of the Ekofisk Area correlation system.

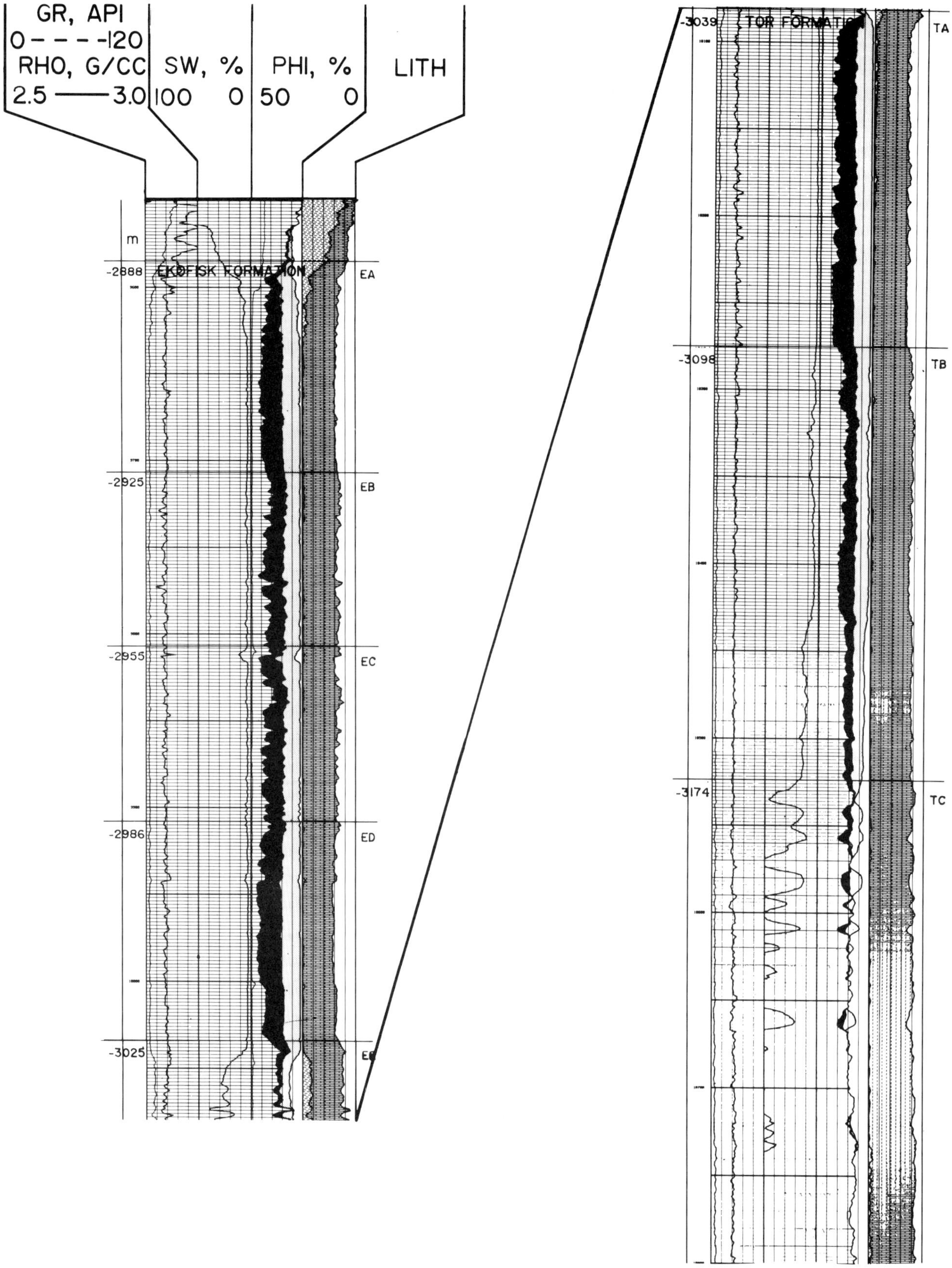

Fig. 8. 2/4C-8 computer-processed interpretation log. GR, API, gamma-ray in standard API units; RHO, G/CC, Matrix density in g/cm^3; SW, %, average percentage water saturation; PHI, %, average percentage porosity; LITH, interpretive lithology percent shale, limestone and pore volume.

logy, predominantly homogeneous biomicrite chalks and laminated chalks with nodular chert horizons. These are interpreted as debris-flows, mud-flows and pelagic deposits. Some matrix shearing and evidence of slumping is present. Flank wells are characterized by an increase in quartz content, mainly due to the nodular chert. The taxa of this layer are predominantly Danian in age but individual debris-flows units may have up to 80% Maastrichtian taxa. This layer varies laterally in both thickness and lithology across the field.

Layer EA, the uppermost Ekofisk Formation layer, is predominantly a homogeneous biomicrite with little internal structure. Average porosities reach 40% in the crestal area of the field. Some individual units can be identified and are commonly capped by a burrowed horizon. These are interpreted as debris-flows and mud-flows. The upper part of this layer contains a greater proportion of laminated, lower-porosity chalks that represent an increased pelagic input. The taxa in this layer are predominantly Danian in age with thin units of mainly Maastrichtian taxa. The upper boundary coincides with the top of the Ekofisk Formation and is marked by an increase in gamma-ray, sonic and density responses representing a change in lithology from chalk to shale.

Differences in these layers are also apparent when mapping the distributions of reservoir properties. Consequently, each of the above described layers have been individually mapped. Figure 9 shows a suite of maps for layer EA as a representative example of the observed general trends. Figure 9(a) is the structure map for layer EA. Figure 9(b) is the gross isopach for the layer and shows characteristic thickening over the crest, which is largely a reflection of increasing porosity preservation higher on the structure, as shown in Fig. 9(c) and Plate 2. Water saturation distribution is shown in Fig. 9(d).

SOURCE

The source for the oil in the Central Graben chalk fields is the Mandal Formation of Late Jurassic age. Van den Bark and Thomas (1981) have shown a convincing correlation between a Jurassic-age shale extract and an Ekofisk crude oil, proving that the Ekofisk oil was sourced by these shales.

Only one well, the 2/4-B19, has been drilled to the Jurassic on the Ekofisk Field. Geochemical analyses from this well indicate that the shales (4072–4358 m) are thermally mature and have fair to very good remaining potential to generate and expel oil and gas. Vitrinite reflectance (R_0 = 0.65–0.88%) and thermal alteration indices (TAI = 3 − to 3 +) indicate that this interval is currently in the main phase of oil generation. This interval has good source-rock potential for oil with the total organic carbon (TOC) values of 1.25–2.91% and oil-prone kerogen values of 72–93%.

A detailed crude-oil geochemical study yielded the observation that molecular maturity indicators generally show a systematic decrease by field along a NW–SE trend across eight Ekofisk area fields. These maturity variations are probably due either to differences in the thermal maturity of the source-rocks along that trend or to thermal alteration in the reservoirs. Also, minor but definite differences in the crude oils can be attributed to localized differences in the source-rocks; i.e. Ekofisk Field crudes have a greater terrestrial input and oxidity than crudes from the Edda and Eldfisk Fields.

HYDROCARBONS

Ekofisk reservoir crude is an undersaturated volatile oil with an initial pressure of 485 bar and bubble-point of 383 bar. It is a low-sulphur, low-asphalt, intermediate, paraffinic crude with a specific gravity of 0.844 g/cm^3.

Drill-stem test results from the four exploratory wells were: 2/4-1AX, 170 Sm^3/day; 2/4-2X, 612 Sm^3/day; 2/4-3X, 602 Sm^3/day and 2/4-4X, 514 Sm^3/day. None of these rates were representative of the full potential of Ekofisk production wells. All these wells were in flank positions and thus did not encounter the best reservoir-zone characteristics. Some tests were limited by surface equipment, and some intervals were not stimulated with acid. When these same four wells were re-entered and completed for early production, per-well average production rates in excess of 1750 Sm^3/day were achieved.

Defining a completely reliable description of formation waters at Ekofisk has been problematic. Characteristics appear to vary areally and samples from most wells have a wide scatter in the data. Recent water samples taken from production wells averaged a salinity of 61 000 ppm NaCl equivalents, specific gravity of 1.038 g/cm^3 and a pH of 6.27.

RESERVES

There was little or no experience in the industry with fractured chalk reservoirs prior

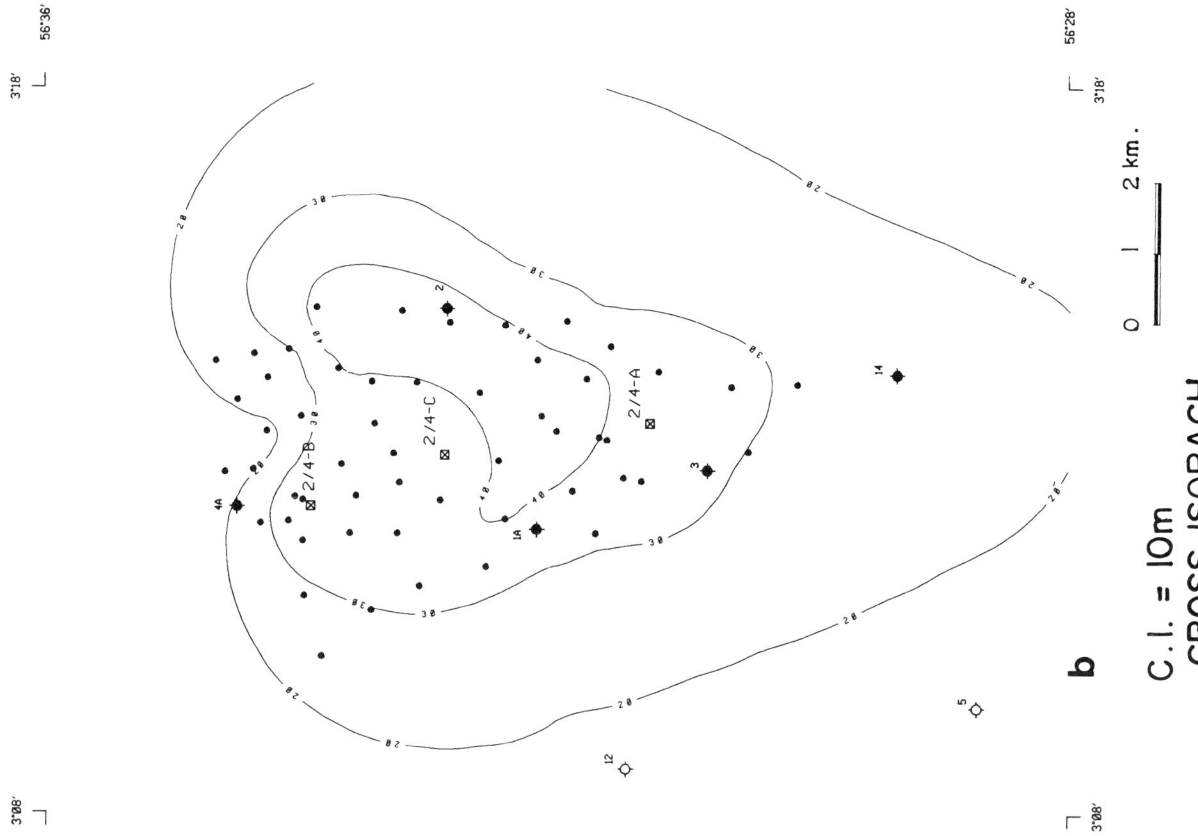
b
C.I. = 10m
GROSS ISOPACH
2/4-A
2/4-B
2/4-C
0 1 2 km.
3°18′
56°36′
56°28′
3°08′

Fig. 9(b).

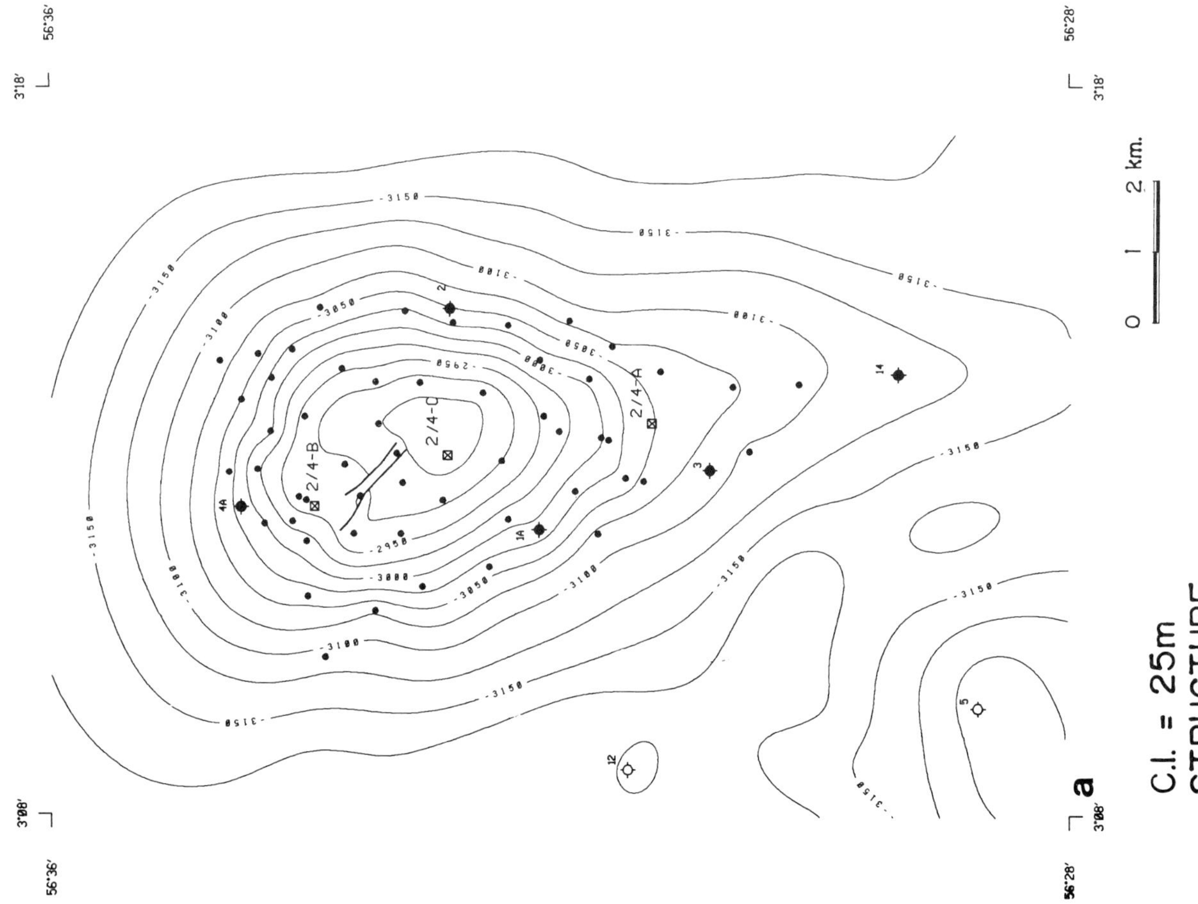
a
C.I. = 25m
STRUCTURE
2/4-A
2/4-B
2/4-C
0 1 2 km.
3°18′
56°36′
56°28′
3°08′

Fig. 9(a).

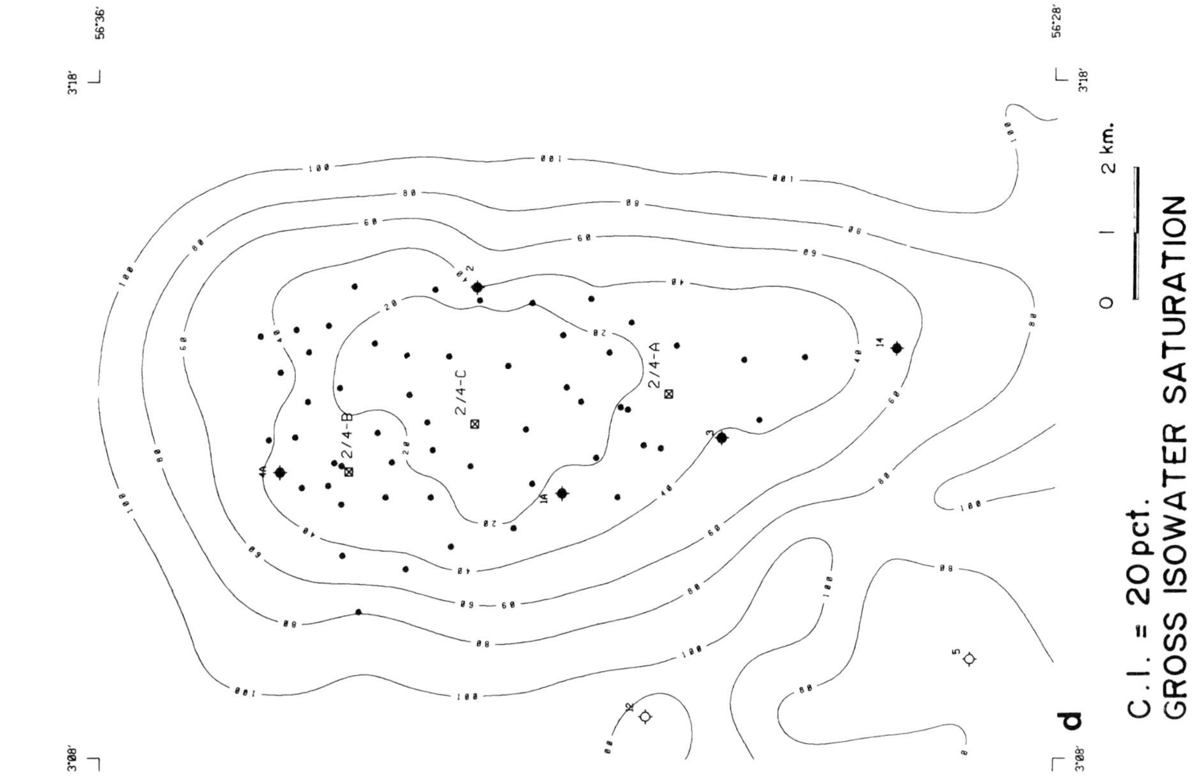

Fig. 9(d).

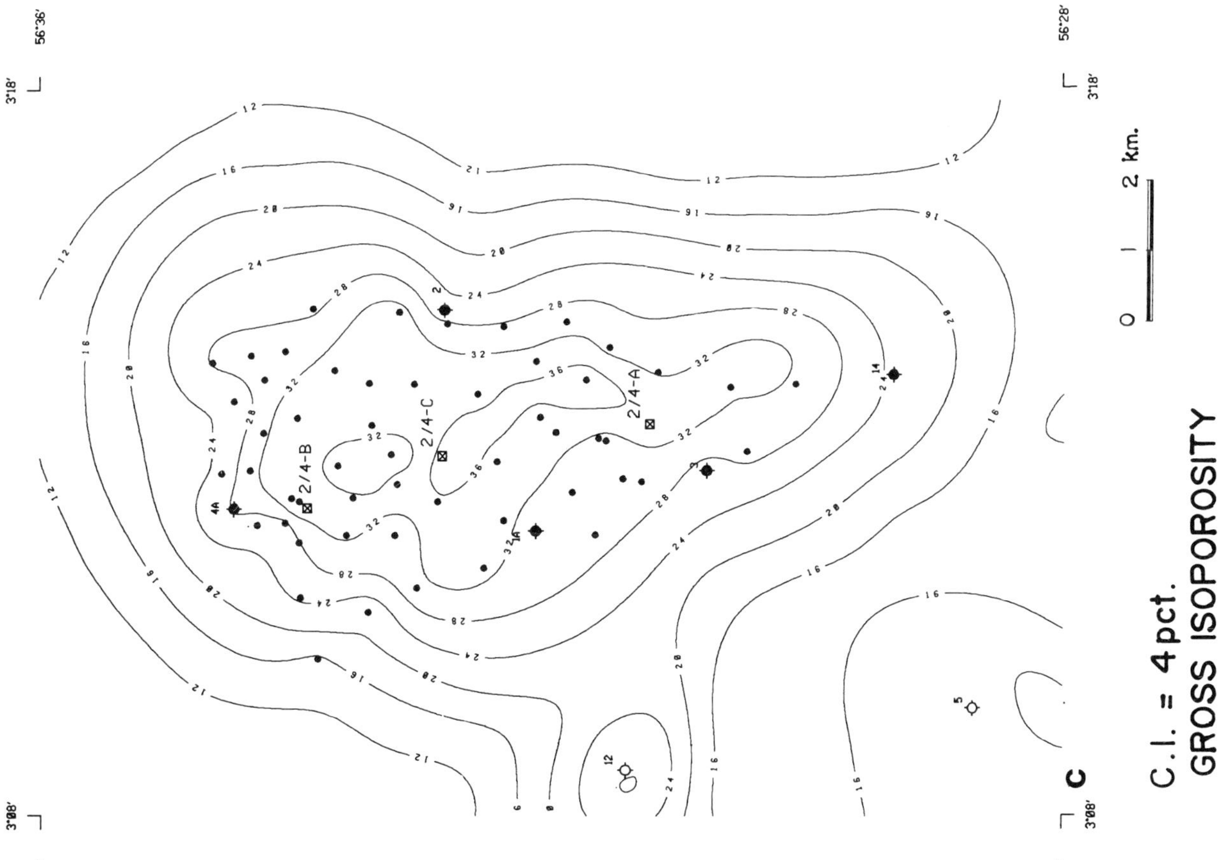

Fig. 9(c).

Fig. 9. Suite of maps describing the Ekofisk Formation layer 25EA.

to the discovery of the Ekofisk Field. Production characteristics and the type of problems that were encountered could not always be anticipated. Hydrocarbon distribution and recovery factors could not be precisely determined. Thus, Ekofisk reservoir management and optimization has been a learning process and remains a particularly challenging task.

As mentioned earlier, the Ekofisk reservoir is naturally fractured. This high-permeability fracture system is critical to the field's productivity and acts as the collecting conduit through which fluids travel from the lower-permeability matrix blocks to the well-bore. To enhance this capability, wells are stimulated with a hydraulic acid fracture treatment.

Water encroachment or increasing water cut in production wells has not been a significant problem. The field does not benefit from an active water drive. Both porosity and permeability are reduced in the non-hydrocarbon intervals. However, repeat formation tester (RFT) pressure data collected from the aquifer zone more than 100 m beneath the pay show partially reduced pressures. The aquifer, therefore, provides limited pressure support to the field.

Production difficulties are most commonly related to the mechanical properties of the chalk. Casing collapse is a relatively common problem. Solids production occurs occasionally and may temporarily reduce or stop a well's flow. Normally, a simple clean-out of the well is all that is required to restore production. As depletion of the reservoir has continued the frequency of these problems has not increased and most wells produce without trouble. Gravel-pack completions and the use of proppant in stimulation jobs are not required.

Down-hole corrosion and scale build-up have been a problem, causing frequent tubing leaks or restrictions. However, successful preventive maintenance programmes have been developed and the effects of these problems have been greatly reduced.

Zones of differential depletion have developed within the reservoir and this has increased drilling difficulties. New casing programmes, accurate selection of casing points and reliable pore-pressure estimates all help to alleviate this problem.

Seabed subsidence has recently been discovered on the Ekofisk Field. This subsidence is a consequence of reservoir compaction caused by the production of reservoir fluids. Compaction of over three metres has been recorded in some wells, most of which occurs in the higher-porosity Ekofisk Formation. The chalks of the Ekofisk Area are abnormally pressured; i.e. the pore fluids support a portion of the weight of the overburden. When this pressure is reduced by fluid production the support of the weight of the overburden is transferred from the fluid to the matrix. This increased stress on the matrix has caused compaction in the reservoir. There has, however, been no observable deterioration in reservoir performance. In order to reduce the rate of compaction, gas injection rates into the Ekofisk Formation have been temporarily increased from an average 3×10^6 Sm^3/day to 10×10^6 Sm^3/day. Modifications are being made to those platforms most affected so as to accommodate the increased water depth.

The productive interval of the Ekofisk Field, like the other chalk fields, is characterized by high-porosity, low-water saturation (see Fig. 9), low permeability, a large volume of oil-in-place and a low recovery factor. Originally oil recovery from pore-fluid expansion had been estimated at 157×10^6 Sm^3 or only 15% of the initial oil-in-place. However, as experience with operating chalk reservoirs was gained, programmes have been enacted to improve this performance.

Gas injection commenced in 1975 and has been credited with increasing oil recovery, more than 17×10^6 Sm^3. Implementation of the Tor Formation water-flood programme is anticipated to improve recovery further. Most recently, the increased injection of gas to counteract compaction and subsidence has reduced gas sales, but oil recovery will again be increased.

The combined effect of these programmes has raised Ekofisk oil recovery estimates to 237×10^6 Sm^3 (Norwegian Petroleum Directorate's estimate). In terms of recovery factor, this represents 22% of the original oil-in-place.

Enhanced-recovery studies continue. Expansion of the Tor Formation water-flood into the southern portion of the reservoir is under consideration. Also, a pilot water-flood field-test is in progress to help determine if the lower interval of the Ekofisk Formation (Reworked Maastrichtian Zone, 25ED) is suitable for full-field water-flood. The evaluation of nitrogen as an injection gas has been in progress for approximately three years and is in an advanced stage. It is anticipated that nitrogen injection may not significantly change Ekofisk oil recovery but would substantially increase recoverable gas reserves. If carried out, the project would by far surpass

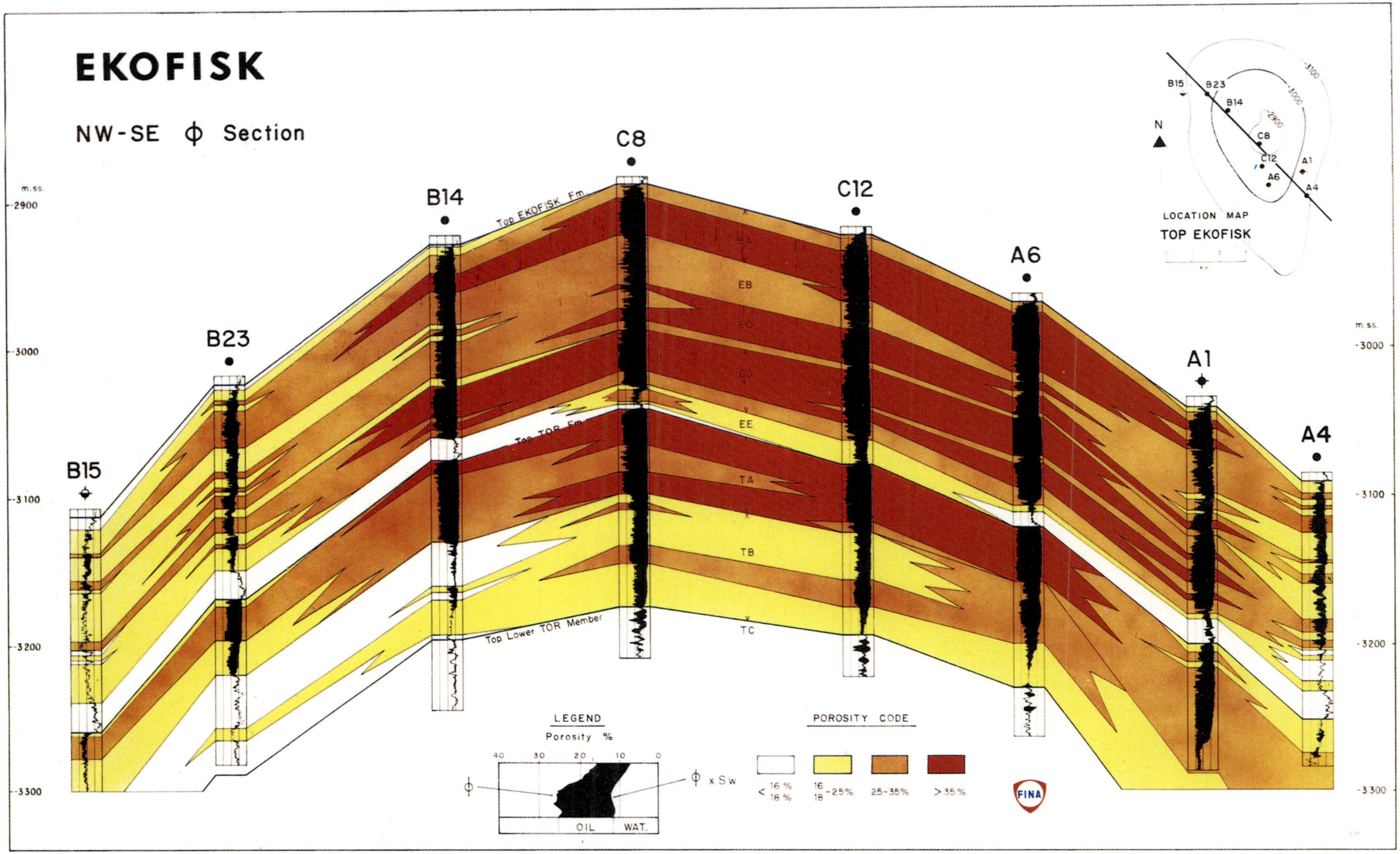

Plate 2. Ekofisk Field structure cross-section showing porosity distribution.

any previous experience in the industry with nitrogen for enhanced recovery.

ACKNOWLEDGEMENTS

The authors express their thanks to Phillips Petroleum Company, Norsk Fina A/S, Norsk Agip A/S, Elf Aquitaine Norge A/S, Norsk Hydro Production AS, Total Marine Norsk A/S and the other members of the Petronord Group by whose permission this paper is published. We are also grateful to Mr M. D'Heur for his assistance and encouragement. The opinions expressed in this paper are those of the authors and do not necessarily reflect those of Phillips Petroleum Company or any of the coventurers of the Ekofisk Field.

REFERENCES

Deegan, C.E. and Scull, B.J. 1977. A standard lithostratigraphic nomenclature for the Central and Northern North Sea, *Report Institute Geological Sciences UK* 77/25.

D'Heur, M. 1984. Porosity and hydrocarbon distribution in the North Sea chalk reservoirs, *Marine Petroleum Geology* **1,** 211–238.

D'Heur, M., de Walque, L. and Michaud, F. 1985. Geology of the Edda Field reservoir, *Marine Petroleum Geology* **2,** 327–340.

Gowers, M.B. and Sæbøe, A. 1985. On the structural evolution of the Central Trough in the Norwegian and Danish Section of the North Sea, *Marine Petroleum Geology* **2,** 298–318.

Hancock, J.M. and Scholle, P. 1975. Chalk of the North Sea, In Woodland, A.W. (Ed.), *Petroleum and the Continental Shelf of North-west Europe*, Vol. 1, Geology, Applied Science, Barking, pp. 413–427.

Hancock, J.M. 1976. The petrology of the chalk, *Proceedings Geologists Association* **86** 499–535.

Hardman, R.F.P. 1982. Chalk reservoirs of the North Sea, *Bulletin Geological Society, Denmark* **30,** 119–137.

Heidbreder, W.L. 1978. The development history of Greater Ekofisk, EUR 95. European Offshore Petroleum Conference and Exhibition, London, 24–27 October, 1978, pp. 229–306.

Kennedy, W.J. 1980. Aspects of chalk sedimentation in the Southern Norwegian Offshore, In *The Sedimentation of the North Sea Reservoir Rocks* (Norsk Petroleums Forening), Article IX; 29 pp.

Kennedy, W.J. 1985. Sedimentology of the Late Cretaceous and Early Paleocene chalk group, North Sea Central Graben, *Chalk Research Program Symposium*, Stavanger, May 1985.

Nygaard, E., Lieberkind, K. and Frykman, P. 1983. Sedimentology and reservoir parameters of the chalk group in the Danish Central Graben, *Geol. Minjb.* **62,** 177–190.

Van den Bark, E. and Thomas, O.D. 1981. Ekofisk: first of the giant oil fields in Western Europe, *American Association of Petroleum Geologists Bulletin* **65,** 2341–2363.

Watts, N.L., Lapré, J.F., van Schijndel-Goester, F.S. and Ford, A. 1980. Upper Cretaceous and Lower Tertiary chalks of the Albuskjell area, North Sea: deposition in a slope and a base-of-slope environment, *Geology* **8,** 217–221.

Watts, N.L. 1983. Microfractures in chalks of Albuskjell Field, Norwegian Sector, North Sea. Possible origin and distribution, *American Association of Petroleum Geologists Bulletin* **67**(2), 201–234.

Ziegler, P.A. 1982. *Geological Atlas of Western and Central Europe*, Shell Internationale Petroleum Maatschappij B.V., The Hague.

Ziegler, W.H. 1975. Outline of the geological history of the North Sea, In Woodland, A.W. (Ed.), *Petroleum and the Continental Shelf of North-west Europe*, Vol. 1, Geology, Applied Science, Barking, pp. 413–427.

6

Eldfisk

Frédérique Michaud

Phillips Petroleum Company Norway, Present address, Elf Italia, 8 Largo L. Mossa, I-00165 Roma, Italy

The Eldfisk Field is located in the southern part of the Norwegian sector of the North Sea. The discovery well, 2/7-1X, was drilled in 1970. Since then, 44 wells or side-tracks have been drilled from two platforms: Alpha (south) and Bravo (north). The reservoir is chalk of the Hod, Tor and Ekofisk formations which are of Turonian to Danian age. It has porosities of up to 42% and permeabilities ranging from 0.01 to 35 mD. The field contains an undersaturated volatile oil. The initial recoverable reserves are estimated to be 48×10^6 Sm^3 of oil plus condensate and 39×10^9 Sm^3 of gas. Production started in 1979 from the two platforms. The present development plan involves draining the reservoir by solution gas drive only, but secondary recovery studies are taking place.

INTRODUCTION

The Eldfisk Field is located in Block 2/7 of the Norwegian North Sea, 330 km south-west of Stavanger (Fig. 1). It is one of the seven producing fields making up Greater Ekofisk. Water depth in the area is approximately 70 m. Eldfisk is situated south of the giant Ekofisk Field and north-west of Amoco's Valhall Field. The field consists of two main structures and a small one on the east. Chalk of Maastrichtian and Danian age constitutes the principal reservoir. Initial oil-in-place is estimated to be 252×10^6 Sm^3.

The field was discovered in 1970, one year after the discovery of the Ekofisk Field.

It received its name through a combination of two practices of the Phillips Group. It begins with 'E' because the Block 2/7 was part of an area designated by the letter E, and Eldfisk (*fisk* = fish in Norwegian) because at that time all Phillips Group prospects in Norway were being named after fish.

HISTORY

Production Licence 018, which covers Block 2/7, was granted in 1965. Phillips Petroleum Company Norway is the operator and the present licence holders are:

Phillips Petroleum Norsk A/S	36.98%
Norske Fina A/S	30%
Norsk Agip	13.04%
Elf Aquitane Norge A/S	8.094%
Norsk Hydro Production A/S	6.7%
Total Marine Norsk	4.047 %
Eurofrep Norge A/S	0.456%
Coparex Norge A/S	0.399%
Cofranord A/S	0.304%

The Eldfisk structure was identified by seismic survey in the late 1960s. But over most of the field the seismic signal was severely degraded and pulled down due to a gas effect in the Tertiary shales.

In 1969 a seismic grid of approximately 2 × 2 km existed in the area. Well 2/7-1X, the

Geology of the Norwegian Oil and Gas Fields © A. M. Spencer *et al.*, eds (Graham & Trotman, 1987) pp. 89–105.

Eldfisk Field data summary

Trap	
Type	dome structure over salt swell
Depth to crest	At top Ekofisk Formation: Alpha 2700 mss, Bravo 2800 mss
Lowest closing contour	2896 mss at top Ekofisk Formation
Oil–water contact	variable
Oil column (average)	380 m
Pay zone	
Formation	Ekofisk, Tor and Hod formations chalks
Age	Danian and Late Cretaceous
Gross thickness (average)	200 m
Net/gross ratio (average; range)	80%; 20–100%
Porosity (averages; range)	32%; 18–40% Ekofisk Formation 28%; 14–42% Tor Formation 18%; 8–29% Hod Formation
Hydrocarbon saturation (average)	85% in the net pay
Permeability (range)	(0.01–35 mD) Formation matrix with effective permeability up to 140 mD (fractures)
Hydrocarbons	
Oil gravity (initial)	0.84 g/cm^3
Gas gravity (initial) (relative to air)	0.70
Bubble point	414 bar (North); 352 bar (South)
Gas/oil ratio (Initial)	377 m^3/m^3 (North); 264 m^3/m^3 (South)
Formation volume factor (Initial)	2.08 m^3/m^3 (North); 1.74 m^3/m^3 (South)

Formation water	
Salinity (average)	52 000 ppm NaCl equivalent
Resistivity (average)	0.042 m at 118 °C
Reservoir conditions	
Temperature	119 °C
Pressure gradient in reservoir	Ranging from 0.156 to 0.163 bar/m
Initial pressure	470 bar
Field size	
Net area	18 km^2
Net rock volume	1963×10^6 m^3
Hydrocarbon pore volume	475×10^6 reservoir m^3
Drive mechanism	Solution gas drive
Recovery factor	18% (North), 22% (South)
Recoverable oil (initial)	48.26×10^6 Sm^3
Recoverable gas (initial)	39×10^9 Sm^3
Production	
Start-up	October 1979 (North: Bravo), August 1979 (South: Alpha)
Development scheme	2 production platforms with 35 production wells
Production rate (October 1985	9208 Sm^3/day oil, 5×10^6 Sm^3/day gas
Cumulative production (October 1985)	26×10^6 Sm^3 oil, 10×10^9 Sm^3 gas

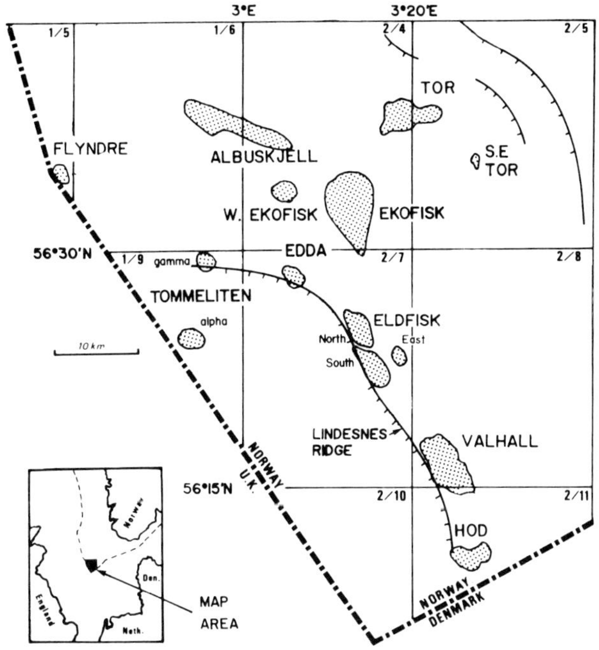

Fig. 1. Location map showing main tectonic elements.

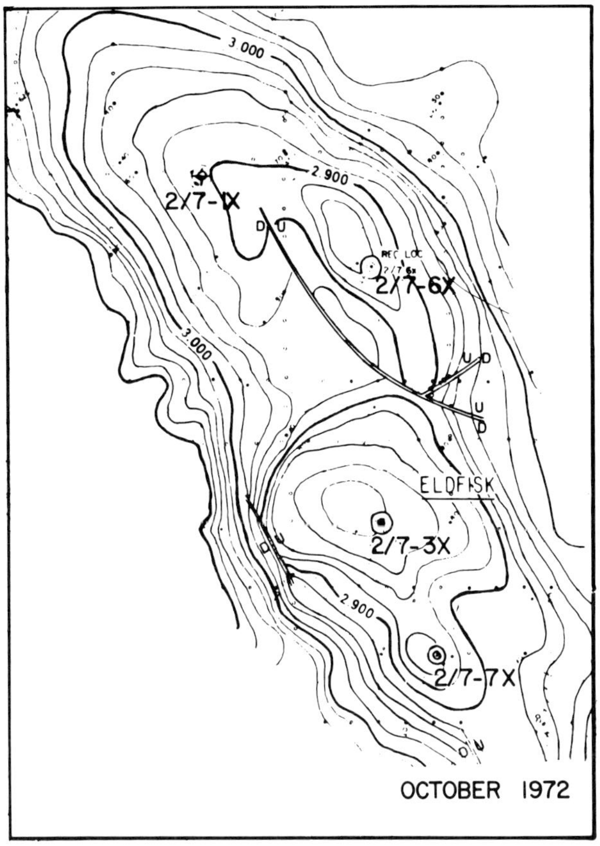

Fig. 2. Top Danian time map of the Eldfisk Field with exploration and appraisal well proposed locations (October 1972).

discovery well, was drilled north-west of the crest of the northern structure and was completed in 1970 with oil shows in the Ekofisk and Tor formations. Drill-stem tests in the Ekofisk Formation were unsuccessful: flow-rates were too low to measure. Attempts to enhance seismic resolution continued throughout the 1970s with re-shooting and reprocessing but little improvement was achieved.

Based on a structural map made in late 1972 (Fig. 2), four appraisal wells were drilled and completed in 1973: 2/7-6X, 2/7-3X, 2/7-7X, 2/7-8X. The Ekofisk and Tor formations gave encouraging results, and measured oil rates were 715-795 Sm^3/day from 2/7-3X and 2/7-6X and 1192 Sm^3/day from 2/7-7X. In addition Wells 2/7-3X and 2/7-7X revealed a good pay zone in the Hod Formation which tested 667 Sm^3/day and 977 Sm^3/day in each well, respectively. Well 2/7-8X discovered the presence of a small separate accumulation to the east of Eldfisk, known today as East Eldfisk. A cross-section made at this time shows rapid changes in thickness and productivity over the field (Fig. 3).

Eldfisk was then declared commercial: three permanent platforms were installed and 49 development wells were planned. Their proposed locations were based mainly on the preliminary pay mapping, derived from the exploration and appraisal wells and the reinterpreted structural map (Fig. 4). One drilling and production platform 2/7 Bravo was set on the northern structure. On the southern structure, two platforms were set, a drilling platform 2/7 Alpha, and a production/process platform (2/7 FTP). The latter is used to process the oil and gas before it is sent to the 2/4 Tank on the Ekofisk Complex.

Development drilling started on 2/7 Alpha in 1978 and on 2/7 Bravo in 1979. On the South Eldfisk structure several dry holes were drilled and disappointing results, with higher water saturations and thinner pay sections than expected, were obtained. The North Eldfisk structure did not experience such disappointments, however. Development drilling proved that the original structural map and simple reservoir interpretation were inadequate to explain the complexity of the field. New maps were then drawn. A total of 15 producing wells were drilled from 2/7 Bravo and 20 from 2/7 Alpha. Two of the wells drilled from Alpha reached the East Eldfisk accumulation, but only one is currently producing. Development drilling is still taking place on both 2/7 Bravo and 2/7 Alpha and studies are currently being made to determine the feasibility for gas lift and gas injection on Eldfisk.

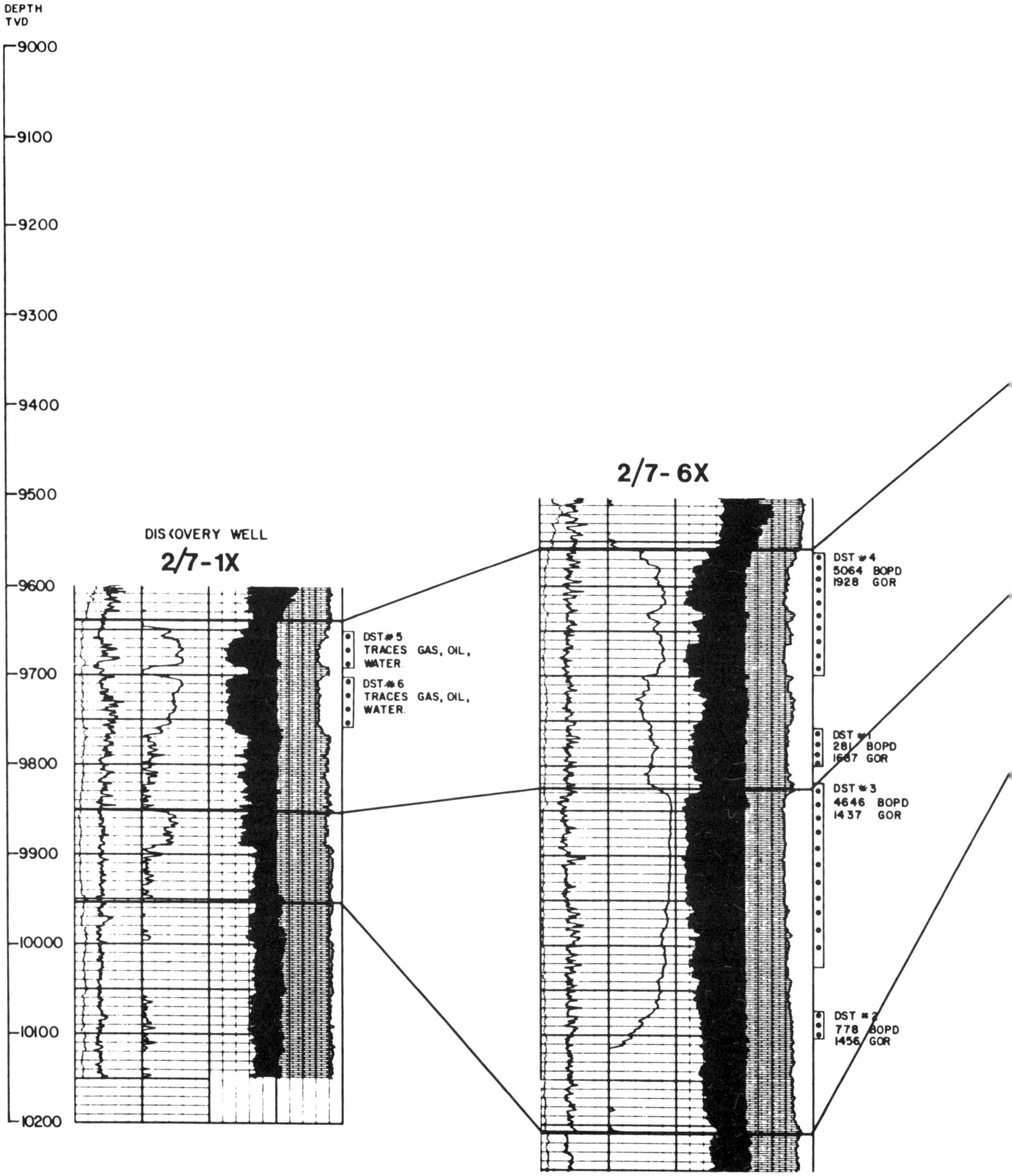

Fig. 3. Stratigraphic section on Eldfisk with available data in 1975.

The wells were completed either in the Ekofisk Formation only, the Tor Formation only, the Hod Formation only, or a combination of these three.

Cumulative production as of October 1985 has been 26×10^6 Sm^3 of oil and 10×10^9 Sm^3 gas and production is expected to continue beyond expiration of the production licence agreement in the year 2011.

GEOLOGICAL SETTING

The Eldfisk Field is situated in the Central Graben. The general geological setting as

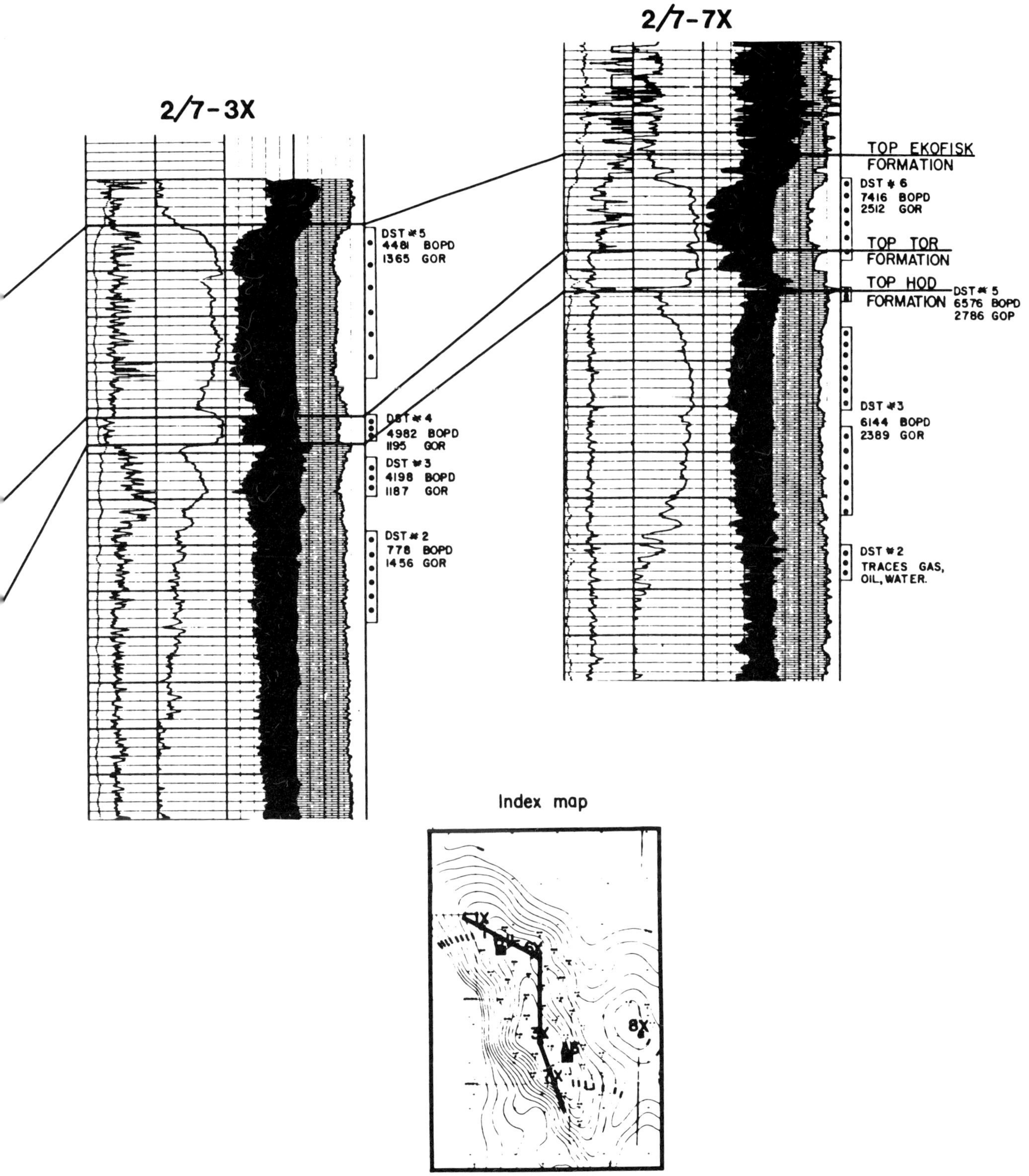

well as the tectonic development of the area has been documented by D'Heur (1984); Ofstad (1983); Skovbro (1983); Ziegler (1982). A recent study by Gowers and Sæbøe (1985) describes the structural evolution of the area from Permian to Tertiary times.

Of particular importance in the Eldfisk area is the Lindesnes Ridge feature, which was initiated in early Jurassic times as a down-to-the-east fault, giving rise to a deep Jurassic basin to the east and north-east. The rapid influx of sediments created by extensional movements and basin subsidence probably initiated halokinetic movements. During

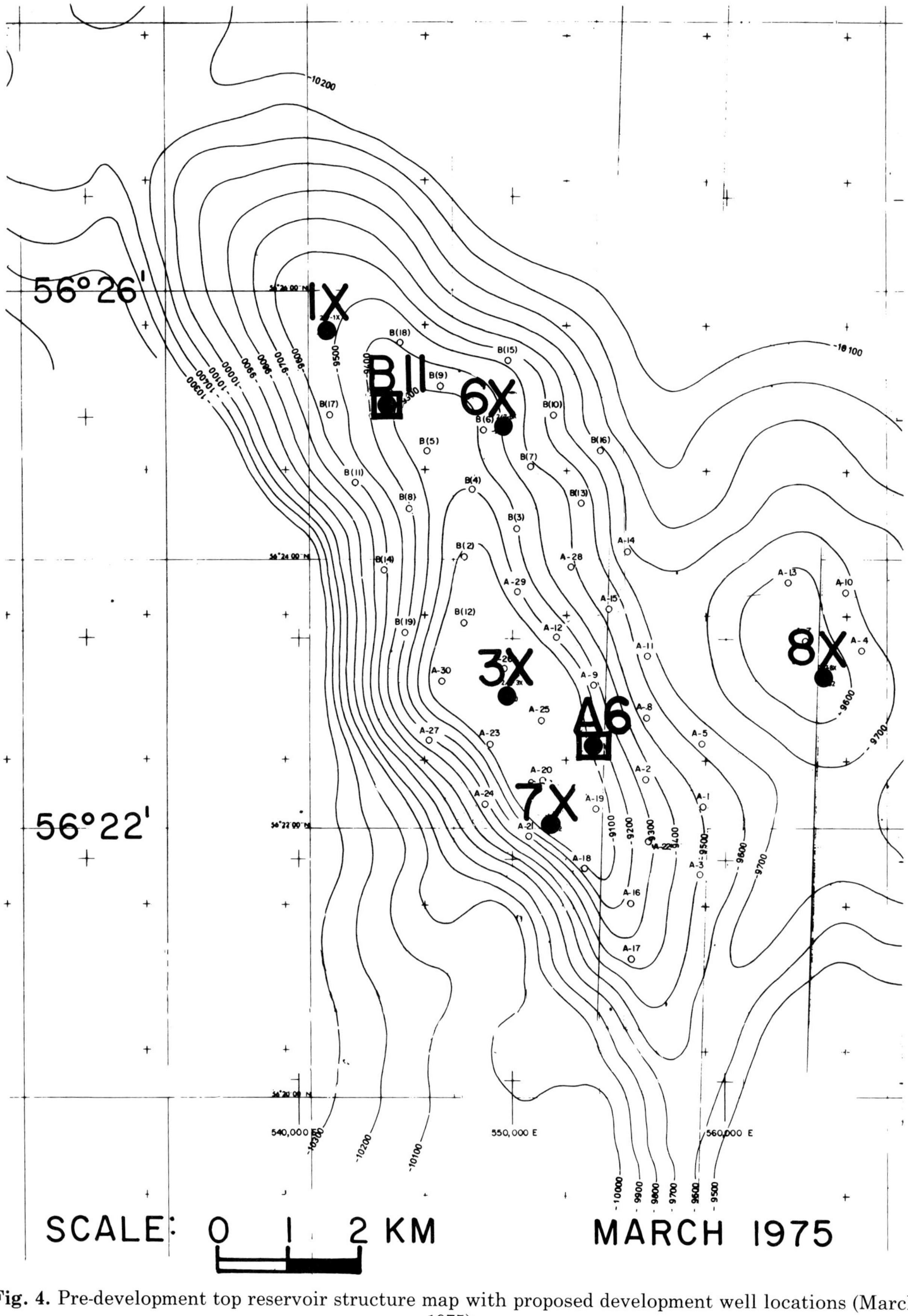

Fig. 4. Pre-development top reservoir structure map with proposed development well locations (March 1975).

earliest Cretaceous times inversion of the Lindesnes Ridge transformed the down-to-the-east normal fault into an apparent down-to-the-west normal fault. Salt diapirism took place along this line of weakness, with antithetic normal faults.

A marine transgression during Late Cretaceous times resulted in the beginning of chalk deposition. The tectonic activity controlled the thickness distribution: the chalk exceeds 1000 m in the western basin area and is as thin as 150 m on the crestal parts of the Lindesnes Ridge where Eldfisk is located (Fig. 5). Tectonics of the area were further complicated by an episode of wrenching in latest Cretaceous times. The resulting *en échelon* structures also influenced the deposition of the chalk in this area. The Hod–Valhall–Eldfisk area was a structural high subject to erosion, whereas the Albuskjell–Ekofisk–Tor area was a depocentre (Fig. 6). Pulses of halokinesis continued sporadically throughout Cretaceous times and were responsible for: syndepositional and post-depositional fault movements, mobilization and redistribution of unstable carbonate sediments, and the generation of traps and fractures and, therefore, creation of the hydrocarbon reservoirs. Chalk sedimentation ended in Late Danian times when a new regional relief generated a rapid and extensive clastic infill. This rapid deposition caused abnormally high pressures in the overburden, producing one of the factors responsible for the high-porosity preservation in the chalk reservoirs.

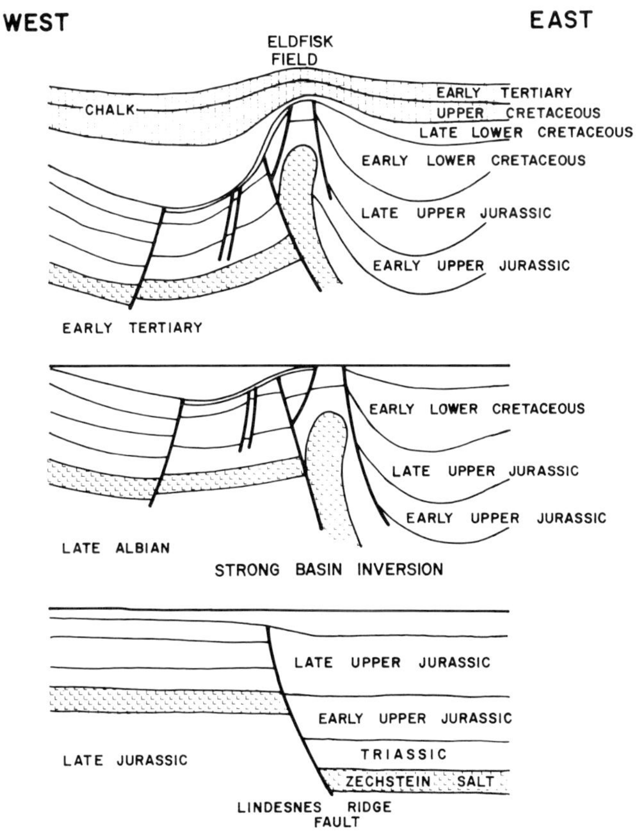

Fig. 5. Structural evolution of the Eldfisk Field area.

The main subdivisions of the Chalk Group employed at Eldfisk conform to the nomenclature defined by Deegan and Scull (1977), i.e. the Ekofisk, Tor, Hod, Plenus Marl and Hidra formations. The Eldfisk reservoir intervals occur in the Ekofisk, Tor and Hod Formations (Fig. 7). Sedimentological descriptions of these formations have been published by Kennedy (1980). The Phillips Group uses further stratigraphic subdivisions based on biostratigraphic correlations, core descriptions and wire-line measurements. This layering can be identified throughout the Greater Ekofisk Area and reflects the consistent pattern of chalk deposition.

GEOPHYSICS

The structure was located by seismic data collected in the late 1960s but most of the seismic shot over the crest of the field was poor because of the gas effect in the Tertiary shales. This caused either dramatic push-down or destruction of the seismic reflectors (Fig. 8). Attempts to enhance seismic resolution continued throughout the 1970s with reshooting and reprocessing, but little improvement was achieved. Remapping and reinterpretation of seismic data led to the production of various maps with a variety of fault patterns affecting the top chalk. The faults were removed or reintroduced to explain the thickness variations in the Tor and Hod formations encountered during development drilling. A suite of top chalk maps produced between 1979 and 1984 showing the various postulated fault patterns were reproduced by Brewster and Dangerfield (1984).

In 1983 the first vertical seismic profile (VSP) was conducted in a straight hole from the Bravo platform, Well 2/7 B-12. Superposition of the VSP on the surface seismic improved the resolution of the data and indicated a fault 600 m east of B-12, downthrowing the base of the Ekofisk Formation to the east. The VSP improved the mapping of the structure in this area. Unfortunately, over most of the field, the area of the gas effect is too wide and this technique is of limited use.

Originally the top Ekofisk Formation was mapped directly from seismic data. The struc-

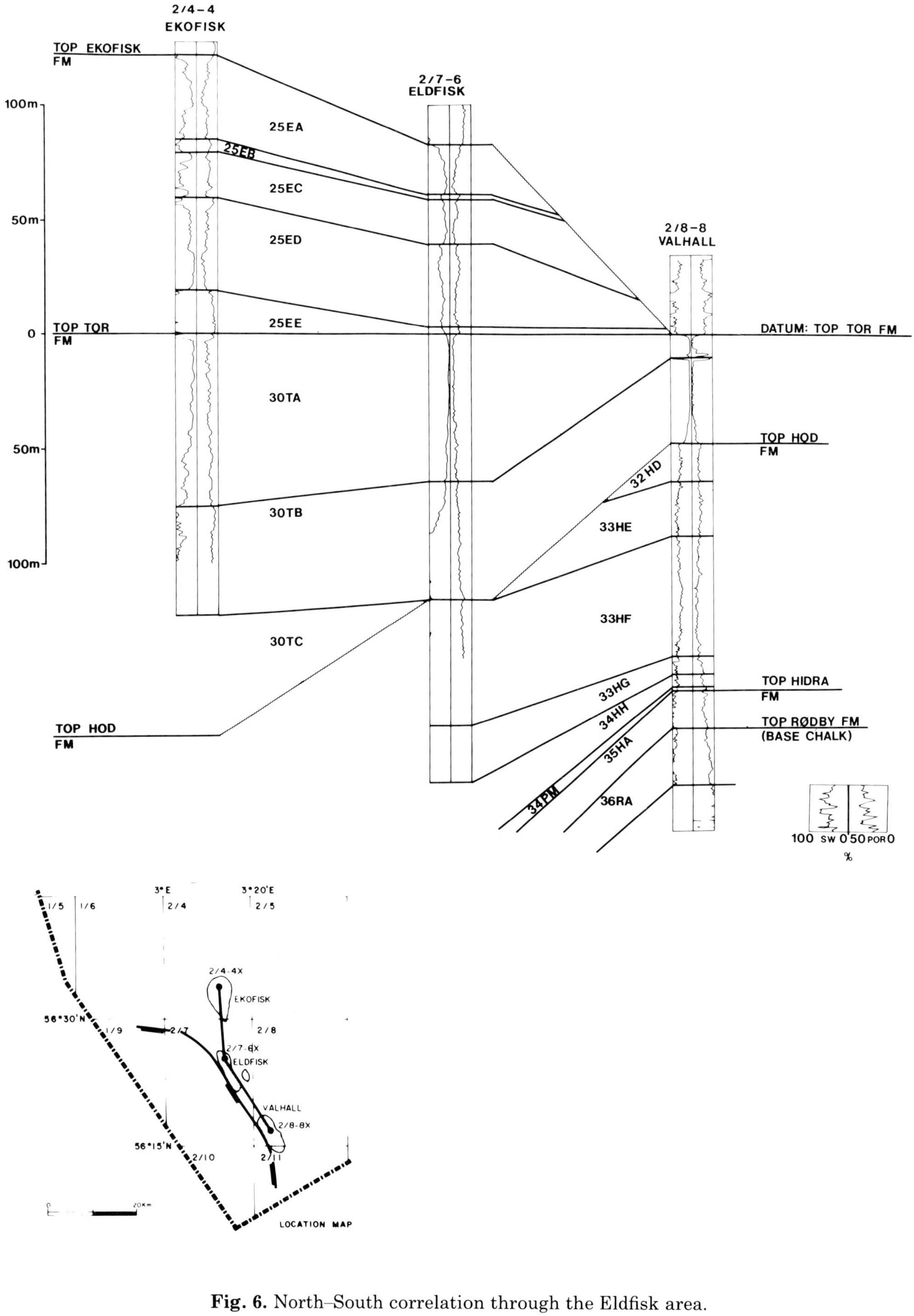

Fig. 6. North–South correlation through the Eldfisk area.

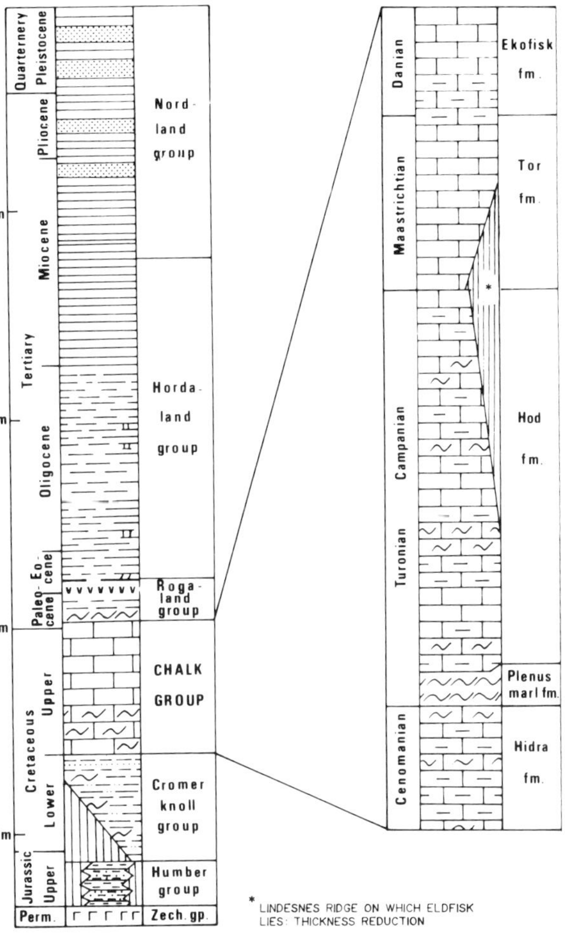

Fig. 7. Lithostratigraphy.

ture map was then improved using well control. The structure maps of the Tor and Hod formations were constructed by subtracting the isopach of each formation.

TRAP

The Eldfisk structure consists of three domes separated by structural saddles. They were created by the interaction of halokinetic movements, inversion, strike-slip faulting and compressional folding. There is a lateral displacement between the northern and southern parts of the field (Fig. 9).

Halokinesis continued sporadically throughout Cretaceous and into Tertiary times providing a four-way dip closure. Salt was encountered at 4192 m in the 2/7-3X crestal well. The hydrocarbon accumulation is controlled by reservoir characteristics, however, and no oil–water contact can be clearly defined. A much thicker transition zone between the oil- and water-bearing sections is observed on the flanks than in the central area (see West Ekofisk, this volume, p. 165). An apparent dextral strike-slip fault, produced by differential salt movement, offset the north and the south structures. It resulted in a present-day difference in their structural elevations and it created a permeability barrier between North Eldfisk and South Eldfisk.

The seal is formed by the thick sequence of Tertiary shales and claystones, which are about 3000 m thick. Gas has escaped from the crest of the reservoir into the Tertiary section and has been trapped in Miocene silty beds.

RESERVOIR

Rock types

Most of the North Sea chalk is composed of skeletal debris of coccolithophoridae (pelagic unicellular algae). These coccoliths are often disaggregated into laths (or minute platelets). Their mineralogy consists of a chemically stable form of low magnesium calcite. Their size varies between 0.5 and 1 μm which creates a low original matrix permeability.

Several sedimentological descriptions of the chalk of the North Sea have been published. Kennedy (1980, 1985), for example, has recognized a number of sedimentary facies which can be classified as follows:

- Autochthonous pelagic chalk, which commonly consists of rhythmic alternations of pale and dark sequences. The lighter units are termed limestones and are often bioturbated. The darker layers are termed marls and are often laminated.
- Allochthonous chalk, resulting from reworking of a previously deposited coccolith ooze. These layers generally correspond with the high-porosity reservoirs.

Most of the chalk reservoirs of the Central Graben appear to be allochthonous, with only thin, interbedded, in-situ pelagic chalks. The alternation of low-porosity pelagic chalk and high-porosity debris-flows reflects alternating episodes of quiescence and tectonic instability. In the Eldfisk Area, the Lindesnes Ridge had a major influence on chalk deposition and redeposition (D'Heur, 1986).

Regional correlation based on petrophysical characteristics and sedimentary facies (Fig. 10, Plate 3), developed from a well database of over 200 wells in the Great Ekofisk area, demonstrate that these units have a common depositional history over hundreds of square kilometres.

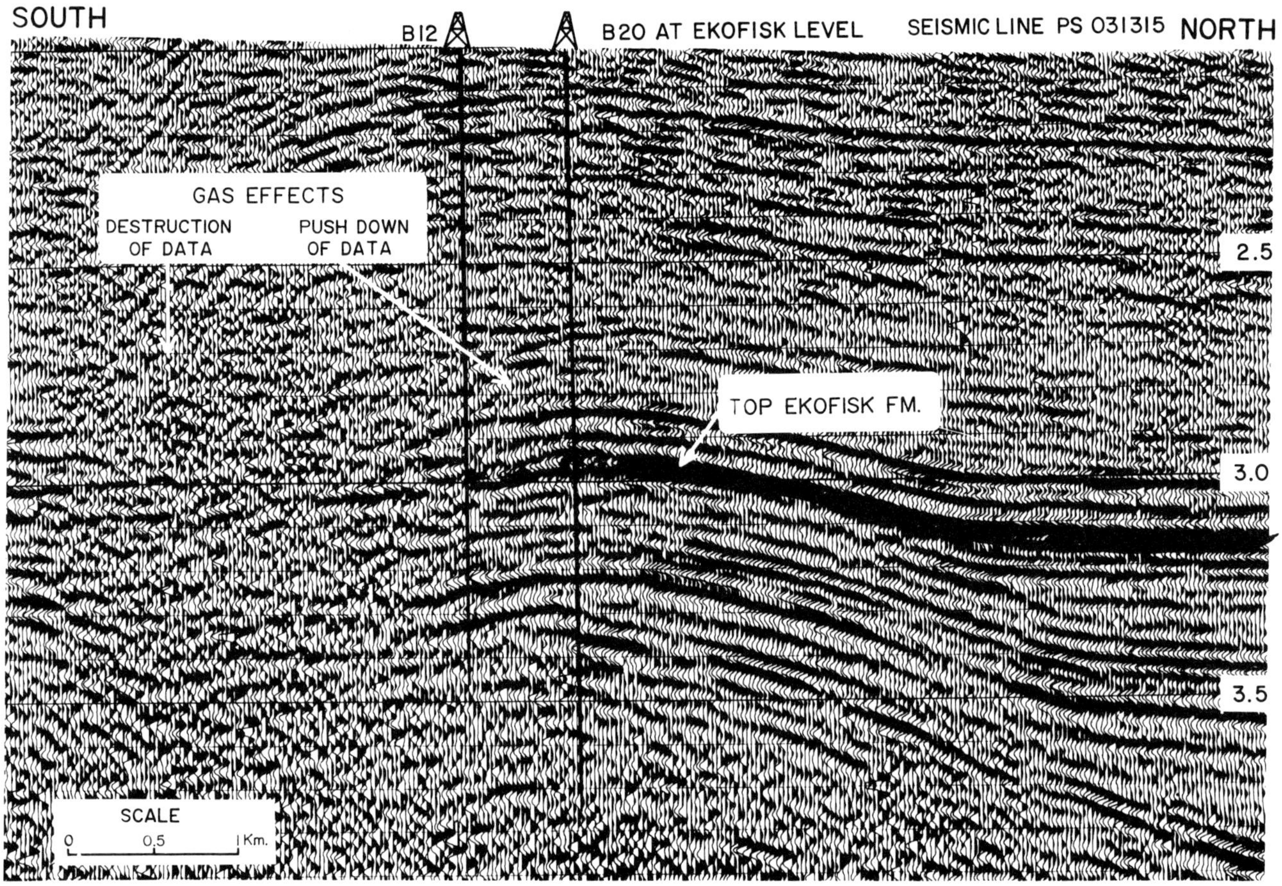

Fig. 8. S–N seismic line PS-031315 through Well 2/7 B-12 showing gas effect.

Reservoir Subdivision

Reservoirs in the Eldfisk Field occur within the Hod, Tor and Ekofisk formations.

Hod Formation

The Hod Formation is of Turonian to mid-Campanian age and subdivisions into Lower, Middle and Upper Members can be recognized on Eldfisk (Ofstad, 1983). A finer subdivision into eight units has also been recognized (Brewster and Dangerfield, 1984). The Lower Hod Member consists of clean chalks. The Middle Hod Member shows alternations of clay-poor layers. The Upper Hod Member is cleaner than the Middle Hod but similar in lithology. The Hod Formation is productive in the South Eldfisk structure. The Hod reservoirs are found mainly in the Upper Hod Member (2/7-3X, 2/7A-19B, 2/7A-25, 2/7A-11, 2/7A-26, 2/7A-16A), but the Lower Hod Member is also hydrocarbon bearing in the structurally highest part of South Eldfisk, where the Upper Hod Member is missing. The boundary between the Hod and the Tor formations is characterized by an angular unconformity which is commonly a hard-ground. This unconformity is less marked in some wells (2/7B-10, 2/7B-6) and on the flanks where a fairly complete sequence of Upper Hod exists. The reservoir performance indicates a permeability restriction at the unconformity.

Tor Formation

The Tor Formation is of Maastrichtian age and is subdivided into three layers.

The Lower Tor Member is recognized only in a few wells in the centre of North Eldfisk, close to Well 2/7B-10. It is characterized by a succession of thin, high-porosity debris-flows and thinner, low-porosity pelagic chalks.

The Middle Tor Member consists of alternations of clean, allochthonous, very poor chalks and shaly marls. The deposition of this member has been influenced by tectonism; chalk eroded from the uplifted axis has been redeposited in the adjacent subsiding area.

The Upper Tor Member is the best reservoir of the field with average porosities of up to 42% and water saturations below 10%. This member is a clean chalk which consists of many allochthonous units in which slumping and plastic deformation are common. The deposition of this member is strongly affected by the palaeorelief and the tectonic move-

2/7 - B17

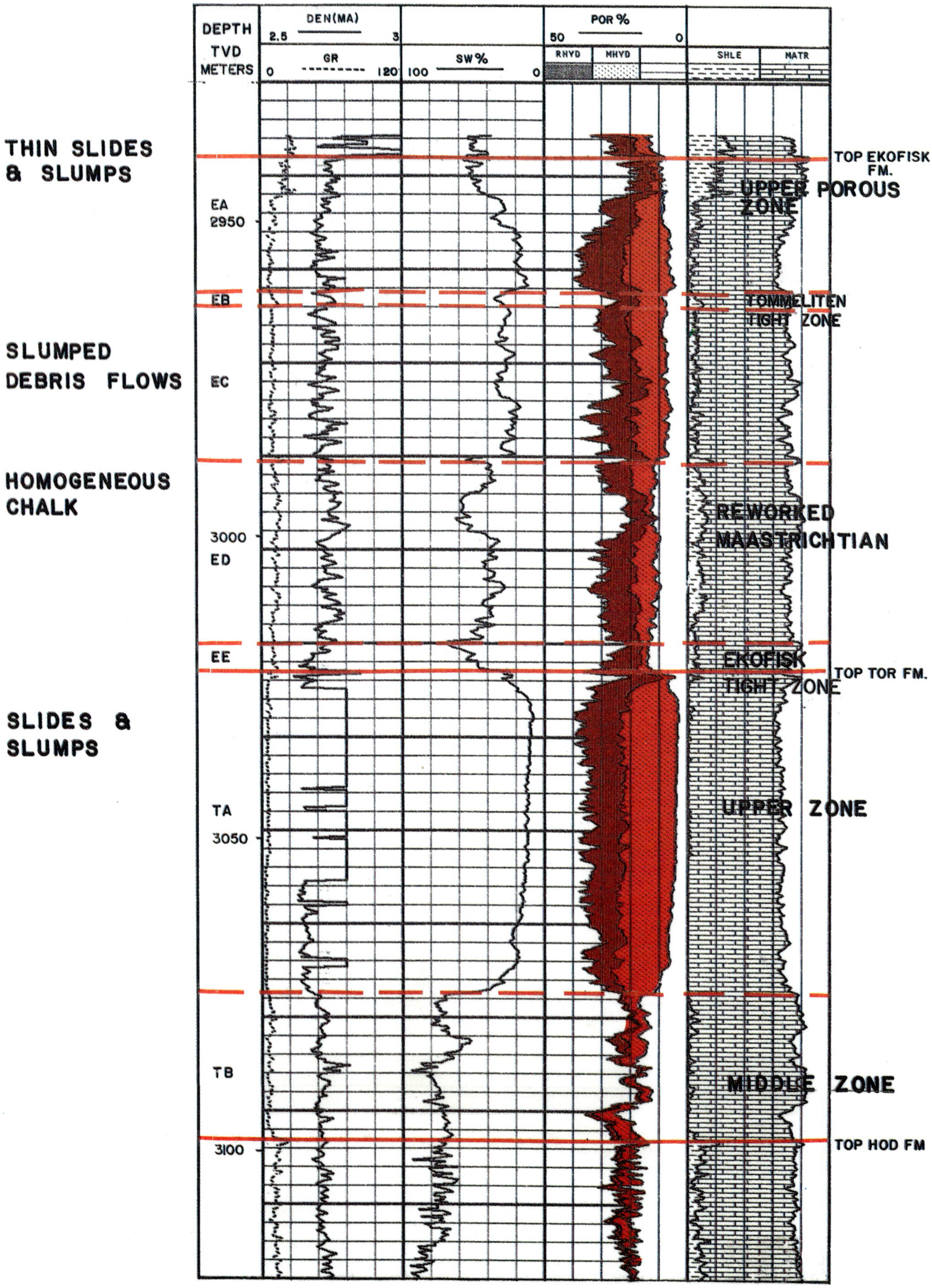

Plate 3. Computer-processed interpretation of Well 2/7 B-17.

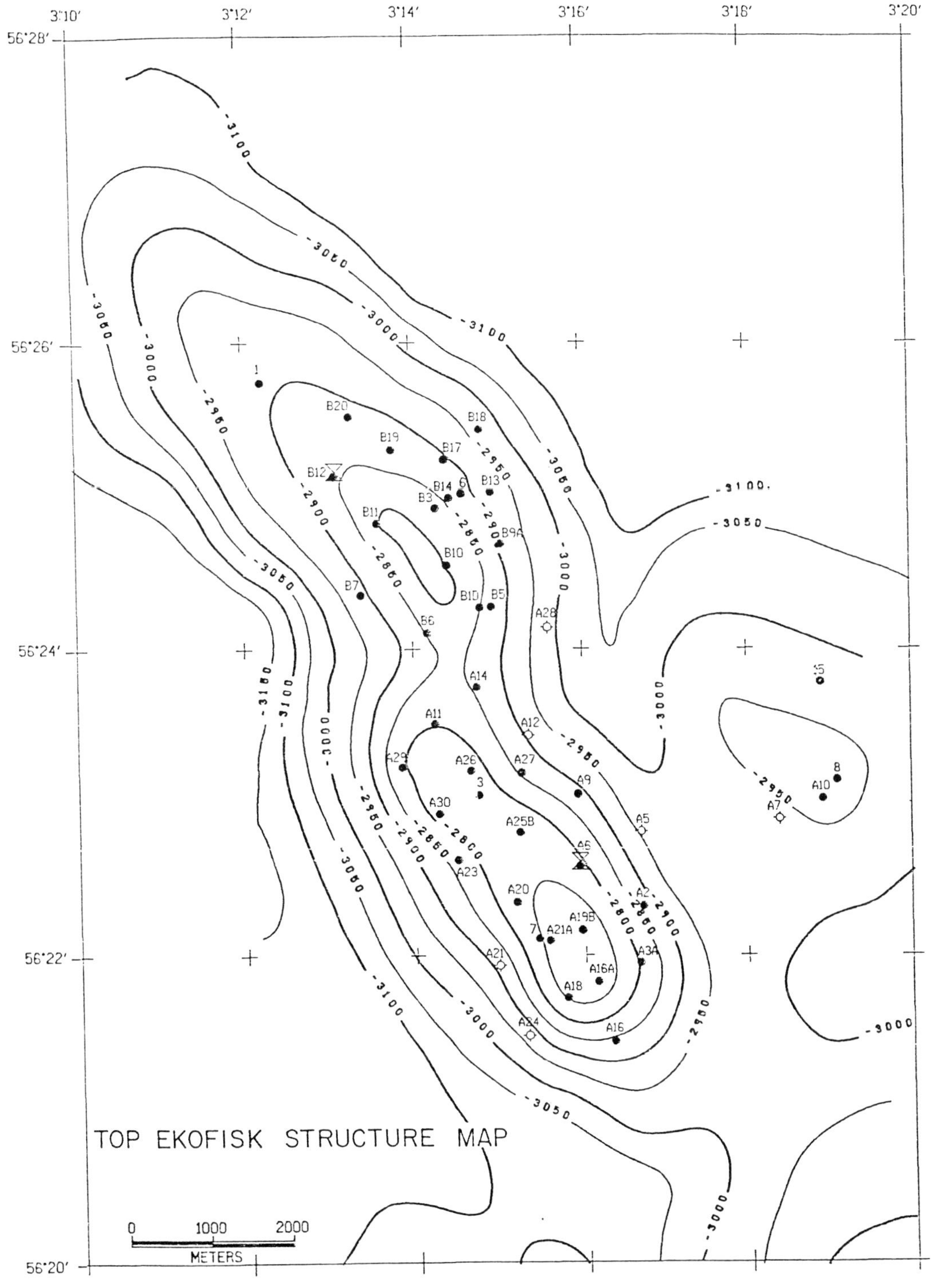

Fig. 9. Present-day structure map of the Top Ekofisk Formation in the Eldfisk Field area. Contour interval 50 m.

ments. The boundary between the Tor Formation and the Ekofisk Formation coincides with the Cretaceous/Paleocene boundary and is usually unconformable. It corresponds with a hard-ground or an erosion surface.

Ekofisk Formation

The Ekofisk Formation is of Danian age. A low-porosity interval occurs overlying the Tor Formation. However, it does not correspond with the typically clay-rich, periodite facies of the Ekofisk Tight Zone of the Ekofisk Field.

The Reworked Maastrichtian Zone (ED, see Plate 3) consists of homogeneous chalk and argillaceous interbeds. Much of the bedding is preserved suggesting the homogeneous deposition of remobilized Maastrichtian sediments without further slumping. Core analyses show poor reservoir qualities.

PALAEOCENE	EKOFISK FORMATION	25EA	UPPER EKOFISK POROUS ZONE
		25EB	TOMMELITEN TIGHT ZONE
		25EC	REWORKED DANIAN ZONE
		25ED	REWORKED MAASTRICHTIAN ZONE
		25EE	EKOFISK TIGHT ZONE
MAASTRICHTIAN TO TURONIAN	TOR FORMATION	30TA	UPPER TOR
		30TB	MIDDLE TOR
		30TC	LOWER TOR
	HOD FORMATION	31HA-32HD	UPPER HOD
		32HE	MIDDLE HOD
		34HF-34HH	LOWER HOD
	PLENUS MARL	34PM	
CENOMANIAN	HIDRA	35HA	

Fig. 10. Phillips Group zonation in the chalk.

Recent repeat formation tester (RFT) results show a transitional pressure gradient, which indicates a possible reduced vertical communication within this zone between the Tor Formation below and the remaining Ekofisk Formation above.

The Reworked Danian Zone (EC—Plate 3), is usually argillaceous and exhibits rather low porosity, but reservoir quality is better than that of the underlying unit. Much of the bedding is disturbed, and thin slumped layers of both shale and chalk are present. This suggests extensive mass flow and slumping.

The Tommeliten Tight Zone (EB—Plate 3), which corresponds to a pause in sedimentation, is poorly represented over the field except in Well 2/7-1X in the north-western area.

The Upper Porous Zone (EA—Plate 3) is a homogeneous clean chalk deposited throughout the field. There is little evidence for reworked material in this zone although extensive bioturbation masks most of the primary depositional features. The upper boundary is gradational, passing from chalky limestone to marl.

Reservoir Distribution and Characteristics

Thickness distribution

The Hod Formation isopach varies from 120 to 730 m with the thickest accumulation to the west and a thickness reduction over the palaeocrest.

The Tor Formation thickness varies from 6 m to more than 200 m. The isopach map (Fig. 11) shows regional thinning on the eastern flank, and thickening on the western flank. There is a local thickness reduction along the crest of the palaeohigh (B-12, A-29, A-30, 7X), and a thickening in a contemporaneous topographic low ('depositional sink') in the B-10 area. Faults in the Tor Formation demonstrate both syndepositional and post-depositional movements in response to repeated halokinetic pulses.

The Ekofisk Formation varies in thickness from 6 m to 100 m. The isopach map (Fig. 12) shows a regional thinning towards the south-east, a regional thickening towards the west and thin deposition over the palaeohigh. A slight topographic low still existed in Danian times in the B-10 area. The thickness distribution of the Ekofisk Formation still shows the influence of the Lindesnes Ridge, but the effects are less pronounced.

It is worth noting that in both structures (North and South) the present axis of folding is 150 m east of the palaeocrest. This is probably due to mechanical reasons; it also shows that the separation and lateral displacement of the two structures took place well after Danian time (D'Heur, 1984).

Porosity Distribution

The present-day porosity distribution on the Eldfisk Field is directly related to both the original porosity distribution and the structural configuration of the reservoir. Excellent average porosities are found in the crestal areas with up to 42% in South Eldfisk (Plate 2) and up to 32% in North Eldfisk and in the 'depositional sink' (B-10 Area). Favourable overpressuring conditions, entry of oil into the reservoir and low magnesium calcite, are the main factors responsible for the abnormally high chalk porosity.

The porosity decreases with depth in each reservoir unit due to progressive infill of the trap by hydrocarbons and subsequent inhibition of diagenesis. This explains why the best porosities are encountered on South Eldfisk which is structurally higher than North Eldfisk. For the same reason, the Hod Formation is a good reservoir in the crestal part of South Eldfisk only. However, the Tor Formation locally retains a high porosity even when not oil bearing (B-18, A-5). This could be explained by late tectonic movements and subsequent lowering of the high-porosity reservoir into the aquifer. Such a structural

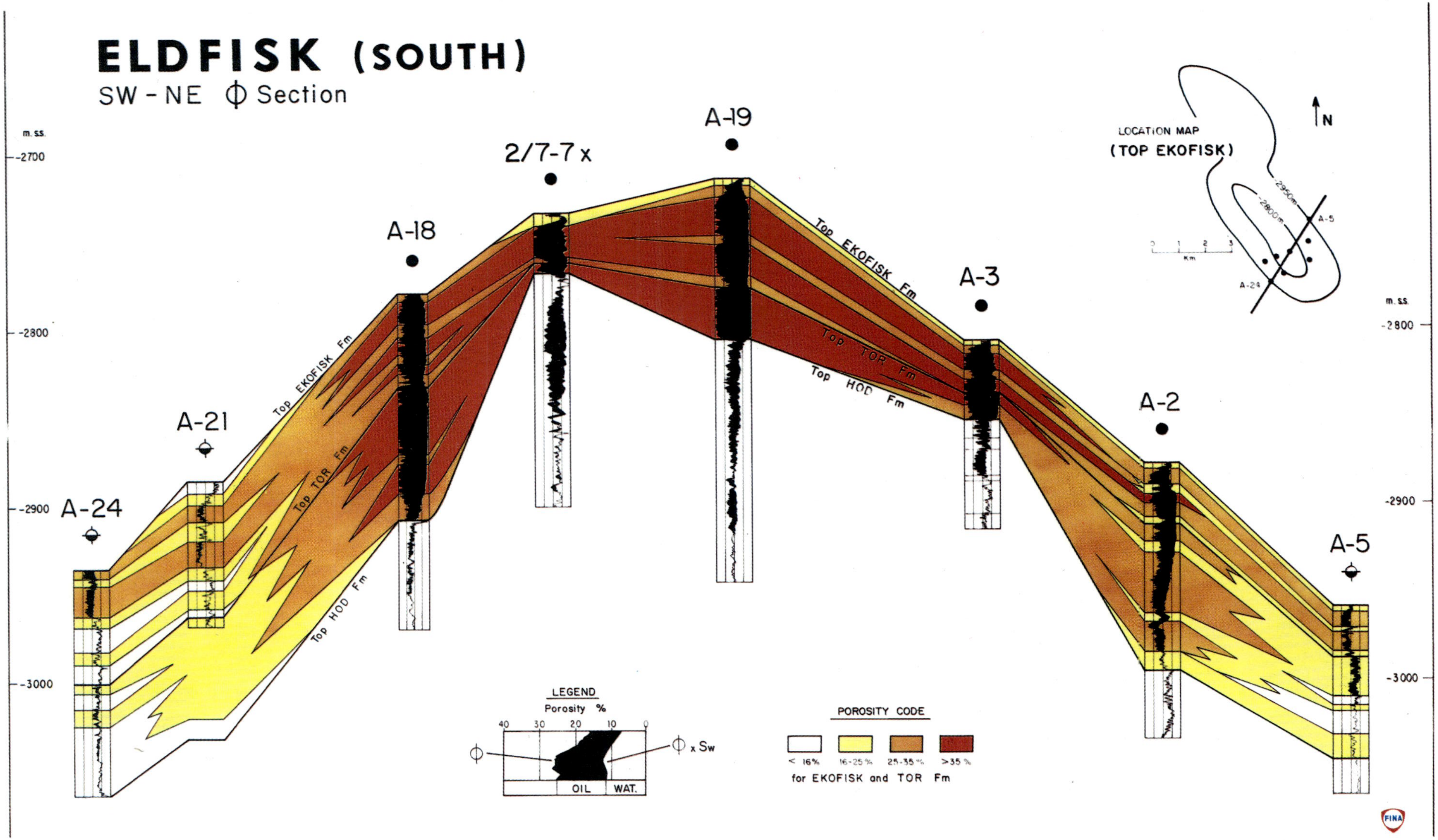

Plate 4. Porosity and hydrocarbon distribution in South Eldfisk (after D'Heur, 1986).

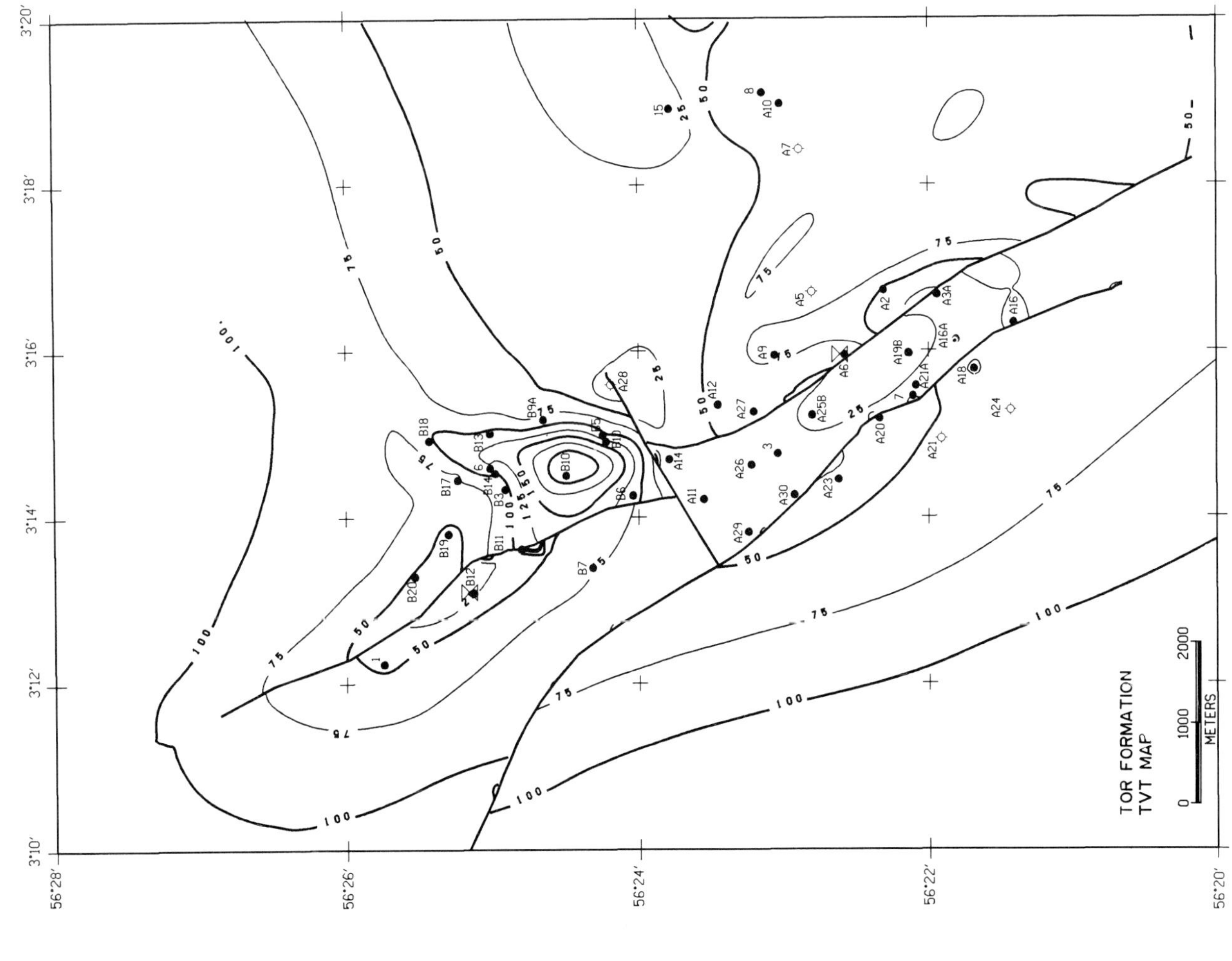

Fig. 11. Tor Formation true vertical thickness map. Contour interval 25 m.

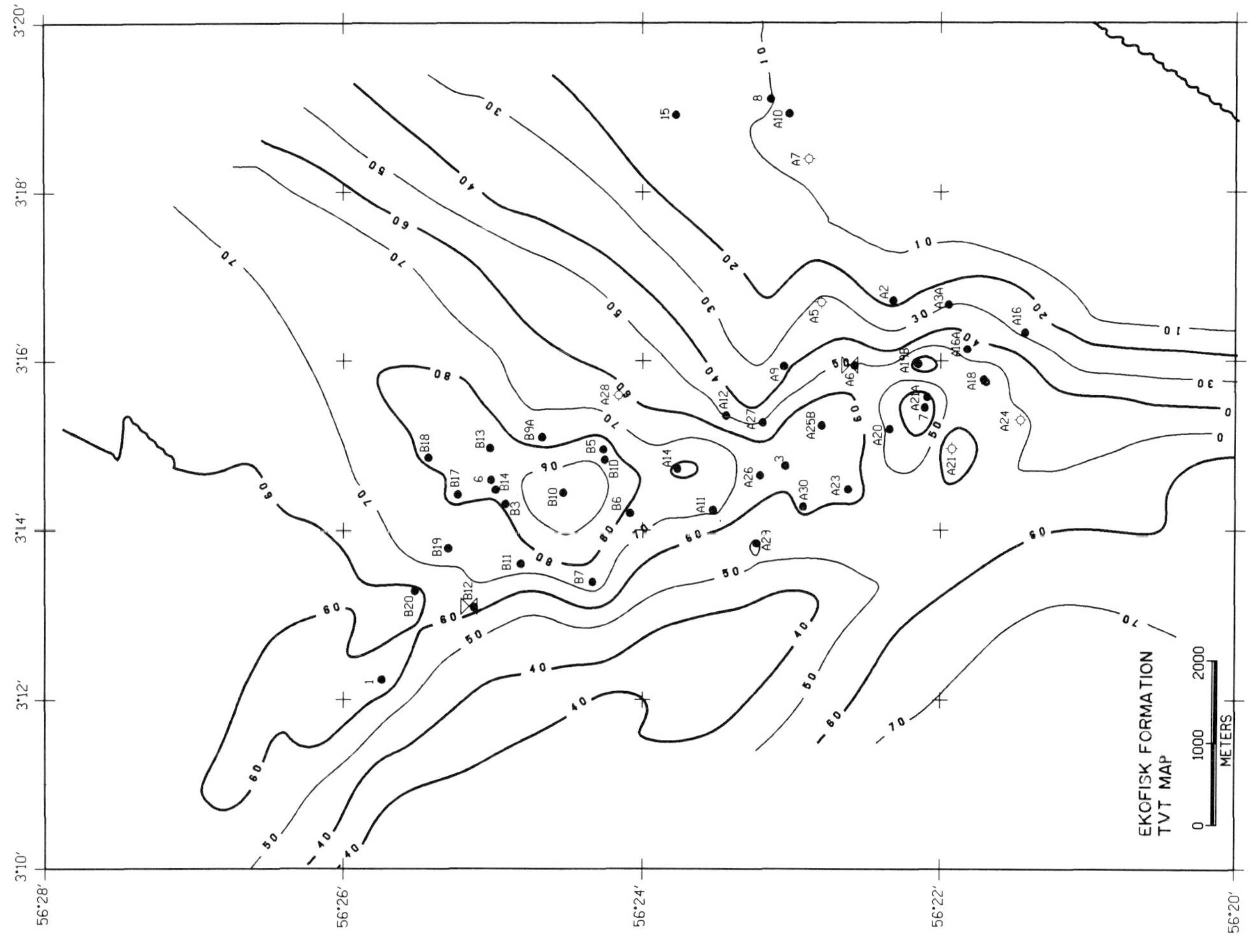

Fig. 12. Ekofisk Formation true vertical thickness map. Contour interval 10m.

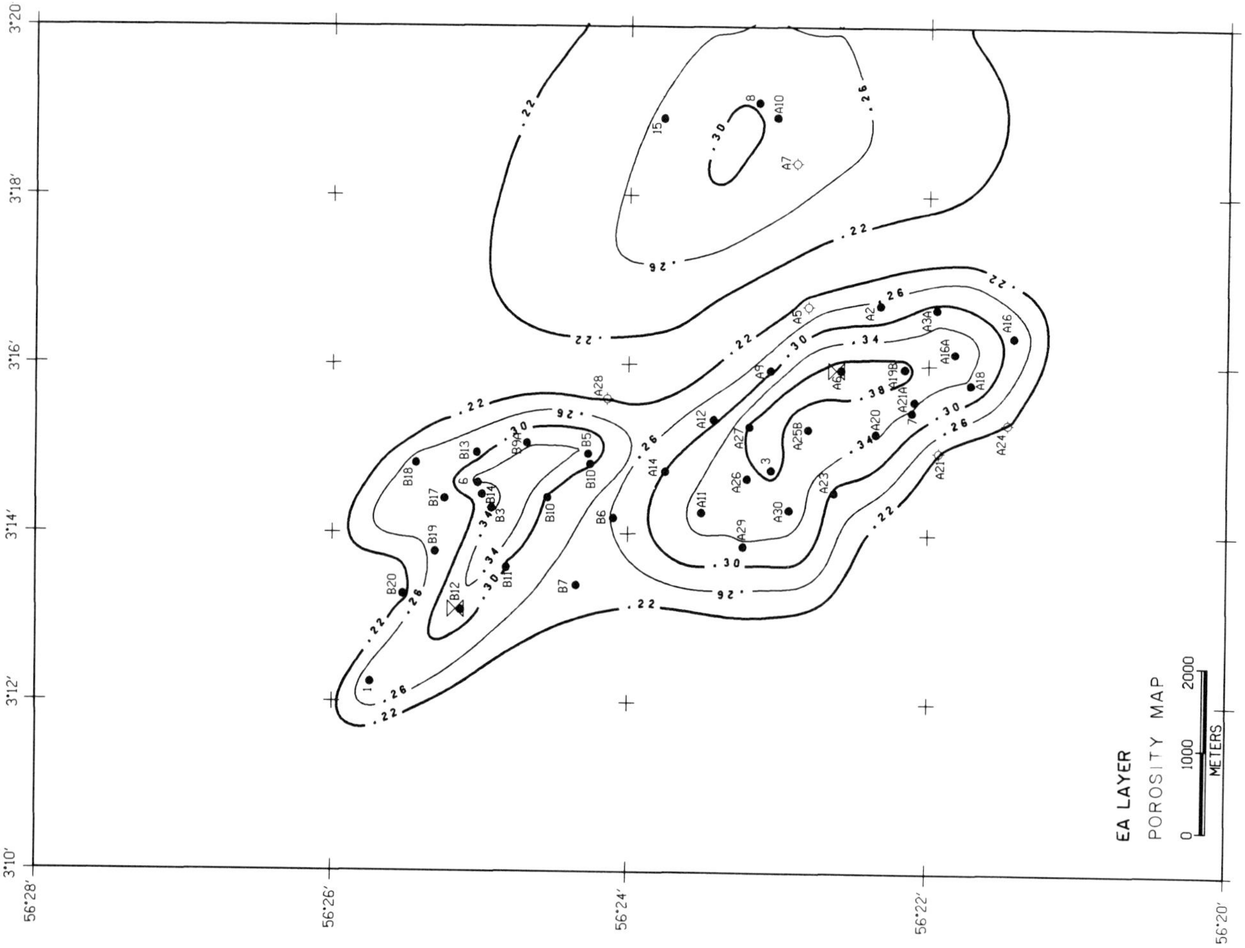

Fig. 14. EA layer average porosity map.

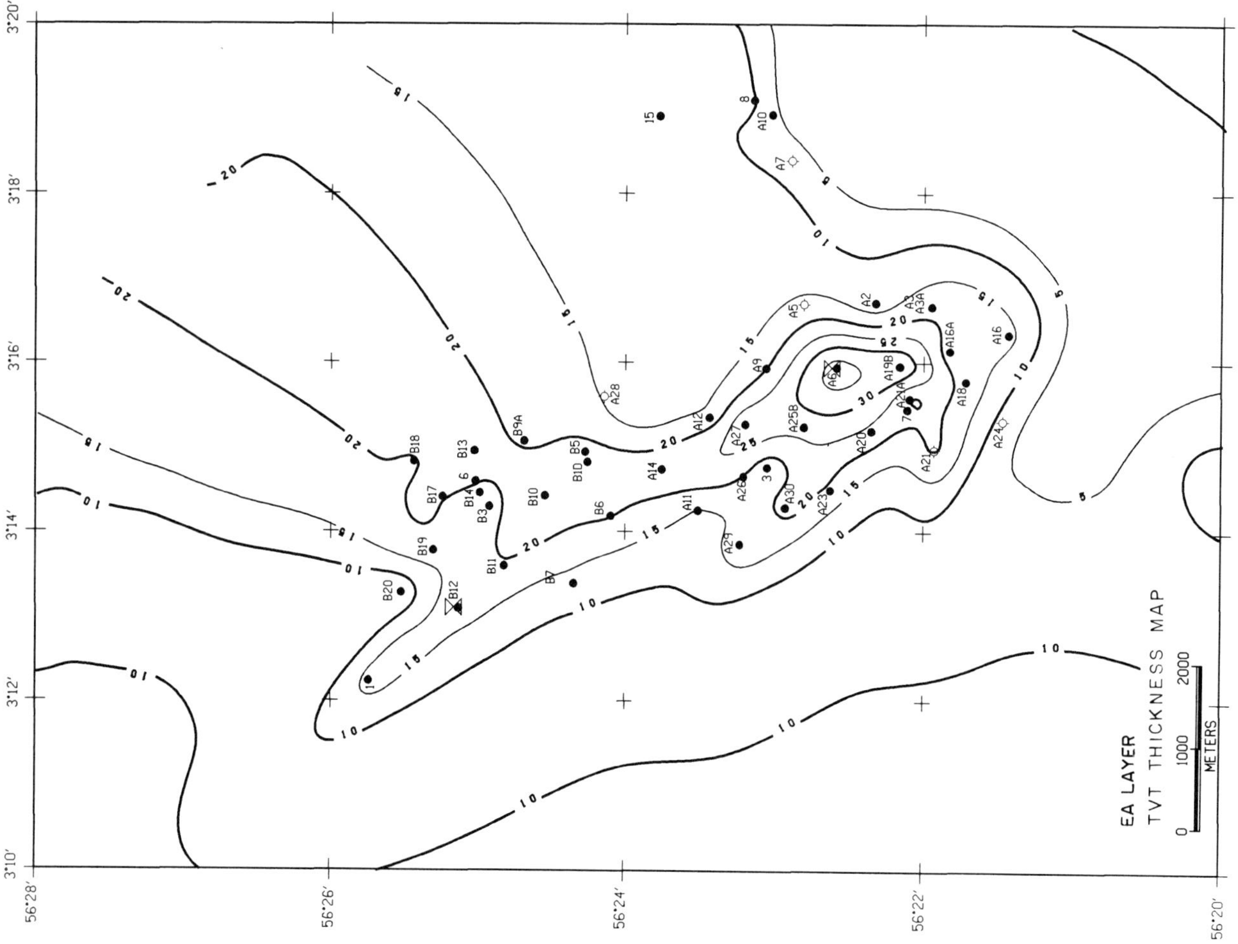

Fig. 13. EA layer thickness map.

tilting could also explain why the porosities are generally higher on the eastern flank than on the western flank.

Permeability Characteristics

The matrix permeability of the chalk varies from 0.01 mD to 35 mD. The average permeability is higher in the Tor Formation than it is in the Ekofisk Formation. Permeability decreases towards the flanks in both the south and the north areas. Matrix permeability seems to be a linear function of porosity on a log–log plot. In South Eldfisk, close to the western fault, the wells have very high permeabilities (up to 140 mD on 2/7A-20A), probably due to extensive fracturing.

The variation in thickness, porosity and water saturation is illustrated in Figs 13, 14 and 15 with an example: the EA layer.

SOURCE

Detailed geochemical analysis has clearly shown that organic shales of Late Jurassic age are the primary source-rock for the Greater Ekofisk Fields (Van den Bark and Thomas, 1980). The timing of hydrocarbon generation, expulsion and subsequent entry into the chalk reservoirs still raises considerable discussions.

A recent geochemical study of Greater

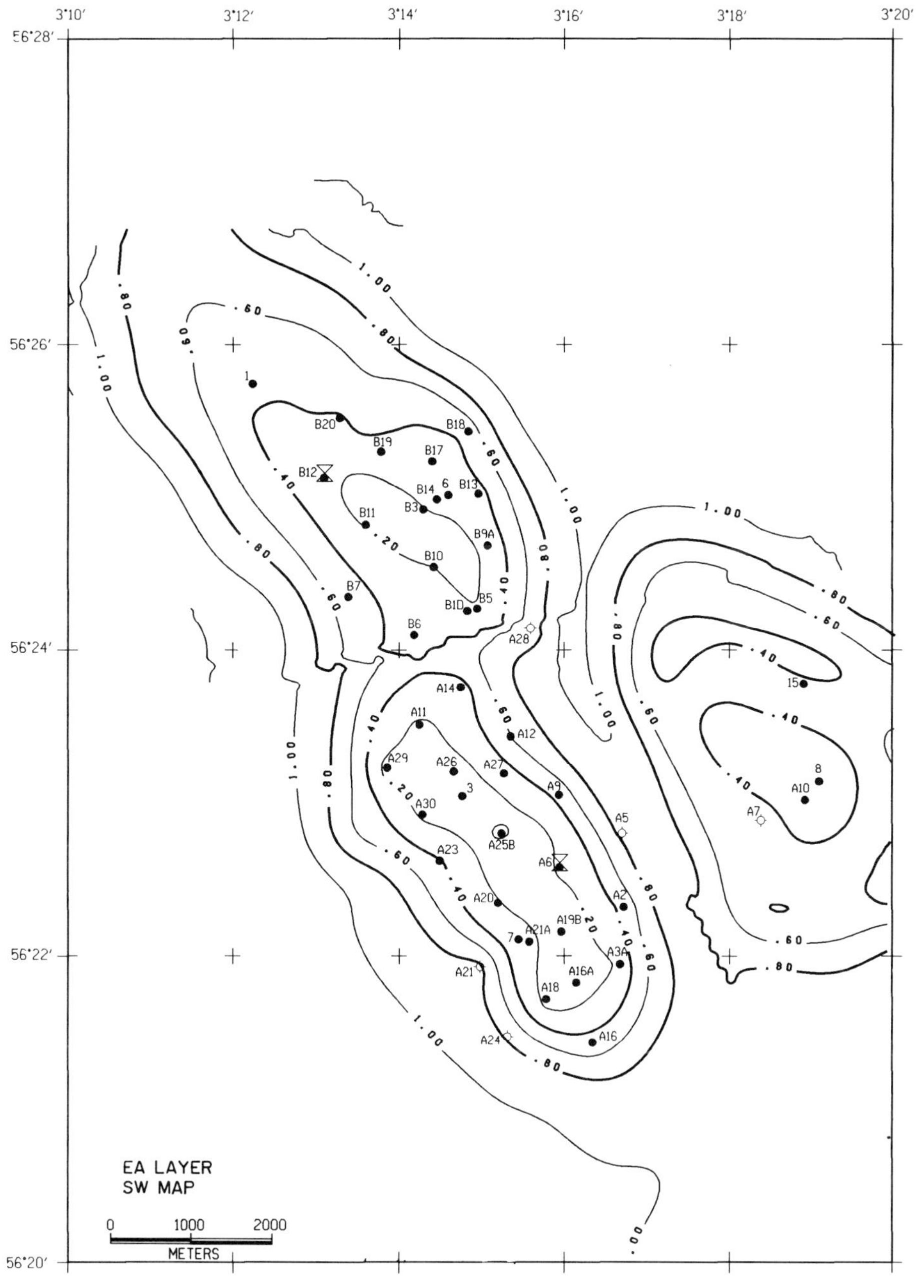

Fig. 15. EA layer average water saturation map.

Ekofisk crude oils undertaken by Phillips Petroleum shows that the oils of the Ekofisk, Tor and Hod formations in the Eldfisk Field are slightly different from each other, possibly due to different source-rock subfacies within the Upper Jurassic. No distinction could be proved between South Eldfisk and North Eldfisk oils.

The migration of hydrocarbons into the Eldfisk Field may have taken place via faults and fractures after the growth of the structure.

The maturity of the Eldfisk Field oil is very similar to that of the Ekofisk Field, even although maturity variations can be observed on a NW–SE regional trend with a decrease in maturity from the Albuskjell Field to the Valhall Field.

HYDROCARBONS

The oil in the Eldfisk reservoir is an undersaturated volatile oil. The initial reservoir temperature is 119 °C; the initial pressure is 470 bar (at 2865 mss), and the initial average gas–oil ratio is 294 Sm^3/Sm^3. The bubble-point pressure of the oil is 414 bar. The crude oil has a specific gravity of 0.84 g/cm^3, 0.20 wt% of sulphur and 0.14 wt% of nitrogen. The asphaltene content is very low. The gas has a specific gravity of 0.70.

The oil in the East Eldfisk reservoir is also an undersaturated volatile oil. The initial reservoir temperature is 123 °C, the initial reservoir pressure is 491 bar (at 2972 mss), and the initial average gas–oil ratio is 134 m^3/m^3. The bubble-point pressure is 242 bar.

RESERVES

The Eldfisk reservoir has a low matrix permeability but is fractured. The fracture system is a critical factor to the productivity of the field. The wells are stimulated with a hydraulic fracture treatment to improve productivity.

There is no evidence of water drive. Most of the non-hydrocarbon-bearing units are dead-end pore volume since both porosity and permeability are drastically reduced by late diagenesis in the aquifer (D'Heur, 1986). Early water production was observed locally on the eastern flank and corresponds with the area of the field with abnormally high porosity, mentioned above. In this area the aquifer might provide limited pressure support.

The present development plans involve draining the reservoir using solution gas drive only. Eldfisk is considered to be a poor candidate for water injection due to the unfavourable wettability and imbibition characteristics of the reservoir rock, but gas injection into the Ekofisk and Tor formations in South Eldfisk is being studied, and the potential for nitrogen injection will also be evaluated.

The effects of compaction and subsidence were first identified in November 1984 in the Ekofisk Field. The seabed movement results from reservoir compaction when pore pressure declines. Pressure maintenance by gas injection is one way to counteract the compaction in the chalk. The Eldfisk Field reservoir has undergone some compaction since production started, but no seabed subsidence has been detected. The Eldfisk platforms are not located on the structural crest of the field where the porosities reach their maximum, and consequently the concern over subsidence is less.

Most of the problems encountered during oil production are caused by solids fill in tubing, solids production or collapsed casing.

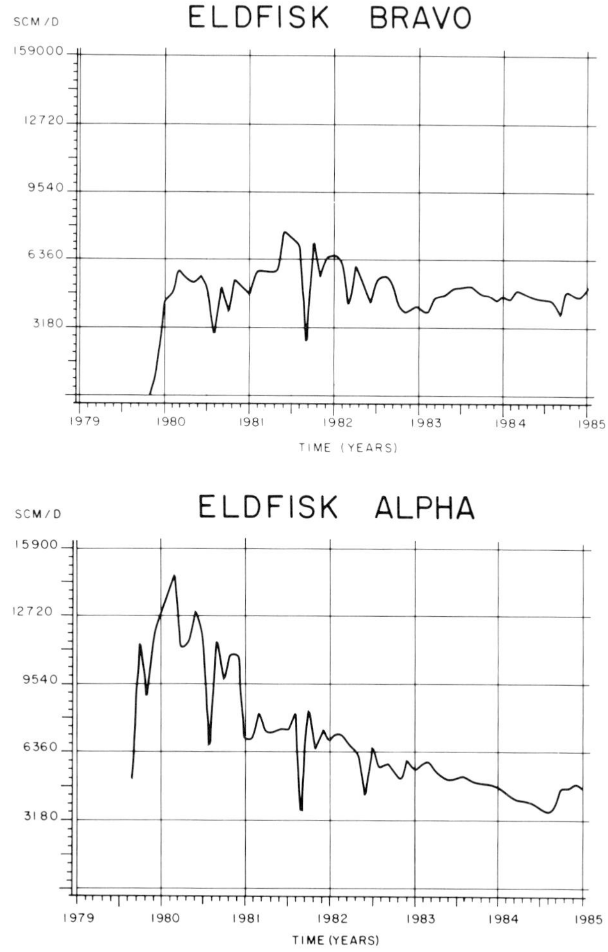

Fig. 16. Eldfisk oil production history.

They result from the stress caused by the fluid drag forces in the formation surrounding the perforations. In order to decrease the fluid transport velocities near the perforations the perforation density was increased and the wells are opened slowly after a shut-down.

Initial recoverable reserves are currently estimated to be 48×10^6 Sm^3 of oil plus condensate and 39×10^9 Sm^3 of gas, utilizing a recovery factor of 18% for North Eldfisk and 22% for South Eldfisk. In October 1985 cumulative production had reached 26×10^6 Sm^3 of oil and 10×10^9 Sm^3 of gas. The production rate was 9208 Sm^3/day of oil and 5×10^6 Sm^3/day of gas.

ACKNOWLEDGEMENTS

The author thanks Phillips Petroleum Norway, Fina, Agip, Elf, Total, Norsk Hydro and the other members of the Group for giving permission to publish this article, and in particular M. D'Heur and J. Brewster for their helpful comments, discussion and constructive criticisms. The opinions expressed are those of the author and may not neccessarily represent those of the Operator or the coventurers in the Phillips Norway Group.

REFERENCES

Brewster, J. and Dangerfield, J.A. 1984. Chalk fields along the Lindesnes Ridge, Eldfisk, *Marine and Petroleum Geology* **1,** (3), 239–278.

Deegan, C.E. and Scull, B.J. (Compilers). 1977. A proposed standard lithostratigraphic nomenclature for the central and northern North Sea, *Norwegian Petroleum Directorate Bulletin* I, *Institute of Geological Science Report* No. 77/25.

D'Heur, M. 1980. Chalk reservoir of the West Ekofisk Field, In *The Sedimentation of the North Sea Reservoir Rocks* (Norsk Petroleumsforening), Article X.

D'Heur, M. 1984. Porosity and hydrocarbon distribution in the North Sea Chalk reservoirs, *Marine and Petroleum Geology* **1** (3), 211–237.

D'Heur M. 1986. The Norwegian chalk Fields. In: Spencer, A.M. *et al.* (Eds), *Habitat of Hydrocarbons on the Norwegian Continental Shelf.* Graham and Trotman, London, pp. 77–89.

D'Heur, M., De Walque, L. and Michaud, F. 1985. Geology of the Edda Field Reservoir, *Marine and Petroleum Geology* **2,** (4), 327–340.

Gowers, M.B. and Sæbøe, A. 1985. On the structural evolution of the Central Trough in the Norwegian and Danish sectors of the North Sea, *Marine and Petroleum Geology* **2,** 298–318.

Illing, L.V. and Hobson, G.D. (Eds). 1980. *Petroleum Geology of the Continental Shelf of Northwest Europe*, Institute of Petroleum, Heyden, London.

Kennedy, W.J. 1980. Aspects of chalk sedimentation in the southern Norwegian offshore. In *The Sedimentation of the North Sea Reservoir Rocks*, Norsk Petroleumsforening, Article IX.

Kennedy, W.J. 1985. Sedimentology of the Late Cretaceous and Early Paleocene Chalk Group, North Sea Central Graben, In *Chalk Research Program Symposium*, Stavanger May 1985.

Ofstad, K. 1981. The Eldfisk Area, *Norwegian Petroleum Directorate, Paper No. 30.*

Ofstad, K. 1983. Geology of the southernmost part of the Norwegian section of the Central Trough, *Norwegian Petroleum Directorate Paper No. 32.*

Skovbro, B. 1983. Depositional conditions during chalk sedimentation in the Ekofisk Area, Norwegian North Sea, *Geologie Mijnbouw* **62,** 169–175.

Van den Bark, E. and Thomas, O.D. 1980. Ekofisk: first of the giant oil fields in Western Europe, In Halbouty, M.T. (Ed). *Giant Oil and Gas Fields of the Decade 1968–78, Memoir American Association Petroleum Geologists* **30,** 195–224.

Ziegler, P.A. 1982. *Geological Atlas of Western and Central Europe*, Shell Internationale Petroleum Maatschappij B.V, The Hague.

7

Hod

I. Norbury

Amerada Hess Norway Limited, Langkaien 1, N-0150, Oslo 1, Norway

The Hod Fields are located in Block 2/11 on the south-western flank of the Central Trough close to the Lindesnes Fault zone and comprise two small oil accumulations in Upper Cretaceous and Early Tertiary chalks of the Hod, Tor and Ekofisk formations. The trap is formed by a combination of structural and stratigraphic factors. The reservoir quality of the chalk is quite variable and is related to its mode of deposition, differential diagenesis and palaeostructure. The Ekofisk and Tor formations are interpreted to be largely redeposited chalks with uniformly very good reservoir quality, whereas the Hod Formation is predominantly pelagic in origin and shows considerable variation in reservoir quality.

The Hod structures are two separate anticlinal features orientated NW–SE. The structural history of the West Hod is closely linked to movements along the Lindesnes Fault zone, whereas East Hod appears to be a salt-induced feature perhaps partly related to compressional movements along a NW-trending strike-slip fault. Recoverable oil resources are estimated to be 6×10^6 Sm^3. In view of the considerable geological uncertainties and in the light of technical difficulties experienced in the operation of the nearby Valhall Field, further drilling on the Hod structures has been deferred and development is not contemplated at the present time.

INTRODUCTION

The Hod Fields are located 370 km south-west of Stavenger and 11 km south of the Valhall Field in the Norwegian sector of the North Sea (Fig. 1). The fields contain small oil accumulations in two separate structural accumulations, informally named West and East Hod. Each has an areal closure of approximately 7 km^2. Water depth is in the order of 70 m. The reservoirs are the Upper Cretaceous and Danian chalks of the Hod, Tor and Ekofisk formations. The trapping mechanism is a complex combination of structural and stratigraphic factors, and there is considerable uncertainty attached to the estimate of recoverable resources of 6×10^6 Sm^3. The field was named from Norse mythology after the blind god Hod.

HISTORY

Block 2/11, Licence PL033, was awarded to the Amoco/Noco Group in 1969 in the second round of licencing. Amoco was appointed operator for the licence. Participating interests are as follows:

Amoco Norway Oil Company (Operator)	25%
Amerada Hess Norway	25%
Texas Eastern Norwegian Corporation	25%
Norwegian Oil Consortium A/S	25%

Statoil has a 10% net profit interest.

All compulsory relinquishments have been completed and the licence expires in the year 2015.

Geology of the Norwegian Oil and Gas Fields © A. M. Spencer *et al.*, eds (Graham & Trotman, 1987) pp. 107–116.

Hod Field data summary

Trap	
Type	combined stratigraphic and structural trap
Depth to crest	2575 mss (W. Hod) and 2650 mss (E. Hod)
Lowest closing contour	2750 mss
Oil–water contact	not adequately established
Oil column (gross proven)	78 m (W. Hod) and 166 m (E. Hod)
Pay zone	
Formation	Ekofisk, Tor, Hod
Age	Danian, Maastrichtian—Campian, Turonian
Net/gross ratio (average; range)	95%; 90–100%
Porosity (average; range)	33%; 24–38%
Hydrocarbon saturation	80% (Ekofisk and Tor) and 40% (Hod)
Permeability (average; range)	1.4 mD; 0.5–6.5 mD
Productivity index	935 Sm^3/day/bar
Hydrocarbons	
Oil gravity	0.68 g/cm^3
Oil type	Low sulphur
Gas/oil ratio	156 Sm^3/Sm^3
Bubble point	270 bar
Formation volume factor	1.405
Formation water	
Salinity	60 000 ppm NaCl equivalents
Resistivity	0.046 ohm m at 88 °C
Reservoir conditions	
Temperature	100 °C
Pressure	178 bar
Pressure gradient in reservoir	0.079 bar/m
Field size	
Area	14 km^2 (7 km^2 for both W. and E. Hod)
Gross rock volume	242.0×10^6 m^3 (W. Hod)
Hydrocarbon pore volume	60.9×10^6 m^3 (W. Hod)
Recovery factor—oil	0.16
Drive mechanism	Water drive
Recoverable oil	6×10^6 Sm^3
Production	
—	

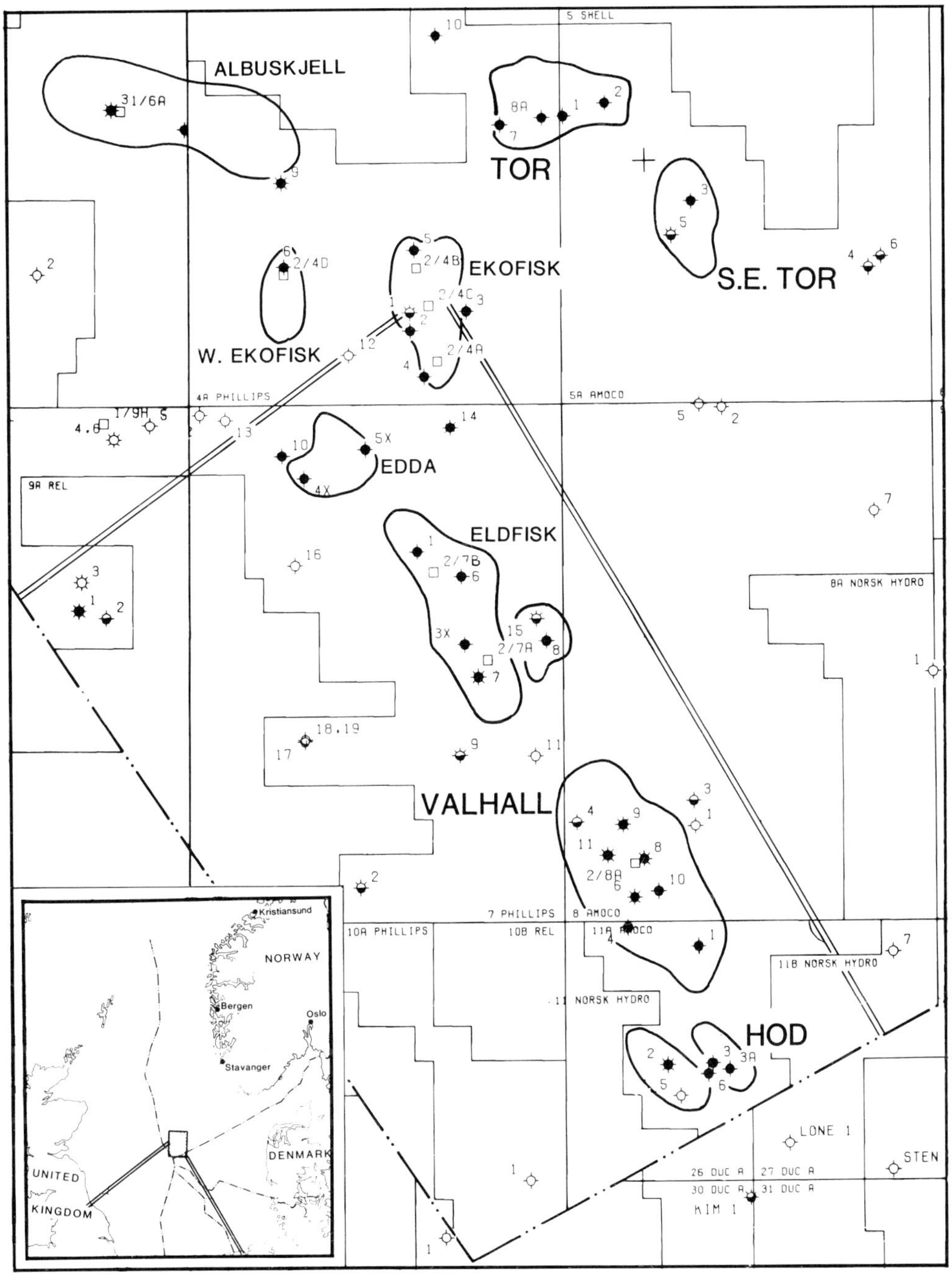

Fig. 1. General location map, Hod Fields.

To date, six wells have been drilled by the Amoco/Noco Group in Block 2/11, four of which were on the Hod structures. The discovery well, 2/11-2, was drilled in 1974 to test the hydrocarbon potential of chalk reservoirs on a structural high on the Lindesnes Ridge. The well proved a 50 m net oil column in the Hod Formation, which in this location is overlain unconformably by Paleocene strata. The reservoir quality was very good. However, subsequent appraisal drilling has given unexpected results. Well 2/11-3 was drilled in 1977, some 2.5 km east of the discovery well, on the western flank of East Hod where seismic data indicated an increased chalk thickness. Top chalk (Hod Formation) was encountered significantly lower structurally than anticipated, which indicated the presence of a structural saddle separating West and East Hod. Potential reservoirs of the Hod Formation were found to be tight and water bearing. The well was side-tracked (2/11-3a) to a target location on the crest of the East Hod, 1 km east-south-east of the original hole, and found 37 m of net pay in the Ekofisk, Tor and Upper Hod formations. The side-track was terminated without penetrating potential reservoirs of the Middle and Lower sections of the Hod Formation. The structural and geological models were re-evaluated in the light of the 2/11-3 and -3a results, and 2/11-5 was drilled in 1979 on the southern flank of the West Hod. The objective was to search for the good reservoirs of the Tor and Ekofisk formations. The well did indeed find the Ekofisk Formation above a thin, somewhat tighter than normal Tor Formation, but only minor hydrocarbon shows were recorded.

A careful re-evaluation was undertaken in 1980. On the basis of this, a twelve-slot drilling template was installed in 1981, located over the saddle between West and East Hod. Well 2/11-6 was drilled through the template during late 1981. The objective was to prove a substantial reserve in all three reservoir levels within the chalk on the East Hod feature. The well was eventually completed as a side-track, 2/11-6ST, having found 78 m net oil column in the Tor and Hod formations.

A further reassessment of the Hod Fields was completed in 1985. This study resulted in a significant reduction of estimated recoverable resources, raising serious doubts about the economic viability of developing the Hod Fields. Because of this, and in view of severe production and completion problems experienced at Valhall, further drilling on Hod has been deferred and development is not contemplated at the present time.

GEOLOGICAL SETTING

The Hod Fields lie on the south-west flank of the Norwegian Central Trough (Fig. 2). The most important structural feature in this part of the basin is the Lindesnes Ridge, with which are associated five oil accumulations, namely Hod, Valhall, Eldfisk, Edda and Tommeliten/Gamma (Brewster and Dangerfield, 1984). The Lindesnes Ridge is a prominent inversion structure, directly east of a major Jurassic or older normal fault, and contains a thick succession of Jurassic strata. The major tectonic episodes affecting the area occurred in Late Jurassic and Early Cretaceous times. However, other tectonic movements took place, of which inversion, strike-slip (transpressional and transtensional) faulting and salt diapirism had an important effect on the character and distribution of the chalk reservoirs. Inversion movements were initiated in Early Cretaceous times and continued periodically through Late Cretaceous to Middle Tertiary times. Tracing the history of movements along and adjacent to the Lindesnes Ridge gives valuable insight into the struc-

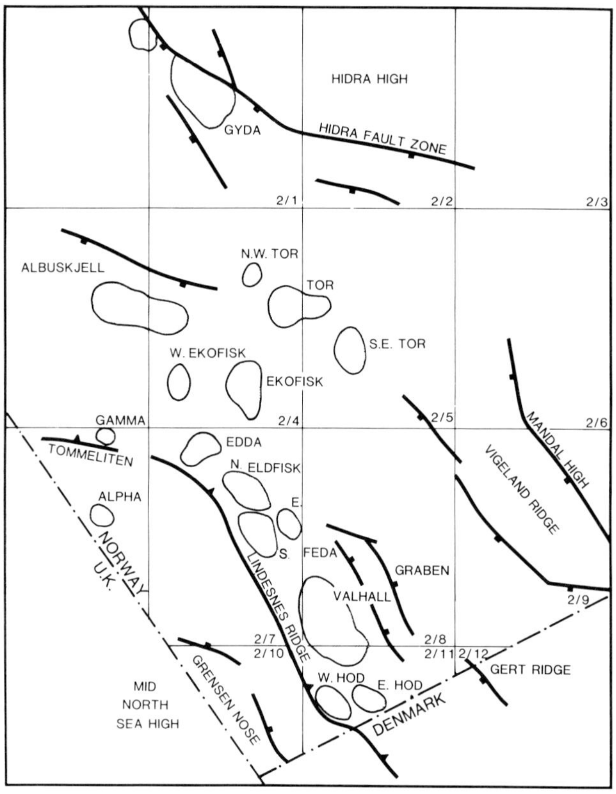

Fig. 2. Location map Valhall/Hod area showing main tectonic elements.

tural and stratigraphic development of the Valhall and Hod structures.

West and East Hod are separate structures with different origins and histories. West Hod directly overlies the Lindesnes Ridge and its history is linked to structural movements along the ridge. East Hod, however, appears to be directly related to halokinetic movement and has a less complex and, in part, more recent history.

A generalized stratigraphic column for the Hod area is given in Fig. 3. The lithostratigraphy adopted is that proposed by Deegan and Scull (1977). No well in the Hod Fields has drilled into strata older than late Early Cretacous. However, Well 2/11-1 on the southern flank of Valhall Field terminated whilst still in Upper Jurassic strata, having drilled 1200 m of bituminous, black marine shales of the Humber Group. These are characterized by a high content of organic matter. The Lower Cretaceous sequence consists of interbedded shales and marls and is of variable thickness over the Hod structures.

The Chalk Group is divided into five formations, the upper three of which (Ekofisk, Tor and Hod) have reservoir potential. The Hod Formation is further subdivided into three informal members, the Upper, Middle and

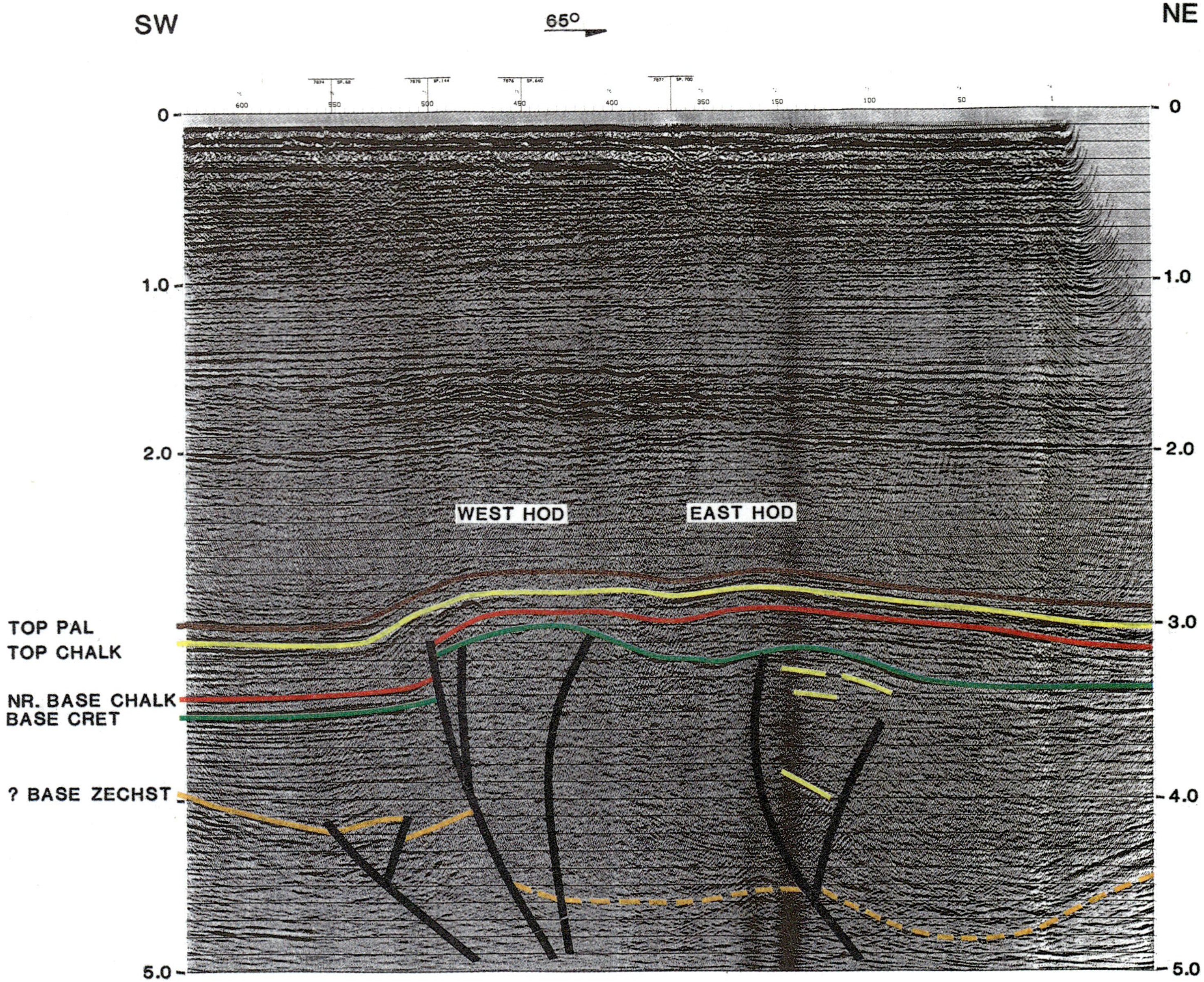

Plate 5. Seismic line, Hod Fields.

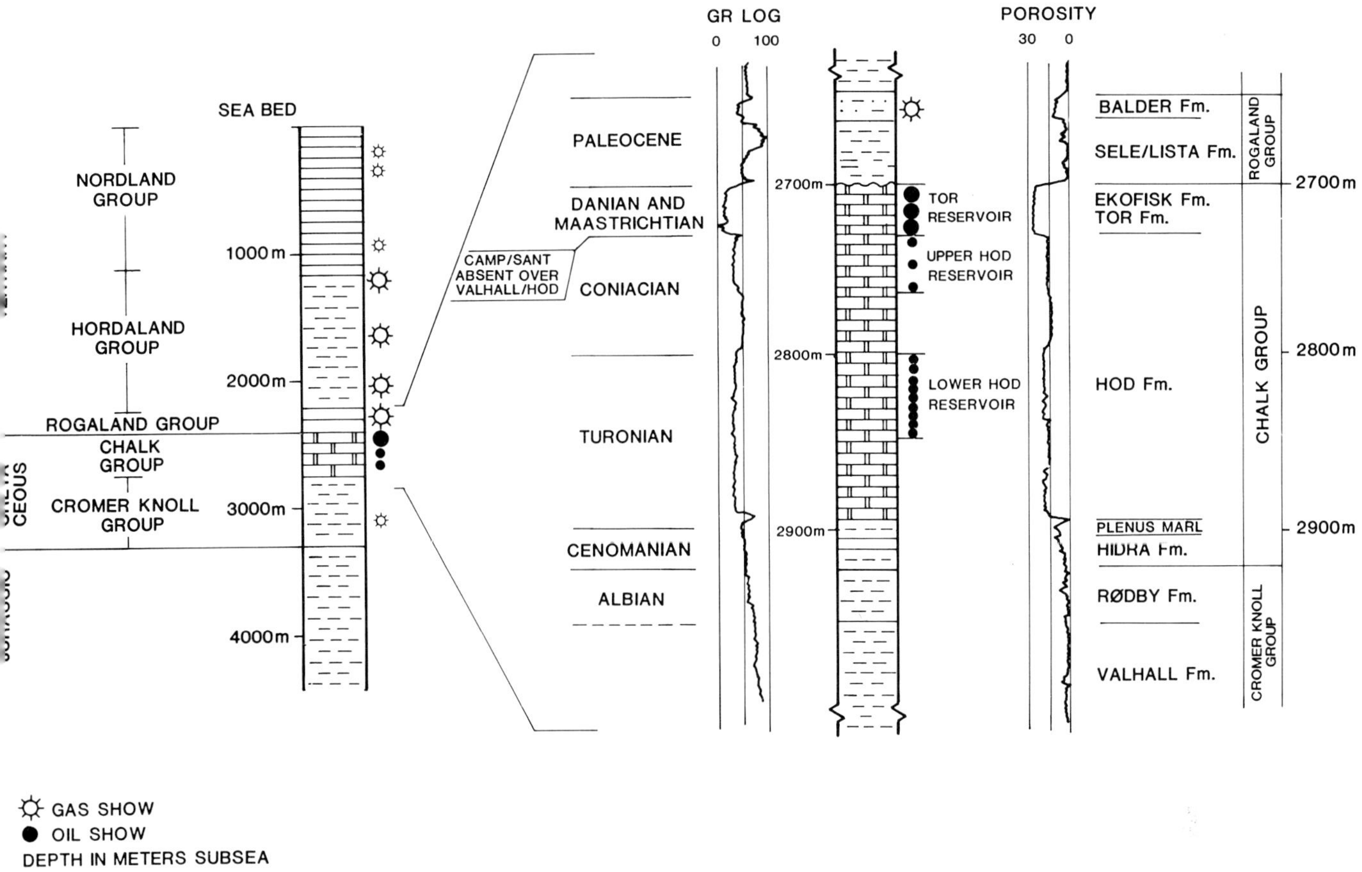

Fig. 3. Stratigraphic column.

Lower Hod. The Tor Formation is a single discontinuous unit and the Ekofisk Formation, where identified, is a thin, condensed deposit.

The petrology and sedimentology of the chalk are well known (Hancock, 1976; Hardman and Kennedy, 1980; Kennedy, 1980 and 1985). The rock consists of the skeletal debris of planktonic algae together with a variable non-carbonate fraction, largely of clay and silica. The non-carbonate elements have a detrimental effect on reservoir quality and are particularly abundant in the Hod Formation in the Hod Fields. The chalks of the Hod Fields are for the most part normal, pelagic deposits, including distal turbidites. In addition, allochthonous components are present, particularly in the Tor and Ekofisk formations, representing slump or debris-flow activity of a semi-regional or, as is perhaps more likely for the Hod area, local derivation.

The Chalk Group is overlain by in excess of 2600 m of clays with thin silts and occasional sands of the Rogaland, Hordaland and Nordland groups. This sequence forms the seal to the chalk hydrocarbon accumulations.

GEOPHYSICS

The primary seismic database for the Hod Fields is a 1981 SSL survey comprising a closely spaced grid of mainly NE-trending lines, although one line trends E–W and six are randomly orientated tie-lines, totalling 260 line-kilometres. The quality of these data is generally fair to good, and provides better resolution than earlier surveys in spite of acquisition problems. In particular, it has been possible to identify intra-chalk events and features. Mapping is possible with a reasonable degree of confidence over much of the Hod area, but gas charge of the Tertiary sediments over West Hod causes severe distortion on seismic records, effectively creating a zone of no data over the structural crest (Plate 5).

The top chalk reflector is generally a clear event, apparently little affected by faulting, and is therefore fairly easily mapped. However, over West Hod it is necessary to extrapolate from the flanks of the structure and to compensate for the time lag on seismic lines. Depth conversion is straightforward over

East Hod, whereas over West Hod the velocity field has to be carefully constructed to account for the severe gas effects.

TRAP

The Hod structures are two separate anticlinal features orientated NW–SE which are believed to have different origins and histories (Fig. 4). West Hod is very similar to the Valhall structure. Its growth is closely related to movements along the Lindesnes Ridge. These movements strongly influenced the thickness and quality of the chalk reservoirs. Growth of the structure appears to have started during Early Cretaceous times and continued, periodically, into Mid-Tertiary times. Uplift of West Hod was particularly severe during Late Cretaceous times. Well 2/11-5, close to the Lindesnes Fault zone, has the thinnest chalk section of any of the Valhall and Hod wells. Furthermore, this well has residual oil in the Tor Formation some 75 m below the base of the present moveable oil column, possibly indicating a shift of the crest of the West Hod during Tertiary times.

East Hod appears to be a salt-induced feature, possibly partly related to compression along a NW-trending strike-slip fault. The major period of growth occurred during the Tertiary, but structural movements also occurred during Cretaceous times. The result is that a sequence containing a reduced thickness of chalk, representing a palaeocrest, lies east-south-east of the present-day structural crest and above an inverted Lower Cretaceous basin.

The overall trapping mechanism is structural, the chalk reservoirs being ultimately sealed by Tertiary claystones. The disposition of hydrocarbons is, however, strongly influenced by two factors: first, the distribution of the better quality reservoirs of the Tor and Ekofisk formations, which is at least partly fault controlled; secondly, by the

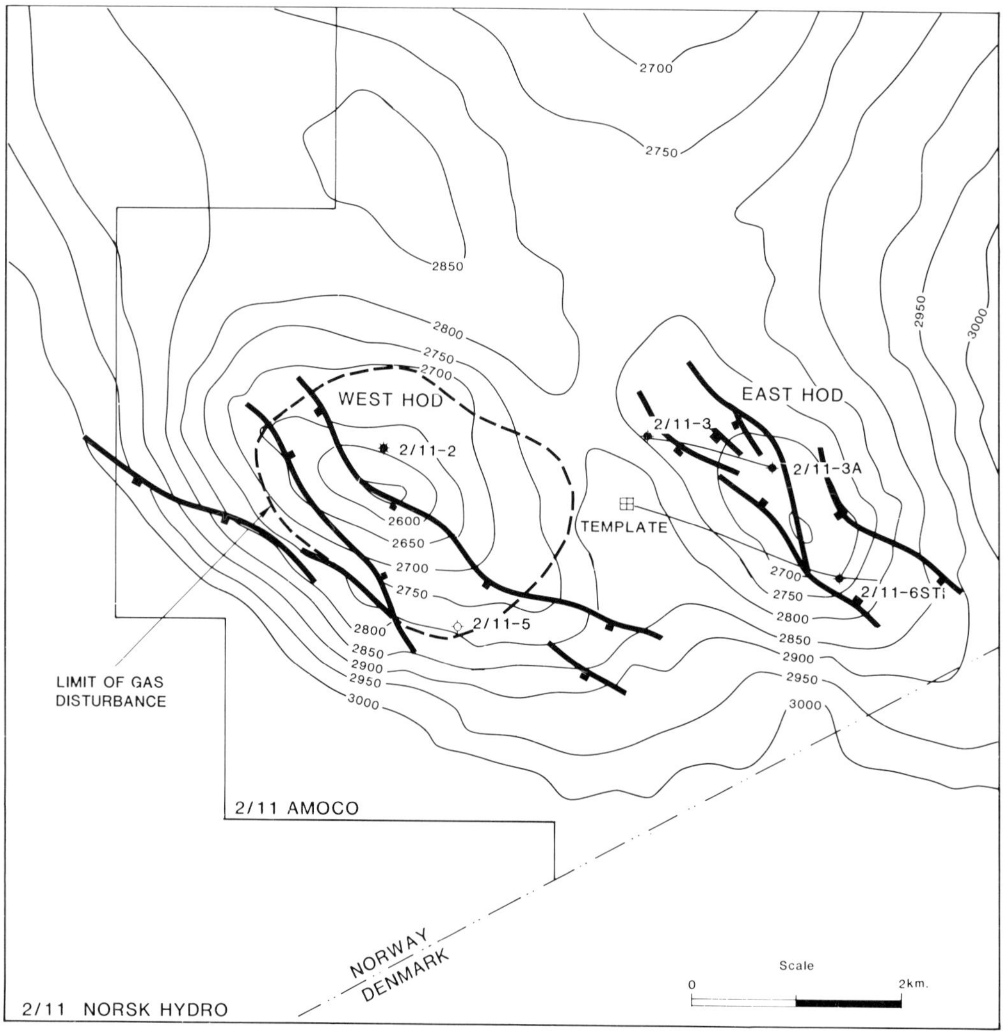

Fig. 4. Top Chalk depth structure map.

lateral and vertical variations in reservoir characteristics, particularly within the Hod Formation. These are related to differences in environment of deposition, diagenesis and to palaeostructural effects.

The Hod structures are largely unaffected by faulting except over their crests. A well-defined graben dissects the crest of the East Hod, the development of which has resulted in the deposition and preservation of the Ekofisk and Tor formations within the graben and, to some extent, along the eastern flank. The western flank is characterized by non-deposition or erosion. The definition of faults is difficult on West Hod; the distribution of the Ekofisk and Tor formations is, however, most likely fault related here also. A schematic geological cross-section across the Hod structures is given in Fig. 5.

West Hod has a vertical relief of approximately 200 m, and East Hod of approximately 100 m, at top chalk level. The area of closure of each structure is about 7 km^2 and the structural spill-points at top chalk level lie at 2775 mss subsea for West Hod and 2750 mss for East Hod.

An effective oil–water contact has not been determined on the Hod Fields. On West Hod, using 80% Sw cut-off, moveable oil is predicted to a depth of 2755 mss in the Hod Formation. On East Hod moveable oil has been found to a depth of 2842 mss, considerably below the mapped structural spill-point at the top of the chalk.

RESERVOIR

The reservoirs of the Hod Field occur in the Late Cretaceous and Early Tertiary chalks of the Hod, Tor and Ekofisk formations. The chalk is quite inhomogeneous and exhibits

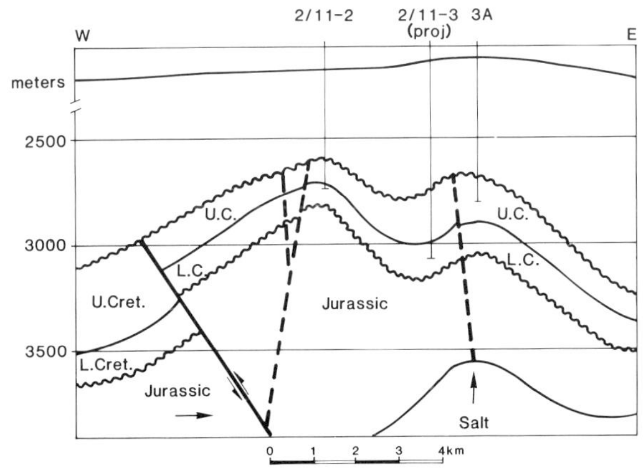

Fig. 5. Schematic geological cross-section.

considerable vertical and lateral variation in rock quality related to mode of deposition, to diagenesis, including mechanical compaction, and to palaeostructure. A detailed discussion of the chalk reservoirs of the Hod Fields is provided by Hardman and Kennedy (1980).

A subdivision of the chalk reservoir section into ten units has been made based on biostratigraphy and log character (Fig. 6). The Ekofisk, Upper Tor (T1) and Lower Tor (T2) units are generally very similar in terms of origin and petrophysical character and, for mapping purposes, may be combined into one unit. Of the six subdivisions of the Hod Formation, units 1 and 4, in the Upper and Lower Hod, exhibit the better reservoir quality. The Tor and Hod Formations are separated by a tight zone (HO) of Santonian to Campanian age.

Tor and Ekofisk formations

The limited core and dip-meter information available suggests that the Tor and Ekofisk formations are largely redeposited chalks which are extensively bioturbated. The reservoir quality is very good and is related both to the large grain size (as a result of relatively rapid deposition) and to the absence of clay or silica. The deposition and preservation of the Tor and Ekofisk formations in the Hod area appears to be closely related to faulting (Fig. 7). These formations are predicted to be present also on the south-west flank of the West Hod and the northern flank of East Hod.

Hod Formation

The Hod Formation is predominantly pelagic in nature and shows considerable variation in reservoir quality related to the presence of non-carbonate constituents. The non-carbonate fraction averages about 20% and consists largely of clay and silica. The percentage varies through time, reflecting changing rates of deposition and hence reservoir quality. Correlation of porosity zones in the Hod Formation is hindered by the fact that not all the wells fully penetrate this section, but there are indications of a general porosity layering, related to variation in the non-carbonate fraction described above. However, the layering has been distorted due to the effects of differential compaction, differential diagenesis and enhanced preservation of porosity by early hydrocarbon emplacement and the development of overpressure within the pore-fluid system. D'Heur (1986) gave an expanded discussion of the effects of

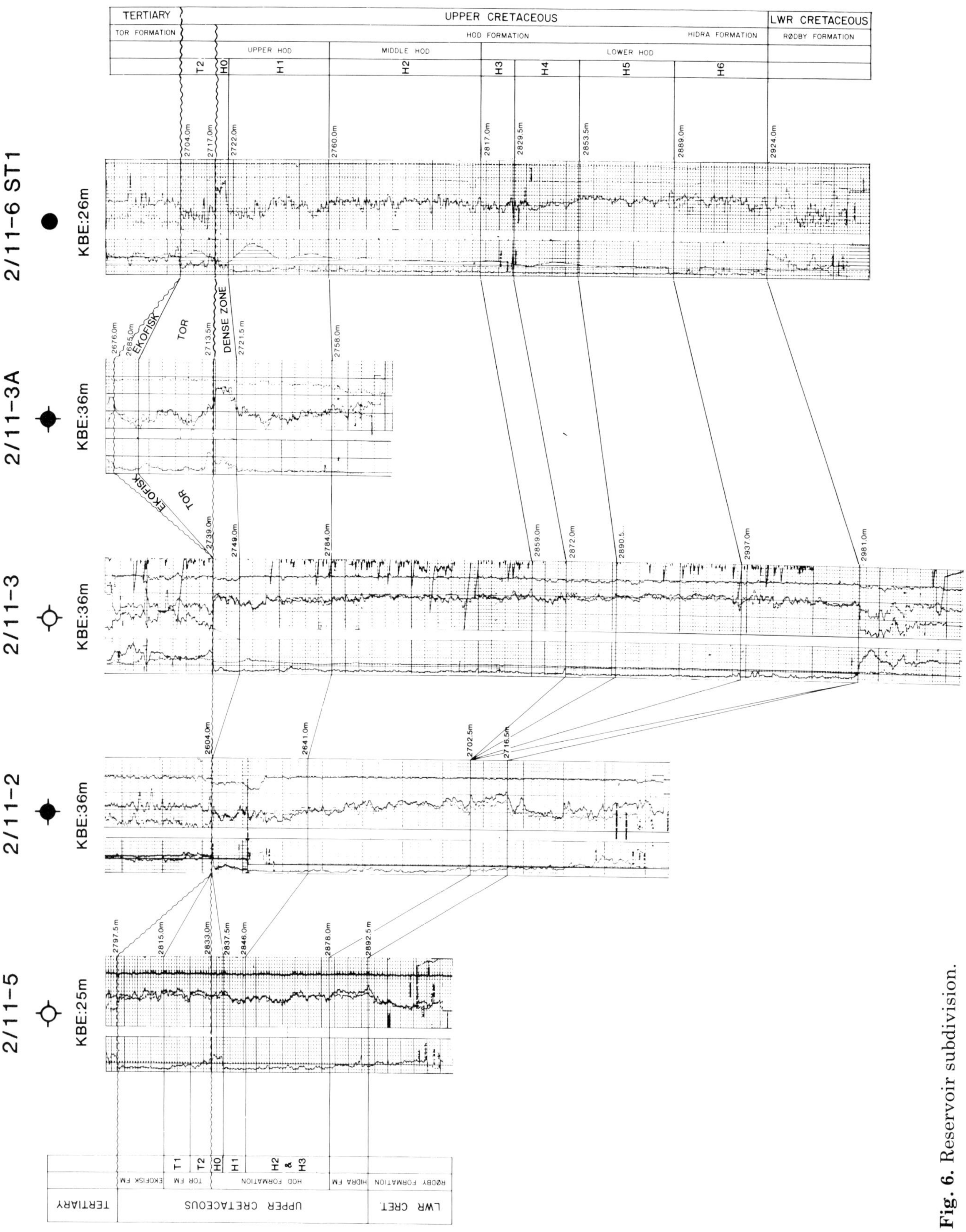

Fig. 6. Reservoir subdivision.

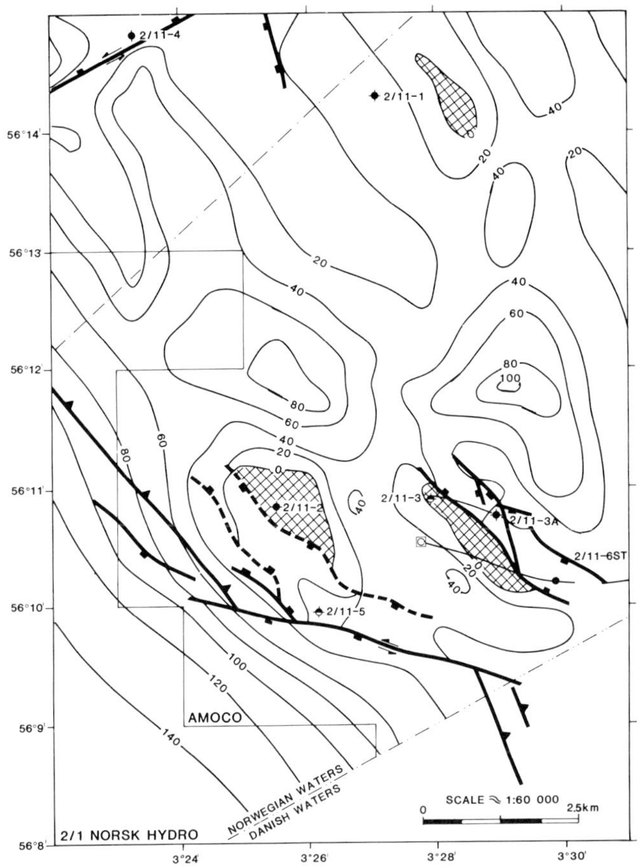

Fig. 7. Tor and Ekofisk Formation isopach map.

these factors for a number of Norwegian chalk fields.

In addition to the factors described above, reservoir quality is affected by the consequences of palaeostructure. The palaeocrests of the Hod structures were characterized by the deposition of thin chalk sequences with the development of hard grounds whereas, at the same time, high-quality chalks were deposited on the flanks of the structures. The palaeocrest of West Hod lay to the south of the present-day crest, and for East Hod the palaeocrest lay to the south-east.

SOURCE

The source-rocks for the oil and associated gas in the Hod Fields are the organic-rich, restricted-marine shales of the Upper Jurassic Humber Group, which directly underly the Hod structures. A vertical migration path through faults and fractures is believed to have existed. Hydrocarbon emplacement began during Early Tertiary times and, together with the development of high pore-pressures, may have contributed to the enhanced preservation of high porosities in the Chalk.

HYDROCARBON POOL

Because of the lack of adequate well control on Hod, knowledge of the hydrocarbon pool remains incomplete. Before Well 2/11-6ST was drilled, a field-wide oil–water contact at 2755 mss was assumed, based on saturation-versus-depth profiles generated from the available data. This depth corresponds closely to the mapped structural spill-points. Well 2/11-6ST, however, proved moveable hydrocarbons to a depth of 2842 mss. The implications of this are not yet fully understood.

West Hod

Well 2/11-2 found 47 m of net pay in the Upper and Middle Hod formations. A decrease in oil saturation occurs with depth in the well, though this coincides with an increase in 'clay matrix' and a corresponding decrease in porosity and permeability. Using an 80% water saturation cut-off, the lowest moveable oil in 2/11-2 is at 2682 mss. Well 2/11-5 found only residual hydrocarbons at a depth of 2798 mss. Saturation-versus-depth profiles suggest a tentative effective oil–water contact at 2755 mss. However, in view of the complex relationship between elevation, porosity and oil saturation in chalks, this figure is considered to be unreliable. The Hod Formation in Well 2/11-2 was tested at a maximum rate of 3000 BOPD after stimulation.

East Hod

Well 2/11-3a found 40 m of net pay in the Tor and Ekofisk formations and 37 m of net pay in the Upper Hod. The well, in fact, terminated in the Middle Hod (H2). Moveable oil is present to a depth of 2749 mss. However, based on the 2/11-6ST results, hydrocarbons may be expected below this depth at the 2/11-3a location, in the unpenetrated Middle and Lower Hod reservoirs. Well 2/11-6ST found 12 m of net pay in the Tor Formation and 66 m of net in the Upper, Middle and Lower Hod. Moveable oil is present to a depth of 2842 mss.

By contrast, Well 2/11-3, the first well drilled on East Hod, found entirely tight, water-bearing chalk (Hod Formation) the top of which was penetrated at 2739 mss. This gives an indication of the complexity of the hydrocarbon pool and of the variation of reservoir quality. One possible explanation is that between 2/11-6ST and 2/11-3 the base of the moveable oil zone rises by in excess of 100 m. Such a peripheral rise of the base of the pay zone has been recognized in West Ekofisk

(D'Heur, 1984). This has been attributed to the effects of a general porosity reduction towards the flanks, possibly augmented by a modification of the pore and pore-throat geometry, resulting in a detrimental alteration of the capillary characteristics of the chalk.

The Tor and Hod formations in Well 2/11-6ST were tested after stimulation at rates up to 6500 BOPD and 2300 BOPD respectively.

RESOURCES

The Hod Fields (and Valhall) are distinguished from the other chalk fields by having a high proportion of their resources in Hod Formation reservoirs. Most likely recoverable oil resources were recently estimated to be 6×10^6 Sm^3. However, this figure is subject to the considerable uncertainty arising both from mapping of oil-in-place, and in selecting an appropriate recovery efficiency.

Many different development concepts have been studied for the Hod Fields. However, due to the small reserve level, the geological uncertainties discussed above, and the technical reservoir difficulties experienced in the operation of the nearby Valhall Field, no acceptable development scheme has been identified.

ACKNOWLEDGEMENTS

This article is in part based on work carried out by Amoco Norway Oil Company as operator for the PL033 licence group. The author wishes to thank Amoco, Texas Eastern, Noco and Amerada Hess for kind permission to publish this paper and is indebted to S. Sørensen for his valuable assistance in its preparation.

REFERENCES

Brewster, J. and Dangerfield, J.A. 1984. Chalk Fields along the Lindesnes Ridge, Eldfisk, *Marine and Petroleum Geology* **1** (3), 239–278.

D'Heur, M. 1984. Porosity and hydrocarbon distribution in the North Sea chalk reservoirs, *Marine and Petroleum Geology* **1,** 211–238.

D'Heur, M. 1986. The Norwegian chalk fields, In Spencer A.M. *et al.* (Eds), *Habitat of Hydrocarbons on the Norwegian Continental Shelf*, Graham and Trotman, London, pp. 77–89.

Hancock, J.M. 1976. The Petrology of the Chalk, *Proceedings Geologists Association* **86** (for 1975), 499–535.

Hardman, R.F.P. and Kennedy, W.J. 1980. Chalk reservoirs of the Hod Fields, Norway, In *The Sedimentation of the North Sea Reservoir Rocks*, Norsk Petroleumsforening, Article XI.

Kennedy, W.J. 1980. Aspects of chalk sedimentation in the Southern Norwegian Offshore, In *The Sedimentation of the North Sea Reservoir Rocks*, Norsk Petroleumsforening, Article XI.

Kennedy, W.J. 1985. Sedimentology of the Late Cretaceous and Early Palaeocene Chalk Group, in North Sea Central Graben, In *Chalk Research Programme Symposium,* Stavanger, May 1985 (proprietary report).

8

Tommeliten

Michel D'Heur* and Lawrence J. Pekot†

** Petrofina, c/o Fina Exploration Norway, Skogstøstraen 37, N-4000 Stavanger, Norway*
† Phillips Petroleum Company Norway, PO Box 220, N-4056 Tananger, Norway

The Tommeliten fields are located in the southern part of the Norwegian sector of the North sea. They consist of two accumulations on which a total of five wells were drilled between 1976 and 1982. Hydrocarbons are contained in chalk reservoirs of the Ekofisk and Tor formations. The distribution and characteristics of the reservoirs result from sedimentary conditions and diagenetic effects. These marginal fields were declared commercial in early 1986, based on recoverable reserves (Statoil estimates) of 21×10^9 Sm^3 of gas and 9×10^3 Sm of oil and NGL.

INTRODUCTION

The Tommeliten chalk fields Alpha and Gamma are located in Block 1/9 in the southern part of the Norwegian sector, 325 km south-west of Stavanger and some 25 km south-west of the giant Ekofisk Field (Fig. 1). They consist of two diapiric structures, 10 km apart, which contain hydrocarbons in chalk of Late Cretaceous and Early Paleocene ages.

The fields were discovered in 1976 and 1977 after the discovery of seven other chalk fields in the same area. They were given the name of Tommeliten after the fairy-tale character who is known as Tom Thumb in English and Tom Pouce in French.

The water depth in the field area is about 77 m. The areal extent is 7 km^2 for Alpha and 4.5 km^2 for Gamma.

HISTORY

Production Licence 044 was granted to Statoil, Phillips, Fina and Agip on August 27, 1976. The licence covers Block 1/9 only, and Statoil is the operator. The present participation interests in the Tommeliten Fields are Statoil (operator 70.64%), Norske Fina A/S (20.23%) and Norsk Agip A/S (9.13%).

Only a few regional seismic lines were available on the block before the first detailed survey was shot in 1974. This 1×1 kilometre seismic grid indicated the presence of four structures. The largest, named Alpha, has a vertical relief of 150 m at Top Chalk level (Figs 2, 3). Beta is a small and low-relief feature located in the western part of the block. Gamma is a near-circular dome with relief in excess of 300 m. The Delta-Epsilon structure is a salt piercement straddling the boundary between Blocks 1/9 and 2/7. The first well, 1/9-1, was spudded in October 1976 near the crest of the Alpha structure. The well reached a total depth of 3681 mss in the Hidra Formation after penetrating 100 m of porous and hydrocarbon-bearing chalk (lithostratigraphy and nomenclature are given in Fig. 4; porosity and hydrocarbon-saturation logs through the reservoir are shown on Fig. 5 and

Geology of the Norwegian Oil and Gas Fields © A. M. Spencer *et al.*, eds (Graham & Trotman, 1987) pp. 117–128.

Tommeliten Field data summary

	Alpha	Gamma
Trap		
Type	dome structure over salt swell	
Depth to crest at top Ekofisk Fm	3000 mss	3025 mss
Lowest closing contour at top Ekofisk Fm	3140 mss	3260 mss
Hydrocarbon contact (lowest)	3180 mss	NA
Gas column	180 m	> 295 m
Pay zone		
Formation	Ekofisk, Tor and Hod formations chalks	
Age	Danian and Maastrichtian	
Gross thickness (average)	125 m	> 165 m
Porosity (averages; maximum)	27%; 42%	24%; 43%
Permeability (average; range)		
Ekofisk Fm	0.2 mD; 0.02–2.0 mD	0.1 mD; 0.01–1.0 mD
Tor Fm	1.0 mD; 0.4–4.0 mD	0.8 mD; 0.04–4.0 mD
Hydrocarbons		
Oil gravity	0.78–0.85 g/cc	0.78 g/cm^3
Gas gravity (relative to air)	0.71	0.68
Formation water		
Salinity	70 000 ppm NaCl equivalents	78 000 ppm NaCl equivalents
Resistivity	0.040 ohm m at 124 °C	0.039 ohm m at 130 °C
Reservoir conditions		
Temperature	124 °C	130 °C
Pressure (at 3100 m)	485 bar	487 bar
Field size		
Net area	7 km^2	4.5 km^2
Recoverable hydrocarbons (Alpha + Gamma)		
gas	21×10^9 Sm^3	
oil + NGL	9×10^6 Sm^3	

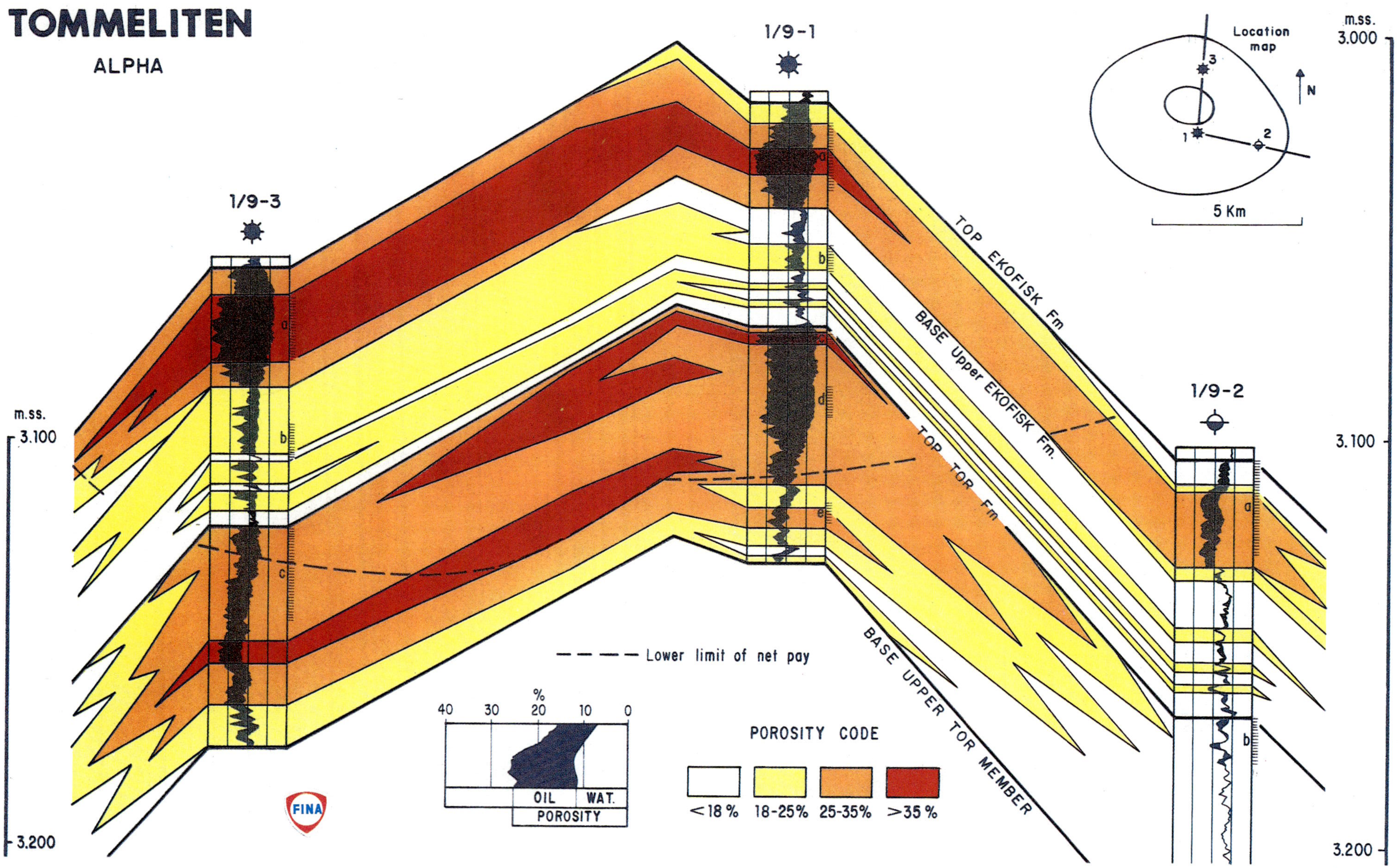

Plate 6. Tommeliten Alpha — porosity cross-section.

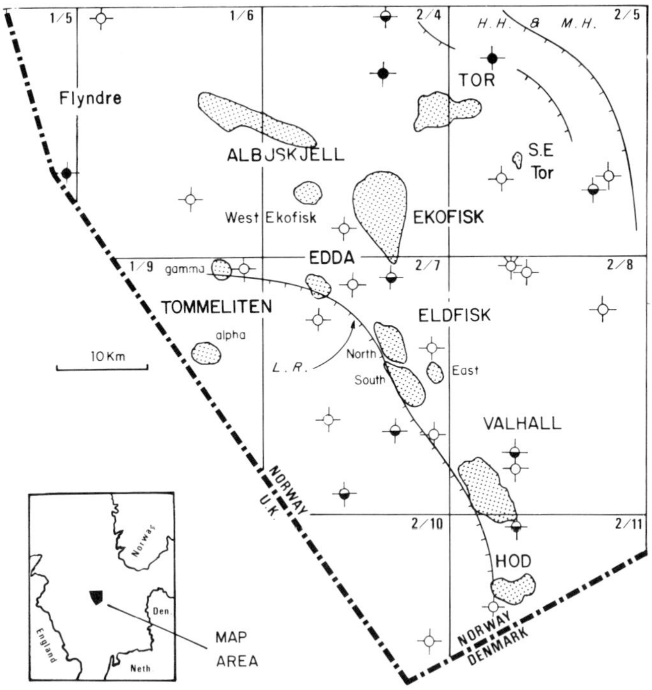

Fig. 1. Location map of the Norwegian chalk fields and finds. L.R. = crest of the Lindesnes Ridge; H.H. are faults of the Hidra and Mandal highs bordering the Central Graben to the east.

Plate 6). Selective production tests were made on numerous intervals of the Ekofisk and Tor formations; the best one flowed oil at 63 Sm^3/day from 13 m of perforations (interval 'd' on Plate 6).

The second well, 1/9-2, was drilled in summer 1977 on the flank of the same structure, 1.5 km east of the discovery well. It was designed to appraise the extent of the field (Fig. 2, 1983). The well reached a total depth of 3434 mss in the Hod Formation. The Ekofisk and Tor formations were present but they were thinner and their porosity and permeability were much lower than in the discovery well. Two intervals were tested but produced only water. Such a reduction in reservoir quality over a short distance from crest to flank of the structure, and the resulting thickening of the transition zone has since been recognized as a characteristic of chalk fields (see West Ekofisk Field, this volume, p. 165).

The third well, 1/9-3, was drilled 1.6 km north of the discovery well immediately after 1/9-2. It encountered the top of the chalk 32 m deeper than 1/9-1 but 53 m higher than 1/9-2. The Ekofisk Formation reservoir was of good quality with a net pay of more than 30 m, but the water saturation was high in the Tor Formation reservoir. Four drill-stem tests were performed. The best Ekofisk Formation test flowed oil at 400 Sm^3/day and gas at 650×10^3 Sm^3/day from an 18-m perforated interval, whereas the Tor Formation produced only water. The well was later deepened to 4535 mss and terminated in the Upper Jurassic without encountering any additio-

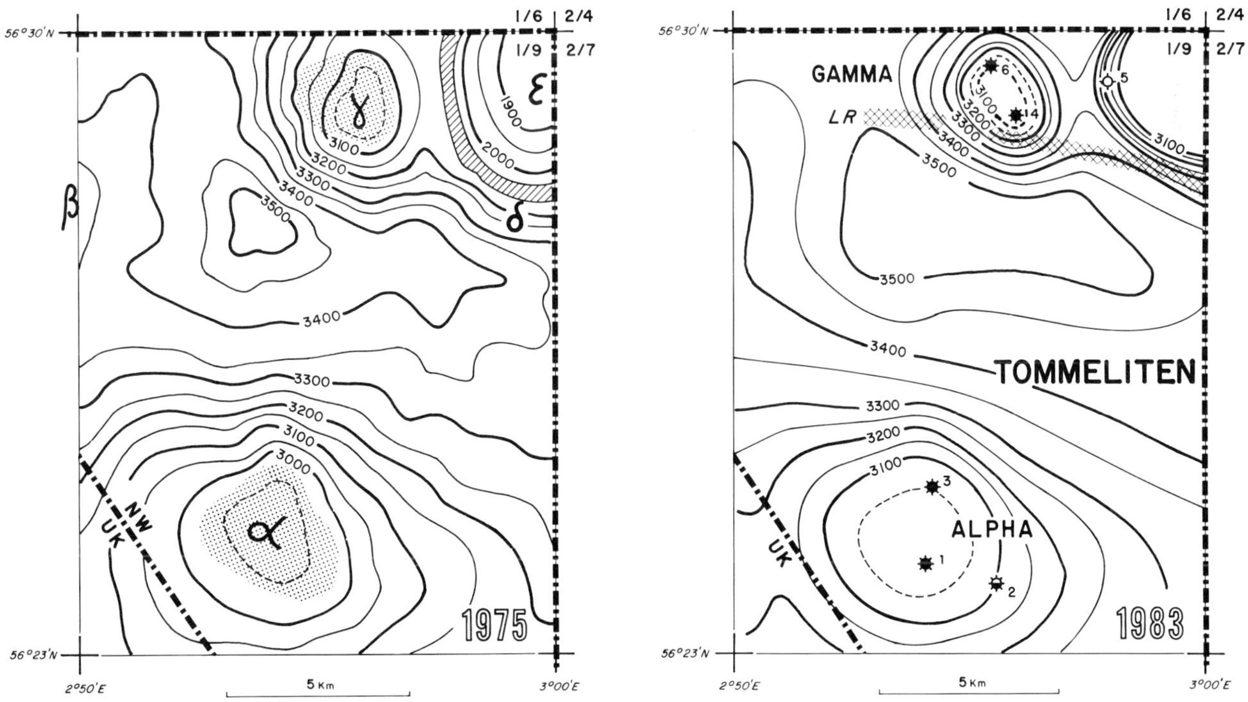

Fig. 2. Structural interpretation of part of Block 1/9 before and after discovery and appraisal drilling of the Tommeliten Fields. Contours show depth of Top Ekofisk Fm. Shaded areas on Alpha and Gamma (1975) are zones of seismic anomaly (see Fig. 3) whereas the hachures around Epsilon represent a salt wall. Crossed shading (LR) on the 1983 interpretation represents the north-western end of the Lindesnes fault zone.

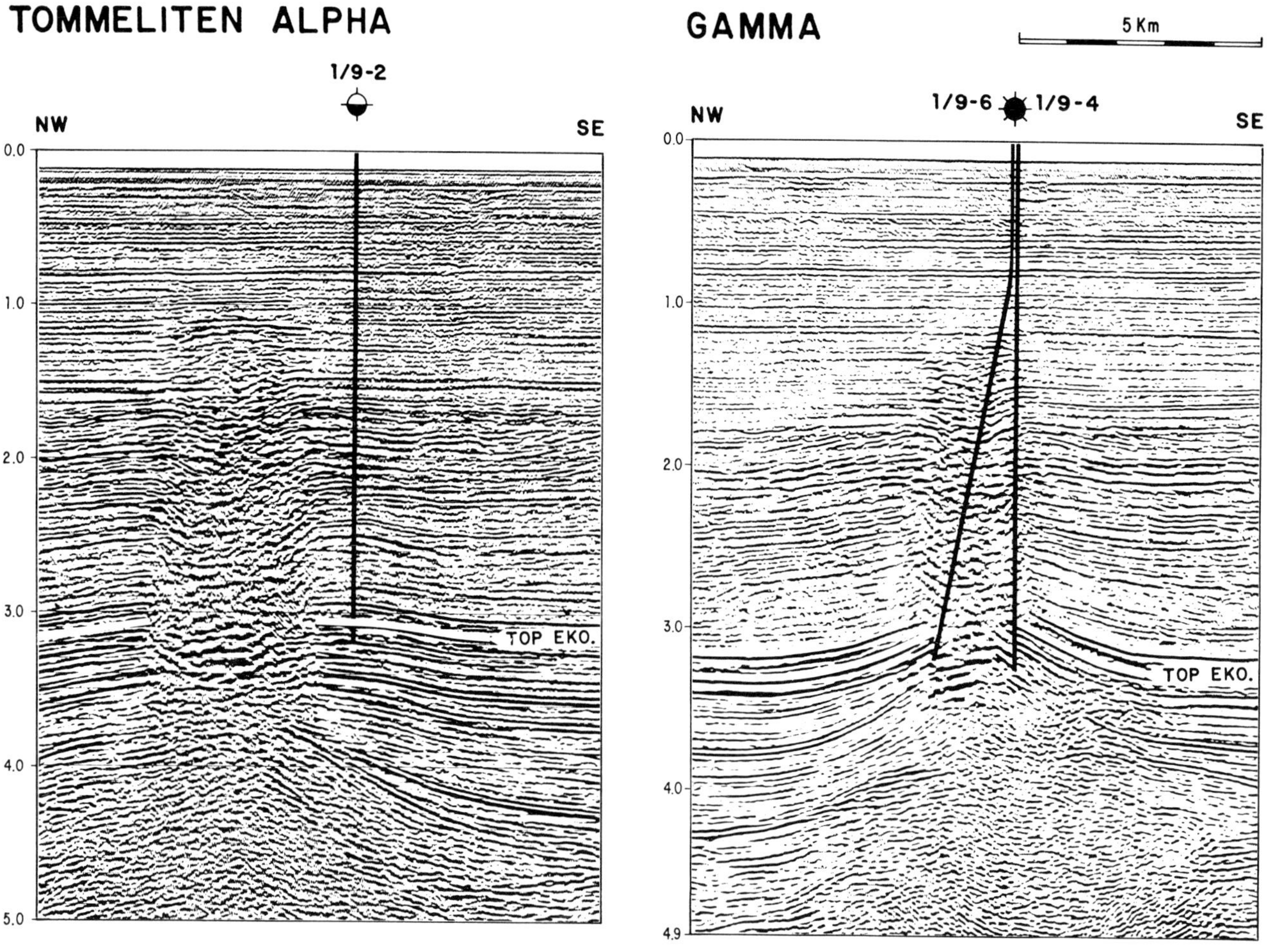

Fig. 3. Seismic lines (NW–SE) across the Alpha and Gamma structures.

nal reservoir. Well 1/9-4 was located near the crest of the Gamma structure; it found a hydrocarbon-bearing chalk reservoir between 3100 and 3300 m and reached a total depth of 3685 mss in Zechstein Salt without encountering any Jurassic or Triassic rocks. The upper reservoir of the Ekofisk Formation (Fig. 5, zone 'a') and the Tor Formation flowed gas at rates of 500 Sm^3/day and 700 Sm^3/day respectively.

Well 1/9-5 was then drilled on the west flank (Delta) of a salt plug. The well reached a total depth of 3415 mss in the Hod Formation without encountering any hydrocarbons. No drilling took place between December 1978 and March 1982. In the meantime, a 135-km radial-pattern seismic survey was shot on both structures.

Well 1/9-6 was drilled in 1982 to appraise the Gamma Field. It was deviated from a sea-bed location close to 1/9-4 to a bottom hole location 1.2 km northwards in the Hod Formation. Tests carried out on the Ekofisk and Tor formations flowed gas at 840 and 860 Sm^3/day respectively. The two wells on Gamma are temporarily abandoned. Development drilling commenced on Tommeliten Gamma in September 1986 from a template installed on the sea floor. The wells will be completed subsea and linked to the Edda platform.

GEOLOGICAL SETTING

Block 1/9 is located in the south-western part of the Central Graben, which was initiated early in Triassic time and persisted until the Early Paleocene. The main tectonic episodes affecting the graben took place during late Jurassic and Early Cretaceous times. Tectonic events within the Graben, as well as instability on the margin slopes and surrounding highs, played a major role in determining the distribution and characteristics of the reservoirs.

In this regard, Tommeliten Alpha and Gamma are different (Fig. 1). Alpha is probably entirely of halokinetic origin and had no marked relief at the time when the reservoir rock was deposited. By contrast, the Gamma area is located at the north-western end of the Lindesnes Ridge, a palaeostructure of half-

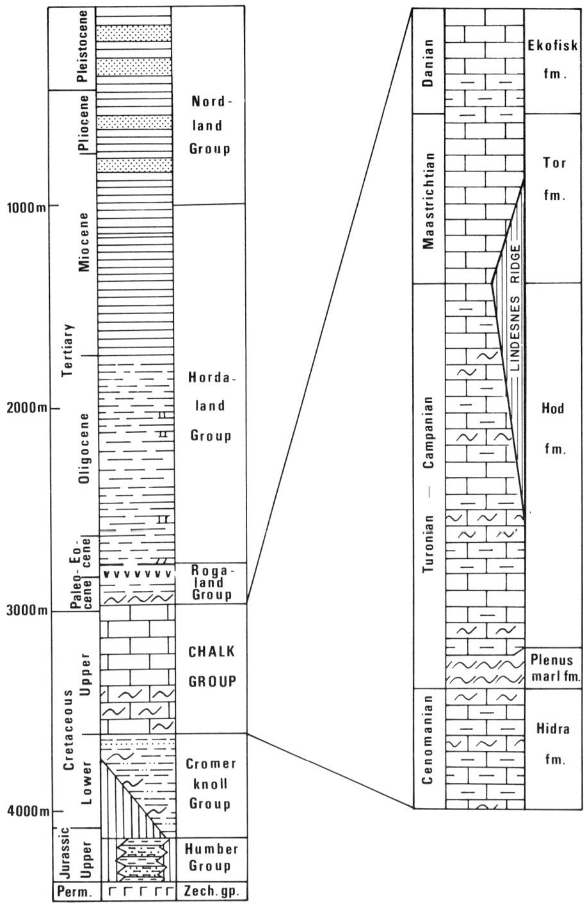

Fig. 4. Lithostratigraphy for the 'Greater Ekofisk Area'.

horst type existing during Maastrichtian and Danian chalk deposition. The north flank of the half-horst was domed by several halokinetic movements of the Permian salt. These differences in structural history influenced hydrocarbon distribution.

Since the end of chalk deposition, the Greater Ekofisk area, including Tommeliten, has remained in the centre of the large Tertiary basin that covers most of the Central North Sea. The general stratigraphy in Block 1/9 conforms to that of the Greater Ekofisk area (Fig. 4). The description below is, however, based on wells drilled on salt structures where some formations are obviously thinner than in the surrounding areas. Synclines between the structures might contain additional strata.

The Chalk Group (500–700 m) is represented by the Ekofisk, Tor, Hod, Plenus Marl and Hidra formations.

The Ekofisk Formation (60–100 m) has a

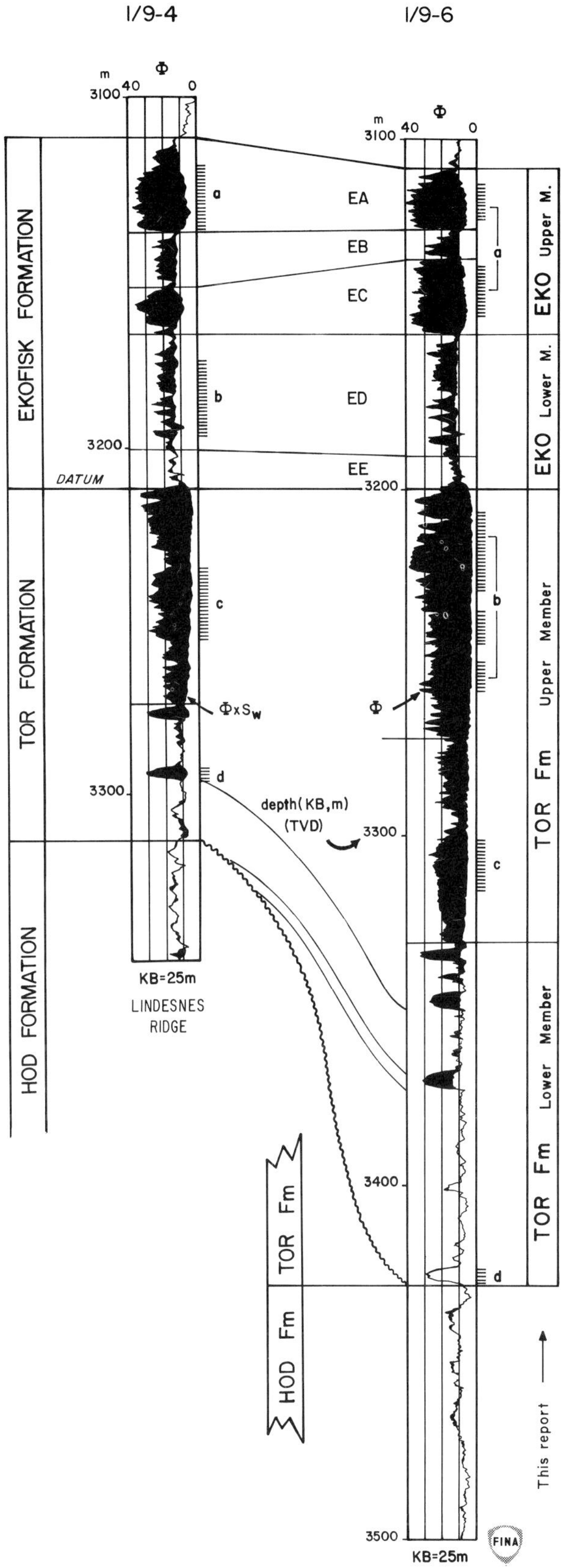

Fig. 5. Tommeliten Gamma—correlations to illustrate thickness variation over the Lindesnes Ridge. The datum is taken at the top of the Tor Formation but since it has been encountered at approximately the same depth in both wells, the section is nearly structural. Note that this diagram does not illustrate the shape of the palaeohigh, which is not yet clearly determined.

gradational upper boundary, passing from marl in the overlying formation to a chalky limestone. The limestone is firm, with stylolites. In the lower part, it becomes more argillaceous, dark grey in colour and is commonly interbedded with dark shale.

The Tor Formation (100–230 m) consists entirely of light grey chalky limestone. It is firm and homogeneous and becomes harder with depth. Thin shale laminae and stylolites are commonly found throughout the whole interval.

The Hod Formation (140–440 m) consists of buff to light grey–brown limestone, becoming more argillaceous and darker lower in the section. The lower part is interbedded with a few thin layers of silty limestone and marl.

The Plenus Marl Formation (10–50 m) is dominated by a light grey to brown marl interbedded with a black silty shale and thin beds of white limestone.

The Hidra Formation (100–150 m) consists of white to light grey limestone interbedded with dark grey shale and light grey marl.

The Cromer Knoll Group (470 m) is composed of some 30 m of pink to red claystone, shale and marls of the Rødby Formation and up to 440 m of Valhall Formation. The latter consists of grey and brown shale with thin interbeds of sand. It includes 120 m of very argillaceous, fine- to coarse-grained sandstone, 70 m of light brown to grey shale, which is often calcareous with thin beds of marl, limestone and sand, and a very argillaceous sandstone 120 m thick.

The Humber Group (300 m) is represented by grey shales of the Kimmeridge Clay Formation interbedded with very argillaceous limestone and thin layers of sand. The Heather Formation has not been penetrated. Triassic red beds have not been encountered but mobilized salt of the Zechstein Group has been penetrated. Older rocks have not been reached.

GEOPHYSICS

Two seismic surveys have been shot in Block 1/9. The first, shot in 1974, consists of a 1 × 1 km grid of seismic lines covering the whole block. The line directions are NW–SE and NE–SW. The second survey was shot in 1981 over the Alpha and Gamma structures in 'star' patterns centred at the top of each structure to optimize configuration for two-dimensional migration. The quality of the seismic data is generally good, except over the central part of the structures, where the seismic signals are heavily delayed and reflector strength destroyed due to abnormal pore-fluid conditions in the Tertiary section. The first strong reflector above the chalk is correlated with the top of the Paleocene, while the second, which runs approximately parallel at about 0.15 s deeper, marks the top of the Ekofisk Formation. Because of the seismic anomaly in the Tertiary over the crest of the structures, neither the top Paleocene nor the top of the Ekofisk Formation can be properly mapped in these areas by geophysical means only. A VSP run in the first development well has unfortunately not helped in solving this problem.

An expression of the Alpha and Gamma structures is also present in the overlaying sediments, since salt movements may have continued at least until early Miocene times.

TRAP

Both Tommeliten structures are circular domes formed by upward swelling of the Zechstein salt. The halokinesis started during Mesozoic times and continued until Early Miocene times. However, as for most chalk fields, the structural relief has been reinforced by differential compaction in the reservoir (see West Ekofisk, this volume, p. 165). The diameters of Alpha and Gamma are respectively 4 km and 3 km, and the vertical closures at top Ekofisk Formation level are 150 m and 350 m.

Although the normal concepts of vertical closure, spill-point and water-table have little significance in chalk fields (D'Heur, 1984), the deepest occurrence of hydrocarbons in each field is approximately equal to the limit of structural closure. Nevertheless, in detail, the base of hydrocarbons is found at different depths in each well, and the base of hydrocarbons surface does not necessarily correspond to the lowest closing contour.

The seal consists of Tertiary shale and claystone about 3000 m thick. The chalk is the only reservoir rock encountered in the Mesozoic and Tertiary.

RESERVOIR

Rock types

The hydrocarbon-bearing chalks of the Ekofisk and Tor Formations are friable to well-cemented biomicrites mainly made up of the skeletal debris of unicellular algae. The exoskeleton (coccosphere) is generally disarticulated into wheel-shaped coccoliths (10–30 μm) or into their minute elements. The latter, called platelets or laths, are monocrystals of low magnesium calcite whose size varies from < 1 μm to > 2 μm. The sediment is chemically stable. The detailed petrography of these chalks is now well documented in the literature (Hancock, 1976). Among the non-carbonate fractions, clay plays a major role as far as diagenesis and permeability are concerned (Hardman, 1982) while silica is of significance for the mechanical behaviour of the rock (da Silva *et al.*, 1985). The main sedimentological characteristic of the Central Graben chalk is the high percentage of allochthonous material it contains.

This material comes from the reworking of a coccolith ooze previously deposited on the surrounding high areas; it was then transported into the deepest part of the graben by gravity flows. A number of autochthonous and allochthonous chalk facies have been described by Kennedy (1980, 1985). Of particular interest is the fact that the allochthonous chalk possesses much better original porosity than the normal pelagic sediment, which was originally more cemented. In addition to the abundance of redeposited material, the most fortunate feature of these chalks is the extraordinarily high level of porosity preservation in the areas of the fields (general discussion in Scholle, 1977). Under normal conditions porosity would have decreased to about 12% (±6%) by compaction under 3000 m of younger sediments, but the chalk of the Ekofisk and Tor formations has porosity of up to 50%, with field averages often higher than 30%. This is due to a number of factors that inhibited the normal process of porosity destruction of a carbonate ooze with increasing overburden (see West Ekofisk, this volume, p. 165).

Reservoir subdivision

The reservoir formations consist of an alteration of more and less porous beds, depending on the depositional mode of the chalk. The porosity contrast is often obvious in the hydrocarbon zone while it is sometimes masked by stronger diagenesis in the aquifer (Fig. 6).

The Tommeliten reservoir has been subdivided into four members. Table 1 is a comparison of these members with the Greater Ekofisk area porosity zonation as used in other chalk fields (see Ekofisk, this volume, p. 165).

The Upper Ekofisk Member, or 'Ekofisk Porous' zone, mainly consists of buff, bioturbated chalk with low clay content. Stylolites are abundant. Most of the material is allochthonous. The reservoir characteristics are generally good, with average porosity higher than 30% in the hydrocarbon zone. In Tommeliten Gamma this member is divided by a 10- to 15-m interval of low-porosity chalk. Its appearance is similar to that of the Lower Ekofisk member, containing buff and light grey alternations with frequent dark clay laminations within the grey intervals.

The Lower Ekofisk Member, or 'Ekofisk Tight' zone (*sensu lato*), consists of a much more argillaceous chalk, which is generally laminated. The typical 'periodite facies' (Kennedy, 1980) with alternations of light, clay-poor and dark, clay-rich streaks representing a significant part of this interval; turbidites and other redeposited sediments are also recorded. A hard-ground is observed at the base of the member in some wells, and a resedimented conglomerate has been found in Well 1/9-2. The reservoir characteristics of this interval are rather poor, with porosity ranging between 8 and 22%, except in Well 1/9-3, where it is higher.

The Upper Tor Member consists of alternations of homogeneous and bioturbated chalks. The upper part has a very low clay content, a high percentage of allochthonous material and good reservoir properties. The lower part is more argillaceous, contains more autochthonous material and has a lower porosity.

The Lower Tor Member is mainly made up of pelagic sediment interbedded with distal allochthonous facies. The porosity is low to very low with the exception of several debris-flows three to seven metres thick, a unit recognized in Gamma and in all of the fields north-east of the Lindesnes Ridge. On Alpha, the Upper Tor Member is equivalent to the TA zone of the Greater Ekofisk area zonation. On Gamma, it is equivalent to zones TA plus TB. This convention simplifies two problems. First, it places essentially the entire Tor Formation productive interval in a single unit. Secondly, it overcomes the difficulty of identifying the contact between zones TA and TB, which is particularly uncertain at Tommeliten.

Reservoir distribution and characteristics (Table 2)

Tommeliten Alpha

The Upper Ekofisk Member contains most, if not all of the Danian hydrocarbons. The 'solid

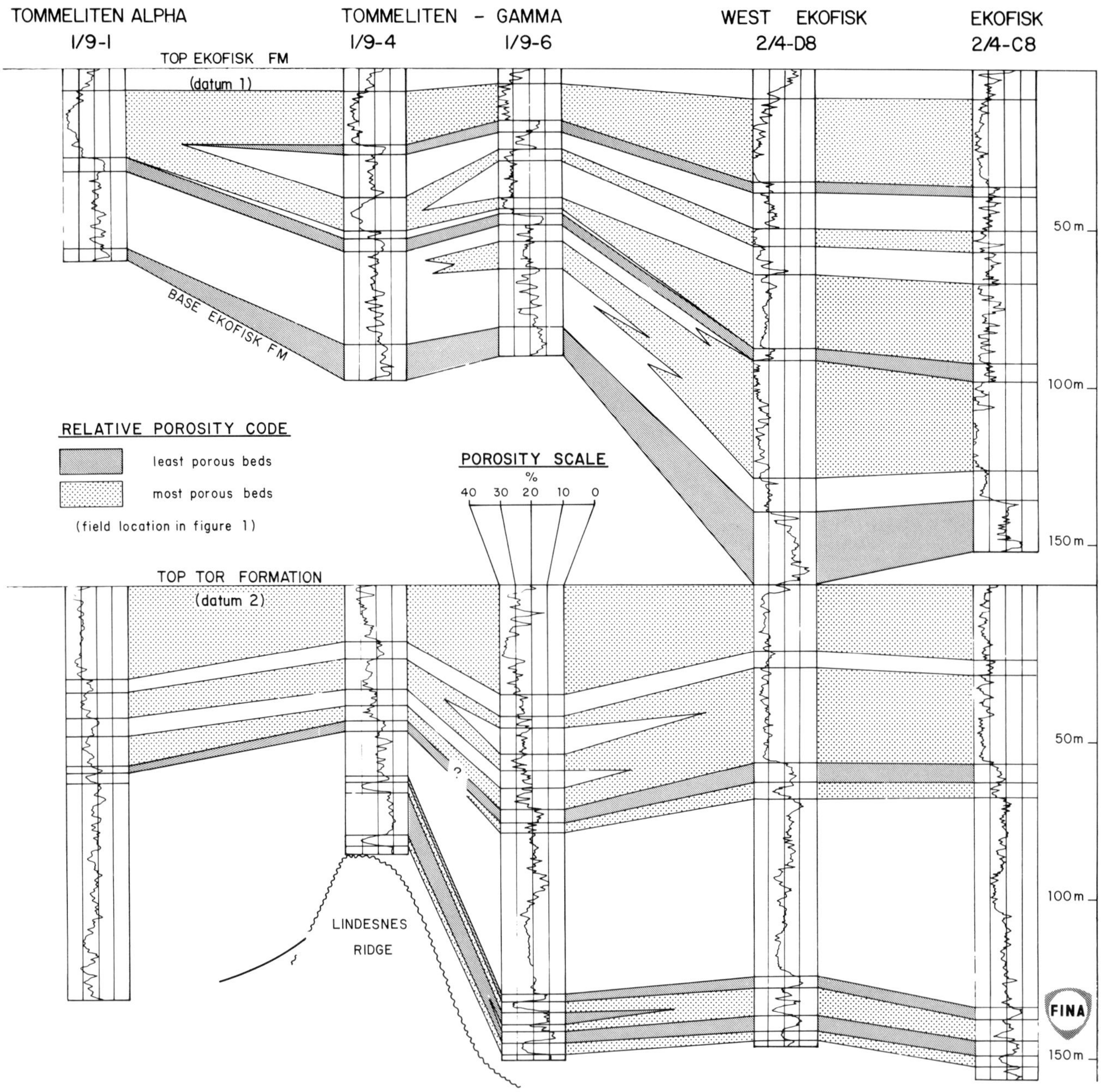

Fig. 6. Regional correlation of the Ekofisk and Tor formations to illustrate the lateral and vertical porosity variation (porosity logs).

thickness' map (Fig. 7) shows a thinning from 23 to 19 m which probably indicates a north-easterly source for the allochthonous chalk of this layer. The isoporosity contours follow the structural shape from 32% in the crestal area to 22% at the field limit. Nevertheless, the porosity maximum is eccentric towards Well 1/9-3, probably because of a slight tilt of the structure since oil emplacement. The gross thickness map of this member shows the effects of both the regional thinning and of differential compaction, corresponding to differential porosity preservation. The rapid porosity decrease interpreted north-east of Well 1/9-3 would result in a slight thinning in the same direction within the field limit. Farther north-east, towards Tommeliten Gamma, the thickness of this member increases slightly. The field boundary corresponds approximately with the contour of 22% average porosity. In low-flank structural positions, the Upper Ekofisk Member has an extremely low permeability even if its porosity still exceeds 20% (D'Heur, 1980 and West Ekofisk, this volume). The Lower Ekofisk Member may contain a small quantity of recoverable hydrocarbons in the very crestal part of the structure.

Table 1. Tommeliten reservoir zonation compared with that of the 'Greater Ekofisk area'.

	Ekofisk area	*Tommeliten* *Alpha* *Gamma*
Ekofisk Fm Upper	EA	Upper Ekofisk member
Middle	EB EC	
Lower	ED	Lower Ekofisk member
Tight zone	EE	
Tor Fm Upper	TA	Upper Tor member
Middle	TB	
Lower	TC	Lower Tor member

Table 2. Reservoir characteristics of Tommeliten and Ekofisk crestal wells. Net pay criteria vary from field to field and are different for the Ekofisk and Tor formations.

	Tommeliten *Alpha*	*Gamma*	*Ekofisk*
Gross pay thickness	90 m	295 m	310 m
Net pay thickness	65 m	140 m	280 m
Average porosity net pay	0.30	0.25	0.31
Average water saturation pay	0.25	0.17	0.10

The Upper Tor Member contains about 40% of the hydrocarbon pore volume in Alpha. The gross interval average porosity map (Fig. 8) shows a marked reduction of porosity, with probably more than 30% (average) in the crestal area down to about 18% in Well 1/9-2, where the rock has always been in the aquifer.

Tommeliten Gamma

Only two exploration/appraisal wells have been drilled. Both encountered the tops of the Ekofisk and Tor formations at similar depths, so the geological interpretation is tentative and based on observation of other fields located on the Lindesnes Ridge (Fig. 9). In addition, since the Tertiary gas effect is very strong between the two wells (see Fig. 3), the structural interpretation in the crestal area is very speculative; faults exist but cannot be mapped.

For the Ekofisk Formation, the limit of the reservoir corresponds approximately with the 22% average porosity contour (upper member), the same as for Alpha. The average

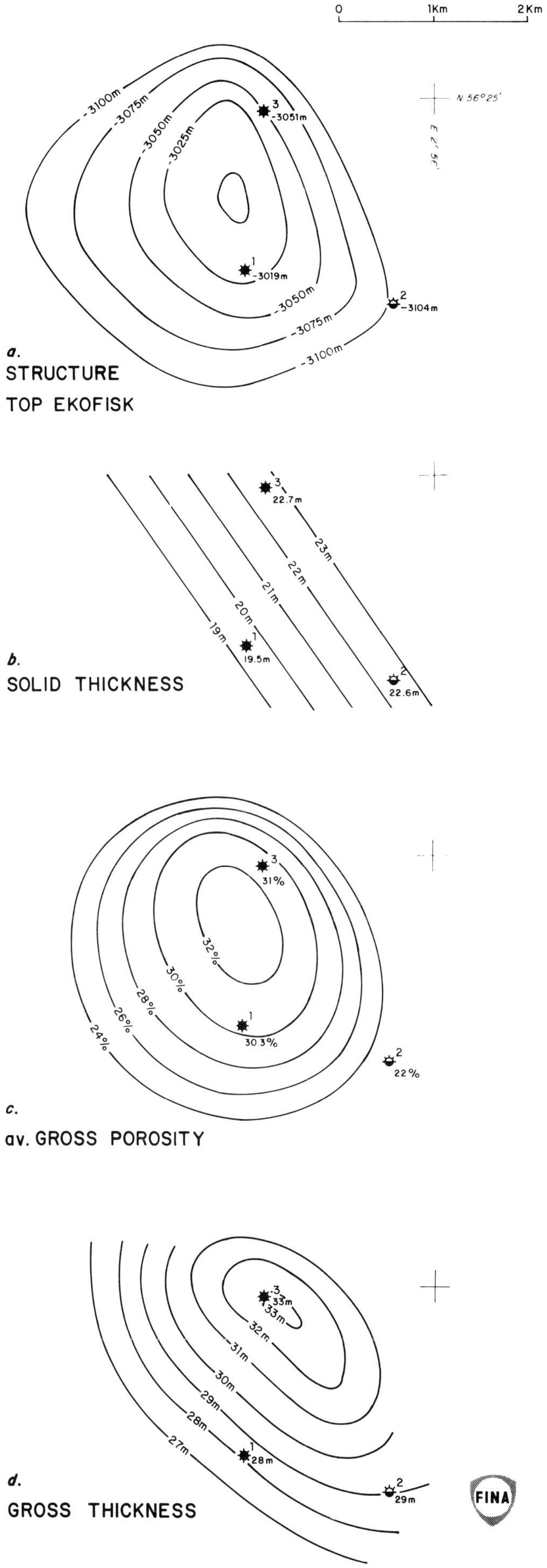

Fig. 7. Tommeliten Alpha—reservoir map of the Upper Ekofisk member. Solid thickness equals true vertical thickness multiplied by (1-ø) where ø is the average porosity for this interval (see West Ekofisk, this volume).

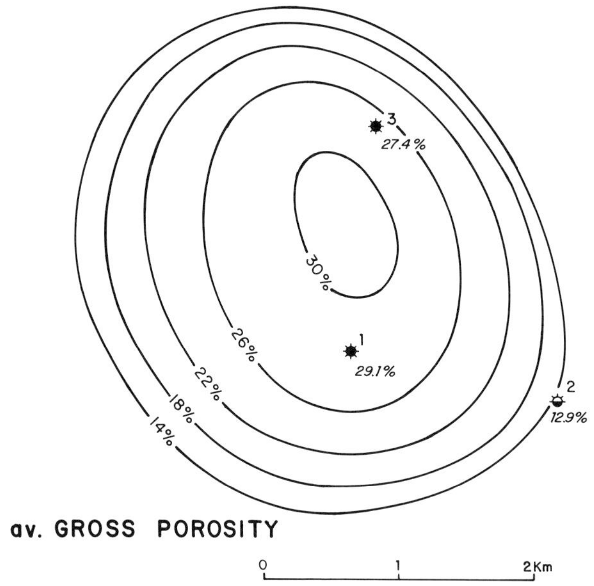

Fig. 8. Tommeliten Alpha—average porosity map of the Upper Tor member.

(net) porosity is approximately 35% in the crestal area and decreases rapidly towards Well 1/9-4 and in the flank.

The gross thickness maps for the Upper Tor Member clearly indicate a thickness reduction in the 1/9-4 area (Fig. 9). This results from the existence of a palaeohigh that strongly influenced the deposition during Maastrichtian times. This high, which is represented on the section in both Figs 5 and 6, is the north-western end of the Lindesnes Ridge (Fig. 1). The precise location of the crest of the palaeohigh is still unknown and therefore, the maps of Fig. 9 correspond to a provisional geological model which will be adapted when development-well data become available. The average porosity map (Fig. 9) shows a slight porosity reduction over the palaeocrest. The limit of the reservoir is controlled by porosity reduction in the flanks as for the Ekofisk Formation.

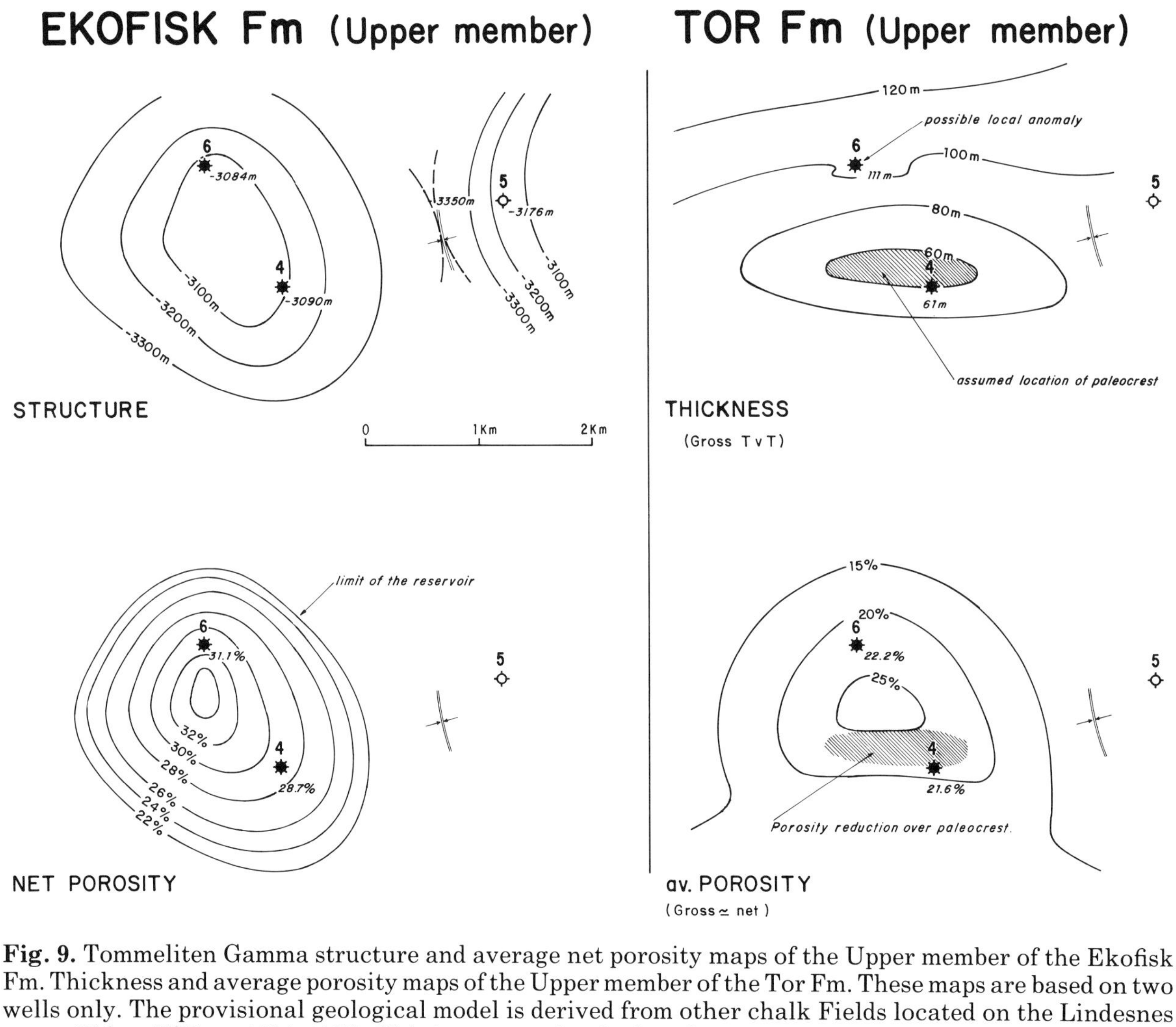

Fig. 9. Tommeliten Gamma structure and average net porosity maps of the Upper member of the Ekofisk Fm. Thickness and average porosity maps of the Upper member of the Tor Fm. These maps are based on two wells only. The provisional geological model is derived from other chalk Fields located on the Lindesnes Ridge (D'Heur 1984, 1985). This interpretation is therefore very schematic and speculative.

The Lower Tor Member reservoir is restricted to only five porous beds, three of which have not been deposited in Well 1/9-4. The net pay thickness, which is highly influenced by the palaeohigh, is arbitrarily limited at 3375 mss because the porous beds have been found water bearing below that depth.

Permeability

Well tests and core analyses indicate that permeabilities from the Upper Ekofisk Member are from five to ten times less than in the Upper Tor Member, even though porosities are higher. The average porosity of the Ekofisk reaches 32% compared with 25% for the Upper Tor. Well tests and core permeabilities generally correlate, indicating a poor natural microfracture system.

The permeability range of the Upper Ekofisk Member is 0.1–0.5 mD with an average of about 0.2 mD; whereas for the Upper Tor Member the range is 0.5–1.0 mD with an average of 0.75 mD.

SOURCE

The Tommeliten fields lie in a deep part of the Central Graben, located in the down-faulted area between the Lindesnes Ridge to the north-east and the graben-margin faults leading up to the Mid-North Sea High in the south-west. The Upper Jurassic Mandal Formation covers the area, and at Tommeliten is considered to be a prolific hydrocarbon source-rock. The top of the Upper Jurassic is often deeper than 5000 m, and the source-rocks are mature enough to have produced the rich gas–condensate of the Tommeliten Fields.

Geochemical analysis suggests that the source-rocks directly beneath the Alpha structure, at the relatively shallow depth of 4300 m, are in the late stages of maturity for oil generation and that some of the section has passed into the stage of condensate and wet gas formation. This degree of maturity is reflected by the presence of an oil layer in Alpha which has not been flushed from the structure by gas.

HYDROCARBONS

The Tommeliten Fields contain a rich gas–condensate. At a reservoir temperature of about 129 °C, the fluid is saturated or slightly undersaturated at initial pressure of 486 bar. The three-stage GOR is about 1350 Sm^3/Sm^3. Condensate gravity ranges from 0.8017 to 0.7796 g/cc. In one well (1/9-IX), which produced from an oil zone, the solution GOR was about 450 Sm^3/Sm^3, with an oil gravity ranging from 0.8498 to 0.8251 g/cm^3.

The distribution of hydrocarbons on the Alpha structure is not entirely clear. Only one well has indicated productive hydrocarbons in the Tor Formation, which included a water–oil transition zone, an oil leg (mentioned above), and a gas–condensate accumulation. Good areal communication seems to exist through the matrix in the hydrocarbon zones, and a single 'base of net pay' has been considered for each field. Its shape is curved, as in the 'West Ekofisk Field geological model', and it has been idealized in Plate 1 for the Alpha structure. In Gamma, the present data only indicate an 'oil down to' corresponding to the tight beds of the basal Lower Tor Member.

RESERVES

The geological evaluation made by the various Tommeliten co-venturers resulted in estimates of total net hydrocarbon pore volume ranging from 120×10^6 to 148×10^6 reservoir m^3 for both Alpha and Gamma together. The mapping presented in this report gives a total hydrocarbon pore volume of 120×10^6 m^3. The high value (Statoil, 1983) corresponds to a total recovery of 21×10^9 Sm^3 of gas and 9×10^6 Sm^3 of oil and NGL. The hydrocarbon pore volume is divided approximately equally between the Alpha and Gamma structures.

ACKNOWLEDGEMENTS

The authors thank Fina, Phillips, Statoil and Agip by whose permission this article is published. The opinions expressed in this paper are those of the authors only and do not necessarily reflect the opinions of any of the Tommeliten co-venturers.

REFERENCES

da Silva, F., Debande, G., Detiege, C., Halleux, L., Monjoie, A., Poot, B. and Schroeder, C. 1985. Mechanical behaviour of chalks, In *Chalk Research Program Symposium*, Stavanger, May 21–22, 1985.

Deegan, C.E. and Scull, B.J. 1977. A standard lithostratigraphic nomenclature for the central and northern North Sea, *Report Institute Geological Sciences London* 77/25.

D'Heur, M. 1980. Chalk reservoir of the West Ekofisk Field, In *Sedimentation of the North Sea Reservoir Rocks*, Norsk Petroleumsforening, Article X.

D'Heur, M. 1984. Porosity and hydrocarbon distribution in the North Sea chalk reservoirs, *Marine and Petroleum Geology* **1,** 211–238.

D'Heur, M., de Walque, L. and Michaud, F. 1985. Geology of the Edda Field reservoir, *Marine and Petroleum Geology* **2,** 327–340.

Hancock, J.M. 1976. The petrology of the chalk, *Proceedings Geologists Association 86* (for 1975), 499–535.

Hardman, R.F.P. 1982. Chalk reservoirs of the North Sea. *Bulletin Geological Society Denmark* **30,** 119–137.

Kennedy, W.J. 1980. Aspects of chalk sedimentation in the Southern Norwegian Offshore, In *The Sedimentation of the North Sea Reservoir Rocks* (Norsk Petroleumsforening), Article IX.

Kennedy, W.J. 1985. Sedimentology of the Late Cretaceous and Early Palaeocene Chalk Group, North Sea Central Graben, In *Chalk Research Program Symposium*, Stavanger, May 1985.

Mapstone, N.B. 1975. Diagenetic history of a North Sea chalk, *Sedimentology* **22,** 601–613.

Scholle, P.A. 1977. Chalk diagenesis and its relation to petroleum exploration: oil from chalk, a modern miracle? *Bulletin American Association Petroleum Geologists* **61,** 982–1009.

9

Tor

Michel D'Heur

Petrofina, c/o Fina Exploration Norway, Skogstøstraen 37, N-4001 Stavanger, Norway

The Tor Field is located in the Central Graben of the North Sea in the southern-most part of the Norwegian sector. The reservoirs are chalks of the Ekofisk and Tor formations which are of Danian and Maastrichtian age respectively. Their very high porosities are due to the combined effects of the abundance of a favourable allochthonous chalk facies and a high degree of porosity preservation, despite the chalk being buried by more than 3000 m of younger sediments. The Tor Field is characterized by areal variation of rock types and, consequently, by lateral variation of primary porosity in each reservoir unit. The present-day porosity variation reflects both the areal changes in facies and the differential 'chemical compaction' that affected the reservoir. Thirteen producer wells have been drilled in Blocks 2/4 and 2/5 from a single platform located near the boundary between the two blocks. The field began production in 1978. The estimated initial recoverable reserves are about 18×10^6 Sm^3 of oil and 12×10^9 Sm^3 of gas.

INTRODUCTION

The Tor Field, named after the Norwegian God of thunder and lightening, lies in the southern part of the Norwegian sector of the North Sea, approximately 310 km south-west of Stavanger and 13 km north-east of the giant Ekofisk Field (Fig. 1). The structure is a faulted anticline of halokinetic origin, 6×3 km in size, mainly contained in Block 2/4, but extending into Block 2/5 to the east. Both blocks were awarded during the first round of Norwegian licencing in 1965. Water depth in this area is about 70 m. Initial recoverable reserves from the chalk reservoir are estimated at about 18×10^6 Sm^3 of oil and 12×10^9 Sm^3 of associated gas.

HISTORY

Production Licence PL018, which includes Block 2/4, was granted to Phillips, Petrofina and Agip in 1965. The present participation interest in Block 2/4 is:

Phillips Petroleum Company of Norway	36.96%
Petrofina (Fina Exploration Norway)	30%
Norsk Agip	13.04%
Elf Aquitaine Norge A/S	8.1%
Norsk Hydro A/S	6.7%
Total Marine Norsk A/S	4%
Others	1.2%.

Block 2/5 is part of PL006 awarded at the same time to the Amoco Group. The present participation interest in Block 2/5 is:

Amoco Norway Oil Company 28.33%
Amerada Hess Norway Limited 28.33%
Texas Eastern Norwegian Incorporated 28.33%
Norwegian Oil Consortium (NOCO) 15%

During the early days of exploration, based on a limited amount of seismic data, a faulted structure was identified north-east of Ekofisk.

Geology of the Norwegian Oil and Gas Fields © A. M. Spencer *et al.*, eds (Graham & Trotman, 1987) pp. 129–142.

Tor Field data summary

Trap	
Type	dome structure above a salt swell
Depth to crest	2872 m
Lowest closing contour	3155 m
Hydrocarbon contact	~3292 (variable)
Oil column	320 m (approximately)
Pay zone	
Formation	Ekofisk and Tor formation chalks
Age	Danian and Maastrichtian
Gross thickness (average; range)	244 m; 213–275 m
Net/gross ratio (average; range)	0.62; 0.1–1.0
Cut-off for N/G	ø = 16%, S_w = 80% Ekofisk Fm ø = 10%, S_w = 90% Tor Fm
Porosity (average; range)	21%; 13–36%
Hydrocarbon saturation	63% net interval
Permeability (average; range)	1.0 mD; 0.04–3.0 mD Ekofisk Fm matrix; 0.6 mD; 0.04–5.0 mD Tor Fm matrix; effective permeability up to 200 mD, dependent upon fracture intensity
Hydrocarbons	
Oil gravity	0.8155 g/cm^3
Gas gravity	0.731
Bubble point	305 bar
Gas/oil ratio	260 Sm^3/m^3
Formation volume factor	1.85
Formation water	
Salinity	117 000 ppm NaCl equivalents
Resistivity	0.037 ohm at 136 °C Ekofisk Fm 0.016 ohm at 136 °C Tor Fm
Reservoir conditions	
Temperature	136 °C
Pressure	492 bar at 3124 m
Pressure gradient in reservoir	0.057 bar/m
Field size	
Area	22 km^2
Net hydrocarbon pore volume	165×10^6 m^3
Recovery factor	0.21
Drive mechanism	Solution gas drive
Recoverable hydrocarbons (initial)	
—oil	18×10^6 Sm^3
—gas	12–15 $\times 10^9$ Sm^3
Production	
Start-up	June 1978
Development scheme	single, steel platfrom with gas and oil oil pipelines to Ekofisk Center
Production rate	1979 Plateau, 12 500 Sm^3/day oil, 3.25×10^6 Sm^3/day gas October 1985; 2000 Sm^3/day oil, 2.4×10^6 Sm^3/day gas
Cumulative production	October 1985; 14.6×10^6 Sm^3 oil, 8.8×10^9 Sm^3 gas

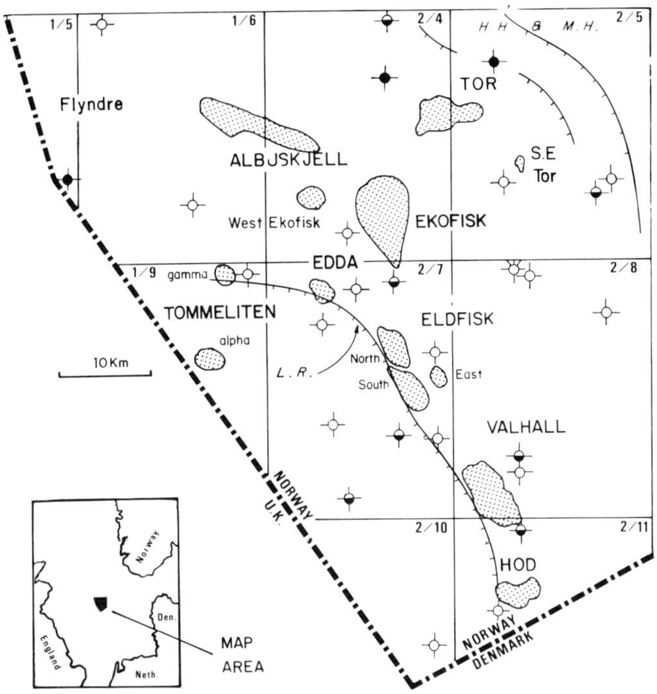

Fig. 1. Location map of the Norwegian chalk fields and finds. L.R., crest of the Lindesnes Ridge; H.H. and M.H. are faults of the Hidra and Mandal Highs complex bordering the Central Graben to the north-east.

A Phillips Group interpretation is shown in Fig. 2; the structure was named 'Ergfisk', as all structures recognized in Phillips Block E (2/4 + 2/7) were given the names of fishes. While the Phillips Group was busy with the appraisal of the Ekofisk discovery, the Amoco Group drilled the first well in Block 2/5 in 1970 and discovered the field they named Tor. The discovery well, 2/5-1X, was nearly crestal and close to the boundary between the two blocks. Oil was found in chalks of Late Cretaceous and Early Paleocene ages, which exhibited high porosity and hydrocarbon saturations beneath a series of Tertiary clays, claystone and shales, 3000 m thick.

The lithostratigraphy of an appraisal well is shown in Fig. 3. The gross pay interval in the discovery well was about 200 m, of which 150 m was net oil-bearing section. The best zone flowed at the rate of 237 Sm^3 oil/day and 70×10^3 Sm^3 gas/day from a nine-metre perforated interval.

Three appraisal wells were drilled in 1971. Computer-processed interpretation logs of the reservoir sections for the discovery and appraisal wells are shown in Fig. 4. Well 2/4-7X delineated the field to the west, whereas Well 2/4-8X, structurally the highest well drilled so far, including all development wells, found the thickest and most porous pay interval. Well 2/5-2X was drilled on an easterly extension of the field which appeared to be separated from the main structure by a slight saddle (Fig. 5(a)). The results of this last appraisal well were disappointing as most of the reservoir was of poor quality.

It was then agreed that the field should be developed and incorporated into the Phillips' Greater Ekofisk development scheme planned for Block 2/4, and an operating agreement was signed by all parties in 1975. Operatorship was assigned to Phillips and the provisional reserves split between the Phillips Group and the Amoco Group was set at 74.5% and 25.5%. It was defined on the basis of the 'net productive hydrocarbon pay' determined from a net HCPV map based on well control and a structure map derived from seismic data. It is in fact remarkable that, despite the paucity of data available at that time and the complexity of hydrocarbon distribution shown by subsequent development drilling, the final equity factors (73.75% and 26.25%) established in a third redetermination in 1984 are very similar to the initial values.

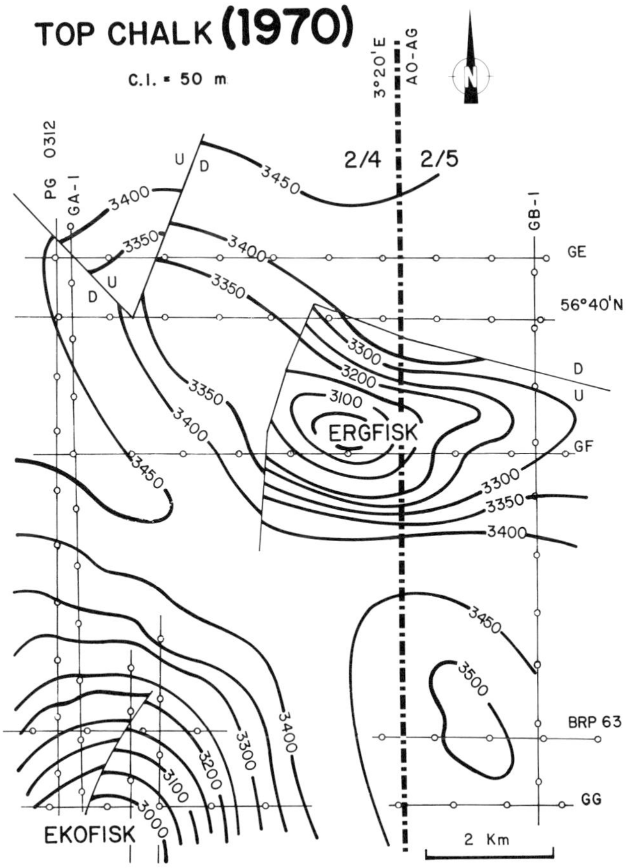

Fig. 2. Structural interpretation of 1970 by Phillips, of the 'Ergfisk' structure (later named Tor) before drilling. Contours in metres subsea on Top Chalk.

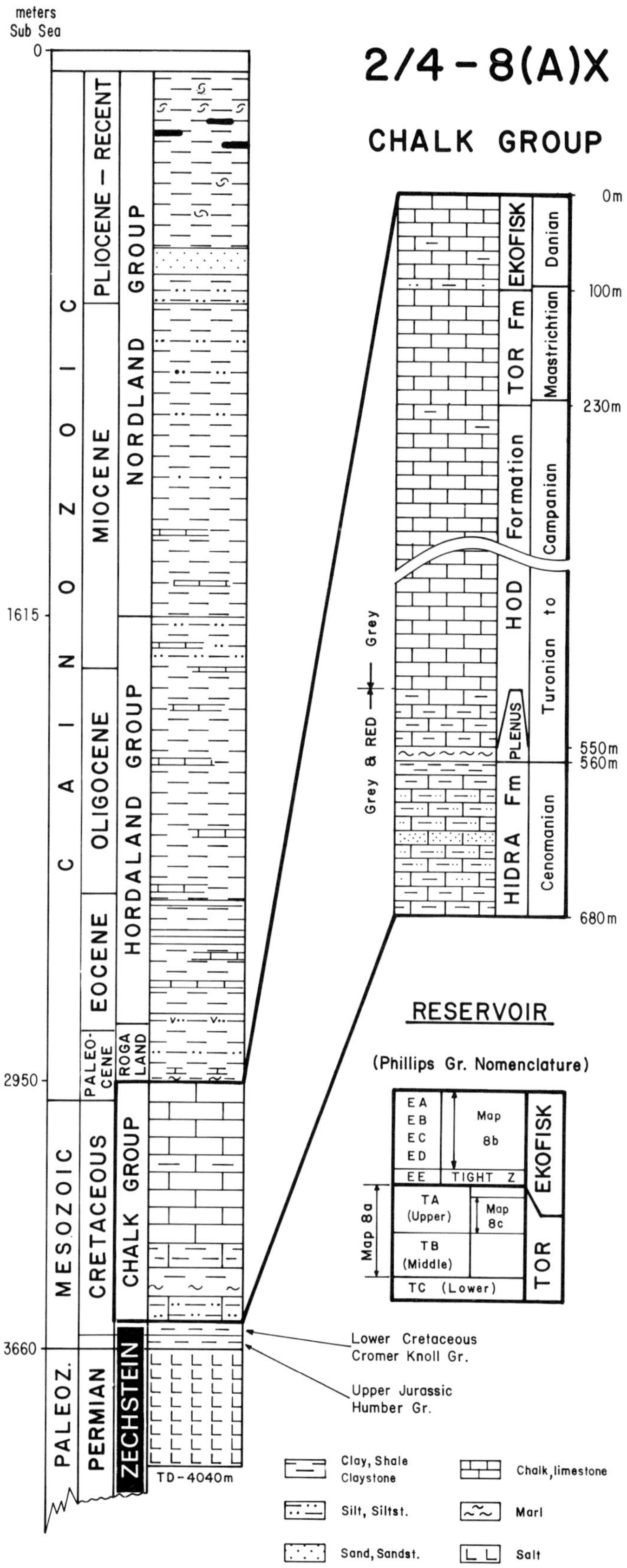

Fig. 3. Lithostratigraphy of Well 2/4-8AX.

The Tor Field platform has 18 well slots and was installed in 1976 in Block 2/4, midway between wells 2/4-7X and 2/5-2X. The development well locations and the 1975 agreed map are shown in Fig. 6(a). The well designation does not reflect the drilling sequence but conforms with the slot numbers. The setting of 30-inch conductors started in November 1976.

The experience from development drilling on the other chalk structures had shown that the best procedure was to drill wells step-by-step from the centre of the field towards the periphery. The explanation for this is that for most chalk fields the porosity generally decreases rapidly from the crestal area towards the flank with a resulting loss of net thickness. Therefore, the drilling sequence at Tor was flexible, each location (except the first three) being contingent on the results of the previous wells. The application of this principle resulted in satisfactory success of the development drilling phase at Tor, whereas several dry development wells were drilled on other fields.

The first development well, E10, was drilled vertically and reached the reservoir at 2990 m subsea in March 1977. Although it was located near the crest of the structure, the results were disappointing. In exploration wells 2/4-8X and 2/5-1X, the main reservoir had average porosities of about 28% whereas it was much less porous in Well E10 (20.5%). The reservoir was also thinner and its top was encountered deeper than expected. This first development well had, in fact, indicated two peculiarities of the Tor Field:

(a) the existence of a palaeohigh during Maastrichtian time (which lay in the vicinity of E10 and was later proved to extend northeastwards to E8);

(b) the presence of a Paleocene crestal graben, in the middle of which E10 (and E5) is located (see Fig. 5(b)).

The second development well, E9, found 110 m of net pay as expected but once more with only poor to fair porosity values.

The third well, E13, drilled to the east found that the main reservoir had very good porosity but that most of the other intervals were of poor quality. It had been previously agreed that the provisional equity split could be subject to revision after the completion of these three wells, but in fact they did not provide sufficient data to permit a precise definition of the reservoir. The field came on production in June 1978 and a further seven wells were drilled by the end of 1979. In 1980 the Amoco Group requested an equity redetermination, but when the parties had failed to reach agreement after three and a half months of joint effort, the matter was referred

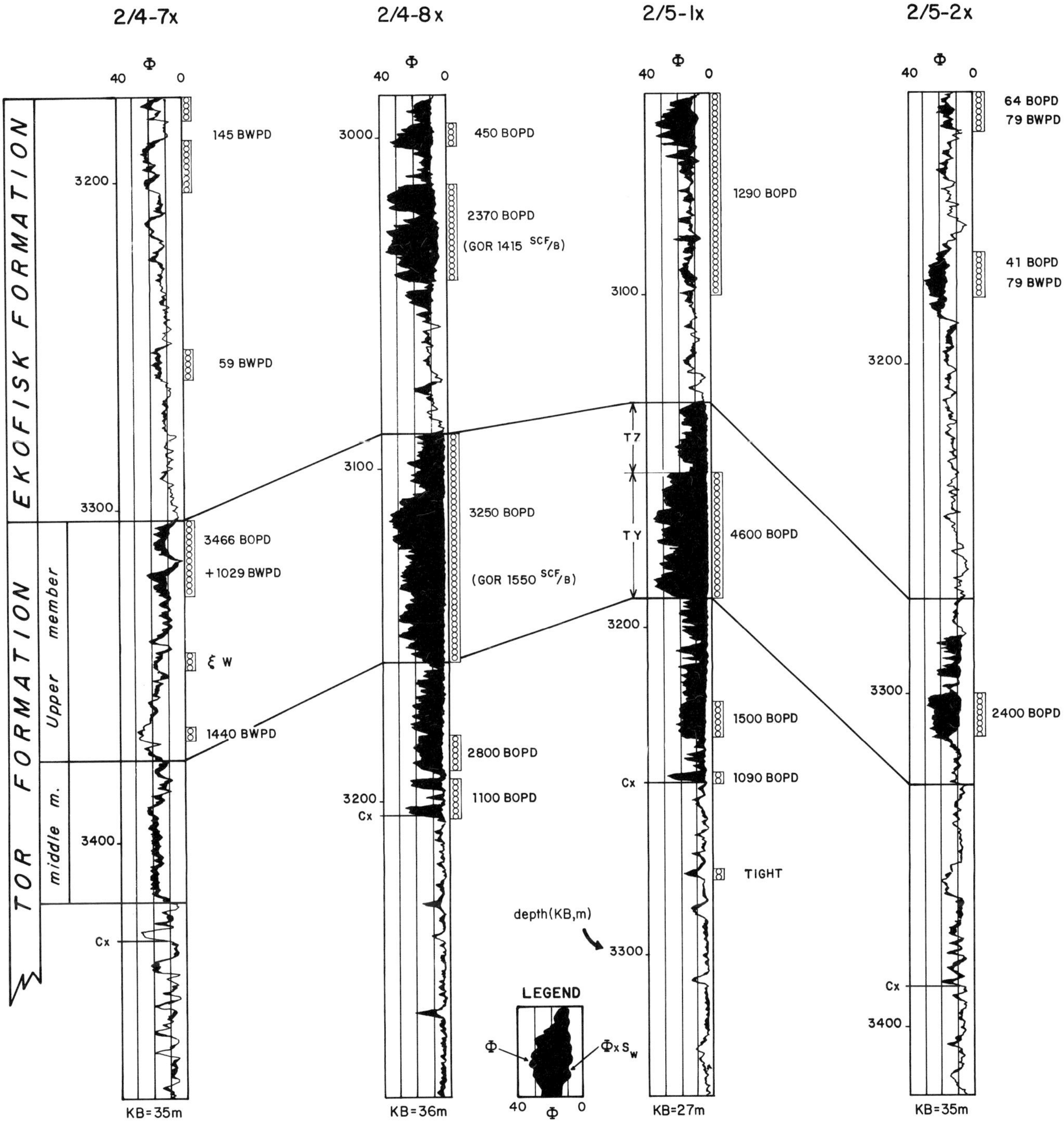

Fig. 4. Computer-processed interpretation of the four discovery/appraisal wells with summary of test results (in barrels of oil or water per day; gas rate omitted).

to arbitration, which increased the Amoco Group's share to 29.2%.

Three additional wells were drilled between 1980 and 1982 in Block 2/5, one of which, E8, was designed to provide information for possible water-flooding of the field. Due to poor economics, this project was cancelled and Well E8, which was located too low on the flank to be a valuable producer, gave very poor results.

Of the two remaining wells, E14 gave information on the saddle between the two culminations and its influence on the porosity and thickness distribution of the reservoir, and Well E7 showed better-than expected results in the south-eastern flank of the adjoining structure.

In 1982, the Phillips Group called for a second equity redetermination which resulted in a reduction of the Amoco Group's share to 24.6%. Although no additional wells had been drilled the Amoco Group called for a

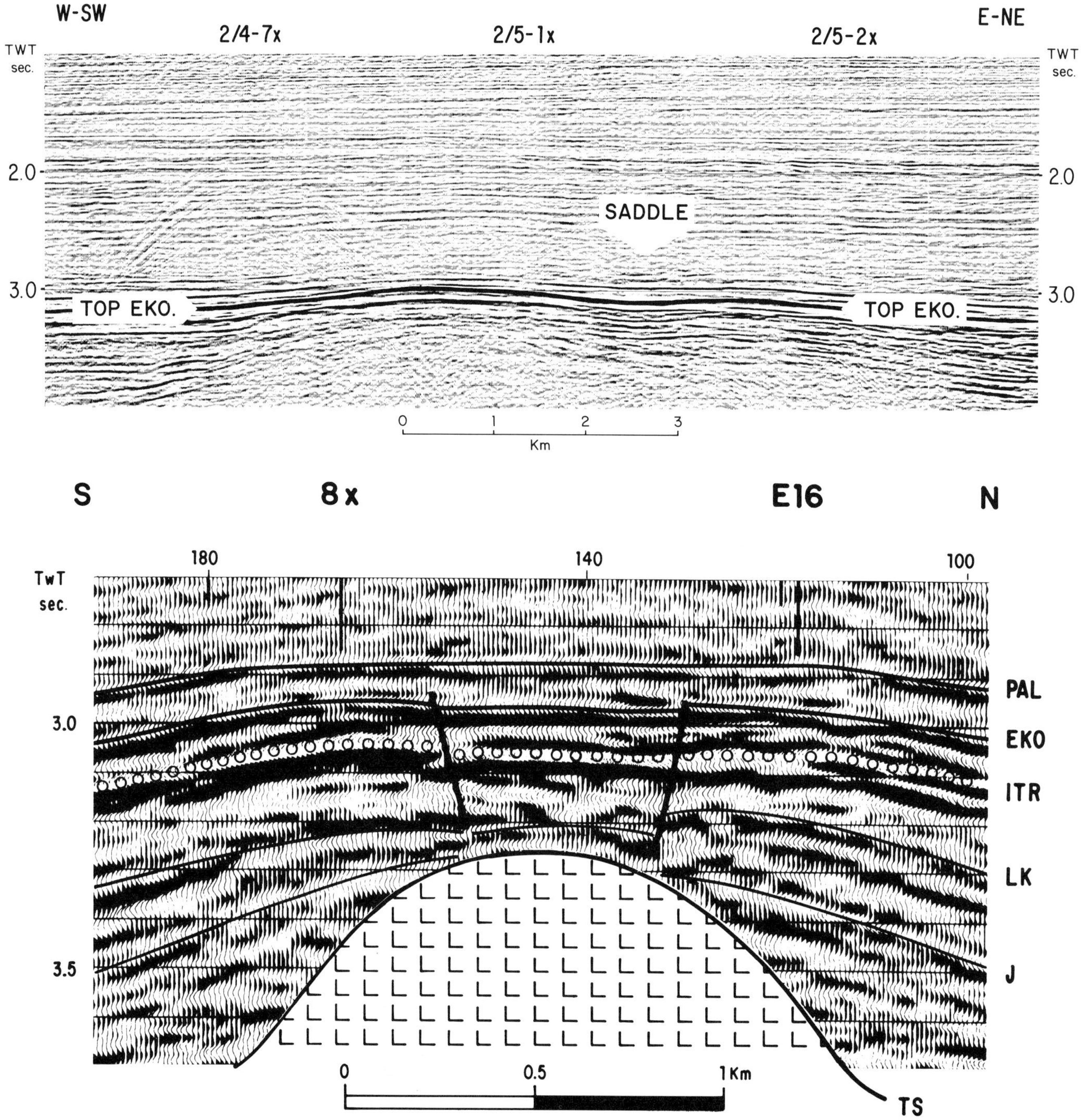

Fig. 5. (a) W–E seismic line. (b) S–N seismic line showing the crestal graben and some seismic markers. PAL, Top Paleocene; EKO, Top Ekofisk Fm; ITR, intra-Tor reflector; LK = top Lower Cretaceous; J, Top Jurassic; TS, Top Salt.

new redetermination in 1984. Their main reasons were based on the field performance, insufficient use of synthetic seismic logs by the arbitrator and dissatisfaction with the methods previously used. At that time, the cumulative production of the field was 14×10^6 Sm^3 of oil and 8×10^9 Sm^3 of gas; this is thought to represent about 80% of the ultimate recovery of the field. The third and probably last equity redetermination increased the Amoco Group's share to 26.2513%. At the end of 1985, the field was producing at a rate of about 2000 Sm^3 oil and 2.8×10^6 Sm^3 gas per day.

GEOLOGICAL SETTING

The Tor Field lies near the north-east margin of the Central Graben.

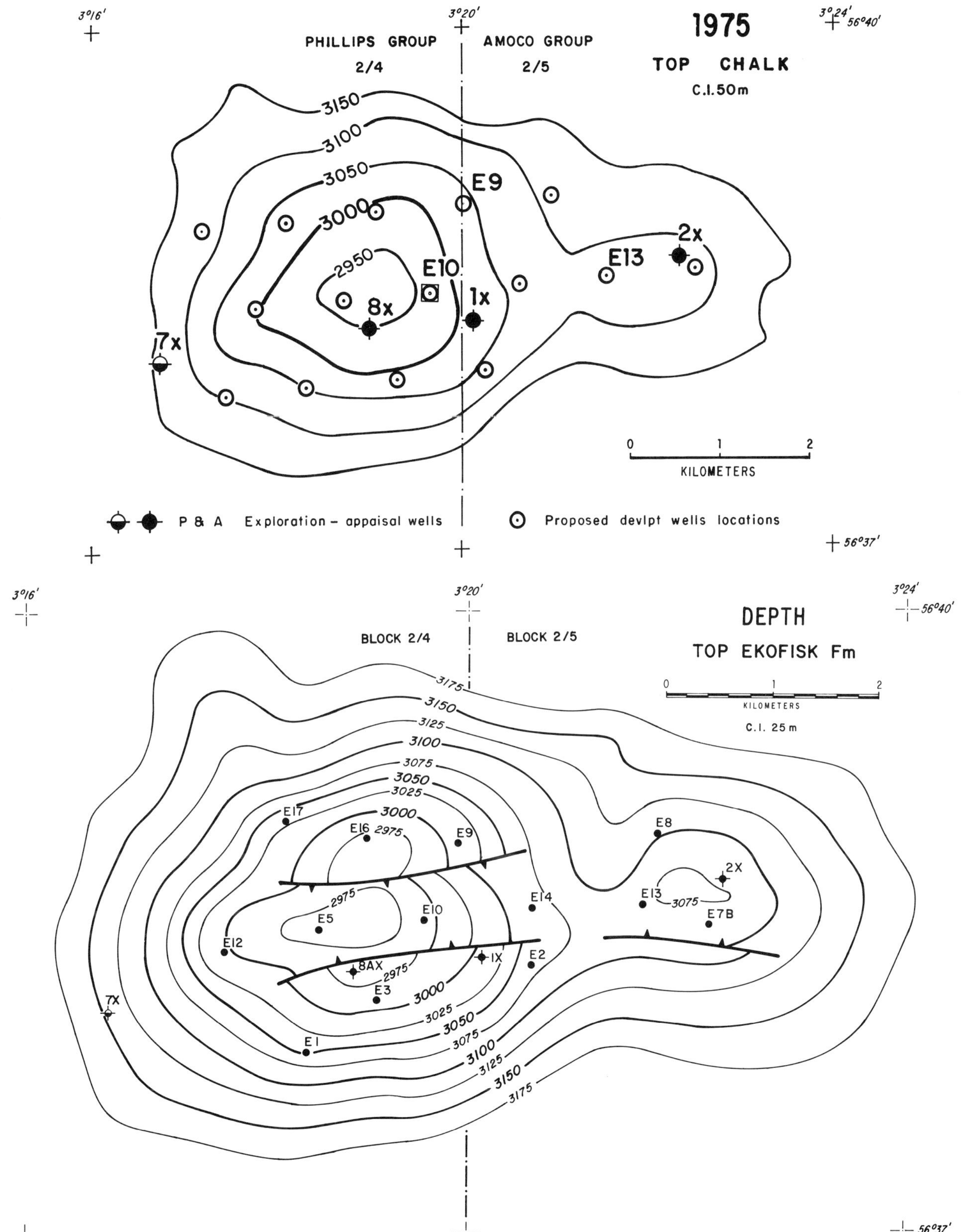

Fig. 6. (a) Proposed development well locations on the 1975 structural interpretation. (b) Structural interpretation of 1985 at Top Chalk level.

Thick Triassic and Jurassic continental to marine strata overlie Permian salt which was subject to halokinesis. After the deposition of a Lower Cretaceous clastic sequence, there was a major marine transgression, which was responsible for restricting the supply of terrigenous sediment into the basin, and a chalk sedimentation regime was initiated. Extensive basinal subsidence prevailed in the graben and the area became a depocentre for both the Chalk Group and for the overlying Tertiary which consists of fine clastic material which forms the seal of the chalk fields.

Locally, the Chalk exceeds 1300 m in thickness. It has been demonstrated that thick allochthonous chalk beds are present and form the best reservoirs (Kennedy, 1980, 1985). These allochthonous chalks result from the remobilization of a calcareous ooze previously deposited on the unstable sea floor of the graben margin slopes or above growing salt domes within the graben itself. The resuspended sediment was transported into deeper areas by gravity flows, and redeposited. Lateral variations in facies and thickness were influenced by current strengths, distance from source, direction of flow and palaeotopography of the sea floor. Distinct distributions of the various allochthonous facies described by Kennedy have been recognized by several researchers and summarized by D'Heur (1986).

Despite their different locations relative to the Graben margin and to the Lindesnes Ridge, which are important elements as far as the chalk facies distribution is concerned, there is a surprising similarity between the Ekofisk and Tor formations encountered in Well 2/4-8 in the Tor Field and Well 1/9-6 in the Tommeliten Fields (Fig. 7).

Sedimentation in the Tor Field was influenced by its proximity to the complex fault zone bordering the Graben to the east. As a result, slumps, slide sheets and widespread debris-flows are common within the chalk. Movement of the Tor structure occurred contemporaneously with deposition, resulting in a variable distribution of allochthonous chalks over the Tor field area. In addition, there was local small-scale reworking in the uppermost part of the Chalk Group. Another important source for allochthonous material was the Lindesnes Ridge in the Valhall area to the south, which was active during Late Maastrichtian and Danian times.

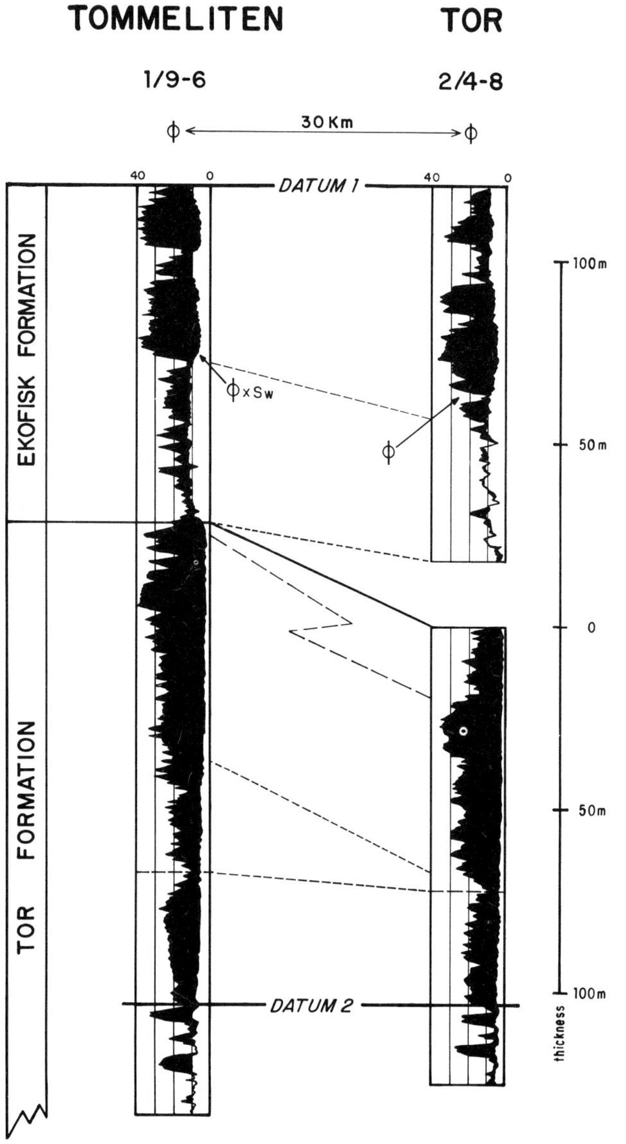

Fig. 7. Comparison of the Ekofisk and Tor formations in the crestal well of the Tor and Tommeliten fields (note that two reference levels have been considered).

GEOPHYSICS

Only sparse seismic data were available before the discovery of the Tor Field (Fig. 2). Today the useful database comprises surveys conducted in 1976 and 1977 (184 km shot by the Amoco Group) and 1979 (131 km shot by the Phillips Group). Velocity surveys and synthetic seismograms were used to identify seismic reflectors at the four exploration wells. Reflector identification was assisted by the synthetic seismograms and acoustic impedance logs generated from sonic and density logs. Synthetic sonic logs were used in conjunction with sonic and porosity data from well logs to interpret porosity development in the Tor and Ekofisk formations beyond well control.

Typical seismic lines across the Tor Field are shown in Fig. 5. The first (upper) prominent reflector around 3.0 s originates from the thin Lower Eocene/Upper Paleocene tuff, while the second one, 100 ms deeper, results from the shale–chalk interface at the top of the Ekofisk Formation. About 100 ms below this reflector is another seismic event designated as the intra-Tor reflector; it arises from the superposition of reflections from the top and the base of the main porous zone within the Tor Formation. Reflections from the top of the Lower Cretaceous and Jurassic are also recognized in the field area. The crestal graben and the salt dome are also represented on the section.

The Top Chalk structure map (Fig. 6(b)) shows the main faults identified from seismic data. Two E–W faults having throws of 15–30 m define the crestal graben and die out on the eastern and western flanks. A third fault in the vicinity of Well E13 may be the continuation of the southern fault of the crestal graben.

The top of the Tor Formation, which contains two-thirds of the field's reserves, cannot be mapped directly from seismic data but may be derived by adding the isopachs of the Ekofisk Formation to the Top Chalk structure map.

An indication of porosity can be gained from the synthetic sonic logs. Considerable discussion arose between the two licence groups about the validity of this method to quantify porosity in the Tor Field.

TRAP

Trapping in the chalk reservoir is the result of structuring caused by halokinetic uplift and sealing by Tertiary shales. The structure is broadly elliptical (8 × 4.5 km) consisting of a large domal feature separated from a small eastern dome by an intervening saddle. Both structures have been faulted at their crests.

A significant growth of the structure occurred before Mid–Late Maastrichtian times as suggested by the isopach map on Fig. 8(a). However, syndepositional movements were responsible for the complicated distribution of the reservoir layers and for local reworking.

The first indication of the development of the saddle was in Mid-Danian times when the old palaeocrest moved southwards from the E10-E8 towards its present locations close to 2/4-8X and E13.

The present-day vertical closure at Top Chalk level is 240 m. The eastern culmination is over 120 m deeper than the main crest. The structural features at the top of the Tor Formation are very similar to those observed at the top of the Ekofisk Formation, except that the eastern fault lies north of E13.

Below the oil accumulation, in the palaeo-aquifer zone, late diagenesis has reduced the porosity below the entry value. An aquifer of limited extent may nevertheless exist in the south-western and south-eastern areas below 3350 mss.

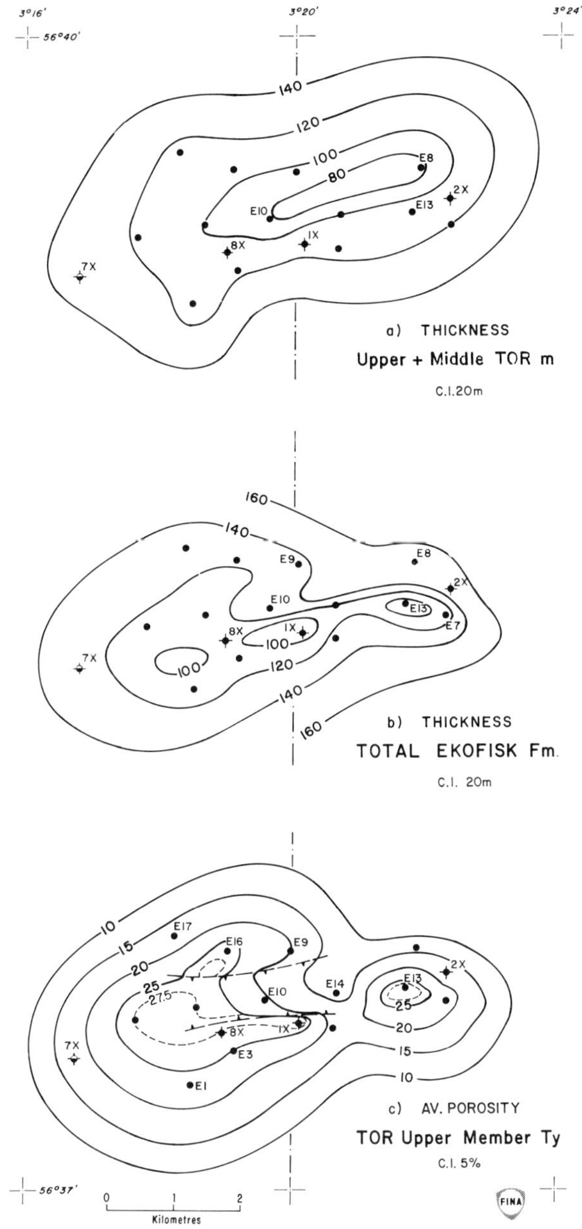

Fig. 8. (a) Thickness map of the Upper and Middle members of the Tor Formation showing the existence of a palaeohigh in the E10–E8 area during or before Maastrichtian times. (b) Thickness map of the Ekofisk Formation suggesting a southwards shift of the palaeohigh during Danian times. (c) Average porosity map of the Upper member of the Tor Formation (excluding the uppermost TZ layer—see 2/5-1X log on Fig. 4 for definition). Porosity decreases from crest to flank but the gradients are different on the northern and southern flanks. There is also a slight porosity reduction in the crestal graben area.

RESERVOIR

The reservoir of the Tor Field comprises chalks of Late Cretaceous (Maastrichtian) and Early Paleocene (Danian) age (Table 1).

The most fortunate anomaly of the Central Graben chalk is its excellent porosity, which results from its very high initial porosity and excellent porosity preservation during more than 60 million years of progressively deeper burial. Note that all porosities are matrix, fracture porosity being negligible.

The very high primary porosity of the deposits is due to the high percentage of rapidly resedimented allochthonous chalk which passed quickly through the water–sediment zone and escaped bioturbation and significant early cementation. The excellent porosity preservation is due to a number of favourable factors (Scholle, 1977):

- low magnesium calcite which is chemically stable;
- low clay content which favoured spot-welding between calcite grains (Mapstone, 1975);
- overpressuring, caused by rapid subsidence and accumulation of chalk, led to a reduction of effective stress;
- entry of oil into the reservoir inhibited pressure dissolution and recrystallization in adjacent pore space, which otherwise is the main cause of porosity desruction.

The present-day porosity distribution in the Tor Field is the result of the combined effects of two factors: the distribution of the various sedimentary facies; and the differential porosity preservation. Both vertical (stratigraphic) and lateral (geographic) variations in facies have been recognized in the Tor Field. The separate members of the Ekofisk and Tor formations have distinct proportions of allochthonous to autochthonous chalks and therefore have distinct 'average' porosities. The growth of the Tor structure during chalk deposition resulted in lateral thickness and porosity variations in the favourable allochthonous facies. For example, the allochthonous facies is generally more abundant in the southern flank of the palaeostructure, this flank being exposed to the Valhall source-area for Maastrichtian (and Danian?) resuspended sediments. On the other hand, less allochthonous material was deposited over the palaeocrest, and the higher proportion of pelagic chalk there resulted in a lower primary porosity. These effects are apparent on the maps on Fig. 8, although differential porosity preservation has altered the initial porosity distribution.

Table 1. Reservoir subdivision, average clay + silica content and average porosity (crestal area)

Fm	*This report*	*Phillips Group*		*Average Ø (crest) (%)*	*Average clay + silica (%)*
Ekofisk	Upper	EA EB EC		22	5
	Lower	ED EE		14	9
Tor	Upper	TA	TZ	19	3
			TY	27	0,5
	Middle	TB		16	1
	Lower	TC			2

Differential porosity preservation within a layer with uniform primary porosity value resulted in decreasing porosities from the crest to the flank of the structure. This can be explained by the progressive invasion of the trap by oil from top to bottom, with corresponding progressive inhibition of chemical compaction at deeper and deeper levels and at lower and lower porosity values. Another explanation, however, is the existence of differential overpressuring in the trap. The relative importance of these two factors for the Tor Field is discussed in the section on source (p. 139).

Tor Formation

The Lower Tor member mainly consists of pelagic chalk containing some low energy allochthonous material. This, together with a high clay content, makes the rock of very poor reservoir quality except for several porous debris-flow beds, 3–5 m thick, which can be correlated over most of the Greater Ekofisk area north-east of the Lindesnes Ridge. They contain oil in the crest of the main structure.

The Middle Tor member is composed of alternating units of autochthonous and allochthonous origin. It has a high content of large bioclasts, which are either solid or, if hollow, commonly spar-filled. Locally the clay content is high. For these reasons this unit has only fair to good reservoir quality compared to the excellent quality of the overlying interval.

The Upper Tor member contains two distinct intervals, TY and TZ. *Unit TY*, the main reservoir of the Tor Field, is composed mainly of stacked debris-flows sourced from the south and the north-east. This unit, as well as the Lower and Middle members of the Tor Formation, correlates from field to field, which suggests uniform sedimentation, although

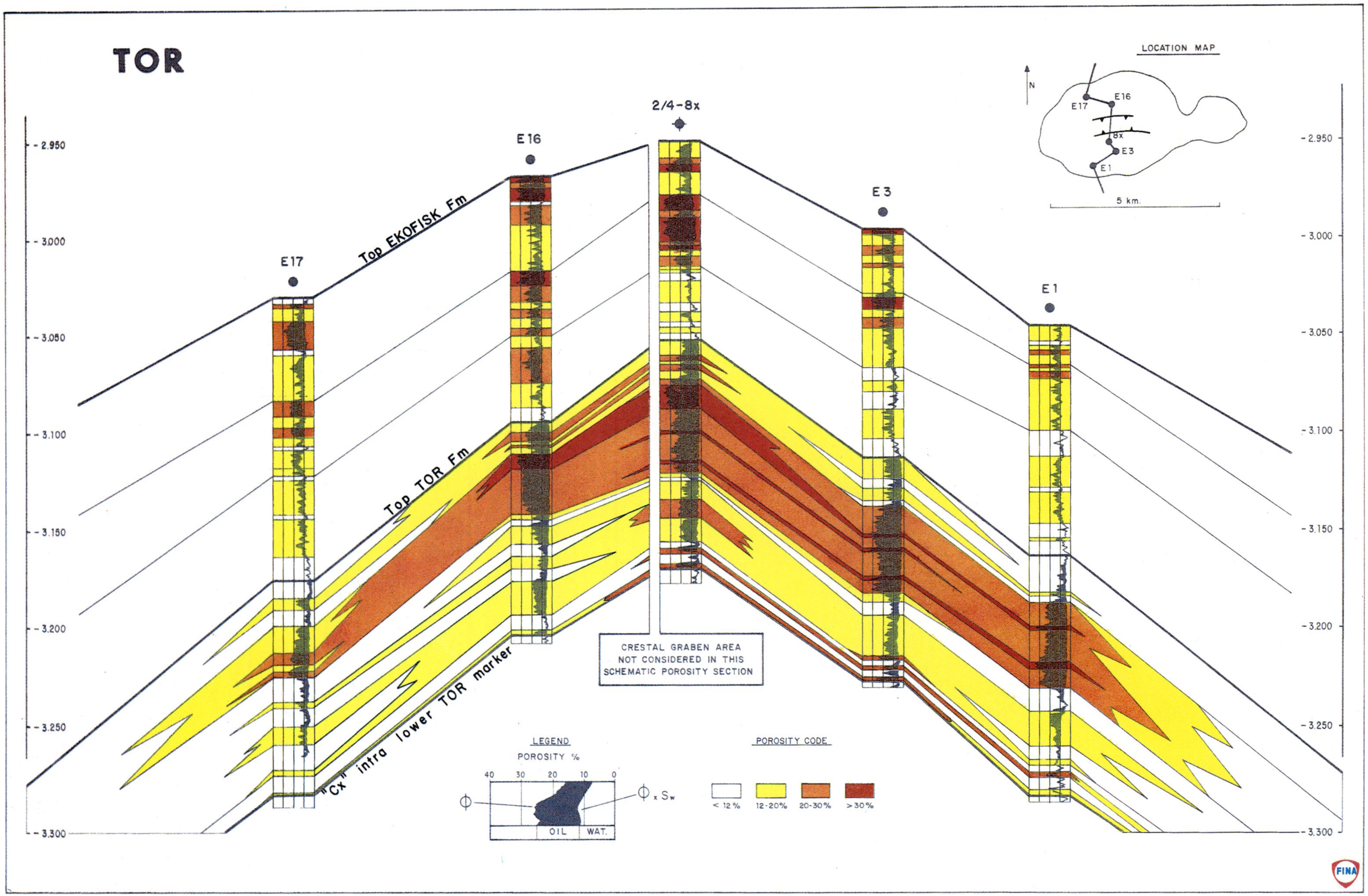

Plate 7. Porosity cross-section from north to south (the crestal graben has not been represented).

local variations are known, suggesting local tectonic activity during deposition. This interval has excellent porosity and hydrocarbon content in all the Ekofisk fields. It has a few large bioclasts, is generally not bioturbated and has a very low clay content. The isoporosity map of this interval (Fig. 8(c)) shows that porosity decreases faster on the northern flank. At the same structural elevation the porosity is much higher in Well E1 than in E17. This is also true for the Middle Tor Member and probably reflects the higher proportion of debris-flow material in the southern area. In addition, a lower than expected porosity characterizes the E10–E8 area. There is probably a causal relationship between this low-porosity zone and the existence of a palaeohigh in this area since early Maastrichtian times, as suggested by the thin isopach (Fig. 8(a)). A higher proportion of pelagic and low-energy chalk was deposited over this palaeocrest with a corresponding lower primary porosity.

Unit TZ has an average thickness of 20 m. It is mainly composed of thin debris-flows and turbidites interbedded with pelagic chalk. The interval has a high oil saturation despite its low porosity. It is thin over the palaeocrest and thickens towards the flanks. 'Thicks' exist on the western and eastern flanks due to local slumping or sliding during a period of structural uplift.

Ekofisk Formation

In the Tor field this formation is much less uniform than are the underlying Tor members. Erosion, truncation, non-deposition, slumping, sliding, faulting and the creation of the crestal graben complicate correlations and mapping. In addition, as usual, differential compaction also influences the present-day thickness. The Maastrichtian palaeohigh became a flank area as uplift occurred near wells E13 and 2/5-1 during Early–Middle Danian times and between wells E3 and 2/4-8X during Late Danian times (Fig. 8(b)).

The Lower Ekofisk member is characterized by normal pelagic, intensely burrowed chalk. It has a high non-carbonate fraction and contains only rare turbidites and occasional debris-flows. The reservoir quality of this member is very poor. Slump structures are very common and reflect the movement of the Permian salt.

The Upper Ekofisk member contains more debris-flows of good reservoir quality. Silica and clay content are variable, being lower than in the Lower Ekofisk member and higher than in the Upper Tor member. Correlations are very difficult and not illustrated on the porosity cross-section (Plate 7). The uppermost beds are absent in several wells, indicating uplift and erosion at the end of Danian times. Porosity generally decreases from crest to flank which means that differential porosity preservation has reversed the trend of primary porosity variation. Core studies reveal considerable fracturing, principally in areas of greatest uplift. There is a probable correlation between high fracture intensity, high porosity and low clay content. There is no clear relationship between poor porosity and abundance of stylolites and they seem less frequent in the shaly interval of the Ekofisk Formation.

SOURCE

Geochemical analyses have established that the hydrocarbons found in the chalk reservoirs have been generated from organic matter in the Upper Jurassic shales. Within this interval is the Mandal Formation, consisting of highly bituminous black marine shales characterized by a high organic content. Its role as an excellent source-rock has been discussed by Van den Bark and Thomas (1981) for the Ekofisk area. The thickness of the Upper Jurassic is not well known because only a few wells have drilled deep enough to fully penetrate the sequence, except over salt domes where the thickness is not representative. In the Tor Field, Well 2/5-1X encountered more than 120 m of Upper Jurassic shales but bottomed without reaching the base of the series. In the Eldfisk area to the south, the Upper Jurassic is more than 1000 m thick.

The timing of oil migration into the chalk reservoir is a key factor in evaluating the relative importance of overpressuring versus the presence of hydrocarbons in pore spaces, in preserving the very high porosities. Phillips Group studies conclude that peak oil generation occurred during Miocene times in the Tor Field area. Studies by Amoco suggest that the source-rock of the same area entered the early mature stage at the end of Middle Miocene times with peak oil generation during Pleistocene to Holocene times depending on location. Thus, in the immediate vicinity of the Tor Field, oil has been generated rather late in geological history, and so its role in porosity preservation would be less important than the role of overpressure. However, oil could have migrated into the Tor reservoir from less proximal areas, much earlier. In the deep zones of the graben, north of Albuskjell (Fig. 9, area 3), significant oil generation took place as early as Paleocene times and migration paths could have developed in the slope area (4) west of Tor.

The relative timing of formation of traps in the Ekofisk area and their size is also of importance. No palaeostructure existed in the West Ekofisk to Ekofisk region during early Paleocene times (D'Heur, 1980) and probably only a slight closure existed during Eocene times.

The opposite case can be seen at Tor, where a palaeostructure existed during Maastrichtian and Danian times and a trap was present as soon as the Paleocene clay effectively sealed the chalk.

The hydrocarbon pore-volume of the Tor Field is very small compared to the amount of oil generated before Miocene time in the deep area to the west, and less than 5% of the Ekofisk field reserves would have been sufficient to fill up the best reservoir (∅ > 30%) of the Tor Field. If the Tor Field oil originated from the deep zones west of Ekofisk, the presence of oil in the reservoir may be the main factor allowing porosity preservation.

In any case, R. Hardman and R. Leonard/F. Jansen (unpublished Amoco Group reports) have demonstrated from wells in Blocks 2/5, 2/8 and 2/9 that overpressure alone can preserve porosity up to at least 26%. However, porosities of 35–50% have only been encountered in reservoirs where oil is or has been present. In several wells, a progressive but rather fast decrease of porosity from more than 40% to about 25% has been observed in the oil zone, whereas a much slower trend of porosity reduction is recorded in the aquifer.

The above discussion concerns only the thick porous interval of the Ekofisk and Tor formations, since it is recognized that the thin debris-flows intercalated in pelagic chalk have variable porosities which bear no relationship to their hydrocarbon content (D'Heur, 1984, Fig. 6).

HYDROCARBONS

The 'net productive hydrocarbons pay' is the criterion used to calculate reserves for the purpose of equity determination. It involves the establishment of cut-off values for porosity and oil saturation. The porosity cut-off to be applied to porosity logs has been established by considering porosity vs permeability correlations, capillary pressure tests and residual saturation core analyses. The values have been checked by the analysis of production tests. Curves for permeability–porosity relationships calculated from five cored wells are given in Fig. 10. As usual, correlations are satisfactory for the pure chalk of the Upper and Middle Tor members but are much less

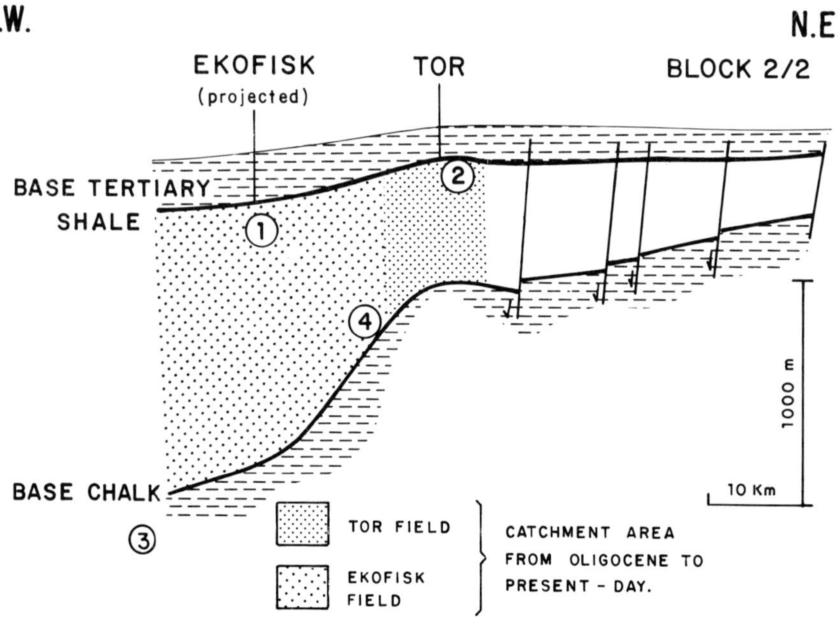

Fig. 9. Schematic reconstruction in the vicinity of Tor during Eocene times. At that time a significant amount of oil had been generated already in the Jurassic shale to the west (area 3). No pronounced palaeorelief existed in the Ekofisk area (1), whereas a trap already existed at Tor (2). Migration paths might have developed along the flexure (4).

satisfactory for the fair and low porosities of the more argillaceous layers of the Ekofisk Formation.

Therefore, the choice of cut-off values of 10% and 16%, respectively, for the Tor and Ekofisk formations (corresponding to a permeability of 0.04 mD) is an oversimplification of the problem, since vertical and lateral variations of rock type have been insufficiently taken into account. In addition, it is believed that the relationship between porosity and permeability is affected by a 'tectonic stress facies' (see West Ekofisk, this volume, p. 165).

The water-saturation cut-off was defined from production test analyses. Significant amounts of oil with low water-cut were produced from intervals where water saturation

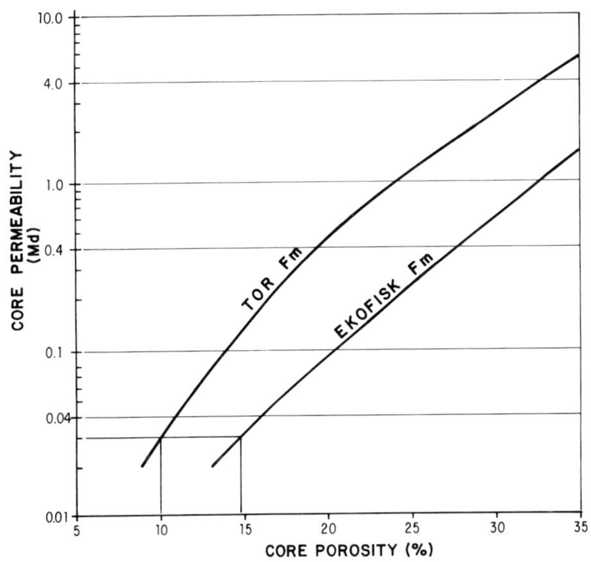

Fig. 10. Porosity/permeability relationship for the Ekofisk and Tor formations.

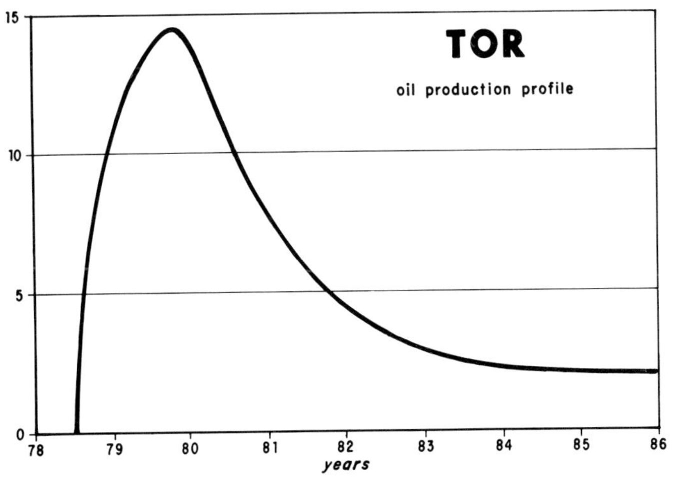

Fig. 11. Production history. Vertical scale in units of 1000 Sm^3/day.

was lower than 80%. The choice of this high value for water-saturation cut-off caused no major disagreement between interested parties, since equity was in fact not sensitive to this factor.

Hydrocarbon distribution: Considerable discussion took place concerning the iso-water-saturation maps needed for establishing the hydrocarbon pore volume in both blocks. As for porosity–permeability correlations, the use of one set of porosity–saturation–depth curves (by porosity classes) for each of the two reservoir formations is an oversimplification, since the effect of lithology (non-carbonate fraction, matrix structure, depositional modes) has been underestimated.

Hydrocarbon properties: the oil in the Ekofisk Formation is similar to that in the Tor Formation, i.e. 0.8155 g/cc, low sulphur, bubble point of about 305 bar and GOR of approximately 260 Sm^3/Sm^3. As in the other Ekofisk area chalk fields, the reservoirs at Tor are overpressured with an initial pressure of 492 bar at 3120 m subsea.

RESERVES

The estimated initial recoverable reserves of the Tor Field are about 18×10^6 Sm^3 of oil and 12×10^9 Sm^3 of associated gas. The Tor Field is linked to the Ekofisk Centre by oil and gas pipelines. At the end of 1985, 83% of the recoverable oil had been produced and the production rate was 2000 Sm^3 oil and 2.6×10^6 Sm^3 gas per day.

ACKNOWLEDGEMENTS

Numerous extensive studies have been made by members of both groups on this unitized field. The author has summarized his own work and has attempted to reflect as much as possible the common ideas about the Tor Field geology. Nevertheless, the opinions expressed in this paper may not necessarily represent those of any of the co-venturers. The author thanks Petrofina, Fina Exploration Norway, Phillips, Agip, Elf, Norsk Hydro, Total, Amoco, Amerada Hess, Texas Eastern and Noco for their permission to publish this paper.

Thanks also to Bridget Bennet for comments on the manuscript and to Dominique Moens for her assistance.

REFERENCES

D'Heur, M. 1980. Chalk Reservoir of the West Ekofisk Field, In *The Sedimentation of the North Sea Reservoir Rocks*, Norsk Petroleumsforening, Article X.

D'Heur, M. 1984. Porosity and Hydrocarbon Distribution in the North Sea Chalk Reservoirs, *Marine and Petroleum Geology* **1,** 211–238.

D'Heur, M. 1986. The North Sea Chalk Fields In Spencer A.M. *et al.* (Eds), *Habitat of Hydrocarbons on the Norwegian Continental Shelf*, Graham and Trotman, London, pp. 77–89.

Hardman, R.F.P. 1982. Chalk Reservoirs on the North Sea, *Bulletin Geological Society Denmark* **30** 119–137.

Kennedy, W.J. 1980. Aspects of chalk sedimentation in the Southern Norwegian Offshore, In *The Sedimentation of the North Sea Reservoir Rocks* (Norsk Petroleumsforening), Article IX.

Kennedy, W.J. 1985. Sedimentology of the Late Cretaceous and Early Paleocene Chalk Group, North Sea Central Graben, In *Chalk Research Program* (Proprietary Report).

Mapstone, N.B. 1975. Diagenetic history of a North Sea chalk, *Sedimentology* **22,** 601–613.

Scholle, P.A. 1977. Chalk diagenesis and its relation to petroleum exploration: oil from chalk, a modern miracle? *American Association Petroleum Geologists Bulletin* **61,** 982–1009.

Van den Bark, E. and Thomas, O.D. 1981. Ekofisk: First of the giant oil fields in Western Europe, *American Association Petroleum Geologists Bulletin* **65,** 2341–2363.

10

Ula

P. C. Home

BP Petroleum Development (Norway) Limited U.A., Postboks 197, 4033 Forus, Norway

The Ula Field was discovered in 1976 by Well 7/12-2, drilled by Conoco, which encountered 154 m of oil-bearing Upper Jurassic sandstone. British Petroleum subsequently became operator and drilled three appraisal wells. The oil occurs in a faulted, four-way dip-closed dome located over a salt swell, with an OWC of 3508 mss. The field has an area of 11 km^2 and a vertical closure of *c.* 160 m. The reservoir is a fine- to medium-grained, shallow-marine sandstone, with a porosity of 15–22% and typical permeability of 650–850 mD. Recoverable reserves are currently estimated to be 25×10^6 Sm^3 of 0.83 g/cm^3 oil. The reservoir pressure is 483 bar at 3450 mss and the reservoir temperature 143 °C at the same depth. The field has been developed using a three-platform PDQ scheme and came on stream in late 1986. The production life of the field is planned to be 11 years, with a three-year offtake plateau of *c.* 11 750 Sm^3/day.

INTRODUCTION

The Ula Field occurs in 70 m of water some 260 km south-west of Stavanger (Fig. 1). It is situated in the Central Graben, 65 km north-west of the chalk oil fields of the Ekofisk complex and 25 km east of the Paleocene Cod gas–condensate field. Recoverable reserves in the Ula Field are estimated to be 25×10^6 Sm^3 of 0.83 g/cc oil and 2.5×10^9 Sm^3 of associated gas, and occur in an Upper Jurassic sandstone. The field was named after a famous Norwegian sea pilot called Ulabrand.

The data presented in this paper are based on a detailed review of the Ula Field by Bailey *et al.* (1981), updated where appropriate.

History

The Ula Field lies entirely within Licence area PL 019A in Block 7/12 (Fig. 1). Current licencees are:

BP Petroleum Development of Norway A/S 57.5%
Svenska Petroleum Exploration A/S15.0%
Norske Conoco A/S 10.0%
A/S Pelican and Company K/S 5.0%
Den norske stats oljeselskap A/S 12.5%

The operator is BP Petroleum Development (Norway) Limited U.A.

Block 7/12 was originally awarded in 1965, in the first round of licencing on the Norwegian continental shelf, to Norwegian Gulf Oil Production Company. It was allocated as part of Licence Pl 019, together with Blocks 2/1 and 11/8.

The first well in Block 7/12 was drilled in 1968, following discoveries in Paleocene reservoirs in the Balder and Cod fields. The Paleocene was the primary objective in Well 7/12-1, but only traces of sandstone and siltstone were found at this level. The well was drilled on through the Chalk Group and was plugged and abandoned in Lower Cretaceous mudstone at approximately 3284 mss as a dry hole. Subsequently, oil was discovered in chalk reservoirs in the Ekofisk area from 1969

Geology of the Norwegian Oil and Gas Fields © A. M. Spencer *et al.*, eds (Graham & Trotman, 1987) pp. 143–151.

Ula Field data summary

Trap	
Type	Dome over salt swell
Depth to crest (at top Ula Formation)	3345 mss
Lowest closing contour	*c.* 3700 mss
Oil–water contact	3508 mss
Oil column	163 m
Pay zone	
Formation	Ula Formation sandstone
Age	Late Jurassic (Oxfordian–Kimmeridgian)
Gross thickness	84–166 m (TVT in wells)
Net/gross ratio (average; range)	0.93%; 0.85–1.0 (Units I–III)
Cut-off for N/G	10%
Porosity	15–22% (Units I–III in oil zone); 12–17% (Units IV–V)
Hydrocarbon saturation	75–97%
Permeability	650–850 mD (Unit II); 0.2–2800 mD (whole reservoir)
Productivity index	120–350 Sm^3/day/bar
Hydrocarbons	
Oil gravity	0.83 g/cm^3
Oil type	Low sulphur (0.1%), high wax (9%)
Bubble point	159 bar at 143 °C
Dew point	Not determined (gas lean)
Gas/oil ratio	89 Sm^3/Sm^3
Formation volume factor	1.33
Formation water	
Salinity	*c.* 175 000 ppm (NaCl equivalents)
Resistivity	0.0145 ohm m at 143 °C
Reservoir conditions	
Temperature	143 °C at 3450 mss
Pressure	483 bar at 3450 mss
Pressure gradient in reservoir	0.067 bar/m
Field size	
Area	11 km^2
Gross rock volume	631.6×10^6 m^3
Recovery factor	*c.* 40%
Drive mechanism	Water
Recoverable oil	25×10^6 Sm^3
Production	
Start-up date	Late 1986
Development scheme	Three (PDQ) steel platforms
Production rate	Expected plateau rate *c.* 11 750 Sm^3/day

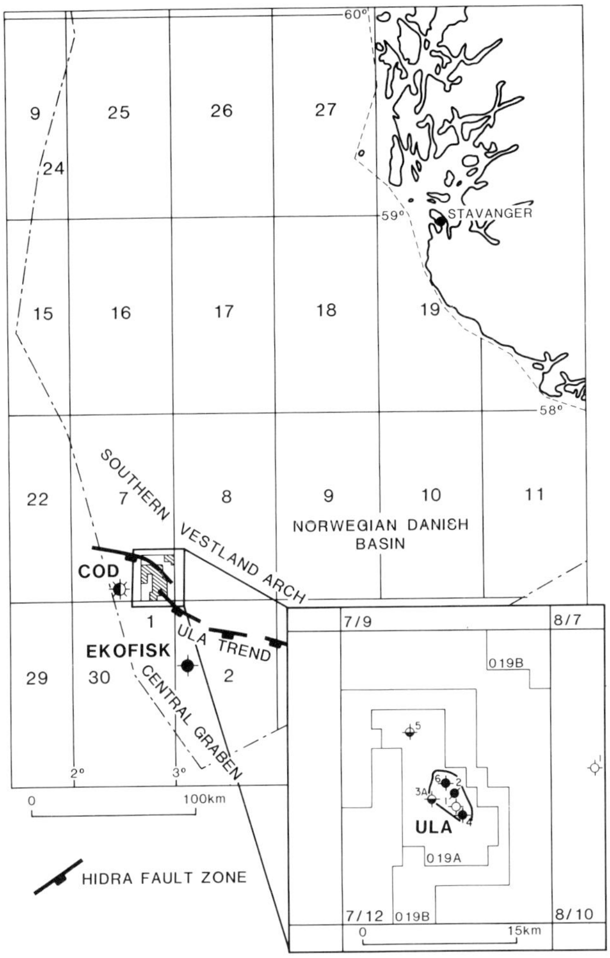

Fig. 1. Location map of the Ula Field.

to 1974, establishing the Central Graben in Norway as a major oil province, at least for Upper Cretaceous reservoirs.

In 1970 Gulf farmed-out an interest in Licence PL 019 to Conoco (operator) and Pelican. Conoco drilled a dry well, 2/1-1, in 1972 under the terms of the farm-out agreement. Subsequently, Gulf withdrew from the licence (July 1975) and early in 1976 BP farmed-in in return for the drilling of one well (7/12-2), although the well location was to be chosen by Conoco and the well was to be operated by Conoco. Well 7/12-2 was drilled in summer 1976 to test the Triassic–Jurassic below the total depth of Well 7/12-1. It penetrated an unexpectedly thick (154 m) Upper Jurassic sandstone sequence with excellent reservoir properties, which was oil bearing throughout. Deeper oil-bearing sandstones were encountered in the Triassic section but had poor porosities, very low permeabilities and produced only small amounts of oil and water when tested.

British Petroleum subsequently took over operatorship of the licence and drilled three appraisal wells, on the west (7/12-3 and -3A, 1977), south-east (7/12-4, 1977) and north-east (7/12-6, 1981) flanks of the field. Well 7/12-3A was deviated up-structure from Well 7/12-3 in the hope of encountering the oil–water contact, after Well 7/12-3 had found the Upper Jurassic reservoir to be water bearing. A small oil column and the oil–water contact were penetrated in 7/12-3A. They proved that the structure was not full of hydrocarbons to the dip-closed spill-point and that the extent of the oil column was probably controlled by faulting.

Early development plans for the Ula Field involved using a jack-up process–accommodation platform linked to an adjacent wellhead tower, with crude production handled by offshore loading and with surplus gas flared. This scheme was unanimously endorsed by the licencees who also agreed that it was an economically marginal prospect. The declaration of commerciality followed in 1979 and a Landing Application was submitted in the same year. Flaring of the associated gas was not acceptable to the authorities and its disposal was then planned to be by pipeline through the Ekofisk system to Emden. Oil export was to be via a separate pipeline to Ekofisk and then to Teesside. However, detailed design plans proved the project to be more expensive than originally estimated. This and other factors reduced the project's economic viability and in 1981 development was deferred before major contracts had been awarded.

Engineering screening studies on alternative concepts were subsequently recommenced. They concluded that any future development should be based on three relatively small, piled steel platforms for process, drilling and quarters (PDQ). This scheme was approved by the licencees in December 1982 and fabrication commenced in early 1984. The jackets were installed in 1985 and first oil production commenced in late 1986. The plateau production rate is expected to be *c.* 11 750 Sm^3/day for three years from nine oil producers, with pressure support being provided by six peripheral water-injector wells. The economic life of the field is anticipated to be 11 years. Oil and gas export will be via Ekofisk as previously planned, although the gas will be transported via the Cod Field.

GEOLOGICAL SETTING

The Ula Field is situated within the Ula Trend (Fig. 1)—'a Jurassic hydrocarbon province located along the faulted north-eastern margin of the oil-rich Central Graben' (Spencer *et al.*, 1986). At Jurassic level the graben is separated from the dry hole province of the Southern Vestland Arch by a NW-trending zone of SW-hading faults, referred to as the Hidra Fault Zone (Gabrielsen *et al.*, in press). The regional structural dip at this level is to the south-west, towards the axis of the Central Graben. The graben margin faults are believed to have been growth faults in Late Jurassic times and to have throws of hundreds of metres (Spencer *et al.*, in press). Shallow-marine sandstones accumulated in topographical depressions on the down-side of the fault scarps and shaled-out basinwards. However, the presence of reservoir sandstones in this 'fairway' is unpredictable, possibly due to contemporaneous and subsequent salt diapirism.

The post-Palaeozoic stratigraphy of the Ula Field is illustrated in Fig. 2. The lithostratigraphic nomenclature is based on that proposed by Deegan and Scull (1977) and modified by Vollset and Doré (1984). The Mesozoic sequence is underlain by a thick section of Zechstein halite which was deposited in the Northern Permian Basin. The Triassic sequence consists of continental red beds and is separated from overlying Jurassic strata by a major angular unconformity. The Lower Jurassic Gassum and Fjerritslev formations consist of shallow-marine sandstone and mudstone respectively, but are only locally present in the field. The Mid-Jurassic Bryne Formation is more widely developed, although thin (20–35 m), and consists of deltaic sandstones, shales and coals. It is disconformably overlain by the shallow-marine reservoir sandstones of the Ula Formation which range in age from Oxfordian to Kimmeridgian. The Ula Formation varies in thickness from 84 m (faulted) to 166 m in the Ula wells and is conformably overlain by over

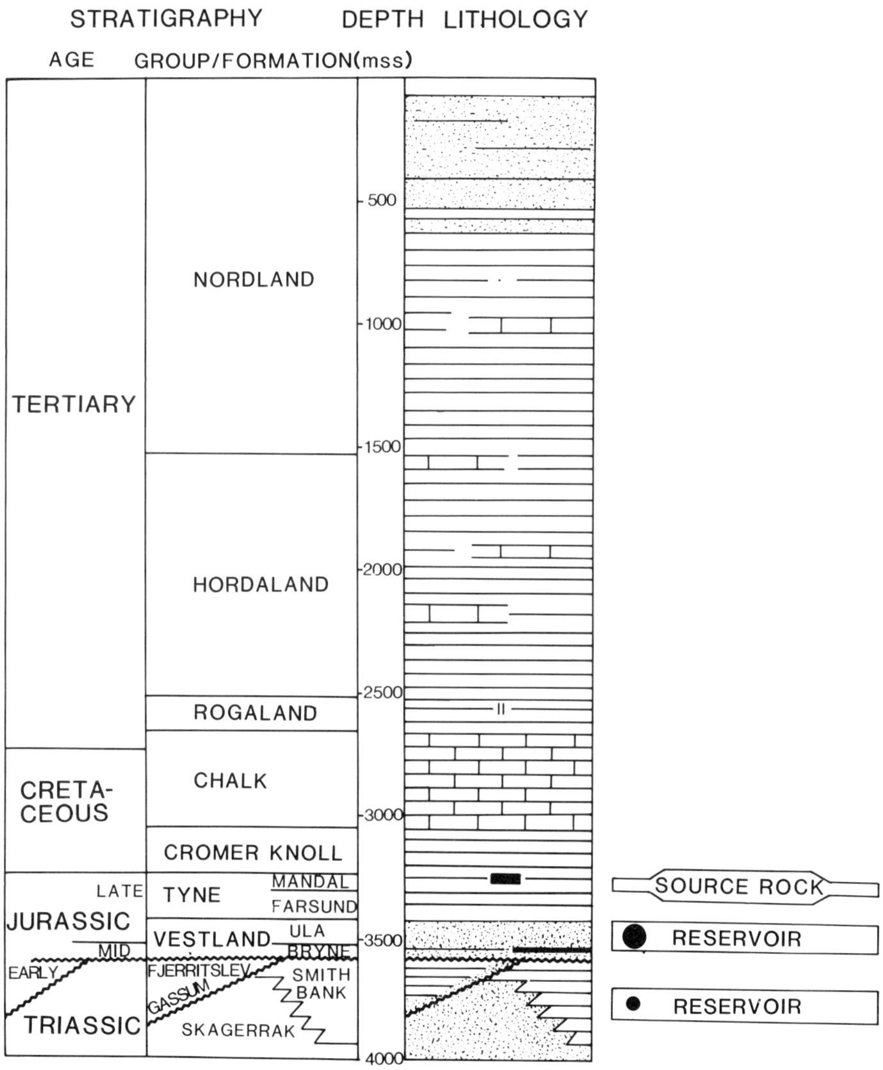

Fig. 2. Generalized stratigraphic column for the Ula Field.

100 m of Upper Jurassic mudstone and shale. The Late Jurassic to Recent section is essentially continuous and contains no prospective reservoir intervals. The reader is referred to Deegan and Scull (1977) for a description of Cretaceous and Tertiary strata in the area.

GEOPHYSICS

Until 1985, structural mapping in the Ula Field was based on a number of two-dimensional seismic surveys shot between 1971 and 1981. Data shot in 1971, 1972 and 1973 formed a one-kilometre grid over the field, which was infilled to a 500-m line-spacing in 1980 and 1981. Thc quality of these data is generally good, except between the Base Cretaceous and Top Zechstein Group reflectors (Fig. 3). In particular, the top of the Ula Formation reservoir is a weak and discontinuous reflector, probably because of its proximity to the high-amplitude Base Cretaceous event. This was confirmed by trace inversion studies, although some improvement in resolution of the Jurassic reflectors was observed. Figure 4 shows the top reservoir map following the completion of Well 7/12-6.

In 1984 GSI shot and processed a three-dimensional grid of seismic data. Two hundred and forty-seven SW-trending lines were acquired, spacing at 50 m intervals. They were shot with a 4075-in^3 air-gun array and a 240-channel, 3600-m streamer, for 60-fold coverage. The total line-kilometres shot was 2826.45 in an area 12×12 km. The quality of the data is good and the resolution of the reservoir significantly improved. The top and base reservoir horizons have been interpreted using the GSI interactive work station. Mapping will be completed in early 1986 and will form the basis for the development drilling.

TRAP

The Ula structure is a four-way, dip-closed domal feature overlying a Zechstein salt swell. The salt is non-penetrative and the structural dips at Jurassic level are gentle

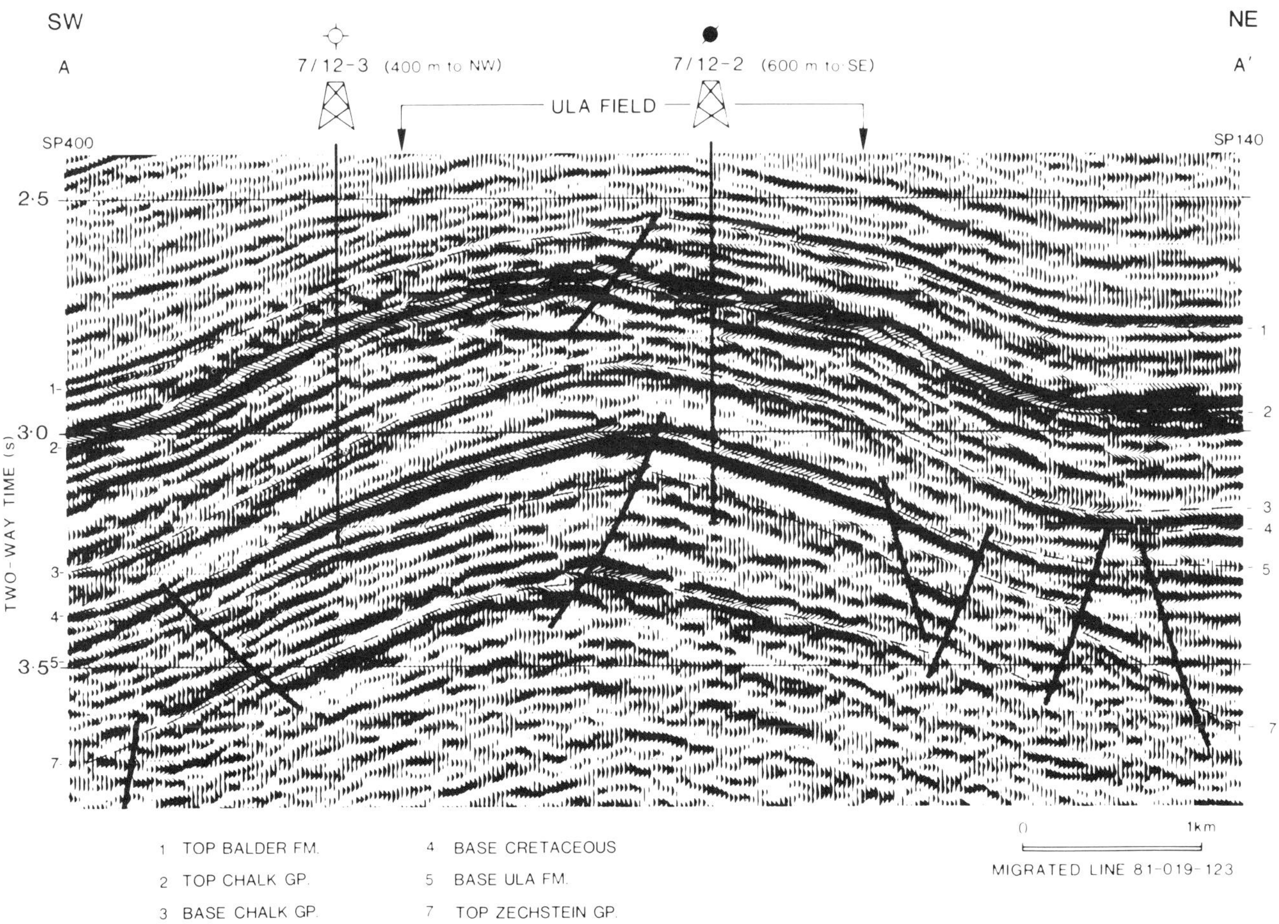

Fig. 3. Migrated seismic section through the Ula Field. See Fig. 4 for location. From Spencer *et al.* (1986).

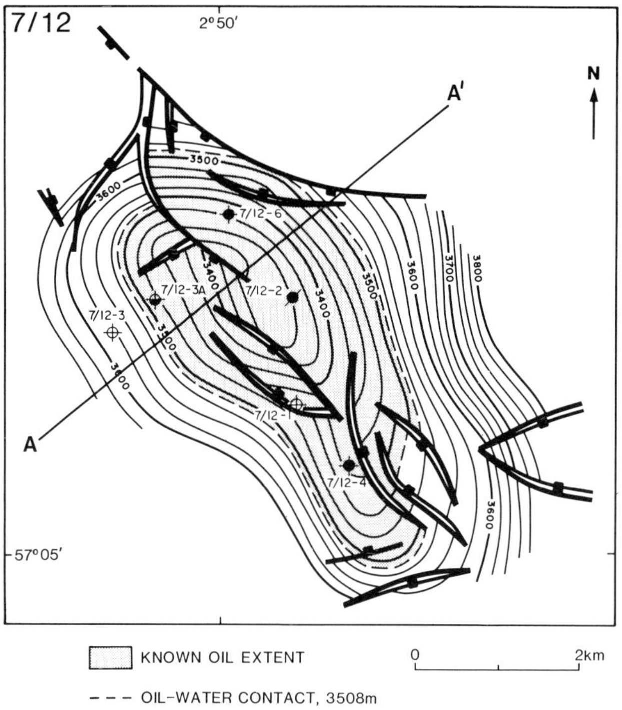

Fig. 4. Structure map of the Ula Field. Contours in depth at top reservoir level; CI 20 m. Modified after Bailey *et al.* (1981).

(< 6°). The trap was formed by major salt doming in Tertiary times and is bounded to the north-east by the Hidra Fault Zone. Figure 5 is a cross-section through the field.

An oil–water contact was not encountered in discovery well, 7/12-2, because of its near-crestal location. However, it was hoped that the reservoir might be hydrocarbon bearing down to its maximum dip-closed spill-point of *c.* 3700 mss. Wells 7/12-3 and 7/12-3A demonstrated that this was not so and indicated an oil–water contact at 3508 mss. This is underlain by a residual oil zone to 3530 mss. The oil–water contact appears to coincide with the intersection of the top of the reservoir with the fault to the north-east of the field. The reservoir is in juxtaposition with Triassic sandstone across this fault.

The oil pool has an area of 11 km^2 and vertical extent of *c.* 160 m. Vertical seal is provided by the mudstones and shales in the Upper Jurassic Tyne Group.

RESERVOIR

The reservoir in the Ula Field is an Upper Jurassic sandstone formally named the Ula Formation (Vollset and Doré, 1984). Detailed petrological and log studies of the reservoir in the wells have produced a division into five informal lithological units (Fig. 6), based on differences in sedimentary structures, grain size, bioturbation, mineralogy and log response (including the HDT). Unit V is a thin basal sandstone which grades upwards into siltstones of Unit IV. The bulk of the reservoir then falls into two major sequences of upwards-decreasing gamma-ray response—Units III and II. The uppermost Unit I shows a constant or upwards-increasing gamma-ray response. Units I and II have been cored in all Ula wells. Units III, IV and V have been cored in 7/12-4 and 7/12-6.

Unit V is a thin, fine-grained grey sandstone with a sharp base and gradational top. The sands are poorly to moderately sorted, micaceous and glauconitic with some carbonaceous debris. Mainly vertical burrows have obliterated any original sedimentary structures.

Unit IV is a thoroughly bioturbated, dark grey siltstone with some grey, fine-grained sandstone. Occasional relics of a wispy lamination are discernible. Vertical burrows predominate, and belemnites, bivalves and sponge spicules occur.

Unit III consists of fine- to medium-grained, moderately to well-sorted, slightly micaceous and glauconitic sandstones. A general coarsening-up trend has been identified, but is superimposed on several smaller coarsening- and fining-up cycles. Details of sedimentary contacts are generally obscured by the pervasive horizontal bioturbation. Some zones do, however, occur, particularly in the lower half of Unit III, in which parallel and low-angle inclined lamination are preserved. Also present are cosets of planar cross-beds with opposing dips.

Unit II is very similar to Unit III, though with less evidence of original sedimentary features. It is, in general, well sorted and apparently the best reservoir unit in the field. The sands have been homogenized by the pervasive horizontal bioturbation. Vertical burrows occur near its base.

Unit I contains finer-grained, less well-sorted sandstones with more vertical burrows than Unit II. Belemnite debris also reappears. Sedimentary features are largely obscured by bioturbation, but thin (5–20 cm) fining-up sequences capped by siltstones can be found. Traces of low-angle cross-bedding, planar bedding and minor ripple structures are also noted.

The sandstones are arkosic to sub-arkosic grainstones, with rounded to sub-rounded grains, becoming less well-sorted and more angular in Units IV, V and the upper part of

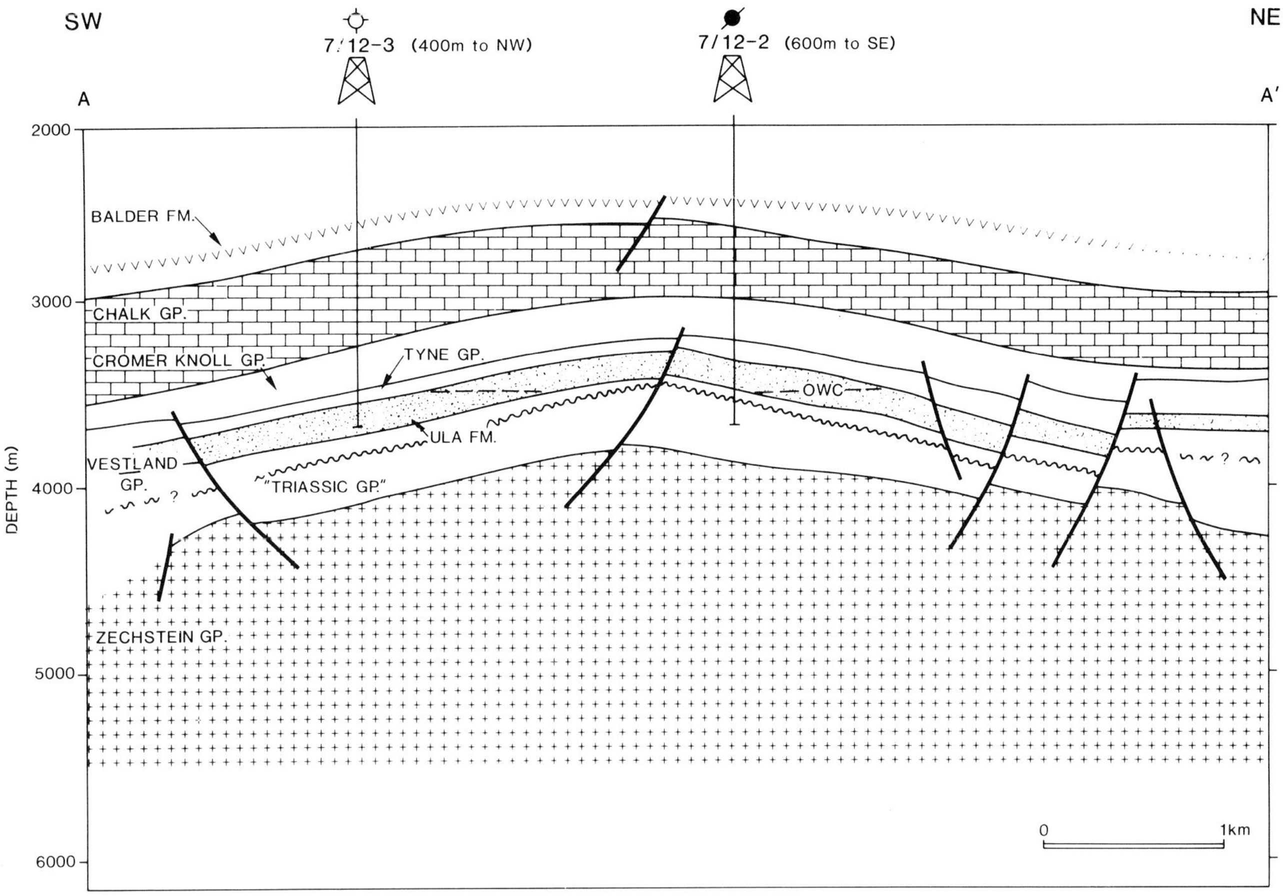

Fig. 5. Cross-section through the Ula Field. Drawn along the line of the seismic section of Fig. 3. No vertical exaggeration. From Spencer *et al.* (1986).

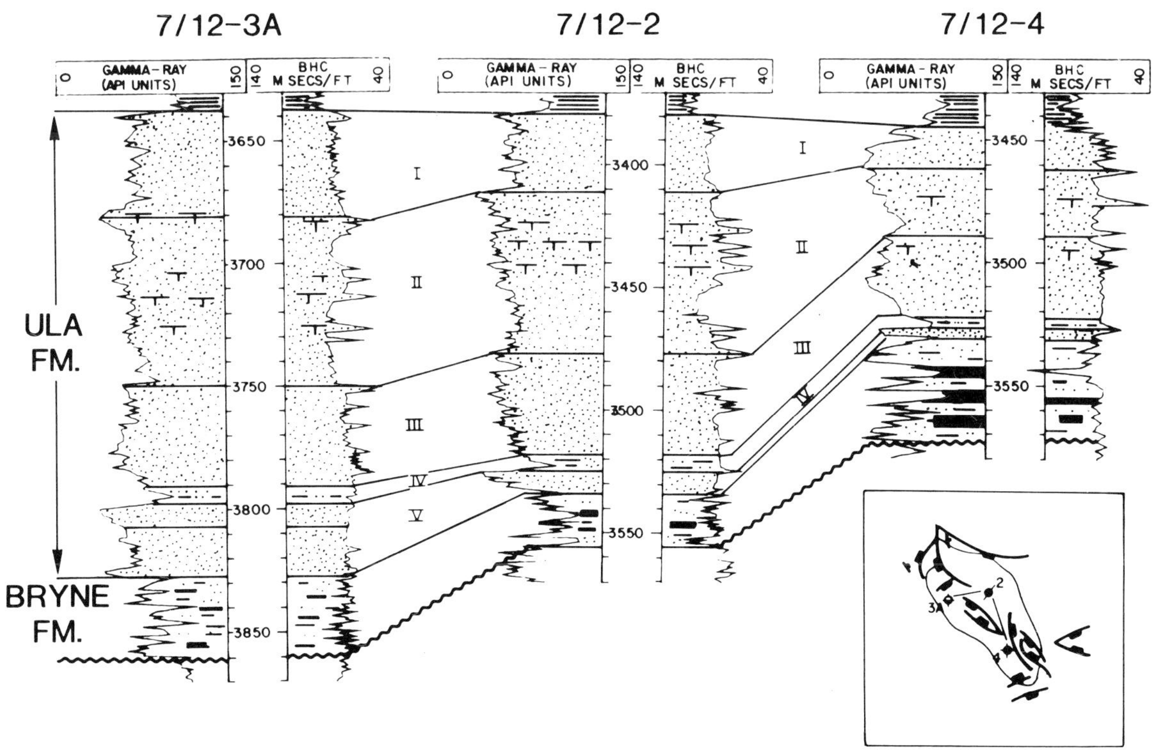

Fig. 6. Stratigraphic correlation of the lithological units within the Ula Formation in the Ula Field. Depths in mbrt. Modified after Bailey *et al.* (1981).

Unit I. Both alkali and plagioclase feldspars occur, including some microcline and rarer perthite. Micas and glauconite are ubiquitous but more common in the finer-grained units. The heavy mineral assemblage of zircons, apatites and rarer tourmalines has been taken to indicate an ultimate granitic provenance, but it is likely that the sands were reworked from Triassic deposits and, perhaps, from the Bryne Formation.

Feldspar and quartz overgrowths occur, whilst wispy illitic clay is especially marked in the finer-grained units. Calcite cements occur throughout, but are particularly concentrated in nodular horizons where there is an association with bioclastic debris. Minor dolomite, pyrite and white mica cements occur with traces of kaolinite, chlorite and siderite (see Larese *et al.*, 1984).

The only discernible secondary porosity (save for localized shell leaching), is provided by the dissolution of alkali feldspars. The sandstones are relatively loosely packed, with little evidence of intergranular pressure solution. Porosities are lower in the water zone, probably due to the greater development of microcrystalline quartz cements.

As far as the deposition of the reservoir is concerned, the microfaunas, the pervasive bioturbation, the shell debris and the presence of glauconite clearly demonstrate a shallow-marine origin for the Ula Formation. However, the unusual thickness of the formation and the scarcity of sedimentary structures are not readily explained. No wholly satisfactory modern analogue of the formation can be suggested.

Vertical grain-size changes suggest that the Unit V–IV–III sequence represents a transgression and subsequent regression. Changes in burrow types and in the ratio of bivalve to belemnite debris are consistent with this interpretation. Unit I is regarded as a transgressive sequence on similar evidence, although its thickness prevents it being interpreted as a typical transgressive sheet sand. Units II and III both show coarsening-upwards trends and are interpreted as broadly regressive sequences.

An early interpretation suggested that Units II and III both represented regressive cycles of offshore bar formation. Synsedimentary faulting was invoked to explain the excessive thickness of each unit in relation to modern bars. Detailed weaknesses of this hypothesis were listed by Bailey *et al.* (1981, p. 8), most significantly the pervasive bioturbation, which is difficult to explain in a high-energy bar crest environment. Instead, they suggested that a tidal sand 'shoal' existed across the Southern Vestland Arch with sand transport towards the Central Graben by low-relief (metre scale?) tidal bedforms (sand-waves or dunes). Migration of these features was frozen by higher subsidence rates and consequent deeper water along the Ula Trend. The model assumes a sufficient time interval between the arrival of successive sand-waves for them to be thoroughly bioturbated prior to being buried. In this model the overall transgressive–regressive–transgressive sequence seen in the Ula Formation is assumed to be the manifestation of background sea-level cyclicity, perhaps eustatically controlled.

Prior to drilling Well 7/12-6 it was believed that the sandstone thickened northwards across the structure towards the Hidra Fault Zone, which was regarded as having an important influence on sedimentation throughout Late Jurassic times. It is now apparent that the reservoir does not simply thicken against the fault scarp (Bailey *et al.*, 1981, fig. 12). The current sedimentary model envisages sands trapped in a general tectonic sink in front of the fault zone, but not directly abutting a synsedimentary fault scarp. Salt tectonics may also have influenced local rates of sand accumulation. Special dip-meter studies of the field and detailed core and correlation work have revealed that the thinner reservoir encountered in Well 7/12-4 is partially explicable by post-depositional faulting.

The lithological units are encountered in all wells and have distinct poroperm characteristics which are of importance in determining reservoir performance. Variations in porosity and permeability appear to be primarily related to variations in textural maturity and grain-size seen in the different depositional units. Average oil zone porosities range from 15 to 22% (Units I–III), decreasing to 12–17% in the finer-grained units. Permeabilities range from 0.2 to 2800 mD, with an average of 650–850 mD in Unit II. Some decrease in both porosity and permeability is recorded in the water zone in 7/12-3A and in the thinner section in 7/12-4. The thin Units IV and V do not constitute effective reservoirs. In Units I to III the overall net-to-gross ratio is 0.93, with a range of 0.85–1.0.

No extensive permeability barriers are recognized in the field. The consistency of the reservoir units suggests that development problems due to significant lateral heterogeneity in the reservoir will not arise.

In addition to the Ula Formation, the Ula structure also contains oil in the Triassic

Skagerrak Formation. A small quantity of oil was produced from this reservoir in Well 7/12-2 and a further oil discovery was made in the Skagerrak in Well 7/12-6. In this well the formation showed a low test productivity.

SOURCE

The source of the oil in the Ula Field is believed to be the 'hot shale' of the Upper Jurassic Mandal Formation (Spencer *et al.*, 1986). This is the only significant oil-prone source-rock in the Ula Trend. It has an average thickness of 36 m in the Ula wells and an S_2 Rock-Eval yield of up to 21 kg/tonne. The Mandal Formation occurs within the of oil window in the drainage kitchen for the Ula Field and generated significant quantities oil and gas from its labile kerogen component in Mid–Late Tertiary times. The gas occurs in solution in the oil pool. The Ula Formation appears to be the principal carrier (as well as reservoir) horizon.

HYDROCARBONS

A total of six drill-stem tests (DSTs) have been conducted on the Ula Formation oil pool with the best flow-rates being achieved in Unit II of the reservoir (1100 Sm^3/day). Drill-stem test 2 in Well 7/12-2 and DST 2 in Well 7/12-4 established dry oil production down to 3507 and 3505 mss, respectively. Drill-stem test 2 in Well 7/12-3A produced essentially 100% water up to a level of 3509 mss. The effective oil–water contact was thus picked at 3508 mss.

Samples of the Ula Formation aquifer water were collected from tests in Wells 7/12-3 and 7/12-3A. It has a high salinity of *c.* 175 000 ppm (NaCl equivalents) and a resistivity of approximately 0.0145 ohm m at reservoir temperature. The pH is 5.4 at 25 °C.

RESERVES

Recoverable reserves in the Upper Jurassic reservoir are currently estimated to be 25×10^6 Sm^3 of oil and 2.5×10^9 Sm^3 of associated gas. The hydrocarbon saturation ranges from 0.75 to 0.97 and the formation volume factor is 1.33. The ultimate recovery is expected to be *c.* 40% of the oil-in-place.

ACKNOWLEDGEMENTS

The author wishes to thank The British Petroleum Company plc, Norske Conoco, Svenska Petroleum, A. S. Pelican and Company and Statoil for permission to publish this paper.

REFERENCES

Bailey, C.C., Price, I. and Spencer, A.M. 1981. The Ula Oil Field, Block 7/12, Norway, In *Norwegian Symposium on Exploration*, Norsk Petroleumsforening, Article 18, pp 1–26.

Deegan, C.E. and Scull, B.J. 1977. A standard lithostratigraphic nomenclature for the Central and Northern North Sea, *Institute Geological Sciences London*, Report 77/25.

Gabrielsen, R.H., Ulvøen, S., Elvsborg, A. and Ekern, O.R. 1985. The geological history and geochemical evaluation of block 2/2 Norwegian sector, North Sea, In B. M. Thomas *et al.* (Eds), *Petroleum Geochemistry in Exploration of the Norwegian Shelf*, Graham and Trotman, London, pp. 165–178.

Larese, R.E., Haskell, N.L., Prezbindowski, D.R. and Beju, D. 1984. Porosity development in selected Jurassic sandstones from the Norwegian and North Seas, Norway—an overview, In A. M. Spencer *et al.* (Eds), *Petroleum Geology of the North European Margin*. Graham and Trotman, London, pp. 81–96.

Spencer, A.M., Home, P.C. and Wiik, V. 1986. Habitat of hydrocarbons in the Jurassic Ula Trend, Central Graben, Norway, In A. M. Spencer *et al.* (Eds), *The Habitat of Hydrocarbons on the Norwegian Continental Shelf*, Graham and Trotman, London, p. 111–127.

Vollset, J. and Doré, A.G. 1984. A revised Triassic and Jurassic lithostratigraphic nomenclature for the Norwegian North Sea, *Norwegian Petroleum Directorate Bulletin* **3**.

11

Valhall

R. C. Leonard and J. W. Munns

Amoco Europe and West Africa Incorporated, 1, Stephen Street, London W1P 2AU, UK

Valhall Field is the southernmost of the currently producing Norwegian Chalk Fields. Hydrocarbons were first encountered on the structure in 1969, but the field was not declared commercial until 1976. Production commenced in 1982.

The Valhall structure was formed by wrench and compressional movements along the Lindesnes Ridge during Late Cretaceous and Tertiary times. The main reservoir, the Tor Formation of Maastrichtian age, is characterized by porosity in excess of 40% over most of the field. The high porosity is due mainly to the extreme overpressure in the reservoir. Reservoir thickness ranges from 0 to 50 m, the variation being due to the influence of Late Cretaceous structural movements and an erosional channel formed during Late Coniacian to Early Maastrichtian times.

The geophysical interpretation of the field is difficult, as the Tertiary overburden is highly gas charged throughout, causing the seismic data to be very poor in quality. The main effects are, first, a marked 'time sag' at the crest of the field, and secondly, major attuenuation of the signal which results in a weak discontinuous reflection from the top of the reservoir. Several unconventional seismic techniques have been used to map the top Chalk horizon and to map thickness variations within the Tor Formation.

INTRODUCTION

The Valhall Field is located about 300 km south-west of Stavanger in the southernmost part of the Norwegian North Sea (Fig. 1). It straddles Blocks 2/8 and 2/11, and lies on the northern part of a large structural closure measuring 280 km^2 at Top Cretaceous level. Water depth is 65–70 in the field area. The field is noteworthy for its complex structural style and for the uneven distribution of the main chalk reservoir, the Tor Formation which is of Late Campanian to Maastrichtian age.

HISTORY

Exploration rights to Block 2/8 were granted to the Amoco/NOCO group (Amoco 28.33%, Texas Eastern 28.33%, Amerada Hess 28.33% and NOCO 15%) in 1965 during the first round of licencing in Norway. Rights to Block 2/11 were granted to the same group in the next round in 1969, each company holding a 25% interest, with a net profit interest to Statoil, the Norwegian State Oil Company.

The initial test of the structure at the Upper Cretaceous (Chalk) level was Well 2/11-1, which was drilled in 1969. It encountered 258 m of Upper Cretaceous Chalk with net pay of 17 m at the top of the section. The well tested 900 barrels of oil per day (BOPD). It was the first oil discovery in the Chalk in the Norwegian North Sea. At that time, the central area of the structure was interpreted to be structurally low due to salt withdrawal. The next two wells, 2/8-3 and 2/8-4, were drilled on the north-east and north-west

Geology of the Norwegian Oil and Gas Fields © A. M. Spencer *et al.*, eds (Graham & Trotman, 1987) pp. 153–163.

Valhall Field data summary

Trap	
Type	Structural/stratigraphic trap
Depth to crest	2400 mss at Top Chalk
Lowest closing contour	*c.* 2700 mss
Oil–water contact	Variable
Oil column	*c.* 200 m, dependent on reservoir quality
Pay zone	
Formation	Tor and Hod formations
Age	Upper Cretaceous
Net/gross ratio	100%
Porosity (ranges)	Up to 45% in crestal Tor Formation; decreases rapidly with
Hydrocarbon saturation	98% in crestal Tor Formation
Permeability (average; range)	Matrix 0.1–10 mD; occasionally enhanced by natural fractur
Hydrocarbons	
Oil gravity	0.845 g/cm^3
Oil type	Paraffinic
Bubble point	324 bar (average)
Gas/oil ratio	231 Sm^3/Sm^3
Formation volume factor	1.65
Formation water	
Resistivity	0.045 ohm m at 90°C
Reservoir conditions	
—	
Field size	
Hydrocarbon pore volume	259 × 10^6 Sm^3
Recovery factor	*c.* 12–15%
Production	
Start-up date	October 1982
Cumulative production to end of 1985	
—oil	6.3 × 10^{6}Sm^3
—NGL	0.3 × 10^6 tonnes
—gas	1.0 × 10^{9}Sm^3

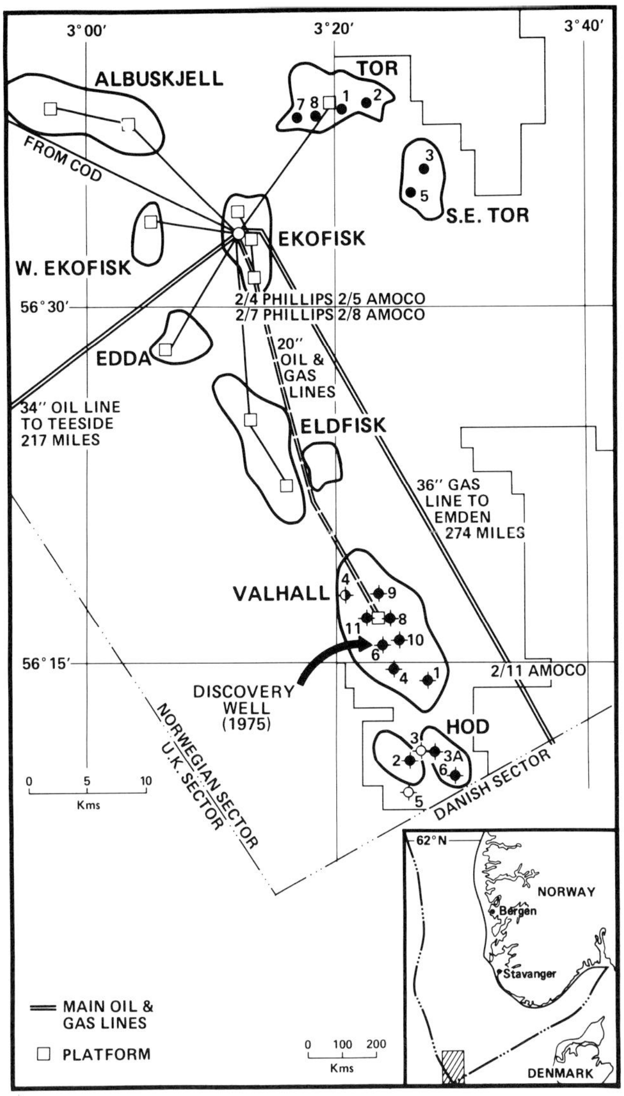

Fig. 1. Location map of Valhall Field.

flanks of the closure and did not encounter significant hydrocarbons.

A reappraisal of the seismic data in 1974 led to the conclusion that the apparent low (in time) in the central part of the field was due to shallow gas anomalies. In early 1975, Well 2/8-6 was drilled in a crestal position on the structure and proved to be 160 m high to 2/11-1 at the top Chalk level. The well encountered over 100 m of hydrocarbon-bearing chalk, and is considered the Valhall discovery well. Four delineation wells were drilled (2/8-8, 2/8-9, 2/8-10 and 2/8-11), following the shooting of a further seismic grid.

In late 1976, the Amoco/NOCO group was granted approval to develop the field. The development plan called for separate platforms for production, drilling, and living quarters, to be located over the crestal part of the field. Development drilling commenced in December 1981. Peak production of 11 000 Sm^3/day oil and $1.8 \times 10^6 Sm^3$/day gas was reached in June 1985.

GEOLOGICAL SETTING

The Valhall Field is located near the south-western flank of the Central Graben, which is an intracratonic rift zone that came into existence in Triassic times. The sediments filling the Central Graben rift are underlain by a Mid- to Late Palaeozoic sequence, laid down after the Hercynian orogeny, that was capped by evaporites of the Permian Zechstein Formation. However, the oldest rocks penetrated in the field are of Late Jurassic age, being reached in Well 2/11-1 (TD 4664 m).

The Valhall structure is associated with the Lindesnes Ridge, a NNW-trending elongate antiformal feature which also contains four other Chalk fields, namely Hod, Eldfisk, Edda and Tommeliten (Fig. 2). The Lindesnes Ridge was formed in Mesozoic times, during a period of active tectonism when the main structural elements in the Central Graben came into existence. The Tertiary, by contrast, was a period of gradual subsidence, when a great thickness of marine shale was laid down in the Central Graben.

Based upon Valhall well control, the stratigraphy can be broken into three main subdi-

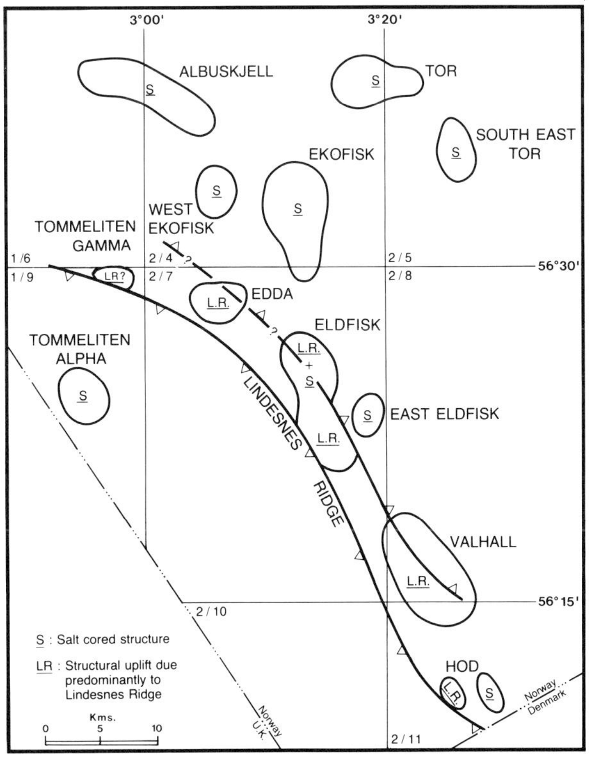

Fig. 2. Structural setting of Valhall Field.

visions, namely: the Upper Jurassic to Lower Cretaceous shales of the Rødby Formation; the Upper Cretaceous Chalk Group consisting of the Hidra, Plenus Marl, Hod and Tor formations; and the Tertiary shales of the Rogaland, Hordaland and Nordland groups.

GEOPHYSICS

The Valhall structure was first mapped seismically in 1967–68. The map showed a central 'no data' zone, which was interpreted as a structural low, due to the withdrawal of an inferred salt piercement. The drilling in 1975 of the Valhall discovery well, 2/8-6, and subsequent appraisal wells, demonstrated that the 'no data' zone in the central part of the field was a time-sag caused by the presence of a low-velocity, gas-charged section in the Tertiary overburden (Fig. 3). Munns (1985) has demonstrated that at least 75% of the velocity pull-down at Top Chalk level is due to a single gas-charged silt unit of Middle Miocene age. Data quality in the crestal part of the field is further degraded both by the severe attenuation of the seismic signal and particularly the higher frequencies, which limit the resolution of the data, and by the degradation of the stack response due to severe ray-path distortion in the overburden.

The most recent seismic data acquired in 1983 resulted in a major improvement in data quality, particularly along the flanks of the gas cloud area, and also gave minor, but significant, improvements in the crestal area of the field.

The geophysical interpretation is primarily concerned with determining both the structural configuration at the Top Chalk level and the isopach variation of the Tor Formation. Due to the data-quality problems previously mentioned, the geophysical interpretation is heavily dependent on data from alternative seismic techniques, including vertical seismic profiles (VSP), flattened seismic sections and SEISLOGS.* This integrated seismic interpretation contributed to the evolution of a new geological model for the structural development of the field, which is considered further in the section covering the trapping mechanism.

The Tor isopach has been mapped primarily by using two geophysical techniques, namely flattened sections and SEISLOGS. Vertical seismic profiles have proved useful in some areas, particularly in the crestal part of the field. These techniques have several advantages over the conventional seismic data:

1. Vertical seismic profiles were acquired in most development wells, and gave much improved data quality in the crestal part of the field. They are particularly helpful in fault identification. Sufficient resolution was achieved to allow resolution of the Tor isopach down to 20 m.
2. Flattened seismic sections, produced by the application of static shifts, enabled the identification of erosional channels that are crucial to the preservation of a thick Tor section. Flattening the Top Chalk event effectively removes the time distortion due to the gas cloud and the structural effects of post-chalk movement.
3. SEISLOGS are pseudosonic logs derived by the inversion and depth conversion of the surface seismic data. They aided the identification of the Tor Formation, due to the marked transit-time differences between the Tor Formation (40–50% porosities) and the Hod Formation (30–40% porosities). This technique gives a Tor resolution of about 25 m.

*SEISLOG is a registered trademark of Teknica Resource Development Limited.

TRAP

The Valhall Field is located within a large structural enclosure measuring 280 km^2 at a level corresponding with the top of the reservoir. The general trapping mechanism is therefore structural. However, because the main reservoir, the Tor Formation, was deposited in a series of erosional channels, there is also an important stratigraphic component.

The Valhall structure is asymmetric, with a steep western flank and a more gently dipping eastern flank (Fig. 4). The Valhall Field is divided into compartments by WSW-trending sinistral strike-slip faults (Fig. 4). A discrete horst-block can be mapped in an axial position at the northern end of the field. This horst is offset to the east at the southern end of the field by lateral movements on strike-slip faults. This axial horst-block influenced the development of the Maastrichtian erosional channel system. The present-day distribution of the Tor Formation clearly shows that a major feeder channel was present to the north of the field. Sea-floor topography caused this main feeder channel to divide into several lobes within which the allochthonous chalk has been preserved.

The structural development of the Valhall Field was closely related to movements along

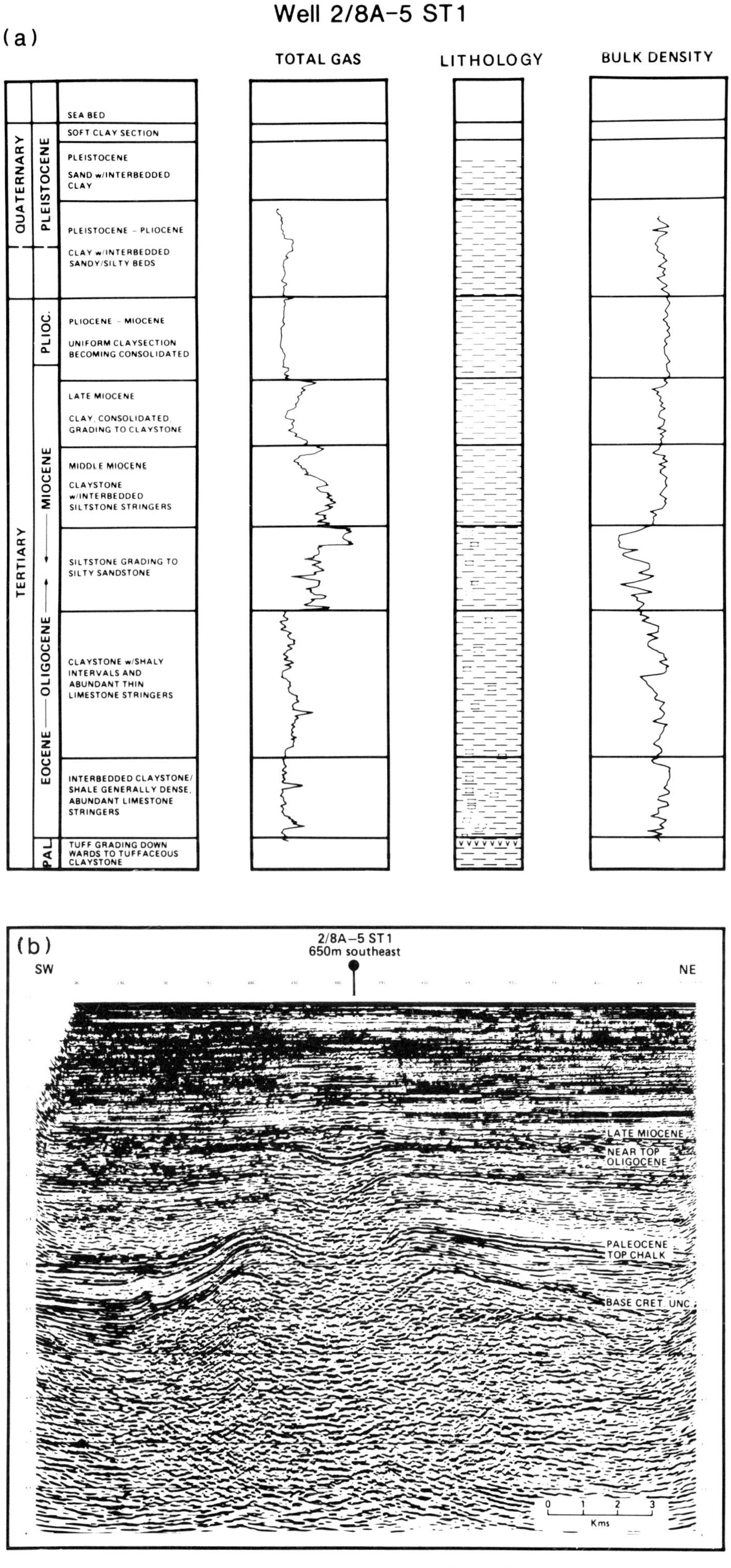

Fig. 3. (a) Mud log of Eocene to Recent section in Well 2/8A-5 ST1. Note the gas-charged section of mudstones and siltstones in the Miocene. (b) Seismic line AN083-02, showing the effect of the gas in the Tertiary on seismic data quality.

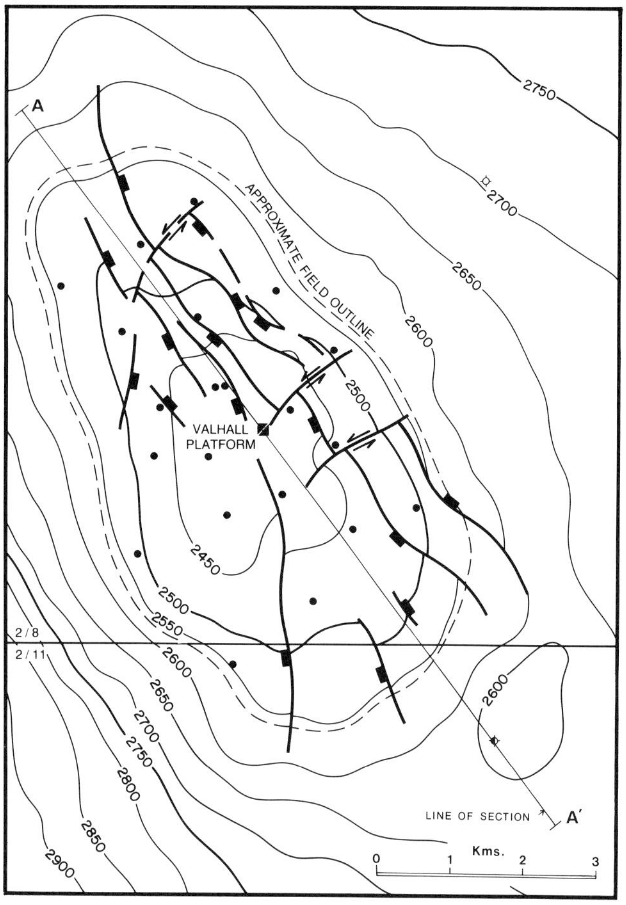

Fig. 4. Top Chalk structure map (depth), Valhall Field.

the Lindesnes Ridge. As illustrated in Fig. 5 the present-day configuration of the ridge is a result of at least three major episodes of deformation.

(i) The first tectonic episode resulted in rapid subsidence in the Upper Jurassic and Lower Cretaceous, allowing the development of a half graben containing several thousand metres of sediment.

(ii) The second episode involved uplift along the western flank of the Lindesnes Ridge, causing localized compressional events resulting in the inversion of the Lower Cretaceous basin. As a consequence, the thinnest Upper Cretaceous section at the present day corresponds with the thickest Lower Cretaceous section deposited prior to the inversion.

(iii) Periodic reactivation of the Lindesnes Ridge during the Eocene and Miocene resulted in the present-day asymmetric nature of the field.

The Chalk reservoirs are sealed by Tertiary shales, but the existence of a gas cloud overlying the Cretaceous reservoirs indicates that sealing conditions are poor with respect to gas. Oil shows are also encountered above the reservoir, particularly in the Paleocene section. They may be due to upward migration

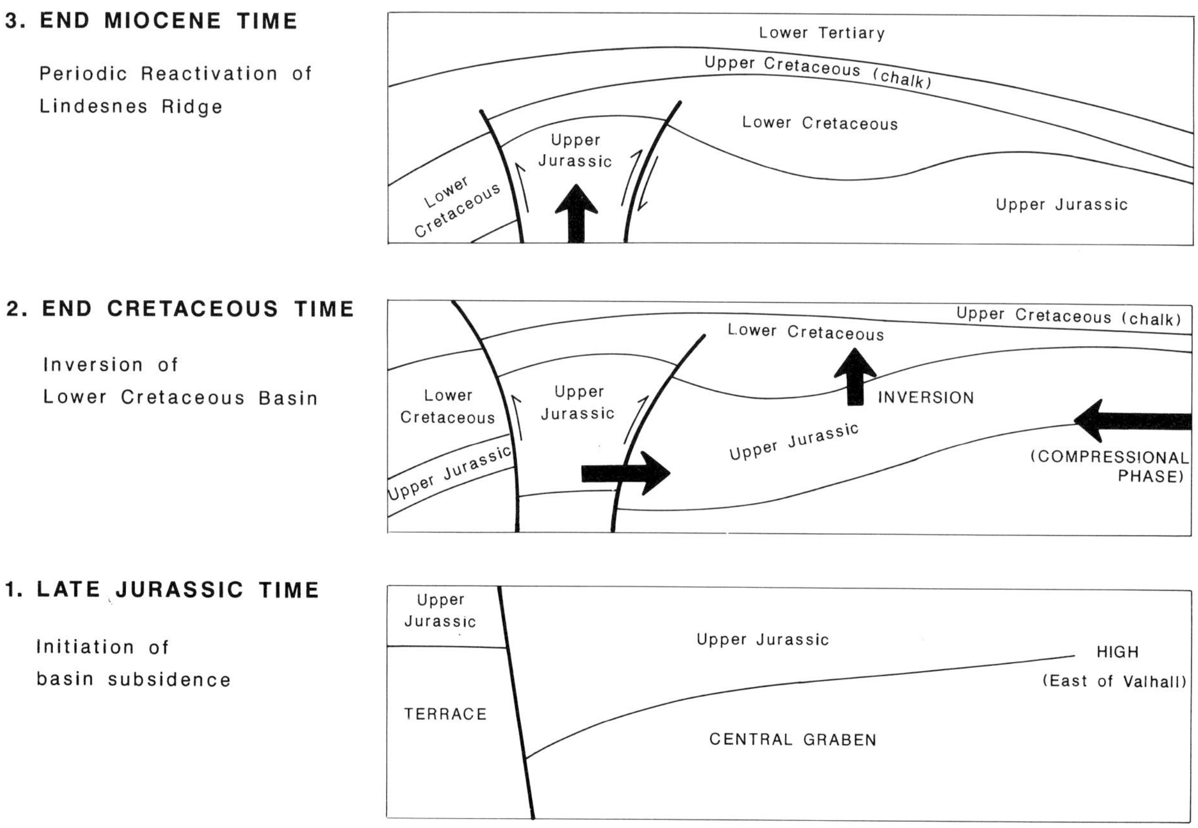

Fig. 5. Structural evolution of Valhall Field.

of oil along faults which can be seen to penetrate into the Paleocene formations.

The crest of Valhall Field lies at about 2400 m subsea, and structural closure extends to about 2760 m, which corresponds with the saddle area between Valhall and the Hod Field to the south. The length of the hydrocarbon column is a function of reservoir quality, and is therefore variable across the field. The lowest known oil is at 2620 m, being found in the Well 2/11-1 to the south. However, values of 100% water saturation have been found as high as 2588 m (Fig. 6). This suggests that saturation levels are a function of reservoir rock quality and that the water level across the field is variable.

RESERVOIR

The hydrocarbon-bearing zones of the Valhall Field occur in chalks of Cenomanian–Maastrichtian age. The stratigraphic section has been divided into eight zones (Fig. 7). At the base lies the Cenomanian Hidra and Plenus Marl formations; above follows the Turonian–Coniacian Hod Formation, which has been subdivided into five zones; and at the top lies the Late Campanian–Maastrichtian Tor Formation, which is divided into two zones. The variations in the reservoir characteristics that distinguish the individual zones are mainly due to depositional environment and the proportion of non-carbonate material in the rock.

Three main lithofacies have been identified within the Chalk Group. These lithofacies can be related to the mechanism of chalk deposition. Two of the lithofacies are allochthonous in origin, and were deposited from slump and debris flows and turbidity currents along erosional channels. The third lithofacies is an autochthonous unit deposited in a low-energy, deeper-water environment.

The slump and debris-flow deposits have the highest porosity and permeability. This is primarily due to high rates of deposition and lack of consolidation and cementation. Four major developments of slump and debris-flow deposits have been identified in the Central Graben, but only one of these flows appears to have reached the Valhall Field, where it is represented by the Upper Tor Member. An intermediate rock type representative of moderate rates of deposition and cementation is the turbidite or channel-flow deposit, found both in the Lower Tor Member and in the middle part of the Lower Hod Member (H4). The pelagic chalks, which occur in the rest of the Hod Formation and the Hidra and the Plenus Marl Formations, were characterized by low rates of deposition and extensive cementation, and have generally poorer porosity and permeability than the other rock types.

A second factor in the variation of reservoir quality is the mineralogy of the chalk. While certain formations, such as the Upper Tor, are composed of almost 100% calcium carbonate, other zones have secondary components, principally clay and silica, that form a significant proportion of the rock. In extreme cases, this fraction can reach 50%.

The major reservoir at Valhall Field is the Tor Formation, containing approximately 70% of the oil in place. There is considerable variation in the Tor Formation thicknesses across the field as indicated in Fig. 8. This is in contrast to the underlying Hod Formation zones which generally show a thinning to a Late Cretaceous palaeocrest located approxi-

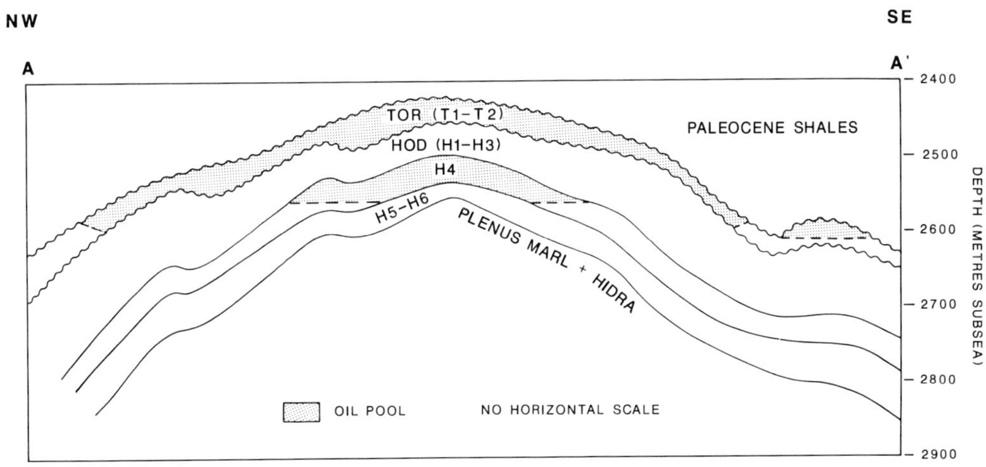

Fig. 6. Schematic cross-section showing major reservoir zones (T1-T2 and H4) with 80% S_w levels. Section located A–A' on Fig. 4.

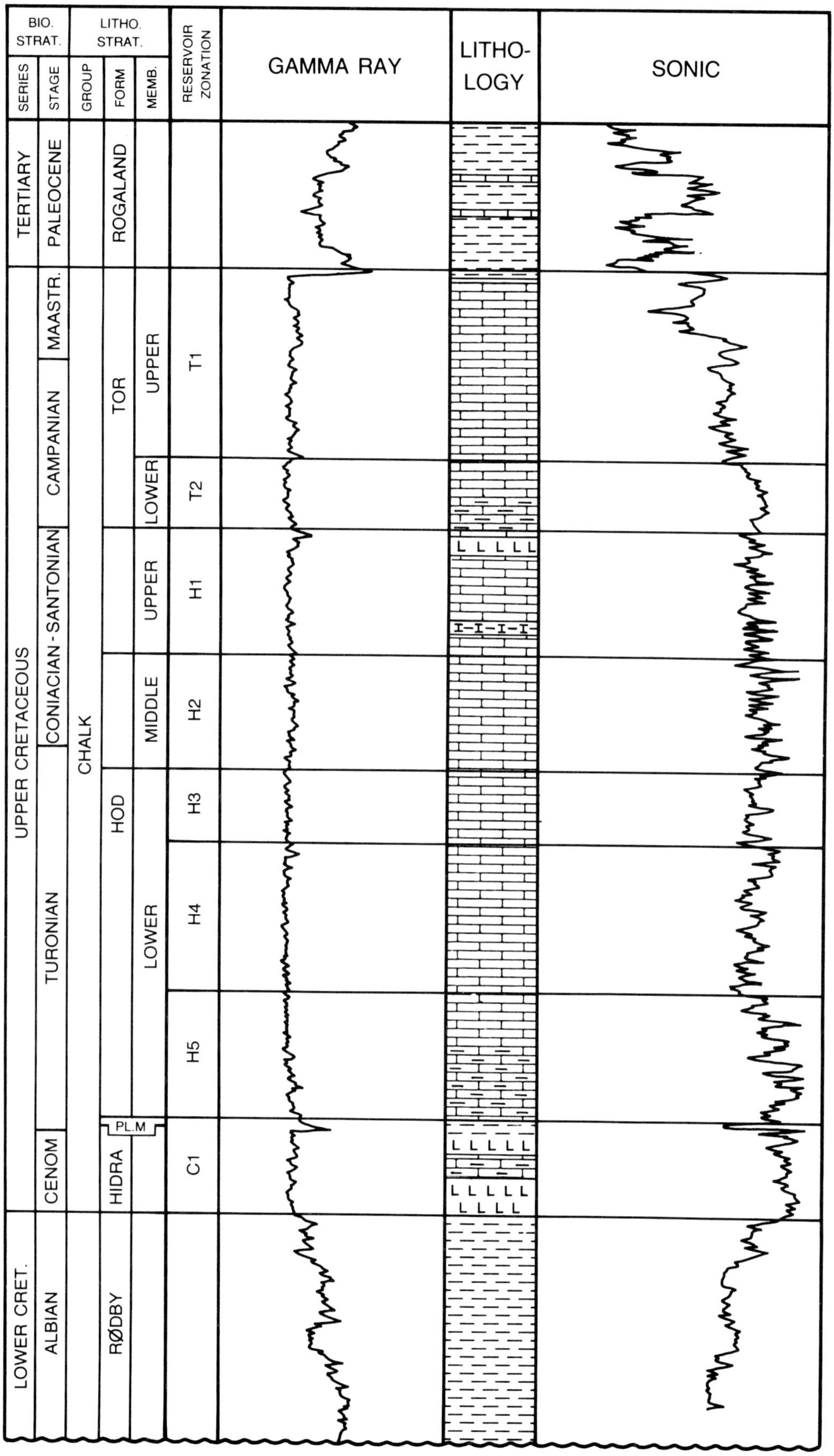

Fig. 7. Reservoir subdivision at Valhall Field.

mately 1.5 km south-west of the Valhall Field platform. During the Late Campanian to Maastrichtian, an erosional channel, originating from the north of the field, cut into the uppermost Hod Formation unit. Sea-floor topography caused this main feeder channel to divide into several lobes. The channel was partially infilled by turbidites and channel deposits of Early (Lower) Tor age, followed by the massive debris flows characteristic of the Upper Tor Formation. Subsequent faulting and erosion resulted in the local removal of the Tor Formation on the up-thrown side of many of the faults that cut the field. Thick sequences of Tor Formation were deposited to the east and west on the flanks of the palaeostructure.

Porosities in the Tor Formation range from almost 50% on the crest of the field to 20% at the edge of the hydrocarbon column. The preservation of high porosities in the Tor Formation is primarily due to the extreme overpressure in the reservoir. For detailed discussion, see Munns (1985).

Porosities in the Hod Formation range from 20 to 40%, with hard-ground intervals at the Coniacian–Maastrichtian unconformity varying from 25% to less than 10%.

Matrix permeability in the Tor Formation generally ranges from 2 to 10 mD, whereas

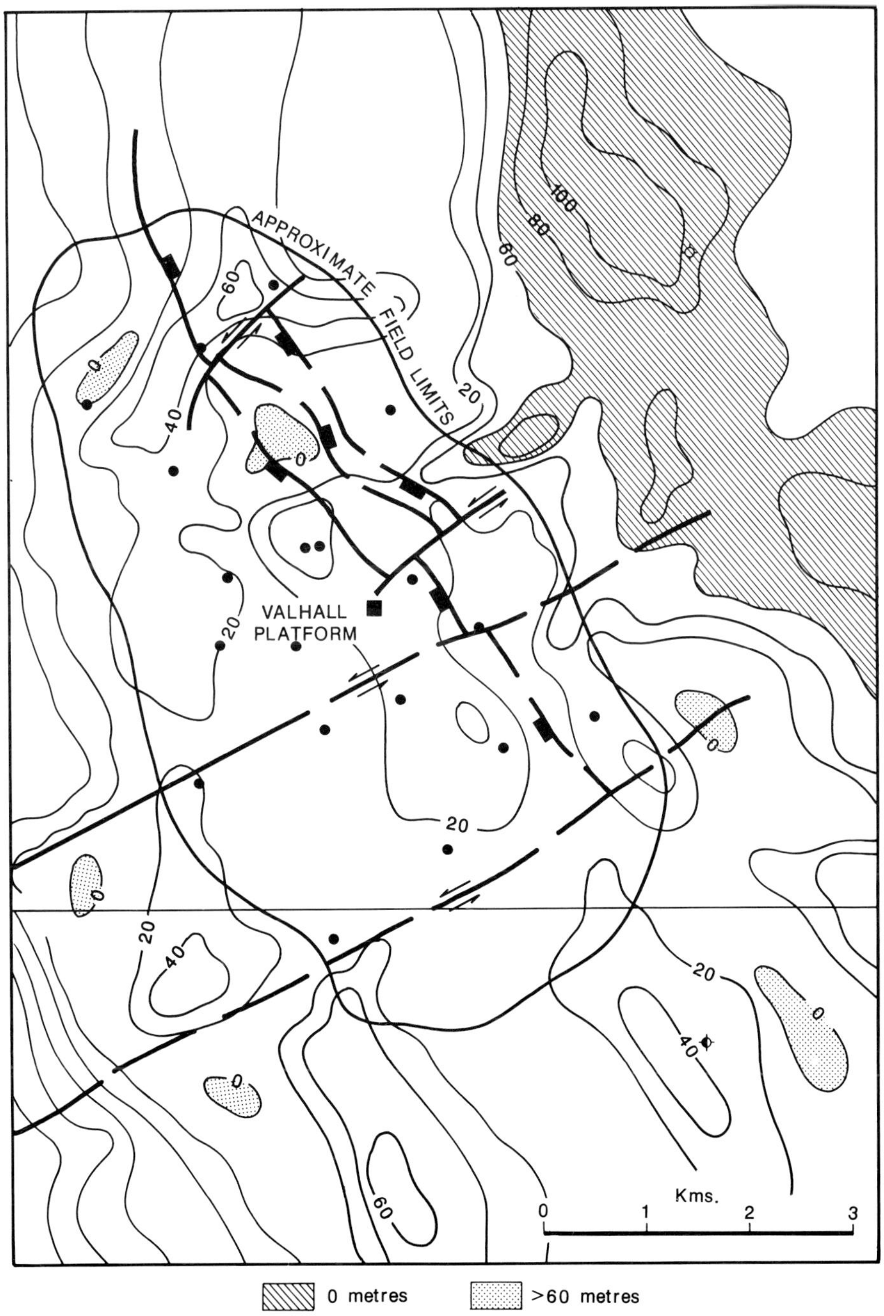

Fig. 8. Tor Formation isopach.

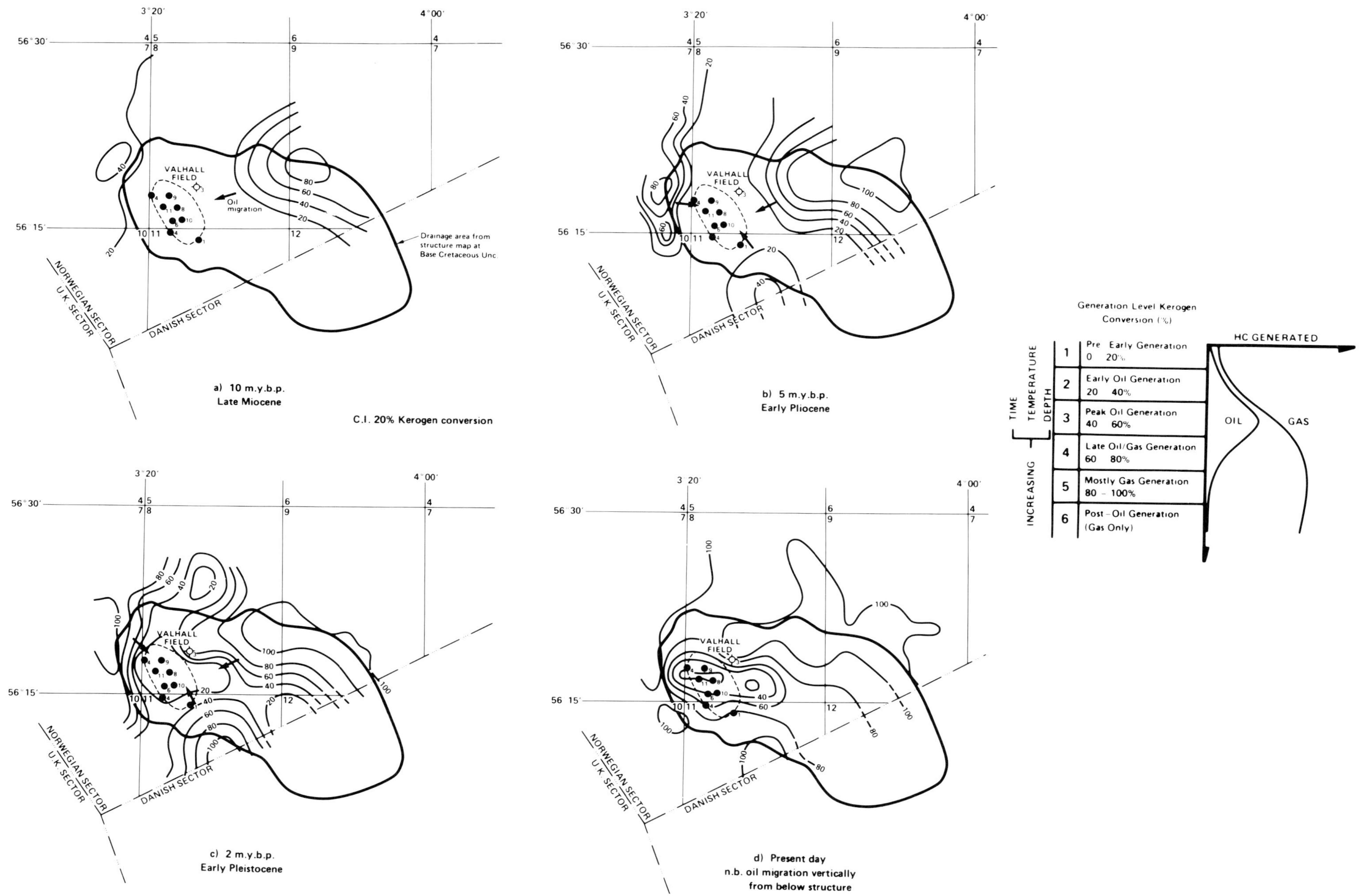

Fig. 9. Hydrocarbon generation and migration, Valhall Field catchment area. The maps are contoured using the percentage kerogen conversion-levels of the hydrocarbon generation curve, which is taken from Tissot and Welte (1978).

typically in the Hod Formation it is less than 2 mD. Permeability calculations from drill-stem test data are in good agreement with matrix permeabilities measured from core analysis in both the Hod Formation and the Tor Formation. However, crestal Tor wells have effective permeabilities of 20–90 mD, indicating the likely contribution of natural fracturing, enhancing the effective permeability by up to a factor of 10.

SOURCE

Geochemical analysis indicates that the source-rock for the oil in Valhall is the Kimmeridge Clay (Draupne Formation) of Late Jurassic age. The presence of organic-rich shales in the Valhall area is shown by the Upper Jurassic section in Well 2/11-1, which encountered 1200 m of shale with an average total organic carbon content of 5–6%. The vitrinite reflectance of the shale varies from 0.57 to 0.67 indicating that the kerogen is at an early stage of hydrocarbon generation.

Based upon regional basin analysis (Leonard, 1984) significant oil generation within the Valhall catchment area began during mid-Miocene times (Fig. 9). Hydrocarbon generation directly under the crest of the field began during Pleistocene times. Approximately $3.5 \times 10^9 Sm^3$ of oil have been generated within the catchment area of the structure. It is estimated that $259 \times 10^6 Sm^3$ oil occur in-place, indicating an expulsion, migration and trapping efficiency factor of about 8%. Migration of hydrocarbons from the Upper Jurassic shale to the Upper Cretaceous chalk probably occurred along normal faults acting as vertical conduits.

HYDROCARBON POOL

The productivity of the Valhall Field wells has been very variable with flow rates varying from 160 to 2000 Sm^3/day oil over extended periods (Munns, 1985).

In common with other North Sea Chalk fields there is not a clearly defined oil–water contact at Valhall. The lowest known oil in the Tor Formation is at 2620 mss in Well 2/11-1 to the south, whereas 100% water saturation has been calculated as high as 2588 mss on the eastern flank. It is apparent that the vertical distribution of the oil is a function of reservoir quality. The structure is not spill-point full.

The crude at Valhall is a paraffinic oil of 0.845 g/cm^3 specific gravity.

RESERVES

During the course of the delineation and development of Valhall Field, estimates of oil-in-place have ranged from 233 to $429 \times 10^6 Sm^3$. The single most important factor affecting the estimates is the distribution of the Tor Formation. Structural elevation of the reservoir is the second most important factor. Accurate assessment of these factors has been difficult due to the poor quality of the seismic data. The most recent estimate gives a figure of $259 \times 10^6 Sm^3$ with approximately 70% occurring in the Tor Formation. It will, however, no doubt be subject to further revisions as the final development wells are drilled.

ACKNOWLEDGEMENTS

The authors thank Amoco Norway Oil Company and its partners, The Norwegian Oil Consortium (NOCO), Texas Eastern Norway and Amerada Hess Norway for their permission to publish this paper.

REFERENCES

Leonard, R.C. 1984, Generation and Migration of Hydrocarbons on the Southern Norwegian Shelf, *AAPG Fossil Fuels of Europe Seminar*, Geneva.

Munns, J.W. 1985, The Valhall Field: a geological overview, *Marine and Petroleum Geology* **2**, 23–43.

Tissot, B.P. and Welte, D.H. 1978. *Petroleum Formation and Occurrence*, Springer-Verlag, Berlin.

12

West Ekofisk

Michel D'Heur

Petrofina, c/o Fina Exploration Norway, Skogstøstraen 37, N-4001 Stavanger, Norway

The West Ekofisk Field, which was discovered in 1970, is located in the Central Graben of the North Sea, in the southern-most part of the Norwegian Shelf, about 7 km west of the giant Ekofisk Field. The reservoirs are chalks of the Ekofisk and Tor formations which are of Danian (Early Paleocene) and Maastrichtian (Late Cretaceous) ages respectively. In the central area of the field, the reservoir thickness exceeds 200 m with a 'net pay' to 'gross interval' ratio of 90%, the average porosity reaches 32% and the average hydrocarbon saturation is about 85%.

The very high degree of porosity preservation is due to the nature of the pore fluid and to the pressure conditions. The porosity and thickness of the reservoir layers decrease from the central area of the field to its periphery from 32 to 20% and from 200 to 150 m, respectively, as a result of differential diagenesis only.

Thirteen wells have been drilled on this small circular structure. The productive area is about 12 km^2. The original recoverable reserves of the field are around 10×10^{6} Sm^3 of condensate and 28×10^{9} Sm^3 of gas.

INTRODUCTION

The West Ekofisk Field is located in the southern part of the Norwegian sector of the North Sea, 315 km south-west of Stavanger and 7 km west of the giant Ekofisk Field (Fig. 1). The structure is a circular dome of halokinetic origin with a diameter of about 4 km. The field, named from its proximity to the Ekofisk Field, is entirely contained in Block 2/4. Water depth is about 70 m. The field is a relatively small accumulation which is expected to produce about 28×10^9 Sm^3 of gas and 10×10^6 Sm^3 of condensate.

HISTORY

Production Licence PLO18, which includes Block 2/4, was granted to Phillips, Petrofina and Agip in 1965, during the first Norwegian round of licencing. The present participating interests in the block are:

Phillips Petroleum Company of Norway	36.96%
Petrofina (Fina Exploration Norway A/S)	30%
Norsk Agip	13.04%
Elf Aquitaine Norge A/S	8.1%
Norsk Hydro A/S	6.7%
Total Marine Norsk A/S	4%
Others	1.2%.

Gas-bearing Paleocene sandstones, which were found by the same group of companies in Block 7/11 (Cod Field), tens of kilometres to the north, were the main objective when exploration drilling began in Block 2/4 in 1969. Well 2/1-1AX was drilled on the Ekofisk structure, the largest in the block, and penetrated the Paleocene without encountering any sandstone, but found an unexpectedly porous chalk which was oil bearing over an interval 180 m thick. The three following appraisal wells clearly established that the Ekofisk Field was a major discovery.

The fifth well, 2/4-5X (sometimes referred to

Geology of the Norwegian Oil and Gas Fields © A. M. Spencer *et al.*, eds (Graham & Trotman, 1987) pp. 165–175.

West Ekofisk Field data summary

Trap	
Type	Dome structure over salt swell
Depth to crest	3065 mss
Lowest closing contour (Top Chalk)	3275 mss
Oil/water contact (lowest)	3320 m
Oil column (maximum)	210 m
Pay zone	
Formation	Ekofisk and Tor Formations (Chalk Group)
Age	Danian and Maastrichtian
Gross thickness (range)	185 m (164–268 m)
Net/gross ratio (central area)	0.86
Porosity (average, range)	30.8%; 18–34%
Hydrocarbon saturation (average)	78%
Hydrocarbons	
Oil gravity (initial)	0.80–0.81 g/cm^3
Gas gravity (relative to air)	0.70
Oil type	Corrosive condensate associated with gas
Bubble point	478 bar
Dew point pressure	444 bar
Gas/oil ratio (initial)	731 m^3/m^3
Formation volume factor (initial)	1.37
Formation water	
Salinity	52 000 ppm NaCl equivalent
Resistivity	0.042 ohm. m at 118 °C
Reservoir conditions	
Temperature (initial)	131 °C
Pressure (initial)	498 bar
Pressure gradient in reservoir	0.022–0.026 bar/m
Field size	
Net Area	12 km^2
Net rock volume	614.6 × 10^6 m^3
Net Hydrocarbon pore volume	148 × 10^6 reservoir m^3
Recovery factor (gas)	0.752
Recoverable condensate	10 × 10^6 Sm^3
Recoverable gas	28 × 10^9 Sm^3
Production	
Start-up	November 1977
Development scheme	One production–drilling platform with 10 producing wells
Production rate (early 1986)	640 Sm^3/day oil, 4.2 × 10^6 Sm^3/day gas
Cumulative production (early 1986)	8.6 × 10^6 Sm^3 oil, 21.7 × 10^9 Sm^3 gas

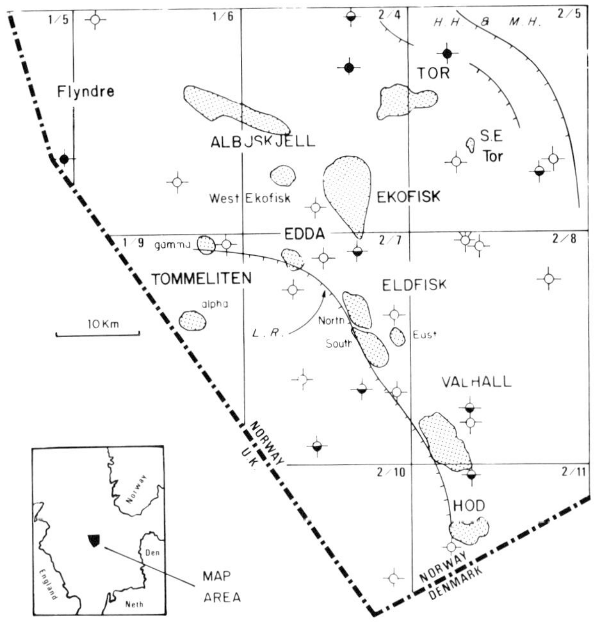

Fig. 1. Location map.

as 2/4-6), was drilled near the crest of a small circular structure 7 km west of Ekofisk, from which it is separated by a saddle at the level of the top of the chalk (Figs. 2 and 3). The well reached the chalk reservoir in October 1970 at a depth of 3090 mss. It penetrated 200 m of very porous and hydrocarbon-bearing chalk and bottomed at 3390 mss in the Maastrichtian. The reservoir interval correlates very well with that of the Ekofisk Field and has similar porosities but a thinner oil column (Table 1). The test results are presented in Table 2 and the corresponding intervals are shown on Fig. 4. The best flow (747 Sm^3 oil/day + 400 × 10^3 Sm^3 gas/day) was obtained from a 27-m perforated interval in the Danian chalk, through a 25 mm choke and after acidizing. Owing to its proximity to the Ekofisk Field, the commerciality of the West Ekofisk Field was declared without any appraisal well being drilled, which may be

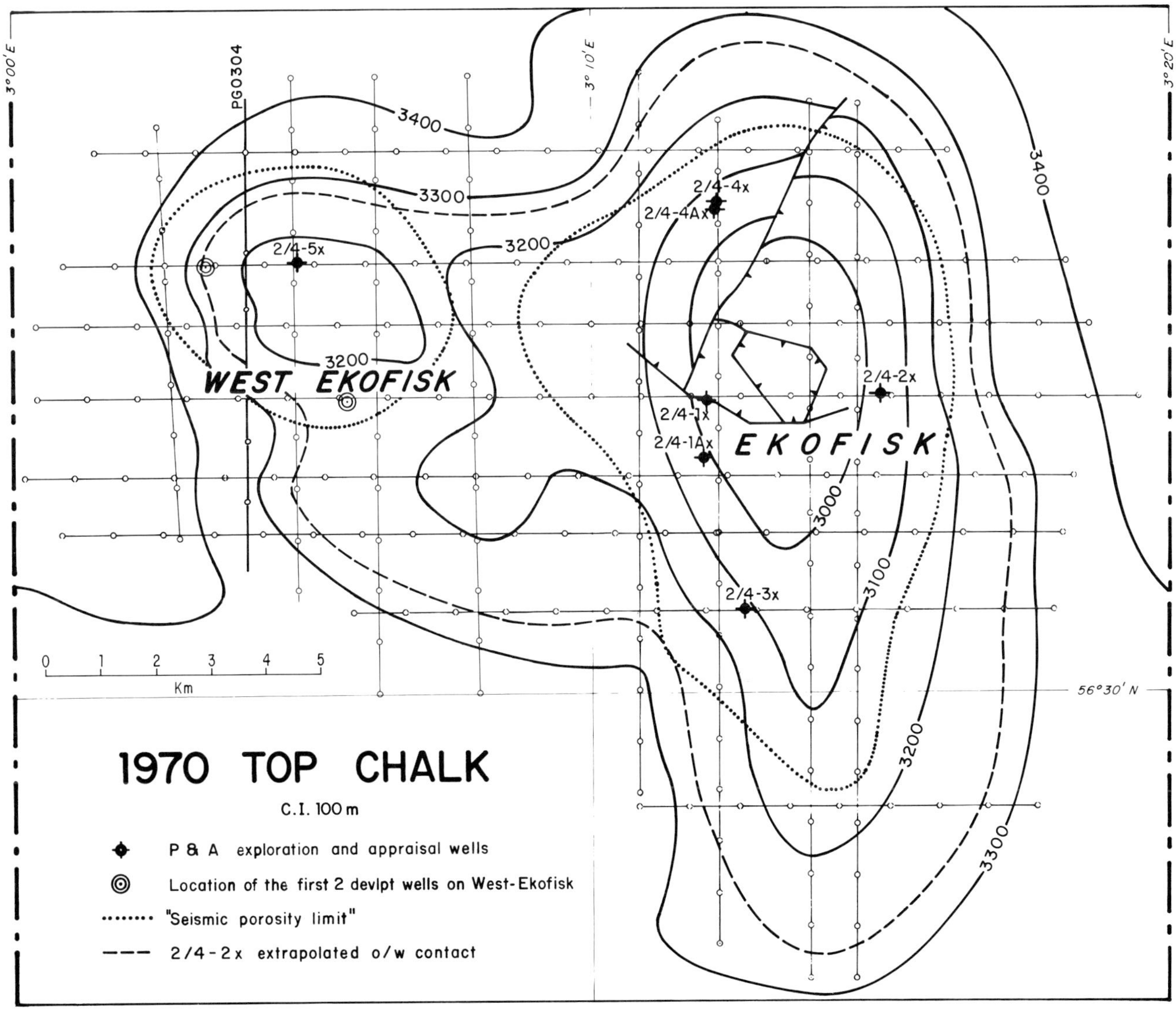

Fig. 2. 1970 structural interpretation of Ekofisk and West Ekofisk.

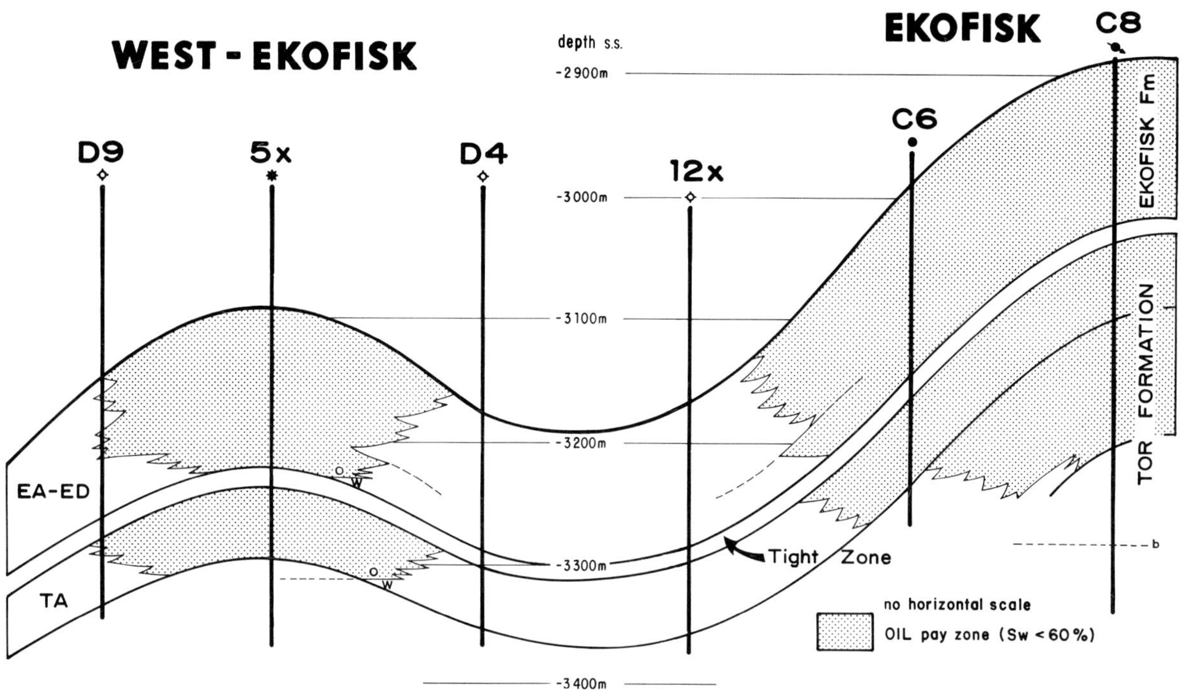

Fig. 3. Diagrammatical cross-section showing the relationship between Ekofisk and West Ekofisk and the location of the first two development wells on West Ekofisk. These wells encountered no pay despite reaching the top of the chalk well above the 'base of pay zone' identified in the discovery well 2/4-5X.

unique for a North Sea field. The lateral extent of the field was assumed on the basis of the so-called 'limit of seismic porosity' which later proved to be imprecise (see Geophysics, p. 170). At that time, the recoverable resources were estimated at 60×10^9 Sm^3 of gas and 30×10^6 Sm^3 of oil, which is more than twice the present estimate of original reserves.

Table 1. Comparison of the total reservoir interval of the crestal wells of Ekofisk and West Ekofisk (net pay cut-off taken at 15–18% for porosity and 50–55% for water saturation).

Crestal wells Net Pay	*Ekofisk*	*West Ekofisk*
Thickness	290 m	186 m
Average porosity	31%	33%
Average water saturation	10%	20%

A single 15-slot drilling–production platform was installed on the field and development drilling was started in October 1974. Two disappointments, out of what became a long series of disappointments during development of the chalk fields, occurred in 1975 when the first two development wells D4 and D9 were drilled on West Ekofisk. Well D4 reached the top of the chalk more than 50 m above the base of the hydrocarbon zone, as identified by the discovery well, but failed to find a productive reservoir. Well D9, drilled on the opposite flank, gave the same negative results. These results led to a drastic reduction in the reserve estimate. It had not then been realized that the best procedure to develop a chalk reservoir was to drill wells step-by-step from the central area of the field towards its periphery, each well being located after careful evaluation of previous results.

Table 2. Summary of test results of the West Ekofisk discovery Well 2/4-5X (intervals shown on Fig. 4).

Interval	*Metres perforated*	*Choke (inch)*	*Oil rate (Sm^3/day)*	*Gas rate (10^3 Sm^3/day)*	*Water rate (Sm^3/day)*
1	—	—	—	—	13
2	30	1	570	260	—
3	21	1	600	200	—
4	27	1	747	400	—

The reason for the failure is that the porosity of chalk reservoir rapidly decreases flankwards and the thickness of the oil–water transition zone increases accordingly. In fact, West Ekofisk became the best field to study these phenomena (see Reservoir, p. 173).

The ten remaining wells were located so as to reach the reservoir at higher structural elevations. They were drilled successfully between 1975 and 1977. Production began in 1977 and reached a peak of 13.5×10^3 Sm^3 oil/day and 15×10^6 Sm^3 gas/day in 1978. At the beginning of 1986, 8.6×10^6 Sm^3 oil and 21.7×10^9 Sm^3 of gas had been produced, and the daily production rate was about 640 Sm^3 oil and 4.2×10^6 Sm^3 of gas.

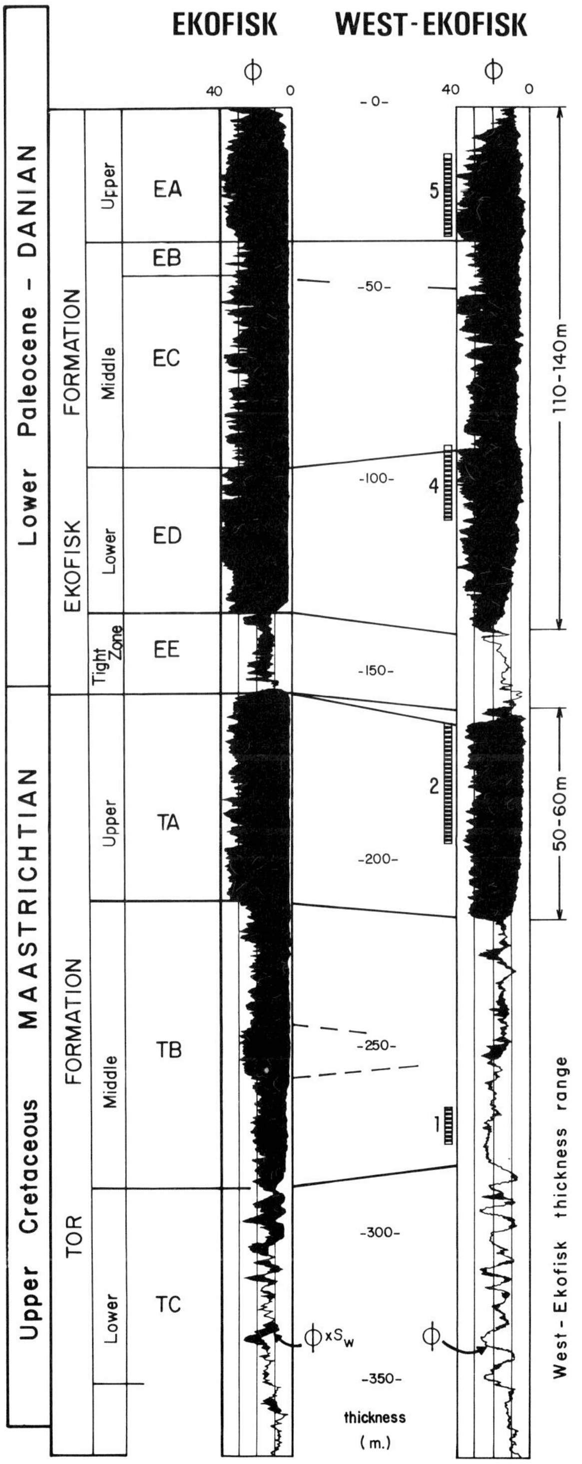

Fig. 4. Comparison of the reservoir intervals in the Ekofisk and West Ekofisk crestal wells (porosity and hydrocarbon saturation logs).

GEOLOGICAL SETTING

The West Ekofisk Field lies in the North Sea Central Graben which originated during early Mesozoic times. Thick Triassic and Jurassic continental to marine strata were deposited over Permian salt which was subject to halokinesis. After the deposition of a Lower Cretaceous clastic sequence, chalk sedimentation persisted for some 60 million years until Early Paleocene times. Because of extensive subsidence, the graben became a depo-centre for the Chalk Group. The structural evolution of part of the Graben has recently been described by Gowers and Sæbøe (1985).

Repeated tectonic pulses, local inversion, halokinesis and instability of the graben margin slopes and of elements within the graben itself all resulted in the resuspension of previously deposited unconsolidated chalks. This material was transported into deeper areas by gravity flows and redeposited as 'allochthonous' chalks.

The various chalk facies in the North Sea have been described by Kennedy (1980–85). In the Ekofisk and West Ekofisk area, the Danian and Maastrichtian sediments consist mainly of allochthonous chalk, deposited principally as debris-flows, which are the most favourable facies as far as chalk reservoir quality is concerned. The relatively thin intervals of interbedded 'normal' authochthonous pelagic chalk have a poor to non-existent reservoir potential.

The chalk is overlain by a Tertiary sequence 3000 m thick which is similar to that of the Ekofisk Field (see Gersib and Pekot, this volume).

GEOPHYSICS

The West Ekofisk structure, like most of the other 'Greater Ekofisk' Fields now developed, was identified in the mid-1960s by reconnaissance seismic work. The seismic grid covering the area when exploration drilling began was widely spaced (Fig. 2). The drilling of the first two development wells, although too far down on the flank to find production, nevertheless provided a good peripheral control.

The reflection quality is generally good (Fig. 5) and the extremely low seismic velo-

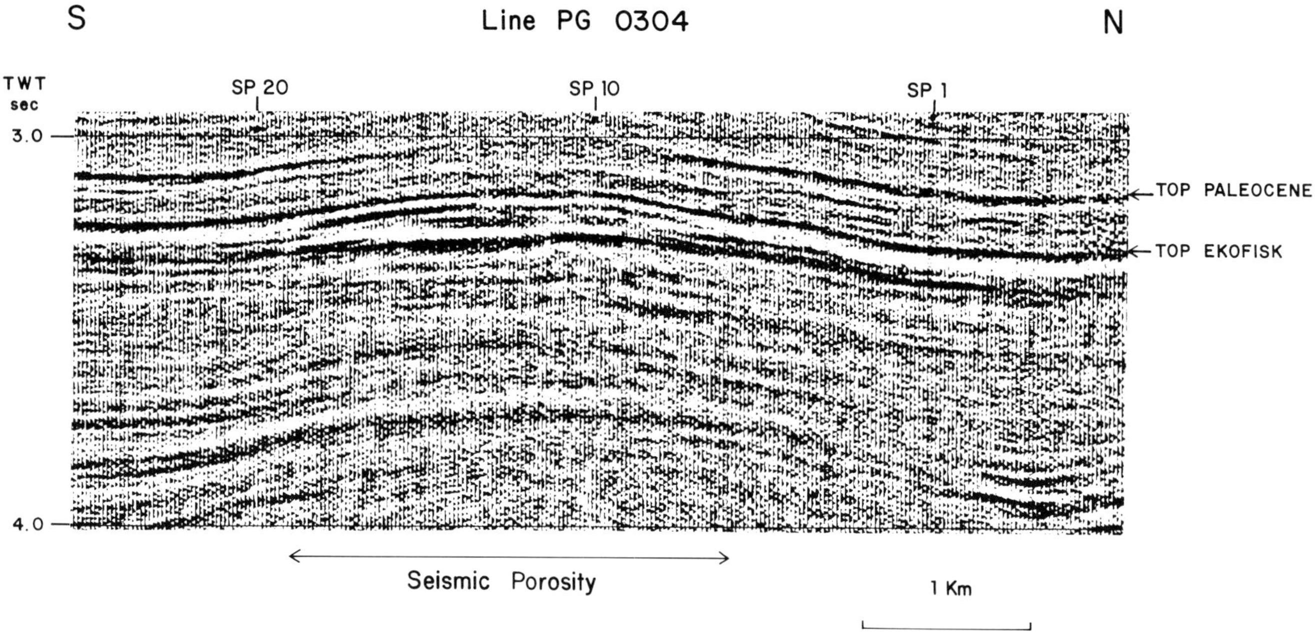

Fig. 5. Seismic line PG 0304 through West Ekofisk (location on Fig. 2).

city, which characterizes the mid-Tertiary section above the central area of Ekofisk, Valhall, Tommeliten and some other structures, is not present above West Ekofisk. The top of the reservoir interval may accordingly be satisfactorily mapped over the field area. After the discovery of the Ekofisk Field, it was recognized that high porosity in chalk creates a marked increase in amplitude.

Based on this observation, a 'seismic porosity limit' was interpreted as shown on Fig. 2, and the locations of the development wells were chosen accordingly. Unfortunately, this method was not precise enough and the first two development wells, although located inside the 'seismic porosity' area, found relatively poor quality reservoirs and were abandoned. West Ekofisk has not been subject to the considerable seismic effort and refinement that has been undertaken over most of the rest of the Greater Ekofisk area.

TRAP

The West Ekofisk dome was created by halokinetic movements of the Permian salt. The phases of upwelling which are responsible for the reservoir shape took place from Eocene to Early Miocene times. No palaeorelief existed during Danian and Maastrichtian times (see Reservoir, p. 174). Tertiary shales seal the top of the reservoir laterally and downwards, the reservoir is sealed by tight chalk as a result of late diagenesis and pressure conditions.

The structure is broadly circular and the productive area has a diameter of 4 km. Vertical closure at top chalk level is 100 m (note that downflank top chalk does not correspond to the top of the reservoir). The vertical relief at Middle Tor Formation level is 65 m. The 35-m difference between these values is the consequence of differential compaction as demonstrated below. This means that about two-thirds of the top chalk relief is due to halokinesis, while the remaining one-third is the result of differential porosity preservation.

The 'Ekofisk Tight Zone' (EE) separates the Ekofisk and Tor Formation reservoirs. The upper reservoir contains oil down to 3220 m, which is 40 m deeper than the closure at the top of the chalk level (Fig. 3). The lower reservoir (TA) is hydrocarbon bearing down to 3320 m.

RESERVOIR

The West Ekofisk Field reservoirs consist of chalk of the Ekofisk and Tor formations which are of Danian (Early Paleocene) and Maastrichtian (Late Cretaceous) ages. The carbonate fraction is essentially made up of extremely fine-grained skeletal debris (coccoliths) of unicellular pelagic algae (coccolithophoridae). The non-carbonate fraction is generally lower than 5% in the Tor Formation but locally reaches 40% in the pelagic chalk interval of the Ekofisk Formation (the

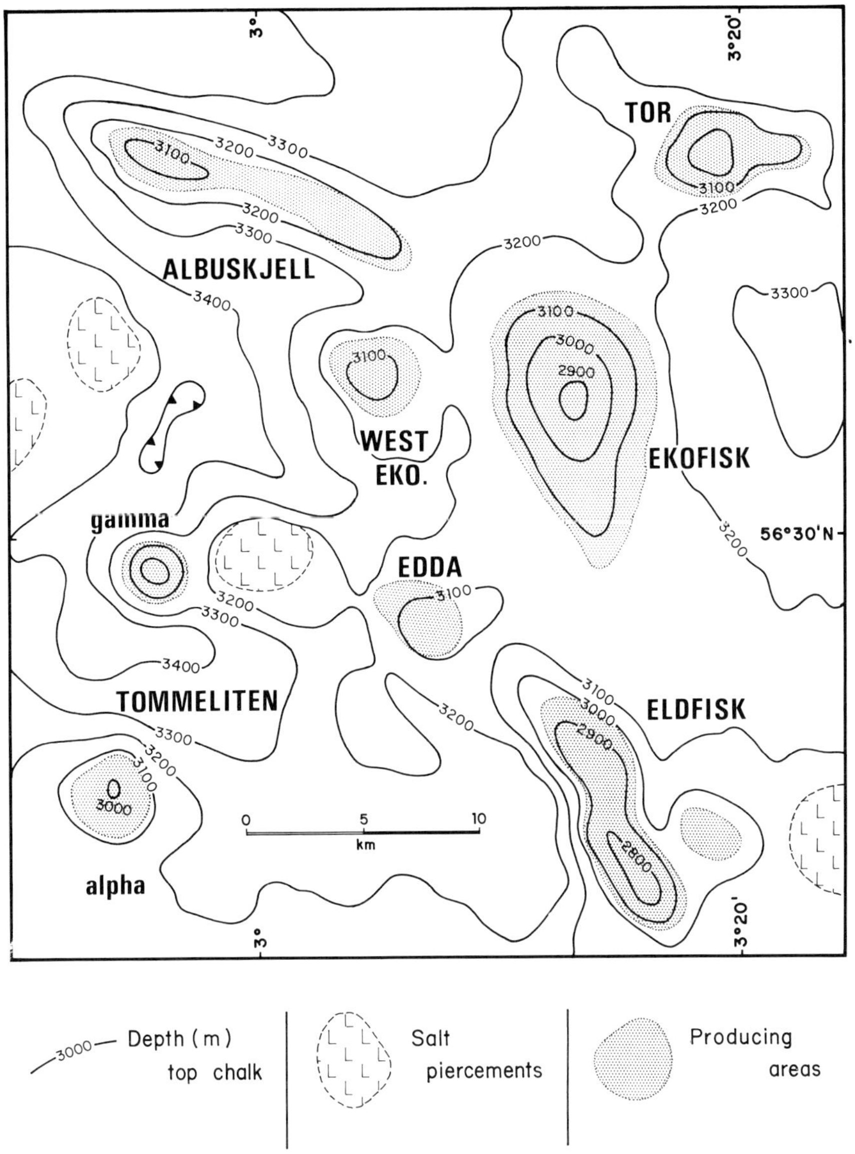

Fig. 6. Present-day structural interpretation of the West Ekofisk reservoir and its relation with the neighbouring fields.

average in the latter Formation being 7%). The non-carbonate components are mainly clay, which reduces porosity and permeability (Hardman, 1982) and quartz, which increases the rock strength (da Silva *et al.*, 1985). The petrology of the North Sea chalk has been described by Hancock (1976) and others.

As mentioned in the section on Geological setting (p. 169), the Central Graben chalk contains abundant reworked (allochthonous) material. The Ekofisk–West Ekofisk area was characterized by thick, massive, debris-flow deposition during Danian and Maastrichtian times. Debris-flows provide by far the most porous chalk sediments which, because of their low clay content, also resist compaction and porosity reduction. The reservoir interval also comprises turbidites, mud-cloud deposits and pelagic (autochthonous) chalk. The autochthonous chalk was deposited slowly, long remained near the water–sediment interface and was burrowed. It consequently has a reduced initial porosity. In addition, it contains more clay, which has an adverse effect on porosity preservation. As a result of variation in the rate and mode of sedimentation, the Ekofisk and Tor formations consist of an alternation of more and less porous chalk respectively allochthonous and autochthonous. The 'porosity stratification' clearly appears in the oil zones where the porosity has been better preserved (see Fig. 4 and Fig. 7).

A reservoir zonation of the Ekofisk and Tor formations has been established for the whole

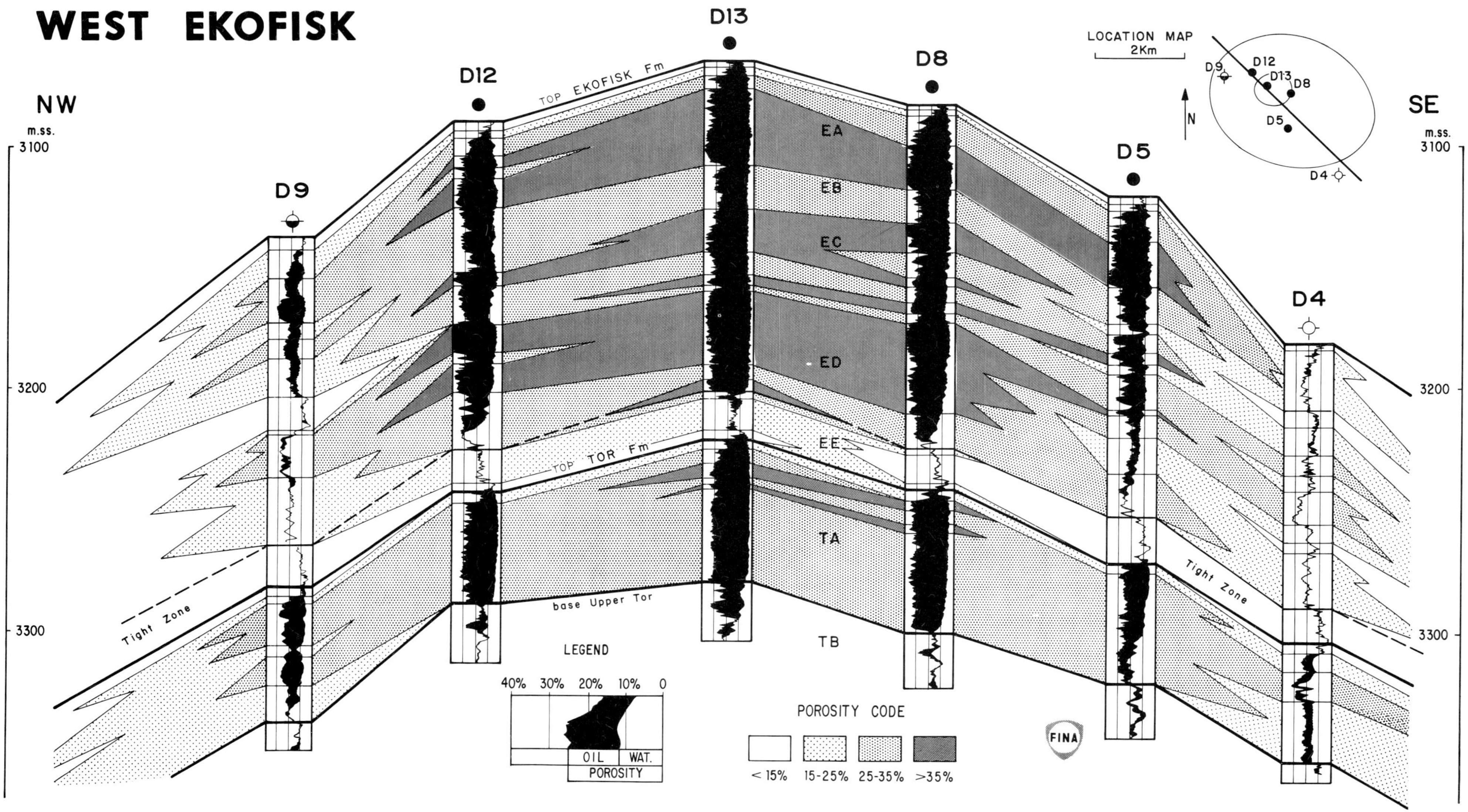

Fig. 7. Porosity cross-section. The initial stratified porosity has been altered by differential diagenesis, which is also responsible for the progressive decrease of porosity from crest to flank in each reservoir zone.

area, on the basis of the 'porosity stratification' which is believed to be isochronous. Eight main units have been recognized (Fig. 4). Each zone has a distinct ratio of allochthonous to autochthonous components. This controlled the initial porosity, which may vary laterally and is often altered (and masked) by diagenesis. For example, the typical 'Ekofisk Tight Zone' (EE) consists mainly of clay/silica-rich pelagic chalk (autochthonous) with interbeds of turbidites and thin debris-flows, while at the other end of the scale, the Upper Tor Member (TA) is essentially composed of massive debris-flows. A detailed description of the reservoir units is given by Pekot and Gersib (this volume, p. 79). It is worth noting that the ED unit of the Ekofisk Formation contains a very high percentage of redeposited Maastrichtian material which gives this interval an abnormally low clay content for a Danian chalk.

It has been estimated that, at a depth of 3000 m, the porosity of the chalk should normally have been reduced to 15% or less by mechanical and by 'chemical' compaction (i.e. pressure solution and reprecipitation of calcite in adjacent pore space). The chalk of West Ekofisk has porosities of up to 40%. The very high degree of porosity preservation is due to several factors which inhibited chemical compaction. These are mainly (1) favourable sedimentary facies (Kennedy, 1980), (2) 'spot welding' at contact points between laths, which may create a rigid framework (Mapstone, 1975), (3) low clay content (Hardman, 1982), (4) nature of pore fluid (Neugebauer, 1974), (5) development of overpressures (Scholle, 1977), and (6) early entry of oil in the reservoir (several authors).

In the West Ekofisk Field, the lateral porosity variation in the various reservoir units is solely the result of differential 'chemical compaction' since chalk facies do not vary in this small area (similarity between porosity logs of the crestal wells in West Ekofisk and Ekofisk).

For each layer the porosity decreases from crest to flank, due to the combined effects of variation of pore pressure with depth, and the progressive infill of the reservoir by oil, from top to bottom, with subsequent inhibition of the 'chemical compaction'. Such inhibition occurred at progressively deeper levels and at lower porosity values. The deeper the reservoir is on the flank, the longer it remains exposed to 'chemical compaction' in the aquifer. The porosity reduction from crest to flank, i.e. from the central area of the field to its periphery, is illustrated on the porosity cross-section (Fig. 7). The similarity between the structural shape and the average porosity variation is shown in Fig. 8 of the total Ekofisk Formation. Average porosity maps of the various reservoir units show the same clear relationship (D'Heur, 1980).

The porosity reduction of the chalk being the result of calcite transfer from sediment grain to pore space, normally corresponds to thickness reduction. The initial thickness variation of a chalk layer may be understood after having removed the effect of differential porosity preservation from the present-day thickness. Assuming that calcite transfer does not occur over long distances, the relationship between the initial (depositional) thickness, D, and the present-day thickness, P, is given by

$$D = P\left(\frac{1-\varnothing_p}{1-\varnothing_D}\right)$$

where $\varnothing_D$ and $\varnothing_p$ are respectively the average

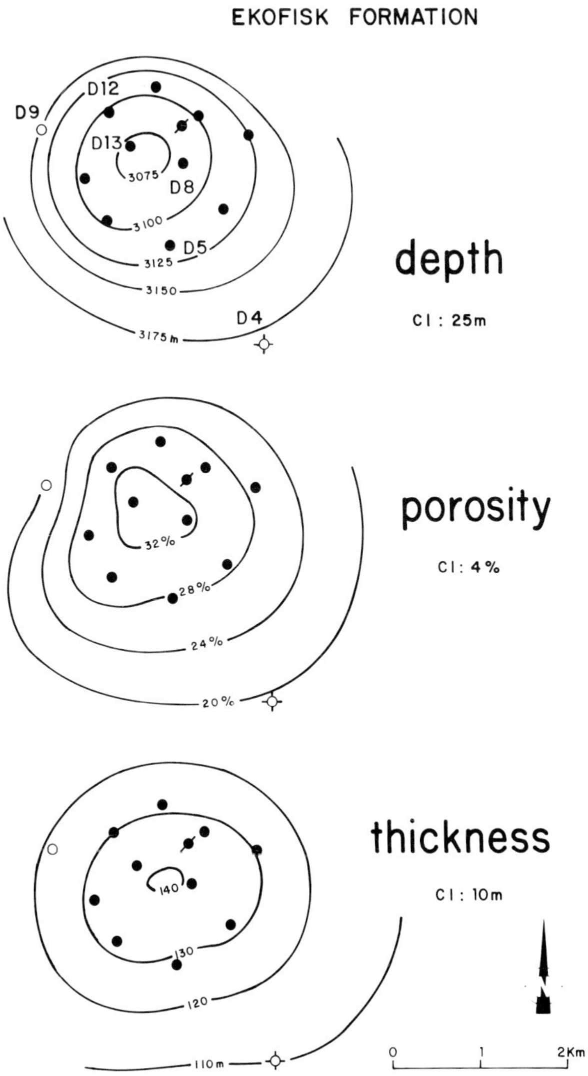

Fig. 8. Ekofisk Formation structure, average porosity and gross thickness maps.

original porosity and present-day porosity of the chalk bed.

Because there was no lateral facies variation and consequently no original porosity variation in the small West Ekofisk area, a constant value of original porosity may be assumed. If it is taken at 70%, then

$$D = P\left(\frac{1-\varnothing_p}{0.3}\right)$$

In the case of West Ekofisk, this value is nearly constant; at least for the main reservoir

in crestal Well D8: $D = 136\left(\frac{1-0.33}{0.3}\right)$

in flank Well D4: $D = 110\left(\frac{1-0.19}{0.3}\right)$
$= 300\ \text{m} \pm 1\%$

Thus, the depositional thickness over the field was constant and the present-day thickness variation is the result of differential porosity-preservation only. This also suggests that the sedimentation was not influenced by structuring in the West Ekofisk area during the Danian times.

The 'gross interval' thickness reduction towards the periphery results in a thinning of the 'net reservoir' in the same direction. This is reinforced by the reduction of porosity from crest to flank.

Another peculiarity of several chalk fields is the thickening of the oil–water transition zone form the central area towards the periphery. This clearly appears in West Ekofisk (Fig. 9) and may be the consequence of variation of pore morphology. For similar porosity values, the hydrocarbon saturation is much higher in the flank than in the central area. Along the axis of folding the chalk may be in extension (upper layers) with resulting pore-throat opening and possible microbrecciation. By contrast, on the flank, where the chalk may be in compression, the pores are flattened and the pore-throat size is reduced ('tectonic facies'). This explains the bad results of wells D4 and D9 despite the fact that they penetrated the chalk layers well above the 'base of hydrocarbons' identified in the exploration well.

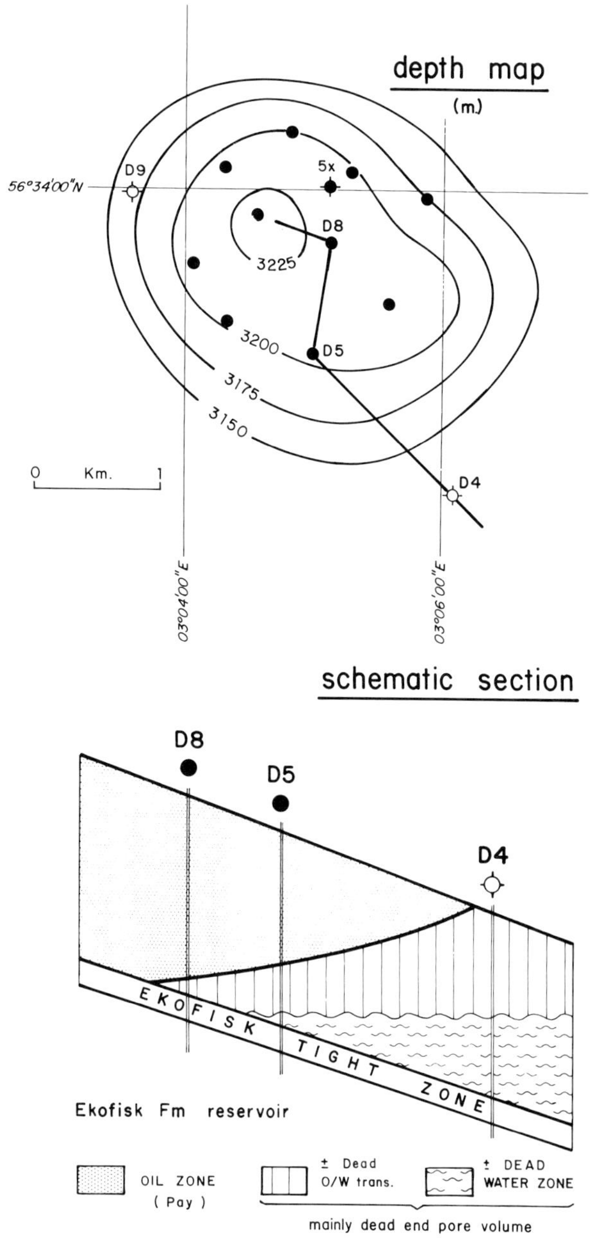

Fig. 9. Map and section showing the geometry of the base of the Ekofisk formation pay zone. It rises up in the flank by more than 75 m in less than 1 km.

SOURCE-ROCK

The Upper Jurassic Mandal Formation, which consists of highly bituminous marine shale, is known to be the source-rock for the Greater Ekofisk fields. West of West Ekofisk, significant oil generation commenced as early as Paleocene times. In that area, the Upper Jurassic is deeper than 5000 m and the source-rock is mature enough to have produced rich gas–condensates.

West Ekofisk is located nearer to the graben axis than is Ekofisk. It is also slightly deeper, and there is only a saddle between the two structures. Before the chalk lost its permeability in the saddle area as a result of late diagenesis, oil may have been flushed by gas from West Ekofisk to Ekofisk. Another point is that in West Ekofisk the Tertiary seal has been perfectly effective, whereas in Eko-

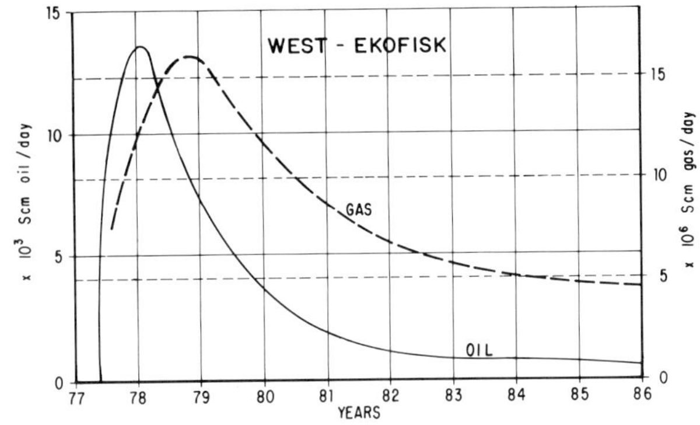

Fig. 10. Production history.

fisk, a significant amount of gas may have dismigrated into the overburden.

HYDROCARBONS

West Ekofisk is a rich gas–condensate field. The initial GOR was 731 Sm^3/Sm^3. It increased to 3200 Sm^3/Sm^3 in 1979 and had reached 7100 Sm^3/Sm^3 by end of 1985. The hydrocarbon distribution within the reservoir is directly related to porosity, structural elevation and distance from the structure axis, as mentioned before.

RESERVES

The reserves of the West Ekofisk Field are now estimated at about 28×10^9 Sm^3 gas and 10×10^6 Sm^3 condensate. The field came on-stream in 1977. The production history is shown in Fig. 10. The best producing well flowed at 2400 Sm^3 oil/day and 3.4×10^6 Sm^3 gas/day from a 70-m, interval through a 1-inch choke. The peak production for oil was 13.5×10^3 Sm^3/day in early 1978 and for gas 15×10^6 Sm^3 day in late 1978, with 10 producing wells. By early 1986, 21.7×10^9 Sm^3 gas and 8.6×10^6 Sm^3 oil had been produced. The field production is transported to the Ekofisk Centre by a 24-inch oil and gas pipeline.

ACKNOWLEDGEMENTS

This article being (alphabetically) the last of a series of papers published by the author in this volume, the author wishes to take this opportunity to thank those who helped: A. M. Spencer, the Editor, spent a considerable amount of time improving the articles; C. J. Campbell (Fina Exploration Norway) and F. da Silva and W. H. Ziegler (Petrofina) made very useful comments. C. Gobeau and J. de Lucena drew all 60 figures. Thanks also to D. Moens and J. M. Questiaux for assistance and to Petrofina, Phillips, Agip, Elf, Norsk Hydro and Total who authorized the publication of the papers.

REFERENCES

da Silva, F., Debande, G., Detiege, C., Halleux, L., Monjoie, A., Poot, B., Schroeder, C. 1985. Mechanical behaviour of chalks, *The Chalk Research Program Symposium*, Stavanger, May 21–22 (proprietary report).

D'Heur, M. 1980. Chalk reservoir of the West Ekofisk Field, In *The Sedimentation of the North Sea Reservoir Rocks*, Norsk Petroleumsforening, Article X.

Gowers, M.B. and Sæbøe, A. 1985. On the structural evolution of the Central Trough in the Norwegian and Danish sectors of the North Sea, *Marine and Petroleum Geology* **2,** (4), 298–318.

Hancock, J.M. 1976. The petrology of the Chalk, *Proceedings of the Geologists Association* **86** (for 1975), 499–535.

Hardman, R.F.P. 1982. Chalk Reservoirs of the North Sea, *Bulletin. Geological Society of Denmark* **30,** 119–137.

Kennedy, W.J. 1980. Aspects of chalk sedimentation in the Southern Norwegian Offshore, In *The Sedimentation of the North Sea Reservoir Rocks*, Norsk Petroleumsforening, Article IX.

Kennedy, W.J. 1985. Sedimentology of the Late Cretaceous and Early Paleocene Chalk Group, North Sea Central Graben. *Chalk Research Program Symposium*, Stavanger, May 21–22 (proprietary report).

Mapstone, N.B. 1975. Diagenetic history of a North Sea chalk, *Sedimentology* **22,** 601–613.

Neugebauer, J. 1974. Some aspects of cementation in chalk, In: Hsü, K.J. and Jenkyns, H.C. (Eds), *Pelagic Sediments on Land and Under the Sea*, Special Publication. International Association of Sedimentologists **1,** pp. 149–176.

Pekot, L.J. and Gersib, G.A. 1987. Ekofisk, this volume, pp. 73–87.

Scholle, P.A. 1977. Chalk diagenesis and its relation to petroleum exploration: oil from chalks, a modern miracle? *Bulletin of the American Association of Petroleum Geologists* **61,** 982–1009.

13

7/11–5 Discovery

J. E. Strand and P. R. Slaatsveen*

Norsk Hydro, P.O. Box 200, N-1321 Stabekk, Norway
**Present address: Deminex (Norge) A/S, P.O. Box 1767 Vika, N-0167 Oslo 1, Norway*

The 7/11–5 discovery is located at the boundary between the Southern Vestland Arch and the Central Trough, 5 km north-east of the Cod Field (Block 7/11) and 20 km west of the Ula Field (Block 7/12). The oil discovery was made in 1982. The reservoir is the shallow-marine Ula Formation sandstone. Depth to the reservoir is 4150 m RKB, and the water depth is approximately 80 m.

An oil–water contact is identifiable in the well, but lies directly above a tight shale horizon, and hence a larger oil column may be present. The recoverable resources of the field are in the range $2–10 \times 10^6$ Sm^3 of oil and $1–4 \times 10^9$ Sm^3 of associated gas dependent on where the true oil–water contact of the field is.

INTRODUCTION

The 7/11-5 structure is located at the boundary between the Southern Vestland Arch and the Central Trough, 5 km north-east of the Cod Field and 20 km west of the Ula Field (Fig. 1.). The structure lies in the central part of Block 7/11. The resources proven by Well 7/11-5 lie entirely within Block 7/11B operated by Norsk Hydro, while a portion of the structure deeper than 4180 mss extends into Block 7/11A, operated by Phillips Petroleum.

The structure covers 26 km^2 down to the structural spill-point and the water depth is 80 m.

HISTORY

Block 7/11 was awarded under Licence 070 in the spring of 1981 to a group consisting of:

Norsk Hydro	25.0%	(operator)
Statoil	50.0%	
Saga	10.0%	
Conoco	7.5%	
Mobil	7.5%	

Licence 070 consists of the relinquished part of Licence 018; the retained part contains the Cod field and is operated by Phillips Petroleum.

To date nine wells have been drilled in Block 7/11, of which four were drilled by the Licence 070 group. The structure was discovered in 1982 by the first Well, 7/11-5, drilled by the new group. The well proved oil in sandstones of the Upper Jurassic Ula Formation. The Phillips group tested the Cretaceous in the same structure in 1969 with Well 7/11-4, 4X, but the well was terminated above the Jurassic.

GEOLOGICAL SETTING

The Rotliegendes volcanics in the Lower Permian, which have been found in several wells in the Central Graben, probably represents the earliest opening of the Northern

Geology of the Norwegian Oil and Gas Fields © A. M. Spencer *et al.*, eds (Graham & Trotman, 1987) pp. 177–183.

11–5 Discovery Field data summary

Trap	
Type	Salt-induced dome
Depth to crest	4040 mss
Lowest closing contour	4260 mss
Hydrocarbon contacts	oil up to 4129 mss
	oil down to 4175 mss
Oil column	> 46 m
Pay zone	
Formation	Ula Unit I
Age	Upper Jurassic
Gross thickness	46 m
Net/gross ratio	0.8
Cut off for N/G	$S_w = 50\%$; $Ø = 12\%$; $V_{sh} = 40\%$
Porosity	17·5%
Hydrocarbon saturation	74.1%
Permeability (average; range)	11 mD; 1–320 mD
Productivity index	3.6 Sm^3/day/bar (7/11-5)
Hydrocarbons	
Oil gravity	0.83 g/cm^3
Oil type	undersaturated
Gas gravity	0.88 (air = 1)
Bubble point	319 bar
Gas/oil ratio	267 Sm^3/Sm^3
Formation volume factor	1.75
Formation water	
Salinity	190 000 ppm NaCl equivalents
Resistivity	0.013 ohm m at 138 °C
Reservoir conditions	
Temperature	161 °C
Pressure	590 bar
Pressure gradient in reservoir	0.051 bar/m
Field size	
Area of closure	26 km^2 (down to spill point)
Gross rock volume	$350–900 \times 10^6$ m^3
Hydrocarbon pore volume	$38–97 \times 10^6$ m^3
Recovery factor — oil	20–35%
— gas	30–50%
Drive mechanism	solution gasdrive/waterdrive
Recoverable oil	$2–10 \times 10^6$ Sm^3
Recoverable gas	$1–4 \times 10^9$ Sm^3 (associated gas)
Production	

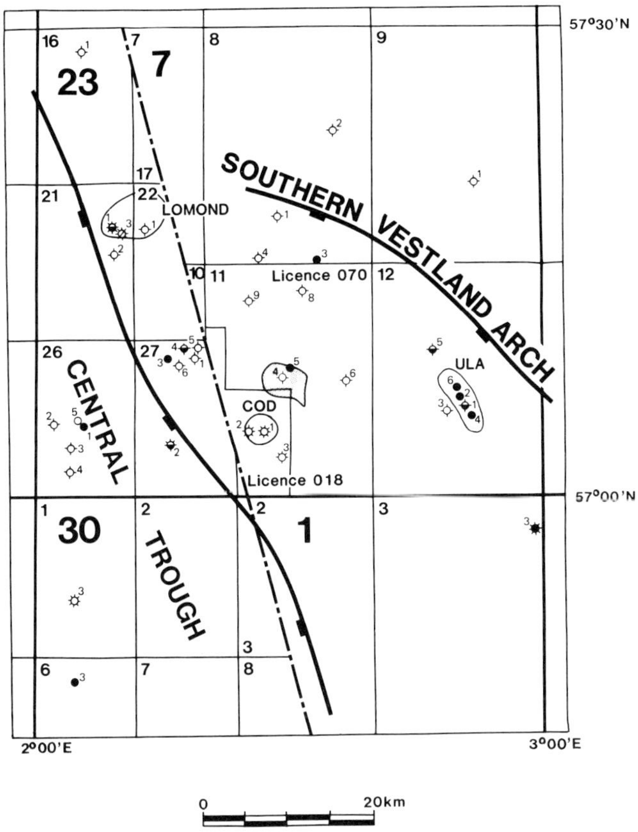

Fig. 1. Location map

Permian Basin. This early rifting was followed by the deposition of thick Zechstein salt in the Late Permian times over the entire area.

In early Triassic times basin subsidence led to deposition of thick sequences of fine-grained sediments. The resulting differential loading, combined with faulting, initiated salt movements. The Triassic strata broke up into large plates that slid on the salt until at some point they rested on the base of the salt. At the plate edges the salt forced its way through the Triassic strata leading to the formation of dome-shaped structures and salt ridges. The first formation of the 7/11-5 structure may have started at this time.

The Triassic lithology is shales, silts and fine-grained sandstones. In some of the Triassic section anhydrites, dolomites and calcites occur. The Triassic depositional environment was continental, in parts tending towards a sabkha environment.

The Triassic structural framework was reactivated by a new tectonic pulse, probably at the end of Middle Jurassic times and the Lower Triassic plates were additionally rotated. The salt was reactivated and the 7/11-5 structure probably underwent additional doming. This tectonic event, together with a marine transgression, resulted in a reworking of much of the Middle Jurassic and some of the Triassic sediments. This was followed by the deposition of Upper Jurassic shallow-marine sandstones, which are thickest on the deeper sides of the rotated plates. As the wells in the 7/11 area are all drilled on structural highs, the contact between the Triassic and the Ula sandstone formation is clearly erosional. Erosional products can be seen as clasts several centimetres in diameter in the lower parts of the Ula Formation. Also, in Well 7/11-5 a sharp erosional contact can be seen. By the end of Jurassic times the transgression had covered the entire area and the organic-rich shales of the Mandal Formation were deposited.

Early Cretaceous times were mostly dominated by a low-energy environment with deposition of marine shales, marls and limestones. The Late Cretaceous was a transgressive period. Subsidence in the Central Trough led to the formation of a chalk depocentre to the south-west of the 7/11-5 structure. Owing to sedimentary loading, the salt was again activated and was squeezed up against major faults.

Paleocene sands, as found on the Cod Field, were almost certainly sourced from the west and cover only the south-western parts of Licence 070. (They pinch out on the flank of the 7/11-5 structure.) Most of the structuring reached its present form during Early Tertiary times due to reactivated halokinesis. This was followed by strong subsidence and the influx of large quantities of clays and siltstones throughout the Tertiary. A generalized stratigraphic column is shown in Fig. 2.

GEOPHYSICS

Several seismic surveys have been shot over the 7/11-5 structure, both by the former operator Phillips and by Norsk Hydro. Two surveys of relatively new date form the basis for the present interpretation. In 1983, Norsk Hydro collected a new seismic survey over the southern parts of Block 7/11 (NH 8302). The data are of excellent quality and provide a grid of roughly 1 × 1 km. Saga Petroleum, in 1980, shot a survey (SG 8048) which is also of excellent quality and provides a very good supplement to the NH 8302-survey.

The upper oil-saturated part of the reservoir in Well 7/11-5 can be tied to the seismic sections and followed down-dip towards the north (Fig. 3). On the other compartments of the structure the identification of the sandstone is difficult. A top Ula sandstone struc-

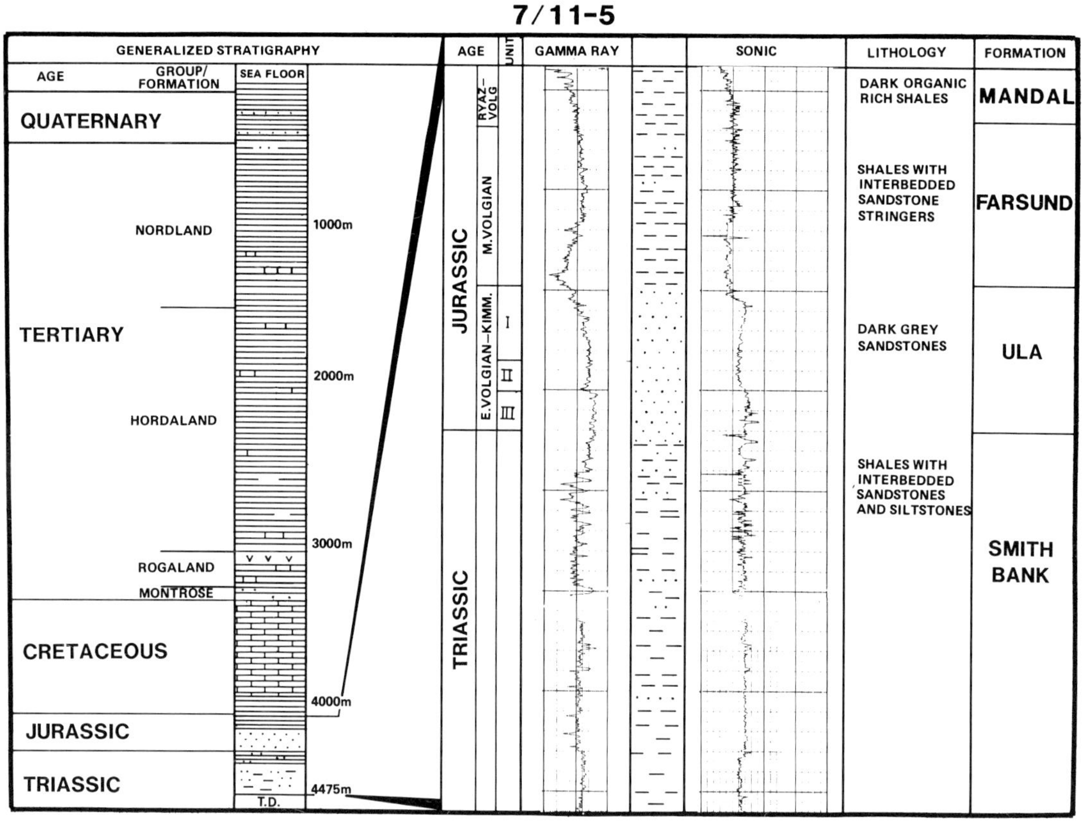

Fig. 2. Generalized stratigraphy of Well 7/11-5

tural time map has been made, however, using information from Wells 7/11-6 and 7. This map was later depth-converted using a linear time-depth relation derived from 11 wells in the area. Figure 4 shows the Top Ula sandstone depth map.

Also, interpretation of the Base Cretaceous 'unconformity' and of an intra-Triassic marker has helped in the understanding both of the structural configuration and of the distribution of the Ula sandstone.

TRAP

The 7/11-5 structure is a salt-induced structure with relief of about 220 m. The crest of the structure is at about 4040 m, while the spill-point is at 4260 m. The structure is divided into three main compartments: A, B and C (Fig. 4).

Well 7/11-5 was drilled on the northern compartment (A). The well encountered 46 m of oil saturated Ula Sandstone. The base of the oil column in this well (4175 m) corresponds to a contact with underlying tight rocks (i.e. it is an oil-down-to level: ODT). This may mean that the true oil–water contact of the field may be as low as the structural spill-point of 4260 m. This compartment of the structure contains about 72% of the recoverable reserves if the 7/11-5 ODT represents the true oil–water contact, and 53% if the structure is filled to structural spill-point.

The second structural compartment is to the south-west (B). It is separated from the northern compartment by a fault of about 40 metres' throw. However, this fault appears not to seal off the migration path, so that this compartment probably has the same oil–water contact as the northern compartment. If this is so the south-western compartment would contain 28% of the recoverable reserves (4175 m ODT) or, if filled to structural spill-point, 26%.

The third and eastern compartment is the most uncertain (C). The compartment is separated from the rest of the field by an important fault, and the sand has probably been eroded on the crest of the compartment. This part of the structure contains only minor reserves above the 7/11-5 ODT level, while if the structure were filled down to the structural spill-point, it would contain 20%.

RESERVOIR

The reservoir of the 7/11-5 structure is a sandstone in the upper portion of the Upper

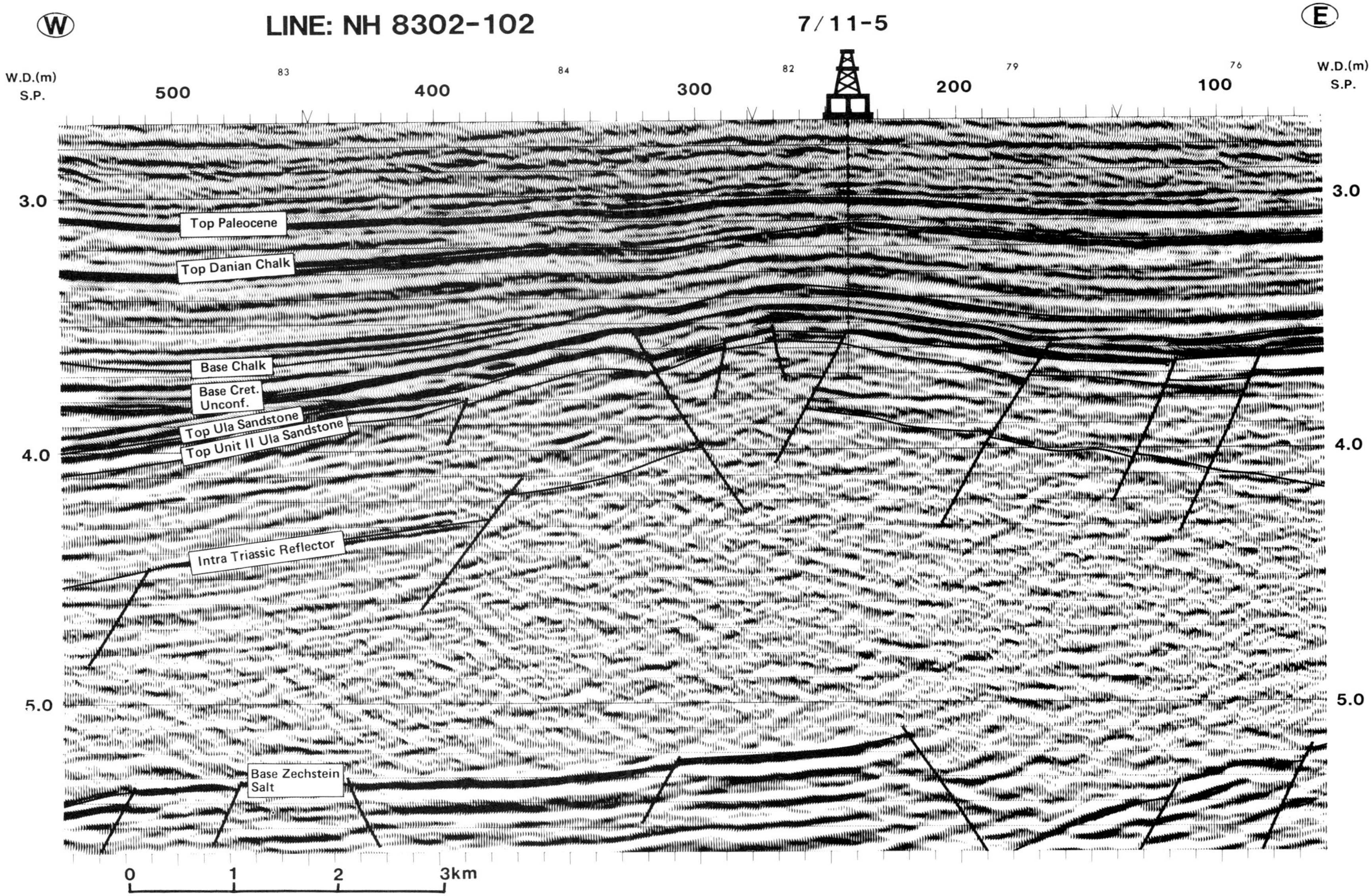

Fig. 3. Seismic cross-section through Well 7/11-5

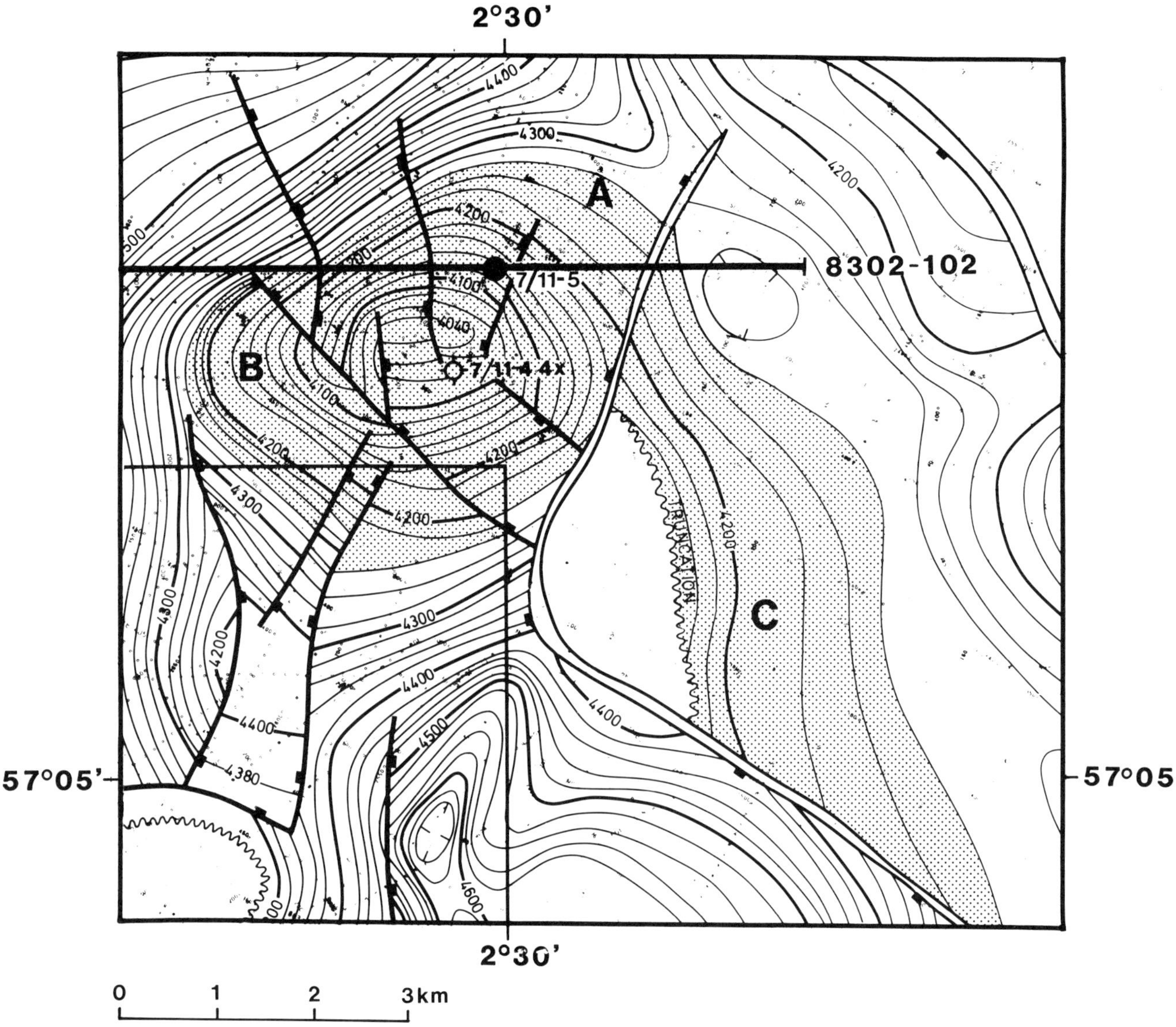

Fig. 4. Structural depth map, top reservoir.

Jurassic Ula Formation. The original sands were shallow marine deposits derived from eroded Middle Jurassic and older sediments and were mainly storm deposits. The formation was laid down in a 'fairway' about 100 km long and 20 km wide. The 'fairway' is delineated to the north-east by the Southern Vestland Arch and to the south-west by a line of salt diapirs.

In a regional study, the Ula sandstone has been divided into three members, which were in turn subdivided into seven units (Fig. 2):

1. The Upper Fine Member — Unit I;
2. The Middle Clean Member — Units II–III;
3. The Lower Silty and Shaly Member — Units IV–VII.

Only the Upper Fine and the Middle Clean Members and their corresponding units I-III are present in Well 7/11-5.

Unit I, which is usually a rather poor reservoir, is the oil-saturated part of the sequence. It has some secondary porosity, most probably due to dissolution of the frequent *Rhaxella* sponge spicules that characterize the formation.

Unit II is separated from Unit I by a thin shale zone. The porosity is severely reduced by silica and/or calcite cementation. Only residual oil seems to be present in this unit.

Unit III has poor reservoir characteristics, and is also water wet.

The oil-filled Unit I consists of very fine- to medium-grained sands and mainly argillaceous sandstones. The sandstones are occasionally micaceous and consist of individual fining-upwards beds in an overall fining-upwards sequence. The unit is severely bioturbated and very few sedimentary structures are preserved. *Rhaxella* sponge spicules are frequent.

SOURCE

The source-rock for the oil in the 7/11-5 discovery is thought to be the Upper Jurassic shales of the Mandal Formation and possibly the Farsund Formation. The oil is most probably sourced from the local basins within Block 7/11. The Upper Jurassic shales reach a thickness of about 150 m in the north-western parts of the block.

HYDROCARBONS

Two drill-stem tests were performed in the reservoir:

- DST-1 was carried out in the interval 4185–4197 m RKB. The flow rate was 470 Sm^3/day of oil and 118 000 Sm^3/day of gas. The oil gravity was 0.83 g/cm^3 and the gas gravity 0.876.
- DST-2 was performed in the interval 4165–4174 m RKB. No hydrocarbons flowed to surface, but a flow rate of 45 Sm^3/day of oil was calculated from the flow/unloading of the water cushion.

RESOURCES

The possible recoverable resources are estimated to be in the range of $2–4 \times 10^6$ Sm^3 (OWC equal to the ODT level of 4175 m in Well 7/11-5). If, alternatively, the oil–water contact lies at a depth between 4175 m (ODT level in Well 7/11-5) and 4260 m (structural spill point), the recoverable resources are in the range of $4–10 \times 10^6/Sm^3$.

The wide range in the resource estimation reflects the uncertainty in locating the OWC and the possible communication with the eastern fault compartment (C).

ACKNOWLEDGEMENTS

The authors would like to thank the partners in Licence 070 for their permission to publish this article. We also would like to thank Per Aure, Jørn Christiansen, Martin B. Gowers and Finn Erik Skaar for valuable advice.

Part III—Egersund Sub-Basin

14

Bream and Brisling

Michel D'Heur and Laurent de Walque

Petrofina, c/o Fina Exploration Norway, Skogstøstraen 37, N-4001 Stavanger, Norway

The Bream and Brisling finds are located in Norwegian Block 17/12, 135 km south-west of Stavanger. The block straddles the boundary between the Sele High and the Egersund Sub-basin. Oil was discovered in the Bream structure, within the Sub-basin, in 1971 and in the Brisling structure, on the High, in 1973. The reservoirs are sandstones of Middle Jurassic age. These are overlain by shales of the Boknfjord Formation which form the seal. The latter formation contains 'hot shales' of Kimmeridgian to early Volgian age identified as the source-rock for both accumulations. Maturity is the main problem in the area, as reflected by the small amount of oil present in the traps. There are no plans to develop Bream and Brisling, owing to their small size and their distance from the nearest commercial fields, pipeline and the coast.

INTRODUCTION

The Jurassic Bream and Brisling oil fields lie in 100 m of water, 135 km south-west of Stavanger. They are entirely contained in Block 17/12 and cover an area of 20 km^2 and 5 km^2 respectively. Although only 14 km apart, their geological settings are very different. Bream is located in the Egersund Sub-basin, while Brisling lies on the Sele High (Fig. 1). The reservoirs are sandstones of Callovian age overlain by Upper Jurassic shales which act both as source-rocks and seal. The combined resources are estimated to be less than 2×10^6 Sm^3 of oil (Norwegian Petroleum Directorate, 1980).

The two discoveries received their respective names through a combination of two practices of the Phillips Group in Norway. Licences or parts thereof were designated by a letter and all prospects in the block receive fish names starting with the said letter. In the case of Block 17/12, the designated letter was B, hence the names Bream and Brisling. There are no plans to develop Bream and Brisling because of the small amount of oil contained in each structure and their distance from the nearest commercial fields, pipeline and the coast.

HISTORY

The Production Licence O16 which contains Blocks 16/11 and 17/12 was granted to Phillips, Petrofina and Agip in 1965 during the first round of licencing in Norway. The participating interests in this licence are:

Phillips Petroleum Company Norway (operator)	36.96%
Norske Fina A/S	30%
Norsk Agip A/S	13.04%
Elf Aquitainc Norge A/S	8.01%
Norsk Hydro A/S	6.7%
Total Marine Norsk A/S	4.05%
Others	1.34%

The first well on the block, Well 17/12-1 was drilled near the crest of a salt induced structure, called Bream, which has an areal closure of 80 km^2 at the Mid-Jurassic reflector

Geology of the Norwegian Oil and Gas Fields © A. M. Spencer *et al.*, eds (Graham & Trotman, 1987) pp. 185–192.

Bream–Brisling Fields data summary

Bream

Trap	
Type	Dome structure over salt swell
Depth to crest	2260 m (top of reservoir)
Lowest closing contour	2400 mss
Oil–water contact	2315 mss
Oil column	55 m
Pay zone	
Formation	Sandnes Fm (sandstone)
Age	Callovian
Gross thickness	51 m (crest)
Net/gross ratio	0.3 (crest)
Porosity (average net)	22%
Hydrocarbons	
Oil gravity	0.855–0.882 g/cm^3
Formation water	
—	
Reservoir condition	
—	
Field size	
Area	20 km^2
Recoverable oil	NPD annual report 1980 classifies this field in the $\leqslant 1 \times 10^6$ Sm^3 category, which is pessimistic
Production	
—	

Brisling

Trap	
Type	faulted anticline
Depth to crest	2120 m (top of reservoir)
Lowest closing contour	2250 mss
Oil–water contact	2138 mss
Oil column	18 m
Pay zone	
Formation	Sandnes Fm (sandstone)
Age	Callovian
Gross thickness	8 m
Net/gross ratio	0.9
Porosity (average net)	20%
Hydrocarbons	
Oil gravity	0.888 g/cc
Formation water	
—	
Reservoir condition	
—	
Field size	
Area	5 km^2
Recoverable oil	NPD annual report 1980 classifies this field in the $\leqslant 1 \times 10^6$ Sm^3 oil category
Production	
—	

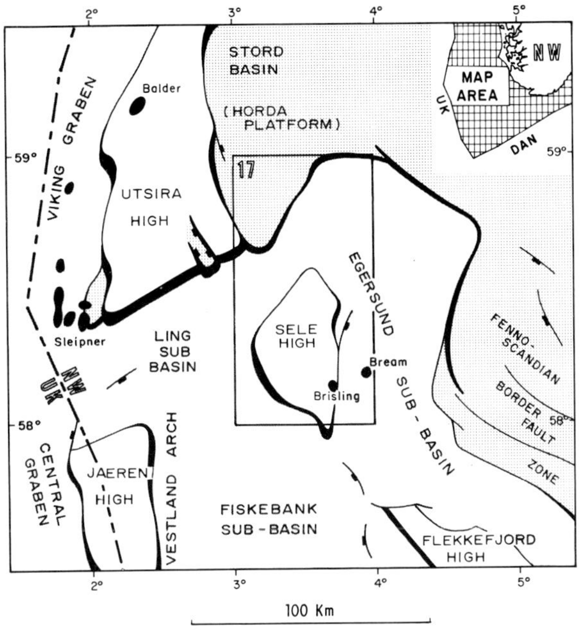

Fig. 1. Location map and structural setting.

level (Fig. 2). Middle Jurassic sandstones were the primary target with Lower Cretaceous and Triassic sandstones being secondary objectives. The well reached the Middle Jurassic sandstone in April 1971 at a depth of 2264 mss, 150 m lower than prognosed. Two sandstone bodies of Callovian age, totalling 15 m of net oil-bearing reservoir, were tested. Drill-stem test number 6 gave the best result: a rate of 160 Sm^3/day of 32.4° API (0.863 g/cc) oil through an 8/64-inch choke from a 15-m perforated interval. The lower sandstone had an oil/water contact at 2315 mss. The older Bathonian to Bajocian sandstones and the Triassic sandstones were found to be water bearing. The well bottomed in Permian salt at a depth of 4271 m.

The second well, 17/12-2, was spudded in August 1973 on a separate structure named Brisling, 14 km west of Bream. The areal closure at Mid-Jurassic level was approximately 40 km^2. The Middle Jurassic sandstone was found at a depth of 2130 mss, was Bathonian to Callovian in age, 13 m thick and had an oil/water contact at 2138 mss. No other Mesozoic sandstones were intersected in this well, and Bajocian strata were found directly above Zechstein carbonates. Below the latter, the well encountered 7.5 m of Lower Permian sandstone and bottomed at 2334 m in strata of presumed Devonian age. An interval of sandstone four metres thick was tested and flowed at a rate of 366 Sm^3/day through a 24/64-inch choke.

The third well, 17/12-3, was drilled in December 1979 to appraise the Bream discovery. The well was located on the western flank of the structure 4 km west of Well 17/12-1. The objective was to investigate the possibility of additional sand development down structure. The Mid-Jurassic reservoir was intersected at a depth of 2352 mss, but the Callovian sandstone was found to be thinner, less porous than in Well 17/12-1, and contained no oil. No test was carried out and the well was abandoned at 2705 mss in Triassic red beds.

GEOLOGICAL SETTING

Bream is located in the south-western part of the Egersund Sub-basin, while Brisling lies near the eastern edge of the Sele High. The Egersund Sub-basin is part of a system of 'inner' basins of Paleozoic to late Mesozoic ages which lie between the stable cratons of Fenno-Scandia and the main Mesozoic rift system of the North Sea. The Tectonic evolution of the area has been discussed by Pegrum (1984) and Hamar *et al.* (1982). The nomenclature used here is that of Deegan and Scull (1977), as revised for the Triassic and Jurassic units of the Norwegian North Sea by Vollset and Doré (1985).

Little is known about the pre-Permian section since most of the Paleozoic sediments have been removed by erosion from much of the area. In the Brisling well, 35 m of sandstone of questionable Devonian age has been penetrated.

After the deposition of continental sediments during Early Permian times (8 m of fine-grained sandstone in the Brisling well), marine incursions in late Permian times gave rise to the cyclic deposition of evaporites. Well 17/12-2 found 20 m of dolomite and limestone on the Sele High and Well 17/12-1, in the Sub-basin, bottomed in the Zechstein salt. It appears that a topographical difference already existed between these two areas before Jurassic times; however, the present-day dissimiliarities between the sections encountered in the Bream and Brisling wells are mainly the result of differential movements and erosion from Mid-Jurassic times onwards (Fig. 3). Whereas Mid-Jurassic sandstones lie on Permian carbonates on the Sele High, more than 1600 m of Triassic alluvial deposits of the Skagerrak Formation are present in Well 17/12-1. This well has been chosen as the type location for the Triassic sequence of the Norwegian–Danish Basin by Jakobsson *et al.* (1980).

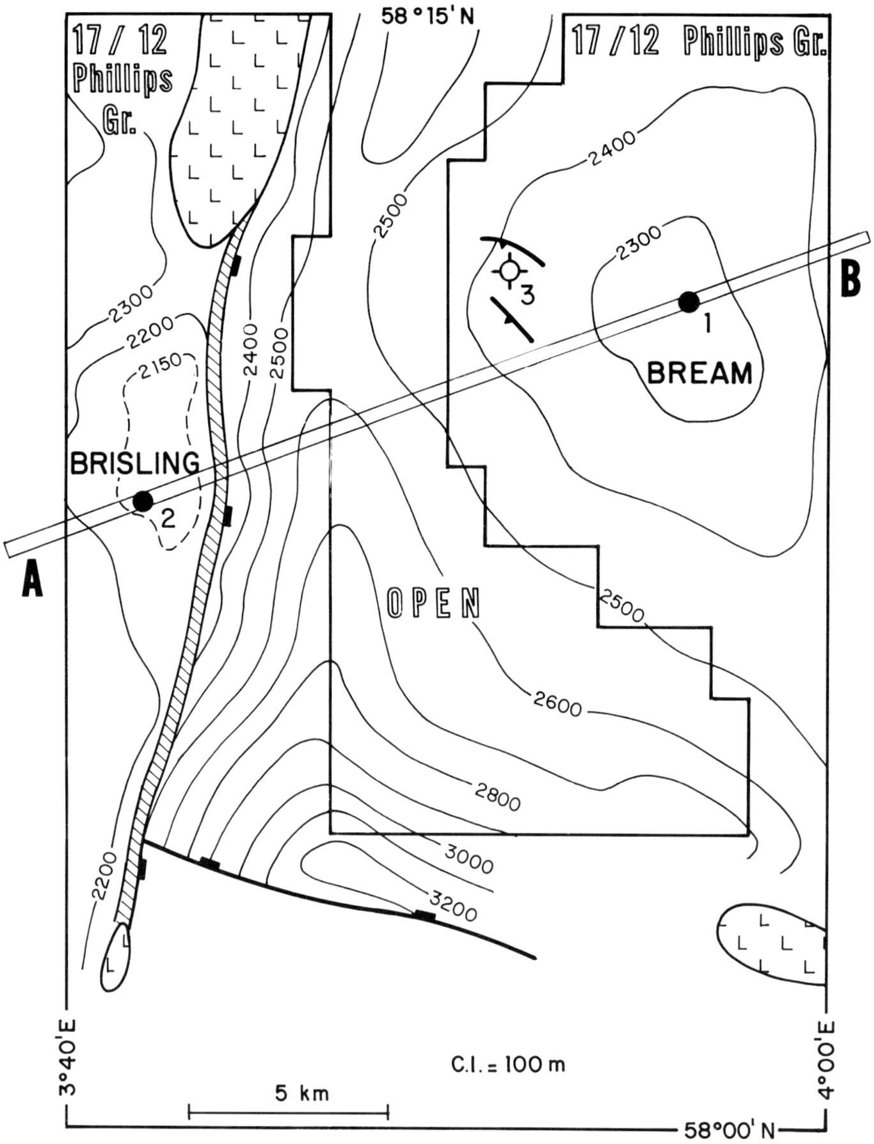

Fig. 2. Depth map at top reservoir level (top Mid Jurassic).

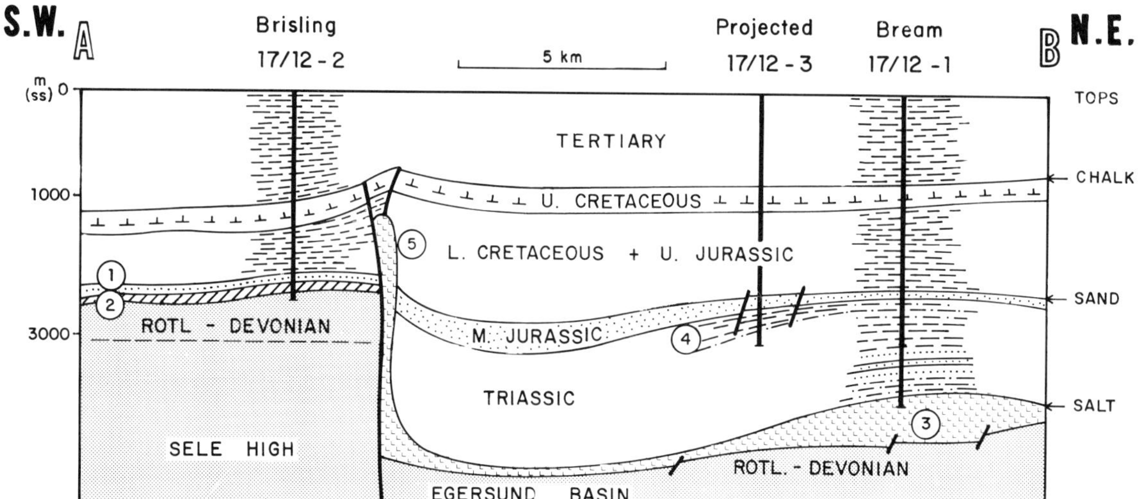

Fig. 3. Geoseismic cross-section from Bream to Brisling. On the Sele High, (1) represents the Jurassic sandstones and siltstones and (2) represents the upper Permian carbonates. In the Egersund Sub-basin, (3) represents the Upper Permian salt and (4) corresponds to the Lower Jurassic. Salt piercement (5) occurs along the Sele High border fault but not necessarily along section AB. See Fig. 2 for location.

In Well 17/12-3, the Skagerrak Formation is overlain by paralic deposits of the Gassum Formation, which pass upwards into marine shales of Early Jurassic age. These shales include the Fjerristlev Formation, of which 170 m has been preserved from erosion at this locality. Uplift of the area resulted in the return to a paralic environment during Aalenian and Bathonian times, while subsidence during Callovian times led to a marine incursion from the North.

More restricted conditions prevailed during late Jurassic times, when dark coloured siltstones and black organic shales of the Boknfjord Group were deposited. Their present day thickness is about 170 m at Brisling and 400 m at Bream. The main phases of Kimmerian tectonism which affected the Central and Viking grabens were felt only mildly in the Egersund Sub-basin. There was, however, a change in environment from anaerobic conditions (Late Jurassic) to open marine conditions (Early Cretaceous). Furthermore, the various sub-basins of the broader Norwegian–Danish Basin became more individualized than previously. The Cromer–Knoll Group, 520–740 m thick, consists mainly of calcareous shales (Valhall Formation) which pass upwards into dark silty shales (Sola Formation) and reddish-coloured claystones and marls (Rödby Formation) (Fig. 4).

Pelagic chalk deposition occurred during Late Cretaceous and Early Paleocene times. The thickness of the Chalk Group is about 450 m. It is overlain by the typical Cenozoic sequence of fine-grained clastic sediments. The Nordland, Hordaland and Rogaland groups total 800–1000 m of shallow-marine clays, silts and minor sands.

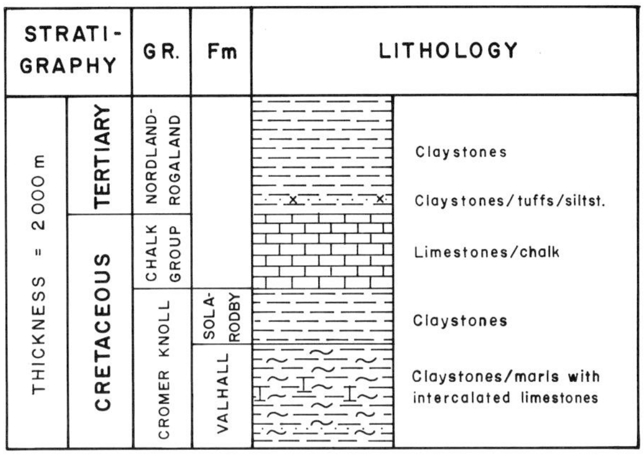

Fig. 4. Outline stratigraphy of the Cretaceous and Tertiary.

GEOPHYSICS

No difficulties have been encountered in identifying the seismic horizons. The top and the base of the Chalk Group are excellent markers due to the velocity contrast with the bounding Tertiary and Lower Cretaceous series above and below. The top of the Jurassic is not identifiable because of the continuity in sedimentation and facies from the Upper Jurassic to the Lower Cretaceous. The top of the Mid-Jurassic sandstones, which represents the primary target, is the first well-defined horizon in the Jurassic.

The base of the Mid-Jurassic sandstone is also a good seismic marker and, together with the top marker, makes up a quite homogeneous doublet which delineates the reservoir. Two weak markers coincide with the top of the Triassic and the top of the Zechstein salt, where these are present. The lowermost horizon occurs at the base of the salts. It marks the top of the Rotliegend, Carboniferous, Devonian, or Basement, depending on the locality.

TRAPS

The Bream structure is a nearly circular, salt-induced anticline, with an area of closure of 80 km^2 at Jurassic sandstone level. The corresponding vertical closure is about 140 m. Small normal faults exist locally on the western flank of the dome.

The Brisling structure is a N–S-trending anticline, 10 km long and 6 km wide. It is cut on its eastern flank by a major fault parallel to its axis. This fault, which is the limit between the Sele High and the Egersund Sub-basin, is thought to mark the eastern limit of the reservoir. Vertical closure is about 130 m and the area of closure is 40 km^2.

The reservoir in both structures is overlain by 700–1000 m of Upper Jurassic and Lower Cretaceous claystone of the Boknfjord and Cromer Knoll groups, respectively. In spite of the thickness of the seal, neither of the two structures is filled to spill-point; the probable reason is that an insufficient amount of oil was generated in the area.

RESERVOIR

The Bryne Formation (Bajocian to Bathonian) and the Sandnes Formation (Callovian) contain the main reservoirs in the area (Figs. 5, 6). The Bryne Formation accumulated on a wide alluvial plain with low relief. It contains mainly siltstone on the Sele High, with less than 5 m of porous sandstone. In the Sub-

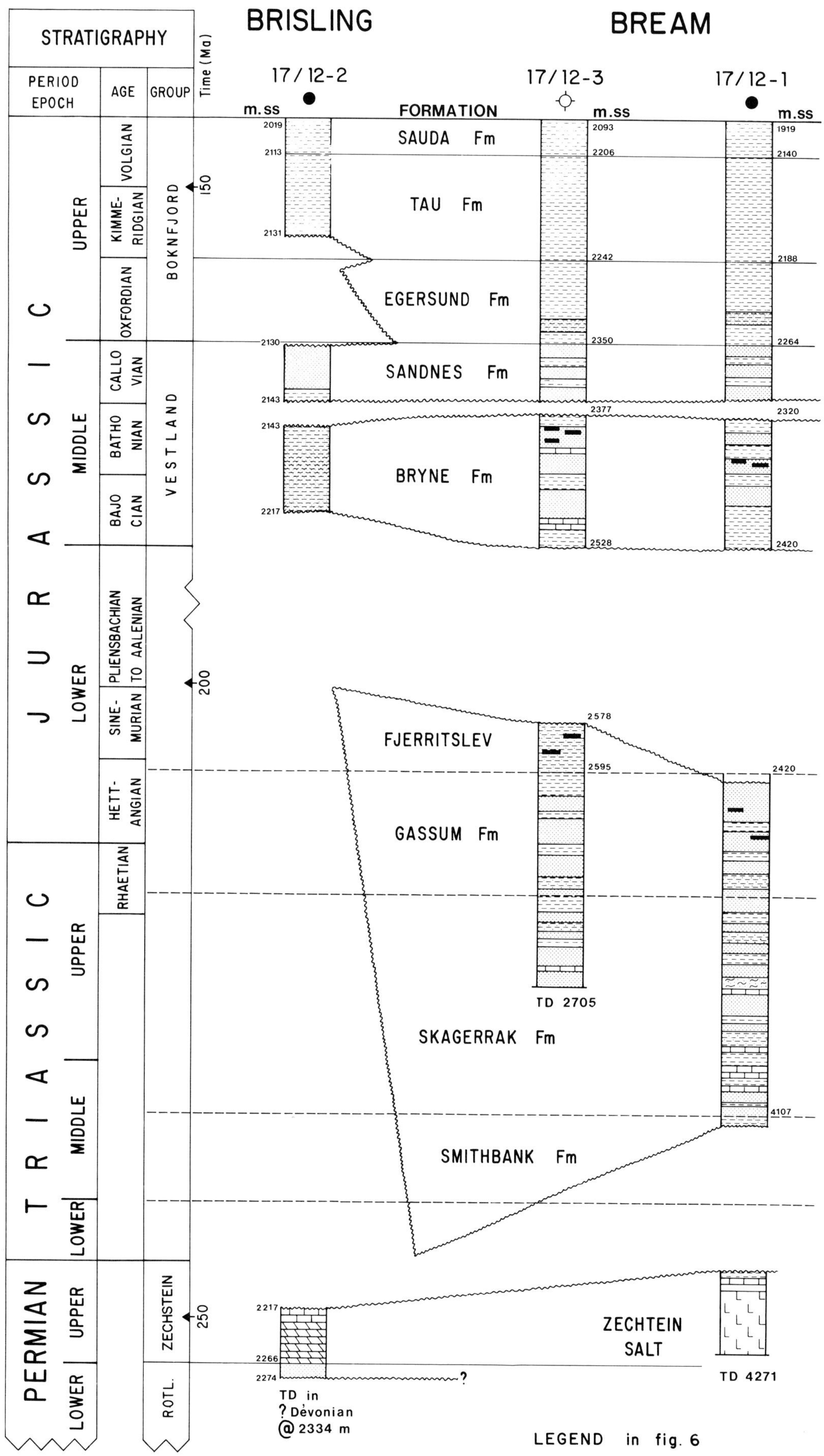

Fig. 5. Stratigraphy and lithology of the Permian to Jurassic sequences. Legend as in Fig. 6. Geological time is used as a vertical scale; consequently, depths and thickness are not to scale.

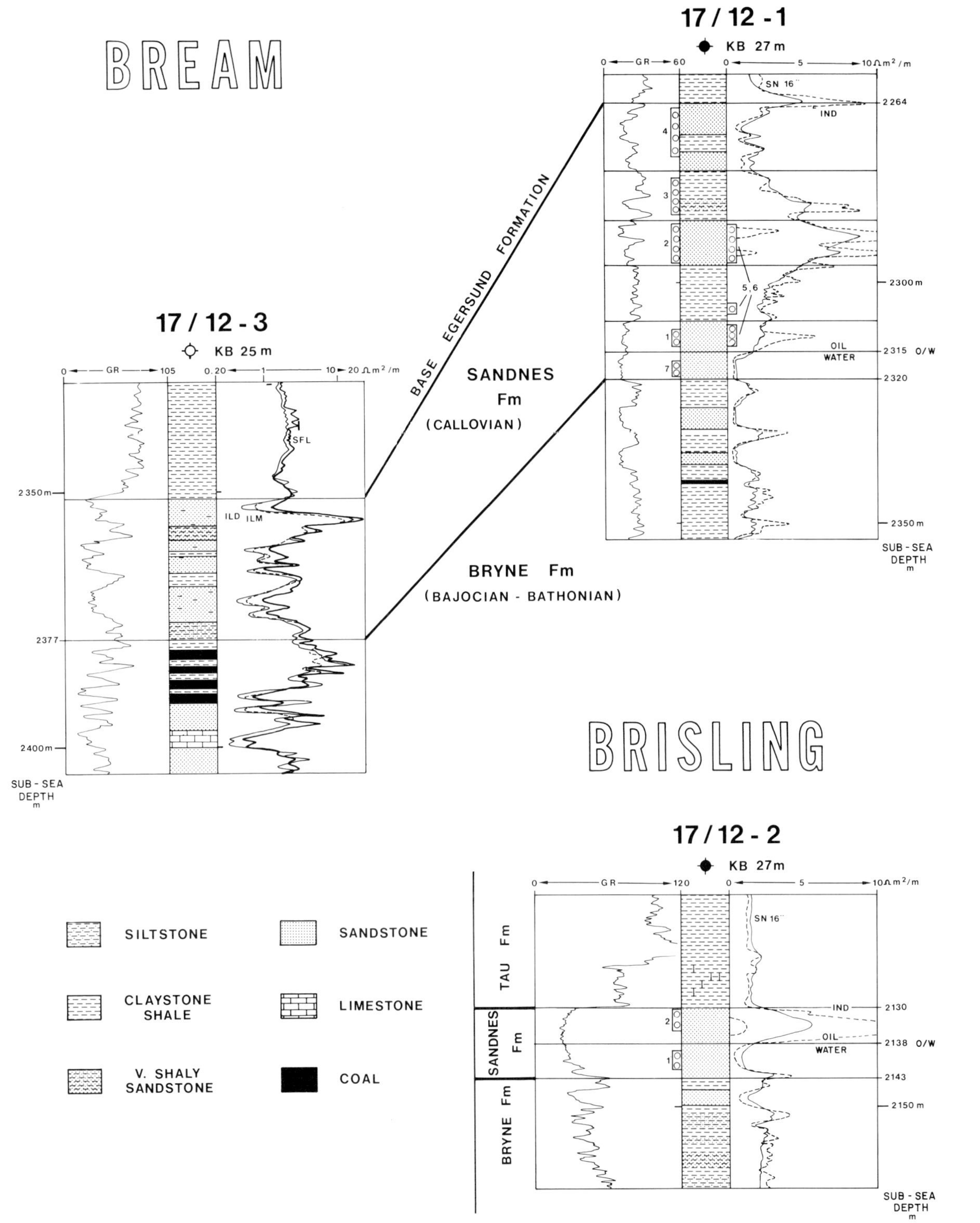

Fig. 6. Reservoir sections of the Bream and Brisling wells.

basin, it includes 170 m of fluvio-deltaic sandstone, siltstone, carbonaceous shale and coal and at Bream 50% of the interval has reservoir potential. At both Bream and Brisling, however, the Bryne Formation lies below the oil/water contact.

The Sandnes Formation consists of transgressive (coastal) glauconitic sandstones interbedded with black shales. It is the equivalent of the Ula Formation of the Central Graben, which is another shallow-marine sand. In the Bream area, the thickness of the Sandnes Formation varies from 27 to 56 m. It contains three sandy units which represent 50% of the 'gross interval'. The upper unit is tight, while the two lower units have average porosities of 19% and 24% in Well 17/12-1. At this crestal location, the lower sandstone unit contains an oil–water contact and the maximum net pay thickness does not exceed 15 m. A combined test of the two porous sand units flowed oil at a rate of 160 Sm^3/day through a 8/64-inch choke.

At Brisling, the Sandnes Formation is represented by a 13-m-thick sandstone unit with an average porosity of 20%. Unfortunately, once again, the reservoir has a shallow oil–water contact, and contains only 7 m of net pay.

SOURCE-ROCKS

Regional well control shows that the Upper Jurassic Boknfjord Group contains satisfactory source-rocks. The characteristics are similar to those of the equivalent formations in the Central and Viking grabens of the North Sea. The Boknfjord Group consists of four formations, two of which, the Tau and Egersund formations, are oil-prone and have satisfactory organic contents. The Tau Formation, which is characterized by a high gamma-ray response, is the best source-rock; its thickness is about 50 m in the Bream–Brisling area, and increases regionally southwards. The 'Hot Shales' of the Tau Formation are older than the 'Hot Shales of the Mandal Formation, which are the prolific source-rocks of the Central Graben (Doré *et al.*, 1984).

Biomarker data indicate that the oils found at Bream and Brisling have been generated in the Tau Formation. The main concern in this part of the Egersund Sub-basin is the degree of maturity of the source-rocks. The thermal gradient found in the Bream wells is 30–32 °C/km and so the source-rocks require an overburden of about 3000 m to reach the hydrocarbon-generation threshold. The Upper Jurassic rocks encountered in the wells are thus too immature to have produced significant amounts of oil.

The right conditions for oil generation occur only in a local depression, in the mid-southern part of Block 17/12. Since there are no other similar mature areas, this depression must be the source-area for the Bream and Brisling accumulations. The main phase of oil generation has probably just been attained. The small size of the catchment area is considered responsible for the limited amount of fill of the traps.

HYDROCARBONS

The hydrocarbons of Bream and Brisling are interpreted to be early mature, non-biodegraded oils. The sulpur content varies from 0.41% (by weight) in Bream to 1.14% in Brisling. The asphaltene content in the sample recovered from DST 7 of Well 17/12-1 is 1.5% (by weight).

RESOURCES

Resource figures of Bream and Brisling have not been published by the operator nor the coventurers. The 1980 annual report of the Norwegian Petroleum Directorate has classified these fields in the $\leqslant 1 \times 10^6$ Sm^3 of oil category, which is probably pessimistic for the Bream Field.

ACKNOWLEDGEMENTS

Several interpretations contained in this short chapter are derived directly from studies of colleague geologists, geophysicist and geochemists of Fina Exploration Ltd and Fina Exploration Norway. The authors thank Fina, Phillips, Agip, Elf, Total and Norsk Hydro who authorized the publication of this article. The opinions expressed in this paper do not necessarily reflect the opinions of the above-mentioned companies.

REFERENCES

Deegan, C.E. and Scull, B.J. 1977. A proposed standard lithostratigraphic nomenclature for the Central and Northern North Sea, *Report Institute of Geological Sciences, London* 77/25.

Doré, A.G., Vollset, J., and Hamar, G.P. 1985. Correlation of the offshore sequences referred to the Kimmeridge Clay Formation—relevant to the Norwegian Sector, In Thomas, B.M. *et al.* (Eds), *Petroleum Geochemistry in Exploration of the Norwegian Shelf*, Graham and Trotman, London, pp. 27–37.

Hamar, G.P., Fjaeran, T. and Hesjedal, A. 1982. Jurassic Stratigraphy and Tectonics of the south-eastern Norwegian offshore, In *Petroleum Geology of the South-eastern North Sea and Adjacent Onshore Areas, Geol. en Mijnb.* 1983, **62,** 103–114.

Jakobsson, H.K., Hamar, G.P., Ormaasen, D.E. and Skarpnes, O. 1980. Triassic facies in the North Sea, North of the Central Highs, In *The Sedimentation of the North Sea Reservoir Rocks*, Norwegian Petroleum Society, Article XV.

Norwegian Petroleum Directorate. 1980. *Annual Report.*

Pegrum, R.M. 1984. The extension of the Tornquist Zone in the Norwegian North Sea, *Norsk Geol. Tidsskrift*, **64** (1), 39–68.

Vollset, J. and Doré, A.G. 1984. A revised Triassic and Jurassic Lithostratigraphic nomenclature for the Norwegian North Sea, *Norwegian Petroleum Directorate Bulletin* 3.

Part IV—South Viking Graben

15

Balder

S. Hanslien

Esso Norge A/S, PO Box 60, 4033 Forus, Norway

The Balder oil field is located on the eastern margin of the Viking Graben in Norwegian Blocks 25/10 and 25/11. The oil is trapped in submarine-fan sands of the Paleocene Heimdal Formation, which were deposited by turbidity currents and originated from the East Shetland Shelf. Secondary reservoir sands of similar origin are present in the Lower Eocene Balder Formation. The sands have excellent reservoir properties with an average porosity of 33% and a permability ranging from 1 to 3 D. A combination of depositional topography and subsequent differential compaction produced a mounded upper surface of the submarine fan, forming the hydrocarbon trap. Individual oil sands are discontinuous and irregular in thickness and they are generally too thin to be mapped with confidence from conventional seismic data. Consequently, even after drilling 17 exploration and appraisal wells, there is considerable uncertainty in the estimate of recoverable hydrocarbons, which ranges from 11×10^6 to 27×10^6 Sm^3.

INTRODUCTION

The Balder oil field is located 190 km west of Stavanger in the Norwegian sector of the North Sea. The field outline covers an area of 44 km^2, primarily in Block 25/11 with a small portion extending into Block 25/10. Water depth in this area is about 125 m. Deep-marine sand deposits of Paleocene and Early Eocene age constitute the reservoir rocks. The field is named after the god Balder, who was the flawless, innocent son of Odin in Norse mythology.

HISTORY

Production Licence 001, which included Block 25/11, was awarded in 1965 as the first offshore production licence in Norway. Licence PL 028, covering Block 25/10, was awarded in 1969. Both licences are wholly owned by Esso but Statoil has a 17 $\frac{1}{2}$% net profit interest in PL 028. All compulsory relinquishments have been completed and the licences expire in the years 2011 and 2015, respectively.

To date, 17 exploration and appraisal wells have been drilled in the Balder Field. The first, Well 25/11-1, was drilled in 1966–7, being the second offshore exploration well in Norway. The objective was to investigate the sequence of strata down to basement in a large Lower Tertiary regional structure. The well encountered 11 m of thin, oil-bearing sands in the Lower Eocene, the first oil found on the Norwegian Shelf. Five subsequent wells were drilled in 1969–70 (25/11–2, -3, -4 and 25/10-1 and -3) and most of them had minor oil shows in the Lower Tertiary section.

In the early 1970s, a careful reinterpretation of the reservoir intervals was carried out using six-fold CDP seismic data and applying the newly established principles of seismic stratigraphy pioneered by P. Vail at Exxon Production Research Company (Vail *et al.*, 1977). Based on this study, Well 25/11-5 was drilled in 1974 and penetrated 24 m of

Geology of the Norwegian Oil and Gas Fields © A. M. Spencer *et al.*, eds (Graham & Trotman, 1987) pp. 193–201.

Balder Field data summary

Trap	
Type	Depositionally controlled structural trap
Depth to crest (Top Paleocene)	1650 mss
Oil/water contact	1760 mss
Gross oil column	*c*. 80 m (maximum 70 m in wells)
Pay zone	
Formation	Heimdal (primary), Balder (secondary)
Age	Paleocene, Early Eocene
Thickness	Average 20–30 m, maximum 56 m
Net/gross	75–100%
Porosity (average; range)	33%; 32–35%
Hydrocarbon saturation	85%
Permeability	1–3 D
Hydrocarbons	
Oil gravity	0.91 g/cm^3 at 15.5 °C/1.0 bar
Oil type	Low sulphur
Gas/oil ratio	53 Sm^3/Sm^3
Bubble point	148–165 bar
Formation volume factor	1.15
Formation water	
—	
Reservoir conditions	
Temperature	77 °C
Pressure	178 bar
Pressure gradient in reservoir	0.80 bar/m
Field size	
Area	44 km^2
Recovery factor—oil	Average 30%
Anticipated production rate	6–9.5 × 10^3 Sm^3/day
Recoverable oil	11–27 × 10^6 Sm^3
Drive mechanism	Aquifer support/water injection
Production	
—	

Paleocene oil sand. This well is designated as the Balder discovery well. Three successful confirmation wells were completed in 1978 and 1979 (25/11-6, -7 and -8). Well 25/11-8 found 56 m of net oil sand and is the best well in the field to date.

A three-dimensional (3D) seismic survey was shot by GSI in 1979. Interpretation of these data led to a depositional model of a submarine fan complex deposited in a series of prograding and channelized suprafan lobes (Sarg and Skjold, 1982). In 1980–81 seven appraisal wells were drilled to confirm the mapping, but gave largely disappointing results. Since then, efforts have been primarily directed towards improving the seismic resolution of the complex reservoir geometry using inverse modelling techniques.

GEOLOGICAL SETTING

The Balder Field is located on the north-western flank of the Utsira High, which is a prominent basement high forming the eastern margin of the southern Viking Graben (Fig. 1). The Utsira High was a positive feature at least as early as Permian times, and the Zechstein evaporites are absent across the crest of the high. When the Kimmerian block-faulting and extensional structuring ceased

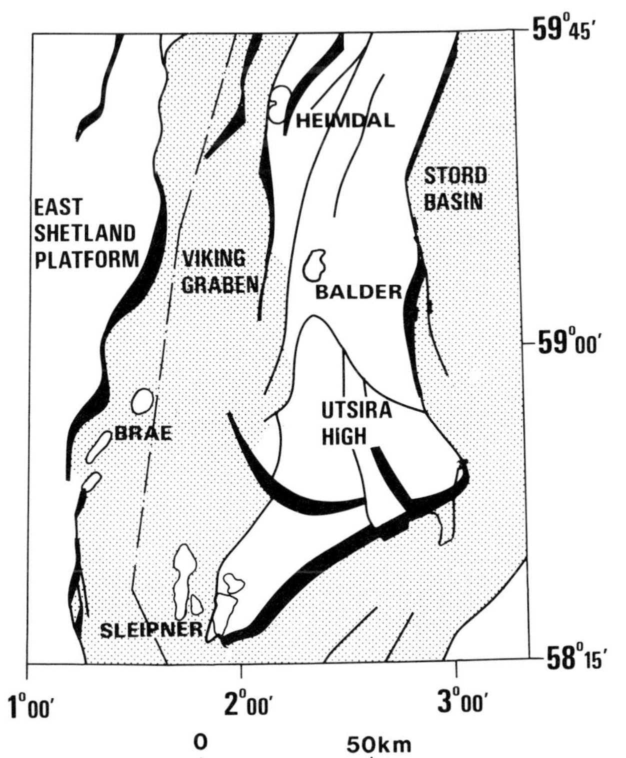

Fig. 1. Location map showing main tectonic elements and major fields.

in Early Cretaceous time, the North Sea basins became the site of passive Upper Cretaceous and Tertiary subsidence. This led to the formation of a N–S-trending Paleocene basin, bounded by shallow-water areas: the East Shetland Platform to the west and the Norwegian Shelf to the east. The Balder Field lies near the centre of this basin. Falls in eustatic sea level during Paleocene time caused an influx of clastic sediments, with sands derived from the East Shetland Platform being shed into the deep water of the basin in the form of large submarine-fan complexes. One such fan complex forms the reservoirs in the Balder Field.

A generalized stratigraphic column is shown in Fig. 2. The lithostratigraphic nomenclature used is that proposed by Deegan and Scull (1977) incorporating the revisions proposed by Vollset and Doré (1984). Well 25/11-1 reached metamorphic basement at approximately 2400 m and this was overlain by Rotliegendes sandstones and conglomerates. Thin carbonates of marginal Zechstein facies are present only in the western part of the field. In most of the field area, the Rotliegendes is directly overlain by Triassic red beds, consisting mainly of shales and siltstones with occasional sandstone interbeds. This sequence is in turn overlain by continental to shallow-marine sandstones and shales of Early and Late Jurassic age. Middle Jurassic rocks have not been encountered in any of the Balder wells. The Lower Cretaceous consists of light grey marls in the lower part and shales in the upper part. The Upper Cretaceous Chalk Group is thin and consists of white to pink limestone with occcasional intercalated shale beds.

The overlying Paleocene clastic interval contains the main reservoir units in the Balder Field and consists of massive deep-marine sands interbedded with shales. This interval ranges in thickness from 80 to 240 m with a sand content varying between 25 and 60%. The main sand units belong to the Heimdal Formation but sands are also present in the older Maureen Formation. Overlying the Heimdal Formation are the hemipelagic shales of the Sele Formation which are 10–60 m in thickness. The top of the Sele Formation coincides with the top of the Paleocene. The Lower Eocene Balder Formation ranges in thickness from 40 m up to 90 m in wells. It consists mainly of tuffaceous siltstone and shales interbedded with stringers of fine-grained sandstone and occasional massive sandstone beds.

The Middle Eocene to Lower Oligocene

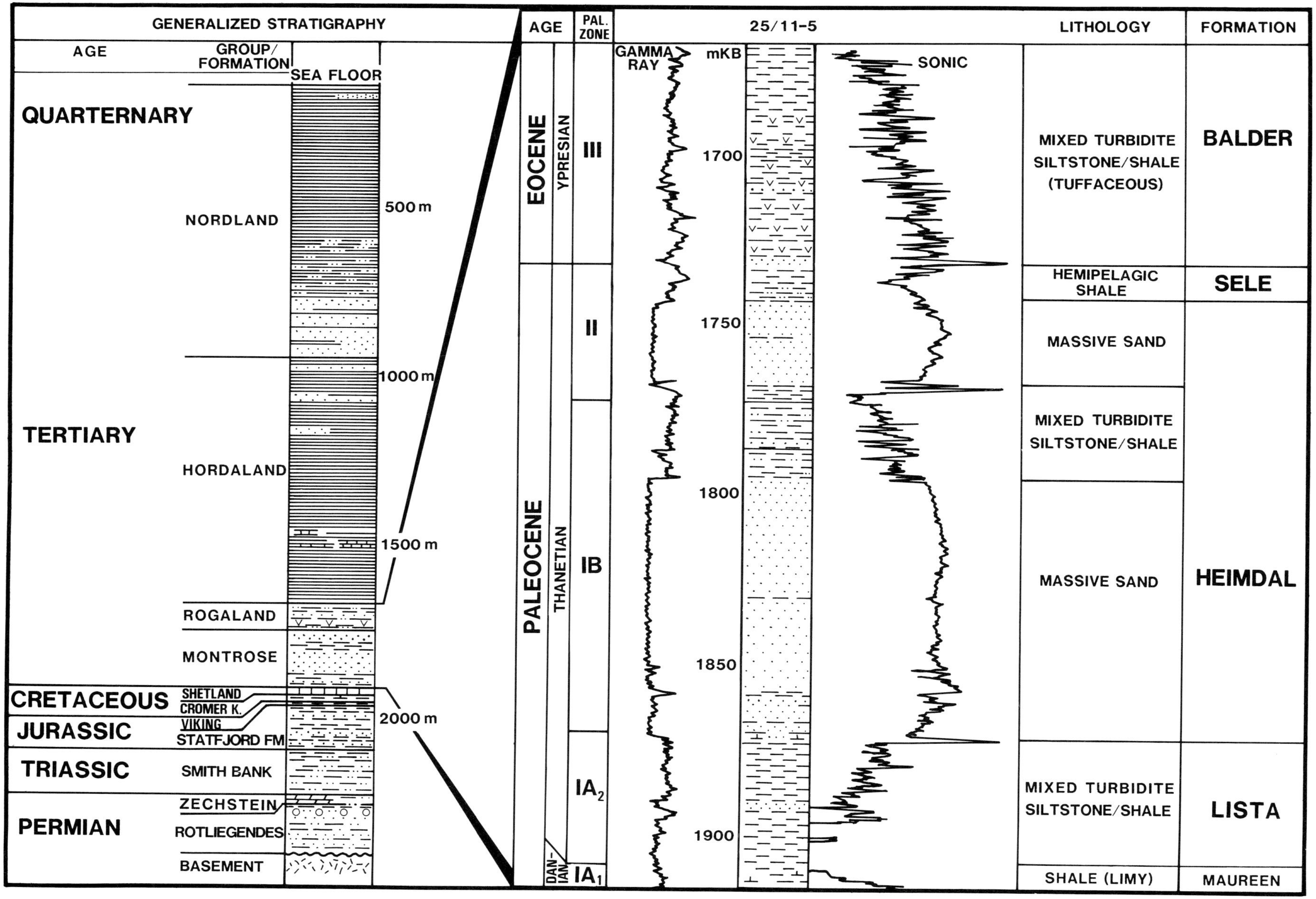

Fig. 2. Generalized stratigraphic column and 25/11-5 type log.

succession is a monotonous sequence of soft gumbo clay belonging to the Hordaland Group, which is overlain by the interbedded sands and shales of the Nordland Group.

GEOPHYSICS

Until 1979 the mapping of the Balder field was based on several vintages of seismic data constituting approximately a 1 × 1 km grid. In 1979 a three-dimensional seismic survey was shot and processed by GSI. This survey consists of 147 E–W-orientated lines 75 m apart and amounts to 1350 km of three-dimensional migrated data.

Interpretation of the three-dimensional seismic survey formed the basis for the seven-well delineation programme in 1980–81. The drilling results demonstrated the need for better seismic definition of the reservoir, and in 1981 the three-dimensional data were reprocessed using inverse modelling techniques in an attempt to resolve better the complex reservoir geometry. The inverse seismic data improved the seismic definition of the oil sands and constitute the basis for the enclosed maps and the current assessment of the oil-in-place.

TRAP

The Balder Field is a stratigraphically induced structural closure resulting from depositional submarine-fan topography enhanced by differential compaction of the Paleocene sands and shales. Most of the faults in the area die out below the reservoir level. The structural relief at the Top Paleocene level (Top Sele Formation) is approximately 80 m with the crest at 1650 mss and the deepest closing contour at 1730 mss (Figs. 3, 4).

The thickest gross oil column in the massive Paleocene sands as seen in the wells is about 70 m (25/11-6 and -8). From seismic mapping, the maximum oil column in the field is predicted to be in excess of 80 m in the area north of Well 25/11-6. It cannot be determined conclusively whether the field-wide oil–water contact at 1760 mss is within the structural closure at the reservoir sand level or if the trapping of the oil is stratigraphically controlled. It appears, however, at least with respect to the upper, more discontinuous sand lobes (Eocene, Zone II and Upper 1B sands, Fig. 5), that the oil is stratigraphically trapped. It is possible, but not directly proved, that the oil–water contact at 1760 mss coincides with the structural spill-point of the widespread Lower 1B sand.

Furthermore, thin sands in the overlying Balder formation have consistently been found to be oil bearing when encountered above 1760 mss. This suggests that these sands are in pressure communication with the massive sands of the Heimdal Formation and that the Sele shale is an inadequate top seal in the field area. Consequently, a lateral stratigraphic change in the Balder Formation is required in order to trap the oil in the field.

RESERVOIR

The primary reservoir sands in the Balder Field belong to the Paleocene Heimdal Formation. Secondary reservoir sands are present in the Lower Eocene Balder Formation. Core and well-log data indicate three submarine-fan lithofacies in the Paleocene (Fig. 2) classified according to the deep-water facies classification of Mutti and Ricci Lucchi (1975) as follows:

1. massive turbidite sandstone;
2. mixed turbidite siltstone–mudstone;
3. hemipelagic shale.

The massive sandstone facies consists of well-sorted, fine- to medium-grained, clean quartz sands with minor amounts of feldspar. The extremely good sorting suggests that the sediments were pre-sorted prior to final transport and deposition. The sandstones are massive bedded and exceed 100 m in thickness in some wells but are more typically 20–40 m thick. X-Ray photographs of the sands show a homogeneous fabric. Rarely, very faint cross-bedding with up to 15° dips and planar bedding can be distinguished. These rocks are classified as massive turbidite sands of the B_1 and B_2 types of the Mutti and Ricci Lucchi scheme. The sands are unconsolidated and have excellent reservoir properties. The net/gross ratio for the individual sand units averages 95%. Porosity averages 33% and absolute permeability ranges from 1 to 3D both horizontally and vertically. Average water saturation is 15%.

A mixed turbidite siltstone–mudstone facies (D_3 of the Mutti and Ricci Lucchi classification) is present in the lower part of the Sele Formation and is the predominant facies in the Balder Formation. It is composed of finely laminated silts and clays.

The hemipelagic shale facies consists of greenish-gray, waxy shale with irregular, dark carbonaceous bands and has very

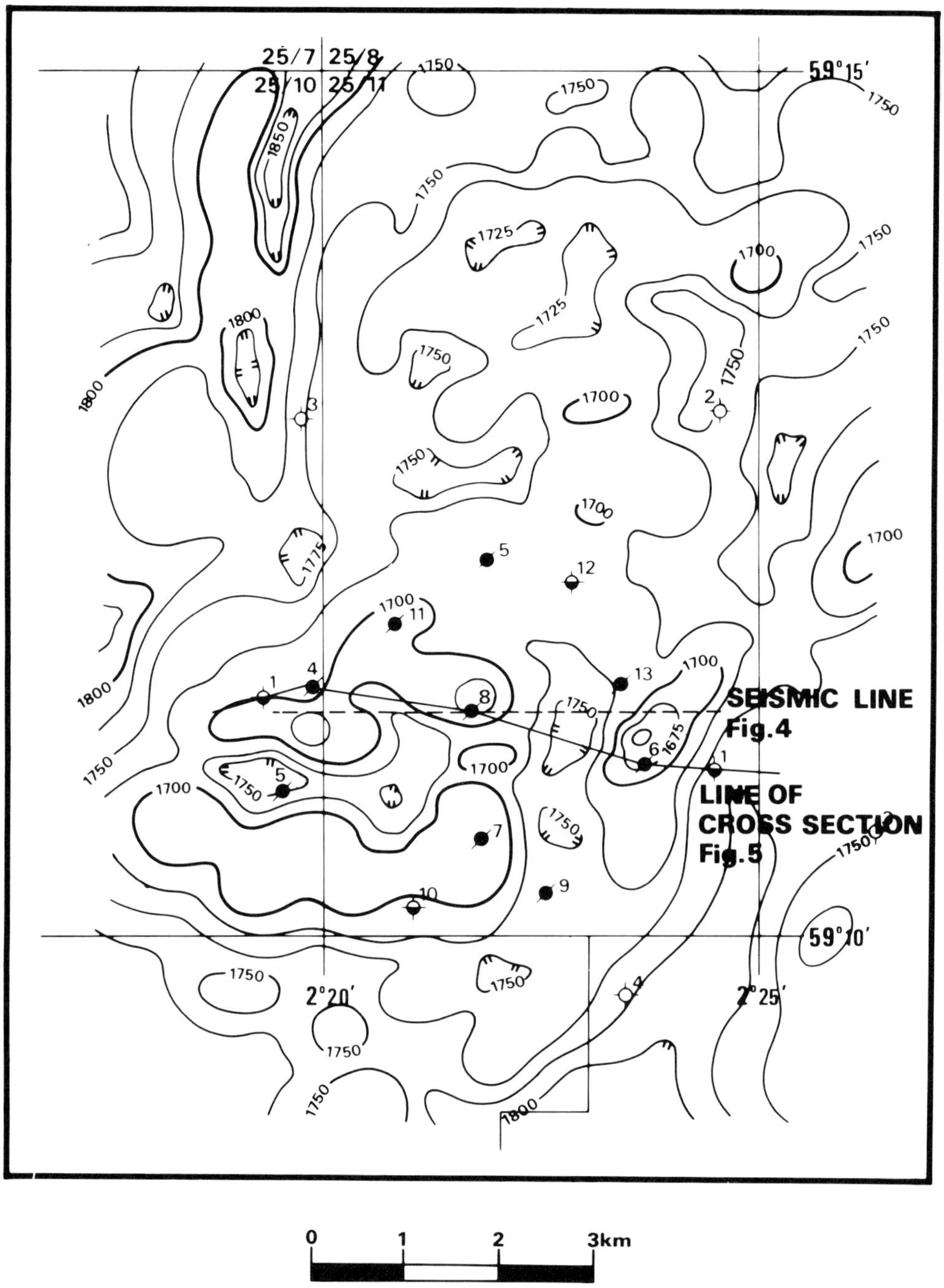

Fig. 3. Top Paleocene (Sele Fm) structure map. Contours are in metres subsea.

poor or no fissility. The gamma ray log character varies between medium and high readings and it is difficult to distinguish this facies from that of the mixed turbidite siltstone/mudstone facies. Beds are generally thinner than 15 m. This lithofacies corresponds to Facies C_7 of the Mutti and Ricci Lucchi scheme.

The Paleocene–Lower Eocene interval has been subdivided into different zones based on palynology (Fig. 2). The oil-bearing sands are all within Palynozones IB, II and III. Four different reservoir sands have been identified from well control and seismic data and they have been named according to their age: Lower IB Sand, Upper IB Sand, Zone II Sand and Eocene Sand. The spatial relationship of the four main reservoir sands is illustrated by the cross-section in Fig. 5 and the plan-view outline of the gross oil sands in Fig. 6. In general, all four reservoir sands are interpreted to be isolated from one another by

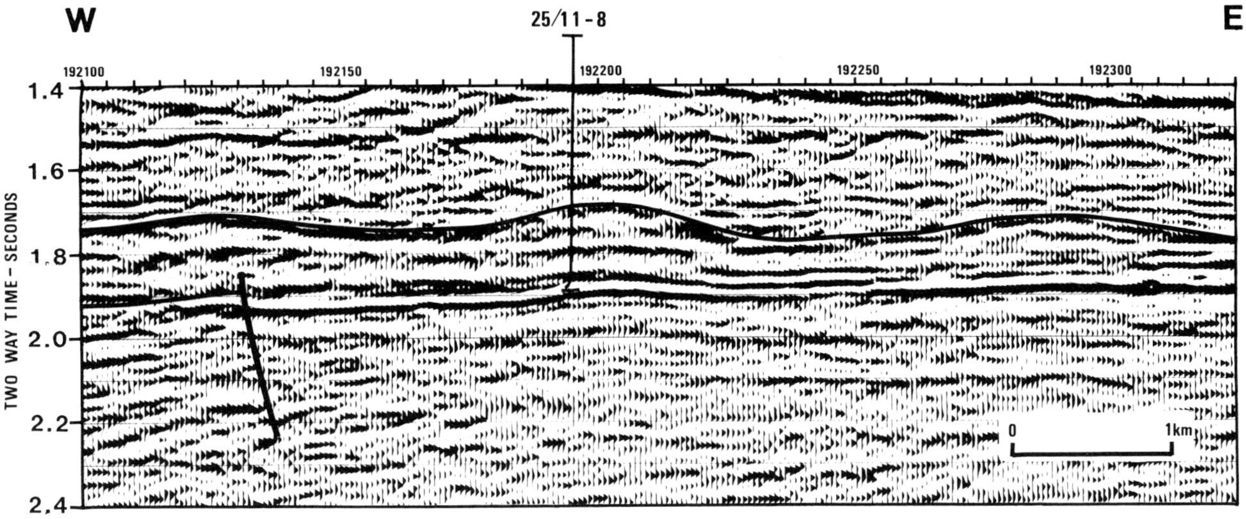

Fig. 4. 3D seismic line 192. For location see Fig. 3.

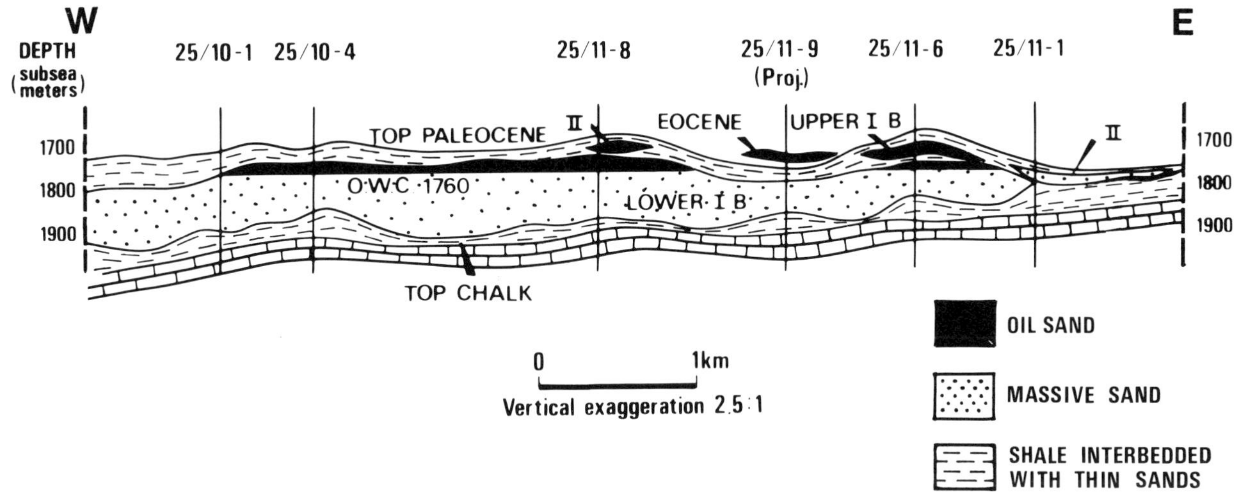

Fig. 5. Structural cross-section. For location see Fig. 3.

shale drapes, although they have a common oil–water contact. Individual sands tend to be discontinuous and lenticular.

SOURCE

The source-rocks for the Balder Field oil are believed to be Upper Jurassic shales in the Viking Graben. Oil was generated and migrated to the Balder reservoirs during Oligocene times.

HYDROCARBONS

The crude in the Balder Field is 'heavy' by North Sea standards, having a specific gravity of 0.91 g/cm^3. The high gravity suggests that the oil may have been water washed along its relatively long migration path from the Viking Graben. The oil is slightly undersaturated, with a gas–oil ratio of 53 Sm^3/Sm^3 and viscosity at reservoir conditions of 3.5 cp. The reservoir temperature is 77 °C and the reservoir is normally pressured with an initial pressure of 178 bars. The bubble point of the oil from various tests is in the range of 148–165 bars. In spite of the relatively viscous nature of the oil, excellent flow rates have been obtained in production tests (up to 485 m^3/day on 14 mm choke).

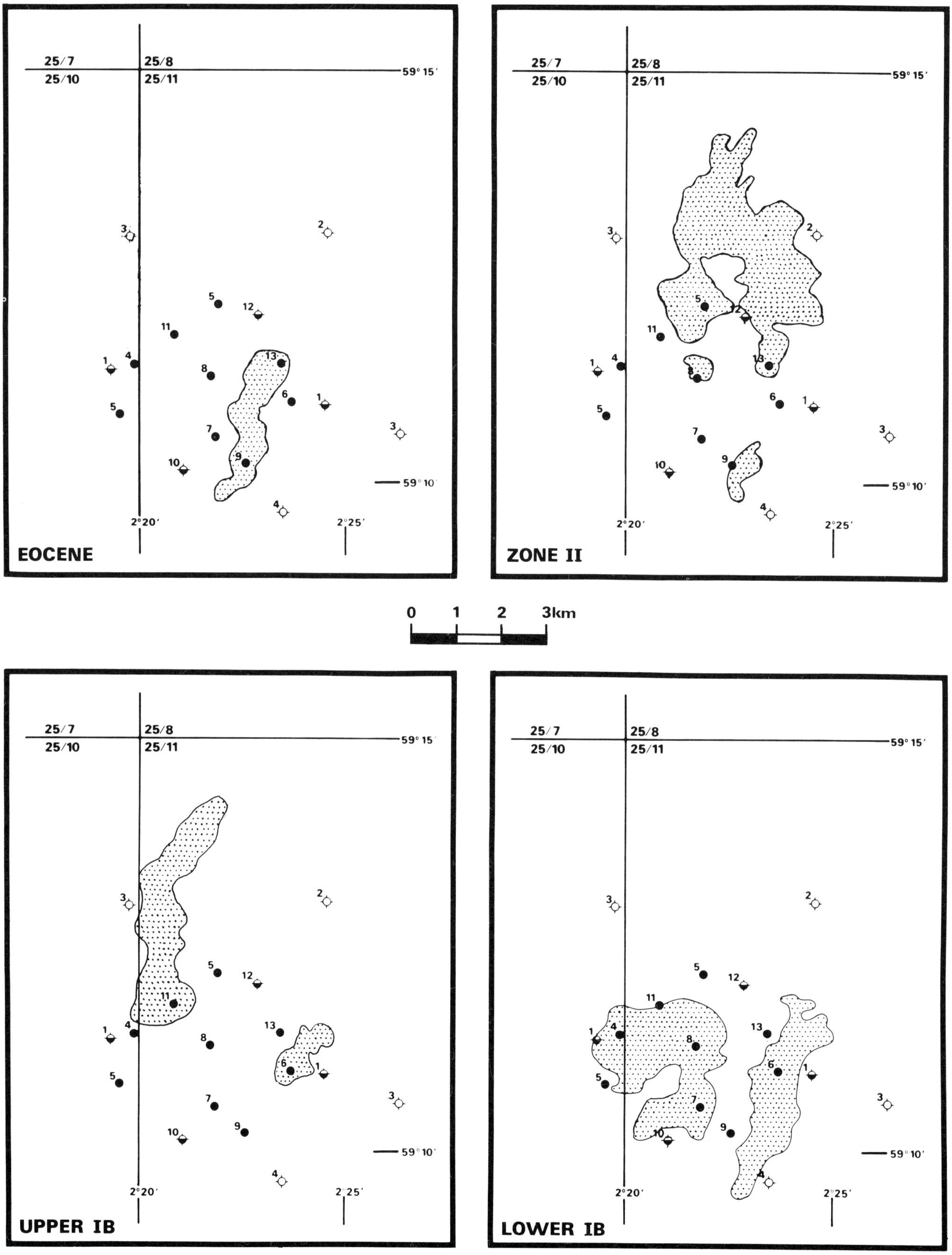

Fig. 6. Gross oil sand outlines for the four main reservoir units.

RESOURCES

Recoverable resources are obviously a function of the type of development facilities and of the reservoir performance. In the Balder Field, recoverable resources are estimated to range between 11×10^6 Sm^3 and 27×10^6 Sm^3. The wide range reflects the uncertainties in mapping the oil-in-place and the difficulty in predicting recovery efficiencies due to the complex reservoir geometry, as well as the various possible development schemes. The individual reservoir sands are discontinuous and thick sands may die out within a few hundred metres laterally. Over large areas the thickness of the oil sands is close to or below seismic resolution. Of the four main reservoir sand units, only the Lower IB sand is believed to have an active water drive. The three other sand units are isolated from the regional aquifer and are separated from each other by shale drapes, and consequently they will require early water injection during production. Water coning and early water cut are anticipated due to the effects of an unfavourable fluid-mobility ratio in combination with thin oil sand. Artificial lift would be required as early as the first year of production in many wells. In addition, the reservoir sands in the Balder Field are unconsolidated, and gravel pack completions would be needed for sand control.

ACKNOWLEDGMENTS

This article is based largely on previously published papers by J. J. Zagar (1984), J. F. Sarg and L. J. Skjold (1982), and L. J. Skjold (1980). D. P. Coney and H. Posamentier, Esso Production Research Company, completed the most recent seismic interpretation of the area and the included maps and cross-sections are based on their interpretation. The author thanks Esso Norge A/S for giving permission to publish this paper.

REFERENCES

Deegan, C.E. and Scull, B.J. (Compilers). 1977. A Standard lithostratigraphic nomenclature for the Central and Northern North Sea, *Institute of Geological Sciences, Report* 77/25.

Mutti, E. and Ricci Lucchi, F. 1975. Examples of turbidite facies and facies association from selected formations of the Northern Apennines, *Guidebook to Excursion no. 11,* 9th International Congress of Sedimentology, Nice, France.

Frodesen, S., Moe, Aa., Ofstad, K., Ormaasen, E., Sjulsen, S.-E. and Ulleberg, K. 1981. The Balder Area, *Norwegian Petroleum Directorate Lithology Paper, no. 28.*

Sarg, F. and Skjold, L.J. 1982. Stratigraphic Traps in Paleocene Sands in the Balder Area, North Sea, *American Association of Petroleum Geologists Memoir 32,* 197–206.

Skjold, L.J. 1980. Paleocene sands of the Balder field, In *The Sedimentation of the North Sea Reservoir Rocks,* Norsk Petroleumsforening, Article VII.

Vail, P. *et al.* 1977. Seismic stratigraphy, In *American Association of Petroleum Geologists Memoir 26* (12 articles).

Vollset, J. and Doré, A.G. (Eds), 1984. A revised Triassic and Jurassic lithostratigraphic nomenclature for the Norwegian North Sea, *Norwegian Petroleum Directorate Bulletin 3.*

Zagar, J.J. 1984. Improved recovery from thin oil sands. Part 3—Balder Field, In *Offshore Northern Seas—Improved Offshore Recovery,* Norsk Petroleumsforening, Article R2.

16

Frigg

E. Mure

Elf Aquitaine Norge A/S, PO Box 168, 4000 Stavanger, Norway

The Frigg Field is the first and the largest gas field in production on the Norwegian continental shelf. It straddles the boundary between Norway and the United Kingdom, in the central part of the Viking Graben, at latitude 60 °N. It lies mostly in Norwegian Block 25/1 and in British Block 10/1. The water depth is about 100 m. The hydrocarbons are trapped at 1785 mss in submarine-fan sands of the Lower Eocene Frigg Formation. In the years 1971–76, nine exploration wells were drilled and a detailed seismic survey were shot to evaluate the size of the 100 km^2 Frigg structure. The original gas-in-place, as estimated by experts in 1976, is 268×10^9 Sm^3 of which 60.8% are in Norway and 39.1% are in the United Kingdom. Development in 1975–79 comprised the installation of five platforms, a flare stack and two pipelines to St Fergus, Scotland, with an intermediate booster platform. Production started in September 1977. New extensive appraisal—six new deep wells and a three-dimensional seismic survey—has been performed in 1984–86 to monitor precisely the water rise over the field. New updated reservoir models, emphasizing the facies heterogeneities of the reservoir, are developed to provide accurate prediction of the decline of the field which should happen in the years 1987–90. Evaluation of the potentially unswept gas pocket is underway. Cumulative production of gas reached 135×10^9 Sm^3 at the end of 1986. Some uncertainty remains in the estimate of the recoverable gas reserves which are presently being revised by the operator.

INTRODUCTION

The Frigg Field is situated astride the boundary between Norway and the United Kingdom in the north central part of the North Sea, some 200 km west of Haugesund, Norway, and 180 km east of the Shetland Islands (Fig. 1). The average water depth is around 100 m. On the Norwegian side, the Frigg Field lies within Petronord Group Block 25/1 with only a small part extending into Esso Block 30/10. On the United Kingdom side, the field lies mostly on Total Oil Marine and Elf Aquitaine Licence Blocks 10/1, 10/6 and 9/10a. A small part of the structure extends into BP Licence Block 9/5 and Ranger Oil Licence Block 4/26.

The field covers an area of approximately 100 km^2. The reservoir section consists of Lower Eocene sands of the Frigg Formation, which were deposited as the final parts of a complex Paleocene/Eocene submarine fan. The official original gas-in-place estimated by 'Experts' in 1976 is approximately 268×10^9 Sm^3. It is probable that a reduced figure will be proposed by the operator after the new appraisal of 1984–86.

The Frigg Field Participating Agreement (13 October 1978) fixed the shares of the production between the participants as follows:

Elf Aquitaine Norge A/S	25.19% (operator)
Norsk Hydro	19.99%
Total Marine Norsk	12.59%
Statoil	3.04%
Elf U.K. Limited	17.41%
Aquitaine U.K. Limited	8.70%
Total Oil Marine Limited	13.06%

The Frigg Field was named after a goddess of

Geology of the Norwegian Oil and Gas Fields © A. M. Spencer *et al.*, eds (Graham & Trotman, 1987) pp. 203–213.

Frigg Field data summary

Trap	
Type	Stratigraphical
Depth to crest	1785 mss
Lowest closing contour	1955 mss
Gas column	160 m
Oil column	9 m
Pay zone	
Formation	Frigg
Age	Lower Eocene (nt IIb palynozone)
Gross thickness	55 m (maximum 170 m)
Net/gross ratio	0.95
Cut off for N/G	Sw < 60%
Porosity (average; range)	29%; 27–30%
Hydrocarbon saturation	91%
Permeability (average; range)	1500 mD; 900–3000 mD
Maximum well rate	3×10^6 Sm^3/day/well
Hydrocarbons	
Gas gravity (relative to air)	0.581
Condensate yield	4.3 $Sm^3/10^6$ Sm^3 gas
Gas expansion factor	197
Oil gravity	0.835 g/cm^3
Oil type	naphthenic
Gas/oil ratio	61 Sm^3/Sm^3
Formation Water	
Salinity	63 000 ppm
Reservoir conditions	
Temperature	61 °C
Pressure	198 bar abs.
Pressure gradient in reservoir	
gas	0.013 bar/m
oil	0.083 bar/m
water	0.103 bar/m
Field size	
Area	100 km^2
Gross rock volume	5.454×10^9 m^3
Hydrocarbon pore volume	1.366×10^9 m^3
Recovery factor—gas	0.72
Recovery factor—oil	0
Drive mechanism	Water drive and depletion
Gas in-place	265×10^9 m^3
Recoverable gas	191×10^9 m^3
Production	
Start-up date	13 September 1977
Development scheme	platforms
Production rate	45×10^6 Sm^3/day (average)
	38×10^6 Sm^3/day (1986)
Cumulative production	135×10^9 Sm^3 (31 December 1986)

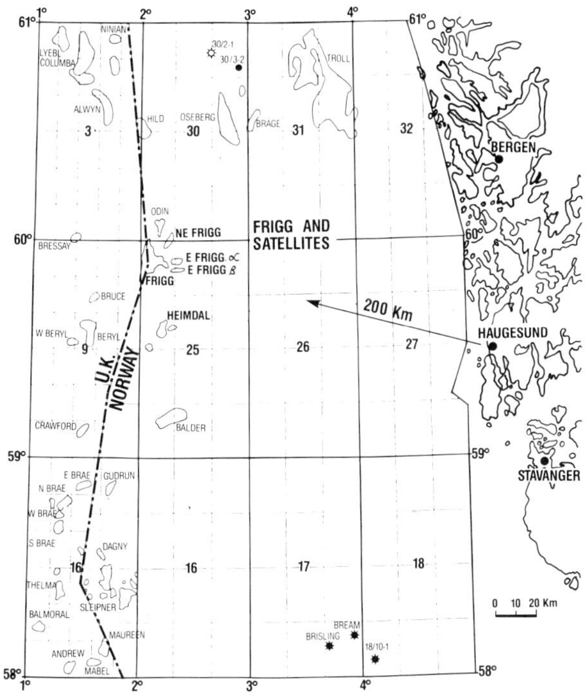

Fig. 1. Location map of the Frigg Field and its satellites.

the Norwegian pantheon, spouse of Odin and divinity of abundance and fertility.

HISTORY

Norwegian Block 25/1 was granted to the Petronord Group in 1969 during the Norwegian second round. In 1970 the United Kingdom Blocks 10/1 and 10/6 were granted by the British authorities to the Total Oil Marine and Elf Aquitaine group. The interpretation of the first seismic surveys with a wide spacing permitted an approximate definition of the Frigg structure. In 1971 the official agreement between the Petronord group and the Norwegian State made possible the drilling of the first exploration well. Well 25/1-1 located at the top of the structural closure entered the Lower Eocene sands at 1820 mss subsea and proved a gas column 129 m thick underlain by a seven-metre thick oil disc. It revealed that a large gas accumulation had been discovered.

From 1971 to 1976, eight appraisal wells were drilled to define the extension of the discovery and in 1973 a detailed seismic survey was shot with a regular grid of 1 × 1 km over the structure.

As the owners of the major portion of the field were unable to reach an agreement on the division of the gas-in-place between the Norwegian and the United Kingdom sectors of the field, De Golyer and MacNaughton were retained to determine the total amount of gas reserves in-place and the division of the gas between the two countries and the various concession blocks. Their report of 20 October 1976 concluded that the Frigg Field reservoir contains about 268×10^9 Sm^3 of gas in-place of which 60.82% are in the Norwegian sector and 39.18% are in the United Kingdom. These figures are the official basis for the 'Frigg Field participating agreement' signed in 1978.

The Frigg Field was declared commercial in April 1972. The first sale contract was signed with British Gas Corporation in 1973 for the British sector and in 1974 for the Norwegian sector. The development of the field was in two phases, one on the United Kingdom sector followed by the other on the Norwegian sector. Owing to the very good reservoir characteristics (high permeability and low viscosity of the gas), the production wells were concentrated in a small area (5 km^2) and two clusters of 24 production wells each were located at the structural culmination. The development plan included five platforms, a flare stack, and 2 pipelines towards St Fergus, Scotland, with an intermediate booster platform (Fig. 2). From 1975 to 1977, the following platforms were installed successively: quarters (QP); concrete drilling (CDP1, UK); oscillating flare; steel drilling (DP2, Norway); treatment (TPI); compression (MCPOI); and the treatment and compression platform (TCP2). All 48 production wells were drilled from 1976 to 1979. Production started on 13

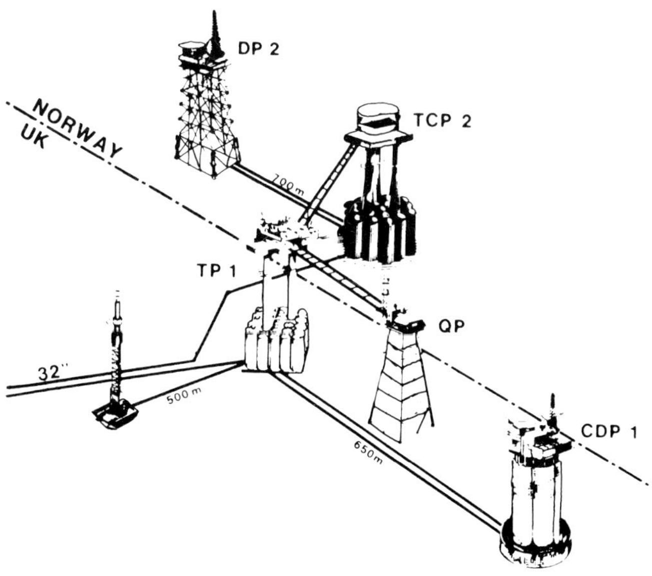

Fig. 2. Installations of the Frigg Field.

September 1977 on CDP1 (UK) and on September 9th, 1978 on DP2 (Norway). More recently, the Frigg treatment installations and pipeline facilities received the gas from the Odin and North East Frigg fields, and later the gas from the East Frigg Alpha and Beta fields will come when these fields are developed.

After five years of production, new appraisal works have been undertaken to monitor precisely the water rise over the field. The aquifer support has proved to be efficient. In 1984 and 1985, three new remote wells and three shallow production wells were drilled down to the aquifer, with continuous coring of the reservoir. A high-resolution three-dimensional (3D) seismic survey was acquired in 1985 and 1986. A total of 4640 km of acquisition lines 100 m apart were recorded providing, after interpolation, a coverage of 4×4640 km of three-dimensional migrated data constituting a 25×25 m grid. This large amount of new geological and geophysical data has been incorporated in the updated reservoir model to provide a better prediction of the decline of the field, which will happen around the years 1987–90. The evaluation of the potentially unswept gas pockets, which could be produced with additional works (new deviated wells—subsea cluster), is under way. Some uncertainty remains in the estimate of the original gas-in-place and the recoverable gas reserves which are presently being revised by the operator.

GEOLOGICAL SETTING

The Frigg Formation Lower Eocene sandstones are the clastic members of a complex Paleocene–Lower Eocene submarine fan, deposited in the central part of the Tertiary Viking Graben basin, east of the structurally higher Shetland Platform (Fig. 3). They are overlain by Middle Eocene open marine shales. The Frigg Field is located in the deepest part of the Viking Tertiary embayment in an area of low structural relief at Base Tertiary level where tectonism was of a very mild intensity. The tilted blocks of the pre-Cimmerian faulting phase are very deeply buried under post-Cimmerian to Cretaceous open-marine strata which are at their maximum thickness in the Frigg area.

The Frigg sands are mainly grain-flow and turbiditic deposits in a marine environment. The majority of the detrital material probably originated from the Shetland Platform to the west.

GEOPHYSICS

The top of the Frigg Formation is clearly defined by a strong seismic marker (Frigg Seismic marker) (Fig. 4) which is caused by the contrast of acoustic impedance between the overlying shales and the Frigg sands. This seismic marker has been picked and mapped on the different seismic surveys with a very high confident level. The initial gas–oil contact gives a very obvious seismic reflection ('flat spot') on the early 1973 seismic. On the more recent seismic no new gas–oil contact has been observed. The amplitude of the original flat spot has decreased with the rise of the water contact all over the field. This attenuation of the original flat spot on the 1985–86 three-dimensional seismic survey is being used to estimate qualitatively and possibly quantitatively the water rise caused by the production of the gas.

The 1976 depth map of Frigg was the result of the migration of a depth map reconstituted from the isochron map with a 'Vo + KP' velocity field calibrated to well data. After the 1985–86 three-dimensional seismic survey a new depth map was produced. The three-dimensional migrated time map was converted to depth using the three-dimensional stacking velocity field calibrated to the wells and to the flat spot limit. The Frigg structure is very flat in time—100 ms of time closure for 100 km^2—and consequently is very sensitive to depth-conversion procedure, i.e. to velocities.

Other seismic markers can be picked within the aquifer but with much less certainty than the top reservoir. A seismic-stratigraphy interpretation of the field is being performed. This interpretation is calibrated to the palynological zonation by Héritier *et al.* (1977), with markers at base nt IIb (Upper Frigg Member), the base of Lower nt IIa (Sele Formation) and the Tuff of the Balder Formation.

TRAP

The structure is mainly formed by the submarine-fan depositional topography and enhanced by the draping and differential compaction of the sands. It is sealed by middle Eocene open-marine shales. The structural relief at the Top of the Frigg sands is about 170 m (Fig. 5). The top of the reservoir is at about 1785 mss; the initial gas–oil contact is at 1948 mss. The structure was filled down to

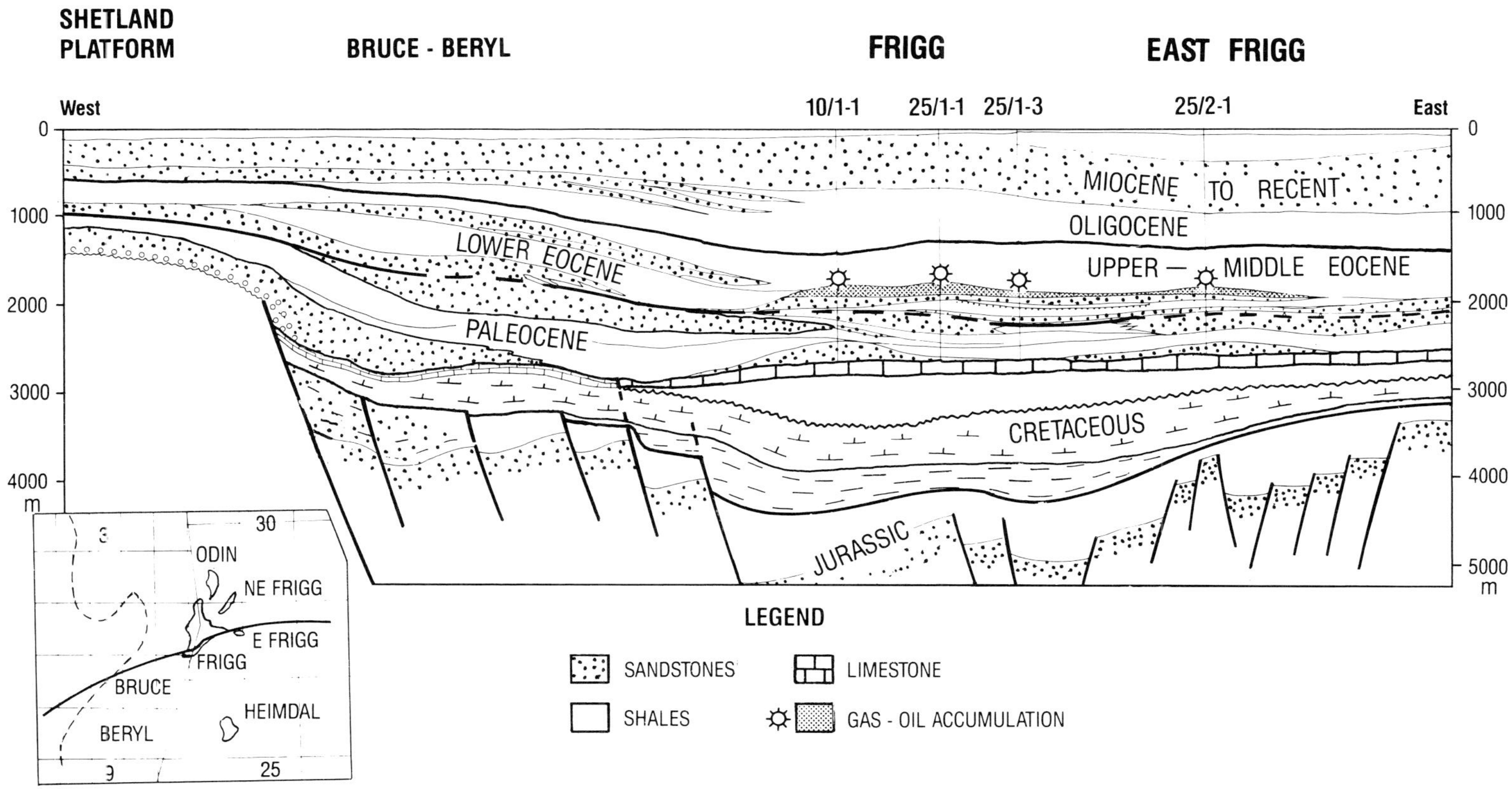

Fig. 3. Geological cross-section from the Shetland Platform to the axis of the Viking Basin (Modified from Héritier *et al.*, 1980).

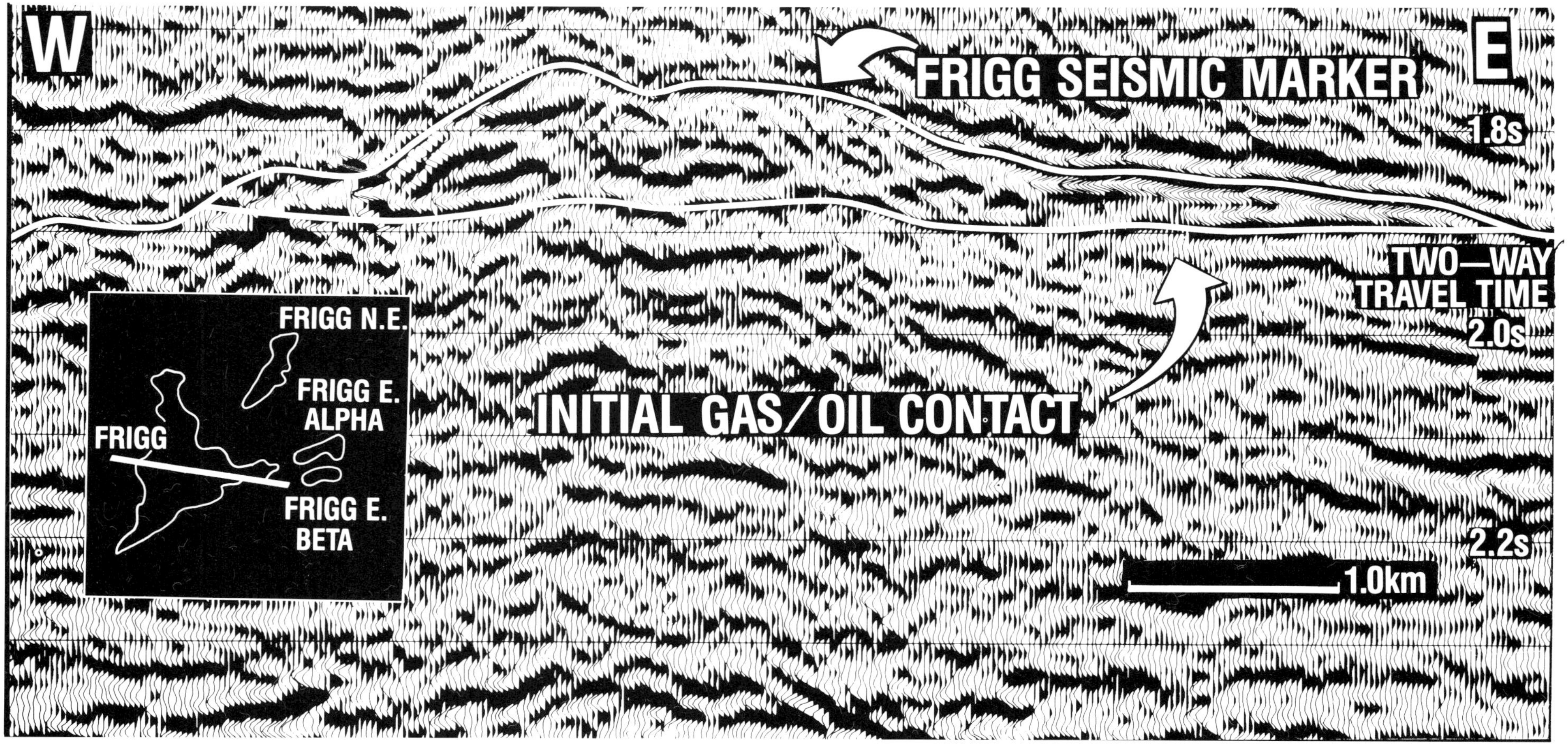

Fig. 4. Seismic line from 1985 3D survey: the Frigg seismic marker is the top of the reservoir.

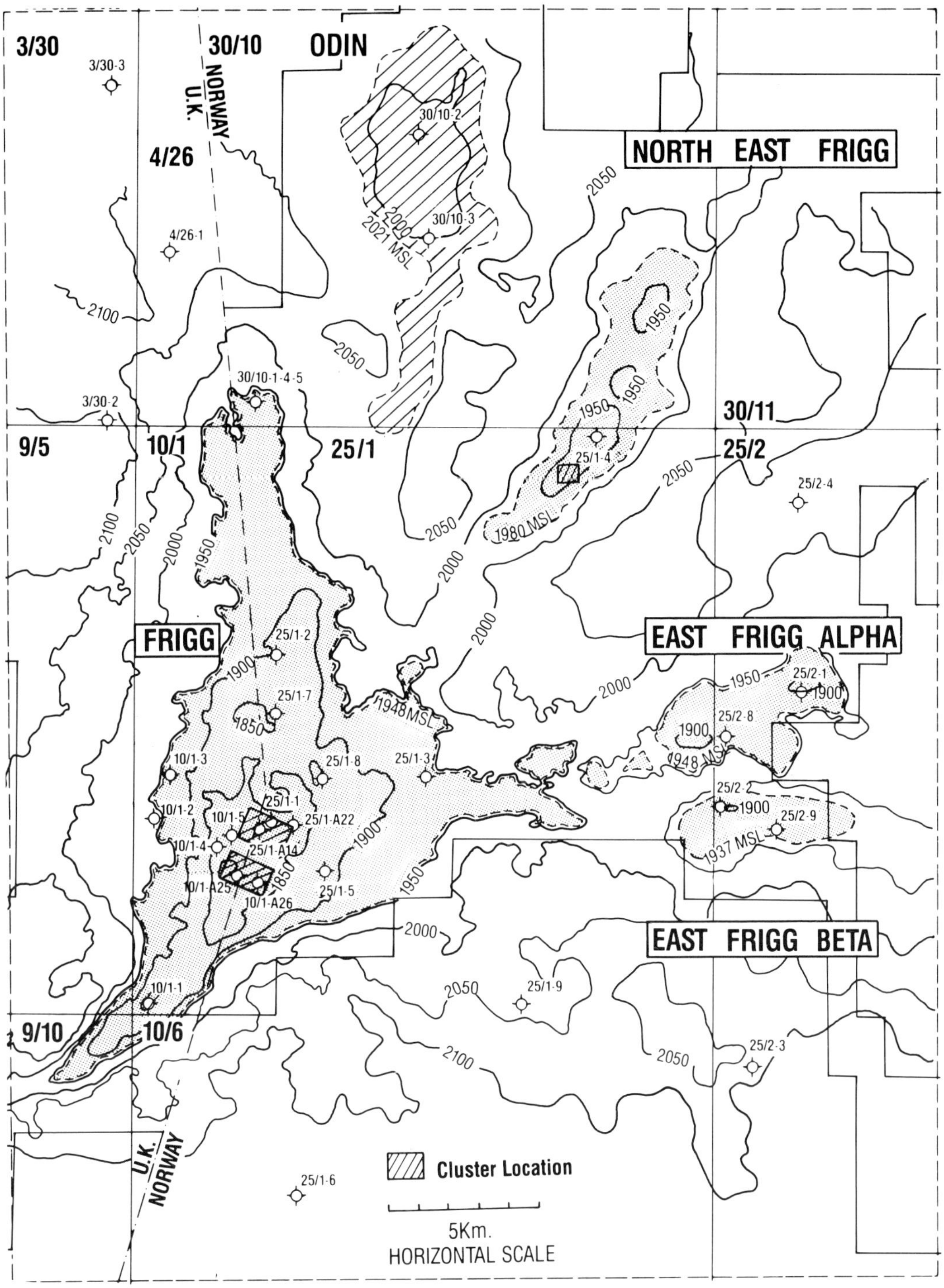

Fig. 5. The Frigg Field and its satellites: structure map of the top of the reservoir (contour interval = 50 m, all depths in mss).

its spill-point at 1955 mss. The structural closure is about 100 km^2.

RESERVOIRS

The gas production comes from sandbodies belonging to a Paleocene–Lower Eocene submarine fan (Figs. 6 and 7). The internal structure of the fan is very complex and was controlled by depositional and/or erosional topography, modified by draping and differential compaction of shales between the sandbodies. The reservoir section has been zoned on the basis of fossil assemblages into series of palynozones which correspond to the formations as shown in Fig. 6.

The Gas Pay Zone is within the nt IIb 1, 2 and 3 palynozones (Frigg Formation Upper Member). The aquifer comprises local Lower

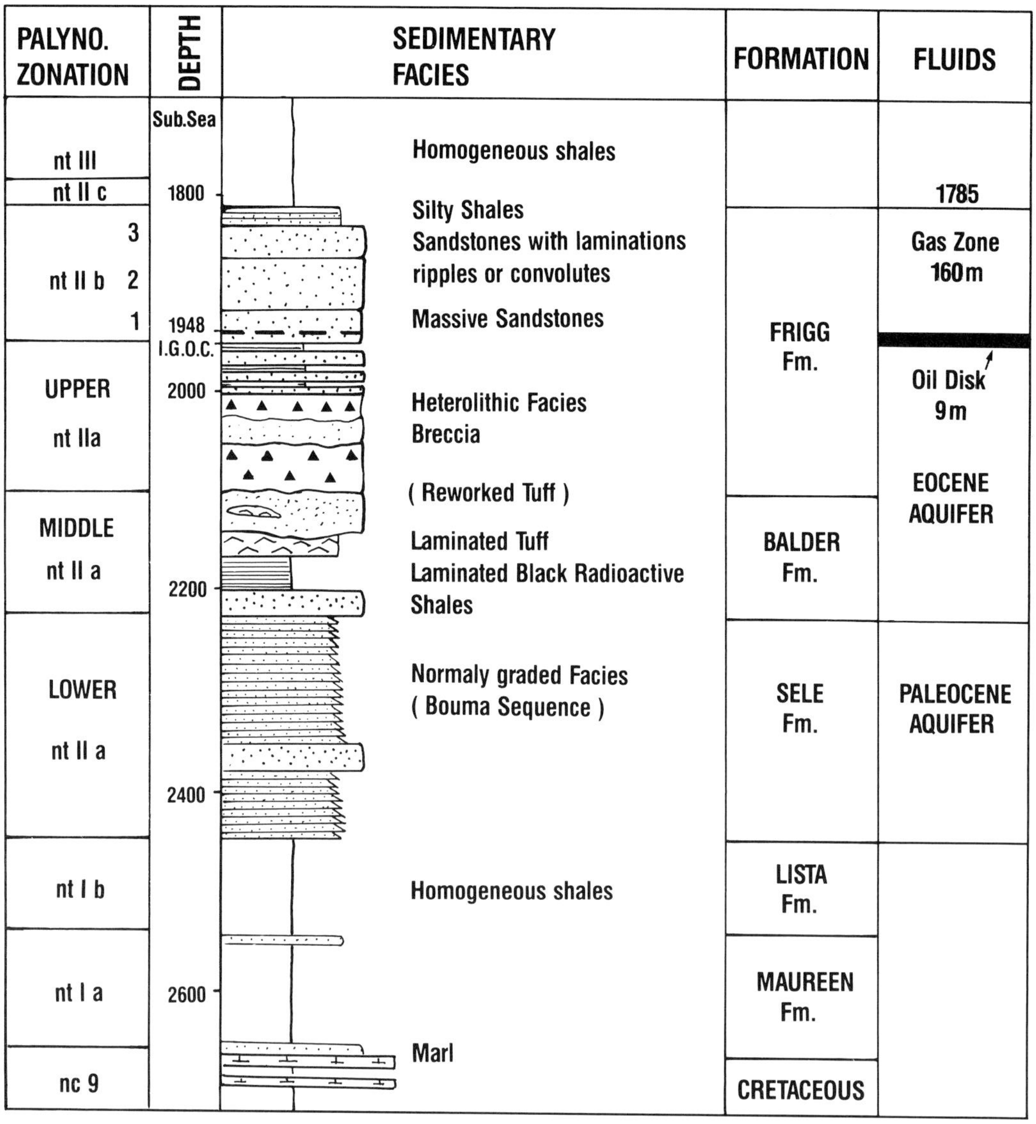

Fig. 6. Composite stratigraphic section of the reservoirs.

Eocene aquifers (Frigg and Balder formations) and a more regional extensive Paleocene aquifer (Sele Formation). The Tuffaceous Balder Formation is mainly represented by thick laminated tuff and black radioactive shales and acts as a partial barrier between the two Eocene aquifers. Nevertheless, in the central part of the field, tuff and shales are completely eroded, so communication is possible between the two aquifers. The Balde and Sele sandstones are also connected to the deeper Paleocene sands of the Maureen and Heimdal formations, which extend to the south-west of Frigg.

The continuous coring of the reservoir in 1984–85 in the new appraisal wells, using special fibreglass core barrels, gave excellent recoveries of the unconsolidated sand and finally allowed representative sedimentological and petrophysical studies. RFT pressure measurements on new wells allowed a good definition of the pressure barriers developed within the aquifer after production.

In the Gas Pay Zone (nt IIb zone), the Frigg reservoir is mainly composed of unconsolidated clean, fine sandstones with excellent reservoir characteristics: porosities range from 25 to 32% and average permeabilities from 0.9 to 3 D. No vertical sequence evolution has been observed; sandstones are organized into amalgamated beds. Locally the shale content increases from metric to decimetric alternations of laminated sandstones, sandstones with ripples and convolutes and very thin centimetric mudstones. These facies are of 'local' lateral and vertical extents but can cause relatively strong dynamic pressure barriers up to a few bar and temporarily stop the rise of the water in local area. These heterogeneities are very difficult to follow

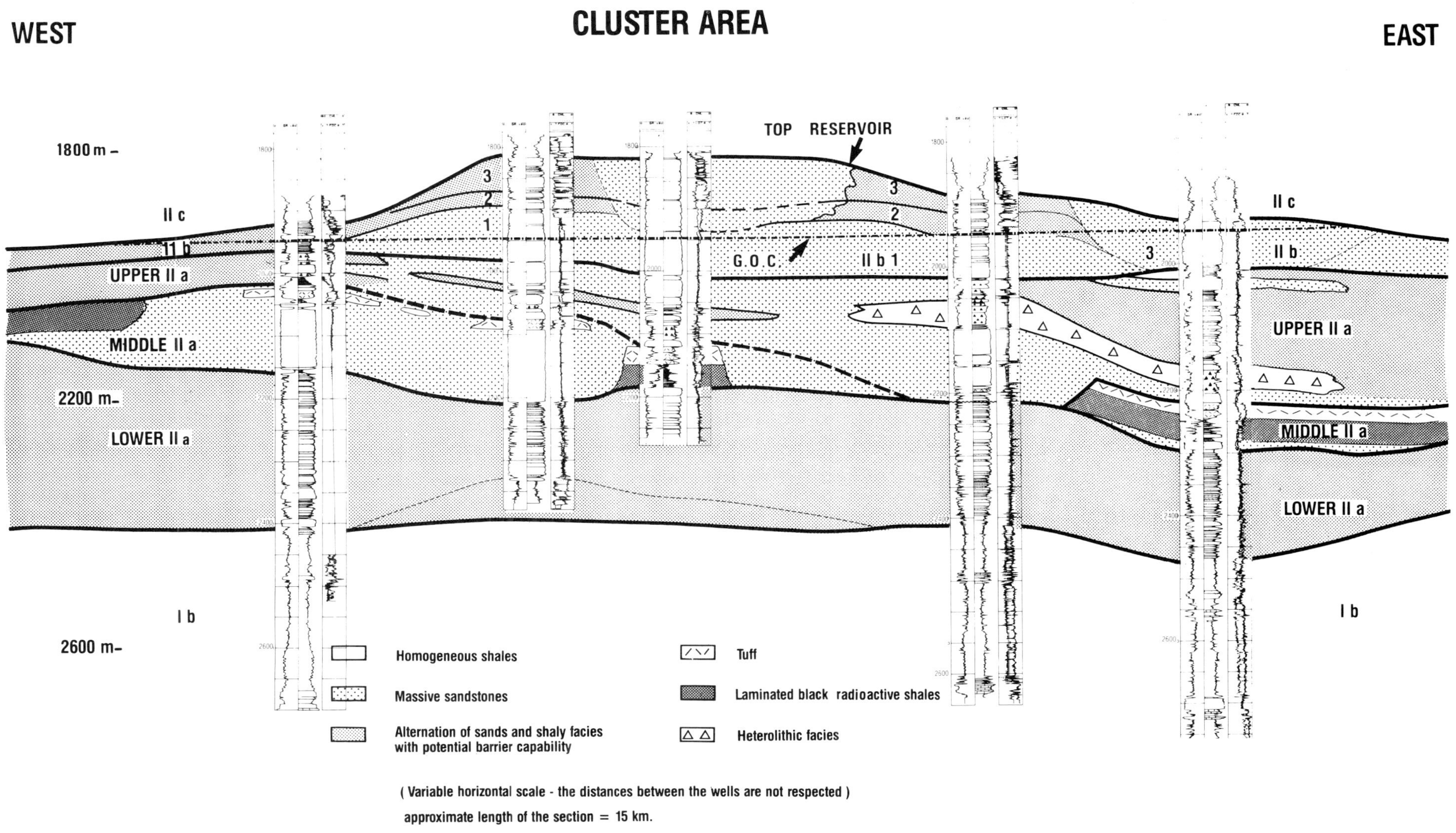

Fig. 7. West–east synthetic geological cross-section of the Frigg Field.

within the sandbodies, even using the new 1985, high-resolution seismic data.

The nt IIb sands are mainly grain-flow deposits, with locally strong erosional contacts which are mainly related to the channelized depositional environment. The Upper nt IIa zone is also mainly sandy but has a large content of heterolithic facies and breccias, related to turbiditic deposits with strong synsedimentary deformations. The lower nt IIa zone is mainly composed of stacked, fining-upwards sandstone (Bouma sequence) more representative of turbiditic deposits in a mid-fan position.

SOURCE

The Frigg gas probably results from a late phase of the maturation of the Jurassic source-rocks (Dogger, Lias), which show vitrinite reflectance values ranging from 1.5 to 1.8. By contrast, Lower Tertiary and Cretaceous source-rocks of the Frigg area are thermally immature gas source-rocks. The deep origin of the Frigg gas is also supported by its carbon-isotope composition which shows a δC^{13} value of -43.3‰, characteristic of a deep gas.

The Frigg oil was probably subject to an earlier phase of migration from the deeply buried Dogger and Lias source-rocks. Geochemical correlations based on pristane/nc^{17} to phytane/nc^{18} ratio can be made between less altered oil of East Frigg and Dogger and Jurassic source-rocks.

HYDROCARBONS

The gas of the Frigg Field is a dry gas with a percentage composition as follows: N_2, 0.4; CO_2, 0.3; C_1, 95.5; C_2, 3.5; C_3, 0.04; C_4, 0.01; $> C_4$ traces. The associated condensate is composed of 86.5% of C_{11+}. At an initial absolute pressure of 199.25 bar at a subsea depth of 1912 mss the gas viscosity is 0.018 cp at a reservoir temperature of 61 °C. The gas compressibility factor (Z) is 0.865.

The oil beneath the gas accumulation has a gravity of 0.835 g/cm^3 and is of a naphthenic composition. It is characterized by a saturate aromatic hydrocarbon ratio of 3:76. The light compounds with a carbon number of $< C_{17}$ represent a minor fraction of the oil, and alkanes are almost absent. The saturated hydrocarbons are mainly represented by cyclanes and isoalkanes.

The accumulation was originally at hydrostatic pressure conditions.

RESERVES

In 1978, experts De Golyer and MacNaughton estimated the gas-in-place to about 268×10^9 Sm^3. Of this total, 60.82% are in the Norwegian sector and 39.18% are in the United Kingdom. The figure given for the recoverable gas reserves was within the range of 227×10^9 Sm^3 down to 215×10^9 Sm^3.

In 1985 MacCord Lewis Energy were retained as experts to update the field reserve after new 1984 appraisal wells. They estimated a lower figure for the gas recovery factor (down to 72%) which gave recoverable reserves of 191×10^9 Sm^3.

The operator is today interpreting the new wells and large amount of seismic data (five wells, three-dimensional seismic survey, aquifer monitoring) acquired from 1984 to 1986. An updated reservoir model is being built, giving a better description of the heterogeneities within the reservoir, emphasizing the potential barrier zones and their extension on the field. A new structural map has been made from the three-dimensional seismic interpretation. The results of the new reservoir model will be known in spring 1987.

Cumulative production of gas reached 135×10^9 Sm^3 in December 1986, with an average production rate of 45×10^6 Sm^3/day, reduced to 38×10^6 Sm^3/day in 1986. The behaviour of the production wells is excellent and no skin or turbulence effects have been noticed up to a production rate of 3×10^6 Sm^3/day per well.

The production mechanisms are depletion and waterdrive, supported by an active aquifer. The oil-in-place was estimated to be 125×10^6 Sm^3. Extensive testing on Well 25/1-3 proved that production of the seven-metre-thick disc of heavy naphthenic oil of the Frigg Field was technically difficult and uneconomic.

ACKNOWLEDGMENT

The author thanks Elf Aquitaine Norge A/S, and the British and Norwegian partners of the Frigg Units (Elf UK, Total Oil Marine, Total Marine Norsk, Norsk Hydro, Statoil) for giving permission to publish this paper.

REFERENCES

Barril, R. 1985. Reservoir management considerations in the Frigg Field, In *Offshore Northern Seas—Advanced Projects Conference—Tech-*

nology Applications Today and Tomorrow, Article A5.

Brewster, J. and Jeangeot, G. 1987. The production geology of the Frigg Field, In Kleppe, J. *et al.* (Eds), *North Sea Oil and Gas Reservoirs*, Graham and Trotman, London.

Goff, J.C. 1983. Hydrocarbon generation and migration from Jurassic source-rocks in the E. Shetland Basin and Viking Graben of the northern Sea, *Journal of the Geological Society of London* **140** (3), 445–474.

Hamre, S. 1984. Improved recovery from thin oil sands. Part 1—Frigg Field, In *Offshore Northern Seas—Improved offshore Recovery* (Norsk Petroleumsforening), Article R2.

Héritier, F.E., Lossel, P. and Wathne, E. 1980. The Frigg gas field, In Illing, L.V. and Hobson, G.D. (Eds), *Petroleum Geology of the Continental Shelf of North-west Europe*, Heyden, London, pp. 380–391.

Héritier, F.E., Lossel, P. and Wathne, E. 1980. Frigg Field—large submarine-fan trap in Lower Eocene rocks of the Viking Graben, North Sea, *Memoir of the American Association of Petroleum Geologists* **30,** 59–79. (Also published in *Bulletin of the American Association of Petroleum Geologists* **63** (11), 1999–2020, 1979.)

Leebeck, A.D. 1987. The Frigg Field reservoir: characteristics and performance, In Kleppe, J. *et al.* (Eds), *North Sea Oil and Gas Reservoirs,* Graham and Trotman, London.

McGovney, J.E. and Radovich, B.J. 1985. Seismic stratigraphy and facies of the Frigg fan complex, *Memoir of the American Association of Petroleum Geologists* **39,** 139–154.

Revoy, M. 1984. Frigg Field: production history and seismic response, In *Offshore Northern Seas—Reduction of Uncertainties by Innovative Reservoir Geo-modelling*, Norsk Petroleumsforening, Article G5.

Skaug, M. and Gundesø, L. 1986. Geological modelling of the Frigg Field with special emphasis on shale mapping, *Society of Petroleum Engineers, European Petroleum Conference Proceedings,* pp. 111–121 (SPE 15859).

Thouand, A., Brewster, J. and Mure, E. 1986. Frigg Field mid-life appraisal programme, In *Offshore Northern Seas—Reservoir Engineering Conference*, (Norsk Petroleumsforening), Article R4.

Ungerer, P., Chiarelli, A. and Oudin, J.L. 1985. Modelling of petroleum genesis and migration with a bidimensional computer model in the Frigg sector, Viking Graben, In Thomas, B.M. *et al.* (Eds), *Petroleum Geochemistry in Exploration of the Norwegian Shelf,* Graham and Trotman, London, pp. 121–129.

17

Øst Frigg—Alpha and Beta

E. Mure

Elf Aquitaine Norge A/S, PO Box 168, 4000 Stavanger, Norway

The Øst Frigg Fields—Alpha and Beta—are twin satellites of the Frigg Field. They are located 15 km east of the Frigg production platform cluster. They straddle the boundary between Blocks 25/1 and 25/2. The average water depth is 100 m. The reservoirs consist of submarine-fan sands of the Lower Eocene Frigg Formation. The two structures are mainly a depositional topography which have been structurally enhanced by draping and differential compaction of the sands. The areas of structural closure are 13 km^2 on the Alpha structure and 7 km^2 on the Beta structure. The reservoirs are at 1900 mss subsea, with a gas column 56 m thick overlain by an oil disc 8 m thick on the Alpha structure, and a gas column 37 m thick and an oil disc 11 m thick on the Beta structure. Recoverable gas reserves are estimated to be 8.9×10^9 Sm^3 on the Alpha structure and 3.7×10^9 Sm^3 on the Beta structure. Declared commercial in 1984, the fields will be developed in 1987 and 1988 with advanced subsea 'Skuld' technology which comprise two subsea clusters, remotely controlled from Frigg Field installations. The gas will be transported to the Frigg treatment and compression platform to be processed and tied into the transportation system of the Frigg Field. Production should start at the end of year 1988.

INTRODUCTION

The Øst Frigg Fields—Alpha and Beta—are twin satellites of the Frigg Field. They are located 15 km east of the Frigg production platform cluster. The average water depth is around 100 m. They straddle the boundary between Petronord Group Blocks 25/1 and 25/2. They belong to Licence 024 (Block 25/1), Licence 026 (Block 25/2) and Licence 112 (Block 25/2). The share of the participating interest is 93% for Licences 024, 026 and 7% for Licence 112. The participating interests in the Øst Frigg Association are:

Elf Aquitaine	40.05% (operator)
Norsk Hydro	31.78%
Total Marine Norsk	20.02%
Statoil	8.15%

Øst Frigg Alpha covers an area of approximately 13 km^2 and Øst Frigg Beta an area of about 7 km^2 (Fig. 1). The reservoirs consist of Lower Eocene sands of the Frigg formation, which were deposited as the final stage of a complex Paleocene–Eocene submarine fan.

The current most likely estimate of original gas-in-place is 21.6×10^9 Sm^3, of which 15×10^9 Sm^3 are in Øst Frigg Alpha and 6.6×10^9 Sm^3 in Øst Frigg Beta.

HISTORY

In 1969 Blocks 25/1 and 25/2 were granted to the Petronord Group (Elf Aquitaine Norge, Norsk Hydro, Total Marine Norsk) during the second Norwegian round. In 1973, two years after the Frigg Field discovery, the first exploration well, 25/2-1, was drilled on the *Alpha structure*, which was interpreted as an extension to the east of the Frigg sands. The well entered the Lower Eocene Frigg sands at 1891 mss subsea and proved a 56 m gas column underlain by an oil disc 8 m thick. The

Geology of the Norwegian Oil and Gas Fields © A. M. Spencer *et al.*, eds (Graham & Trotman, 1987) pp. 215–222.

Øst Frigg data summary

	Alpha	Beta
Trap		
Type	structured stratigraphic	structured stratigraphical
Depth to crest	1891 mss	1900 ms
Lowest closing contour	1955 mss	1950 mss
Hydrocarbon contacts	GOC 1948 mss, OWC 1956 mss	GOC 1937 mss, OWC 1947 mss
Gas column	56 m	37 m
Oil column	8 m	11 m
Pay zone		
Formation	Frigg	Frigg
Age	Lower Eocene (nt IIb palynozone)	Lower Eocene (nt IIb palynozone)
Gross thickness	40 m	25 m
Net/gross ratio	0.987	1.0%
Cut off for N/G	$V_{sh} < 40\%$	$V_{sh} < 40\%$
Porosity (average; range)	28.5%	28%; 27–30%
Hydrocarbon saturation	92.5%; 90–94%	90%; 86.5–92.5%
Permeability	1000 mD (700 $< k <$ 1500 mD)	1200 mD (400 $< k <$ 2000 mD)
Well rate	1.5×10^6 Sm^3/day/well	1×10^6 Sm^3/day/well
Hydrocarbons		
Oil gravity	0.83 g/cm^3	0.83 g/cm^3
Gas gravity (relative to air)	0.586	0.578
Condensate yield	5.1 $Sm^3/10^6$ Sm^3 gas	5.1 $Sm^3/10^6$ Sm^3 gas
Gas expansion factor	196	196
Formation water		
Salinity	55 000 ppm NaCl equivalents	55 000 ppm NaCl equivalents
Reservoir conditions		
Temperature	63 °C	63 °C
Pressure	197 bar at 1930 mss	197 bar at 1930 mss
Pressure gradient in reservoir		
gas	0.013 bar/m	0.013 bar/m
oil	0·083 bar/m	0.083 bar/m
water	0.103 bar/m	0.103 bar/m
Field size		
Area	13 km^2	7 km^2
Gross rock volume	295×10^6 m^3	132×10^6 m^3
Hydrocarbon pore volume	77×10^6 m^3	34×10^6 m^3
Recovery factor—oil	0	0
Recovery factor—gas	0.59	0.61
Drive mechanism	depletion and water-drive	depletion and water-drive
Recoverable gas	8.9×10^9 Sm^3	3.7×10^9 Sm^3
Production		
Start-up date	1988	1988
Development scheme	subsea—'skuld'	subsea—'skuld'

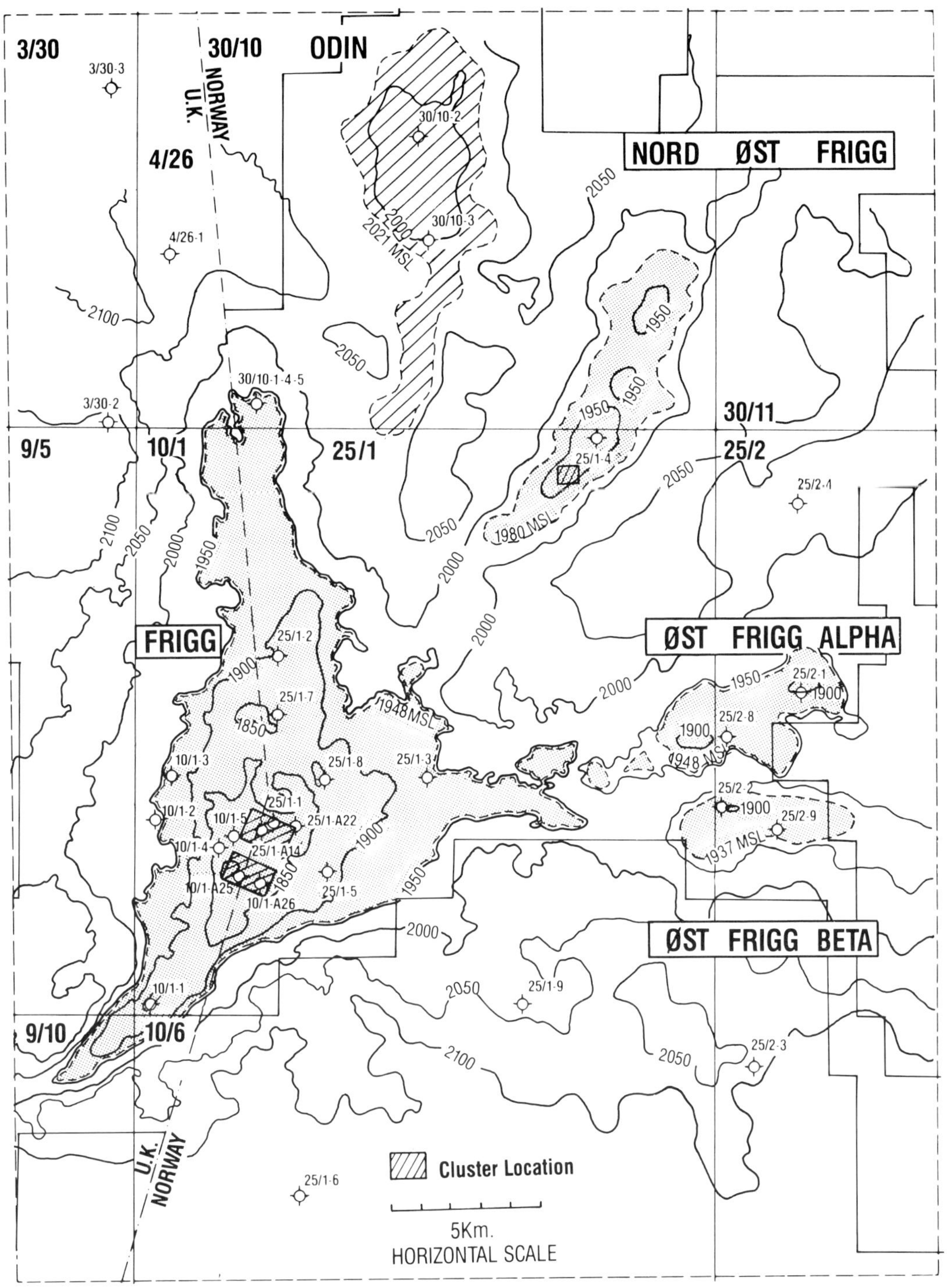

Fig. 1. The Frigg Field and its satellites. Structure map of the top of the reservoirs (contour interval = 50 m, all depths in mss).

original gas–oil contact was at the same depth (1947 mss) as on Frigg Field. In 1974, Well 25/2-2 was drilled on the *Beta structure*. The well entered the Lower Eocene Frigg sands at 1922 mss with a gas column 15 m thick underlain by an oil disc 11 m thick. The initial gas–oil contact at 1937 mss is 10 m above the gas–oil contact at Øst Frigg Alpha. Further seismic interpretation suggested that Well 25/2-2 was on the northern flank of the structure. The maximum gas-column thickness is estimated to be 37 m on the culmination. New high-resolution seismic surveys have been shot every three years on the Øst Frigg Satellites, extending over the saddle with Frigg Field. Two new exploration wells were drilled in 1984 and 1985: Well 25/2-8 on Øst Frigg Alpha and Well 25/2-9 on Øst Frigg

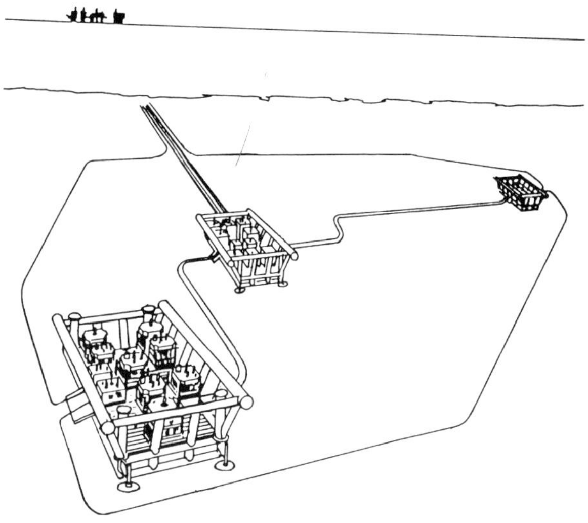

Fig. 2. Future installations of the Øst Frigg Fields.

Beta, with continuous coring of the reservoirs.

The fields were declared commercial in 1984. The gas will be sold to the British Gas Corporation within the existing Frigg sales agreement. Production should start around October 1988 for a production period of approximately 13 years.

The development will be based on advanced subsea 'Skuld' technology (Fig. 2). Development plans include two identical subsea production systems, one on each structure, remotely controlled from Frigg. A central manifold will tie in the two systems, from where the gas will be transported by pipeline to the Frigg treatment and compression platform (TCP2). There the gas will be processed and tied into the transportation system of the Frigg Field.

GEOLOGICAL SETTING

The Øst Frigg fields are part of the complex Paleocene—Lower Eocene Frigg submarine fan. All general geological discussions made in the Frigg Field chapter are still valid here.

GEOPHYSICS

Over the entire Frigg area, the top of the Frigg Formation is clearly defined by a strong seismic marker (Frigg Seismic Marker), which is caused by the contrast of acoustic impedance between the overlying shales and the Frigg sands (Figs. 3 and 4). The initial gas–oil contact is also a very obvious seismic marker ('flat spot') on both the Alpha and Beta structures, but on Beta, it appears to be more discontinuous, probably due to interference caused by lithological changes.

Depth maps have been calculated by two methods: by reconstruction from the migrated time maps with velocity field calibrated on wells and by the 'DHI' method, from a time isopach map of the gas zone.

TRAP

As in the Frigg Field, the Øst Frigg Fields are stratigraphical traps. The two structures are on the top of E–W elongated, large sand-bodies, which have been structurally enhanced by draping and differential compaction of shales. The structural relief is about 64 m for the Alpha structure and 50 m for the Beta structure The structures are filled to their spill-points. The gas column is 56 m on Alpha and 37 m on Beta.

RESERVOIRS

The Øst Frigg Reservoirs are very similar to those of the Frigg Field. The same zonation has been defined within the reservoir section based on palynological studies (Fig. 5 and 6). No obvious regional evolution of the sedimentary facies is observed within the Frigg formation. The gas pay zone is within the Nt IIb palynozones. The reservoirs consist of unconsolidated clean fine sandstone with excellent characteristics: high average porosities (30%) and permeabilities in the range of one Darcy. Locally the shale content increases from metric to decimetric scale with alternations of laminated sandstones, sandstones with ripples and convolutes and very thin centimetric mudstones. These facies have much lower vertical and lateral permeabilities. The Frigg sands are mainly grain-flow and turbiditic deposits in a channelized environment. Easy lateral and vertical communications are possible within the Eocene aquifers of the Frigg and Balder formations.

SOURCE

The hydrocarbons of Øst Frigg are very similar to the gas and oil of the Frigg Field and their origin is the same.

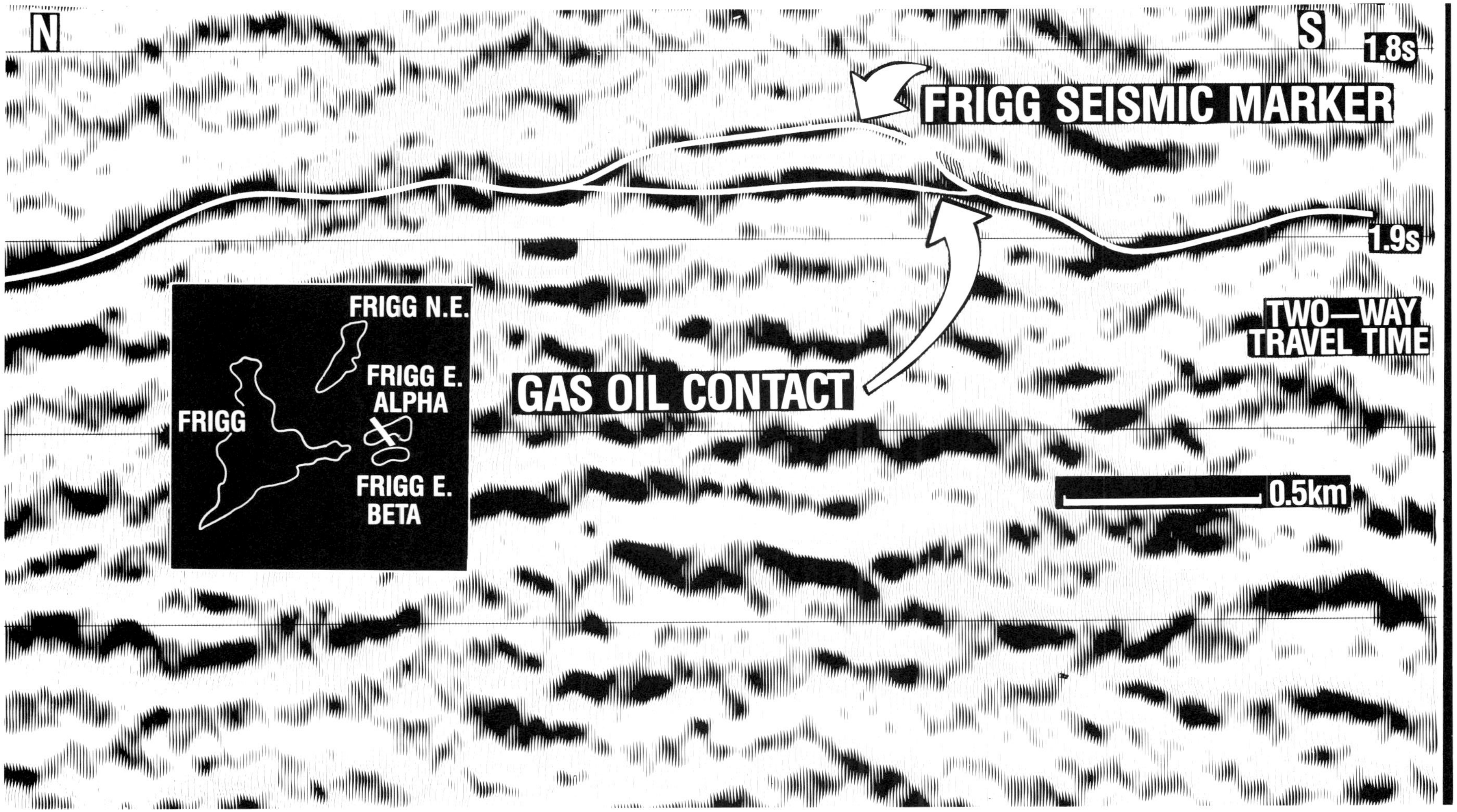

Fig. 3. Seismic line of 1984 across Øst Frigg Alpha.

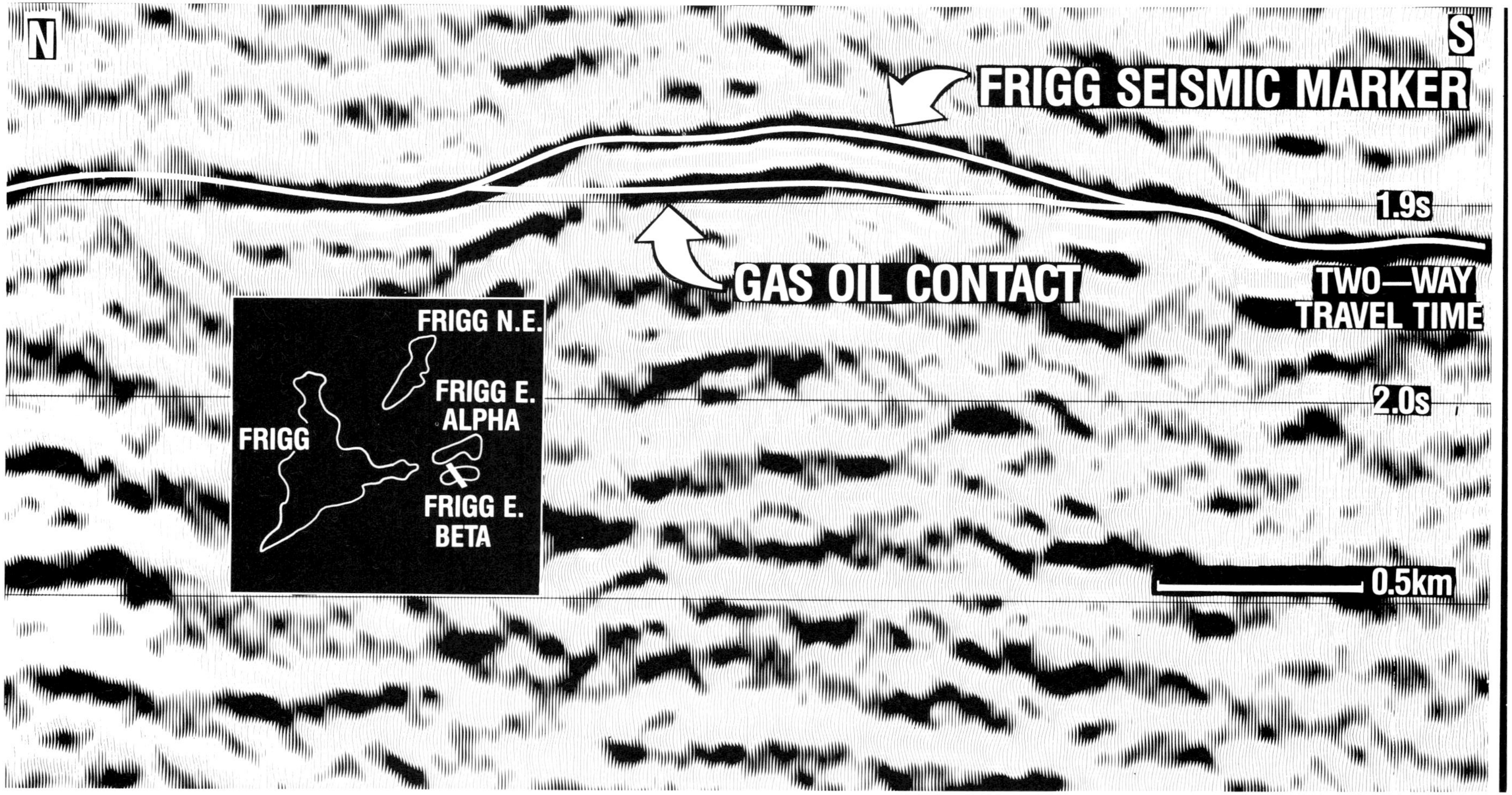

Fig. 4. Seismic line of 1984 across Øst Frigg Beta.

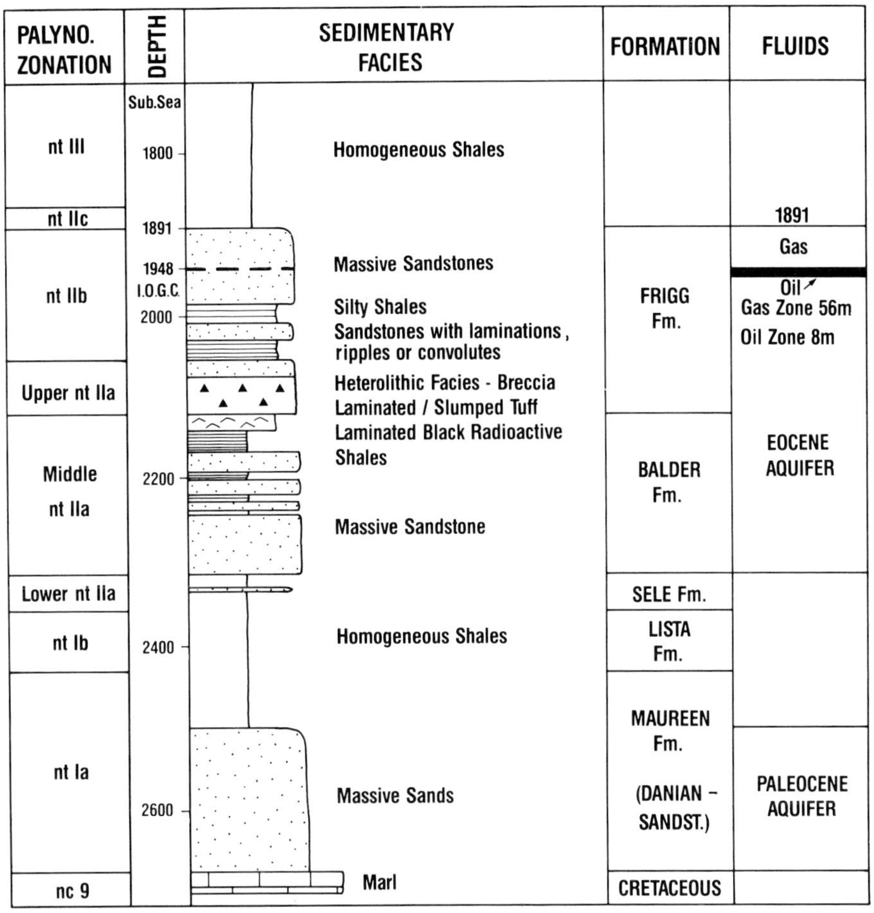

Fig. 5. Composite stratigraphic section of the reservoirs, Øst Frigg Alpha.

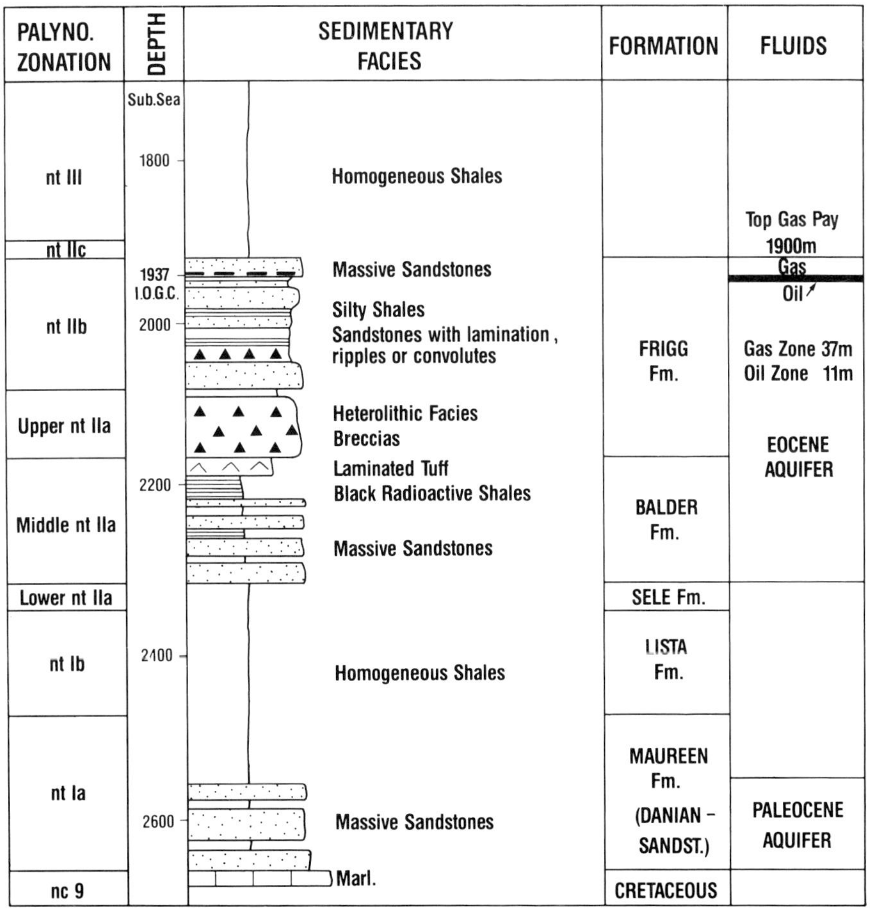

Fig. 6. Composite stratigraphic section of the reservoirs, Øst Frigg Beta.

HYDROCARBONS

The gas of Øst Frigg is a dry gas. Its percentage composition is as follows: Øst Frigg Alpha: N_2, 0.7; CO_2, 0.3; C_1, 94.9; C_2, 3.9; C_3, 0.08; C_4, 0.02; $> C_4$ traces and Øst Frigg Beta: N_2, 0.6; CO_2, 0.05; C_1, 95.6; C_2, 3.6; C_3, 0.07; C_4, 0.01; $> C_4$ traces.

The oil beneath the gas accumulation has a gravity of 0.83 g/cm^3 and is of a naphthenic composition. The oil is much less altered on Øst Frigg than on the Frigg Field. The accumulation was originally at hydrostatic pressure.

RESERVES

The current most likely estimate of original gas-in-place is 21.6×10^9 Sm^3: 15×10^9 Sm^3 on the Alpha structure and 6.6×10^9 Sm^3 on the Beta structure. The gas-recovery factor is estimated to be around 0.59 on the Alpha structure and 0.61 on the Beta structure. These low figures are due to the very flat shape of the accumulation. The recoverable reserves are 8.9×10^9 Sm^3 on Alpha structure and 3.7×10^9 Sm^3 on Beta structure.

ACKNOWLEDGEMENT

The author thanks Elf Aquitaine Norge A/S, and the partners of the Norwegian Licences 024 and 026 for giving permission to publish this paper.

18

Nord-Øst Frigg

E. Mure

Elf Aquitaine Norge A/S, PO Box 168, 4001 Stavanger, Norway

The Nord-Øst Frigg Field is a satellite of the Frigg Field. It is located 15 km north-east of the Frigg production platform cluster. It straddles the boundary between Blocks 25/1 and 30/10. The average water depth is 100 m. The reservoirs consist of Lower Eocene sands of the Frigg Formation sealed by Middle Eocene open-marine shales. The structure has mainly a submarine-fan depositional topography which has been enhanced by draping and differential compaction of the sands. The area of structural closure is 19 km^2. The reservoirs were proved at 1923 mss with a 57 m thick gas column underlain by a 3.4 m thick oil disc. Recoverable gas reserves are estimated to be 11×10^9 Sm^3. The development plan is based on one subsea cluster remotely controlled from an articulated column and from the Frigg Field installations. The gas is treated on the Frigg Field installations and tied into the Frigg Field transportation system. Production started in December 1983. Cumulative gas production had reached 5.6×10^9 Sm^3 at the end of 1986.

INTRODUCTION

The Nord-Øst Frigg Field is a satellite of the Frigg Field. It is located 15 km northeastward from the Frigg production platforms (Fig. 1). The average water depth is in the range of 100 m. The field straddles the boundary between Petronord Group Block 25/1 and Esso Block 30/10. The share of the accumulation between the two blocks was estimated in 1984 to be 42% in Block 25/1 and 58% in Block 30/10. The participating interests after Nord-Øst Frigg unitization are as follows:

Elf Aquitaine Norge	17.39% (operator)
Esso	58%
Norsk Hydro	13.8%
Total Marine Norsk	8.69%
Statoil	2.1%

Nord-Øst Frigg covers an area of approximately 19 km^2. The reservoirs consist of Lower Eocene sandstones of the Frigg Formation which were deposited as the final part of a complex Paleocene–Eocene submarine fan. The current most likely estimate of original gas-in-place is around 18×10^9 Sm^3.

HISTORY

Discovery Well 25/1-4 was spudded in 1974, three years after the Frigg Field discovery. It was located at the culmination of a NE–SW-elongated anticlinal structure, which proved to be an extension to the north-east of the Frigg Formation sands. The prospect showed direct hydrocarbon indicators on the seismic in the form of a 'flat spot'. The well entered the Lower Eocene Frigg sands at 1942 mss and proved a gas column 38 m thick underlain by a oil disc 3.4 m thick. The initial gas–oil contact at 1980 m was 33 m below the Frigg Field gas–oil contact. The field was declared commercial in 1980. The develop-

Geology of the Norwegian Oil and Gas Fields © A. M. Spencer *et al.*, eds (Graham & Trotman, 1987) pp. 223–228.

Nord-Øst Frigg Field data summary

Trap	
Type	Structure stratigraphical
Depth to crest	1923 mss
Lowest closing contour	1990 mss
Hydrocarbon contacts	GOC initial 1980 mss
	OWC initial 1983.5 mss
Gas column	57 m
Oil column	3.5 m
Pay zone	
Formation	Frigg
Age	Lower Eocene (nt IIb palynozone)
Net/gross ratio	0.9
Cut off for N/G	$V_{sh} < 33\%$
Porosity	28%
Hydrocarbon saturation	85%
Permeability	1000 mD (750 $< k <$ 1500 mD)
Well rate	2.5×10^6 Sm^3/day/well
Hydrocarbons	
Oil gravity	0.896 g/cm^3
Gas gravity (relative to air)	0.585
Condensate yield	7.3 $Sm^3/10^6$ Sm^3 gas
Gas expansion factor	195
Formation water	
Salinity	65 000 ppm NaCl equivalents
Reservoir conditions	
Temperature	61.5°C
Pressure	201 bar at 1965 mss
Pressure gradient in reservoir	
gas	0.013 bar/m
oil	0.083 bar/m
water	0.102 bar/m (hydrostatic)
Field size	
Area	19 km^2
Gross rock volume	$420–450 \times 10^6$ m^3
Hydrocarbon pore volume	$90–96 \times 10^6$ m^3
Recovery factor—oil	0
Recovery factor—gas	0.6
Drive mechanism	Depletion and waterdrive
Gas in-place	18×10^9 Sm^3
Recoverable gas	11×10^9 Sm^3
Production	
Start-up date	December 1983
Development scheme	Subsea template with articulated platform
Production rate	6×10^6 Sm^3/day
Cumulative production to date	5.65×10^9 Sm^3 (January 1987)

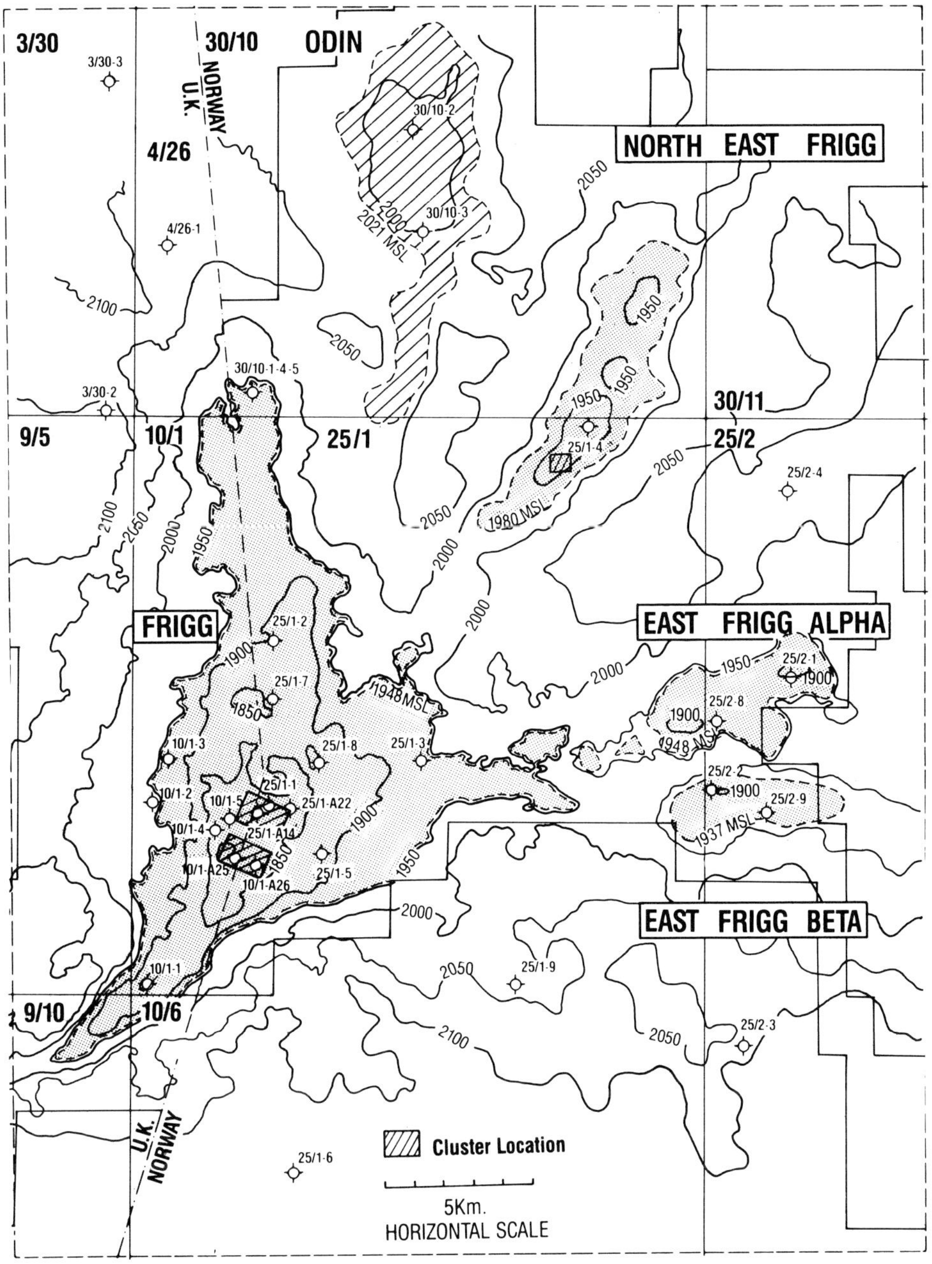

Fig. 1. The Frigg Field and its satellites. Structure map of the top of the reservoir (contour interval = 50 m, all depths in mss).

ment plan was based on advanced subsea technology (Fig. 2). The production wells were concentrated in a small area, with one subsea cluster of six production wells located at the structural culmination. The subsea cluster is remotely controlled from an articulated column and/or directly from the Frigg Field installations. The gas is pipelined and treated at the Frigg installations. Production started in December 1983, but the sale of the gas started in 1981, with the gas being supplied by the Frigg Field. The 'reimbursement' will be accomplished during the production of Nord-Øst Frigg.

GEOLOGICAL SETTING

The Nord-Øst Frigg Field is part of the complex Paleocene–Lower Eocene Frigg submarine-fan. All the general geological discussions of the Frigg Field article are still valid here.

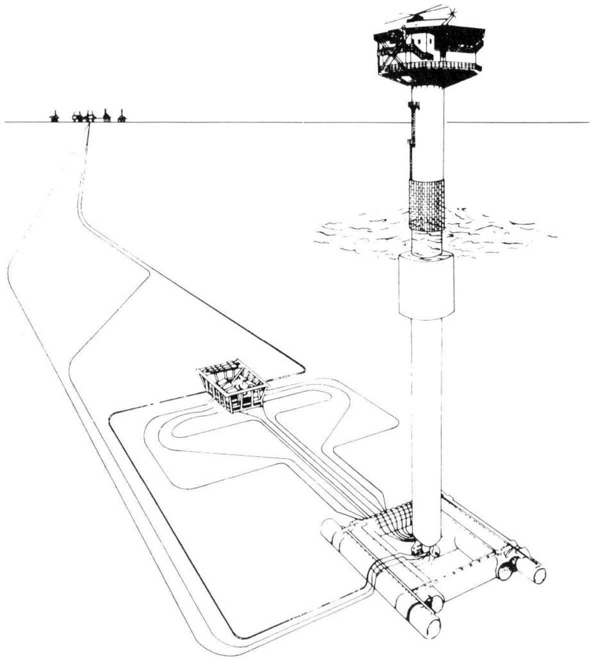

Fig. 2. Installations on Nord-Øst Frigg Field.

GEOPHYSICS

The top of the Frigg Formation is clearly defined by a strong seismic marker (Frigg Seismic Marker) which is caused by the positive contrast of acoustic impedance between the sands and the overlying shales (Fig. 3). The initial gas–oil contact is also a very obvious seismic marker ('flat spot'). The Nord-Øst Frigg structure is very flat in time. There is less than 30 ms of TWT Relief over a 19 km^2 area, and consequently the structure is very sensitive to depth-conversion procedure. The structural map was obtained by conventional depth conversion. The DHI method, using a time isopach between the Frigg seismic marker and the flat spot was applied in the gas zone. The seismic expression of the saddle between the Nord-Øst Frigg and Frigg structures was studied in order to evalute the effect of the production on the initial gas–oil contacts of the two fields.

TRAP

The structure is formed by submarine-fan sediments sealed by Middle Eocene open-marine shales. The structural relief, which is accentuated by differential compaction, is about 60 m at the top of the Frigg sands. The structural closure is approximately 19 km^2. The structure was filled down to its spill-point.

RESERVOIRS

The reservoirs are very similar to those of the Frigg Field. The same zonation has been defined based on palynological studies (Fig. 4). No obvious regional evolution of the sedimentary facies is observed within the Frigg Formation. The gas pay zone is within the Nt IIb palynozone. The reservoirs consist of unconsolidated, clean, fine sandstones with excellent characteristics: a high average porosity (28%) and a permeability of around one Darcy. The shale content increases locally from metric to decimetric scale with alternations of laminated sandstones, sandstones with ripples and convolutes and very thin centimetric mudstones. These facies have low vertical and horizontal permeabilities. The Frigg sands are mainly grain-flow and turbiditic deposits in a channelized environment.

Lateral and vertical communications are possible within the Eocene/Paleocene aquifers (Frigg, Balder and Sele formations).

SOURCE

Hydrocarbons of Nord-Øst Frigg are very similar to gas and oil of the Frigg Field, and their origin is the same.

HYDROCARBONS

The gas is a dry gas with a condensate/gas ratio of only 7.3 $Sm^3/10^6$ Sm^3. Its percentage composition is as follows: N_2, 0.67; CO_2, 0.3%; C_1, 94.2%; C_2, 4.65%; C_3, 0.07%; C_4, 0.02%; $> C_4$, traces. The oil has a gravity of 0.89 kg/cm^3. It is of naphthenic composition. The accumulation was originally at hydrostatic pressure.

RESERVES

The current most likely estimate of original gas in-place is 18×10^9 Sm^3. With an estimated gas-recovery factor of 60%, the recoverable reserves are around 11×10^9 Sm^3. The cumulative production reached 5.65×10^9 Sm^3 at the end of 1986.

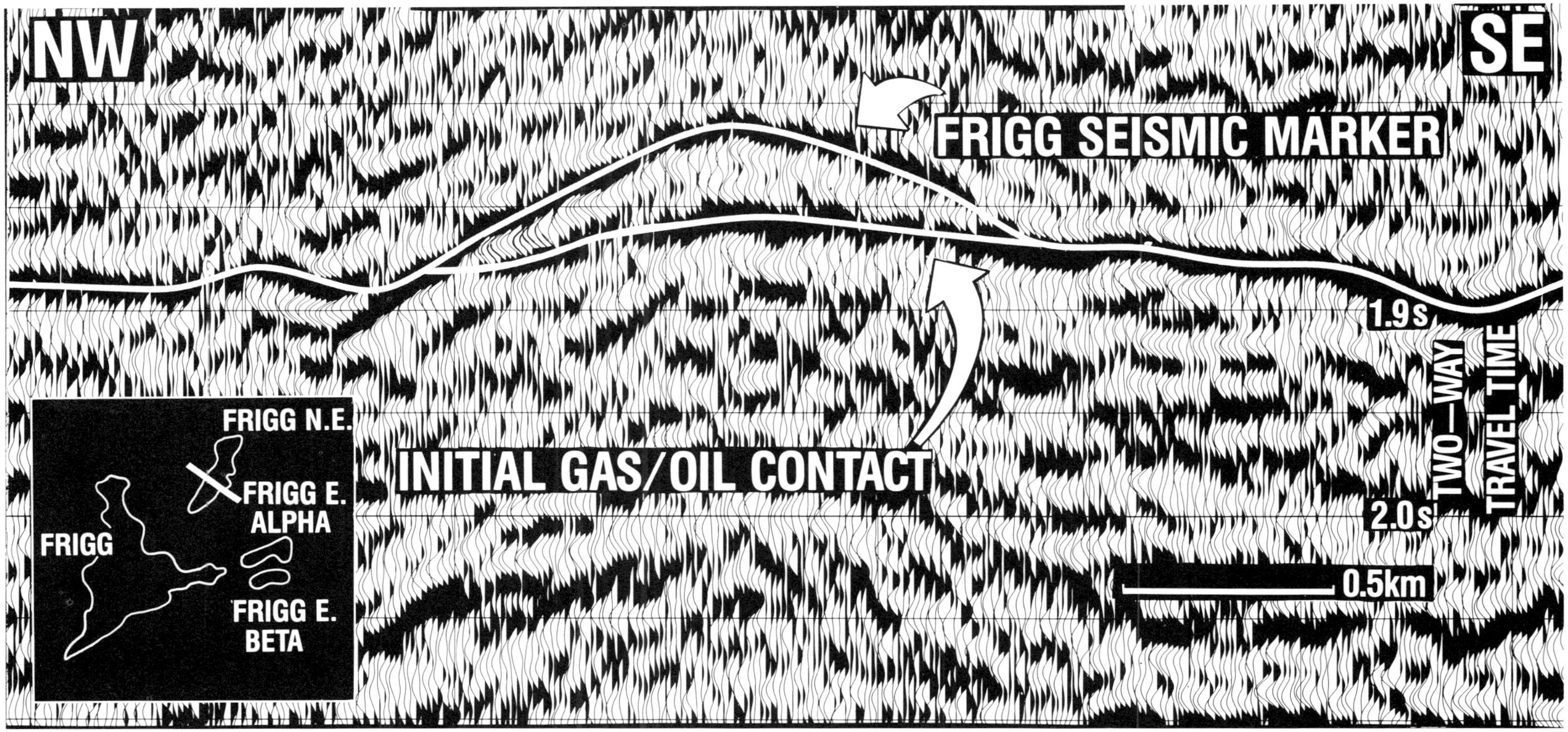

Fig. 3. Seismic line from 1979 across Nord-Øst Frigg.

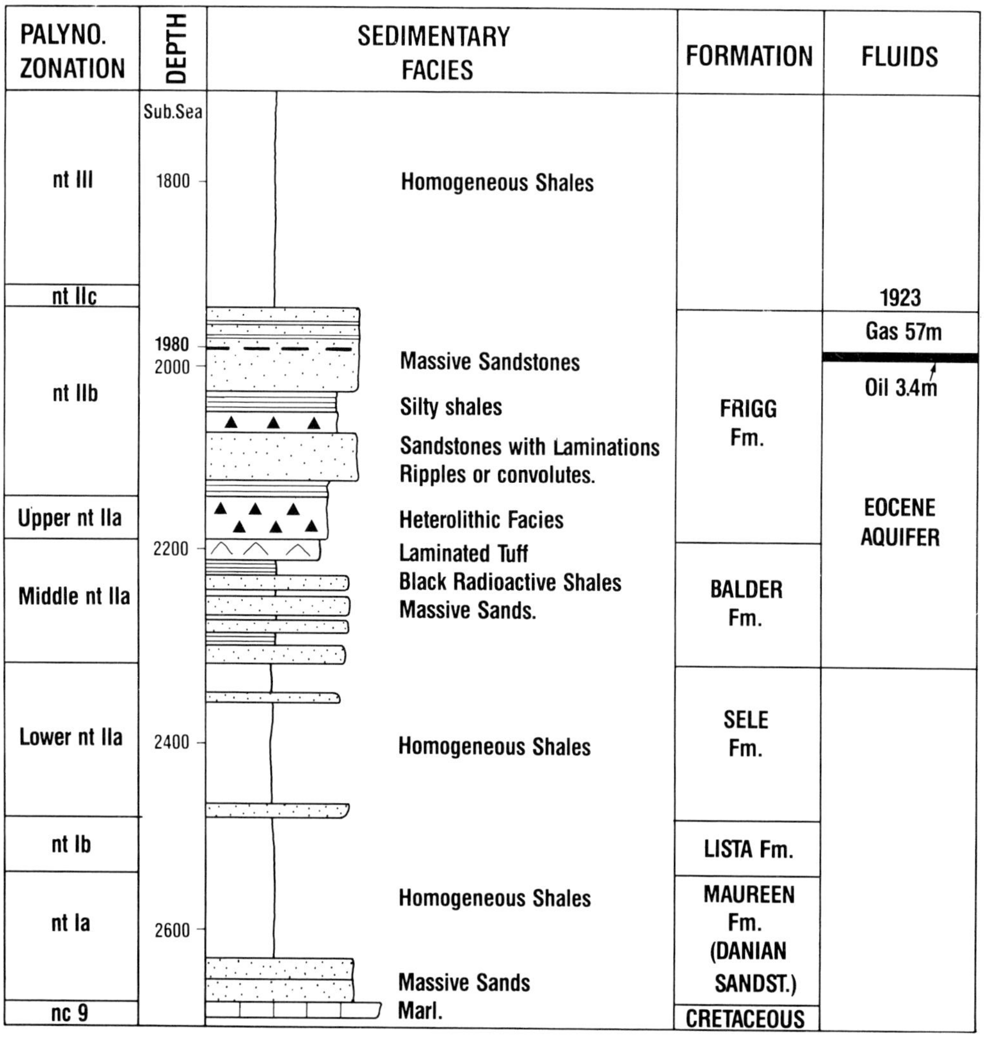

Fig. 4. Composite stratigraphic section of the reservoirs.

ACKNOWLEDGEMENT

The author thanks Elf Aquitaine Norge A/S, and the Norwegian partners of the Nord-Øst Frigg Unit (Esso, Total Marine Norsk, Norsk Hydro and Statoil).

19

Heimdal

E. Mure

Elf Aquitaine Norge A/S, PO Box 168, 4001 Stavanger, Norway

The Heimdal Field is located on Block 25/4, 35 km south of the Frigg Field. The water depth is 120 m. The trap is formed from sands of the Paleocene Heimdal Formation, structured by differential Cenozoic subsidence, above a deep Jurassic high. The field covers an area of 33 km^2. The reservoirs are submarine-fan sands. A 135 m wet-gas column, underlain by an oil leg five metres thick is proved at 1923 mss. The gas-in-place is estimated to be 50×10^9 Sm^3. Recoverable reserves are estimated to be 35.6×10^9 Sm^3 of wet gas. The development of the field is based on a single steel platform which integrates drilling, production, processing and accommodation. The gas is piped to the Statpipe system at Riser Platform 1 and then to Emden (Germany). The condensate is piped to the Brae Field in the British Sector and then to Cruden Bay (Scotland). The platform was installed in 1984. The production started in February 1986. Cumulative production to January 1987 is 2.24×10^9 Sm^3 of dry gas and 352 216 Sm^3 of condensate.

INTRODUCTION

The Heimdal Field lies in Norwegian Block 25/4, 35 km south of the Frigg Field. The average water depth is around 120 m. The Heimdal Field covers an area of approximately 33 km^2 (Fig. 1). The reservoir consists of sands of the Heimdal Formation which are part of a Paleocene submarine fan. The original gas-in-place is estimated at 50×10^9 Sm^3. Recoverable reserves are in the range of 35.6×10^9 Sm^3 of gas and condensate. The field is named after the god Heimdal who was the guardian of the gods and the father of humanity in Norse mythology.

HISTORY

Production Licence 036, which includes Block 25/4, was awarded in 1971, during the Norwegian second round, to the Pan Ocean Group (Pan Ocean Oil Norge A/S, operator; Sunningdale Oil Norge A/S; Syracuse International Norge A/S; K/S Femogtyvefire Norsk A/S). In 1972 after an agreement between the Pan Ocean Group and the Petronord Group, the operatorship was transferred to Elf Aquitaine Norge. In 1975, after the discovery, Statoil exercised its option of 40% on the Heimdal field ('inside deposit'). K/S Femogtyvefire Norsk transferred its interests to Pan Ocean, Elf Aquitaine and Total Marine Norsk in 1978 then to Saga and Ugland Rederi in 1980. Syracuse changed its name to Bow Valley in 1973 and Pan Ocean to Marathon Petroleum Norge A/S in 1981.

Today, the participating interests in the Heimdal Association are as follows:

Elf Aquitaine Norge A/S	9.639% (operator)
Statoil	40%
Marathon	23.798%
Bow Valley	8%
Norsk Hydro	6.228%
Total Marine Norsk	4.820%
Sunningdale	3.875%
Saga	3.471%
A/S Ugland Rederi	0.169%

Geology of the Norwegian Oil and Gas Fields © A. M. Spencer *et al.*, eds (Graham & Trotman, 1987) pp. 229–234.

Heimdal Field data summary

Trap	
Type	Structural
Depth to crest	2010 mss
Lowest closing contour	2150 mss
Hydrocarbon contact	GOC 2145 mss, OWC 2150 mss
Gas column	135 m
Oil column	5m
Pay zone	
Formation	Heimdal
Age	Paleocene
Net/gross ratio	0.79
Cut off for N/G	$V_{sh} < 33\%$
Porosity	24.5%
Hydrocarbon saturation	89%
Permeability	1 D
Well rate	1.5×10^6 m^3/day/well
Hydrocarbons	
Gas gravity (relative to air)	0.7
Condensate gravity	0.72 g/cm^3
Dew point	210 bar
Condensate yield	156 $Sm^3/10^6$ Sm^3 gas
Formation volume factor	208
Formation water	
Salinity	70 000 ppm NaCl equivalents
Reservoir conditions	
Temperature	76.2°C
Pressure	217.5 bar at 2123 mss
Pressure gradient in reservoir	
gas	0.020 bar/m
oil	0.080 bar/m
water	0.102 bar/m
Field size	
Area	33 km^2
Gross rock volume	1372×10^6 m^3
Net OGIP	50×10^9 sm^3
Recovery factor—oil	0
Recovery factor—gas	0.71
Drive mechanism	Depletion and waterdrive
Recoverable	
hydrocarbons	35.6×10^9 Sm^3 wet gas
condensate	4.7×10^6 Sm^3
dry gas	33.9×10^9 Sm^3
Production	
Start-up date	February 1986
Development scheme	production-treatment-drilling platform
Production rate	9.54×10^6 Sm^3/d (wet gas)
Cumulative production to 1 January 1987	2.24×10^9 Sm^3 dry gas; 353 216 Sm^3 condensate

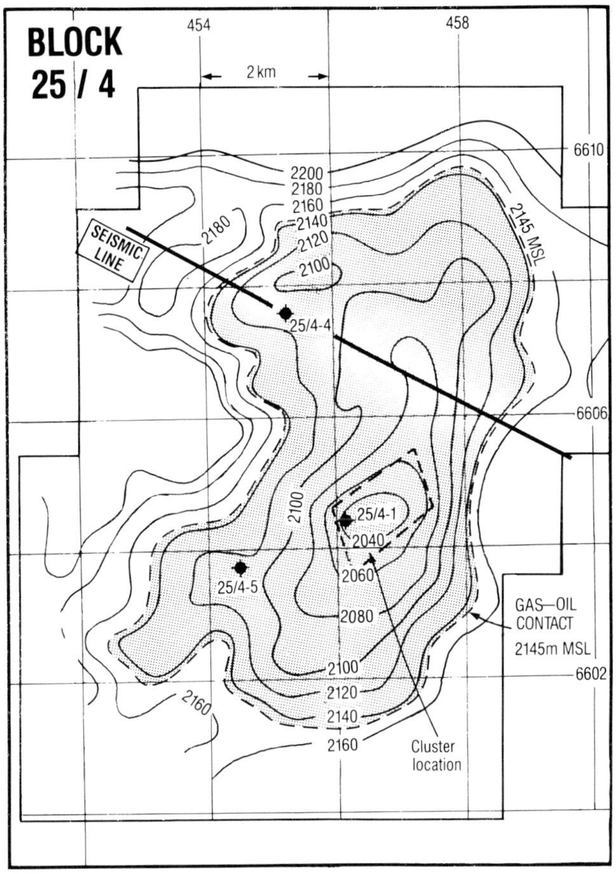

Fig. 1. Heimdal Field. Structure map of the top of the Heimdal Sands (contour interval = 20 m; all depths in mss).

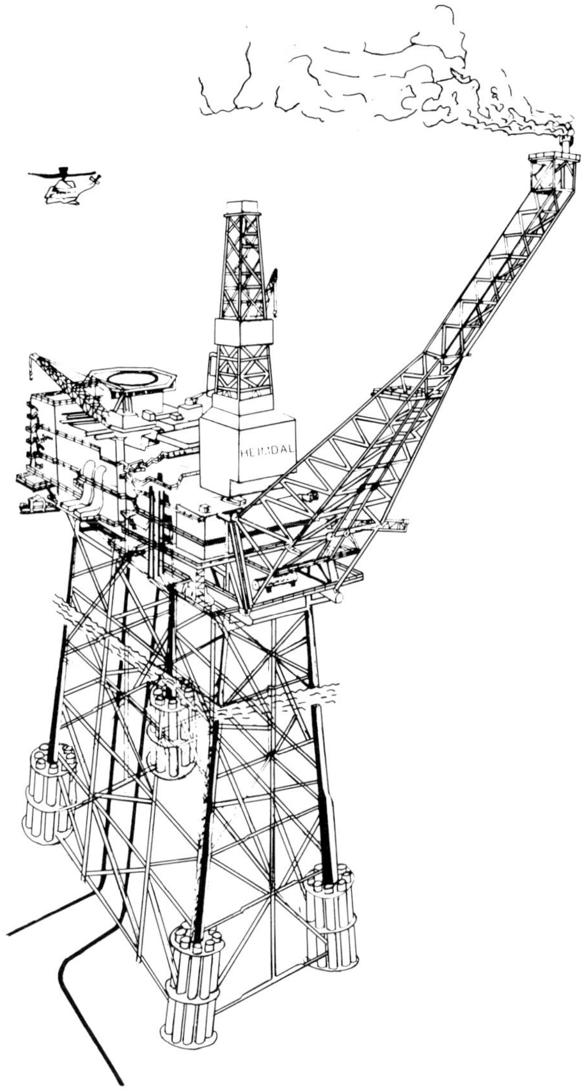

Fig. 2. Installations of the Heimdal Field.

In 1972, the initial discovery well, 25/4-1, was spudded at the culmination of the Paleocene structure but with Jurassic reservoirs as the main objectives. The well entered the Paleocene sands at 2039 mss and proved a gas column 125 m thick underlain by a five-metre-thick oil leg. The field was declared commercial in 1974. Two other exploration wells were drilled in 1975 and 1981 on the flanks of the structure. The development plan is based on a single-jacket based platform, housing drilling, production, treatment and quartering functions (Fig. 2). The platform was installed in 1984. Ten production wells were drilled in 1985 and 1986. Production started in February 1986. The gas is piped to the Statpipe system at Riser Platform 1 and then to Emden, Germany. The condensate is piped to the Brae Field in the British Sector and then to Cruden Bay, Scotland.

GEOLOGICAL SETTING

The sandstones of the Heimdal Formation are the last clastic members of a Paleocene submarine fan, deposited in the northern part of the Tertiary South Viking Graben, half-way between the Shetland Platform to the west and the Utsira High to the east. They are overlain by late Paleocene open-marine shales of the Sele Formation and by Eocene shales and tuffs of the Balder Formation. The Heimdal Field is located above a Jurassic structural high, near the deepest part of the Viking Tertiary embayment in an area of low structural relief at the base Tertiary level. The tilted blocks of the pre-Cimmerian faulting phase are deeply buried below Upper Jurassic and Cretaceous open-marine strata.

GEOPHYSICS

The structure of the Heimdal Field has been defined by several vintages of two-dimen-

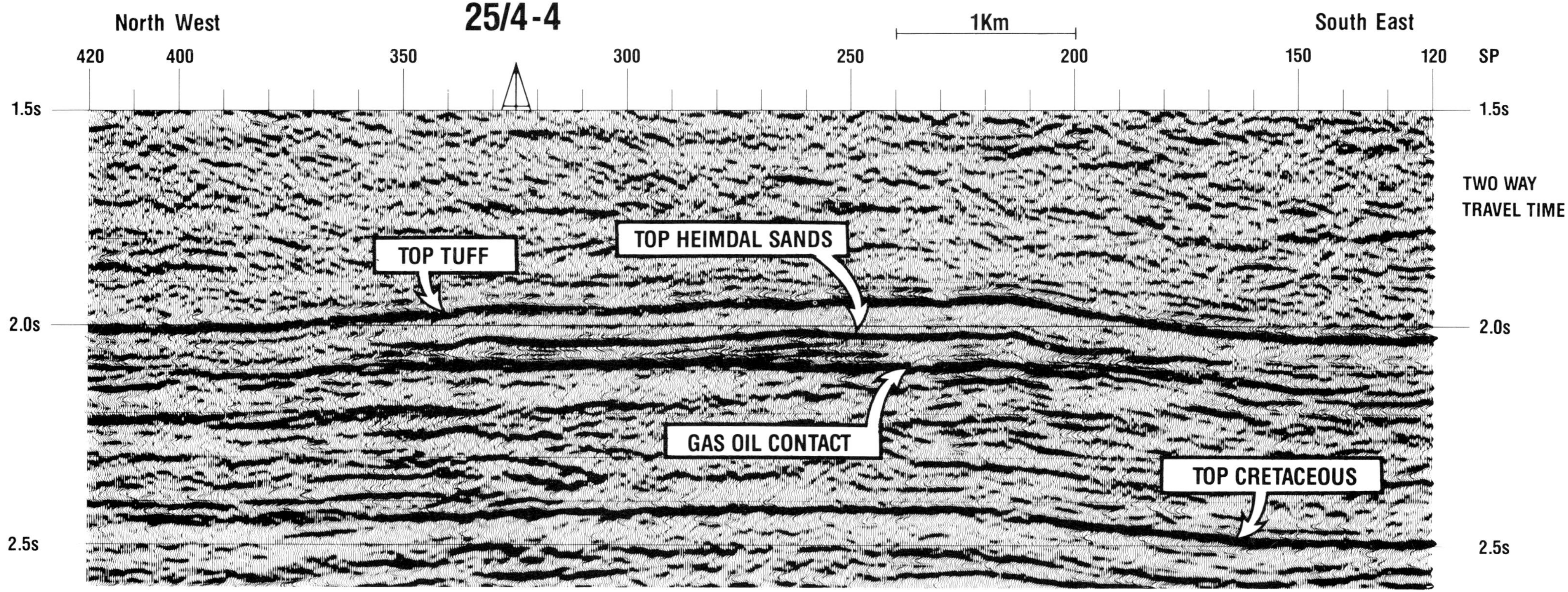

Fig. 3. Heimdal Field: NW–SE seismic line through Well 25/4-4. See Fig. 1 for location.

sional seismic surveys, shot from 1972 to 1983 providing an irregular grid of approximately 500 × 500 m. A three-dimensional (3D) survey was shot in 1986 amounting to 4200 km. The processed three-dimensional seismic data give a subsurface coverage of 37.5 m between E–W lines.

The top of the Heimdal Sands is fairly well defined by a strong seismic marker underlying the characteristic tuff seismic marker (Fig. 3). Direct hydrocarbon indicators have allowed the gas–oil contact and the limits of the field to be mapped. A new seismic evaluation of the reservoirs is under way following the 1986 three-dimensional seismic survey and the new production-well data.

TRAP

The Heimdal Field is a structural trap, mainly formed by differential Cenozoic subsidence above a deep Jurassic high and also by diffential compaction between Paleocene sands and shales. The structure is a NNE-trending anticline. The structural relief at the top of the Heimdal Formation is about 140 m with a culmination at 2010 mss. The closed area is about 33 km^2. The trap is filled to its spill-point.

RESERVOIRS

Heimdal reservoirs are Paleocene sands, dated Nt Ib by palynological studies, on the basis of marine dinoflagellate assemblages (Fig. 4). They are sealed by Paleocene marine shales of the Sele Formation (lower Nt IIa palynozonc).

The reservoir comprises alternations of massive sandstones and turbiditic sequences. The massive sandstones are heterogeneous, fine to coarse, poorly sorted, quartz sands with minor amounts of glauconite, pyrite and organic materials. The turbiditic sequences of

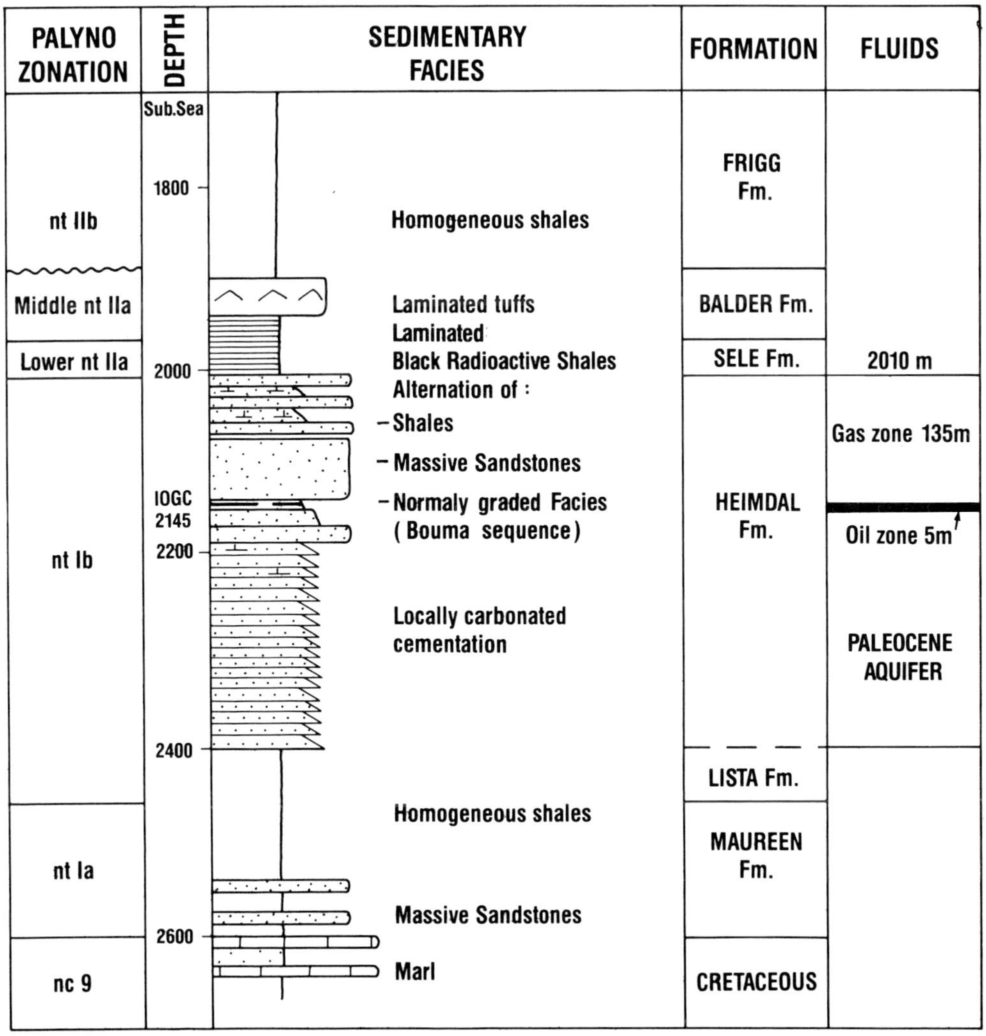

Fig. 4. Heimdal Field: composite stratigraphic section of the reservoir.

Bouma type are amalgamated and/or truncated fining-upwards sequences, showing from base to top the following facies: coarse to fine graded sandstones, sandstones with laminations and/or current ripples, argillaceous sandstones, silts and shales.

The facies association suggests a submarine fan depositional environment with deposits related to channel and/or supra-fan area.

The sands have good reservoir characteristics: average porosity of 25%, permeability in excess of one Darcy. Porosities and permeabilities are much lower in the shaly facies, which comprise up to 25% of the total lithologic column. Discontinuous carbonate cementations occur within the gas-bearing sands, as nodules of metric scale or locally, layers two to three metres thick. These carbonate concretions are formed by an early phase of diagenesis with dolomitization, followed by dedolomitization and calcitization. They reduce drastically the pore volume within the reservoir but to date have no major effect on production. The gas-bearing section of the reservoir is divided into six units on a basis of facies correlations. The Heimdal aquifer is connected with the regional Paleocene/Eocene aquifer of the Bruce, Beryl and Frigg area.

A comprehensive study of the reservoir is in progress. It will incorporate data from the 1985–1986 production wells, continuously cored through the reservoir section, and new seismic data from a 3D seismic survey shot in 1986.

SOURCE

The hydrocarbons of the Heimdal Field are believed to originate from the Jurassic source-rocks (Draupne and Heather formations, Brent and Dunlin groups) of the Viking Graben. In the deepest part of the basin, oil and gas were generated as early as Cretaceous times. The accumulation formed from the generation and migration of oil and gas as soon as the trap was constituted during early Eocene times.

HYDROCARBONS

The Heimdal gas is a wet gas with a condensate yield of 156 $Sm^3/10^6\ Sm^3$ gas. It is free of non-combustible gases. Its percentage composition is as follows: N_2, 0.9%; CO_2, 0.4%; C_1, 86%; C_2, 6%; C_3, 3%; C_4, 1%; C_{4^+}, 2%.

The reservoir was originally at hydrostatic pressure conditions.

RESERVES

Original gas-in-place is estimated at 50×10^9 Sm^3. With a gas-recovery factor of 71%, the reserves are around $35.6 \times 10^9\ Sm^3$ of wet gas, comprising $33.9 \times 10^9\ Sm^3$ of dry gas and $4.7 \times 10^6\ Sm^3$ of condensate. The underlying oil disc is negligible for production. Production started in February 1986. Cumulative production to 1 January 1987 is as follows: $2.24 \times 10^9\ Sm^3$ of dry gas and 353.216 Sm^3 of condensate.

ACKNOWLEDGEMENT

The author thanks Elf Aquitaine Norge A/S, and the partners of the Norwegian Production Licence 036, 'inside deposits', for giving permission to publish this paper.

20

Odin

H. M. Nordgård Bolås

Esso Norge AS, P.O. Box 60, N-4033, Forus, Norway

The Odin gas field is located in Block 30/10 of the Norwegian North Sea, north of the Frigg gas field. Both fields have reservoirs in the Frigg Formation submarine-fan sandstones of Lower Eocene (Ypresian) age. The mounded, lobate fan complex, creating a depositionally controlled structural trap, is situated near the deepest part of the Viking Graben, but not far from the Shetland Platform source area for the reservoir sands. The reservoir properties of the sands are excellent, with an average porosity of 29.5%, and with permeabilities exceeding 1000 mD. Two exploration wells and 11 production wells have been drilled in the Odin Field. Production started in October 1984 from a fixed platform with the gas being piped to the Frigg TCP-2 platform for processing. The Odin field is expected to produce gas until 1995, with a plateau rate of 385 Sm^3/day for about three years. The recoverable gas reserves are estimated to be in the range of 32–36 × 10^9 m^3.

INTRODUCTION

The Odin gas field is located in Block 30/10, 175 km south-west of Bergen in the Norwegian sector of the North Sea (Fig. 1). The field covers a total area of approximately 40 km^2. The water depth in the area is in the order of 100 m. The gas has accumulated in Frigg Formation sandstones, deposited as part of a Lower Eocene submarine-fan complex. The field is named after Odin, the foremost of the Norse gods, who was the god of poetry and occult wisdom.

HISTORY

Production Licence 030 in Block 30/10 was awarded to Esso on 30 May 1969, in the second Norwegian licensing round. The licence is held by Esso with Statoil having 17.5% net profit interest.

The Odin prospect was first mapped on a reconnaissance seismic grid in 1966. The discovery and appraisal wells were located on a 1973 seismic grid, which is approximately a 3 × 3 km grid.

A total of five exploration wells were drilled in PL 030 during the period May 1973 to April 1975, and two of these wells were located on the Odin Field (Fig. 2). The first well drilled in Block 30/10-1 was a Frigg Field extension well. The next, Well 30/10-2, was the Odin discovery well. It was spudded in December 1973 and found about 70 m of good porous sand, 53 m of which were gas bearing and underlain by an oil leg 4 m thick. Later in the same year, an appraisal well 30/10-3 was drilled on the southern flank of the field, to confirm the presence of the Frigg sand and the contact seen in Well 30/10-2. Well 30/10-3 encountered 20 m of gas-bearing sand, and the contact was successfully confirmed. The last two exploration wells on PL 030 were located outside the Odin Field. Well 30/10-4 was intended as a deep Jurassic evaluation well but was abandoned due to operational problems. The well was respudded as 30/10-5, and was completed at 5161 m subsea in the Triassic section. The well was plugged and aban-

Geology of the Norwegian Oil and Gas Fields © A. M. Spencer *et al.*, eds (Graham & Trotman, 1987) pp. 235–242.

Odin Field data summary

Trap	Stratigraphically induced structural trap
Depth to crest	1964 mss
Lowest closing contour	2025 mss
Gas–oil contact	2021 mss
Oil–water contact	2025 mss
Gas column	57 m (max)
Oil column	4 m
Pay zone (gas zone)	
Formation	Frigg
Age	Lower Eocene (Ypresian)
Thickness	Maximum 57 m
Net/gross ratio (average; range)	93.0%, 81–100%
Porosity (average; range)	29.5%, 28–31%
Hydrocarbon saturation (average; range)	81.0%, 76–86%
Permeability	> 1000 mD
Productivity index	1×10^6 Sm^3/day/bar
Hydrocarbons	
Oil gravity	0.83 g/cm^3
Oil type	low sulphur
Gas gravity	0.60 (relative to air)
Condensate yield	9.54 $Sm^3/10^6$ Sm^3
Dew point	83–103 bar
Gas expansion factor	201
Formation water	
Salinity	50 000 ppm NaCl equivalents
Resistivity	0.075 ohm m (at 65°C)
Reservoir conditions	
Temperature	65°C
Pressure	206 bar at 2000 mss
Pressure gradient in reservoir	0.029 bar/m
Field size	
Area (Limit: GOC)	40 km^2
Gross rock volume	1026×10^6 m^3
Hydrocarbon (gas) pore volume	230×10^6 m^3
Recovery factor—gas	0.70–0.80
Drive mechanism	pressure depletion/limited water drive
Well production rate	1.0×10^6 Sm^3/day (average)
	1.7×10^6 Sm^3/day (max)
Recoverable gas	$32–36 \times 10^9$ Sm^3
Production	
Start-up date	October 1984
Cumulative production to end 1985 —gas	5.8×10^9 Sm^3

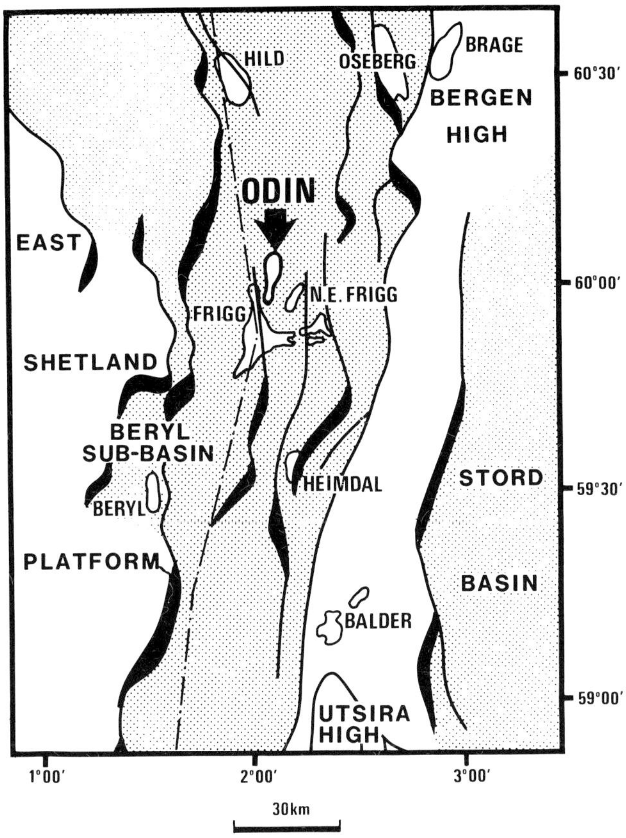

Fig. 1. Odin Field location map showing main tectonic elements and major fields.

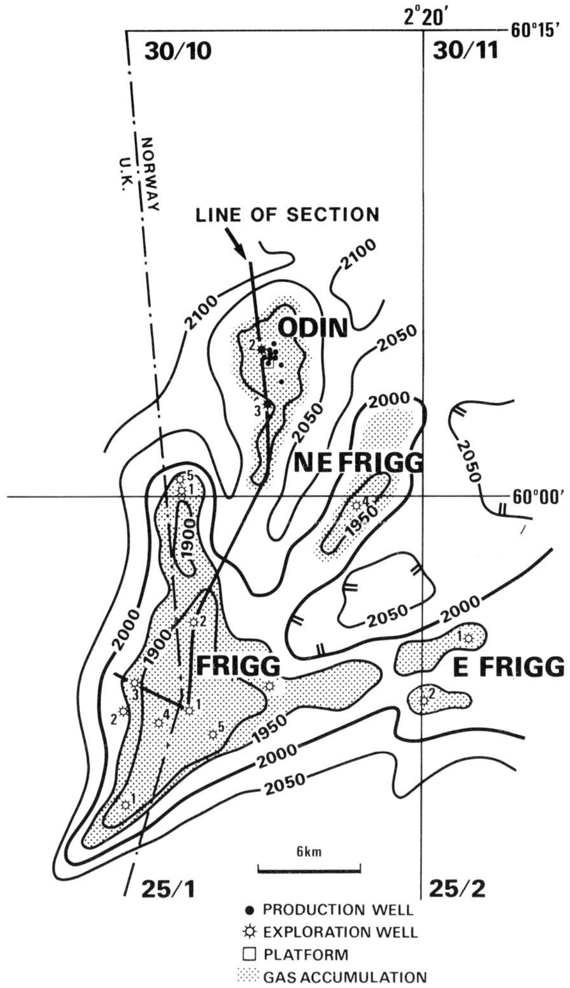

Fig. 2. Top Frigg Formation structure map of the Frigg and Odin field areas. Contours are in metres subsea.

doned as a dry hole with respect to the Jurassic formations, but the discovery of a few metres of gas column in the Frigg sand confirmed the northern limit of the Frigg field.

A detailed, high-resolution seismic survey, 1 × 1 km, was acquired in 1977, and this survey remains the basic geophysical data set for the licence area.

The Odin Field was declared commercial in February 1980, and a platform was installed in July, 1983. The drilling of the 11 production wells was done from December 1983 to January 1985. Production of gas started in October 1984.

GEOLOGICAL SETTING

The Odin Field is located in the central part of the Viking Graben between the East Shetland Platform to the west and the Bergen – Utsira High to the east (Fig. 1). When the block-faulting and extensional tectonics of the Late Kimmerian phase ceased in Early Cretaceous time, the North Sea Basin became the site of passive Upper Cretaceous and Tertiary subsidence. The Odin Field is situated slightly west of the deepest part of this basin.

The generalized stratigraphic column shown in Fig. 3 is based on Well 30/10-A4, which was drilled to a total depth of 2258 m subsea in Paleocene sediments (Heimdal Formation). The discovery Well, 30/10-2, penetrated the entire Paleocene section, and found about 40 m of Danian detrital skeletal limestone and about 60 m of Upper Cretaceous micritic limestone (Fig. 4). Palaeontological data suggest that both lithologies were deposited in a deep shelfal environment.

'Laramide' tectonism and associated eustatic sea-level drop caused profound palaeogeographic changes in the North Sea area. Chalk deposition gave way to a post-Danian clastic regime. Further Paleocene and Eocene subsidence of the Viking Graben and the rise of the Shetlands – Orcadian belt in the west, gave rise to the deep basin environment in which the basinward progradation of Tertiary clastic sediments took place. The Danian Chalk is

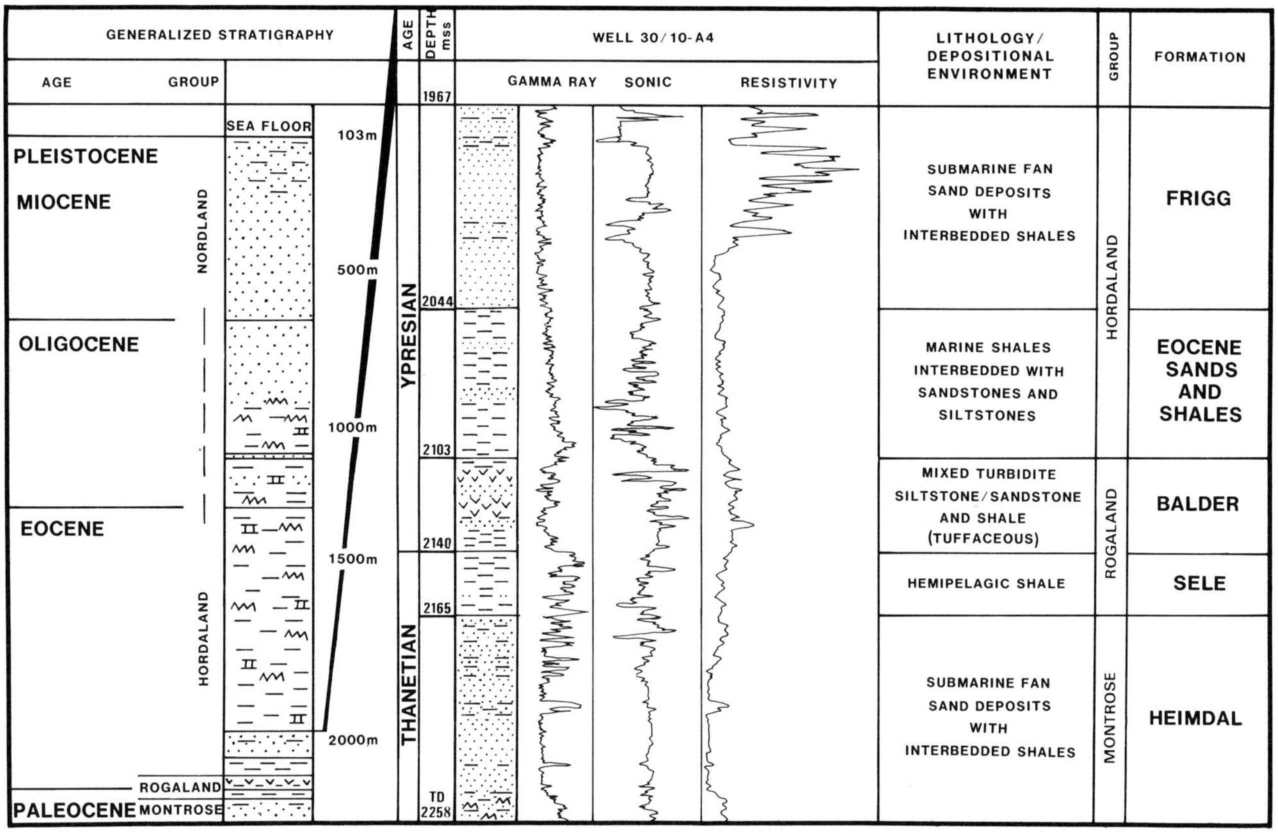

Fig. 3. Generalized stratigraphic column and well 30/10-A4 type log.

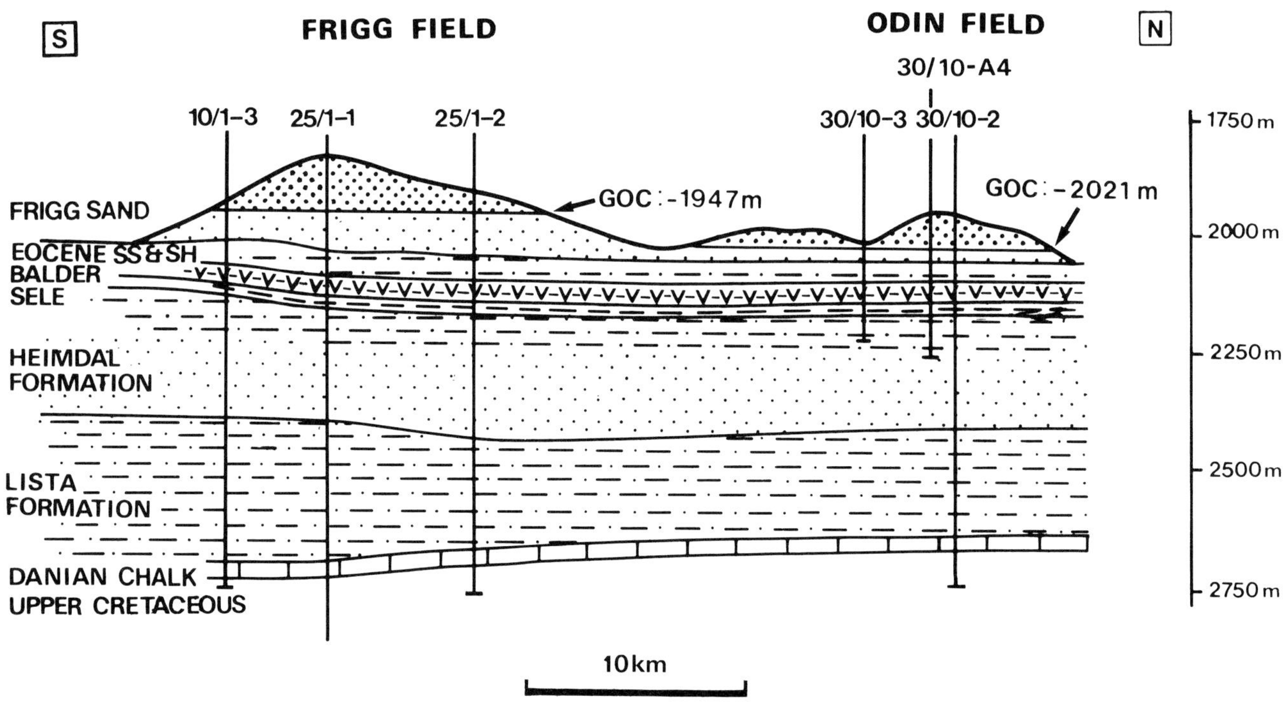

Fig. 4. Structural cross-section of the Frigg Formation Sands and underlying formations, showing the Frigg and Odin fields. For location see Fig. 2.

overlain by the Lista Formation, which in turn is overlain by the Heimdal Formation. These last two formations have total thicknesses of 220 and 200 m, respectively, in Well 30/10-2.

The Lista Formation consists mainly of shale and clay with stringers of sandstone and limestone. In contrast, the Heimdal Formation is represented by thick sandstones in the lower part, and interbedded sands and shales in the upper part. The sands of the Heimdal Formation were derived from the relatively high areas to the west and were deposited in the basin as part of turbidity current-fan complexes during low stands of sea level. The overall rise in sea level eventually drowned the clastic source-area, resulting in the deposition of 25 m of the shaly Sele Formation.

The Lower Eocene Balder Formation is about 40 m thick and consists of interbedded shales, thin sandstones, and siltstones interbedded with stringers of tuffaceous material. This tuff very probably originated from the western Hebridean volcanic belt as part of a major episode of volcanic activity resulting from the opening of the North Atlantic. The Eocene sands and shales below the Frigg Formation have a thickness of about 60 m, and are characterized by shales interbedded with thin siltstones and calcite-cemented sandstones. During Ypresian time, sands were shed from the East Shetland platform into the deep water of the basin during subsequent sea-level low-stand periods, and large submarine-fan complexes like the Frigg and Odin fans were developed. The Beryl subbasin south of the Frigg and Odin fields (Fig. 1) was apparently a zone of weakness along the Shetland Escarpment, and provided the fundamental control for the deposition of clastics in this part of the basin. The Frigg Formation is nearly 80 m thick in the 30/10-A4 Well, and consists of sandstones with some thin shaly zones and a few thin calcareous streaks. During Eocene times, the influx of clastics from the Shetland Shelf and the Scottish Highland gradually diminished, and consequently the Ypresian sand-rich submarine-fan sediments are depositionally succeeded by Upper Eocene shales, clays and interbedded siltstones.

The Oligocene is also represented by deep-marine shales in the lower part, but sandy deposits became more frequent upwards due to the fall in sea level in middle Oligocene times. Similar shallow-marine sands characterize the regressive Miocene period, with shallow-marine shales becoming more abundant during Pliocene times.

GEOPHYSICS

A detailed high-resolution seismic survey 1×1 km, (2 ms sample rate) was shot in 1977, and is still the basic geophysical data set.

The two exploration wells confirmed that the gas-oil interface was coincident with a good seismic reflector. Detailed modelling, seismic reprocessing and interpretation on large-scale (40 cm/s) sections enabled mapping of the gas-sand interval, using direct interpretation of the bright-spot caused by the gas-oil contact. A top Frigg sand-depth structure map was constructed by adding a depth-converted, seismic time, gas-sand thickness to the horizontal gas-oil contact. This method compensates for gas 'pull-down' or 'sag' velocity effects, both related to the gas-sand itself and to local velocity variations in the overburden.

An interpreted seismic line across the Odin Field is shown in Fig. 5.

TRAP

The Odin Field is a depositionally controlled structural trap. The structure is related mainly to the original depositional submarine-fan topography, but has been accentuated by differential compaction of sands and muds (Heritier *et al.*, 1979). The reservoir of mounded Lower Eocene, Ypresian sands is capped by impermeable Upper Eocene shales. The trap is illustrated by the cross-section in Fig. 4. The trap is filled to the structural spill-point, and the vertical closure is 61 m.

RESERVOIR

The reservoir rocks of the Odin field, which have a total thickness of nearly 80 m in Well 30/10-A4, belong to the Lower Eocene Frigg Formation and consist of sandstones interbedded with relatively thin shale layers and a few calcite-cemented streaks. The sandstones often show a gradual upwards increase in silt and clay content, resulting in several fining-upwards sequences ranging in thickness from 0.25 m to about 6 m. Cores from the reservoir interval in 30/10-A4 contain five lithofacies: homogeneous sandstones, graded sandstones, silty shales, clays and calcite-cemented siltstone.

There is no abrupt boundary between the two sandstone facies. The homogeneous sandstones are fine to medium grained, often with

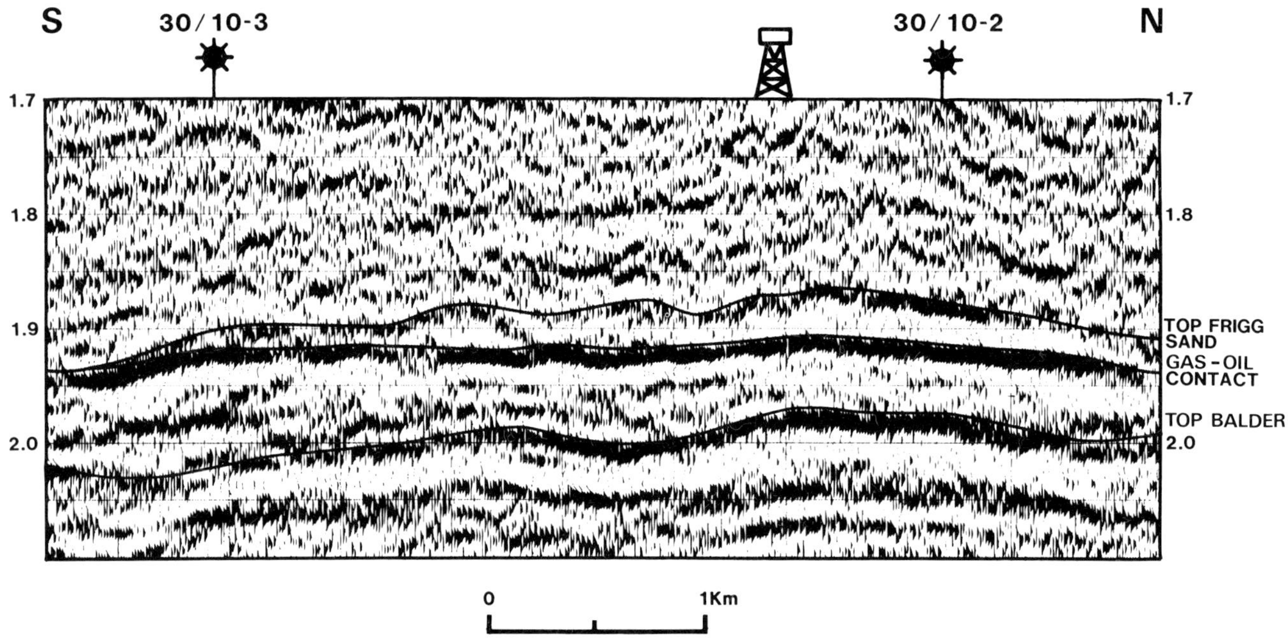

Fig. 5. Three-dimensional seismic line OGD 79–12 A. For location see Fig. 2.

very coarse quartz grains at the base, while the graded sandstones fine upwards from fine to medium grained at the base to very fine grained at the top, with the silt content increasing upwards in the sequence. Both sandstone facies are light grey to light olive grey, very friable to unconsolidated, and the grains are subangular to subrounded and generally moderately sorted. The most common sedimentary structures are load casts, scoured bases, wavy and planar laminations and microfaults. Mineralogically the sands are classified as subarkoses with 80–95% quartz, some feldspar and muscovite, and minor amounts of pyrite, zircon, biotite, glauconite, kaolinite, chlorite and smectite–illite. The porosities range from 28 to 31% and the permeabilities are of the order of 1000 mD or more.

The silty shales range in thickness from 1 cm to 3 m, but are more typically 20–30 cm. Thin to very thin laminations and lenses of very fine-grained silty sandstone, together with flame structures, wavy non-parallel lamination, lamination deformations, dish structures and slump features, are common in the lower parts of these intervals. The silt content and the amount of sedimentary structuration usually decrease towards the top of the intervals.

The hemipelagic/pelagic clay layers are thinner than 20 cm and occur only near the top of the reservoir in Well 30/10-A4. This clay is greenish coloured and soft. There are also a few thin layers of calcareous cemented siltstones. The irregular shape of the cement suggests large calcareous concretions in a matrix of clayey siltstone rather than a totally cemented horizon.

The homogeneous, non-graded sandstones (correlative to facies B of Ricci Lucchi, 1975) were probably deposited by fluxoturbidity currents in the channelized setting of the middle fan. The graded sandstones and associated silty shales (correlative to facies C or E) may have been deposited by high and low concentration turbidity currents or traction flow, as upper parts of channel sequences or in non-channelized settings such as the middle fan fringe. The thickest layers of silty shales are very rich in slump-like structures (correlative to facies F) and may have been formed by gravity slumping and sliding along channel margins of the middle-fan environments. The hemipelagic/pelagic clays (correlative to facies G) probably covered the total fan area as suspension deposits, but may have been locally deposited by dispersal mechanisms in interchannel settings or as fill in abandoned channels.

The topography of the Frigg sands is illustrated by a three-dimensional view in Fig. 6.

SOURCE

No geochemical studies have been carried out specifically on the Odin Field hydrocarbons, but it is assumed that they originated from overmature Upper Jurassic Draupne Formation source-rocks.

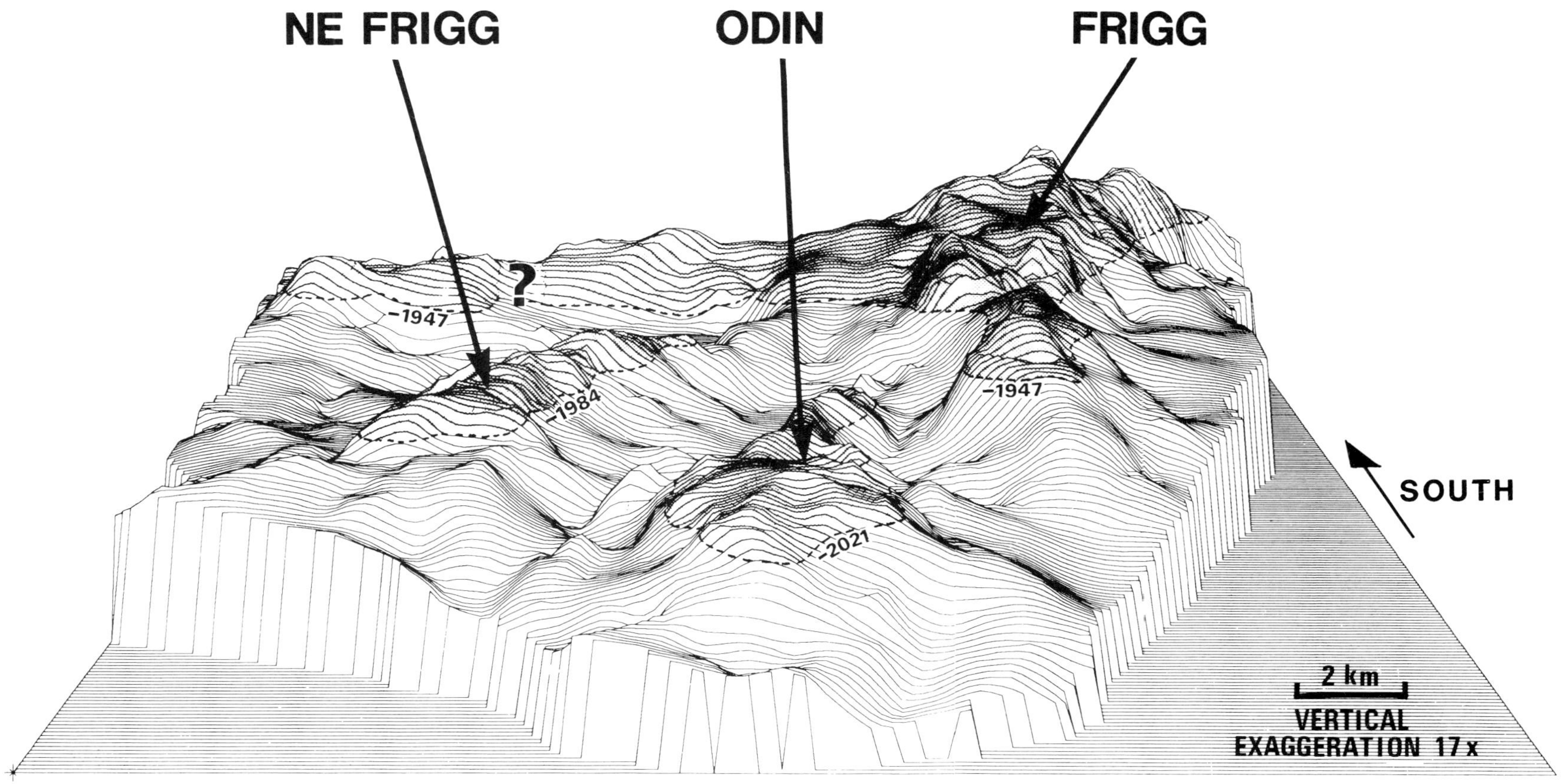

Fig. 6. Perspective view looking south of the structure on top of the Lower Eocene submarine fans of the Frigg, North-east Frigg and Odin fields.

HYDROCARBONS

The Odin gas is dry, with methane and ethane contents of 95.3% and 3.4%, respectively. It has a gas gravity of approximately 0.6 relative to air, and a condensate yield of 9.54×10^6 Sm^3/Sm^3. The average well production rate is 1.0×10^6 Sm^3/day, and the maximum rate is 1.7×10^6 Sm^3/day. The reservoir temperature is 65 °C, and the initial reservoir pressure in the discovery well was 206 bars at 2000 m subsea. The four-metre-thick oil leg under the gas, which is not economically recoverable, has a specific gravity of 0.83 g/cm^3 and is classified as a low sulphur crude.

RESERVES

Recoverable gas reserves in Odin are estimated to be in the range $32–36 \times 10^9$ Sm^3, assuming a recovery factor of 0.70–0.80 with limited water drive. The non-recoverable oil volume is estimated to be 13×10^6 m^3 STOOIP.

ACKNOWLEDGEMENTS

This article is based mainly on Esso in-house work, but the publications listed below have been very useful. S. Vaagene, Esso Norge AS, has carried out the most recent seismic interpretation, on which the attached map and cross-sections are based. The author thanks Esso Norge AS for giving permission to publish this paper.

REFERENCES

Deegan, C.E. and Scull, B.J. 1977. A standard lithostratigraphic nomenclature for the Central and Northern North Sea, *Institute of Geological Sciences, Report* 77/25.

Heritier, F.E., Lossel, P. and Wathne, E. 1979. Frigg Field — large submarine fan trap in Lower Eocene rocks of North Sea Viking Graben, *Bulletin of the American Association of Petroleum Geologists* **63,** (11), 1999–2020.

Ricci Lucchi, F. 1975. Depositional cycles in two turbidite formations of northern Apennines (Italy), *Journal of Sedimentary Petrology* **45,** 3–43.

21

Sleipner Øst

O. J. Østvedt

Esso Norge A/S, PO Box 60, N-4033 Forus, Norway

The Sleipner Øst gas and condensate field is located in Block 15/9, on the eastern margin of the South Viking Graben. The hydrocarbons are entrapped within two different reservoirs. The primary reservoir is provided by submarine fan sandstones of the Paleocene Heimdal and Maureen Formations. The sandstones were deposited by sediment gravity flows sourced from the East Shetland Platform. The trap mechanism is a combination of stratigraphic pinch out of the sandstones along the southeastern flank, and structural closure caused by doming in Mid-Tertiary times. The reservoir properties of the sandstones are very good with an average porosity of 26 % and permeabilities ranging from 200 mD to 1500 mD. Secondary reservoir sandstones are present in the Callovian Hugin Formation. These marginal-marine sandstones were deposited during an overall transgressive phase in Middle to Late Jurassic times.

The field was discovered in 1981 and to date a total of four wells have been drilled and a three-dimensional seismic survey has been acquired. Recoverable wet gas is estimated to be 60×10^9 Sm^3. A statement of commerciality was issued in 1984 and the production start-up is planned for 1993. The gas will be transported to the European Continent.

The ideas and interpretation expressed are those of the author, and it is recognized that Statoil, the operator of the field, has a different interpretation of the seismic data in certain areas and of details in the geological depositional model.

INTRODUCTION

Two major gas–condensate fields, Sleipner Øst and Sleipner Vest, are located in Block 15/9 in the Norwegian sector of the North Sea. Sleipner Øst is located in the eastern part of the block, 230 km west-south-west of Stavanger. The hydrocarbons are trapped in submarine-fan sandstones of Paleocene age, and in marginal-marine sandstones of Middle–Late Jurassic age. Recoverable reserves are estimated to be 60×10^6 Sm^3 of wet gas. The field is named after Sleipner, the eight-legged horse of the god Odin in Norse Mythology.

HISTORY

Production Licence 046, which includes Blocks 15/9 and 15/8, was awarded in 1976 to Statoil (50% and operator), Esso Exploration and Production Norway (40% and technical assistant) and Norsk Hydro (10%).

During the period from 1977 to 1981 the exploration activity was concentrated on the Sleipner Terrace in the western part of Block 15/9 (Fig. 1). A gas discovery had previously been made in the southern part of Block 15/6, on a structure that was believed to extend southwards into Block 15/9. Gas was discovered in the first wells drilled in Block 15/9 and confirmed the southern extension of the structure. The gas was trapped in sandstones of Middle–Late Jurassic age. By 1981, 8 wells had been drilled and the outline of the gas accumulation, later named Sleipner Vest, had been established.

During the summer of 1981 Well 15/9-9 was

Geology of the Norwegian Oil and Gas Fields © A. M. Spencer *et al.*, eds (Graham & Trotman, 1987) pp. 243–252.

Sleipner Øst Field data summary

	Paleocene Heimdal and Maureen Fm	Jurassic Hugin Fm
Trap		
Type	Combined structural and stratigraphic	Structural
Depth to crest	2260 mss	2645 mss
Lowest closing contour	2417 mss	2800 mss (approx)
Gas–water contact	2417 mss	2800 mss (approx)
Gas column	maximum 157 m	maximum 155 m
Pay zone		
Formation	Heimdal and Maureen	Hugin
Age	Paleocene	Callovian
Thickness	0–120 m	35 m (average)
Net/gross ratio (average)	0.95	0.78
Porosity (average)	26%	22%
Hydrocarbon saturation	67%	78%
Permeability	400 mD (average)	100–400 mD
Hydrocarbons		
Gas gravity (relative to air)	0.93	0.94
Condensate yield	560 $Sm^3/10^6$ Sm^3 gas	490 $Sm^3/10^6$ Sm^3 gas
Gas expansion factor	208	234
Formation water		
—		
Reservoir conditions		
Temperature	93 °C	104 °C
Pressure	244 bar at 2417 mss	300 bar at 2800 mss
Pressure gradient in reservoir		
Gas	0.023 bar/m	0.024 bar/m
Water	0.099 bar/m	0.1 bar/m
Field size		
Area	59 km^2	22 km^2
Hydrocarbon Pore Volume	371×10^6 m^3	85×10^6 m^3
Drive mechanism	Volumetric expansion plus water influx	Volumetric expansion
Recoverable gas	53×10^9 Sm^3	7×10^9 sm^3

Production

The two reservoirs will be produced simultaneously and the field will be depleted in 10–12 years. Forecast plateau production is about 7×10^9 Sm^3 of wet gas per year.

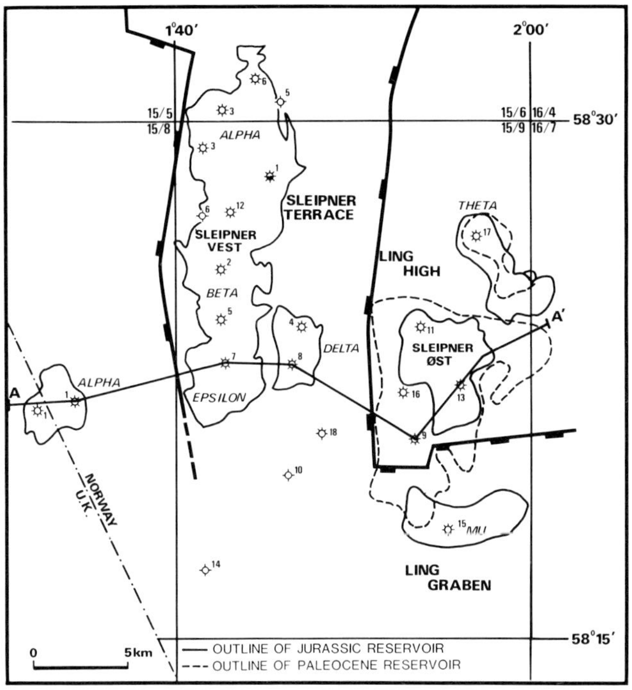

Fig. 1. Outline of fields in Block 15/9 including main structural elements.

drilled on the Ling High in the eastern part of the Block (Fig. 1). The main target for this well was Middle–Upper Jurassic sandstones. The well penetrated 100 m of gas bearing Paleocene sandstones, assigned to the Heimdal and Maureen Formations. No Jurassic sands were encountered.

This was the first time hydrocarbons had been encountered in sandstones of Paleocene age in the Sleipner area, and Well 15/9-9 is the Sleipner Øst discovery well. The three subsequent wells, 15/9-11, 13 and 16 drilled in 1981 and 1982 confirmed the presence of a thick Paleocene gas-bearing sandstone with good reservoir properties. Middle–Upper Jurassic sandstones were present in only two of the wells, 15/9-11 and 15/9-13. These sandstones were gas-bearing with a gross thickness of about 35 m and represent the secondary reservoir in the Sleipner Øst Field.

The encouraging drilling results made it desirable to better define the reservoir, and the seismic coverage was extended from a two-dimensional (2D) to a three-dimensional (3D) seismic grid in 1982. This three-dimensional grid remains the basic geophysical data set.

Field development studies of the entire Sleipner Field area were initiated in 1979, and in February 1984 a statement of commerciality was issued for the Sleipner Øst Field. In the first field development plan, forwarded in 1985, it was suggested to develop Sleipner Øst and Sleipner Vest concurrently. Changes in the gas market made it necessary to revise the plan, and in August 1986 a new field development plan was forwarded which includes Sleipner Øst only. The field will be developed with one fully integrated platform with a capacity of 24×10^6 Sm^3/day. Gas and condensate will be transported to buyers on the Continent as part of the Troll/Sleipner sales agreement. Production is planned to start in 1993.

GEOLOGICAL SETTING

Sleipner Øst is situated on the Ling High which is a south-western extension of the Utsira High (Figs. 2, 3). The Ling High is limited to the west by a N–S-trending fault zone along which the Sleipner Terrace is down faulted. To the south, the Ling Graben is downfaulted along an E–W-trending fault zone.

Late Permian marginal-marine carbonate and evaporite strata of the Zechstein Group were penetrated in Wells 15/9-16 and 15/9-13 (Fig. 1). They are about 50 m thick, which is very thin compared to the thick salt accumulations identified on seismic data on the Sleipner Terrace. This indicates that the Ling High was a structurally elevated area in Late Permian times. During Triassic times the Ling High was subaerially exposed and continental deposits characterize this period. Argillaceous sediments assigned to the Smith Bank Formation (Deegan and Scull, 1977) constitute most of the 300–500 m thick Triassic interval (Fig. 4).

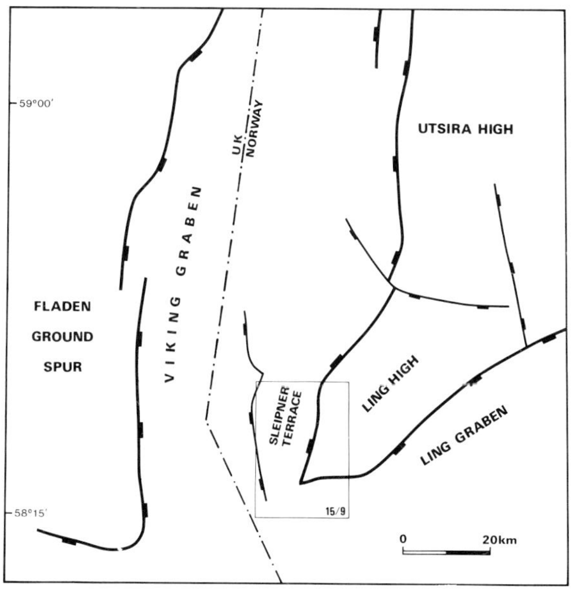

Fig. 2. Regional structural elements. Block 15/9 is outlined.

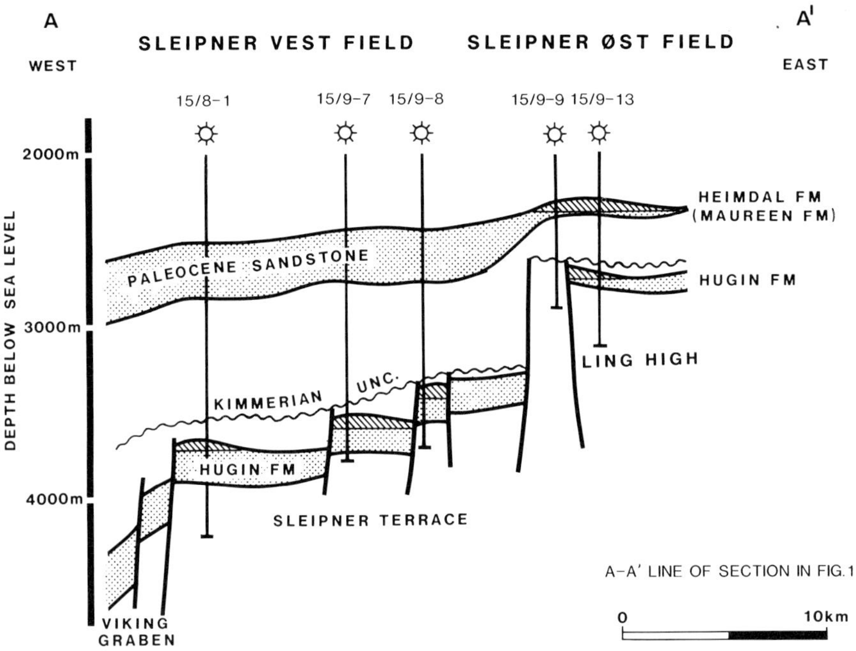

Fig. 3. Structural cross-section.

These sediments were probably deposited in a lacustrine or muddy floodplain environment. The overlying sandy Skagerrak Formation (Norian–Carnian) is thinner and in some places absent. This formation is believed to represent continental deposits of braided-stream/floodplain origin. The Ling High then remained above sea level until late Callovian times, when a major transgression occurred from the north. The transgression was associated with a main phase of extensional rifting in the Viking Graben. During the transgression, sands ascribed to the Hugin Formation were deposited along the margin of the Ling High. These sands are interpreted as shallow-marine shoreface deposits, which accumulated during pulses of westward regression during the overall marine transgression towards the southeast.

The Callovian Hugin Formation is 35 m thick in Well 15/9-11 but thins towards the south, either due to depositional thinning or erosion associated with the Late Jurassic unconformities. The sands are overlain by open-marine shales of the Heather Formation and Draupne Formation (Vollset and Doré, 1984), which cover the entire area and represent the culmination of the Late Jurassic transgression.

The Late Jurassic phase of rifting had a great influence on the final structural configuration of the Sleipner area. Faults trending NNE–SSW were formed along the margin of the Viking Graben where the dominant extensional stress resulted in mostly normal faulting. A set of faults trending NNW–ESE were also developed. Pegrum and Ljones (1984) suggested that some of these faults were displaced by left lateral strike-slip movements resulting from transtensional rejuvenation of pre-Permian 'basement faults' along the continuation of the Tornquist Zone. Later episodes of faulting in late Early Cretaceous and Middle Tertiary times occurred mostly as reactivation of the Late Jurassic faults.

The Lower Cretaceous is represented by a thin succession consisting predominantly of marine shales, marls and silts. In Well 15/9-16 this section is completely absent indicating that the Sleipner Øst area was elevated, probably as a result of the Mid-Cretaceous compressive phase (Pegrum and Ljones, 1984). A complete Upper Cretaceous chalk sequence suggests tectonic quiescence during Late Cretaceous time. In Early Paleocene times a N–S-trending basin was formed as the East Shetland Platform and Fladen Ground Spur (Fig. 2) became subaerially exposed. During rapid falls in eustatic sea level, sediments supplied from the shelf edge formed large submarine fans. Sandstones belonging to the Maureen Formation and the Heimdal Formation were deposited in a submarine fan environment, in the Sleipner area. The maximum recorded thickness of the Maureen Formation is 22 m (Well 15/9-9), while the thickest accumulation of the Heimdal Formation is 119 m (Well 15/9-11). Both formations thin rapidly towards the south-east (Fig. 5). The Heimdal sands are overlain by the shaly Lista and Sele forma-

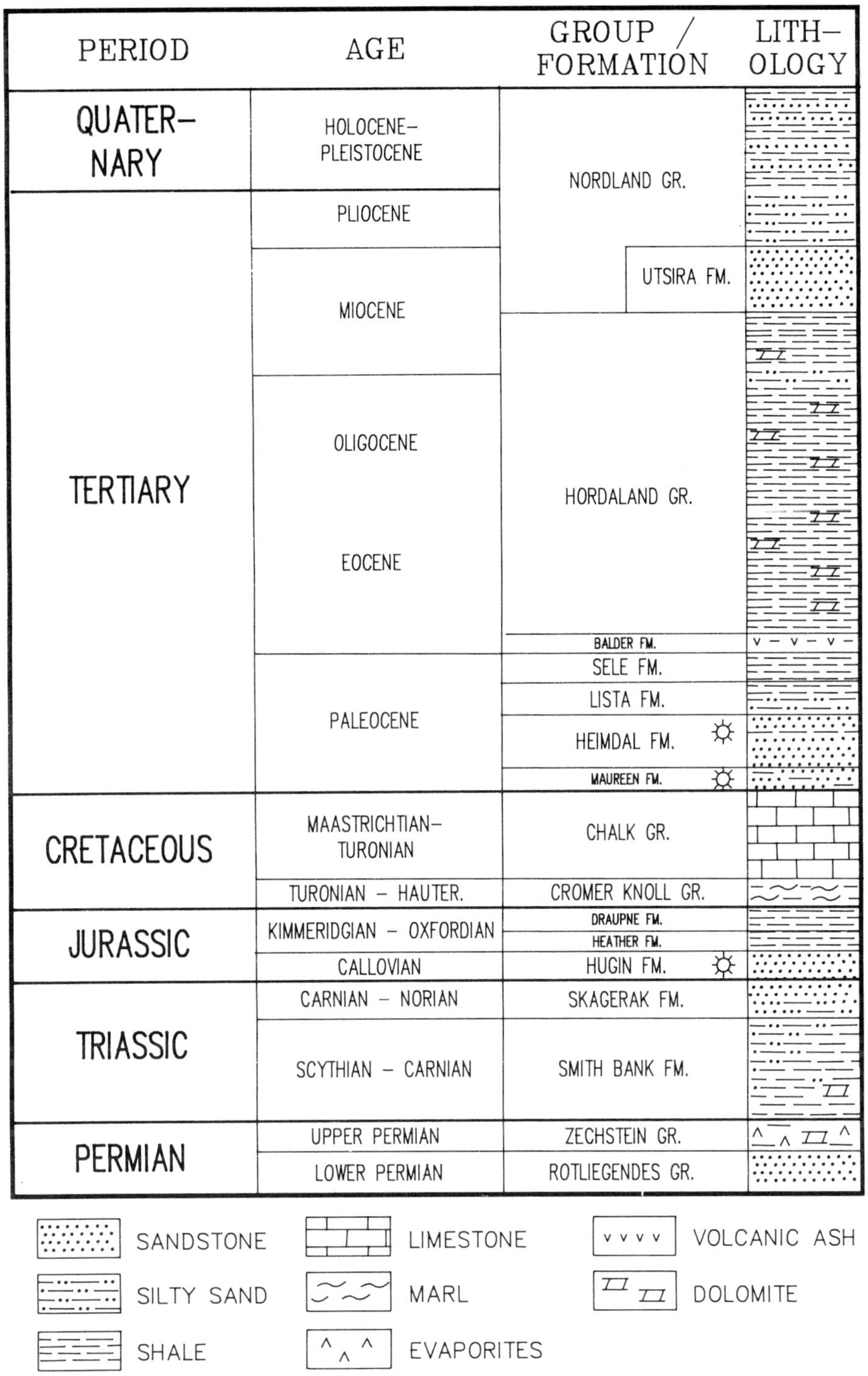

Fig. 4. Generalized stratigraphy in Sleipner Øst.

tions. The overlying Balder Formation is characterized by volcanic ash bands, which can be readily correlated from well data, form an excellent marker bed and produce a strong regional seismic reflector.

In Mid-Tertiary times, transpressional stress (Pegrum and Ljones, 1984) resulted in reactivation and inversion of older faults, and gentle folding of the rocks into domal structures. The Eocene to Lower Miocene interval consists of marine shales with thin carbonate stringers and interbeds of sand and silt. The Upper Miocene Utsira Formation is represented by about 250 m of shallow-marine sands. Alternating clays and sands make up the Pliocene and Pleistocene sequences.

GEOPHYSICS

A 1 × 1 km survey shot in 1975 was infilled to a 1/2 × 1/2 km grid in 1981. In 1982 the whole

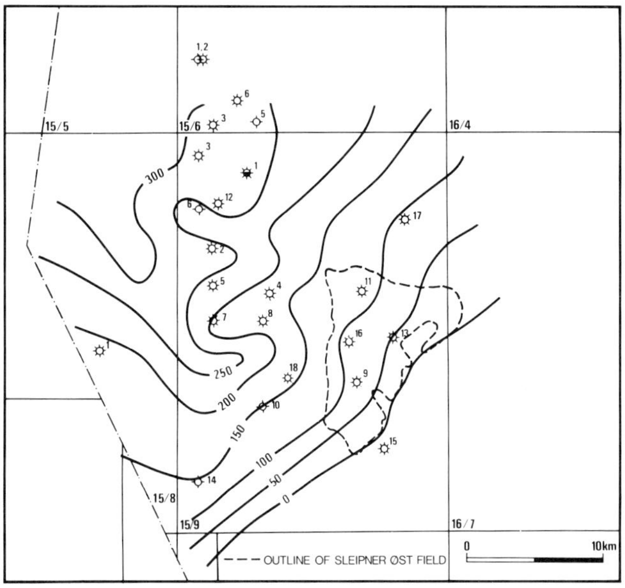

Fig. 5. Isopach map; total thickness of Heimdal and Maureen formations. Contours in metres.

structure was covered by a three-dimensional seismic grid. The survey was conducted by Geco and the lines were shot in a N–S direction, 75 m apart.

The following horizons have been mapped: Top Paleocene, Top Heimdal Sand, Top Chalk, Top Lower Cretaceous, Base Cretaceous and Top Jurassic/Triassic Sand (Fig. 6). Top Paleocene and Top Chalk are the two strongest reflectors. In between these two, the Top Heimdal Sand horizon suffers from interference and gas effect, and is of intermediate quality. However, the pinchout of the Top Heimdal Sand can be seen as a down-lapping event on the Top Chalk reflector and can be mapped with fairly high confidence. The Top Jurassic/Triassic Sand is a weak reflector with many faults and discontinuities and is rather difficult to map.

A direct hydrocarbon indicator (flat spot) can be recognized at about 2.35 s in parts of the Paleocene reservoir, and is believed to arise from the gas–water interface.

TRAP

The Paleocene reservoir is a combined structural and stratigraphical trap. To the west and north the sands have structural closure. To the southeast, the extent of the field is partly determined by the depositional limit of the sandstones, which form a stratigraphic trap (Fig. 7). Shales of the overlying Lista, Sele and Balder formations constitute a good seal, and the trap is filled to the structural spill-point at 2417 mss.

The Middle–Upper Jurassic reservoir sandstones are faulted and gently folded into a domal structure, sealed by Upper Jurassic shales. Although the reservoir is limited to the west by erosional truncation or sandstone pinch-out, the trap is basically of structural origin. The gas–water contact is estimated at 2800 mss. The contact is in a shaly sequence in both wells and is therefore difficult to accurately determine.

RESERVOIR

Sandstones of the Paleocene Heimdal and Maureen Formations form the primary reservoir in Sleipner Øst. The Heimdal Formation, which is the thicker of the two, constitutes the major part of the reservoir. Middle–Upper Jurassic (Callovian) sandstones assigned to the Hugin Formation represent a secondary reservoir (Fig. 8).

Interpretation of core and log data indicates that the sediments of the Maureen and Heimdal formations were deposited in a submarine-fan environment. Three main lithofacies, namely massive sandstones, interbedded sands and shales, and shale–claystones are recognized in the cores.

1. *Massive sandstones* are well sorted, fine to medium grained, and commonly occur in units from 2–3 m up to about 20 m thick. Very few beds can be distinguished within the units. Dish structures and sand pillars formed by escaping pore water are the most common sedimentary structures. In some of the beds it is possible to recognize a faint parallel and inclined stratification. Shale rip-up clasts are abundant in the lower parts of beds associated with shales. The sandstones are interpreted to be deposited from high-density turbidity currents and other types of sediment gravity-flows. The large thicknesses of the beds are due to stacking and amalgamation of several flows, probably related to infilling of structural and depositional lows.

2. *Interbedded sands and shale* occur in units with thicknesses up to 10 m. The thickness of the individual shale and sand beds is from 5–10 cm up to about 1 m. Soft sedimentary deformation is a common feature. The sandstones are fine to medium grained and generally massive. Shale rip-up clasts are commonly present in the basal parts of the beds. The shales are often marly, occasionally sandy with a dark green–grey to black colour. The sand–shale units are interpreted as channel overbank and distal fan deposits.

3. *Shale–claystones* are the least abundant

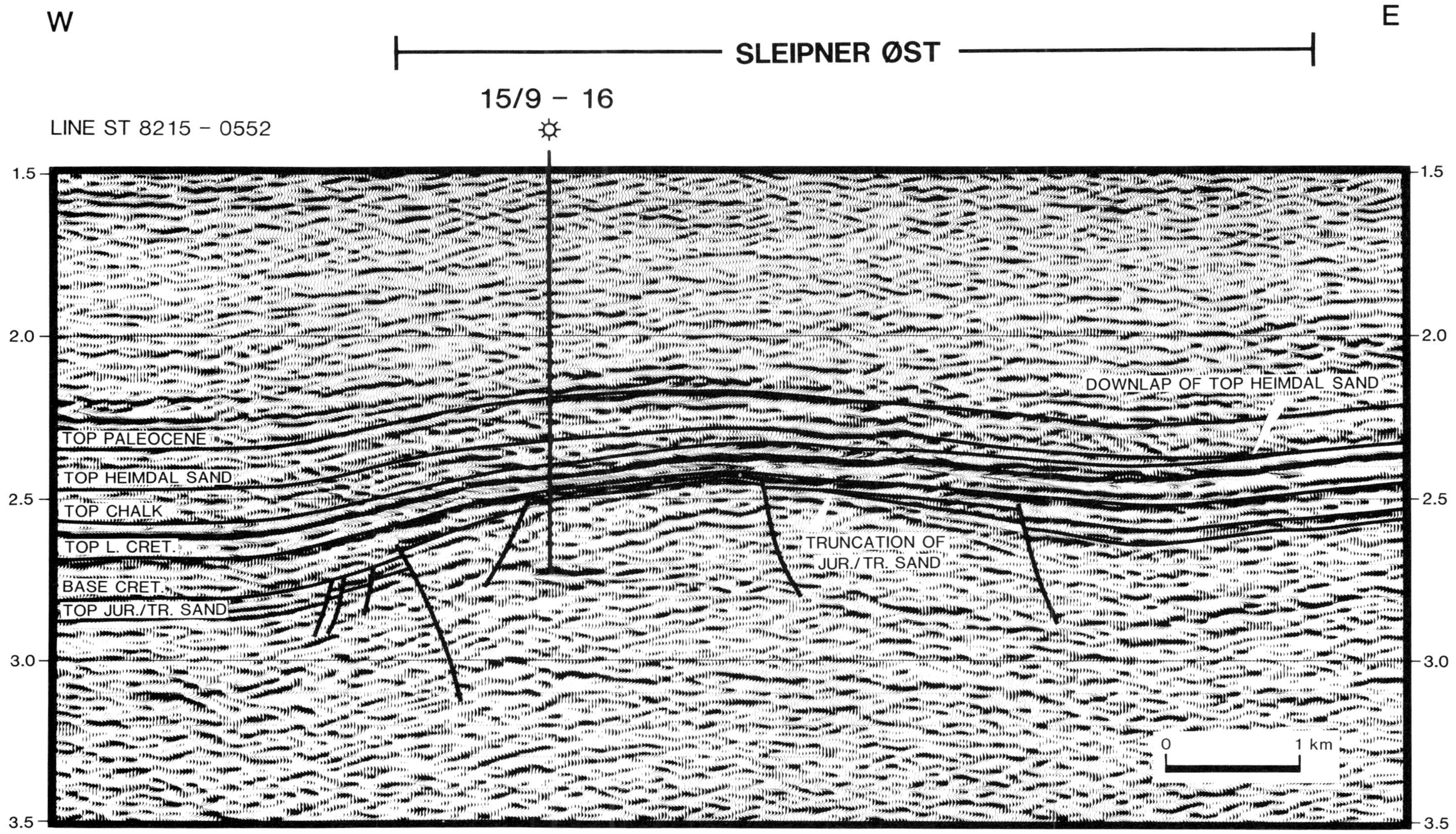

Fig. 6. East–west seismic line 552.

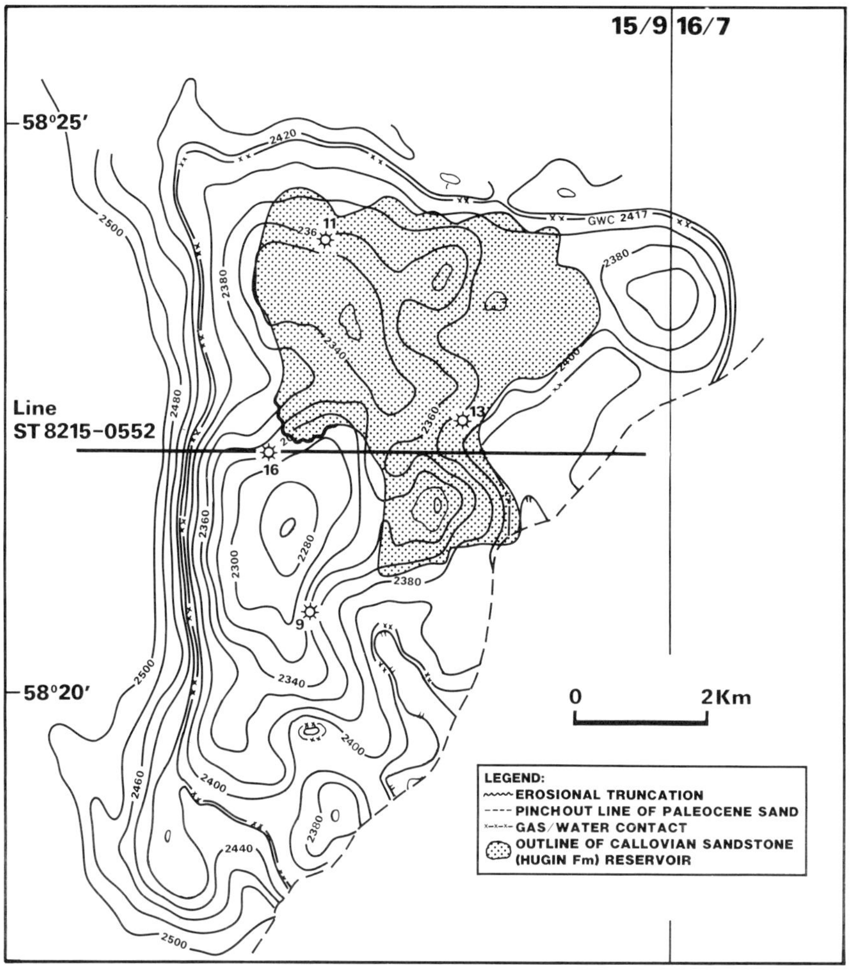

Fig. 7. Structure map; Top Heimdal Sand. Contours in metres subsea.

lithofacies. These fine-grained sediments occur as dark grey to dark green beds between thicker units of massive sands. Deposition was by settling from suspension during temporary abandonment of the fan-lobe.

The Heimdal Formation

This consists dominantly of thick beds of massive sandstones, interbedded with thin beds of shale–claystones and thin sand–shale units. The depositional sequence is interpreted to represent several stacked mid-fan lobes (see Mutti and Ricci Lucchi, 1972 and 1978) that spread out from a depocentre located north-west of the Sleipner Øst Field. The shaly units represent a shift of the active lobe or periods of fan recession when only very fine-grained sediments were deposited.

A shaly unit in the upper part of the formation can be correlated between the wells and separates the Heimdal Formation into an Upper and a Lower zone. This shale unit is not believed to represent a complete permeability barrier, as pressure measurements in the wells indicate that both zones belong to the same pressure system. The common occurrences of shale rip-up clasts in the sandstone beds overlying the shales is evidence of erosion. Locally the erosion may have been strong enough to disrupt the shales and provide permeability pathways. However, in terms of production the shale unit will probably behave as an impermeable layer. The reservoir properties of the sandstones are very good. Net gross ratio and porosity average 0.95 and 26% respectively, and the permeability ranges from 200 mD to 1500 mD.

The Maureen Formation

This formation consists mostly of massive sandstones, except in Well 15/9-11 where interbedded shales and sands dominate. Isopach maps indicate that the depocentre of the Maureen Formation was towards the west and only the distal part of a fan-lobe reached into the Sleipner Øst area. Calcite cement and

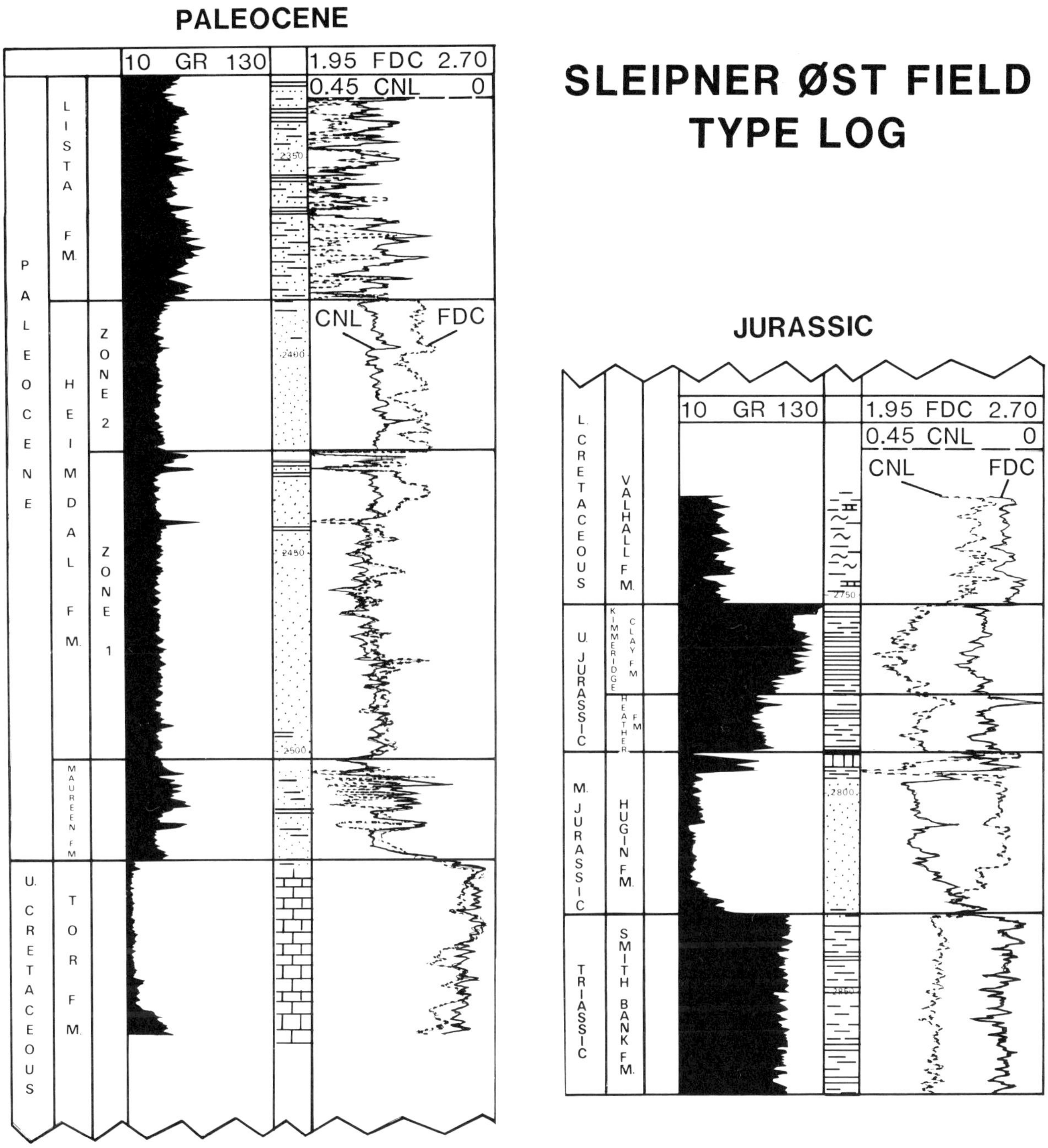

Fig. 8. Sleipner Øst Field type log.

pyrite concretions are present in parts of the sandstones. The average net gross ratio is 0.66 and the average porosity is 27%.

The Hugin Formation

This sandstone reservoir has been penetrated by only two wells within the Sleipner Øst Field. This formation is interpreted to represent shallow-marine shoreface sands interbedded with coastal plain deposits. A facies change occurs between Well 15/9-13 and Well 15/9-11, and makes it difficult to correlate the sands and to predict field-wide reservoir properties. The estimated average net gross ratio and porosity are 0.78 and 22% respectively. Permeability measurements in the wells were fairly good and indicate a range from 100 to 400 mD in the reservoir, although faults and sand discontinuities may represent permeability barriers.

SOURCE

Geochemical studies suggest that both the Jurassic and Paleocene reservoirs were sourced by Jurassic shales and coals in the

Viking Graben. The hydrocarbon gases have probably migrated through permeable sands and up fault zones and fractures to successively shallower reservoirs. The hydrocarbons in the Paleocene and Jurassic reservoirs are very similar in composition except for the CO_2 content. Loss of CO_2 by dissolution during migration is probably the reason for a much lower CO_2 content in the Paleocene reservoir than in the Jurassic (see Field data summary).

HYDROCARBONS

Sleipner Øst is a gas–condensate field. The gas in the Paleocene reservoir has a methane content of 71% and a gravity of 0.93 relative to air. Flow rates for tests in the Heimdal Formation were excellent. The best test gave 890×10^3 Sm^3/day with a condensate yield of 420 Sm^3 liquids per 10^6 Sm^3 of gas, using a 31.8 mm (80/64-inch) choke. The test interval was 2370–2390 mss.

The best test of the Jurassic Hugin Formation was from a four-metre interval in Well 15/9-13 at a depth of 2740-2769 mss. The test produced 800×10^3 Sm^3/day of gas through a 25.4 mm (64/64-inch) choke, with a condensate yield of 320 Sm^3 liquids per 10^6 Sm^3 of gas. The wet gas has a gravity of 0.94 relative to air. The hydrocarbons in the two reservoirs are very similar in composition, except for the CO_2 content (see Field data summary).

RESERVES

Total recoverable reserves for the Sleipner Øst Field are estimated at 60×10^9 Sm^3 of wet gas which is converted to 51×10^9 Sm^3 of dry gas and 23×10^6 tons of condensate after processing. The majority, 88% of the recoverable gas (53×10^9 Sm^3) is in the Paleocene reservoir. A strong water drive is expected for this reservoir, and the estimated recovery factor is about 70%. Recoverable wet gas in the Jurassic reservoir is estimated at 7×10^9 Sm^3.

ACKNOWLEDGEMENTS

This chapter has been compiled mainly from studies and interpretations performed by geologists and geophysicists of Esso Norge AS. The ideas and interpretations expressed herein are those of Esso Norge AS and do not necessarily represent those of the other partners. The author thanks the PL 046 partners—Statoil, Norsk Hydro and Esso Norge A/S—for permission to publish this chapter.

REFERENCES

Deegan, C.E. and Scull, B.J. 1977. A standard lithostratigraphic nomenclature for the Central and Northern North Sea, *Institute of Geological Sciences, Report* 77/25.

Mutti, E. and Ricci Lucchi, F. 1978. Turbidites of the northern Appenines; introduction to facies analysis, *International Geology Review* **20,** (2), 161–199. (Translated by T.H. Nilsen from: Le torbiditi dell' appeninno settentrionale: Introduzuone all' analisi di facies, *Mem. Soc. Geol. Italiana*, 1972, 161–199.)

Pegrum, R.M. and Ljones, T.E. 1984. 15/9 Gamma gas field offshore Norway; a new trap type for North Sea basin with regional implications: *American Association Petroleum Geologists Bulletin* **68,** (7), 874–902.

Vollset, J. and Doré, A.G. 1984. A revised Triassic and Jurassic lithostratigraphic nomenclature for the Norwegian North Sea, *Norwegian Petroleum Directorate Bulletin* No. 3, September, 1984.

22

Sleipner Vest

H. K. A. Ranaweera

Statoil, PO Box 300, N–4001 Stavanger, Norway

The Sleipner Vest Field is located at the southern end of the Viking Graben in Norwegian Blocks 15/9 and 15/6. The field was discovered in 1974 by Well 15/6-3 and later drilling has proved hydrocarbons in four more structures in Block 15/9. The reservoir consists of marginal-marine sandstones, 100–170 m thick, of the Callovian Hugin Formation, which have an average porosity of 18% and permeabilities ranging from 10–430 mD. The structural pattern is dome shaped and was formed by pre-Cretaceous and Late Eocene movements. The two main fault systems run in N–S and E–W directions. The reservoir sequence is capped by Upper Jurassic shales which seal the reservoirs. Thick coal beds of the Middle Jurassic Hugin and Sleipner formations and radioactive shales of the Upper Jurassic Draupne Formation have sourced the hydrocarbons. Two different pressure systems exist in the Sleipner Vest reservoir. Estimated recoverable resources are 126×10^9 Sm^3 of gas and 29×10^6 tonnes of condensate. The field is currently not under development, but a production plan shows that 30–45 production wells could drain the reservoir.

INTRODUCTION

The Jurassic Sleipner Vest Gas–Condensate Field lies in the Norwegian sector of the North Sea, approximately 240 km west-southwest of Stavanger at latitude 58° 22′ N and longitude 01° 45′ E. The field as currently defined is situated mainly in Block 15/9 with a small portion extending northwards into Block 15/6. Water depths in the field area are approximately 110 m. Marginal-marine sandstones of Callovian age form the main reservoir in the Sleipner Vest Field. Estimated recoverable reserves are 126×10^9 Sm^3 gas and 29×10^6 tonnes condensate. The field is named after the eight-legged horse of the God Odin in Norse mythology.

HISTORY

Licence PL046, comprising Block 15/9 and the neighbouring 15/8 Block, was first awarded in 1976 to the following group:

Statoil (operator)	50%
Esso Norge A/S	40%
Norsk Hydro Produksjon a.s.	10%

Previous exploration in the neighbouring block to the north, 15/6 (PL029 — Esso, 100%), had discovered gas in Jurassic sandstones in Wells 15/6-2 and 15/6-3, drilled in 1974. Well 15/6-3 encountered a gas–condensate column 74 m thick, and subsequent seismic mapping indicated that the structure extended southwards into Block 15/9 (Fig. 1). The discovery was informally designated as the Sleipner Field.

Early seismic interpretation in Block 15/9 defined a major N–S-trending fault system which separated the Sleipner Terrace in the west from the Utsira High fault-block in the east. In addition, Jurassic strata were thought to be thin or absent on the Utsira High and in the southern half of Block 15/9.

Geology of the Norwegian Oil and Gas Fields © A. M. Spencer *et al.*, eds (Graham & Trotman, 1987) pp. 253–264.

Sleipner Vest Field data summary

Trap	
Type	dome structure induced by salt movement
Depth to crest	3375 mss (Epsilon), 3450 mss (Alpha)
Lowest closing contour	3648 mss
Gas–water contact	3628 mss (Alpha NW), 3581 mss (Alpha NE), 3637 mss (Alpha S/Beta), 3648 mss (Epsilon), 3538 mss (Delta)
Gas column	250 m
Pay zone	
Formation	Hugin Formation sandstone
Age	Middle Jurassic Callovian
Gross thickness	54–203 m
Net/gross ratio	0.80
Porosity	18–23%
Hydrocarbon saturation	79%
Permeability	10–430 mD
Hydrocarbons	
CO_2 content	9%
Gas gravity (relative to air)	0.76
Gas type	dry
Condensate gravity	0.79 g/cm^3
Condensate type	paraffinic
Condensate yield	3754 Sm^3/10^6 Sm^3
Gas expansion factor	285
Formation water	
Salinity	70 000 ppm NaCl equivalent
Resistivity	0.031 ohm m
Reservoir conditions	
Temperature	123°C
Pressure	440 bar at 3637 mss
Pressure gradient	0.0315 bar/m at 3637 mss
Field size	
Area	80 km^2
Gross rock volume	6960×10^6 Sm^3
Hydrocarbon pore volume	688×10^6 Sm^3
Recovery factor	0.75
Recoverable gas	126×10^9 Sm^3
Recoverable condensate	45×10^6 Sm^3
Production	
—	

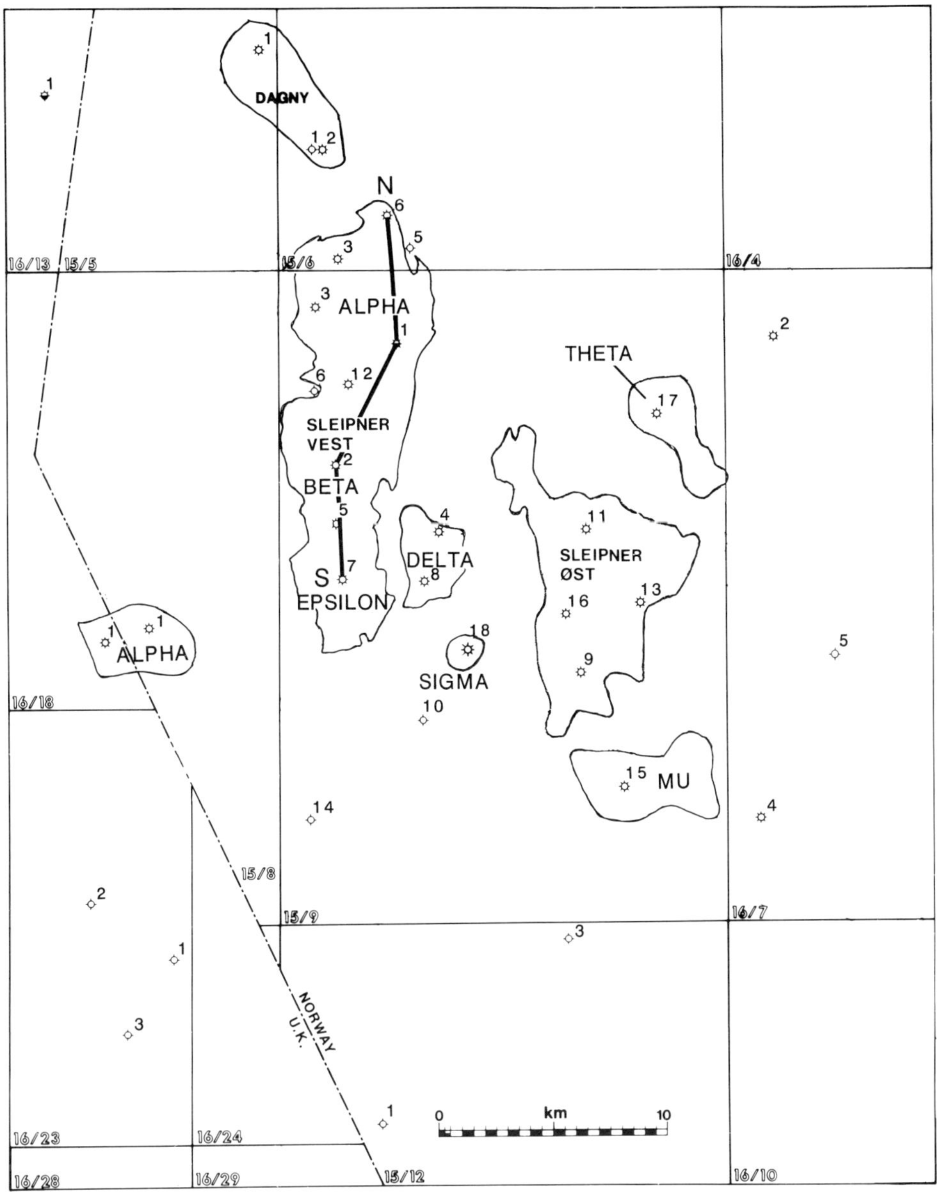

Fig. 1. The Sleipner Fields.

For these reasons early exploration was directed towards the known gas–condensate-bearing sandstones of the Sleipner Terrace.

After the 1976 award the first well, 15/9-1, was spudded in 1977 on the Sleipner Alpha prospect (Fig. 1). It encountered a gas–condensate column of about 100 m which was underlain by thin oil-bearing sandstones. Reserve estimates for the Alpha structure were, however, reduced in late 1977 after Well 15/6-5 proved to be dry. In 1978 Well 15/9-2 proved the presence of gas–condensate in the Beta structure. Reserves on the Alpha structure were again decreased after the drilling of Well 15/9-3. Consequently, further exploration was directed towards the other Jurassic structural culminations south of the Alpha structure. The next two wells, 15/9-4 and 15/9-5, drilled in 1979, proved the presence of hydrocarbons in the Delta and Beta structures respectively.

Well 15/9-6, drilled during 1980 between the Alpha and Beta structures, located a down-faulted area and was dry. The results of the initial phase of exploration have been described by Larsen and Jaarvik (1981). At this time it was apparent that the 1975 seismic data were of too poor a quality for detailed reservoir mapping. Consequently, 2250 km of three-dimensional seismic data on a 75-m grid were shot on the Alpha structure during 1980. A further 3050 km were shot during 1981 to cover the remaining area of Sleipner Vest. Interpretation of these seismic data is the foundation for the present reservoir mapping. In the meantime, Well 15/9-9 was drilled in

1981 on the Gamma structure (now designated Sleipner Øst), a high structural culmination on the up-thrown Utsira Block, to test for hydrocarbons in possible Permian/Triassic and Jurassic intervals. Although no hydrocarbons were discovered in these intervals, Well 15/9-9 did, however, penetrate a significant gas-bearing interval within porous Paleocene sandstones which had a gross pay thickness of 103 m (Pegrum and Ljones, 1984). Following this gas discovery, the eastern half of Block 15/9 was remapped and an additional seismic programme was shot in September 1981. The extent of the Sleipner Øst Field was proven by Well 15/9-11 in 1981 and Wells 15/9-13 and 15/9-16 in 1982. Wells 15/9-13 and 15/9-11 proved the presence of additional hydrocarbons in Middle Jurassic sandstones.

During this period of exploration of Sleipner Øst, activity continued on Sleipner Vest. Early in 1982, Well 15/9-12 confirmed the new seismic interpretation that no structural closure existed between the Alpha and the Beta structures. Well 15/6-6 confirmed the presence of gas–condensate in the northern area of the Alpha structure, although the gas–water contact proved to be higher than expected.

During this period, drilling activity also extended to previously unexplored areas. Wells 15/9-10 and 15/9-14 south of Sleipner Vest proved to be dry, whilst hydrocarbons were located in Well 15/8-1. Hydrocarbons were also proved in Wells 15/9-15 and 15/9-17, drilled on the Mu and Theta structures respectively. The latest well, 15/9-18 (Fig. 1), located south-east of Sleipner Vest, found only a minor accumulation. In 1982 an additional three-dimensional seismic survey was shot over the remaining area of Block 15/9.

GEOLOGICAL SETTING

Sleipner Vest is situated near the southern end of the Viking Graben where several major structural elements converge. These include the Utsira High and Ling Graben to the east, the East Shetland Platform and the Fladen Ground Spur to the west, and the Andrew Salt Ridge and Jæren High to the south (Fig. 2). The structural pattern within the Sleipner area is dominated by two major basement-controlled fault-blocks, namely the Gamma High to the east and the Sleipner Terrace to the west. The westerly down-thrown Sleipner

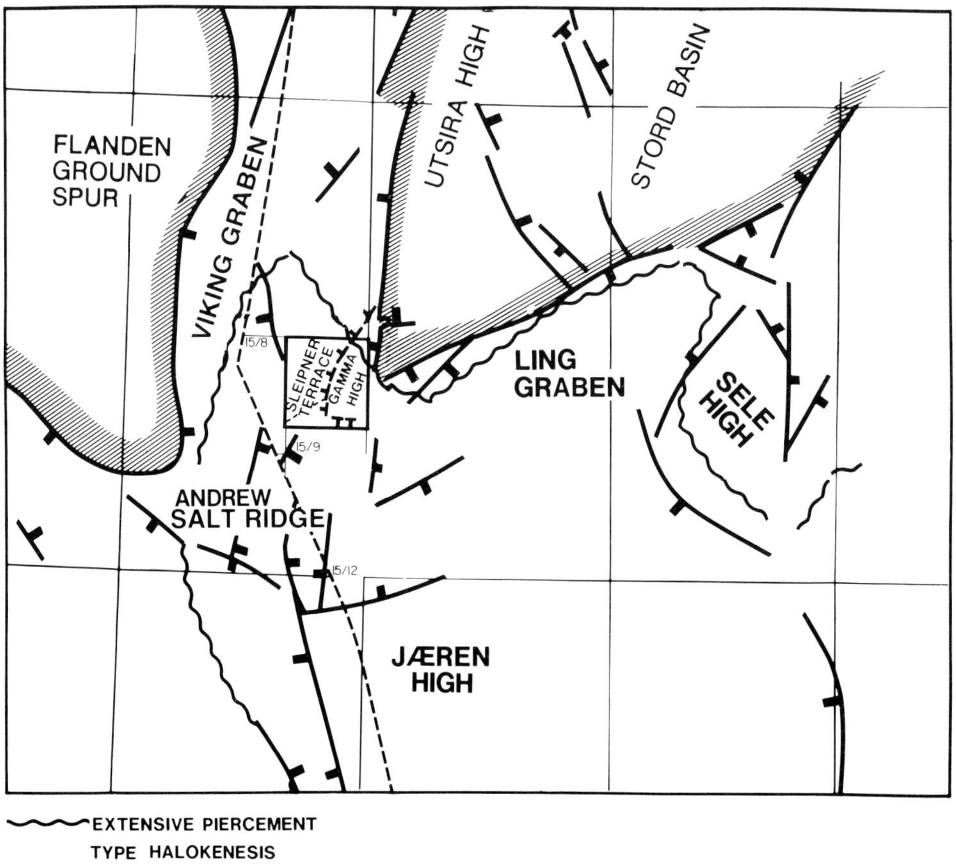

Fig. 2. Tectonic Map.

Terrace preserves a cover of Jurassic rocks of variable thickness. The structural configuration of the Sleipner Terrace is further complicated by the presence of a thick sequence of Zechstein evaporites. Extensional rifting affected the 15/9 area periodically from Early Permian until Late Jurassic times. Extensional rifting reached a maximum in Late Jurassic times, with differential rates of rifting leading to localized compression and strike-slip fault movements and, perhaps, rejuvenation of basement faults (Fig. 3). This extensional rifting in Late Jurassic times could have triggered movement in the underlying Zechstein evaporites, deforming the Triassic and Jurassic cover in the Sleipner area. Reversal of normal faults, due to compressional movements in early Mid-Cretaceous times, resulted in further structuring at Jurassic/Triassic levels. A period of relative tectonic quiescence followed this in Late Cretaceous times reflecting a major change in the regional stress field, due perhaps to the initial opening of the North Atlantic Ocean.

Rift movements within the North Sea had essentially ceased by Early Cretaceous times. At the end of Late Cretaceous times the Laramide orogenic event resulted in an uplift and stripping of the East Shetland Platform. This event put coarse clastics into the Sleipner Area in the form of Paleocene submarine-fan deposits. A further rejuvenation of the fault system occurred in Block 15/9 during Mid-Tertiary times.

A generalized stratigraphic column is given in Fig. 4. The lithostratigraphic nomenclature used is that proposed by Deegan and Scull (1977) as revised by Vollset and Doré (1982).

Pre-Triassic rocks have not been penetrated on the Sleipner Terrace in Block 15/9. In Sleipner Øst, Well 15/9-9 drilled the Rotliegend, which comprised interbedded red sandstones and breccias deposited in a terrestrial to fluvial/lacustrine environment. The Zechstein sequence in this well was composed of interbedded dolomites and anhydrites with a basal, thin dark-coloured, radioactive shale, ascribed to the Kupferschiefer Formation.

Triassic strata are present throughout Block 15/9 but their stratigraphy is poorly known. Most of the wells drilled on the Sleipner Terrace have terminated after drilling a few tens of metres of Triassic red-bed shales, siltstones and sandstones. Well 15/9-9 penetrated palaeontologically barren sandstones of the Skagerrak Formation, underlain by thick, red shale-beds of the Smith Bank Formation.

Jurassic strata are well developed in Sleipner Vest as proved in Well 15/9-4. The oldest Jurassic strata encountered belong to the Middle Jurassic Bathonian–Bajocian Sleipner Formation, comprising shales, siltstones and thick coal-beds with minor sandstones deposited in delta top/coastal plain environments. The Callovian Hugin Formation, which forms the main reservoir, comprises shallow-marine shelf sandstones and shales, barrier-bar sandstones, back-barrier sandstones, siltstones and shales, deposited

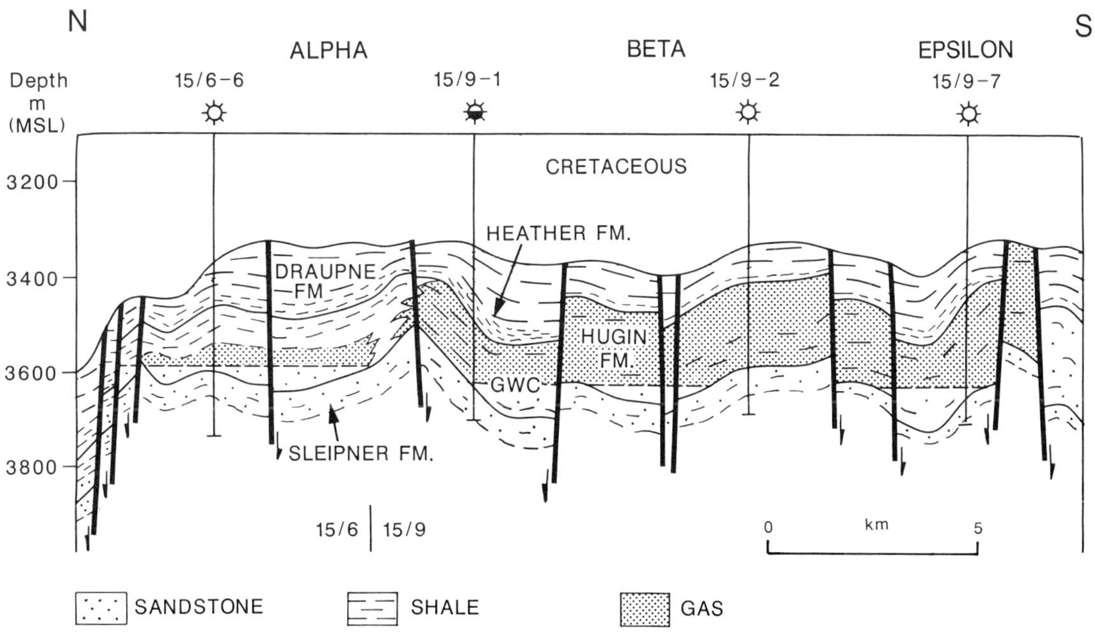

Fig. 3. Cross-section N–S through the Sleipner Vest Field.

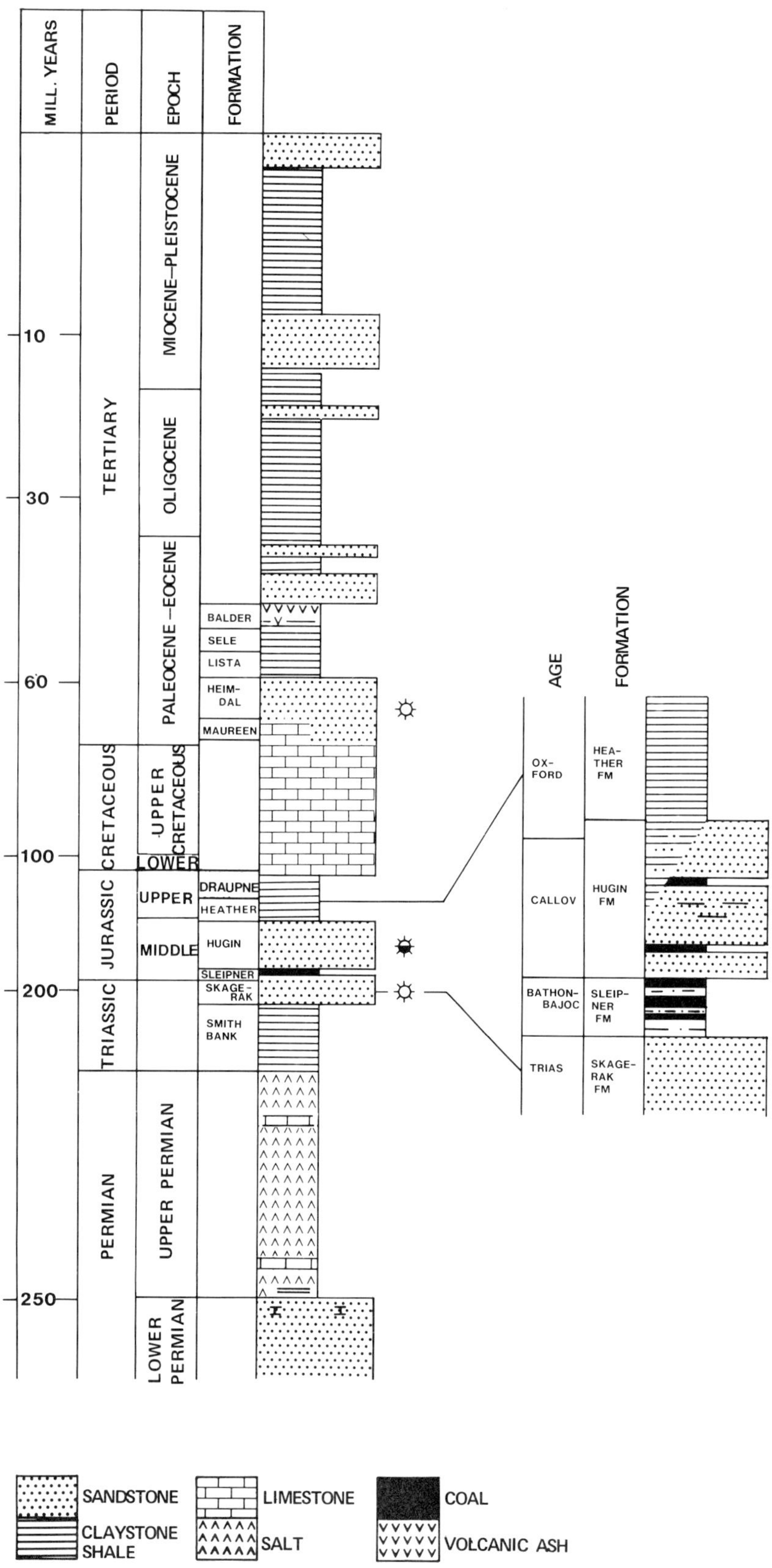

Fig. 4. Generalized Stratigraphic column.

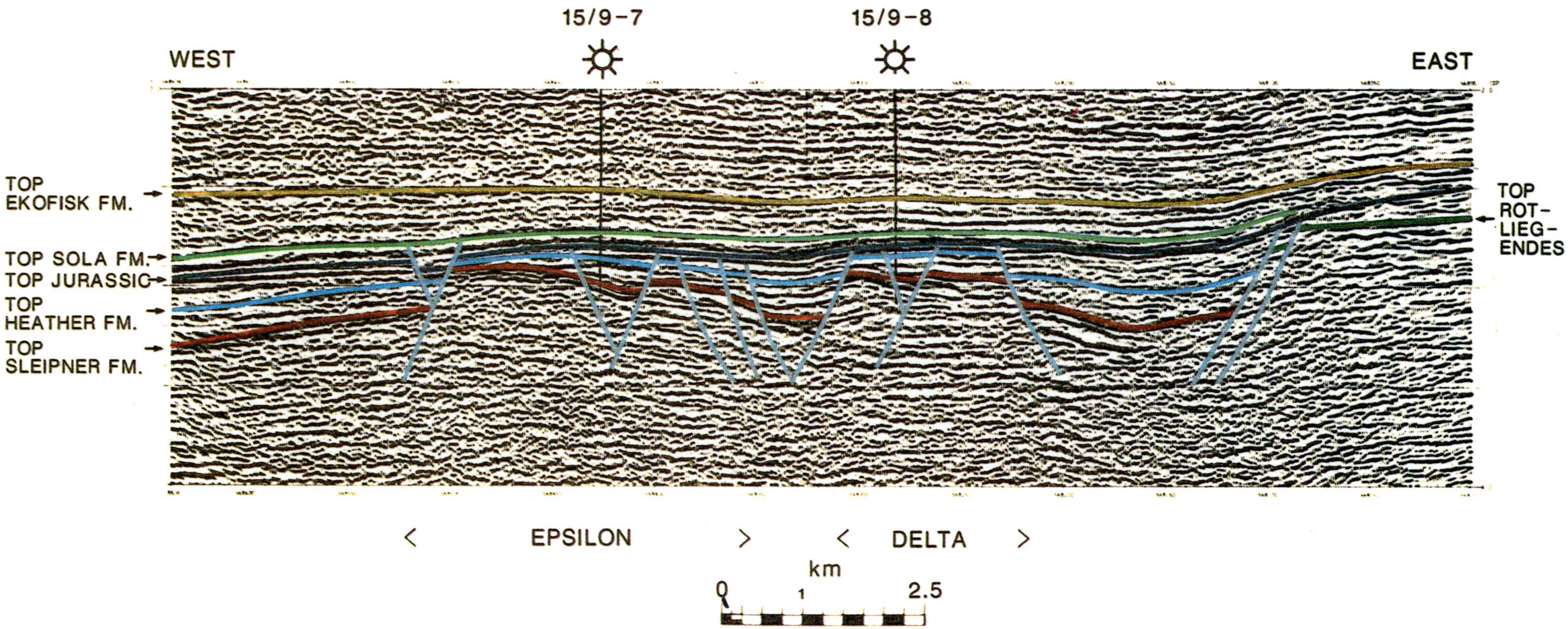

Plate 8. Three-dimensional seismic line close to Wells 15/9-7 and 15/9-8.

as a result of three transgressive cycles. The Upper Jurassic strata encountered in Sleipner Vest are the Oxfordian Heather shales and the Kimmeridgian and Volgian Draupne clays.

Lower Cretaceous strata in the 15/9 area are represented by the Rødby Formation, the Sola Formation and the Valhall Formation and comprise marine shales, marls and siltstones. A thick Upper Cretaceous Chalk Group is present (Cenomanian to Maastrichtian).

A complete sequence of Lower Paleocene is encountered in the Sleipner Vest Field and is represented by the Maureen, Heimdal, Lista and Sele formations. The Balder Formation straddles the Paleocene–Eocene boundary. The Lower Paleocene Heimdal Formation forms the reservoir in the Sleipner Øst Field. The overlying Eocene and Oligocene sequence consists of monotonous sections of clays and silts with stringers of sand and limestones. The Miocene sequence in Block 15/9 comprises a lower group of marine shales and an upper group of shelf sands. The latter corresponds to the Utsira Formation.

GEOPHYSICS

The discovery of the Sleipner Vest field in 1974, by Well 15/6-3, apparently extending into neighbouring Block 15/9, led to the acquisition of a detailed seismic survey over the area in 1975. A 1 × 1 km seismic grid covering the field area was acquired in 1975 and 1976. Six good reflectors were interpreted: base Utsira Formation, top Paleocene, top Ekofisk Formation, top Sola Formation, top Jurassic and one intra-Jurassic reflector. These seismic data provided the basic data until 1980 when it became apparent that the quality of the data was too poor for detailed reservoir mapping.

This led in 1980, to the first three-dimensional (3D) seismic survey to cover the Alpha structure. It was acquired and processed by Geco, and consisted of 170 lines, totalling 2250 km, with an E-W shot-line orientation. The shot-point interval was 25 m and the line-spacing 75 m. The three-dimensional migration was done in the FK domain, both for the row migration (in line) and the column migration (cross-line). Six horizons were interpreted: top Ekofisk Formation, top Sola Formation, top Jurassic, top Heather Formation, top Sleipner Formation and top Triassic. The mapping showed that the Alpha structure is broadly dome shaped. Two main fault-trends, in N–S and E–W directions, are recognized. The post-Jurassic and top Jurassic reflectors are broken by a few small faults. The top Heather and top Sleipner Formation horizons are much more faulted.

The second three-dimensional survey, in 1981, covered the Beta, Delta and Epsilon structures. The data were acquired and processed by GSI. The survey consisted of 179 lines with a total length of 3046 km. The direction and the spacing used was the same as for the previous survey. Six horizons were again picked (Plate 8).

The interpretation confirmed the extensively faulted dome-shaped structure of the Alpha culmination (Fig. 5). It also showed a N–S, elongated dome-shaped structure with a fairly flat top (Beta), a steeply dipping half-dome-shaped structure (Epsilon) and a horst structure (Delta).

A third three-dimensional survey was shot in 1982 and covered the Sleipner Øst Field and the southern part of Block 15/9. It consists of 5979 km of fully three-dimensional migrated rows.

TRAP

The structural pattern of the Sleipner Vest Field comprises several dome-shaped culminations formed by pre-Cretaceous movements. The extensional rifting and localized strike-slip movements in Late Jurassic times are believed to have triggered the movement in the underlying Zechstein evaporites, deforming the Triassic and Jurassic cover in the Sleipner area. The Zechstein salt acted as a zone of structural décollement on the Sleipner Terrace, the post-Zechstein strata being deformed independently of the Pre-Zechstein basement. Finally, Late Eocene compressive movements could have rejuvenated the Mesozoic strike-slip faults, inducing salt movement and forming a series of en-échelon domal structures (Pegrum and Ljones, 1984). The Upper Jurassic shales cap the reservoir rocks and act as a seal.

RESERVOIR

The main reservoir at Sleipner Vest, the Hugin Formation, is divided into three units, A, B and C, (Fig. 6) corresponding to three major phases of sediment accumulation. These three phases were the consequence of three episodes of marine transgressions separated by three regressive, progadational pulses.

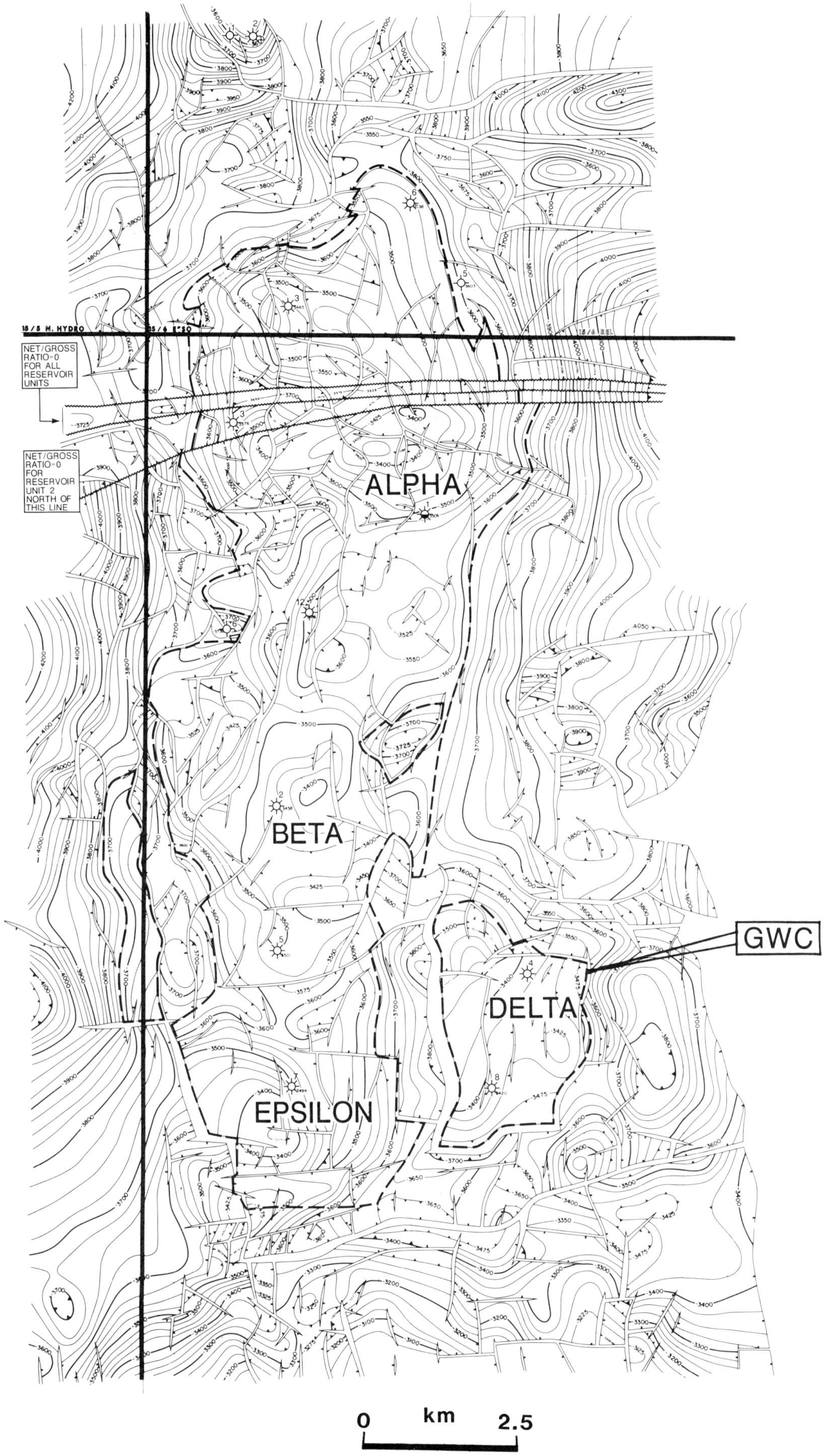

Fig. 5. Top reservoir depth map of the Sleipner Vest Field.

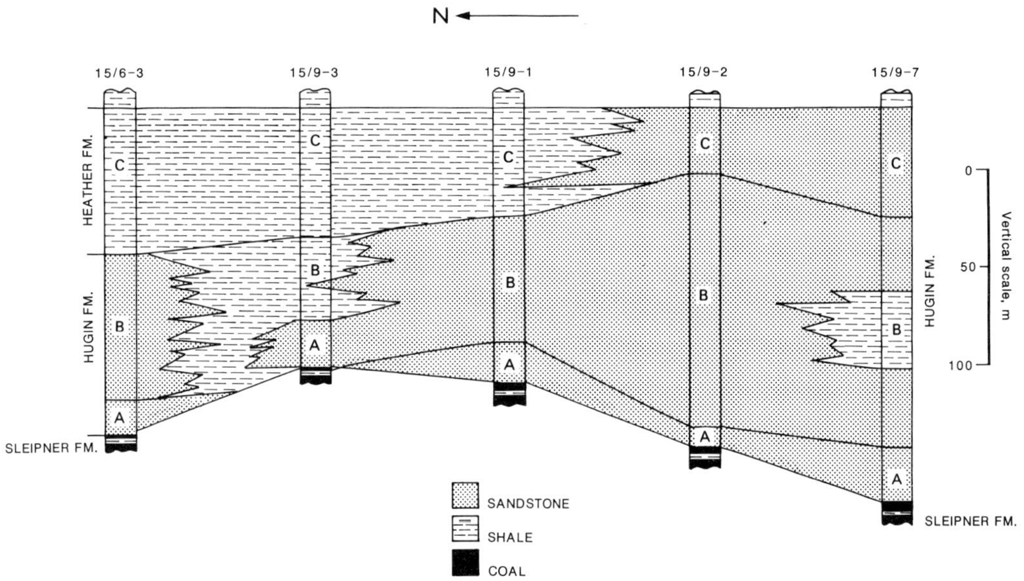

Fig. 6. Correlation of reservoir with major depositional units.

Unit A is the basal member and has a sharp, planar contact with the underlying Sleipner Formation. It comprises fine- and medium-grained, graded sandstone units of storm origin, which are overlain by very fine-grained, argillaceous, burrowed, shoreface sandstones. These pass up into coarsening-upwards fine- and medium-grained, locally pebbly sandstones characterized by fine parallel laminations and low-angle cross-bedding. This sequence is capped by a thick, drifted coal.

Unit B is the middle member of the Hugin Formation and includes a number of different facies. It is composed of a basal, medium- and fine-grained, occasionally pebbly, graded unit of bedded and cross-bedded (15°) carbonaceous sandstones, separated by scour surfaces. This is overlain by a coarsening-upwards, argillaceous, carbonaceous, burrowed sandstone. Above this is an interbedded, thoroughly burrowed and bioturbated, organic-rich siltstone and mudstone which becomes progressively less argillaceous and bioturbated upwards.

Unit C comprises restricted, moderately silty burrowed, marine shales and fine sandstones and is characterized by coarsening-upwards grain-size trends and relatively sharp tops. At least three or more such sequences are recognized in unit C, which is capped by marine shales of the Heather Formation.

Quartz is the most abundant detritalmineral in the Hugin Formation sandstones, usually followed by feldspar. Normal unstrained quartz is the most abundant type and strained and polycrystalline quartz are less common. Both orthoclase and microcline feldspars occur, with some plagioclase and perthites. Minor, more scattered detrital grains include mica, rock fragments, ooliths, altered glauconite and heavy minerals including tourmaline and zircons. The detrital heavy mineral composition, combined with the presence of granitic and extrusive volcanic rock fragments, suggests that the source-rocks for the sandstones were largely igneous. Evidence of diagenesis is widespread. The main diagenetic phases are quartz overgrowth development, siderite, pyrite, vitrinite gel, kaolinite, dolomite, illite, leucoxene, anatase and feldspar dissolution. Some of these exhibit more than one phase of development, depending upon variations in the diagenetic history.

Interpretation of the available geological evidence suggests that the Hugin Formation developed in a fairly stable, marginal-marine environment in which major facies changes resulted from changes in sea level. The major sand development was the aggradation of the nearshore barrier–bar complex, centred on an ENE–WSW-orientated linear locus through Well 15/9-2 (Fig. 7), with estuarine facies developed to the south and shoreface facies to the north.

In Early Callovian time the sea transgressed the area from the north, reaching south to the area of Wells 15/9-10 and 15/9-14. A progradation of the shoreline towards the north followed, depositing shoreface sandstones and the associated back-barrier lagoonal and coastal marsh deposits. The second transgression covered approximately the same area as the first, and was followed by the development of an aggrading barrier–bar complex with shoreface and estuarine facies developed to the south and offshore shelf facies to the north in Block 15/6 (Fig. 7).

In Late Callovian times a third transgressive phase eventually flooded the entire Sleipner area. This general transgression was interrupted by some regressive phases, due to substantial sediment accumulation, resulting in a well-developed shoreface facies in the southern part of the field. No sands were deposited north of Well 15/9-12.

SOURCE

The condensates in Sleipner Vest were derived from both the Draupne and Hugin/Sleipner Formations. The contribution from the Draupne Formation increases signifi-

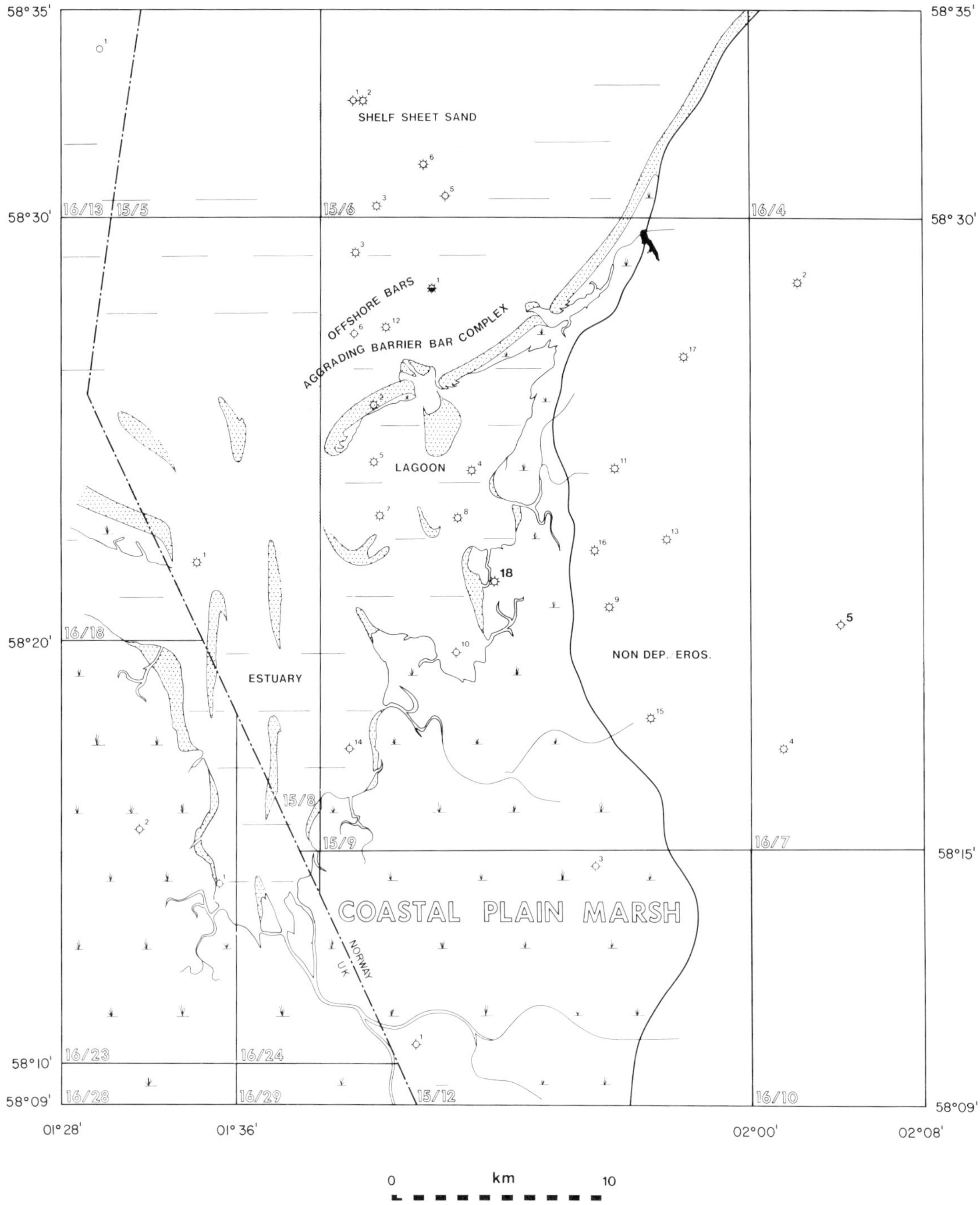

Fig. 7. Palaeogeographic reconstruction of Hugin Formation.

cantly to the north into Block 15/6. In Block 15/9 the accumulated hydrocarbons were generated from high-maturity source-rocks, with late gas from the deepest Hugin/Sleipner formations. Early and mid-mature hydrocarbons from Draupne Formation were generated in the down-faulted areas between Sleipner Vest and the Utsira High in Block 15/9 and migrated into Wells 15/9-1 and 15/9-3. This explains the early oil in 15/9-1. Early and mid-mature condensate generated in the Hugin and Sleipner formations, migrated into the adjacent Alpha structure in Block 15/8. Late condensates and gas generated from the Hugin/Sleipner formations in the graben bypassed the 15/8 Alpha structure and migrated into the Sleipner Vest area.

HYDROCARBONS

The pressure in the Sleipner Vest reservoir is 70 bar higher than the hydrostatic pressure (Fig. 8). Pressure distribution in the reservoir is complex, as indicated by varying pressure measurements from the different wells. Some of the pressure variations can be attributed to imprecision of measurement, but real differences are thought to exist and compartmentalized pressure regimes can be inferred.

For resources-calculation purposes the pressure model shown in Fig. 8 has been used. The model indicates the existence of two separate pressure systems in the Alpha structure. Pressure measurements from Well 15/6-6 support the definition of a separate pressure regime in the northern part of the Alpha structure. The Delta structure has its own pressure regime (Fig. 8).

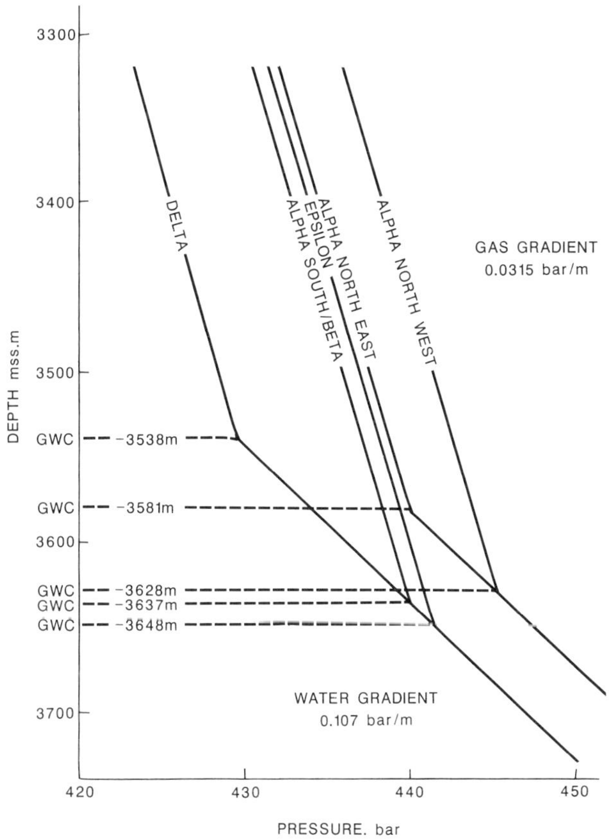

Fig. 8. Pressure model for the Sleipner Vest Field.

RESOURCES

Recoverable resources in Sleipner Vest are estimated to be 126×10^9 Sm^3 gas and 29×10^6 tonnes condensate. Owing to the complexity of the reservoir, the production wells will be spread to cover most of the field. Between 35 and 45 wells will be needed to drain the reservoir. The field is assumed to have a very small aquifer and therefore will be produced by gas depletion. The final development plan for the Field is not yet decided and will depend on the market possibilities for the sale of the gas.

ACKNOWLEDGEMENT

This article is based on the results of contributions made by many people from Statoil who have worked on the Sleipner Vest Field over the past few years. The author thanks Statoil, Norsk Hydro and Esso Norge AS for giving permission to publish this paper.

REFERENCES

Deegan, C.E. and Scull, B.J. 1977. A standard lithostratigraphic nomenclature for the Central and Northern North Sea, *Report Institute of Geological Sciences* 77/25.

Larsen, R.M. and Jaarvik, L.J. 1981. The geology of the Sleipner Field Complex, In *Norwegian Symposium on Exploration*, Norsk Petroleumsforening, Article 15.

Pegrum, R.M. and Ljones, T.E. 1984. 15/9 Gamma gasfield offshore Norway, new trap type for North Sea basin with regional implications, *American Association of Petroleum Geologists Bulletin* **68,** (7) 874–902.

Vollset, J. and Doré, A.G. 1984. A revised Triassic and Jurassic lithostratigraphic nomenclature for the Norwegian North Sea, *Norwegian Petroleum Directorate, Bulletin 3.*

23

15/5-1 Discovery

J. B. Christiansen* and J. E. Strand

Norsk Hydro, PO Box 200, N-1321 Stabekk, Norway

The discovery named after Well 15/5-1 is located on the western edge of the Sleipner Terrace, 20 km north-west of the Sleipner Field. The discovery was made by Esso in 1974 by the Well 15/6-2 and confirmed in 1977 by the Well 15/5-1. Gas–condensate was proven in sandstones of Callovian to Bathonian age (Hugin Formation). The reservoir is a sandstone, approximately 60 m thick, which was deposited in a shallow-marine/restricted environment between the Utsira High and the East Shetland Platform. The depth to the reservoir is 3400 m. The field is covered by a three-dimensional seismic grid shot in 1980. The in-place resources are estimated to be about 8.75×10^9 Sm^3 gas. No development plans exist today. The discovery has been evaluated as a possible satellite to a Sleipner Field development.

INTRODUCTION

The 15/5-1 discovery is located at the western edge of the Sleipner Terrace on the eastern side of the southern Viking Graben, 20 km north-west of the Sleipner Field (Fig. 1). The discovery covers an area of 8.3 km^2. It lies in the south-eastern part of Block 15/5 and stretches into the neighbouring Block 15/6 (Esso), where the discovery Well 15/6-2 is located.

The water depth is approximately 120 m. The Hugin Formation of Callovian to Bathonian age constitutes the reservoir in this relatively small gas field.

HISTORY

Licence PLO48 was awarded in February 1977 to the following group:

Norsk Hydro AS (operator)	17.300%
Elf Norge A/S	14.535%
Total Marine Norsk A/S	10.900%
Aquitaine Norge A/S	7.265%
Statoil	50.000%

Originally the Production Licence consisted of the two Blocks 15/2 and 15/5 adjoining the median line. In February 1983 Block 15/2 was totally relinquished and Block 15/5 was retained.

To date four exploration wells have been drilled within the licence area. The first well, 15/5-1, was spudded in November 1977 and tested the structure in the south-east of Block 15/5. The well tested gas–condensate in a Callovian sandstone reservoir, the Hugin Formation.

In 1978 a development concept study was undertaken to evaluate the reservoir potential and possible development of such a small field: it was treated as a possible satellite to the neighbouring Sleipner development.

In 1980 a three-dimensional (3D) seismic survey was shot over the field in order to obtain better seismic mapping of the relatively complex structure.

*Present address: *3DC Consultants, Brynsveien 96, N-1352 Kolsås, Norway*

Geology of the Norwegian Oil and Gas Fields © A. M. Spencer *et al.*, eds (Graham & Trotman, 1987) pp. 265–272.

15/5-1 Field data summary

Trap	
Type	structural (faulted dome)
Depth to crest	3380 mss
Lowest closing contour	3670 mss
Hydrocarbon contacts	no GWC in well
Gas column	58 m observed
Pay zone	
Formation	Hugin
Age	Upper Jurassic
Gross thickness (average)	58 m
Net/gross ratio (average)	0.75
Cut-off for N/G	$S_w = 50\%$; $Ø = 9\%$; $V_{sh} = 40\%$
Porosity (average)	14%
Hydrocarbon saturation	86%
Permeability (average; range)	28 mD; 1–180 mD
Productivity index	25×10^3 Sm^3/day/bar
Hydrocarbons	
Gas gravity (relative to air)	0.73
Dew point	471 bar
Condensate yield	544 Sm^3/10^6 Sm^3 gas
Gas expansion factor	256
Formation water	
Resistivity	0.038 ohm m at 127°C
Reservoir conditions	
Temperature	127°C
Pressure	474 bar
Pressure gradient in reservoir	0.038 bar/m
Field size	
Area	8.3 km^2
Gross rock volume	376×10^6 m^3
Hydrocarbon pore volume	30.4×10^6 m^3
Recovery factor — condensate	22.0%
— gas	76.5%
Drive mechanism	natural depletion
Recoverable NGL/condensate	0.9×10^6 Sm^3
Recoverable gas	6.7×10^9 Sm^3
Production	
—	

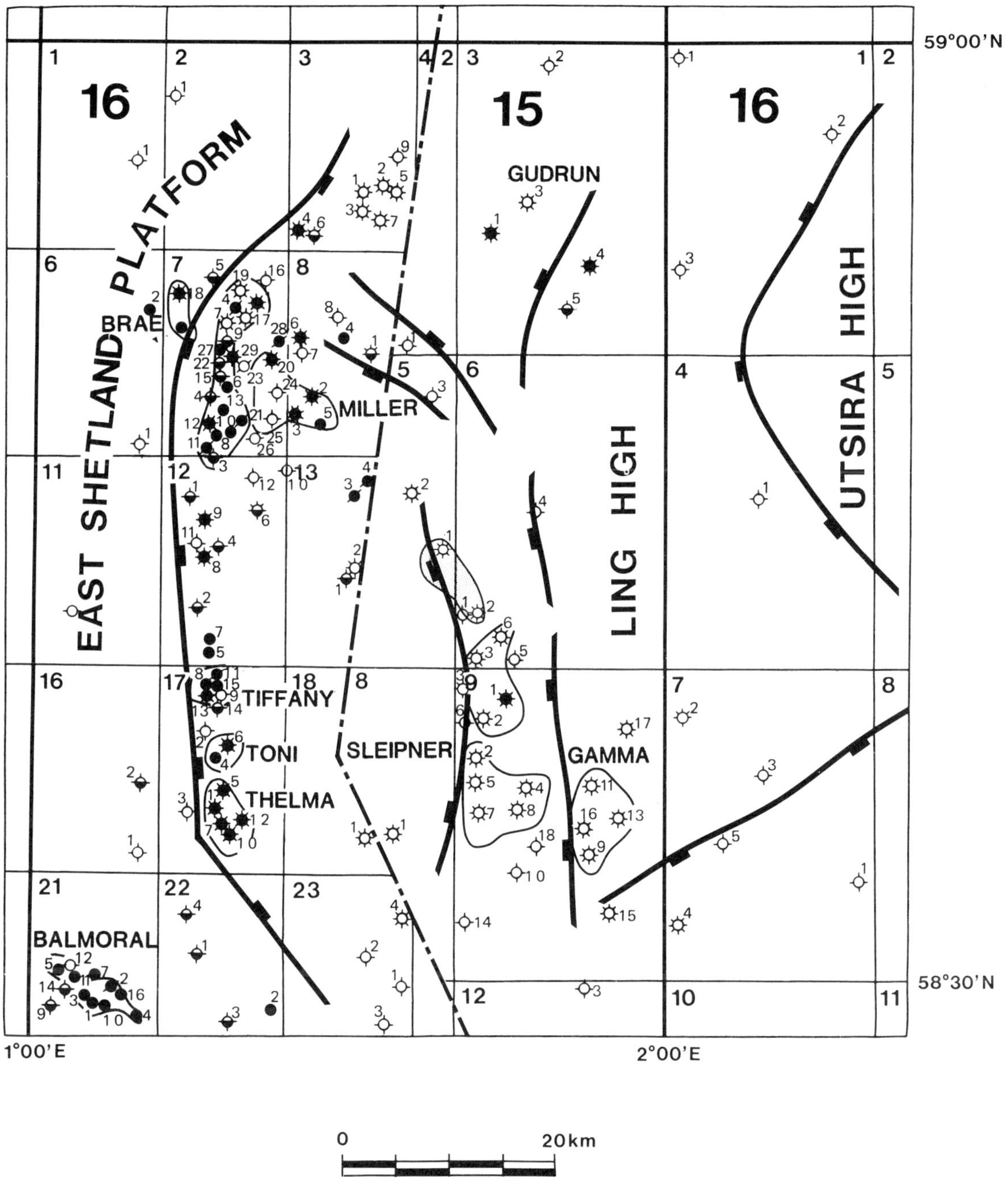

Fig. 1. Location Map.

The conclusions of the development concept study and the detailed three-dimensional (3D) seismic interpretation are that the proven in-place gas resources (8.75×10^9 Sm^3) are at present too small to justify a stand-alone development, but that as a satellite to the Sleipner Field the structure may become interesting.

GEOLOGICAL SETTING

The 15/5-1 discovery is situated at the western edge of the Sleipner Terrace, on the eastern side of the Southern Viking Graben. The Sleipner Terrace started to develop during Permian times, and was a part of the northern Zechstein basin, bounded on the west by faults down-throwing to the Viking Graben and on the east by faults up-throwing to the Utsira High. The Permian development of the Graben System saw thick sequences of Zechstein evaporites deposited on the Sleipner Terrace. In addition to the N–S normal faulting connected with the rifting of the Viking Graben, the Sleipner Terrace is intersected by NW–SE transverse faults of a regional nature.

Unlike the northern part of Block 15/5, the area of the 15/5-1 structure does not seem to have been influenced by early salt movements during Triassic time. Well 15/6-2 penetrated 890 m of Triassic strata before entering anhydrites of Zechstein age. Although top Zechstein is not a seismic event, 800–1000 m of Triassic is regarded as an average thickness for the area.

A widespread seismic reflector corresponding with the Middle Jurassic coals implies there was a flat-lying and tectonically stable base for the deposition of the westward-prograding Callovian reservoir sands, which were deposited in shallow-marine conditions in a restricted basin between the Utsira High and the East Shetland Platform.

The development of the structure started during Oxfordian time and was strongly influenced by rifting in the southern Viking Graben. Westward tilting of Triassic–Jurassic strata caused gravity gliding on the salt–Triassic interface. This resulted in the development of transverse faults on the south and north sides of the structure and two large slump faults on the western side. On the east side of the structure collapse features provide the closure. The two slump faults magnified the salt growth and resulted in the development of two separate highs on the structure. The crests of these highs were exposed to erosion, and the Callovian sands were partly eroded. The structure was finally capped by the Draupne Formation.

Early Cretaceous time was a period of subsidence in the Viking Graben which resulted in further westerly tilting. A tectonic phase during early Tertiary times gave the final contribution to the structuring by a slight doming. A generalized stratigraphic column is shown in Fig. 2.

GEOPHYSICS

In June 1980 a 1200-km, three-dimensional seismic survey was acquired over the structure. Until then the mapping of the field had been based on several vintages of two-dimensional seismic data. The three-dimensional seismic survey consists of 168 E–W lines, rebinned from 75 m to 50 m. This line spacing does not seem to be small enough to avoid spatial aliasing on the steeply dipping north and south flanks of the structure. It is not possible to interpret the reservoir interval, either on the two-dimensional or on the three-dimensional seismic data (Fig. 3).

There are, however, two seismic reflectors in the zone of interest: the Base Cretaceous unconformity and the Middle Jurassic coal marker at the base of the reservoir sand. The top reservoir map (Fig. 4) has been constructed by adding a geologically derived sand isopach to the top of the coal marker. These maps have been the basis for the reserve calculations.

TRAP

The 15/5-1 discovery is a structural trap with a relief of about 290 m (Fig. 4). The crest of the structure is at 3380 m and the spill-point in the east at 3670 m. The gas–water contact (GWC) has not been determined from either of the two wells, 15/5-1 or 15/6-2. For the purposes of reserve estimation the GWC is assumed to be at the structural spill-point. All the faults interpreted in the post-Zechstein strata intersect the reservoir level, and are of such magnitude that they could cause production barriers. The seal of the structure is provided by the Draupne Formation.

RESERVOIR

The reservoir sandstone belongs to the Hugin Formation, which is dated as of Callovian to Bathonian age. The Hugin Formation was deposited during a transgressive pulse in Callovian time. This pulse affected the whole of the Sleipner Terrace and has been encountered in several wells both on the Norwegian and the British side of the southern Viking Graben.

The sandstone was deposited in a fluctuating-marine (prodelta) to marginal-marine (mainly delta-front) and non-marine (lower delta-plain) environment in close proximity to both the Ling High in the east and the Shetland Platform in the west. The topographic elevation between the delta-front and the prodelta environment was of the order of a few metres. This, together with changes in the sea level and variation of the sedimentation rate, led to rapid and major facies changes during Callovian time.

The sandstone is light to medium grey, fine grained, well sorted, friable to moderately hard and in places calcareous cemented. Varying amounts of laminated mica and thin laminae of coal characterize the sandstone which also contains traces of pyrite. Coarse-grained layers of sand stringers are also observed. The porosity varies from fair to poor.

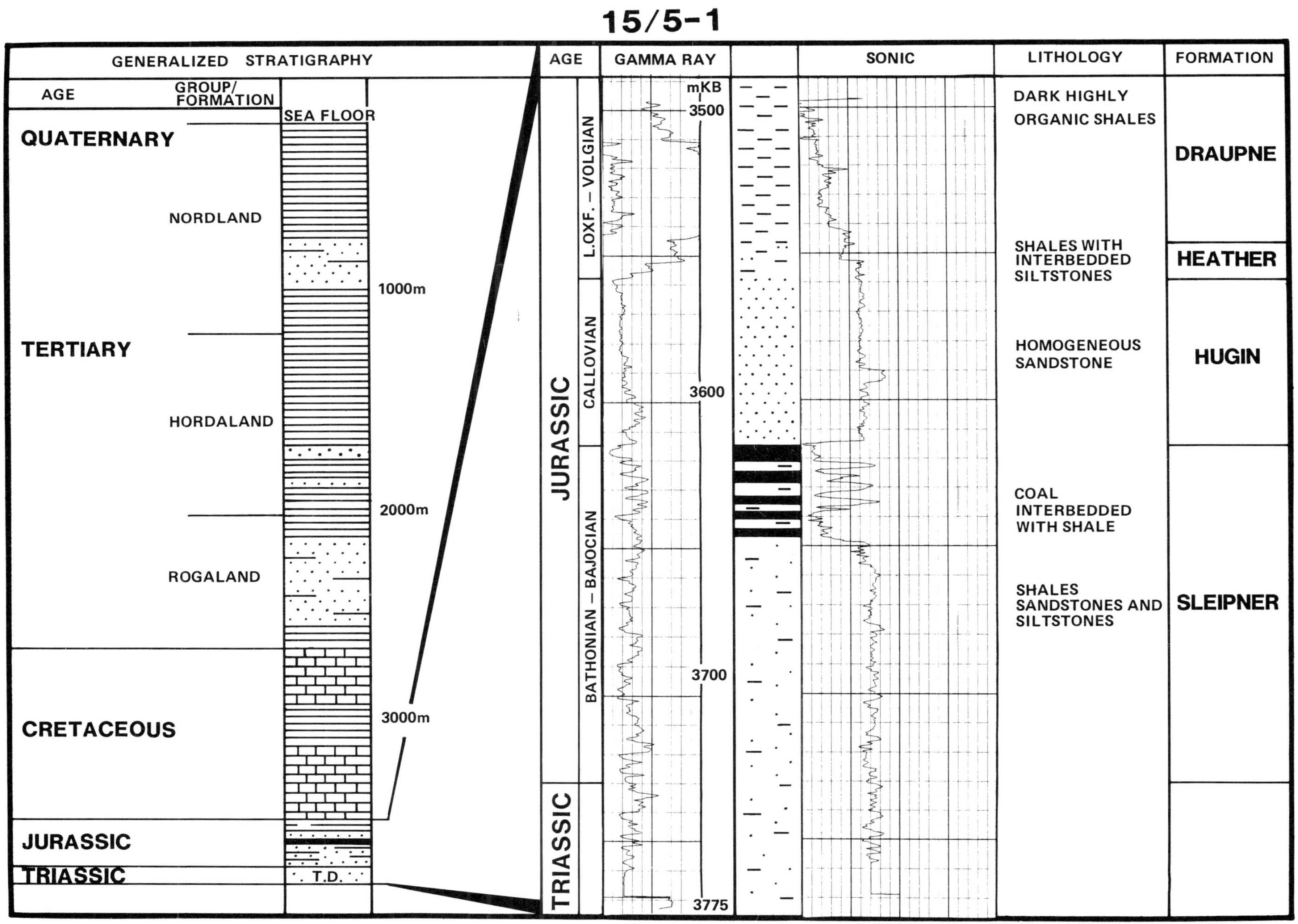

Fig. 2. Generalized stratigraphy Well 15/5-1.

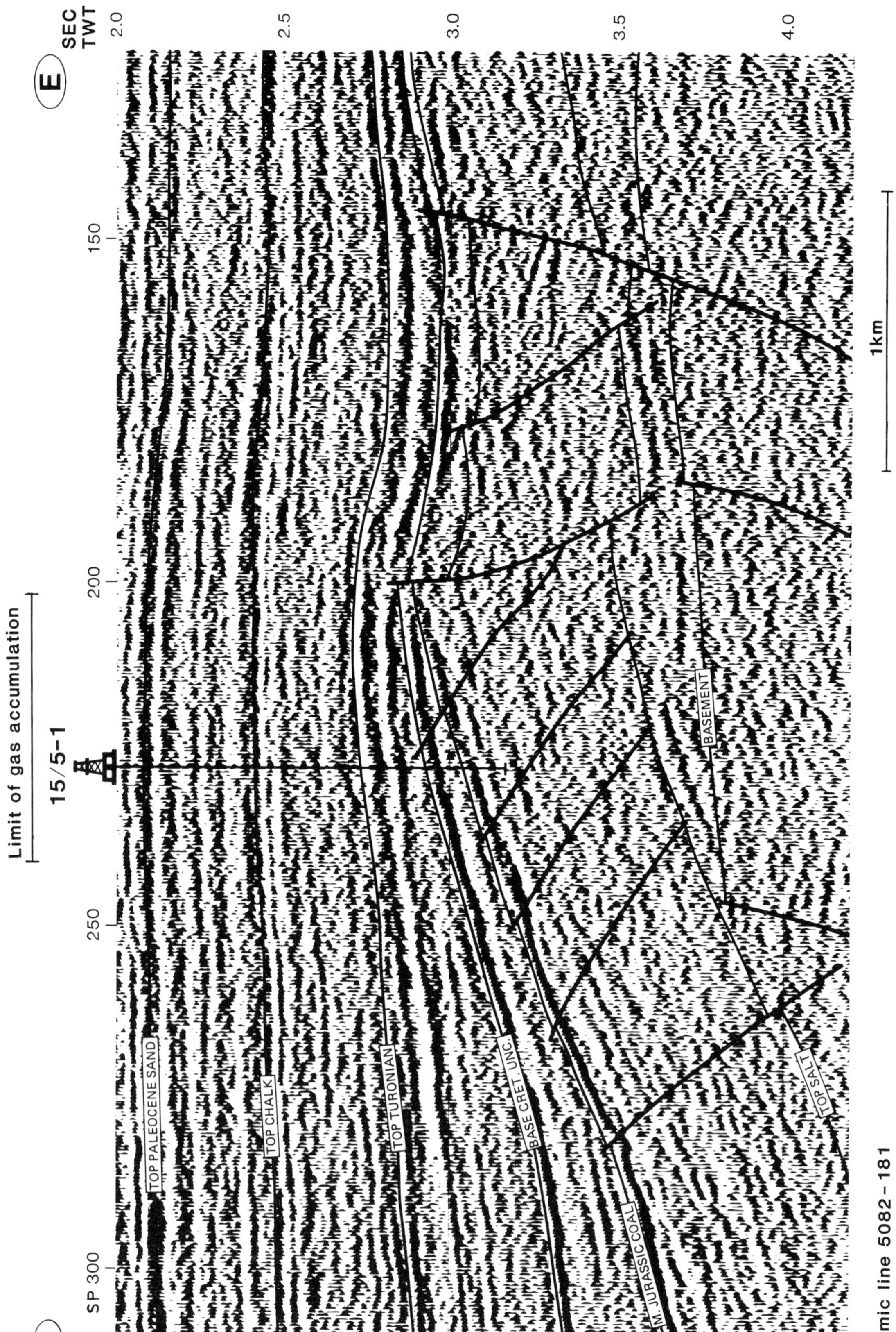

Fig. 3. Seismic cross-section through Well 15/5-1.

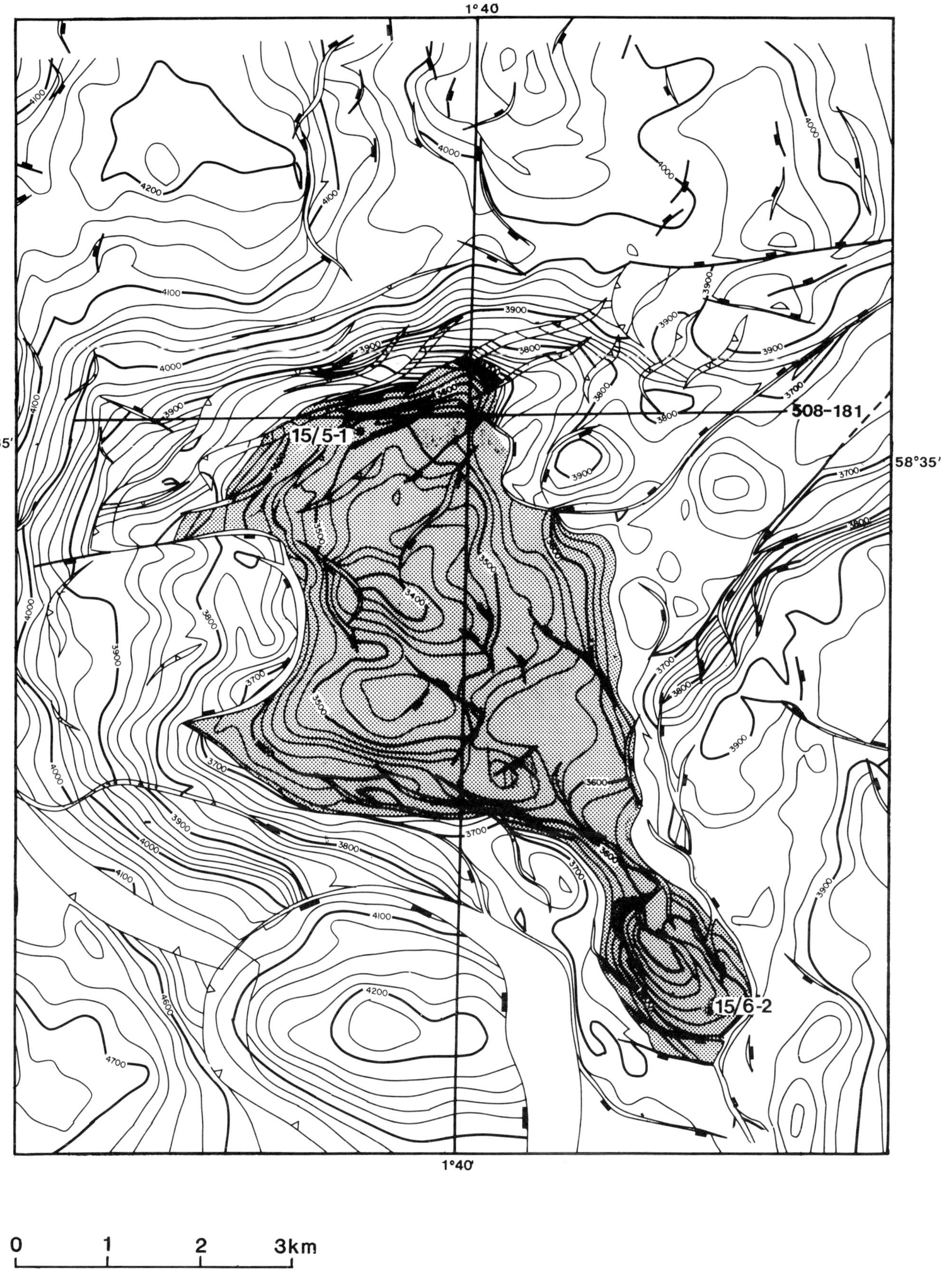

Fig. 4. Structural depth map, top reservoir. Contours are in metres subsea.

SOURCE

The main hydrocarbon source-rock is the Upper Jurassic shale. Thick sequences of this shale have been encountered in the southern Viking Graben where it has been proved to be hydrocarbon generating. The hydrocarbons have migrated from the main graben into the structure and been trapped.

HYDROCARBONS

Two drill-stem tests (DSTs) have been performed in Well 15/5-1 in the reservoir horizon. The flow was not stabilized in the first DST. The second DST, however, yielded a maximum flow rate of 620×10^3 Sm^3/day gas and 470 Sm^3/day oil (gravity 0.809 g/cm^3).

These test results and the log interpretation indicate that all of the Hugin Formation in 15/5-1 is gas filled from 3533 to 3591 mss.

RESOURCES

The gas resources are estimated to be 8.75×10^9 Sm^3 in-place with approximately 43% in Block 15/5.

ACKNOWLEDGEMENTS

The authors thank the PLO48 partnership for giving permission to publish this paper. We also acknowledge the contributions from several colleagues in Norsk Hydro, in particular A. Sæbø and P. Aure.

Part V—North Viking Graben-West

24

Gullfaks

T. Erichsen, M. Helle, J. Henden and A. Rognebakke

Statoil, P.O. Box 1212, N-5001 Bergen, Norway

Gullfaks is an oilfield located in Block 34/10 in the Norwegian sector of the North Sea. Exploration started in 1974 and in 1978 Block 34/10 (PL 050) was awarded to Norsk Hydro, Saga Petroleum and Statoil. Statoil is the operator, with Conoco as technical assistant. Gullfaks was declared commercial in 1980. The reservoirs are sandstones in the Brent Group, the Cook Formation and the Statfjord Formation and contain 493×10^6 m^3 STOIIP. The field is divided into several small rotated fault blocks. The presence of gas in post-Jurassic strata, in addition to a complex structural pattern, requires detailed and high-quality seismic surveys. The field will be produced from three gravity-based platforms and a number of satellite wells. The first platform started production from three satellite wells in December 1986. Seawater will be injected in order to maintain reservoir pressure. The maximum production from the three platforms is estimated to be 23.8×10^6 Sm^3/year.

INTRODUCTION

The Gullfaks Field is located 175 km north-west of Bergen in the Norwegian sector of the North Sea (Fig. 1). The Gullfaks area includes several interesting hydrocarbon-bearing structures (Fig. 2), amongst which the main Gullfaks Field is the largest. This structure covers an area of 47 km^2 entirely confined within Block 34/10. Water depth varies from about 135 m to 220 m increasing north-eastwards. The reservoirs of the main Gullfaks Field are Jurassic sandstones which contain reserves in the order of 493×10^6 m^3 STOIIP.

The field name is from a Norwegian fairy-tale where Gullfaks was a sailing vessel with the capacity of sailing both on sea and on land. The name is also connected with the fact that the licence (Block 34/10) was earlier designated the Golden Block.

HISTORY

Block 34/10 (PL 050) was awarded in 1978 to a group comprising Statoil as operator (85%), Norsk Hydro (9%) and Saga Petroleum (6%). Conoco is technical assistant for the operator.

The exploration of Block 34/10 started in 1974 when conventional seismic with a 1-km spacing was collected across the block. In development of the first well prognosis Statoil used information from Statfjord and nearby UK fields. In contrast to the Statfjord Field, structures in Block 34/10 have suffered more deformation expressed by intense faulting and steeper structural dip. Further, geophysical data indicated that the prospective reservoirs on the Gullfaks structure were structurally shallower than the Statfjord Field.

The first well in Block 34/10, drilled in July 1978 just after concession award, was a success (Fig. 3). The Brent Group in Well 34/10-1 was encountered about 50 m higher than predicted. As expected, the formation was oil productive. Well 34/10-1 also encountered the deeper Cook and Statfjord formations, but they were found to be waterbearing. The entire Brent Group was cored. Three produc-

Geology of the Norwegian Oil and Gas Fields © A. M. Spencer *et al.*, eds (Graham & Trotman, 1987) pp. 273–286.

Gullfaks Field data summary

	Brent gp	Cook fm	Statfjord fm
Trap			
Type	Structural	Structural	Structural
Depth to crest	1740 mss	1785 mss	1865 mss
Hydrocarbon contacts	1947 mss	2090 mss	2043 mss
Oil column	207 m	305 m	178 m
Pay zone			
Group	Brent	Cook	Statfjord
Age	Middle Jurassic	Lower Jurassic	Lower Jurassic
Cut-off for N/G	ø < 12% and V shale > 40%	ø < 12% and V shale > 40%	ø < 12% and V shale > 40%
Porosity	20–34%	24–36%	19–30%
Hydrocarbon saturation	85%	75%	85%
Permeability	20–5000 mD	5–1000 mD	20–3000 mD
Productivity index	100–200 Sm^3/day/bar	100–200 Sm^3/day/bar	100–200 Sm^3/day/bar
Hydrocarbons			
Oil gravity	0.88 g/cm^3	0.85 g/cm^3	0.83 g/cm^3
Gas gravity (relative to air)	0.70	0.82	0.83
Bubble-point	244 bar at 1825 mss	236 bar at 1835 mss	234 bar at 1981 mss
Gas/oil ratio	100 Sm^3/Sm^3 at 1825 mss	155 Sm^3/Sm^3 at 1835 mss	162 Sm^3/Sm^3 at 1981 mss
Formation volume factor	1.266 at 1825 mss	1.498 at 1792 mss	1.507 at 1981 mss
Formation water			
Salinity (NaCl equiv)	25 000 ppm at 21.6 °C	25 000 ppm at 21.6 °C	25 000 ppm at 21.6 °C
Resistivity	0.168 ohm m at 21.6 °C	0.168 ohm m at 21.6 °C	0.168 ohm m at 21.6 °C
Reservoir conditions			
Temperature	72.2 °C at 1850 mss	76.0 °C at 1850 mss	80.0 °C at 1998 mss
Pressure	310 bar at 1850 mss	317 bar at 1827 mss	325 bar at 1930 mss
Pressure gradient in reservoir	0.075 bar/m at 1850 mss	0.079 bar/m at 1827 mss	0.058 bar/m at 1930 mss
Field size			
Area	35 km^2	15 km^2	10 km^2
Hydrocarbon pore volume	503×10^6 Rm^3	64×10^6 Rm^3	68×10^6 Rm^3
Recovery factor—oil	0.445	0.26	0.42
Drive mechanism	water injection	water injection	water injection
Recoverable oil	179×10^6 Sm^3	12×10^6 Sm^3	19×10^6 Sm^3
Production			
Start-up date	22 December 86		
Cumulative production to date (23 February 87)	324×10^3 Sm^3		

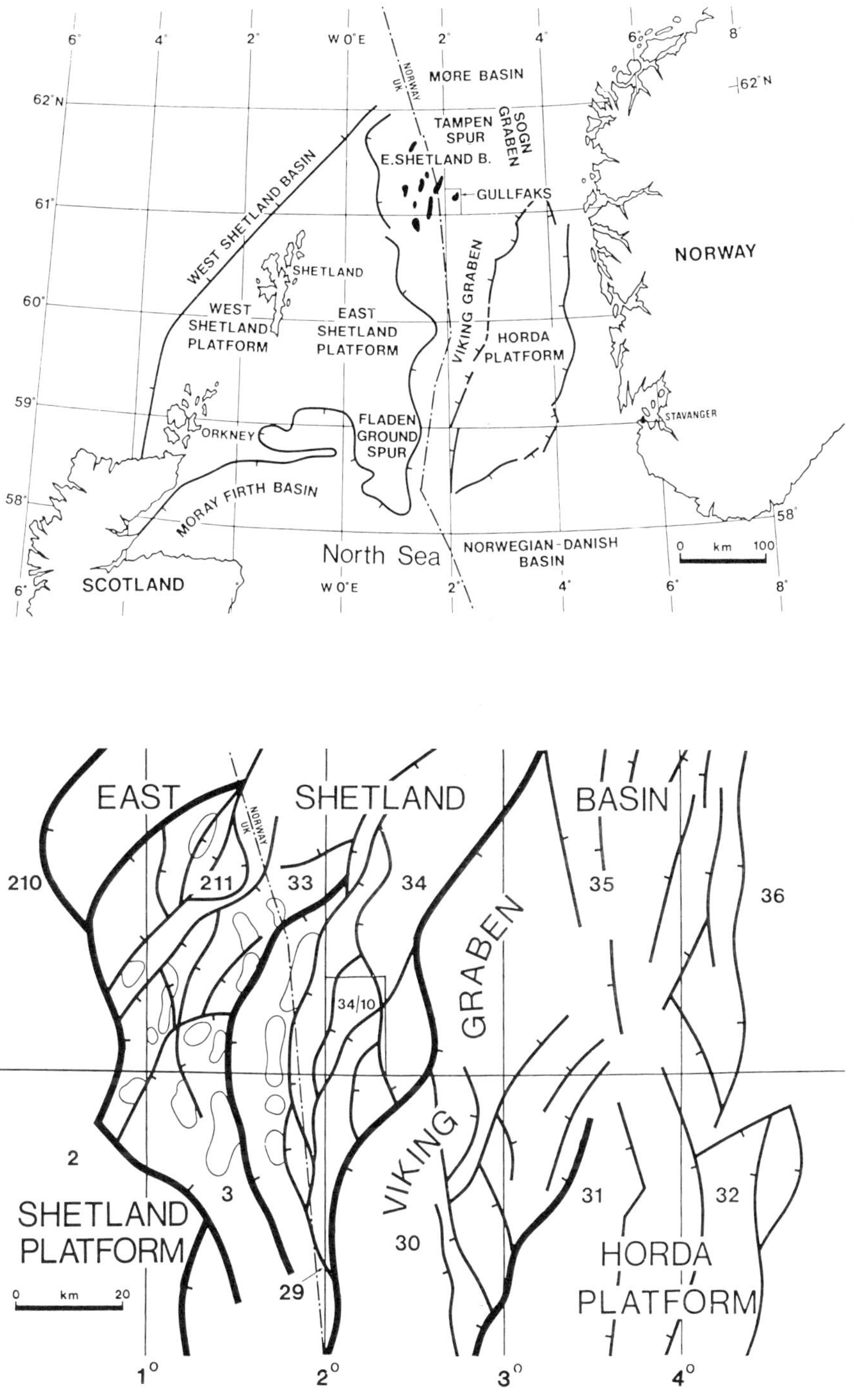

Fig. 1. Tectonic map of the Northern North Sea.

tion tests were performed in the oil zone providing information about fluid properties, rock permeability and reservoir pressure. As the depth of the Brent Group is quite shallow, 1800 m below sea level, the rock is loosely consolidated. The likelihood of the formation producing sand together with oil was extensively evaluated during these tests, and later in other wells drilled on the structure. The second well in Block 34/10 was drilled on the Gullfaks South structure (Fig. 2). This well encountered gas in the Brent Group and oil in the Statfjord Formation.

After these two wells, the exploration/appraisal activity was concentrated on the Gullfaks structure and at times two locations were drilled simultaneously. It was soon realized that Gullfaks was a large oilfield. Hence, a three-dimensional seismic survey was carried out in 1979 across the entire structure to improve the interpretation. Despite this, several geophysical surprises occurred dur-

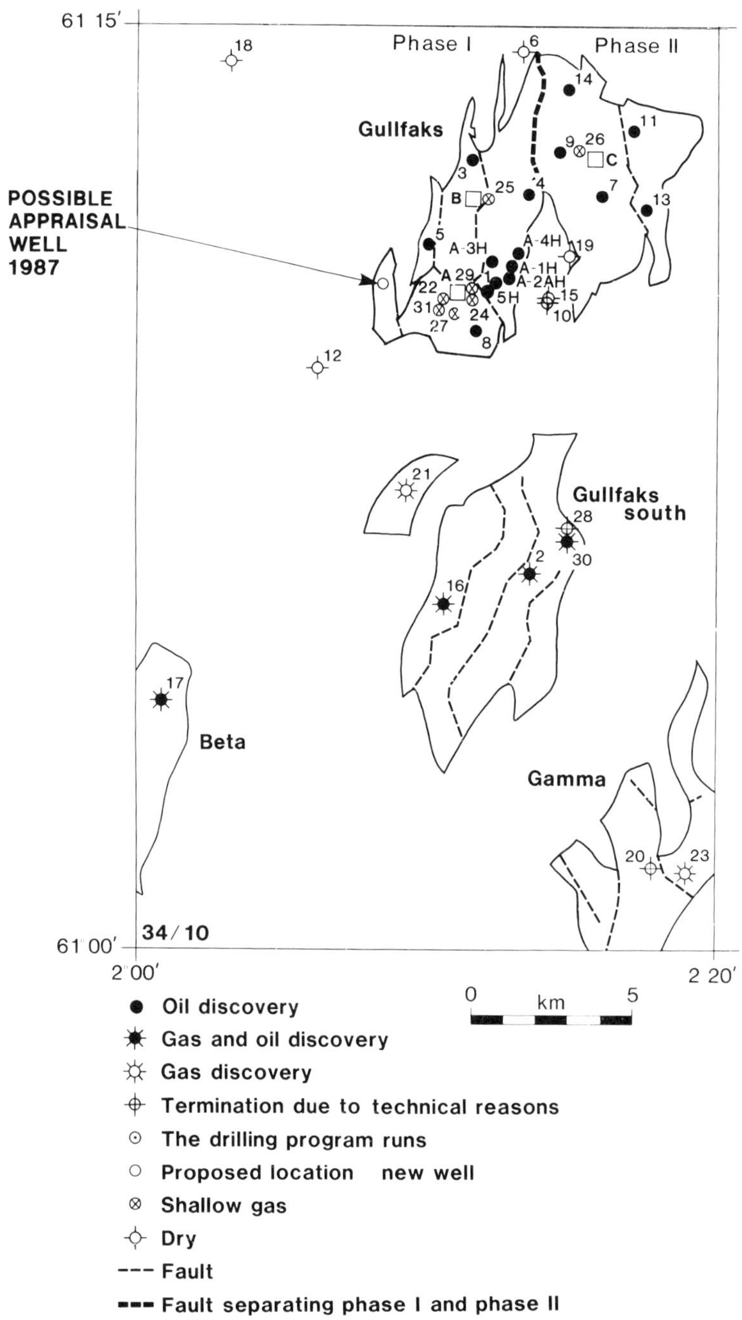

Fig. 2. Structures and wells in Block 34/10.

ing the appraisal phase. The eastern part of the structure has been especially difficult to interpret.

In the eight years since the concession was awarded there have been 13 exploration and appraisal wells drilled through the potential oil reservoirs of the Gullfaks Field. Ten of these encountered hydrocarbons and three were dry holes. One well had to be abandoned before reservoirs were reached because of problems with shallow gas and high pore pressure.

Statoil has conducted an additional three-dimensional seismic survey on the Gullfaks structure (summer 1985), using the most modern and recently developed equipment. The interpretation of the 1985 three-dimensional survey is in progress and forms the basis for prognosing ongoing and future development drilling on the structure.

Development work to date comprises the drilling of four satellite production wells, which were drilled and completed during 1985–86, and one water-injection well which is currently being completed. Production from these wells started on 22 December 1986 on the Gullfaks A-platform, which was installed in June 1986. Development drilling is planned

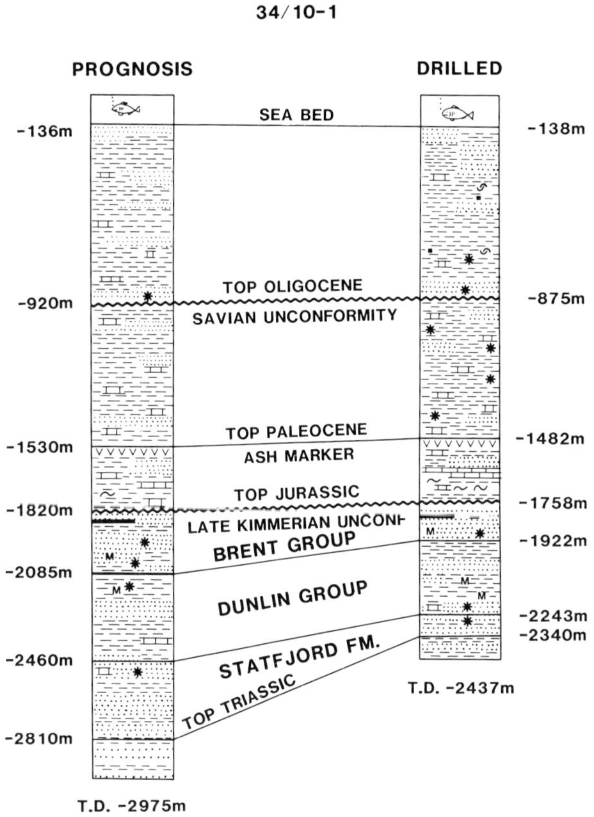

Fig. 3. Well 34/10-1, prognosis and drilling results.

to start during the second quarter of 1987. The Gullfaks B-platform is scheduled to be on site during the summer of 1987 and in production at the end of 1988. The B-platform will be a satellite platform to the A-platform, which means that after preliminary separation, the oil and gas will be pumped in separate pipelines to the A-platform for further separation, storage and transportation. The Gullfaks C-platform is planned to be on stream in 1990 in the phase II area of the structure.

The A- and C-platforms are both fully integrated PDQ-platforms (process, drilling, quarters); the B-platform is a DQ-platform (it has only one-stage separation). The total process capacity of the A-platform (12.1×10^6 Sm^3/year) will be needed in 1989 and the total production from the whole field (phase I and phase II) will reach a maximum in 1993 (23.8×10^6 Sm^3/year). Figure 4 shows the production platforms and flowlines. The production capacity is about 33 000 tons crude oil and 3.5×10^6 Sm^3 gas per day for both the A- and the C-platform, and 20 000 tons oii for the B-platform.

The natural water drive is believed to be negligible. Therefore seawater will be injected from all the platforms in order to maintain the reservoir pressures above the bubble point. Owing to the fact that the reservoirs are highly faulted, injectors must be drilled in order to achieve pressure maintenance.

GEOLOGICAL SETTING

Block 34/10 is situated in the central part of the East Shetland Basin (Fig. 1). The Gullfaks structure lies in the north-east part of Block 34/10 and represents the most positive structural element of the Tampen Spur.

The structural genesis of the Gullfaks Field took place in Late Jurassic to Early Cretaceous times in connection with regional tectonic movements. The structure is a sloping high with a generally westerly stratal dip. The dip gradually decreases towards the east. The major part of the structure has been extensively block faulted and rotated, but fault-block 1 (Fig. 5) has been less affected and stands up as a horst block. The N–S-striking faults split the field into several rotated fault-blocks. In the up-dip parts of the fault-blocks and in the central parts of the eastern structures, the Brent and the Cook sandstones have been partially or totally eroded (Fig. 5).

The Triassic/Lower Jurassic coarse clastic sediments of the Statfjord Formation were deposited in a rift basin initiated in Late Triassic times. The lowest part represents an early alluvial facies, while the deposits above represent a northward-draining fluvial system. The top section of the Statfjord Formation contains shallow-marine coastal sandstones representing a marine transgression.

The lower parts of the Dunlin Group and the Amundsen and Burton formations, have an increased marine character and contain predominantly shale/claystone. A temporary regression ended the major transgression (Pliensbachian) and sand and silt were deposited farther out in the basin, building up the Cook Formation.

The Cook Formation is divided into five different facies. The lower four facies represent normal, open-marine setting, while the uppermost facies was deposited in a restricted-marine environment, probably with anoxic bottom conditions.

The general transgression continued through Late Pliensbachian and Toarcian times and the Drake Formation was deposited as marine clay with a variable silt content.

Through Bajocian and Bathonian times a shallow-marine, deltaic coastal-plain developed in the northern North Sea Basin, forming the Brent Group. It is possible to identify

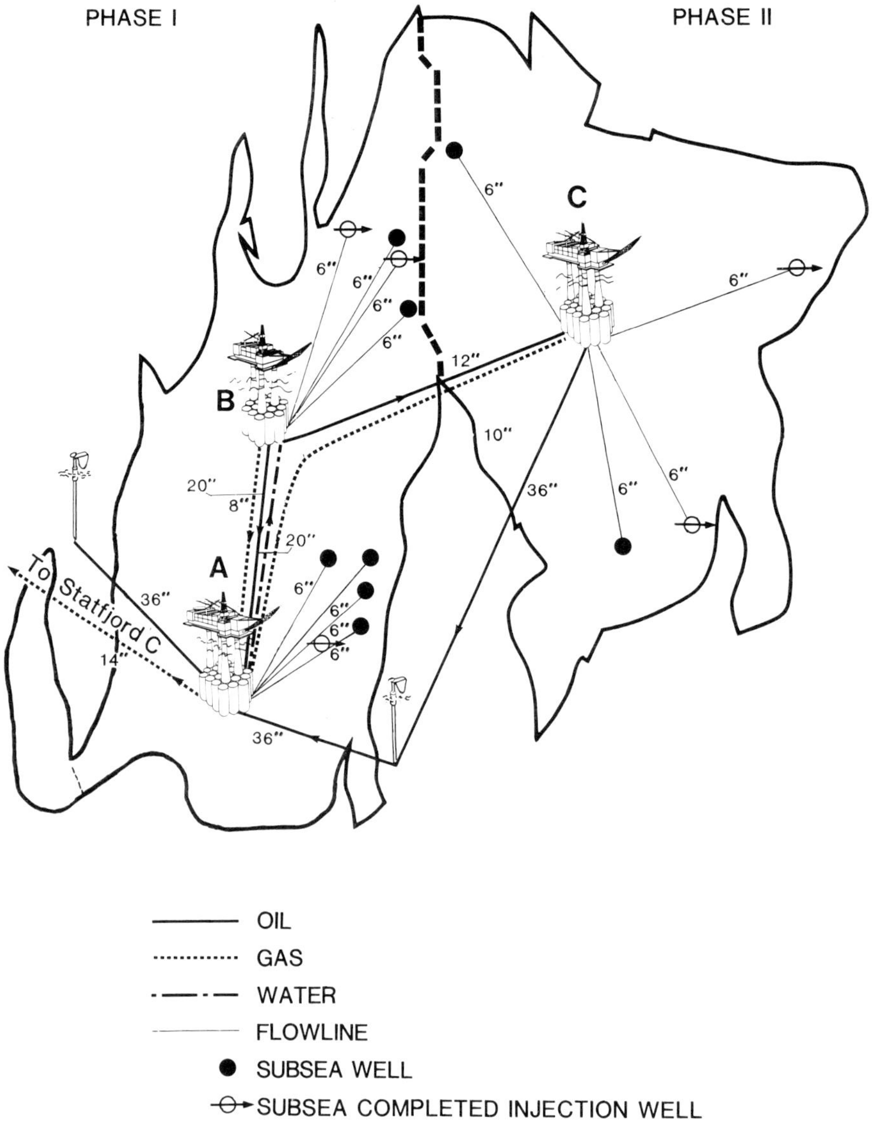

Fig. 4. The Gullfaks Field, development layout.

two separate deltaic facies: (1) the lower Brent delta is interpreted as a wave-/fluvial-dominated system; (2) the upper Brent delta resembles a tide-/fluvial-dominated delta. The lowest unit, the Broom Formation, consists of prodeltaic clay-, silt- and sandstones. Progradation continued through Middle Bajocian and Earliest Bathonian times, depositing the lower Brent delta represented by: (a) the Rannoch Formation, a prograding delta-front shoreface; (b) the Etive Formation, an upper delta-front complex; and (c) the Ness Formation, with delta-plain sediments. In Bathonian times a major transgression moved the coastline southwards, depositing a tide-/fluvial-dominated delta, the Tarbert Formation (the upper Brent delta).

The boundary between the Middle and Upper Jurassic strata coincides with the main Kimmerian tectonic event, and within the Viking Group there are several unconformities. Extensive erosion took place on structural highs due to the Late Kimmerian tectonic phase. The deposits of the Upper Jurassic consist mainly of claystones and siltstones. Both the Heather Formation and the overlying Draupne Formation were deposited in a strongly reducing environment with restricted circulation. Early Cretaceous sedimentation was still influenced by the Late Kimmerian tectonic phase, and deposits are lacking on structural highs. In Late Cretaceous times the tectonic movements ended and the whole East Shetland Basin was under

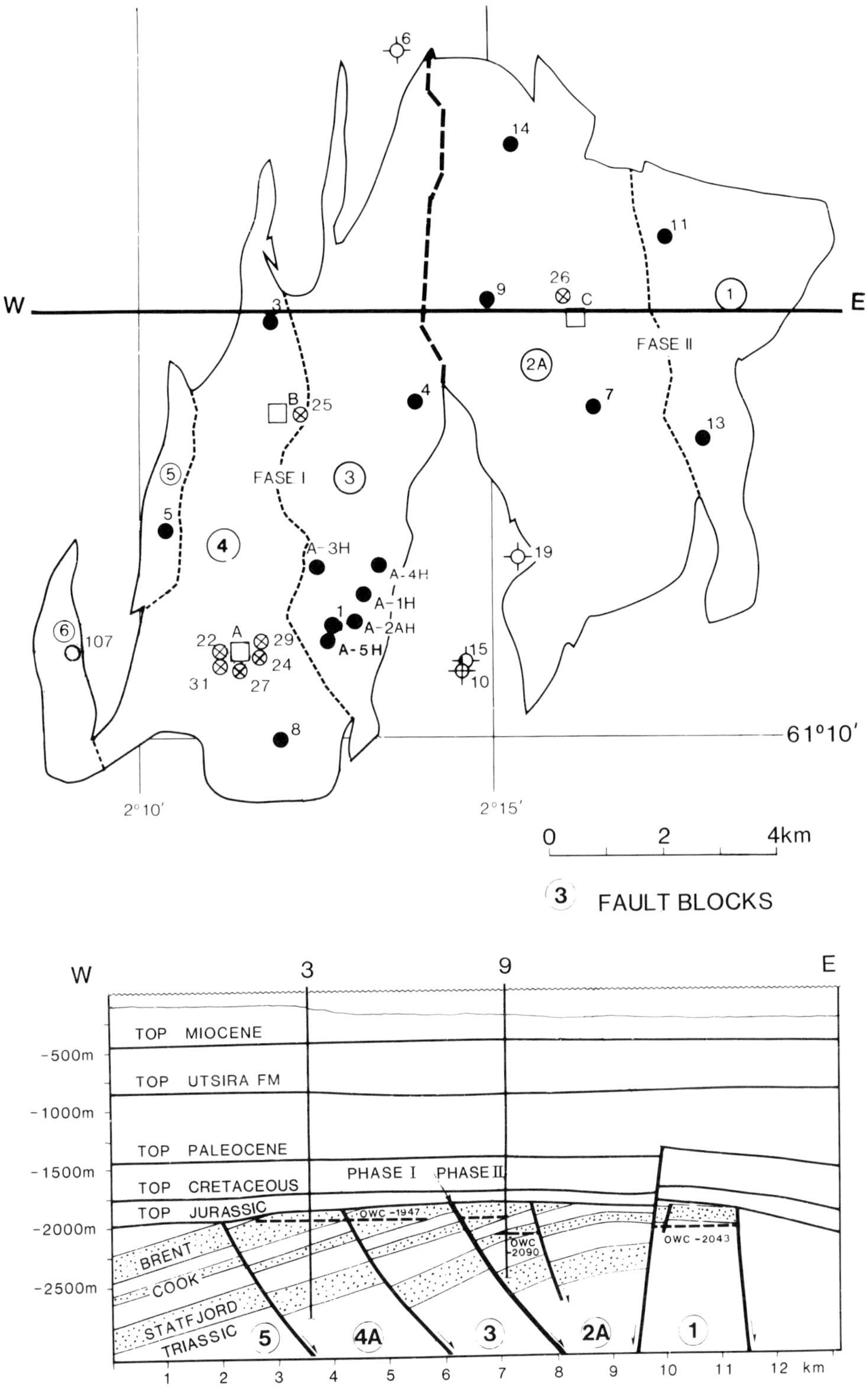

Fig. 5. The Gullfaks Field, map of fault-block compartments.

the influence of general subsidence. Minor tectonic movements seem to have continued until Early Eocene times.

GEOPHYSICS

It is difficult to obtain good-quality seismic data from the Gullfaks Field. The most important reasons for this are:

- the presence of gas in the post-Jurassic section results in high absorption and hence, poor signal penetration;
- the two horizons top Paleocene and top Cretaceous are very strong reflectors, and

therefore a low proportion of the original seismic energy penetrates the pre-Cretaceous section;

- these two reflectors also generate strong multiples, which interfere with the pre-Cretaceous reflections;
- the poor seismic quality in the pre-Cretaceous section may also be the result of low seismic reflectivity of these layers.

Several sets of seismic data have been acquired over the Gullfaks Field. A conventional two-dimensional survey was shot between 1974–78. In order to improve the data quality a three-dimensional seismic survey was carried out in 1979 (Fig. 6). The quality of the data was still considered to be low, and in 1985 acquisition began of a new three-dimensional data set. The new survey consists of 8700 km with seismic data divided into 550 lines with a line-spacing of 25 m.

The processed data from the last survey have proved to be of a higher quality than the data collected in 1979. In particular, the cross-line sections are improved, and it is easier to correlate and map the reflectors.

At reservoir level the following horizons have been mapped:

- base of the Cretaceous;
- the top of the Brent Group in the western part of the field;
- base of the Brent Group;
- top of the Statfjord Formation.

The improved data quality has resulted in a more accurate and complete structural interpretation.

TRAP

The Gullfaks structure represents the most positive element of the Tampen Spur, which here limits the Viking Graben along its western and north-western edge.

The seismic interpretation shows the Gullfaks Field as a westerly dipping fault-block system consisting of small fault-blocks 1–2 km in width orientated mainly N–S. We may characterize the trap as a structural trap, comprising many faulted, rotated and later partly eroded fault-blocks.

The main reservoirs are Jurassic sandstones with the Upper Jurassic shales as source-rocks. Sealing is provided by Cretaceous shales. The oil is trapped mainly in Brent sandstones which are 250 m thick in the uneroded part of the phase I area (Fig. 5). The oil–water contact in the Brent sandstones is 1947 mss. In the phase II area the oil–water contact is 2090 mss in the Cook Formation and 2043 mss in the Statfjord Formation.

In addition, gas-bearing sands have been encountered at depths of 300–450 mss in strata of Pliocene age. The thickness of the sands varies from 1 to 5 m and the areal extent seems to be 6–8 km^2 for the most important interval. In 1980, penetration of these gas-bearing sands caused a blowout and abandonment of the well 34/10-10. The so-called 'shallow-gas' wells (Figs 2 and 5) have been drilled in order to document the existence of shallow gas at platform sites, test the gas and drain it out from under the platforms.

RESERVOIR

The main reservoir of the Gullfaks Field lies in the Brent Group of Early Bajocian–Earliest Bathonian age, but older reservoir systems in the Cook Formation of the Dunlin Group and in the Statfjord Formation are also important.

Statfjord Formation

The main oil reservoir of the Statfjord Formation (81–224 m) is in fault-block 1 in the phase II area (Fig. 5). Due to poor core recovery, data available are sparse. The formation is normally divided into three units (Fig. 7) and represents a gradual upwards transition from an arid alluvial environment with episodic sheet-flood deposits (Unit 1), to a poorly drained alluvial plain with channelized flows (Unit 2, lower Unit 3). Finally, a transgressive low-energy shoreline advanced across the area (upper Unit 3).

Unit 1 consists of interbedded sandstones, siltstones and shales. The interbedding appears random, with no obvious systematic change in grain size or bed thickness. Sand bodies up to 16 m in thickness have been observed. Units 2 and 3 consist of interbedded sandstones, siltstones and shales which generally define fining-upwards sequences normally 3–10 m thick. Each sequence consists of smaller sub-sequences grading from conglomerate and very coarse sandstone at the base to medium-grained sandstones at the top. The primary textures are well preserved, normally. Locally the sandstones are cemented. In the sandstones the porosities and permeabilities average 19–30% and 30–3000 mD respectively.

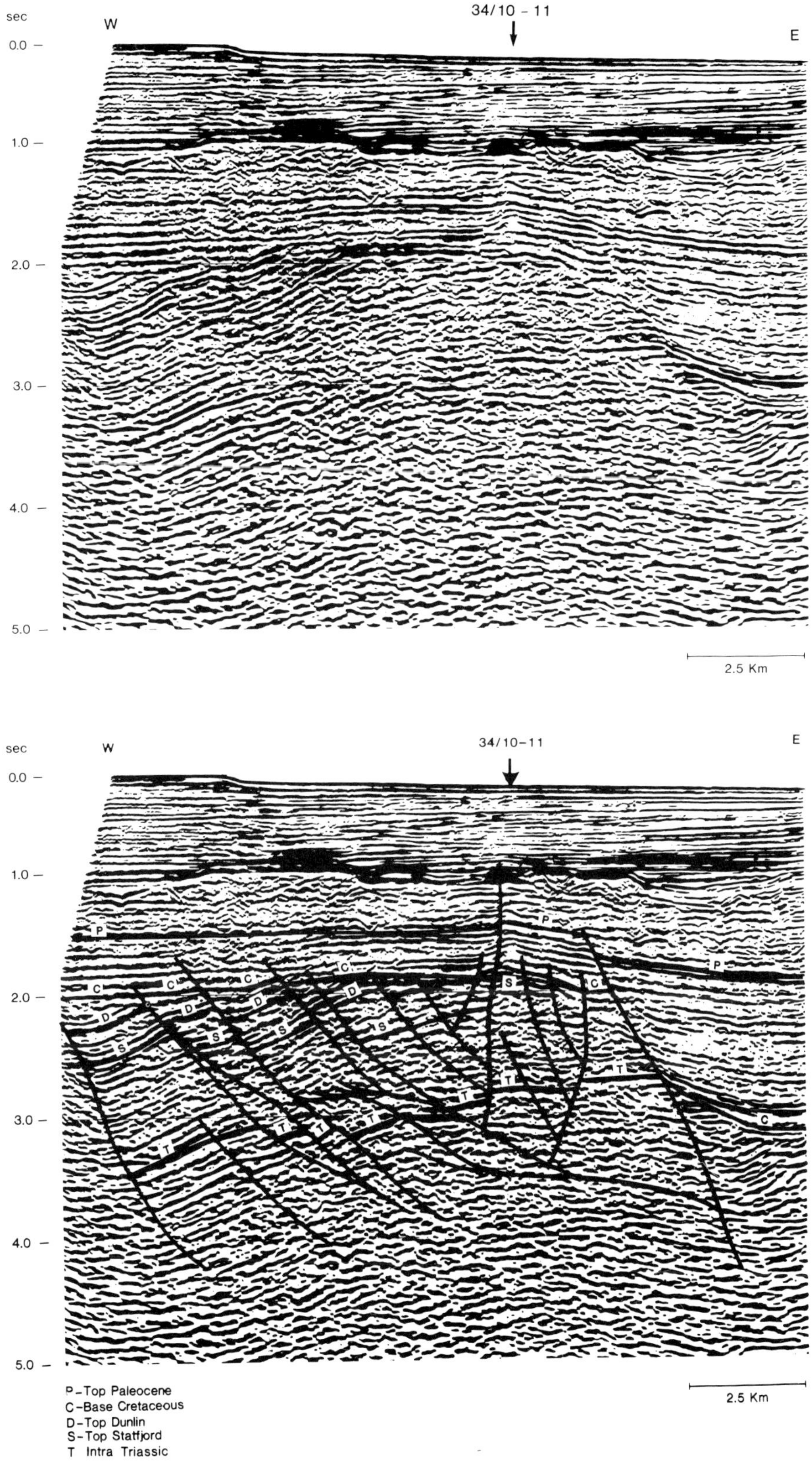

Fig. 6. East–west three-dimensional seismic line through Well 34/10-11.

Cook Formation

The Cook Formation (106–150 m in wells) plays an important role as a reservoir in the phase II area. It belongs to the Dunlin Group, a Lower Jurassic, argillaceous, marine sequence. The formation, which is normally divided into three units (Fig. 8), consists of shales, siltstones and sandstones deposited under both open-marine conditions (Unit 1

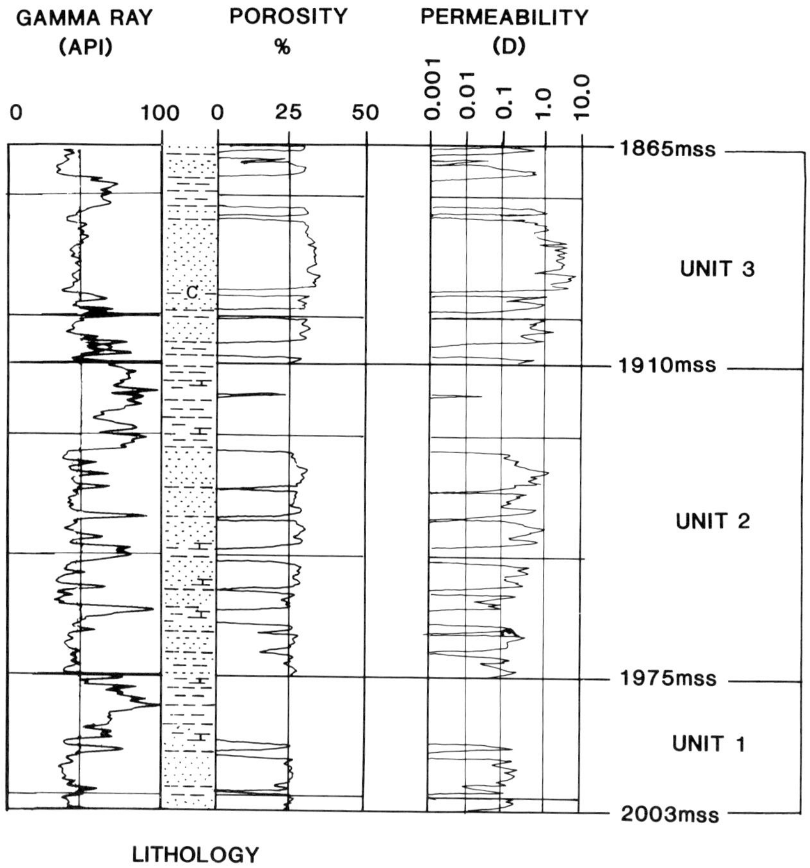

Fig. 7. Statfjord reservoir, Well 34/10-11

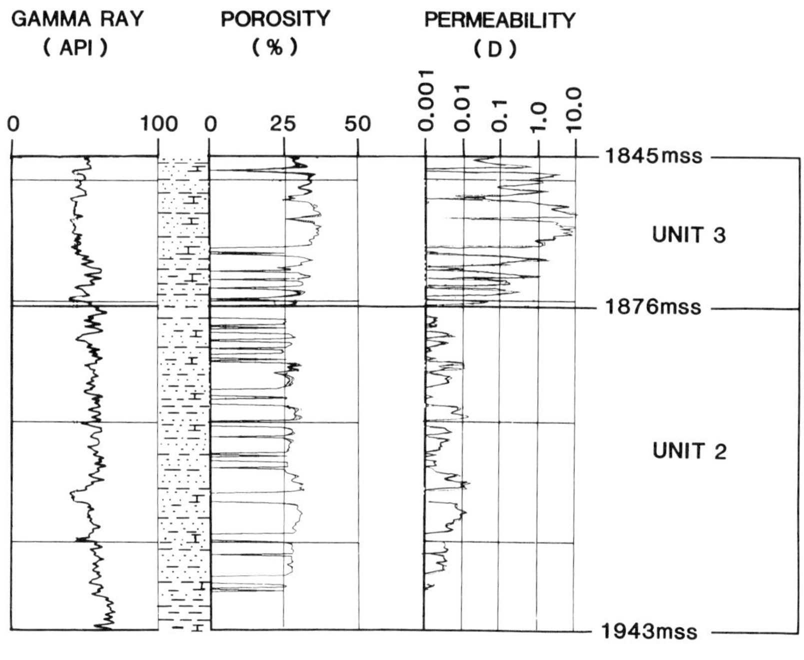

Fig. 8. Cook reservoir, Well 34/10-5

and 2) and restricted-marine conditions (Unit 3). The formation has its greatest thickness in the southern part of the Gullfaks Field.

Four sedimentary facies have been recognized within the formation: (1) laminated sand and shale, consisting of interlaminated very fine sandstone and shale/siltstone, moderately bioturbated; (2) bioturbated muddy sand, consisting of very fine- to fine-grained sand in a clay matrix; (3) hummocky sand—very fine hummocky, cross-stratified sands with minor shale; (4) interbedded sand and shale—generally fine- to medium-grained sandstones (locally coarse to conglomeratic) interbedded with shales or siltstones.

The sandstones of the Cook Formation are mostly arkosic in composition. Carbonate-cemented layers may be important in producing the oil from the Cook Formation as they form laterally continuous, impermeable zones. Despite this, the overall effect of diagenesis on the poro/perm properties is believed to be minor. The porosities are 20–36% and the permeabilities are in the range of 1–1000 mD.

Brent Group

The Brent Group is divided into five different formations of which the Tarbert, Ness, Etive and Rannoch formations are important as reservoirs. The reservoir sandstones are generally arkosic in composition.

Rannoch Formation

The Rannoch Formation (48–91 m in wells) is interpreted as a delta-front deposit, comprising an overall upwards-coarsening sequence which thins to the south. The formation is subdivided into three reservoir units: R1, R2 and R3 (Fig. 9). Unit R1 is a muddy micaceous sandstone. Units R2 and R3 are dominated by very fine to fine, well-sorted micaceous sandstones, with R2 as the most homogeneous zone with respect to grain size.

Carbonate-cemented beds are commonly found in the R1 and R2 units. Increasing content of diagenetic kaolinite southwards and downwards in the formation is believed to be an important modifier of the permeability. The average porosity is 30%, with permeabilities in the range of 50–800 mD.

Etive Formation

The Etive Formation (13–38 m in wells) is interpreted as an upper delta-front complex. The formation is subdivided into two units: E1 and E2 (Fig. 9). The lower part, Unit E1, commonly contains upwards-fining sequences which may suggest channel-fill deposits, while the well-sorted sandstones of the upper part (E2) are interpreted as beach deposits. The formation generally thins to the south. The formation has a superb reservoir quality and consists of predominantly medium- to coarse-grained sandstones with minor mica and clay matrix.

The average porosity is 32% and permeabilities are in the range of several Darcies. A slight southward increase in the content of diagenetic kaolinite, similar to that described in the Rannoch Formation, is observed in the Etive Formation.

Ness Formation

The Ness Formation (75–109 m in wells) is subdivided into three units (Fig. 9): N1, N2 and N3 which represent a delta-plain depositional environment. Units N1 and N2 represent the lower delta-plain facies, and N3 the upper delta-plain facies. The lower part of N1 consists of coal-beds with rootlets and interbedded sandstones and mudstones, while the upper part, in addition, can consist of thicker sandstone beds, interpreted as fluvial channel-fill deposits. The N2 Unit is interpreted as an interdistributary bay-fill deposit, and consists of three major upwards-coarsening sequences with generally clean, highly porous and permeable sandstones on top of each sequence. Unit N3 marks the end of the lower Brent delta progradation in the Gullfaks area. Major characteristics of this unit are soil profiles, coal beds and siderite. Floodplain deposits, fluvial channel-sands and lacustrine deposits are recorded. The fine- to medium-grained sandstones of the fluvial channel deposits have porosities above 25% with permeabilities up to one Darcy.

Tarbert Formation

The Tarbert Formation (46–75 m in wells) is subdivided into two units: T1 and T2 (Fig. 9), which represent a transgressive event building up the upper Brent delta. The two units are interpreted as a tide-/fluvial-dominated delta system. Unit T1 consists of two or three coarsening-upwards sequences ranging from bioturbated clay- and siltstones at the bottom, to medium- to coarse-grained sandstones at the top. The lower part of T2 consists of a coarsening-upwards sequence, in general more sandy than T1. The upper part of this sequence is coarse grained with similarities

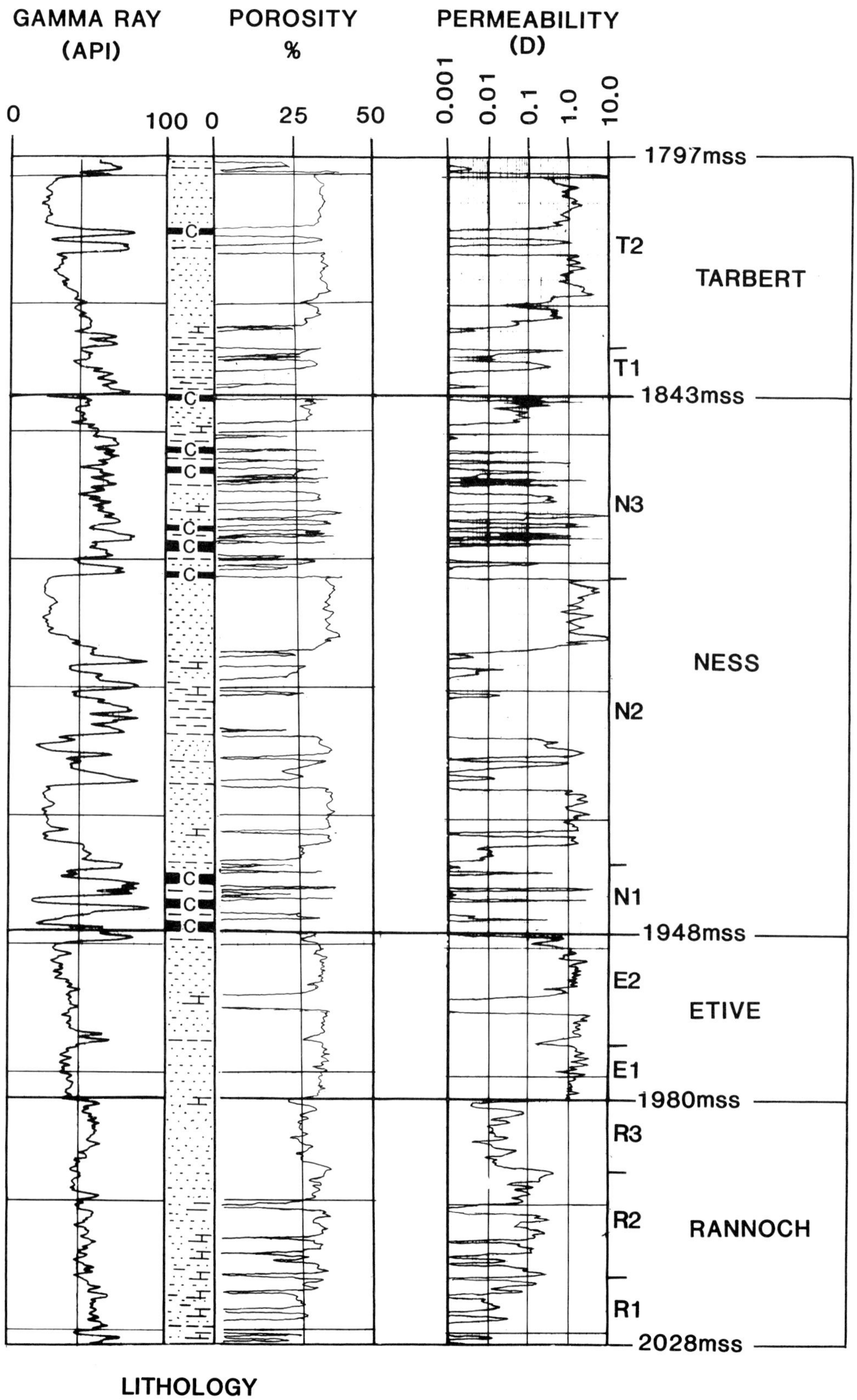

Fig. 9. Brent reservoir, Well 34/10-8

to the Etive Formation. Both T1 and T2 sandstones contain considerable amounts of secondary kaolinite. Despite this, the sandstones have good to excellent reservoir properties.

SOURCE

Geochemical analyses have been performed on many wells in the area and indicate that the main source-rock in the area lies within the Draupne Formation. This formation is partly eroded on the structural highs, but is about 200–300 m thick in the Viking Graben area. The total organic content is variable, but shows values of 5–10% TOC in the Viking Graben area. Other potential source-rocks are the shales and coals within the Brent Group, but their volumes are restricted. Another possible source-rock is the dark marine shales of the Dunlin Group. Generally the maturity is relatively low, but it is anticipated that the Dunlin Group has generated hydrocarbons in the deeper parts of the basin. The Dunlin shales have a limited oil potential, but a good gas potential.

Direct migration of hydrocarbons from the source-rock to the reservoir has probably only partly contributed to the oil-filling of the Gullfaks structure, because of the major faults towards the east and south. The major contribution of hydrocarbons to Gullfaks seems to have taken place by the spilling of oil from other structures in the area, most probably from the structures Snorre 'Vest' and 33/9 Alpha (Statfjord Øst).

HYDROCARBONS

In all three reservoirs the oil is undersaturated. The gas–oil ratios and the bubble-point pressures of the three reservoirs decrease with increasing depth in a roughly linear way. A similar trend is valid for the formation volume factor (see field data summary). The geometry of the accumulations is indicated in Fig. 5 which shows a cross-section through fault-blocks 1 to 6. Reserves of the Statfjord Formation are confined to fault-block 1, and Cook reserves are found mainly in fault-block 2. Brent reserves are present in fault-blocks 2, 3, 4 and 5, and possibly in the, as yet undrilled, fault-block 6.

RESERVES

Total hydrocarbon-pore-volume on Gullfaks has been estimated to be 635×10^6 m^3, of which 10% is in the Cook Formation and 11% is in the Statfjord Formation. Of the oil reserves initially in-place, the distribution is 84.5%, 7.6% and 7.8% in the Brent, Cook and Statfjord units respectively.

Recovery of these reserves has been estimated to be highest in the Brent Group and the Statfjord Formation. A relatively low recovery (26%) is expected from the Cook Formation due to very low permeabilities occupying large parts of this reservoir.

The reservoirs of the Brent Group have demonstrated a common OWC at 1947 mss in wells 34/10-3, -5 and -8 in the western part and in well 34/10-14 in the eastern part of the field. Although a slight difference in bubble-point pressure has been noticed in these two parts, pressure communication cannot be disregarded.

Deeper OWC's for both the Cook and Statfjord formations are evident from wells 34/10-9 and -11.

As seen from the field data summary, the liquid properties of the three reservoirs are different. From west to east there is a decrease in oil gravity, a decrease in bubble-point pressure and an increase in gas–oil ratio. Much speculation as to the reason for these changes has been made, but the most common opinion now is that at least two source-rocks have contributed to the filling of the Gullfaks reservoirs.

ACKNOWLEDGEMENT

We would like to thank Norsk Hydro, Saga Petroleum and Statoil for permission to publish this paper. Special thanks to H. Borgen for typing the text and K.G. Jakobsen for constructive comments.

REFERENCES

Hazeu, G.J.A. 1981. 34/10 Delta structure geological evaluation and appraisal, In *Norwegian Symposium on Exploration*, Norsk Petroleumsforening, Article 13, 36 pp.

Helgøy, T. 1984. Prediction techniques from logging tools: are the results confirmed by testing? In *Offshore Northern Seas—Reduction of Uncertainties by Innovative Reservoir Geo-modelling*, Norsk Petroleumsforening, Article G4, 24 pp.

Karlsson T. and Gausland I. 1986. Application of new 3-D technology to recent field development in Norway, In *Offshore Northern Seas—Exploration Conference*, Norsk Petroleumsforening, Article G4, 16 pp.

Kemp, A.C. and Grinstad, K.C. 1981. Computer modelling of a faulted reservoir, *51st Annual Meeting, Society of Exploration Geophysicists*, Vol. 5, pp. 2935–2951.

Saeland, G.T. and Simpson, G.S. 1981. 3-D seismic data help to delineate subunconformity trap in Norwegian North Sea, *Oil & Gas Journal* **79,** (45), 381–398

Saeland, G.T. and Simpson, G.S. 1982. Interpretation of 3-D data in delineating a subunconformity trap in block 34/10, Norwegian North Sea, *Memoir of the American Association of Petroleum Geologists* **32,** 217–236.

Skarpnes, O., Briseid, E. and Milton, D.I. 1982. The 34/10 Delta prospect of the Norwegian North Sea: exploration study of an unconformity trap, *Memoir of the American Association of Petroleum Geologists* **32,** 207–216.

25

Hild

K. Rønning, C. D. Johnston, S. E. Johnstad and P. Songstad

Norsk Hydro A/S, Exploration Department, Bergen, Norway

The small gas–condensate finds of the Hild Field are located along the British–Norwegian median line at 60°30′ N. The Middle Jurassic Tarbert Formation constitutes the reservoir. It was broken up into numerous fault-block traps during Late Jurassic times. Five wells have five proven different gas–water contacts in different fault compartments. This fact, combined with poor seismic quality not allowing the structures to be confidently defined, makes reserve calculations uncertain. Gas-in-place is estimated to be about 30–40 × 10^9 Sm^3. The reservoir is buried to depths greater than 3600 m, having average porosities in the range of 13–20% and pressures above 740 bar. Hydrocarbon generation commenced in late Paleocene times, presumably from the intra-Brent coal layers.

INTRODUCTION

The Hild Field is the general name for a group of small gas–condensate finds located in the northern North Sea along the British–Norwegian median line at 60°30′ N some 180 km west of Bergen, Norway (Fig. 1). The extents of the hydrocarbon pools are not fully known, hence the name is generally used for the area covered by the three-dimensional seismic survey (see Fig. 2), which was shot in 1981. Water depth in the area is in the range of 100–110 m. The Hild Field takes its name from the old nordic mythology: 'Hild, daughter of Hogne, was kidnapped by Hedin Hjarrandason and his accomplices. Hild's father and his men tracked them down and an eternal battle commenced. Every day the two sides kill each other and each successive night Hild goes out on the battle field and revives them . . .' The exploration history of the Hild Field has been a similar battle, hence this well-founded name.

The main pay zone is the Tarbert Formation (nomenclature after Deegan and Scull, 1977 revised by Vollset and Doré, 1984) of the Middle Jurassic Brent Group. The Hild Field is structurally very complex, consisting of numerous small fault compartments which apparently have separate gas–water contacts as deduced from well results (Fig. 3). Resource estimates are thus uncertain and problematic to assess.

HISTORY

The Hild Field possibly extends into three licence areas (Fig. 2). Gas has been proved within Blocks 30/7 and 29/9 (Licence 040, Norsk Hydro operator/Statoil/Elf/Total) and Blocks 29/6 and 30/4 (Licence 043, BP operator/Statoil) on the Norwegian side. The general structural complex also extends into the British Blocks 3/15 (Licence 090, Total operator/Elf) and 3/20, but no wells have yet been drilled in that area. The Norwegian blocks were awarded in 1975 and 1976, re-

Geology of the Norwegian Oil and Gas Fields © A. M. Spencer *et al.*, eds (Graham & Trotman, 1987) pp. 287–294.

Hild Field data summary

Trap	
Type	structural, tilted fault-blocks
Depth to crest	3600 mss
Lowest closing contour	unknown
Hydrocarbon contacts	variable
Gas column	245 m (maximum)
Pay zone	
Formation	Tarbert
Age	Middle Jurassic (Bathonian)
Gross thickness	100 m
Sand/shale ratio	0.85
Net/gross ratio	0.4–0.7
Porosity (average; range)	13–20.5%; 0–32%
Hydrocarbon saturation	54–87%
Permeability	0.01–2000 mD
Hydrocarbons	
Oil gravity	0.80–0.82 g/cm^3
Oil type	low sulphur
Gas gravity	0.657–1.067
Dew point	630.1–736.5 bar
Gas/oil ratio	1264–4660 Sm^3/Sm^3 (DST)
Formation water	
—	
Reservoir conditions	
Temperature	126–145 °C
Pressure	730–775 bar
Field size	
Area	120 km^2 (three-dimensional area)
Estimated gas—in-place	30–40 × 10^6 Sm^3
Production	
—	

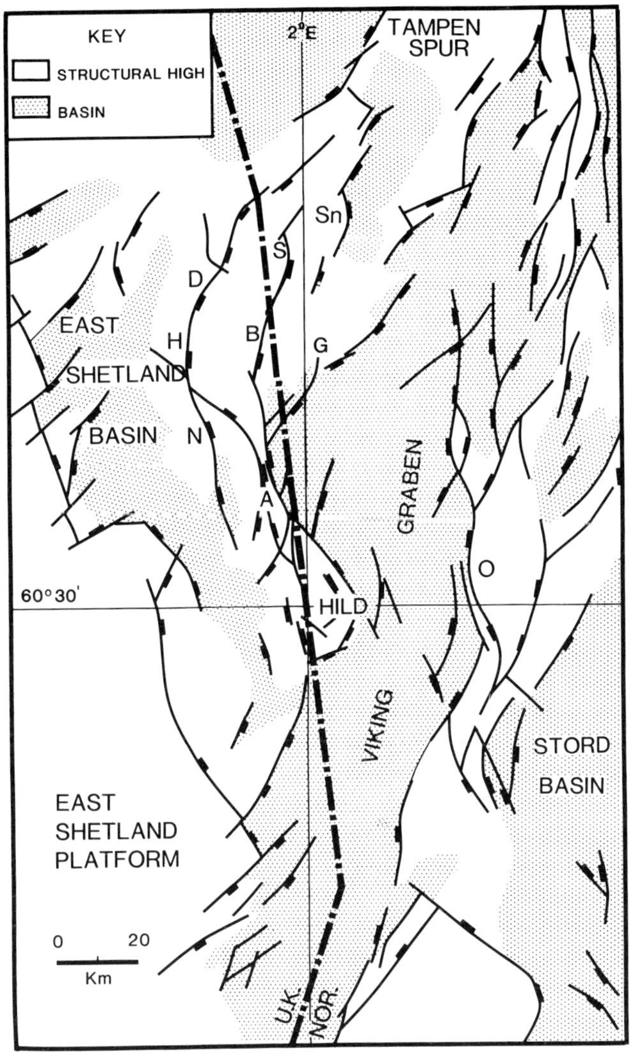

Fig. 1. Regional setting of the Hild Field. The main fault-blocks of the East Shetland Basin are indicated by letters D–H–N (Dunlin–Heather–Ninian), S–B–A (Statfjord–Brent–Alwyn), and Sn–G (Snorre–Gullfaks). The letter O is Oseberg to the East.

spectively. Since then a total of nine wells have been drilled within the Hild area. However, three of these were abandoned at depths less than 1000 m due to technical problems. The first successful well (30/7-2) tested a Tertiary Frigg sand prospect with uneconomic results.

The first Jurassic test (30/7-6) was spudded in February 1977 and reached the target in May 1978 after having been sidetracked and re-entered by another drilling vessel having the proper equipment to handle the abnormally high reservoir pressures (±750 bar) typical of the area. Well 30/7-6 found a well-developed Brent sequence and a hydrocarbon column of 100 m. Subsequent wells (30/4-2, 30/7-8, 29/6-1) have all found gas–condensate accumulations within the Tarbert reservoir with, however, different gas–water contacts and different pressure regimes. The hydrocarbon columns in these wells are in the range of 25–100 m. The 29/9-1 well missed the Tarbert reservoir due to a fault but found gas–condensate within the lower Brent section (Etive Formation).

GEOLOGICAL SETTING

The Hild Field is located on a structural nose that projects over the western flank of the Viking Graben at the southern end of a Base Cretaceous structural high trend responsible for the Statfjord–Brent–Alwyn fields. This trend is generally characterized by Jurassic fault-blocks with westward dips. However, in the Hild area the fault-blocks have counter, regional eastward dips.

These structures formed as a result of differential collapse and subsidence along N–S-orientated normal faults, related to a failed incipient rift (see Fig. 2) which developed during the Late Jurassic North Sea rifting phase. The fairly uniform Brent sequence was thus broken up into a number of fault-blocks. It is overlain by shales of the Viking Group, which are of variable thickness reflecting the syndepositional faulting activity (Rønning *et al.*, 1986).

A general lithostratigraphic column is shown in Fig. 4. The deepest layer penetrated by wells is the upper part of the Triassic Hegre Group. These continental sands and shales grade up into the more sandy and more marginal-marine Statfjord Formation. The thick, Lower Jurassic, Dunlin Group marine shales and intercalated sandstones (Cook Formation) show evidence of minor tectonic movements in the uppermost part as they abruptly grade up into the 300-m-thick, sandy, Brent Group deltaic deposits. The Tarbert Formation of the Brent Group is about 100 m thick in the area and consists of laterally extensive, permeable, coarse-grained sandstones alternating with fairly tight, dirty, finer-grained sandstones, with thin lagoonal shales and with coal beds (Fig. 5). This sequence is abruptly overlain by the marine Viking Group shales (Heather and Draupne formations), which show marked thickness variations (70–400 m) reflecting the syndepositional, Late Jurassic faulting activity.

During Early Cretaceous times the area remained a high, and no sediments of this age are preserved. However, rapid subsidence commenced in Albian times and gave rise to a thick Upper Cretaceous succession of marine

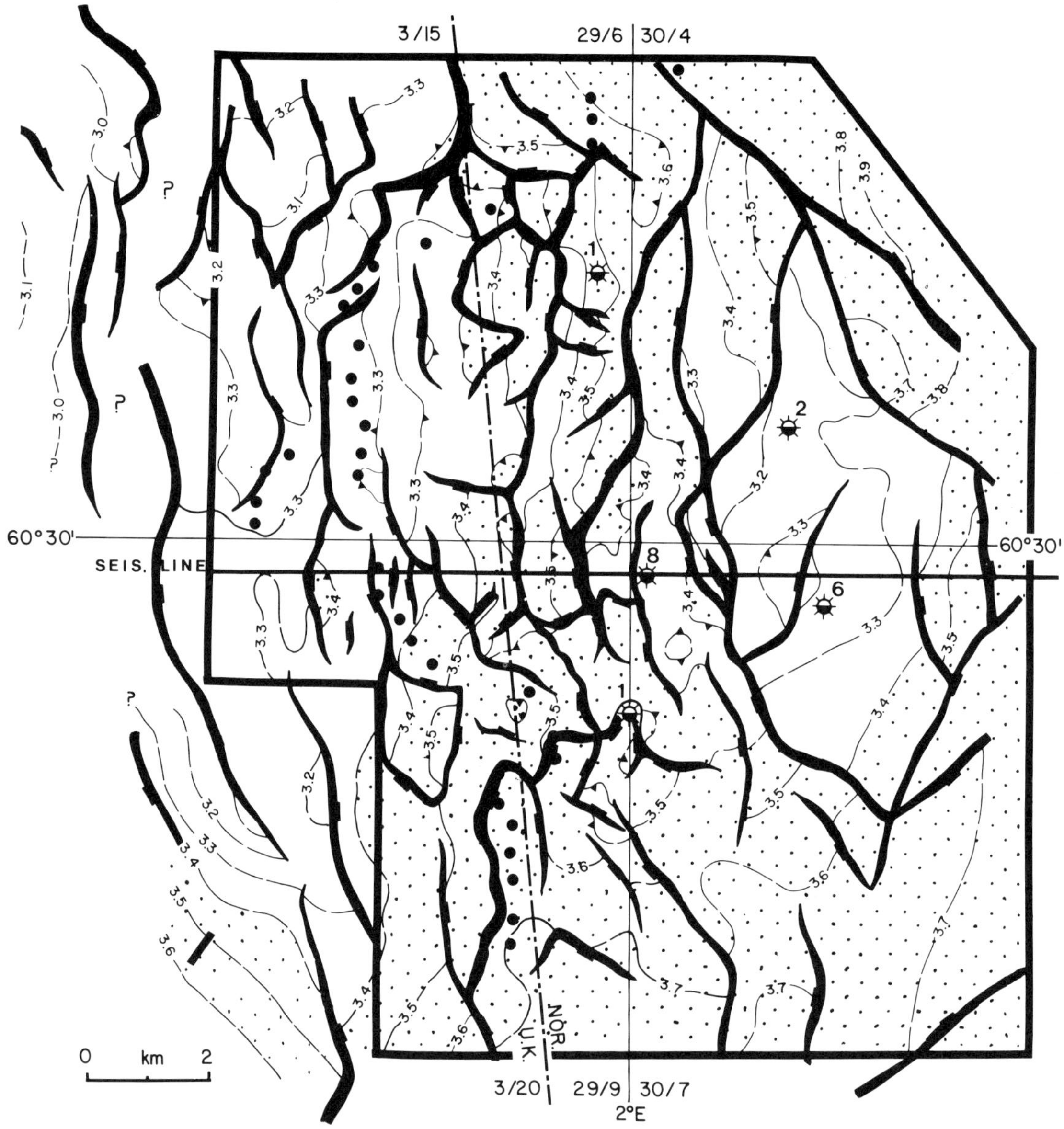

Fig. 2. Time-structure map of the intra-Brent reflector (approximately base Tarbert Formation) and outline of the area covered by three-dimensional seismic. Fault trends are conjectural. Dotted areas are deeper than 3.4 s. Heavy dots outline the trend of the 'incipient rift axis'.

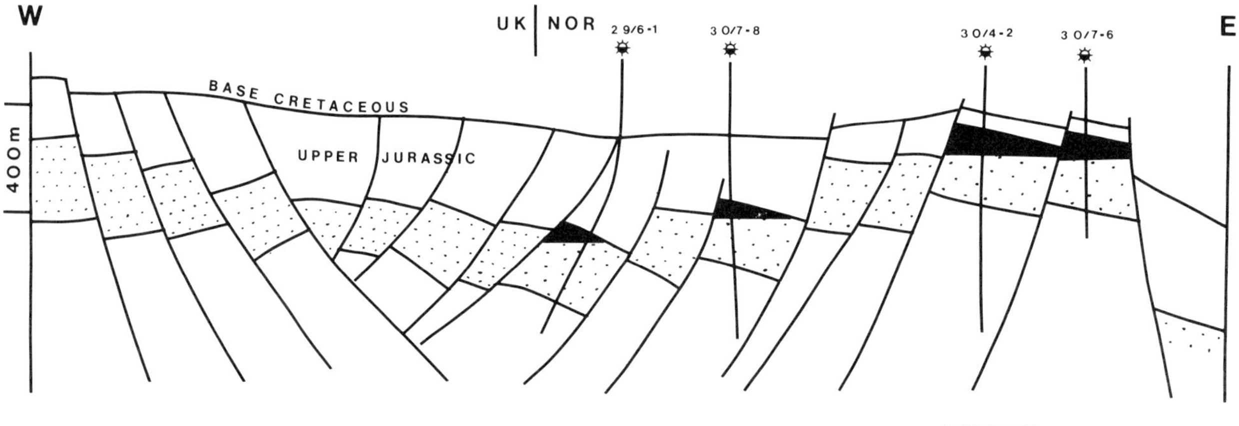

Fig. 3. Generalized structural E–W cross-section across the Hild structural complex based on the three-dimensional seismic. The four wells shown found gas–condensate (black) in the upper part of the Middle Jurassic Brent Group (dotted). Note different gas–water contacts in the different fault-blocks.

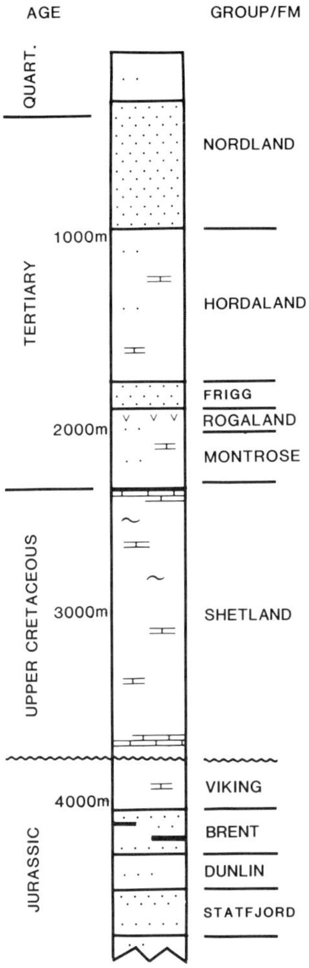

Fig. 4. General lithostratigraphy of the Hild Field.

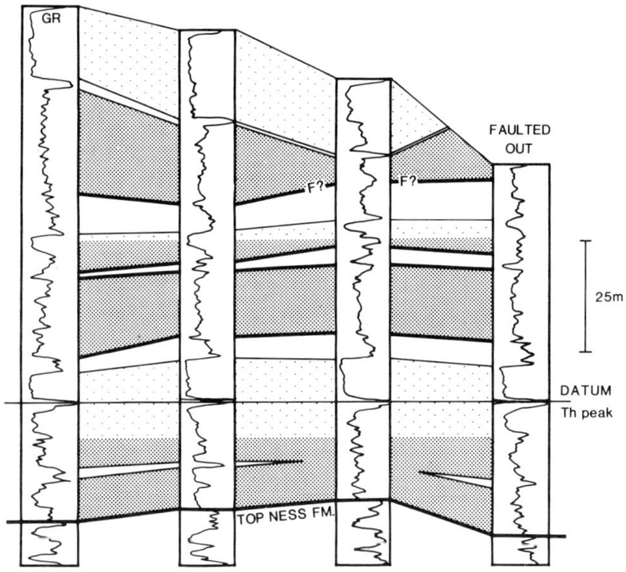

Fig. 5.' Gamma-ray envelope correlations, displaying the Tarbert Formation stratigraphy in the Hild area. Note the prominent zoning of this reservoir and the tabular geometries of the various facies. Open shading denotes permeable-barrier sandstones, dark shading shows tight, back-barrier sandstones, white represents shaley zones, and heavy lines coal beds. Datum is a gamma-ray peak from a zircon-enriched layer. Distances between wells range from 2.8 to 4.8 km. Missing sections are due to post-depositional faults.

shales, marls and nodular limestones. During Tertiary times more than two kilometres of shales and sandstones accumulated in deep-water submarine fans (Frigg Formation) and in shallow-marine settings. Tuff layers occur in the Lower Eocene section (Sele and Balder formations).

GEOPHYSICS

The major exploration problem in the Hild area is the poor quality of the seismic data (Fig. 6). There are several reasons for this: the low acoustic contrasts in the pre-Cretaceous section compared with the strong onlapping basal Cretaceous reflectors; the structural complexity of the area, with numerous small fault-blocks which are often smaller than the Fresnel zone in extent; the small amount of energy which penetrates beneath the Base Cretaceous; and a prominent 'gas chimney' over the eastern-central area.

An additional specific problem is that only the migrated sections of the 2000 km of two-dimensional seismic acquired in the years 1975–80 (one survey each year) can be utilized, as most unmigrated sections fail to show pre-Cretaceous events. The migrated versions of lines shot with different orientations do not tie the same reflector, due to the often drastically changing dips of the Jurassic strata and of the faults. For the same reasons, the faults are almost impossible to correlate between seismic lines. There were wells drilled prior to the three-dimensional seismic interpretation which found the reservoir about 200 m deeper than predicted, due to this poor seismic information and an incorrect geological model inferred from these data.

In order to obtain a consistently acquired and processed set of data it was decided to shoot a three-dimensional seismic survey over the Hild structure, involving all three licence groups. This survey, shot in 1981 and totalling 3000 km, did improve the data quality, but not sufficiently to allow a confident mapping of the area (Fig. 6). The Jurassic reflectors (Fig. 7) vary laterally in quality in relation to their dips relative to the basal

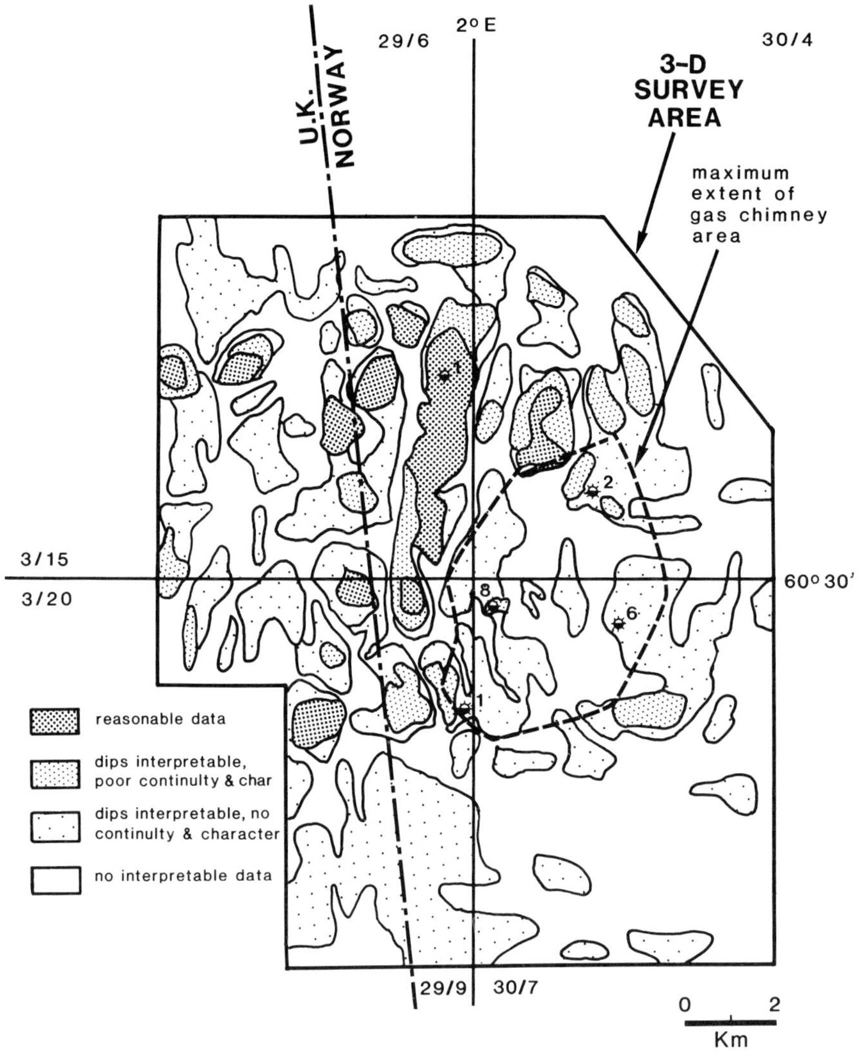

Fig. 6. Seismic data quality map after the three-dimensional seismic. More than half of the area has no reliable data.

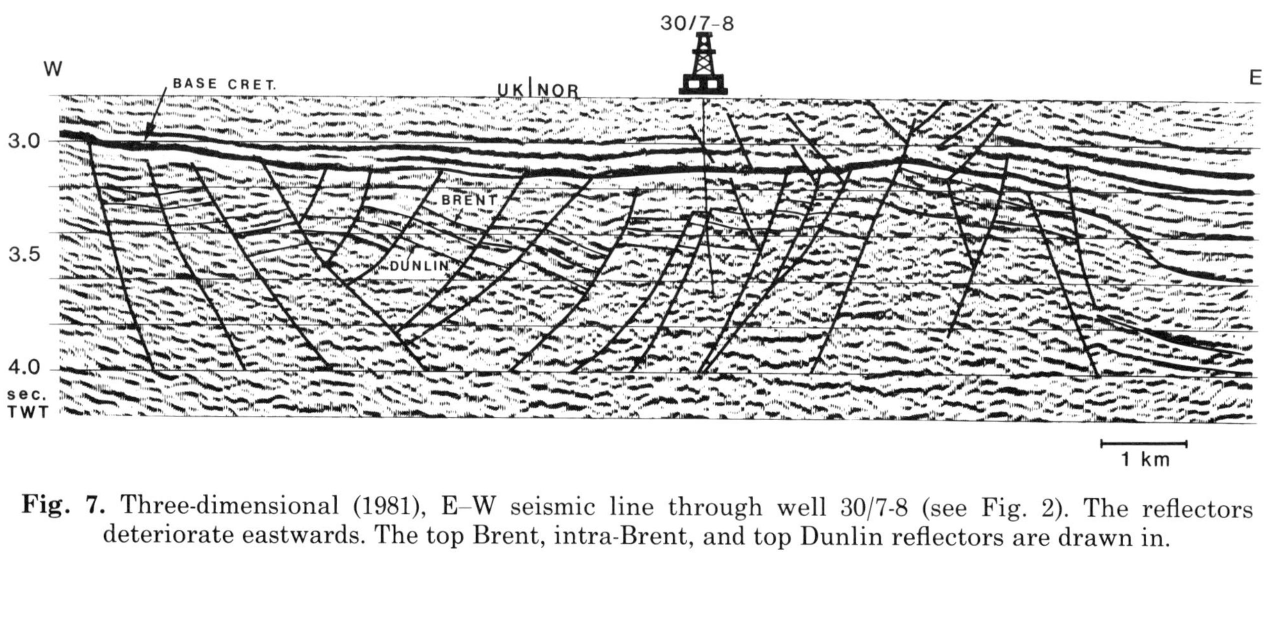

Fig. 7. Three-dimensional (1981), E–W seismic line through well 30/7-8 (see Fig. 2). The reflectors deteriorate eastwards. The top Brent, intra-Brent, and top Dunlin reflectors are drawn in.

Cretaceous reflectors. The best data occur where the Jurassic reflectors are steeply dipping and thus not sub-parallel to subtle reverberations/multiples from the strong Cretaceous horizons. However, the reservoir-level reflectors cannot be correlated continuously for more than a few kilometres. The only viable way to map this area was by trying to character-correlate all of the discontinuous Middle/Lower Jurassic reflectors with the seismic data through the one well that had by far the best data quality. However, the results of the fifth Jurassic well demonstrated that it was incorrect to continue trends into bad data areas. This well found only the lowermost Brent, due to a major fault not evident on the seismic lines, but seen on the displays from the three-component walkaway VSP acquired in the well (Ahmed *et al.*, in press; Rønning *et al.*, 1986).

The Hild complex has been mapped using an intra-Brent reflector (Fig. 2), which appears to be the most continuous one on the seismic sections.

TRAP

The Hild Field consists of a number of rotated fault-block traps of Late Jurassic age, sealed by Upper Jurassic and Cretaceous shales. The highest structure (about 3600 mss) is found in the eastern and western parts with successive down-stepping by normal throws on faults towards the central area (see Figs 2 and 3). Each fault-block apparently has its own gas–water contact, and they are apparently not filled to their spill-points.

RESERVOIR

The Tarbert Formation constitutes the main reservoir in the Hild Field. It can be divided into eight prominent stratigraphic zones based on lithological and depositional criteria (Fig. 5). This zonation is caused by the vertical stacking of three main depositional facies:

1. coarse-grained, permeable-barrier sandstones,
2. finer-grained, micaceous and argillaceous, tight, back-barrier sandstones, and
3. lagoonal shales, siltstones and coal beds.

The reservoir zonation is further emphasized by the fact that the porosities, and in particular the permeabilities, are facies dependent. Due to the higher matrix content in the form of mica, illitized clay and authigenic kaolinite, the back-barrier sandstones are fairly tight as far as production is concerned. The barrier sandstones are coarser grained, cleaner, and have permeabilities that may reach 10 darcies.

Porosities are to a large extent of secondary origin after feldspar and calcite cement dissolution (Lønøy *et al.*, in press), and on average range from 13 to 20% but may be as high as 32%. Porosity and permeability decrease with depth, a trend apparently related to the late-stage diagenetic dissolution/reprecipitation process affecting the calcite cement (Rønning *et al.*, 1986).

Gas–condensate was also encountered in the lower Brent section (Etive Formation), in the 29/9-1 Well. This is a poor reservoir with permeabilities of less than 1 mD and with an average porosity of 14%.

SOURCE

The Hild condensates have a sterane composition typical of terrigenous kerogen (Rønning *et al.*, 1986) and it is assumed that they are derived from the Brent coals. The high (3–4%) CO_2 content of the gas supports this view. Hydrocarbon generation started in Late Paleocene times.

HYDROCARBONS

The waxy Hild condensates have an API gravity ranging from 41.3 to 45°. The gas/oil ratios derived from drill-stem tests are about 1270 in the deeper compartments and about 4650 in the shallow ones. Reservoir temperatures range from 126 to 143 °C and the abnormally pressured reservoirs have pressures of more than 740 bar.

RESOURCES

The Hild resources are difficult to assess due to the inability to confidently define and map the numerous fault-blocks, combined with the fact that they have different gas–water contacts. However, estimates of the total in-place resources lie in the range of 30–40 $\times 10^9$ Sm^3.

ACKNOWLEDGEMENT

Norsk Hydro would like to thank the partners

of Licence 040, Elf, Total and Statoil, for their permission to publish this paper. We also thank the BP/Statoil group for permission to include data from their part of the Field.

REFERENCES

Ahmed, H., Dillon, P., Johnstad, S.E. and Johnston, C.D. 1987. Northern Viking Graben multilevel three-component walkaway VSP—a case history, *First Breake*, in press.

Deegan, C.E., and Scull, B.J. (Compilers). 1977 A standard lithostratigraphic nomenclature for the Central and Northern North Sea, *Rep. Inst. Geol. Sci. No. 77/25; Bull. Norw. Petrol. Direct., No. 1.*

Lønøy, A., Akselsen, J. and Rønning, K. 1987. Diagenesis of a deeply buried sandstone reservoir: Hild Field, Northern North Sea, *Clay Minerals*, in press.

Rønning, K., Johnston, C.D., Johnstad, S.E. and Songstad, P. 1986. Geology of the Hild Field, In Spencer, A.M. *et al.* (Eds), *Habitat of Hydrocarbons on the Norwegian Continental Shelf*, Graham and Trotman, London, pp. 199–206.

Vollset, J. and Doré A.G. (Eds). 1984. A revised Triassic and Jurassic lithostratigraphic nomenclature for the Norwegian North Sea, *Bulletin Norwegian Petroleum Directorate* 3.

26

Murchison

Nils Engelstad

*Mobil Exploration Norway Incorporated, PO Box 510,
4001 Stavanger, Norway*

The Murchison oil field straddles the UK/Norway border in the North Sea and lies in the East Shetland Basin on the west side of the Viking Graben. The field is entirely within UK Block 211/19 and Norwegian Block 33/9. It was discovered by the Conoco Well 211/19-2 in August 1975 and is operated by Conoco UK Limited. The structure is a tilted fault-block. The reservoir sands are of Middle Jurassic age and of progradational coastal deltaic origin. The oil-bearing Brent Group sands vary in thickness from 100 m to 150 m in the main part of the field (thicker in the south flank). Porosities average 22% and permeabilities 500–1000 mD in the producing zones. Original recoverable reserves are estimated at 51×10^6 Sm^3 oil. A total of 26 separate wells have been included in the main development programme. The field is developed from a single, steel-jacket platform. The production peaked in February 1983 at 20 000 Sm^3/day and the field life is expected to be 18 years (1980–98).

INTRODUCTION

The Murchison oil field is located approximately 250 km west of Bergen and 10 km northwest of the Statfjord Field (Fig. 1). Currently 75% of the reserves are in UK Block 211/19 and 25% in Norwegian Block 33/9. The field has a square outline covering approximately 17 km^2 and is in a water depth of 156 m. It is a medium-sized oil field with approximately 51×10^6 Sm^3 original recoverable oil and with associated gas reserves estimated at 3.9×10^9 Sm^3. Coastal deltaic progradation during Middle Jurassic times resulted in deposition of the Brent Group reservoir sands. The main field is structurally simple, whereas the south flank is complexly faulted. Seismic quality is fair to good over the main field, but deteriorates over the south and east flanks. The field is named after the Scottish geologist Sir Roderick Impey Murchison (1792–1871) who is best known for his contribution to the determination of the Palaeozoic sequence.

HISTORY

The Murchison Field was discovered in August 1975 by the Conoco–BNOC–Gulf Well 211/19-2 in Block 211/19, within the UK Production Licence P 104 area. Two delineation wells were drilled, 211/19-3 and -4, the latter establishing an oil–water contact (the level below which water saturation is essentially 100%) at 3080 mss. Geophysical interpretations using data from the UK and Norwegian (PL 037) licences showed that a section of the Murchison Field extended into the Norwegian Block 33/9. The Murchison Field is therefore binational with the following owners:

UK Licence P 104—Block 211/19

Conoco (UK) Limited	33.3%	(operator)
Britoil Limited (formerly BNOC)	33.3%	
Chevron (formerly Gulf Oil Corporation)	33.3%	

Geology of the Norwegian Oil and Gas Fields © A. M. Spencer *et al.*, eds (Graham & Trotman, 1987) pp. 295–305.

Murchison Field data summary

Trap	
Type	Tilted fault-block
Depth to crest	2904 mss, average 2956 mss
Lowest closing contour	3200 mss
Hydrocarbon contacts	3080 mss (main field)
Gas column	—
Oil column	176 m (main field)
Pay zone	
Formation	Brent
Age	Mid-Jurassic
Gross thickness (average; range)	130 m; 75–220 m (S flank inclusive)
Net/gross ratio (average; range)	0.7; 0.45–0.80
Porosity (average; range)	22%; 16.5–26.5%
Hydrocarbon saturation (average; range)	70%; 65–82%
Permeability	500–1000 mD; maximum 10 D
Productivity index	160 Sm^3/day/bar
Hydrocarbons	
Oil gravity	0.83 g/cm^3 at 15 °C
Oil type	low sulphur crude
Bubble point	131 bar at 110 °C
Gas/oil ratio	96.2 Sm^3/Sm^3
Formation volume factor	1.313
Formation water	
Salinity	14.000 mg/litre
Resistivity	0.312 ohm m at 15 °C
Reservoir conditions	
Temperature	110 °C
Pressure, initial	444 bar
Pressure gradient in reservoir	0.070 bar/m
Field size	
Area	17 km^2
Gross rock volume	*c.* 1400×10^6 m^3
Hydrocarbon pore volume	96×10^6 m^3
Recovery factor—oil	41%
Drive mechanism	water injection, aquifer support
Initial recoverable hydrocarbons	
oil	51×10^6 Sm^3
NGL	2×10^6 tonnes
gas	3.9×10^9 Sm^3
Production	
Start-up data	Drilling June 1980, production September 1980
Development scheme	Single platform, 28 slots + 3 subsea completions
Production rate	Peak 20 200 Sm^3/day (1983), May 1985: 15 250 Sm^3/day
Cumulative production (31 May 1985)	26.5×10^6 Sm^3

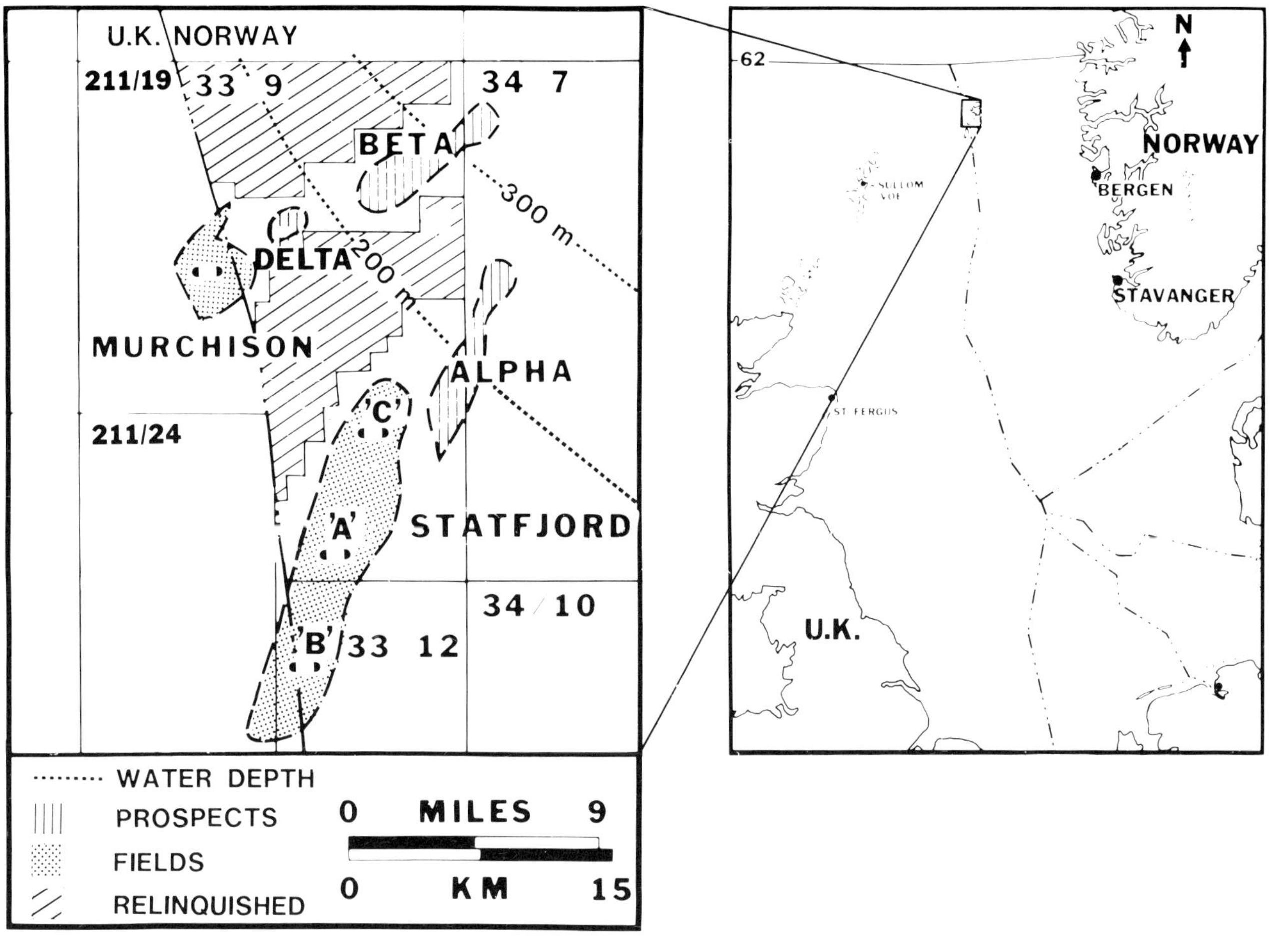

Fig. 1. Location of the Murchison oilfield relative to other discoveries in adjacent blocks.

Norwegian Licence PL 037—Block 33/9

Den norske stats oljeselskap AS	50.000%
Mobil Exploration Norway Incorporated	15.000%
Conoco Norway Incorporated	10.000%
AS Norske Shell	10.000%
Esso Exploration and Production Norway Incorporated	10.000%
Saga Petroleum AS	45/24%
Amoco Norway Oil Company	25/24%
Amerada Petroleum Corporation of Norway	2/24%
Texas Eastern Norwegian Incorporated	25/24%

A seismic map dated June 1971 gave the first hint of the structure. Two years later another seismic map indicated the trap, which is a horst-block, fault-bounded on three sides. The southern boundary fault was thought to control the Brent Group and Statfjord Formation thicknesses. The location of the 211/19-2 Well was selected off the structural crest to avoid a presumed truncation of the Brent sandstones. Subsequent drilling showed that this assumption was wrong in that the Brent was down-faulted rather than truncated (Well 211/19-1, the first well drilled in the 211/19 Block, was a Thistle Field delineation well, and was located west of what became the Murchison Field).

The field discovery well 211/19-2 encountered 127 m gross and 91 m net of oil-bearing Brent Group sandstones below Upper Jurassic shales. After drilling through 122 m of Dunlin shales and a thin Statfjord calcareous sandstone, the well reached its total depth 73 m into the Triassic. The Statfjord and Triassic sandstones were water wet, and subsequent drilling has not proved them otherwise. Two drill-stem tests in the 211/19-2 Brent Mica sandstones flowed 214 Sm^3/day and 890 Sm^3/day oil, respectively.

Several seismic surveys were undertaken between 1973 and 1982, and three-dimensional data shot over the field in early 1985 are currently being interpreted.

The original unit equity was negotiated in 1979 and gave Block 33/9 an equity of 16%. This equity share was increased to 25% by the equity redetermination effective from November 1983, which was initiated after the fifteenth platform well had been drilled. A second equity redetermination started in August 1984 and is currently in progress.

Following the decision to develop the field in 1976, a steel-jacket platform with fully integrated production, drilling and accommodation facilities was installed in August 1979. Production started September 1980. In addition to the three subsea wells, 26 platform wells have been drilled to date. The platform has a total of 28 platform well slots. Seven of the wells have been side-tracked or deepened. The status in May 1985 was: 14 producers; 9 water injectors; 1 gas injector; 10 suspended/abandoned wells; 34 well locations in total.

An infill drilling programme of five wells is planned starting in late 1985. The main objective is to increase production from the crestal and south flank areas, and, depending upon the three-dimensional seismic interpretation, to prove reserves in the east flank of the field.

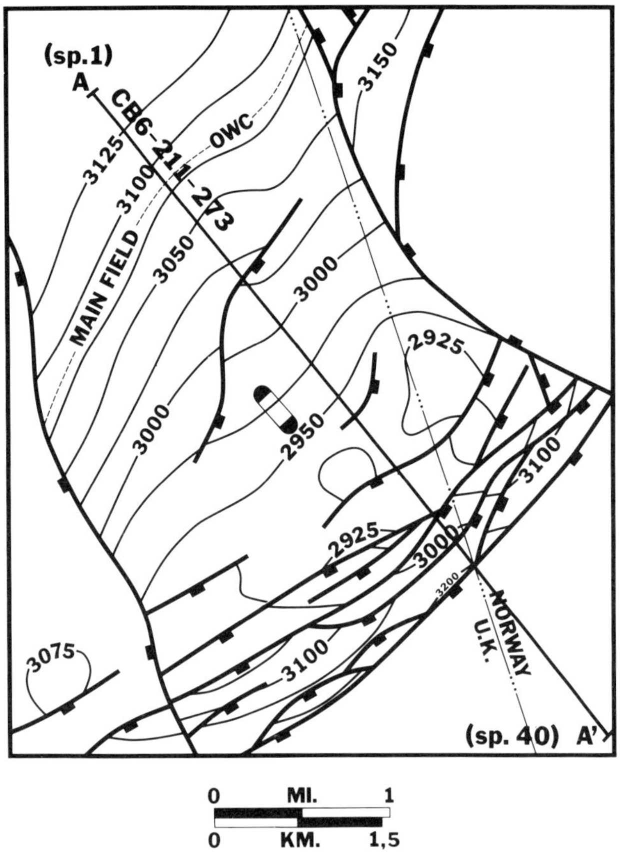

Fig. 2. Mobil-interpreted top Brent Group structure map. Contour interval is 25 m.

GEOLOGICAL SETTING

The Murchison Field is located in the East Shetland Basin on the west side of the Viking Graben, within the same depositional and faulting trends as the Statfjord, Thistle and Dunlin Fields.

A Permian intracratonic basin in the northern North Sea was subjected to Permo-Triassic rifting which established the East Shetland Basin and initiated an extensional stress regime. Sedimentation in the basin lasted from Permo-Triassic times through the Mesozoic and Tertiary, but was interrupted during Jurassic and Early Cretaceous times by Kimmerian tectonic activity, resulting in phases of erosion and marine transgression. The main structural element of the Murchison Field is a large square-shaped, north-west-dipping fault-block, resulting from Middle–Late Jurassic block-faulting, tilting and erosion. It is bounded to the east and west by NW-trending normal faults. The crest in the southern part of the block borders the complexly faulted southern flank, where the faults trend ENE (Figs 2, 3).

The Middle Jurassic Brent Group represents a major terrigenous clastic wedge and provides the main oil and gas reservoir of the East Shetland Basin. Tectonism within the area of the field is interpreted as being mild prior to Brent deposition. The tectonic movements which created the Murchison trap occurred during the deposition of the upper Brent. The amount and intensity of faulting during early–mid Brent deposition are uncertain.

Block faulting within the south flank continued throughout post-Brent time, and a reactivation of the E–W faults delineating the south flank is believed to have occurred in post-Jurassic–Early Cretaceous time.

The Brent sequence was deposited following a marine transgression represented by the Lower Jurassic Statfjord Formation–Dunlin Groups (Fig. 4). It was succeeded by the transgressive Upper Jurassic Humber Group, which partly truncates the upper Brent Tarbert Formation over most of the field. The Brent sequence was laid down in a progradational coastal deltaic depositional environment, sourced from the south-south-east. Its base is a shallow-water sublittoral sheet-sand, followed by pro-delta clays and then by more proximal deltaic deposits, prograding barrier sands, delta-top facies, and an interdistributary bay deposit. This is overlain by a deltaic sequence with interbedded marine shales. The Brent Group therefore represents three distinct phases of sedimentation (Deegan and Scull, 1977; Vollset and Doré, 1984):

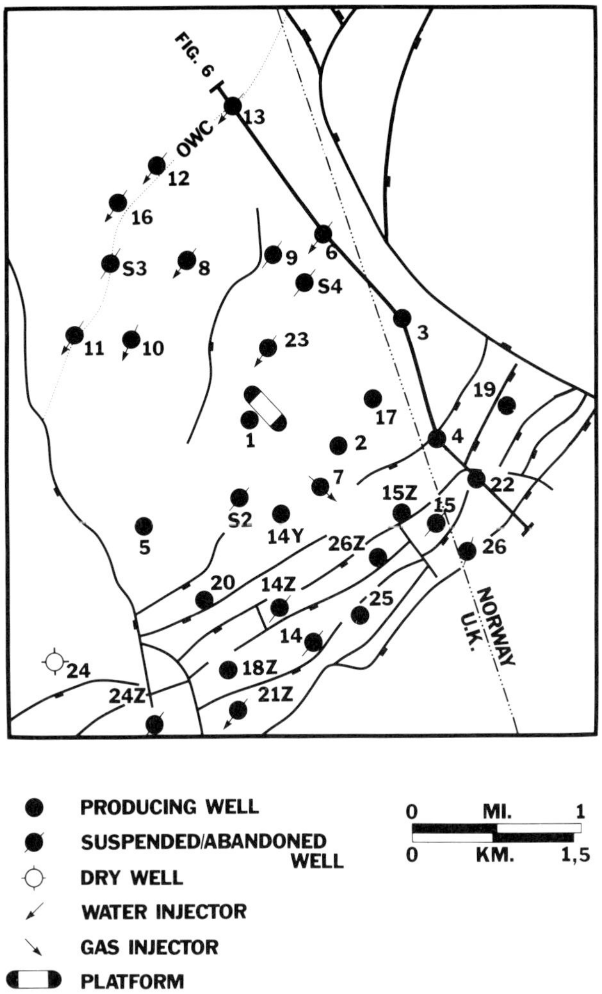

Fig. 3. Outline of the major tectonic elements of the Murchison Field showing wells drilled through May 1985. Complexly faulted south-eastern flank pattern is Mobil interpreted.

1. abandonment, represented by the transgressive facies of the Tarbert Formation;
2. shoreface and coastal plain progradation, comprising the Rannock, Etive and Ness formations;
3. at the base of the Brent, the Broom Formation is interpreted as a sublittoral sand-sheet deposited by an estuarine channel–shoal system, genetically separate from the two facies groups above.

Major uplift and faulting occurred simultaneously, causing a relative subsidence of the south flank and preservation of a thick upper Brent sequence. The Brent envelope averages approximately 130 m thickness in the main field, subdivided into seven zones, of which five are reservoir zones. The facies and thickness distributions indicate a north-westerly direction for the Brent Group progradation. Synsedimentary tilting, controlled by the major faults, influenced the progradation and deposition. The unconformably overlying Upper Jurassic and Lower Cretaceous shales form the cap-rock.

GEOPHYSICS

Following the UK Third Round Licence award of Block 211/19 in 1970 a seismic map was prepared in 1971, based on a three-kilometre reconnaissance seismic grid. The first hint of a closed structure was provided by a segment of the southern major fault interpreted from this map. Infill seismic, shot in 1972, provided a closer grid (1 1/2 km) and confirmed a small closure.

The Murchison trap or trapping potential was first seen on a seismic map based on lines from the 1973/74 MNG (Mobil Norway Group) survey and the 1974 Conoco-operated CNG survey. This map defined the western and eastern boundary faults and recognized the conjugate fault pattern which characterizes the East Shetland Basin (the north-east Caledonide and north-north-west Kimmerian fault lineaments).

The CNG 1974 survey was the first non-regional survey and provided a detailed seismic grid leading to construction of the last map before the 211/19-2 discovery well was drilled in 1975. The MNG 1974 survey was shot over the whole Norwegian PL037 Licence area and the Murchison Field in a 1 × 1 km grid. The CNG survey covers only the Murchison Field in a triangular grid with 2–2.5 km line-spacing.

The second Conoco-operated survey was the CBG/211 survey from 1977 (Fig. 5), shot in a strike-and-dip grid with 1–2 km line-spacing. The quality was poor to fair at the objective Jurassic horizons.

The data from all these surveys, in addition to some lines from the 1976/77 Mobil-operated MNG surveys, were reprocessed by Seismograph Service Limited (SSL). The reprocessed data improved the Top Brent reflector, but the final maps produced for equity determination purposes in the summer of 1982 were inconclusive. Major disagreement existed on the eastern and south-western flank interpretation.

In September 1982 Conoco shot the 206 km C/211-82 survey, which was also processed by SSL. This survey represented the only new seismic data to be used in the current 1984/85 equity redetermination. It did not solve the interpretation problems over the south and east flanks, especially for the Top Dunlin reflector. The owners decided to shoot three-dimensional seismic over the whole field.

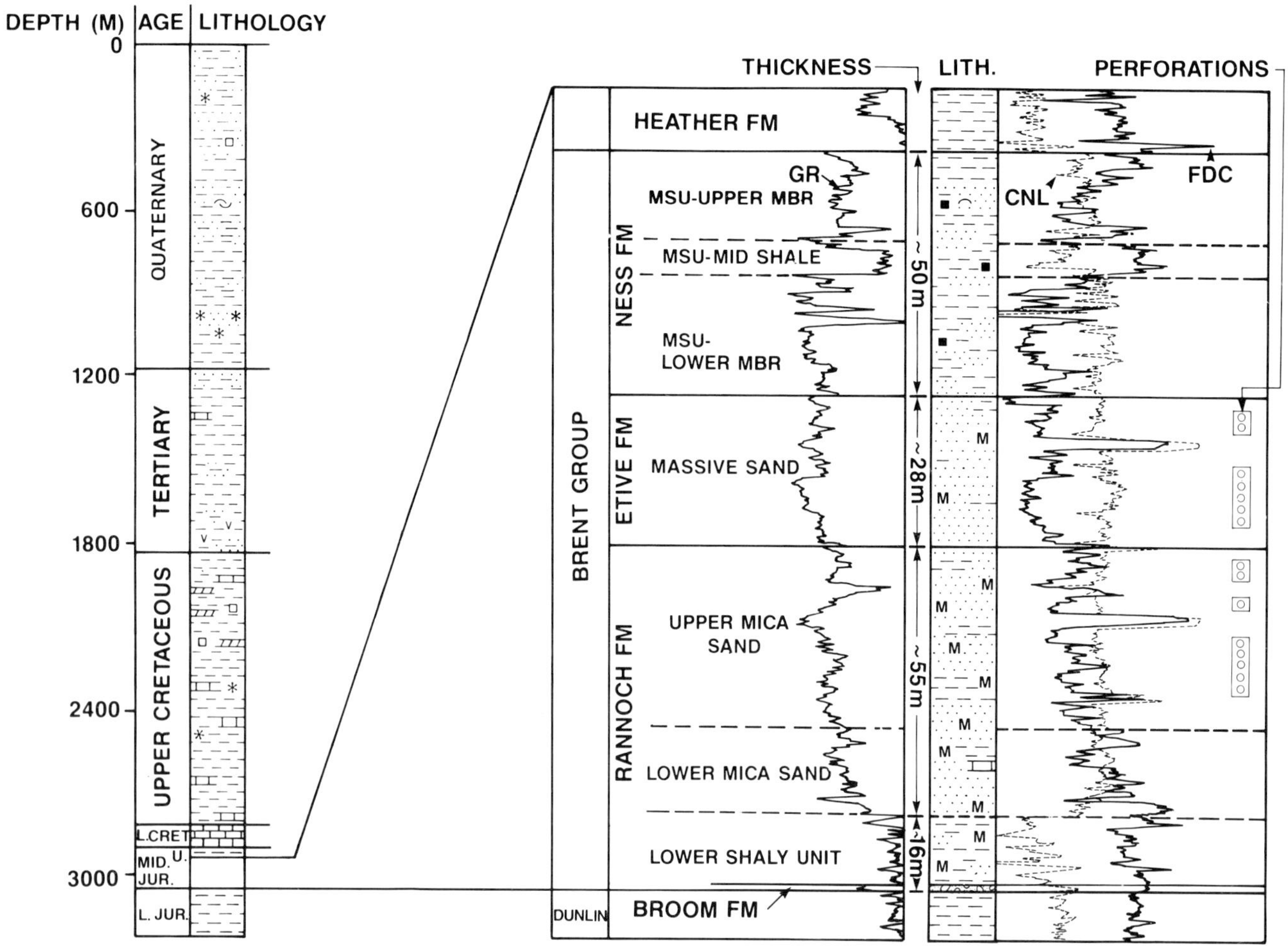

Fig. 4. Representative stratigraphic section for the Murchison Field (well MO3).

During the winter of 1984–85, GECO shot 2600 km of three-dimensional lines with 25-m line spacing. These data are expected to be interpreted in 1986. The interpretation should contribute to deciding optimal locations for infill drilling, mainly over the crestal area of the field, and will, it is hoped, produce structure maps that can be agreed on by all the geophysicists working the field.

The following is a summary of the seismic surveys leading to the current structural interpretation of the Top Jurassic, Top Brent and Top Dunlin horizons:

Reconnaissance Survey	1971
Infill seismic	1972
MNG (Mobil Group)	1973
MNG	1974
CNG (Conoco Group)	1974
MNG	1976
MNG	1977
CBG/211 (Conoco Group)	1977
C/211-82 (Conoco)	1982
3D (Murchison Unit)	1985

TRAP

The Murchison Field oil accumulation occurs in a structural trap resulting from Late Kimmerian tensional movements, which started in Late Bajocian times and continued with intermittent fault-block rotation throughout the remainder of Jurassic times. The accumulation is limited to the north-west by an oil–water contact intersecting the gently north-westerly dipping Brent zones (Fig. 6).

The roughly parallel field-bounding fault systems strike south-east, have throws of up to 150 m and limit the accumulation to the west and east. To the south and south-east occur a series of fault-block terraces with fault throws of 20 to 140 m and dips mainly to the south-east. The southern field fault throws in the same direction but has a magnitude of 100–250 m. Three wells drilled on the down-thrown side of this fault have yielded oil shows. The deepest well penetrated top Brent at 3370 mss.

The trap is delineated by the following

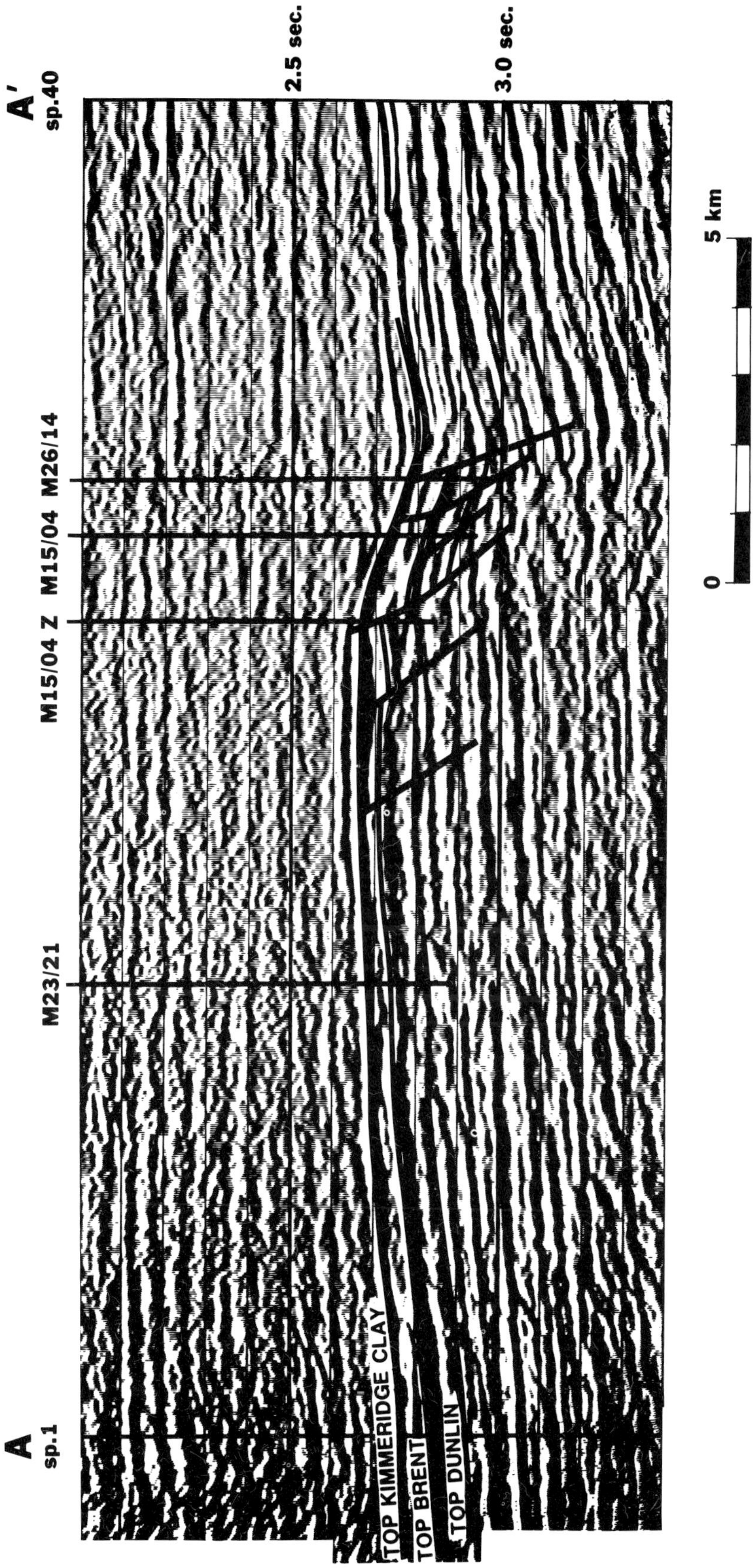

Fig. 5. Dip-orientated seismic line across the Murchison Field from the 1977 CBG/211 survey (Mobil interpretation). The location of the line is shown in Fig. 2.

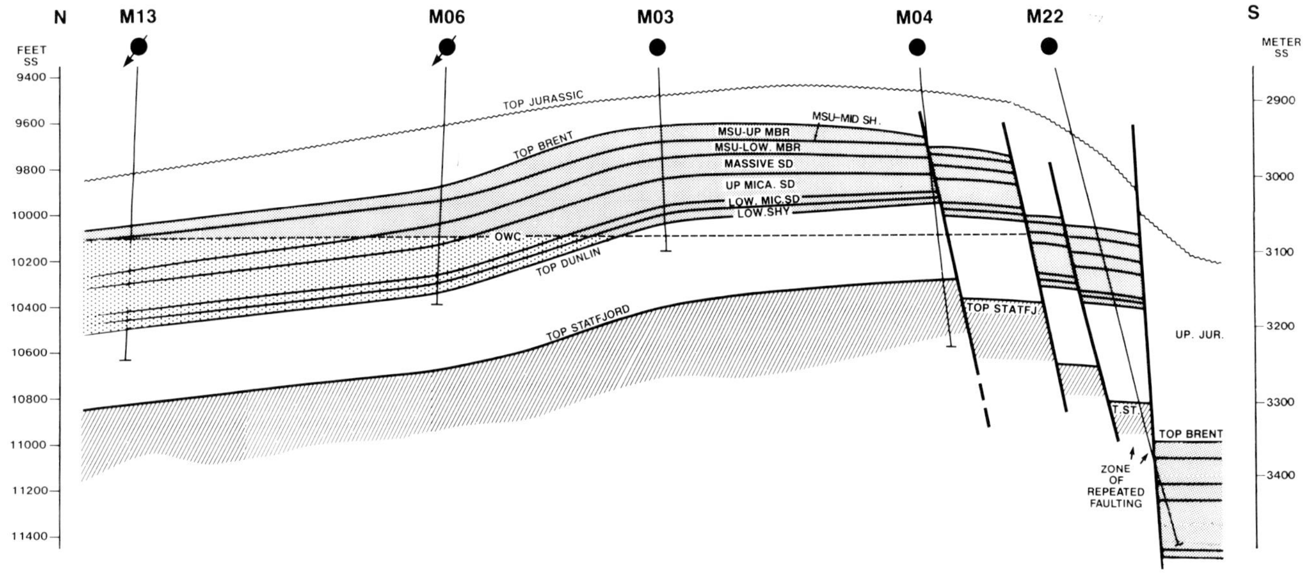

Fig. 6. Dip-orientated structural cross section through the eastern portion of the Murchison Field. Location of this section is shown on Fig. 3.

elements: the oil–water contact in the north; the lithologically controlled oil levels within the south flank fault-blocks; the completely or partially sealing field-bounding faults; and the Upper Jurassic Heather marine shales which truncate the Tarbert Formation. The main field oil column is 176 m high (crest at 2904 m, oil–water contact at 3080 mss). Oil-down-to levels in the south fault terraces deepen the lowest known oil inside the southern-most field fault by more than 300 m.

RESERVOIR

The reservoir sandstones in the Murchison Field occur in the Brent Group of Middle Jurassic (Aalenian–Bathonian) age. Primary depositional characteristics seem to have controlled porosities and permeabilities in the Brent reservoir zones above approximately 3350 m. Below this depth, secondary mineralization and diagenesis have reduced the porosity and permeability. The main field hydrocarbons are found between the depths of 2900 and 3080 m, and so primary sedimentary facies have been dominant in establishing reservoir properties in the oil column. Reservoir zonation corresponds with the stratigraphic breakdown of the Brent Group and consists of seven subunits, of which five are reservoir zones (Fig. 4). A short geological and reservoir property description of the seven-unit zonation is given below. The reservoir parameter values indicated on Fig. 7 are average or approximate and refer to the main section of the field.

The *Lower Shaly Unit* is the basal Brent zone. It includes the Broom Formation sublittoral sand-sheet as its lowermost unit. Cores from this unit and the one above show storm-generated ripples interpreted as of lower shoreface–offshore environment, with quiescent periods between storms allowing for bioturbation and mud deposition. The one-metre-thick Broom sand layer is overlain by approximately 15 m (TVT) of marine pro-delta clays. The unit is not a reservoir unit.

The *Lower Mica Sand Unit* represents the lower shoreface of the progradational delta sequence. It has a gross thickness of 15–18 m and consists of fine mica sands of which 2–4 m are considered reservoir sands. Porosities are 15–16% and oil saturation is 55%. One percent of the oil-in-place in the field is contained in this unit, which has not been perforated to date.

The *Upper Mica Sand Unit* is a middle shoreface facies of the delta front, where shallow-water depths are indicated by large-scale, wave-generated structures. Like the unit below, it consists of fine, micaceous sands but has better porosity (20%) and a higher oil saturation (75%). The gross thickness is 35–60 m. Eleven wells are currently producing from this unit, which is estimated to contain 29% of the oil-in-place. The two Mica Sand Units have poorer reservoir quality than the overlying zones. The units within the Mica Sands coarsen upwards and their

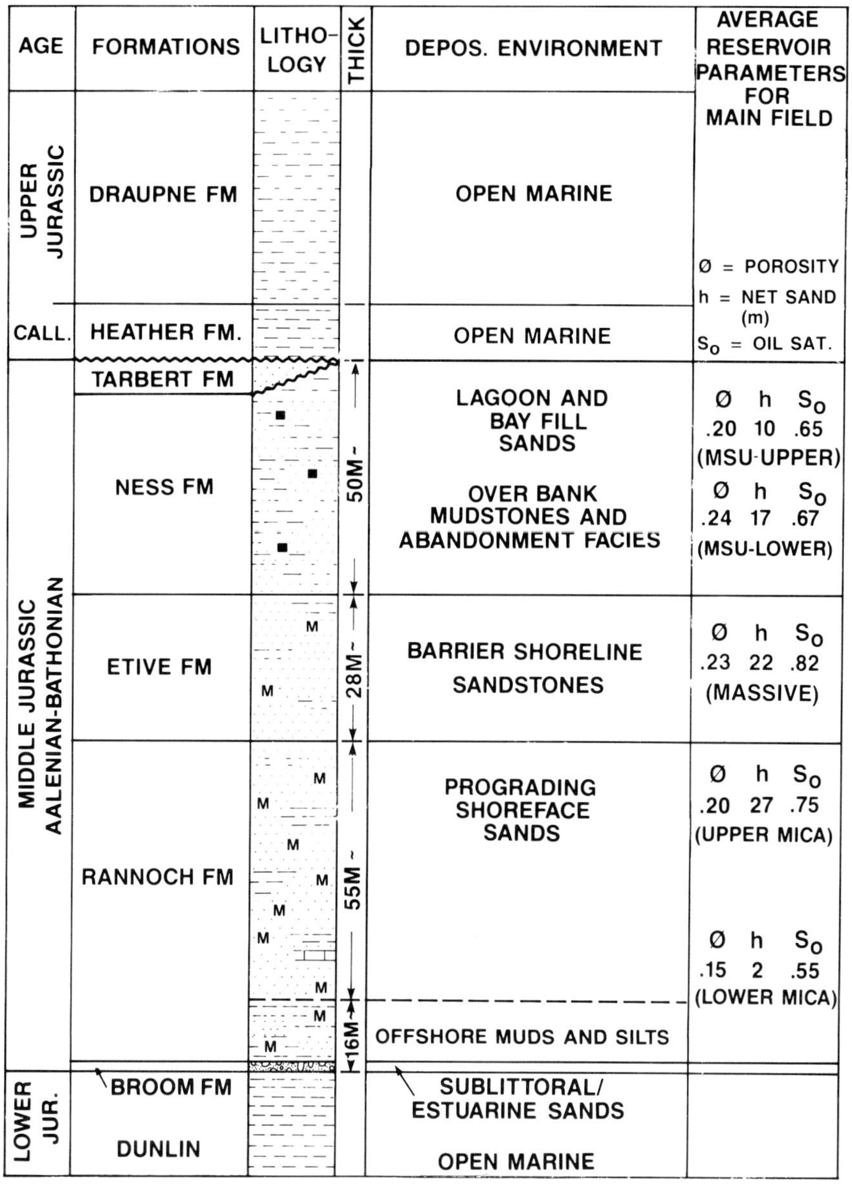

Fig. 7. Formal Jurassic lithostratigraphic nomenclature for the Murchison Field including lithology, interpreted depositional environment and average reservoir parameters.

tops act as preferential permeability conduits. Horizontal permeability trends confirm the diagenetically associated porosity/permeability reduction below the oil–water contact. The Mica Sand Units here correspond to the Rannoch Formation elsewhere in the northern North Sea.

The *Massive Sand Unit* consists of fine- to medium-grained sands showing weak cross-bedding with some erosional contacts. They are interpreted as a prograding barrier bar–sand complex mixed with barrier beach sands and tidal channel deposits. Total thickness is 20–33 m, average porosity 23% and oil saturation 82%. The lower zone of the unit, made up of basal channel fills and tidal inlet deposits, has good porosities (21–26%) and permeabilities of more than four Darcies. An extremely good permeability streak is found at the base with permeabilities of up to 10 D. The unit is currently producing in three wells and is estimated to have 32% of the oil-in-place. It corresponds to the Etive Formation.

The *Middle Shaly Unit–Lower Member* is a typical delta-top facies consisting of multi-storey distributary-channel sands, mouth bars, crevasse splays, overbank deposits and interdistributary muds and coal-swamp deposits. The lithology, consequently, varies from mudstones to coarse-grained channel sands. Within and proximal to the channels the poroperm characteristics are good but become poorer with distance from the channels. The unit has an average thickness of 25–

30 m (maximum 60 m), an average porosity of 24%, an oil saturation of 67% and permeabilities in the 500–1000 mD range. The unit corresponds to the lower Ness Formation. It is estimated to contain 23% of the oil-in-place. Currently, six wells are producing from this unit.

The *Middle Shaly Unit–Mid-Shale* (corresponding to Mid-Ness shale) was laid down during a temporary halt in the process of continental progradation. It is interpreted as a huge interdistributary bay deposit with very little sand influx, and consequently is not a reservoir unit. It is approximately five metres thick and acts as an effective vertical fluid/pressure barrier between the Upper and Lower Members of the Middle Shaly Unit.

The *Middle Shaly Unit–Upper Member* reflects a resumption of deltaic processes along with increasing marine influence. Shallow-marine sands were transgressed and partially reworked as marine shales, terminating the deltaic depositional cycle. Except for the marine transgressions, the lithology and poroperm conditions are similar to the MSU–Lower Member. This zone averages 15 m in thickness, 20% porosity and 65% water saturation. With an estimated 14% of the oil-in-place attributed to the MSU–Upper Member, production is planned from this zone from late 1986/early 1987. It is the equivalent of the Upper Ness Formation and Tarbert Formation where it is not eroded. The Tarbert units have no distinct permeability trends, and only the basal subunit has good reservoir quality.

The Brent sequence was followed by deposition of the thick Heather transgressive sequence, which lies on an erosion surface cutting into the Tarbert by as much as 20 m. Syn- and post-depositional faulting, which was most significant in the extreme south, gave rise to increased and (within fault zones) repeated sediment thicknesses and poorer reservoir properties in the south-flank fault-blocks.

SOURCE

The source- and cap-rock for the Brent hydrocarbons are the overlying Upper Jurassic Draupne carbonaceous clays. The hydrocarbon properties correspond well with those in other North Sea oil fields that undoubtedly have the same source. During Late Cretaceous to Middle Tertiary times the source-rock matured and the oil migrated into the Murchison trap, most likely from the north. Reactivation of the bounding and south flank faults in the Cretaceous could have caused oil to migrate from the south, as suggested by oil shows encountered by wells drilled off the field, having the same quality as the main field hydrocarbons but at greater depths. A possible spill-point in the south-west corner, together with evidence of pressure communication between the south flank and the main field, as shown by reservoir pressure depletion due to production, suggest that migration into the main field may still be taking place.

HYDROCARBONS

Two drill-stem tests in the Brent Mica sands of the 211/19-2 discovery well flowed 890 Sm^3/day and 215 Sm^3/day oil on a 32/64-inch choke. Subsequent drill-stem tests with the same choke setting gave approximately 800 Sm^3/day. The first development well, MO1, reached a peak production of 4300 Sm^3/day on a 140/64-inch choke after 17 months on-stream, from the Massive and MSU–Lower Member sands.

RESERVES

The operator's initial, recoverable reserve estimate for the total field is 51×10^6 Sm^3 oil, 2 million tonnes NGL and 3.9×10^9 Sm^3 dissolved gas. The original oil-in-place was estimated at 125.5×10^6 Sm^3 in 1983. Of this 94×10^6 Sm^3 or 75% is within the UK tract and 31.4×10^6 Sm^3 or 25% is within the Norwegian tract. The three lower Brent zones—MSU–Lower Member, Massive Sand and Upper Mica Sand—are estimated to hold 84% of the reserves.

From production start-up at the end of September 1980 until December 1981 the field reached a production rate of 16 000 Sm^3/day. This was the result of unusually fast and efficient drilling of platform wells (the first 24 wells were drilled in 27 months).

Aquifer influx and injected water are the main drive mechanisms. Platform capacity is 25 400 Sm^3/day and production peaked in February 1983 at 20 200 Sm^3/day. Current (May 1985) rates are 15 300 Sm^3/day, with 14 wells on stream, nine injecting water and one injecting gas. Cumulative oil production is 27×10^6 Sm^3 or 52% of estimated ultimate recovery.

Daily gas production rates average 1.5×10^6 Sm^3/day and cumulative produced gas is 2.6×10^9 Sm^3. Water breakthrough occurred one year after production start-up. Aquifer water with some injected water is encroaching through high-permeability layers. The highest rates occur in the western part of the main field, where water cut has now reached 28%. The development scheme is to perforate the lower Brent sands first, in order to reduce the degree of water override in the high-permeability shallower sands, as permeability contrasts are large. The operator predicts that water production rates will surpass oil production by late 1987, and that the field will produce oil economically until 1998. A reservoir simulation study shows that increased water handling capacity and infill drilling will increase the recovery from 41% to 45%.

Oil and NGL off-take is via the Dunlin and Cormorant Fields, then by the Brent pipeline to the Sullom Voe terminal on the Shetland Islands. Gas sales started July 1983 and the gas goes through the Northern Leg Gas Pipeline and FLAGS systems to St Fergus in Scotland.

ACKNOWLEDGEMENTS

Personal communications and contributions, for which I am grateful, came from W. T. Gans, Conoco (UK), London, A. J. Machell, Conoco (UK), London, K. O. Bakken, Mobil Exploration Norway Incorporated, and M. Wiik, Mobil Exploration Norway Incorporated.

REFERENCES

Davies, E.J. and Watts, T.R. 1977. The Murchison oil field, In *Mesozoic Northern North Sea Symposium*, (Norsk Petroleumsforening), Article 15.

Deegan, C.E. and Scull, B.J. (Compilers). 1977. A proposed standard lithostratigraphic nomenclature for the Central and Northern North Sea, *Bulletin Norwegian Petroleum Directorate No. 1; Inst. Geol. Sci. Rept* 77/25.

Handley, R.L. and Gray, I. 1980. The integration of geological and seismic data in the Murchison oilfield by the use of synthetic seismograms and seismic modelling, In *Offshore North Sea—Exploration—Geological and Geophysical Highlights*, Norsk Petroleumsforening, Article E-III/6.

Massie, I. *et al.* 1985. Murchison—A Review of the Reservoir Performance During the First Five Years (in press). 60th Annual Technical Conference and Exhibition of the Society of Petroleum Engineers, Las Vegas, September 22–25, 1985.

Morton, A.C. and Humphreys, B. 1983. The Petrology of the Middle Jurassic sandstones from the Murchison Field, North Sea, *Journal of Petroleum Geology*, **5** (3), 245–260.

Vollset, J. and Doré, A.G. (Eds). 1984. A revised Triassic and Jurassic lithostratigraphic nomenclature for the Norwegian North Sea, *Norwegian Petroleum Directorate. Bull. 3.*

27

Snorre

N. B. Hollander

Saga Petroleum a.s., Box 9, N-1322 Høvik, Norway

The Snorre Field is located on the Tampen Sub-platform in the Norwegian Blocks 34/4 and 34/7. The reservoir section includes 1300 m of strata of the Upper Triassic Lunde Formation and the Upper Triassic to Lower Jurassic Statfjord Formation. These strata are arranged in a series of tilted fault-blocks dipping to the north-west. Their crests have been truncated by the Kimmerian Unconformity, which has resulted in a relatively flat trap structure. The upper part of the Lunde and Statfjord formations contains the best reservoir zones, with porosities of 19–29% and permeabilities up to several Darcies. The most likely estimate of oil-in-place is 476×10^6 Sm^3. The potentially recoverable resources for the entire field are estimated to approximately 110×10^6 Sm^3 of oil.

INTRODUCTION

The Snorre Oil Field lies in the southern part of Block 34/4 and the northern part of Block 34/7, some 240 km north-west of Bergen, 12 km north-north-east of the Statfjord and 20 km north of the Gullfaks oil fields (Fig. 1). The water depth ranges from 295 m in the south-west to 380 m in the north. The field is elongated NNE–SSW and covers an area of approximately 105 km^2. The reservoir section consists of the Lunde Formation of Late Triassic age and the Statfjord Formation of Late Triassic to Early Jurassic age. The current most likely estimate of oil originally in-place is 476×10^6 Sm^3.

Saga Petroleum has chosen 'norrøna' names for hydrocarbon fields, i.e. names from the old Norwegian sagas. The best-known story-teller was Snorre, who wrote the saga of the Norwegian kings.

HISTORY

Block 34/4 was awarded as Production Licence No. 057 on 6 April 1979. The acreage was originally 494 km^2 which was reduced to 246 km^2 after the first relinquishment in 1984. The licence members are:

Company	*Interest* (%)
Saga Petroleum a.s. (operator)	15
Amoco Norway a.s.	10
Amerada Hess (Norway) Limited	5
Den norske stats oljeselskap a.s. (Statoil)	50
Deminex (Norge) A/S	15
Texas Eastern Norway A/S	5

The first well was drilled in 1979 on the north-eastern flank of what is now called the Snorre Field. It encountered oil in poor-quality reservoir sandstone of the middle member of the Lunde Formation of Late Triassic age. An oil column of 78 m was penetrated. The well had a total depth of 2961 mss in strata of possible Middle–?Early Triassic age. A southerly extension of the discovery was established by the second well on the structure (34/4-4), drilled in the winter of 1982–83. It proved a gross oil column of 161 m in sandstones of the

Geology of the Norwegian Oil and Gas Fields © A. M. Spencer *et al.*, eds (Graham & Trotman, 1987) pp. 307–318.

Snorre Field data summary

	Statfjord Formation upper–middle units	Upper Member Lunde Fm, D/E/F units
Trap		
Type	tilted fault-block	tilted fault-block
Depth to crest	2300 mss	2375 mss
Lowest closing contour	2400 mss	2400 mss
Oil–water contact	2595 mss	2561, 2574 or 2595 mss
Oil column	295 m	220 m
Pay zone		
Age	Hettangian–Sinemurian	Early Norian–Rhaetian
Gross thickness (average; range)	70 m; 24–73 m	570 m; 400–600 m
Net/gross ratio (average; range)	0.53; 0.40–0.63	0.49; 0.35–0.68
Porosity (average; range)	25%; 19–29%	24%; 19–29%
Hydrocarbon saturation	89%; 79–94%	78%; 57–86%
Productivity index	70–175 Sm^3/day/bar	30–90 Sm^3/day/bar
Permeability	1.3–2.0 D	320–535 mD
Hydrocarbons		
Oil gravity	0.711–0.764 g/cm^3	0.674–0.701 g/cm^3
Oil type	undersaturated	undersaturated
Bubble-point	90–126 bar	179 bar
Gas/oil ratio (average-range)	58–85 Sm^3/Sm^3	132 Sm^3/Sm^3; 91–137 Sm^3/Sm^3
Formation volume factor (average; range)	1.24; 1.19–1.27	1.40; 1.32–1.52
Formation water		
Salinity	34 000	42 050
Resistivity	0.196 ohm m	0.175 ohm m
Reservoir conditions		
Temperature	90 °C	93 °C
Pressure	383 bar	383 bar
Pressure gradient in reservoir	0.070–0.075 bar/m	0.068 bar/m
Field size		
Area	30 km^2	50 km^2
Oil-in-place*	103×10^6 Sm^3	224×10^6 Sm^3
Recovery factor—oil (average; range)	0.38; 0.32–0.42	0.35; 0.21–0.41
Drive mechanism	water injection	water injection
Recoverable oil	39 (35–46) $\times 10^6$ Sm^3	77 (63–102) $\times$ Sm^3
Production		
Start-up date	1992	1993
Development scheme	tension leg platform (TLP)	TLP or subsea

* Excluding 149×10^6 Sm^3 in Lower Statfjord, Upper Lunde A, B/C, Middle and Lower Lunde members.

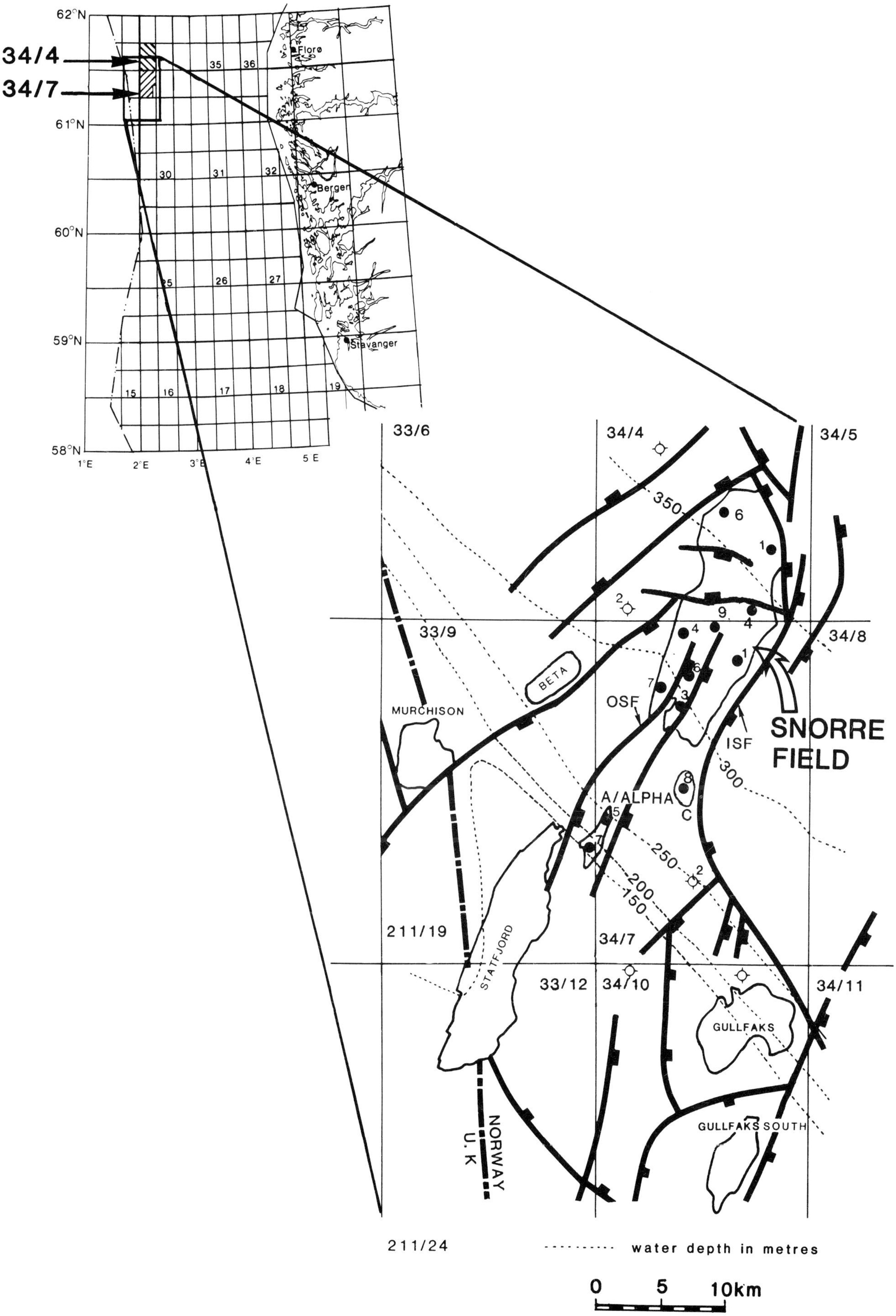

Fig. 1. Location of the Snorre Field, Blocks 34/4 and 34/7, Norway. ISF, Inner Snorre Fault; OSF, Outer Snorre Fault.

upper member of the Lunde Formation near the southern block boundary. The northern extension was tested in early 1986 by Well 34/4-6 which encountered a thin oil column.

Production Licence No. 089, awarded on 9 March 1984, covers 446.8 km^2 in Block 34/7. The participants are:

Company	*Interest (%)*
Saga Petroleum a.s. (operator)	10
Den norske stats oljeselskap a.s. (Statoil)	50
DNO	1
Deminex (NORGE) A/S	4
Elf Aquitaine Norge A/S	8
Esso Norge a.s.	15
Norsk Hydro Produksjon A/S	12

The extension of the 34/4 Lunde discovery into Block 34/7 was confirmed in the spring of 1984 by Well 34/7-1. Well 34/7-3 established the presence of an oil-bearing Late Triassic to Early Jurassic Statfjord Formation reservoir. Subsequently Wells 34/7-4, 6, 7 and 10 have been drilled to define the extent of this oil accumulation. Well 34/7-9 was drilled to assess further the Lunde reservoir.

GEOLOGICAL SETTING

The Snorre Field is located on the north-western, uplifted, rotated and eroded shoulder of the Viking Graben. The area consists of a stable platform, the Tampen Spur, and a less stable sub-platform, the Tampen Sub-platform. This area was a part of a larger sedimentary basin during most of Permian, Triassic and Jurassic times. During Late Jurassic times the Tampen Spur part of the basin was uplifted and formed a regional high through the Early Cretaceous. During Late Cretaceous and Tertiary times the basin subsided and the Snorre Fault Block was buried and became a part of a larger North Sea basin.

The Tampen Sub-platform is divided into a western high and an eastern low area separated by a N–S-trending fault system that is recognized on the eastern side of the Gullfaks Field. This main fault, informally referred to as the Inner Snorre fault, strikes N–S and forms the eastern limit of the Snorre Field. It is a steep, eastward-facing fault or set of faults. The throw is approximately 1500 m down to the east. The main movements took place in Late Jurassic–Early Cretaceous times. The Inner Snorre Fault coincides with the Snorre escarpment which is partly an erosional surface. Thus, the upper part of the fault plane is not known.

The Snorre Field consists of a series of rotated fault-blocks dipping 5–9° to the north-west and located west of the Inner Snorre Fault. The dip is more gentle in the northern, and steeper in the southern part of the field. The major phase of the tilting of the major fault-blocks took place after the deposition of the Brent Group, i.e. in Bathonian times. In the western part of the Snorre Field there is another fault, informally called the Outer Snorre fault, which was first active during Early Triassic times. The movement persisted to Early Cretaceous times.

After Early Cretaceous erosion and reactivation of the Inner Snorre Fault with considerable subsidence of the area to the east, the Snorre Fault Block was a high topographical feature. Some of the faults have, however, been active into Late Cretaceous times.

The Middle Triassic to Pleistocene stratigraphic column is discontinuous in the area, e.g. no Jurassic strata younger than the Early Pliensbachian Amundsen Formation have been encountered in the Snorre Field. The oldest sedimentary rocks penetrated in Blocks 34/4 and 34/7 are the Lower Lunde sandstones and claystones (Fig. 2). These are overlain by marginal marine and fluvial sandstones and shales of the Lunde Formation (middle and upper members). The Statfjord Formation, which is the youngest sandstone reservoir of the Snorre Field, is present in the western and south-western areas. The Cretaceous section consists of claystones with limestone and marls at the base (the Lower Cretaceous Cromer Knoll Group) and calcareous claystones (the Upper Cretaceous Shetland Group). The youngest strata belong to the Rogaland, Hordaland and Nordland groups.

GEOPHYSICS

Blocks 34/4 and 34/7 are well covered with seismic data. Saga's regional data from 1980 and the Tampen Spur Group Shoot from 1981 reach into the area. Early in 1983 Saga, Norsk Hydro and Statoil acquired additional two-dimensional data. Geco–Esso shot a three-dimensional survey over the northern half of Block 34/7 in 1983. The following year PL 089 acquired a three-dimensional survey over the southern half of Block 34/7 and PL 057 acquired another three-dimensional over the southern part of Block 34/4. Thus, the entire Snorre Field is covered by three-dimensional data with a line spacing of 50 m.

Three main horizons have been identified

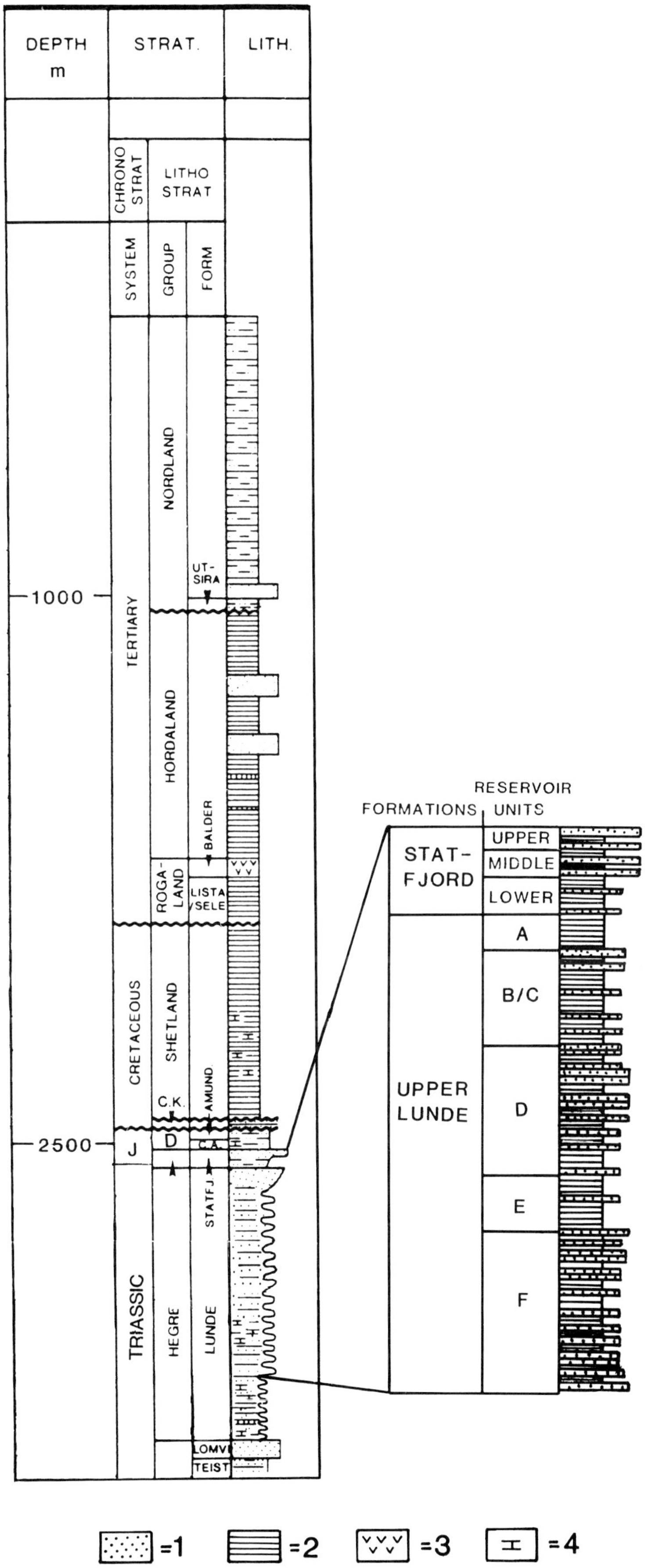

Fig. 2. General stratigraphic column. 1, sandstone; 2, shale; 3, tuff; 4, calcareous. C.K., Cromer Knoll; J, Jurassic; D, Dunlin Group; Amund, Amundsen Formation; C.A., Calcareous Amundsen; Statfj., Statfjord Formation.

seismically and used for mapping the reservoir: base Cretaceous, top Statfjord and top unit B/C. The last represents a predominantly shaly sequence of the upper member of the Lunde Formation. The amplitude of the base Cretaceous reflector strongly depends on the thickness of the overlying Cromer Knoll Group. Typical thicknesses of this marly sequence are 5–15 m. The structure map on the top reservoir (Fig. 3) is a composite map

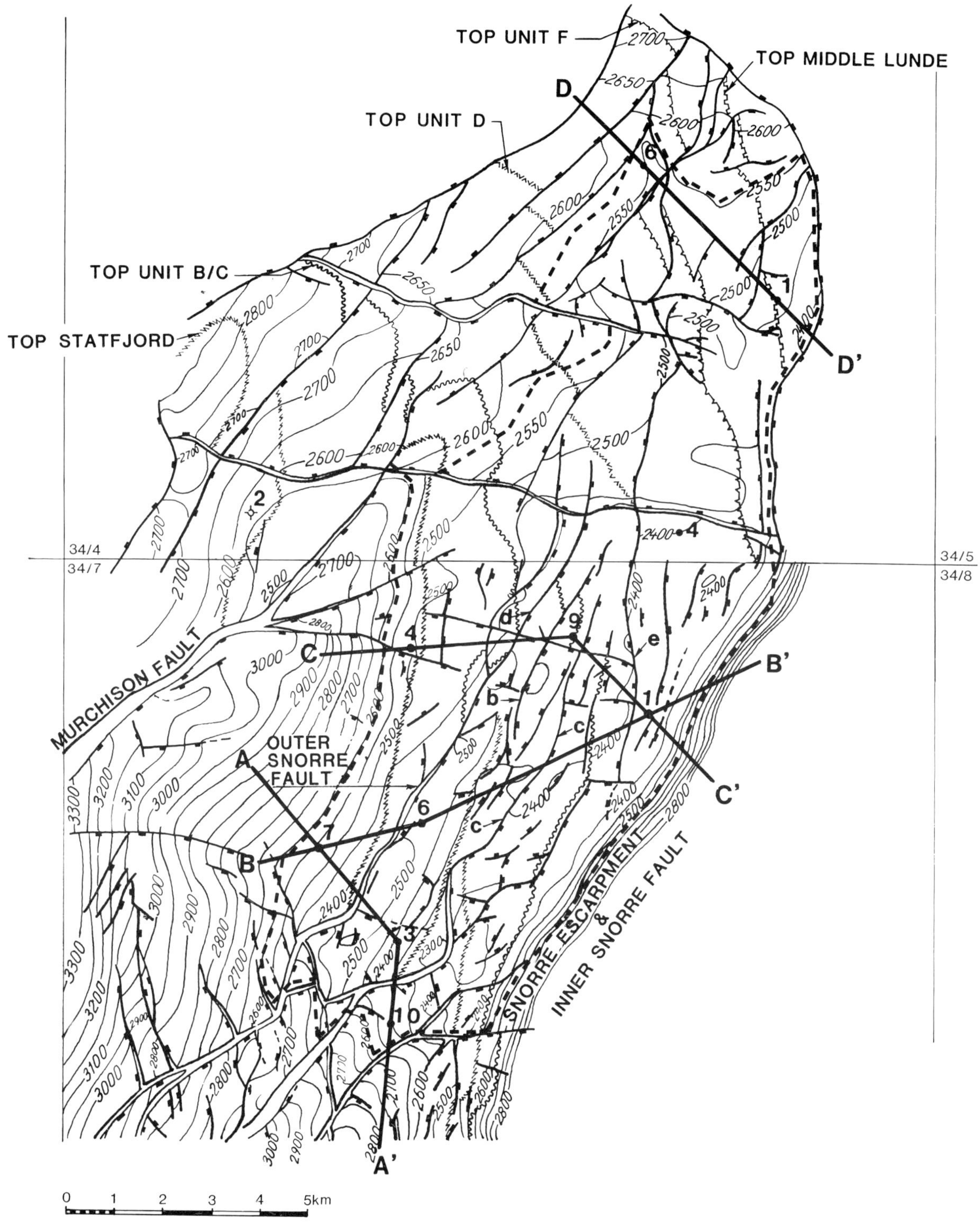

Fig. 3. Top reservoir structure map of the Snorre Field showing the top Statfjord Formation in the south-west and progressively older beds subcropping towards north and east. For convenience the smaller faults have been named b–e to facilitate the correlation with Fig. 4. The lines mark the positions of the profiles in this figure. The subcrop lines show where the top of the Statfjord Formation, B/C, D and F units as well as the middle member of the Lunde Formation have been truncated. The thick, stippled line marks the oil–water contact. Contours in metres subsea.

drawn on the top Statfjord Formation where the top Statfjord is present and on the base Cretaceous Unconformity surface where the top Statfjord Formation has been removed by erosion. Depth conversion has been carried out using a four-zone layer-cake velocity model based on data from the wells. It should be noted that because of the generally flat structure of the Snorre Field the relatively small changes in velocities used for depth conversion can have a significant effect on gross rock-volume estimates.

The top Statfjord is also a composite event. It is created by interference between the positive reflection coefficient at the top of the calcareous Amundsen Member and the negative at the lithologic boundary between the Dunlin Group and the Statfjord Formation.

The seismic pick for the top unit B/C is complex and the uncertainty in this pick is slightly greater than for the Top Statfjord and Base Cretaceous reflectors.

TRAP

The Snorre Field has an elongated shape with a 20 km long main axis stretching in a NNE–SSW direction. The width varies between 5.5 and 8 km. The Statfjord reservoir is located in the S and SW of the field, while the Lunde reservoir is in the E and NE. These strata are arranged in a series of tilted fault-blocks dipping to the north-west. The crest of the faulted reservoir has been truncated by the Kimmerian Unconformity resulting in a relatively flat structure with the highest point at 2300 mss (Fig. 3). The northern part of the field has been more deeply eroded. As a result, progressively older beds subcrop below the Base Cretaceous Unconformity to the east and to the north.

The Snorre Field is dominated by a NE–SW series of faults with throws up to 250 m in the southern part. The faults partly have a listric geometry. These faults play an important role in the distribution of reservoir units by repeating and protecting parts of the section from erosion. In the northern field area, the fault pattern has a similar trend with throws up to 150 m but with displacements dominantly down to the west.

The structural framework is further complicated by the other suites of faults:

(a) faults with small displacements down to the west, which are probably antithetic to the larger faults;
(b) a series of E–W-trending faults, two of which form a graben feature of small displacement in the south of Block 34/4. These are younger than the NE–SW-trending faults;
(c) low-angle, possibly listric faults on the eastern flank of the field.

The trap was first formed in Late Jurassic times and appears to have been tilted to the north during the subsidence of the Møre Basin during Cretaceous–Tertiary times. Seals to the reservoir are provided by the Dunlin shales overlying the Statfjord Formation and by Cretaceous marls and claystones overlying eroded Jurassic and Triassic reservoir rocks at the crest of the structure.

The shale of the upper Lunde unit A may also form a part of the trapping mechanism by separating the Statfjord and Lunde reservoirs.

Three oil–water contacts (OWCs) associated with different gas contents have been interpreted. These are:

1. 2561 mss interpreted in Lunde units E and F in Wells 34/4-4, 34/4-6 and 34/7-1. The oil associated with this contact has a GOR of 91–137 Sm^3/Sm^3.
2. 2574 mss interpreted in Lunde unit B/C in the recently drilled 34/7-9.
3. 2595 mss interpreted in Lunde unit B/C in Well 34/7-3 and in the Statfjord Formation of Wells 34/7-4, 34/7-7 and 34/7-10. The oil associated with this contact has a GOR of 85 Sm^3/Sm^3 to the east of the Outer Snorre fault and a GOR of 58–65 Sm^3/Sm^3 west of the same fault.

As the aquifer for the Statfjord Formation has been encountered only in poor sands in Wells 34/7-4 and 34/7-7, an exact water gradient cannot be defined for the Statfjord Formation. However, oil-down-to values from oil tests, samples and saturation data support the assumption that the oil–water contact at 2595 mss observed for unit B/C in Well 34/7–3, is also valid for the Statfjord Formation across the field.

An understanding of the areal and stratigraphic distribution of the contacts is important for the estimation of the amount of oil-in-place. Data are insufficient to define fully the OWC. It is possible that the variations in the oil properties across the field are consistent with an inclined OWC. However, a step-wise OWC variation across the field has been assumed for oil-in-place calculation. The steps rely for their separation on a combination of fault/offset and stratigraphy.

Thus the 2561 mss contact is restricted to

the northern and eastern and the 2574 mss contact to the western part of the Lunde Formation. The 2595 mss OWC is confined to the Statfjord Formation. The only exception to this model is the 2595 mss OWC interpreted in the unit B/C of 34/7-3. This could be explained by the juxtaposition of unit B/C against the Statfjord Formation across the fault located 750 m to the east of 34/7-3, section AA in Fig. 4.

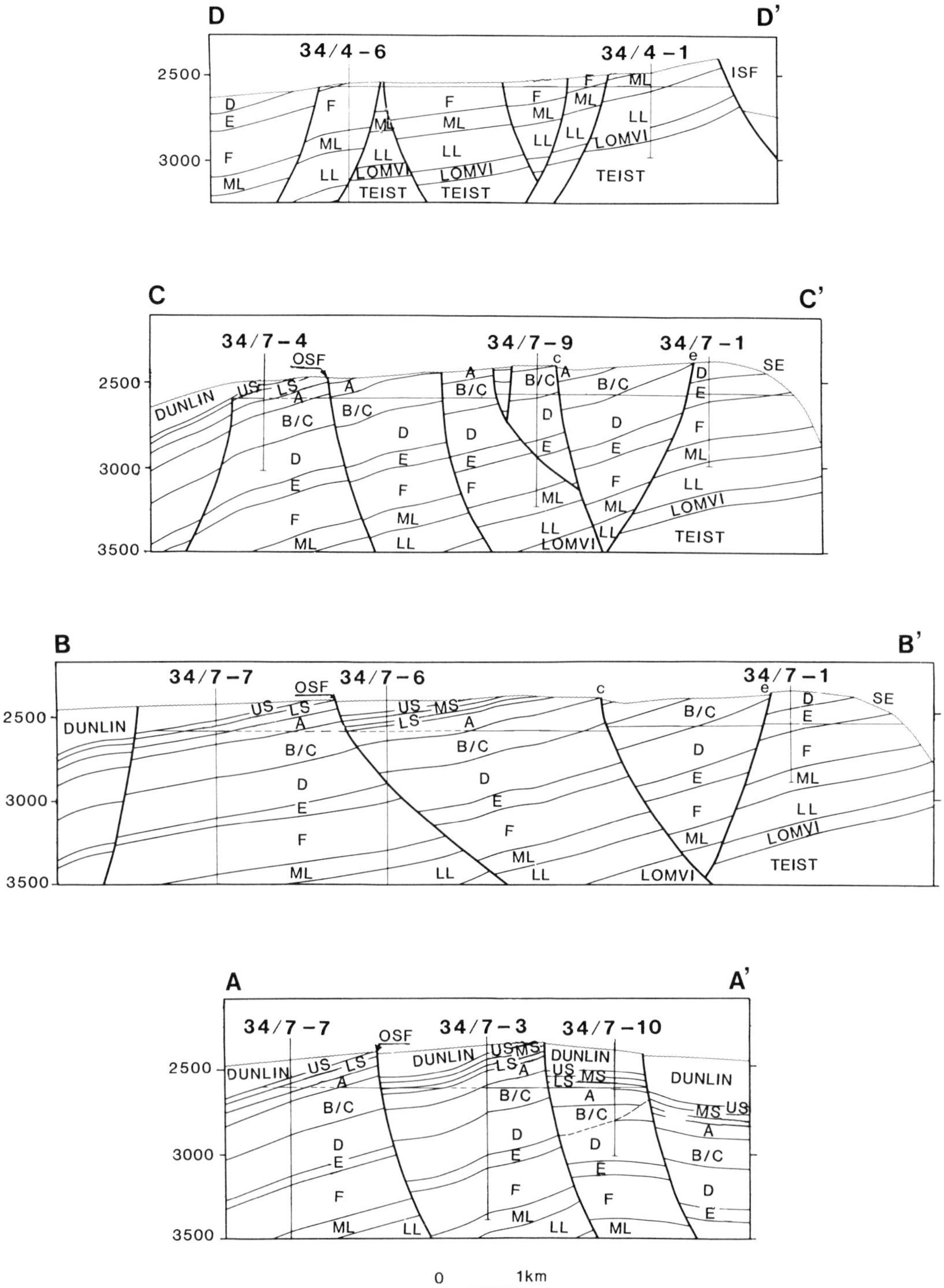

Fig. 4. Structural cross-sections through the Snorre Field below the Base Cretaceous Unconformity. US, MS and LS represent upper, middle and lower units of the Statfjord Formation; A, B/C, D, E and F are the different units in the upper member of the Lunde Formation; ML, middle member, Lunde Formation; LL, lower member, Lunde Formation; OSF, Outer Snorre Fault; ISF, Inner Snorre Fault; SE, Snorre Escarpment, b–e are smaller faults. The horizontal lines mark the OWC. Depths in mss are given on the left side. The locations of the cross-sections are given in Fig. 3.

RESERVOIR

The reservoir section of the Snorre Field includes some 1300 m of strata (Fig. 5) belonging to two major formations. These are:

1. The Lunde Formation of Triassic age, which consists of lower, middle and upper members. The bulk of the oil-in-place in the field is contained in the upper Lunde, which is further subdivided into five reservoir units based on sand content: A, B/C, D, E and F.
2. The Statfjord Formation, which is divided into upper, middle and lower units and contains approximately 26% of the total oil-in-place.

Lower and middle members of the Lunde Formation

These are the oldest reservoir units in the Snorre Field and are restricted to the north-eastern parts, mostly in Block 34/4. The lower Lunde member is a sequence of interbedded fine-grained sandstones and silty claystones which has been interpreted as representing either fluvial or estuarine channel sandstones. Fossils suggest a Ladinian–Carnian age. The total section thins to the south-west from 217 m in 34/4-1 to 154 m in 34/7-6. Reservoir quality is moderate with a net-to-gross sandstone ratio of 0.32 and porosities of 21%. The lower member of the Lunde Formation has not been tested in an oil-bearing location.

The middle member consists of interbedded, fine-grained sandstones and grey shales of Norian age, deposited in a fluvial to marginal-marine environment. The total section is 109–146 m thick over the Snorre area. Reservoir quality is generally moderate with a net-to-gross sandstone ratio of 0·28 and a porosity of 18%. The middle member was tested in an oil-bearing location in Well 34/4-1.

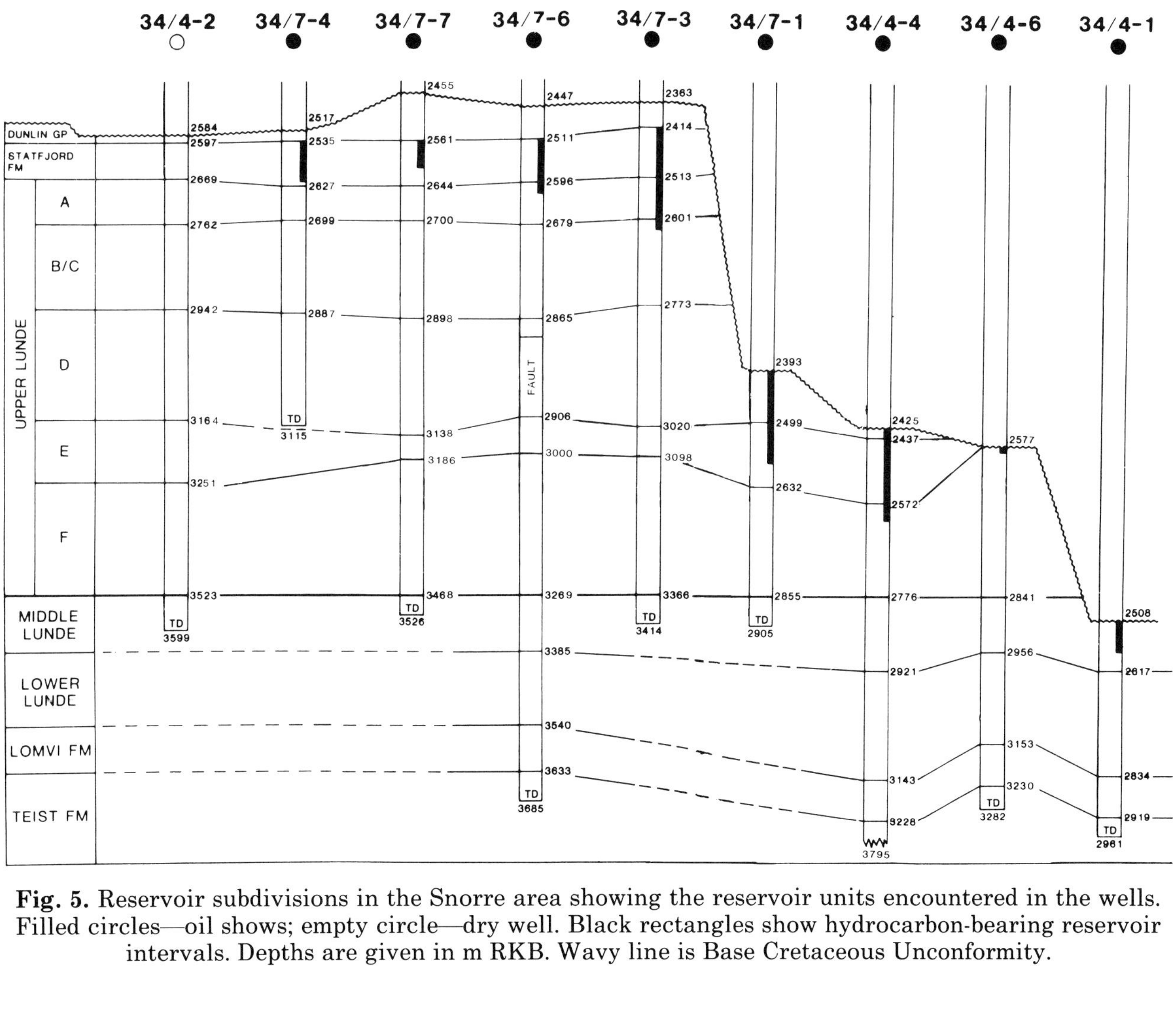

Fig. 5. Reservoir subdivisions in the Snorre area showing the reservoir units encountered in the wells. Filled circles—oil shows; empty circle—dry well. Black rectangles show hydrocarbon-bearing reservoir intervals. Depths are given in m RKB. Wavy line is Base Cretaceous Unconformity.

Upper member of the Lunde Formation

This consists of a repetitive sequence of fluvial channel sandstones and non-channelized sands and silty overbank shales of Norian to Rhaetian age. The geological model for the upper member of the Lunde Formation is based on a depositional environment which was dominated by rivers flowing across a low relief semi-arid alluvial plain (Fig. 6). Regional seismic data show a general thinning of the upper member to the west. However, over the Snorre Field area this member has a relatively constant thickness of 825–850 metres.

The differing properties of the fluvial reservoir sandstones can be explained in terms of their energy of deposition. A range of fluvial sequences is recognized varying from higher- to lower-energy channel and non-channel settings. The higher-energy channel sandstones of better reservoir quality typically have a basal, frequently calcareous cemented, intraformational conglomerate. This is usually less than one metre thick and grades rapidly into massive, fine- to medium-grained, well-sorted sandstones with good porosity (25%) and permeability (200–3000 mD). Clay content is generally low (< 5%). These channel sandstones are usually 2.5–3 m thick and occur as single sandstones or as stacked bodies up to 15 m thick. As the energy of deposition decreased the sandstones became finer grained, laminated and clay rich with a corresponding decrease in porosity (to 21%) and permeability (to 50–200 mD). These sandstones are interbedded with non-channelized sands and red–brown, silty overbank shales. Sandstones within the overbank shales are thinner, less than 2 m thick, and have reservoir properties similar to the upper portions of the channel sands.

The upper member of the Lunde Formation has been subdivided into five units based on varying sandstone content through the section:

- Unit A (57–90 m). This uppermost unit consists of red–brown shales and siltstones with rare, low-quality sandstones. Net-to-gross ratio averages 0.03.
- Unit B/C (170–200 m). Reservoir quality is moderate and variable with average net-to-gross ratio of 0.25 and porosity of 24%. The unit has been encountered oil bearing in 34/7-3 and 34/7-10.
- Unit D (220–250 m). Good reservoir properties; average net-to-gross ratio 0.48 and

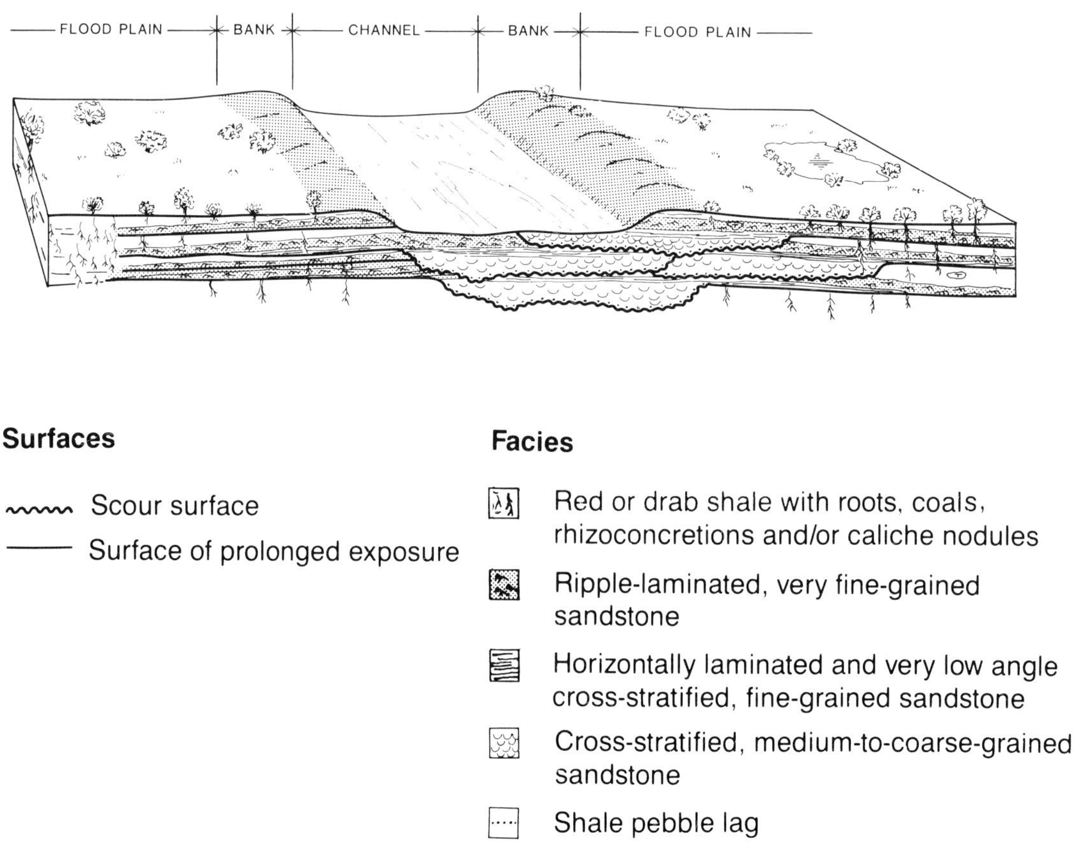

Fig. 6. Depositional model for Statfjord and Upper Lunde formations, Snorre Field.

porosity 24%. Well 34/7-1 encountered an oil-bearing but eroded unit D 106 m thick.

- Unit E (50–135 m). Moderate reservoir quality; average net-to-gross ratio 0.42 and porosity 22%. This unit, which has a high proportion of lower-energy channels, has been encountered oil bearing in 34/7-1 and 34/4-4.
- Unit F (200–240 m). Good reservoir quality; average net-to-gross ratio 0.61 and porosity 25%, with a large proportion of stacked, higher-energy sands. This is the most consistent unit in the Snorre area. A thickness of 16 m has been encountered oil bearing in 34/4-4 and of 11 m in 34/4-6.

Statfjord Formation

The Statfjord Formation is the youngest reservoir of the Snorre Field and is 85–100 m thick. It contains approximately 26% of the total oil-in-place, almost all of which is found in Block 34/7. The Statfjord Formation is transitional into the underlying shales of unit A. The Statfjord Formation consists of sequences of fluvial-channel sandstone and overbank shales, and has for oil-in-place calculations been subdivided into three units based on differing sandstone types and sandstone contents. The upper and middle units are equivalent to the Eiriksson Member of the Statfjord Field, while the lower unit and unit A of the Lunde Formation are considered to be equivalent to the Raude Member.

The lower unit of the Statfjord Formation (25–69 m) consists of fine- to medium-grained channel sandstones, very similar to those in the Lunde Formation, interbedded with mottled red and grey silty shales. The sandstones, which increase in abundance upwards, are generally thin (1–3 m) although they may occur as stacked bodies up to 8 m thick. The reservoir quality is generally poor to moderate (net-to-gross ratio average 0.18; porosity average 20.8%) and although permeability in the thicker sands is good, test results indicate moderate reservoir continuity compared to the upper unit.

The upper unit of the Statfjord Formation (24–73 m) consists of coarse-grained, poorly sorted, stacked fluvial sandstones, interpreted as low-sinuosity braided-stream deposits, interbedded with grey or red silty shales and occasional coals. Individual channels average three metres in thickness, but tend to occur as stacked sand-bodies up to 10 m thick. Reservoir quality is excellent. In Wells 34/7-3, 34/7-6 and 34/7-10 a middle reservoir unit is also recognized. This unit has slightly lower reservoir quality than the upper one. In the western fault-block a middle unit has not been identified.

Wells 34/7-3 and 34/7-6 encountered a thicker section of the upper unit of the Statfjord Formation than Wells 34/7-4 and 34/7-7, thus suggesting growth across the Outer Snorre fault. The main period of faulting, which reached a maximum during Late Jurassic times, probably started during the deposition of the Statfjord Formation. These early fault movements could have caused the rivers to be preferentially located in topographically lower areas on the downthrown sides of the faults.

SOURCE

The most important source-rock for oil and gas in the Tampen Spur area is the Upper Jurassic Draupne Formation. The Heather and the Lower Jurassic Drake formations are poorer source-rocks. All these formations are absent over the Snorre Field due to erosion on the structural highs, but increase in thickness towards the Møre Basin and Viking Graben.

Modelling studies of burial history and maturity evolution tentatively indicate that oil generation in the Draupne Formation started in the Møre Basin about 80 million years ago and in the basin immediately north-east of Snorre 55 million years ago. Gas generation became significant about 55 and 15 million years ago respectively.

The Snorre structure could have been filled by hydrocarbons migrating from two areas: from the Statfjord Field to the west or from the above-mentioned areas north and east of the structure.

An observed GOR gradient across the field, decreasing from a high of 164 Sm^3/Sm^3 in the north-east to 58 Sm^3/Sm^3 in the south-west could be caused by a late migration of oil with a high GOR from the deep drainage basin to the east of the Snorre Field.

HYDROCARBONS

A total of ten wells have tested abnormally pressured, undersaturated oil in the Snorre Field. Five of these wells have penetrated the Statfjord Formation. Permeabilities vary from a few millidarcies up to several darcies in the good-quality channel sands, and short-term production tests indicate high productivity. The pressure build-ups for the Lunde Formation tests indicate boundaries in the vicinity of the wells which may affect flow through the reservoir. The production tests have resulted in excellent flow rates: up to 1750 Sm^3/day on 14.3-mm choke in the upper unit of the Statfjord Formation and 1750 Sm^3/day on a 17.5-mm choke in the Upper Lunde Formation.

The oils in the Jurassic and Triassic reservoirs in the Snorre Field are geochemically similar. However, the oil density, formation volume factor and gas–oil ratio vary systematically across the field. The lower-density, most gas-rich oil occurs in the north-east, and there is a gradual increase to the heaviest oil with the lowest GOR in the south-west. Furthermore, there is a nearly linear relationship between the formation volume factor and oil density in the various wells. Generally, greater variations in the fluid parameters are observed in the Lunde Formation than in the Statfjord Formation.

RESOURCES

Estimates of oil originally in-place vary from 408×10^6 to 557×10^6 Sm^3 with a most likely value of 476×10^6 Sm^3. Approximately 125×10^6 Sm^3 occur in the Statfjord Formation and 350×10^6 Sm^3 in the Lunde Formation. The estimated volume of associated gas-in-place for the most likely case is 51×10^9 Sm^3.

The recovery efficiency has been estimated from reservoir simulation studies. The potential recovery efficiencies range from 42% in the upper to 12% in the lower Statfjord unit. The upper members of the Lunde Formation range from 17% in unit B/C to 41% in unit F. Middle and lower members are expected to be 14 and 20% respectively.

The potential recoverable resources for the entire field are estimated to be approximately 140×10^6 Sm^3. The Statfjord Formation in the southern portion of the field and the upper Lunde D, E and F units in the central part of the field contain 86% of these resources. Considerations of platforms and subsea configurations and economic criteria could limit the produceable resources to approximately 110×10^6 Sm^3 of oil.

ACKNOWLEDGEMENTS

During the preparation of this work the author has drawn freely on the work and experience of his colleagues in the Snorre Team Organization and Saga's geological research team. The author greatly acknowledges H. O. Augedal, K. Bøyum, G. W. Diesen, F. Eide, R. Gabrielsen, K. Jorde, R. Knarud, E. S. Mo, B. G. Pryce, G. R. Ramsayer, N. Raestad, K. O. Stanley, C. P. Stockbridge and S. Vågene for their contributions to the results on which this paper is based.

28

Statfjord

J. D. Roberts, A. S. Mathieson and J. M. Hampson

Mobil Exploration, Norway Incorporated, Statfjord Divisjonen, PO Box 620, 4001 Stavanger, Norway

The Statfjord Field, the largest oilfield in the North Sea, is located on the Norway/UK boundary between 61°10′ and 61°23′ N. It lies within the East Shetland Basin to the south-west of the Tampen Spur. As in many of the fields within the East Shetland Basin, the trap comprises a tilted fault-block of Jurassic–Triassic rocks sealed by unconformable Upper Jurassic and Cretaceous shales.

Most of the hydrocarbon reserves are contained in two separate sandstone reservoirs: the deltaic Middle Jurassic Brent Group and the fluviatile Upper Triassic–Lower Jurassic Statfjord Formation. Both Brent and Statfjord reservoirs exhibit excellent reservoir properties with porosities ranging from 20 to 35% and permeabilities ranging up to several Darcies. The two reservoirs are not in communication and are being developed separately. A third hydrocarbon accumulation is found within the Lower Jurassic marine-shelf sandstones of the Dunlin Group Cook Formation.

Recoverable reserves for Statfjord Field are currently estimated at 482×10^6 Sm^3 of oil and 88×10^9 Sm^3 of gas. The oil is undersaturated and no natural gas-cap is present. Production from the Field commenced in 1979 and to date (1 July 1986) 79 of the planned 97 development wells have been drilled.

INTRODUCTION

The Statfjord Field is located approximately 220 km north-west of Bergen, Norway. The field lies primarily within Norwegian Petroleum Production Licence 037, Blocks 33/9 and 33/12, but extends across the Norway/UK boundary into UK offshore Licence 104, Blocks 211/24 and 211/25 (Fig. 1).

Water depths across the field average 145 m, increasing slightly from south to north. The field comprises a westerly tilted, NE-trending fault-block, which suffered severe faulting and erosion during Late Jurassic and Early Cretaceous times. The field is 24 km in length and averages 4 km in width, with a productive area of around 80 km^2.

The etymological origin of the field name is not clearly defined.

HISTORY

Norwegian Petroleum Production Licence 037 was granted by a special award in August 1973, to a consortium of oil companies including Den norske stats oljeselskap (Statoil), Mobil (designated the field operator), Conoco Norway, Esso Exploration Norway, Norske Shell, Saga, Amoco Norway, Amerada Hess and Texas Eastern. The UK Blocks 211/24 and 211/25 had already been licensed to Conoco North Sea Incorporated (operator), Gulf Oil Corporation and BNOC (Exploration) Limited (later changed to Britoil). Table 1 outlines the current ownership of the Statfjord Field. The Field Equity was last redetermined in 1979, which resulted in a tract participation of 84.09% for the Norwegian Partners and 15.91% for the UK partners. At

Geology of the Norwegian Oil and Gas Fields © A. M. Spencer *et al.*, eds (Graham & Trotman, 1987) pp. 319–340.

Statfjord Field data summary

	Brent reservoir	Dunlin reservoir	Statfjord reservoir
Trap			
Type	Structural/stratigraphic	Structural/stratigraphic	Structural/stratigraphic
Depth to crest	2360 mss	2475 mss	2575 mss
Oil–water contact	2585.8 mss	2600 mss (approximate)	2805.7 mss
Oil Column	225.8 m	110.8 m	230.7 m
Pay zone			
Formation	Brent sandstone	Cook sandstone (Dunlin II)	Statfjord sandstone
Age	Middle Jurassic	Lower Jurassic	Upper Trias–Lower Jurassic
Gross thickness (average)	156 m	70 m (estimated)	123 m
Net/gross ratio (average)	0.75	0.40 (estimated)	0.58
Porosity (average)	27%	23% (estimated)	22%
Hydrocarbon saturation	84%	40% (estimated)	64%
Permeability (average)	2500 mD	150 mD (10–300 mD)	470 mD
Productivity index	230.5 m^3/day/bar	2–65 m^3/day/bar	207.5 m^3/day/bar
Hydrocarbons			
Oil gravity	0.82 g/cm^3	0.86 g/cm^3	0.83 g/cm^3
Oil type	low sulphur crude	low sulphur crude	low sulphur crude
Gas gravity	0.76	0.75 (estimated)	0.78
Bubble-point	275 bar	248 bar	196 bar
Gas/oil ratio	190 Sm^3/Sm^3	140 Sm^3/Sm^3	155 Sm^3/Sm^3
Formation volume factor	1.58	1.444	1.54
Formation water			
Salinity—total chlorides	14 000 ppm NaCl equivalents	—	13 600 ppm NaCl equivalents
Resistivity	0.11 ohm m at 77 °C	—	0.128 ohm m at 83 °C
Reservoir conditions			
Temperature	89 °C	92 °C	100 °C
Pressure	383 bar at 2469 mss	383 bar at 2469 mss	404 bar at 2701 mss
Pressure gradient in reservoir	0.064 bar/m	0.064 bar/m	0.065 bar/m
Field size			
Area	80 km^2	26 km^2 (estimated)	35 km^2
Gross rock volume	6.65×10^9 m^3	1.4×10^9 m^3 (estimated)	4.2×10^9 m^3
Hydrocarbon pore volume	1130×10^6 m^3	33×10^6 m^3 (estimated)	343×10^6 m^3
Original oil-in-place	787×10^6 Sm^3	23×10^6 Sm^3 (estimated)	226×10^6 Sm^3
Drive mechanism	water injection and aquifer	solution gas/water injection	gas injection and aquifer, later water injection
Production			
Start-up date	November 1979	1993 (estimated)	November 1979
Development scheme	3 platforms—SPM tanker loading	recompletion of existing field wells	3 platforms—SPM tanker loading
Cumulative production to 1/7/86	oil—83.7×10^6 Sm^3 gas—19.0×10^9 Sm^3	— —	oil—45.7×10^6 Sm^3 gas—10.8×10^9 Sm^3

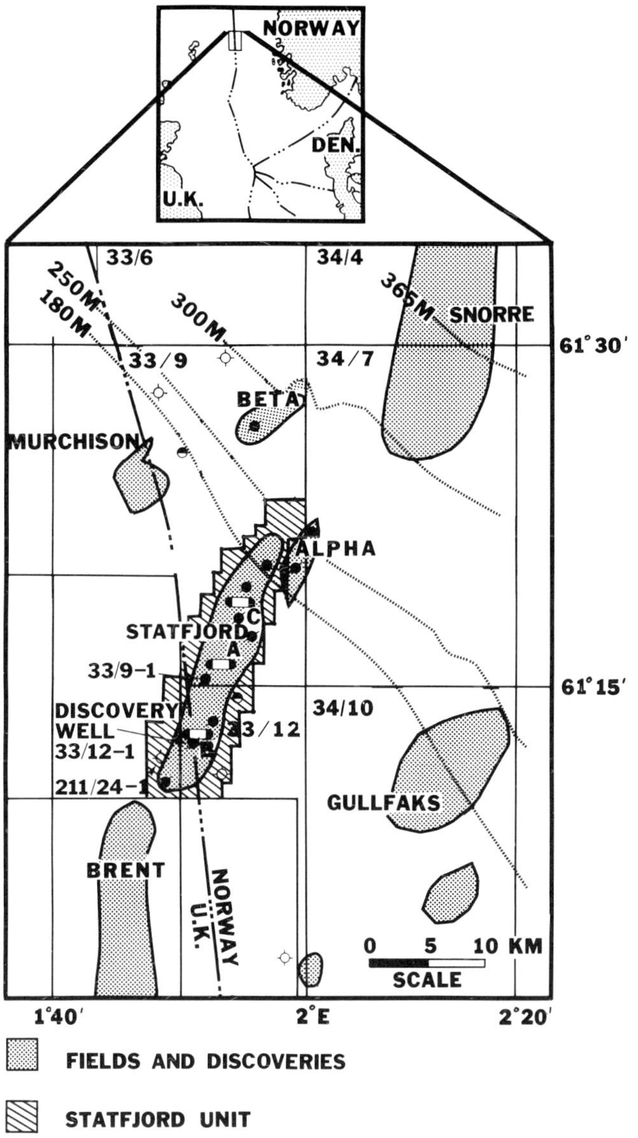

Fig. 1. Statfjord Field location map.

the time of writing, the tract participation is being redetermined. This latest redetermination should be completed by late 1987.

By the time the Statfjord Field was discovered, a number of major hydrocarbon finds had already been made in neighbouring UK blocks within the prolific East Shetland Basin. The first of these discoveries was the Brent Field in 1971, followed in rapid succession by the Thistle, Dunlin, Ninian and Hutton fields.

The prospect was originally mapped as a NE–SW structural anomaly at Base Cretaceous level which lay on-trend with the established Brent Field, 20 km to the south-west in the UK sector. The first well on the Statfjord Field structure was 211/24-1, drilled in the UK sector in 1973. This well encountered reservoir quality, water-wet Brent and Statfjord sandstones separated by Lower Jurassic marine shales of the Dunlin Group. Well 33/12-1 was spudded in December 1973 to the west of the structural crest in order to penetrate the fullest possible prospective section beneath the Base Cretaceous unconformity. The entire 161 m of Brent section was oil bearing but the Lower Jurassic Statfjord sandstones were waterbearing. The well was plugged and abandoned in April 1974 after testing to the capacity of the separator at 1590 Sm^3/day.

Well 33/9-1 was drilled in 1974 in a structurally lower position than the 33/12-1 Well and established an oil–water contact in the Brent section at 2585 mss. This determination of the oil–water contact confirmed the isolation of the Statfjord Field Brent Group oil accumulation from that of the Brent Field. After

TABLE 1 The Statfjord unit owners

Names of participating companies	*Participating Interest (%)*[a]
Mobil Exploration Norway Incorporated (Operator)[b]	12.614 00
Den norske stats oljeselskap AS	42.046 61
Esso Exploration and Production Norway Incorporated	8.409 32
AS Norske Shell	8.409 32
Conoco Norway Incorporated	8.409 32
Conoco (UK) Limited	5.302 26
Britoil P.l.c.	5.302 26
Gulf Oil Corporation	2.651 13
Gulf (UK) Offshore Investments Limited	2.651 13
Saga Petroleum AS	1.576 74
Amerada Petroleum Corporation of Norway	0.875 97
Amoco Norway Oil Company	0.875 97
Texas Eastern Norwegian Incorporated	0.875 97
	100.000 00

[a] Determined in 1979, currently being redetermined.
[b] On January 1 1987 Statoil replaced Mobil as operator.

testing the 33/9-1 Well, a declaration of commerciality was made in August 1974 and the accumulation was named the Statfjord Field. The 33/12-2 Well, also drilled in 1974, established oil-bearing Statfjord Formation but did not encounter the oil–water contact which occurred within a lower Statfjord Formation shale. Well 211/24-4, drilled in 1975, penetrated 56 m of Brent Group above the oil–water contact and thereby established the southern extension of Statfjord Field hydrocarbons into the UK sector. The last of 11 exploration/appraisal wells on the Statfjord structure, Well 33/9-9, was drilled in 1977.

The Statfjord Unit Owners initiated a development programme utilizing three Condeep concrete gravity-base platforms, each with a Single Point Mooring for tanker loading. Figure 2 shows the configuration of the field's platforms, articulated loading points and pipelines.

Development took place in three phases and, following agreement on field unitization, the first phase of development began with oil production from the 'A' platform in November 1979. The second development phase was initiated with the first production of oil from the 'B' platform in November 1982. The final production phase saw oil production commencing on the 'C' platform in July 1985.

By July 1986, 79 of the planned 97 development wells had been drilled and cumulative production had reached 129.4×10^6 Sm^3 of oil and 29.8×10^9 Sm^3 of gas.

GEOLOGICAL SETTING

The Statfjord Field lies in the central part of the East Shetland Basin, as defined by Rønnevik *et al.* (1975) and Hamar and Hjelle (1984). The basin is bounded by the East Shetland Platform to the west and south-west, the Møre Basin and Tampen Spur to the north and north-east and the Viking Graben to the south and south-east (Fig. 3).

The East Shetland Basin owes its origin to the extensional stress regime, initiated in Late Permian/Early Triassic times, which resulted in the formation of the Viking–Central Graben system by the end of Jurassic times. Eustatic sea-level changes, rifting, relative uplift, fault-block rotation and differential subsidence dominated basin development and depositional regimes during Late Triassic, Jurassic and Early Cretaceous times (Fig. 4). Sediments deposited during this time contain the excellent reservoir rocks of the Statfjord Formation and Brent Group and the prolific source-rocks of the Draupne Formation (Fig. 5). The establishment of the Viking Graben tilted fault-block regime was essen-

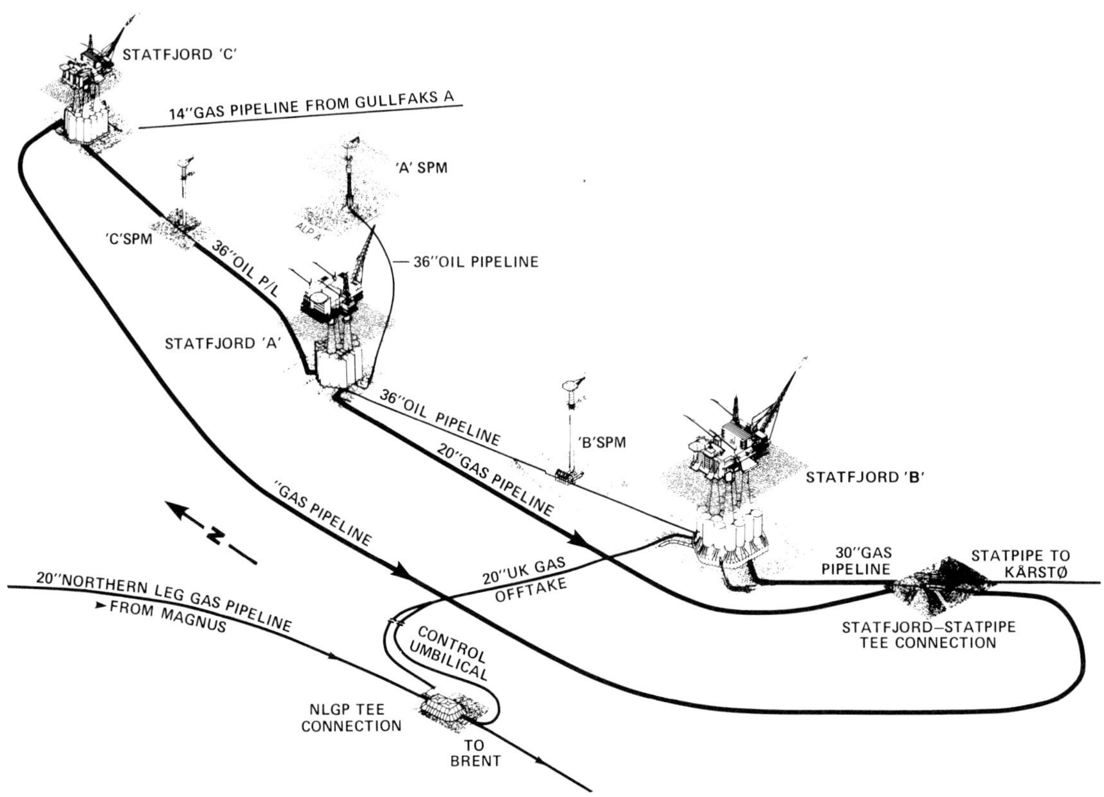

Fig. 2. Platforms, loading points and pipeline configuration.

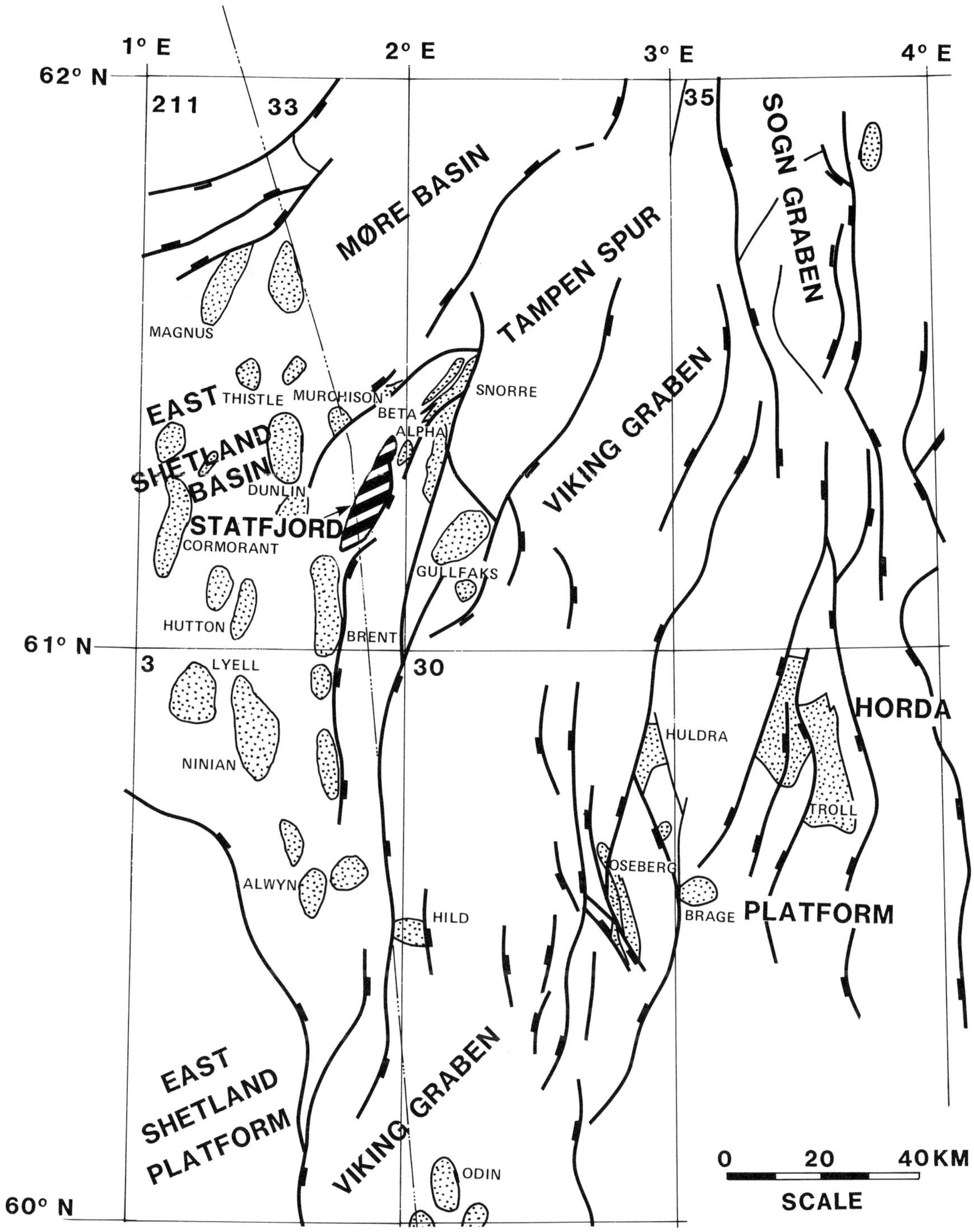

Fig. 3. Statfjord Field—structural setting.

tially completed by the end of Jurassic times in this area.

Triassic–Hegre Group

The oldest strata encountered in the Statfjord Field are in Well 33/12-5 which reached total depth in sandstones and shales of Triassic Carnian age.

The most striking feature of Triassic deposition in the area was the extensive development of thick red-bed sequences. These non-marine sequences range from coarse conglomerates to basin-filling fluvial sandstone and lacustrine deposits. In the Statfjord area, this period of deposition is represented by siliciclastics comprising the Hegre Group. A complete Hegre section has not been penetrated in the Statfjord Field. Well 33/12-2, however, penetrated 1403 m of the group before reaching TD at 4329 mss.

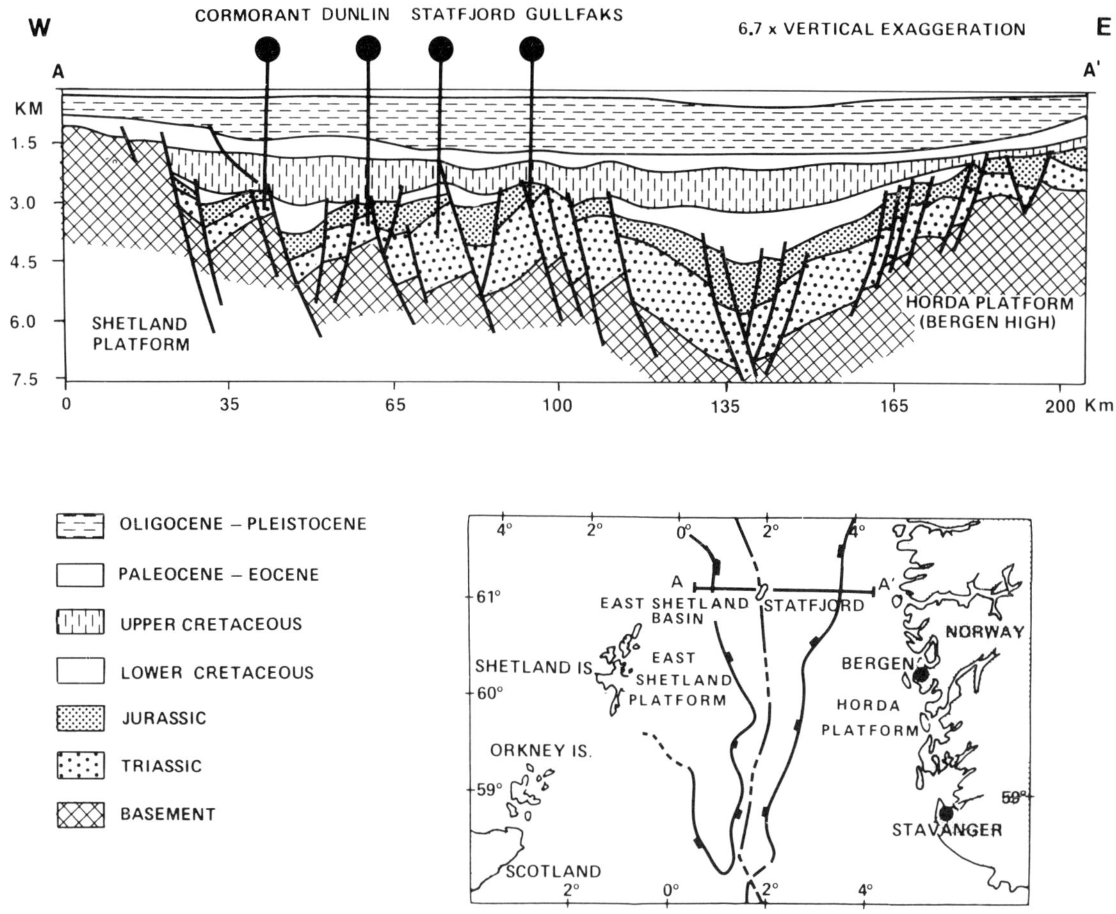

Fig. 4. Generalized cross-section of the northern North Sea (modified after Kirk, 1980).

Triassic/Lower Jurassic–Statfjord Formation

The Statfjord Formation spans the Late Triassic/Early Jurassic boundary, ranging in age from Rhaetian to Sinemurian. It lies with apparent conformity on the underlying Hegre Group continental red-bed deposits. In the Statfjord Field, the Statfjord Formation has an average gross thickness of just over 200 m and shows a slight thinning up-structure. The formation comprises a sequence of sandstones, siltstones and shales, with calichés. A widely distributed, texturally mature, clean sandstone, termed the Nansen Member of the Statfjord Formation, marks the transition into the marine shales of the overlying Dunlin Group.

Late Triassic topography was characterized by extremely low relief. Early Cimmerian movements initiated the transitional phase from Triassic continental deposition to the shallow-marine conditions of the Jurassic. One such Early Cimmerian movement resulted in a major regression during latest Triassic (Rhaetian) time. It affected large areas of the North-west European rift system, accentuating positive structural elements and shedding clastic sediments into intervening lows. The Statfjord Formation represents such a phase of sedimentation and was deposited on a wide alluvial braid-plain and mud-plain. A marked change from an arid to a more humid climate took place during the deposition of the Statfjord Formation, resulting in the increasing sand content of the formation from base to top.

Continued subsidence, coupled with rising sea level, resulted in a marine transgression across the area and the deposition of the Nansen Member. Slight thinning of the formation up-structure indicates early movement on the field-bounding fault system and localized tectonic control of sedimentation, with thickening into structurally low areas.

Lower Jurassic–Dunlin Group

The Dunlin Group conformably overlies the Statfjord Formation. It comprises a series of shallow-marine shales and sandstones ranging in age from Pliensbachian to Toarcian,

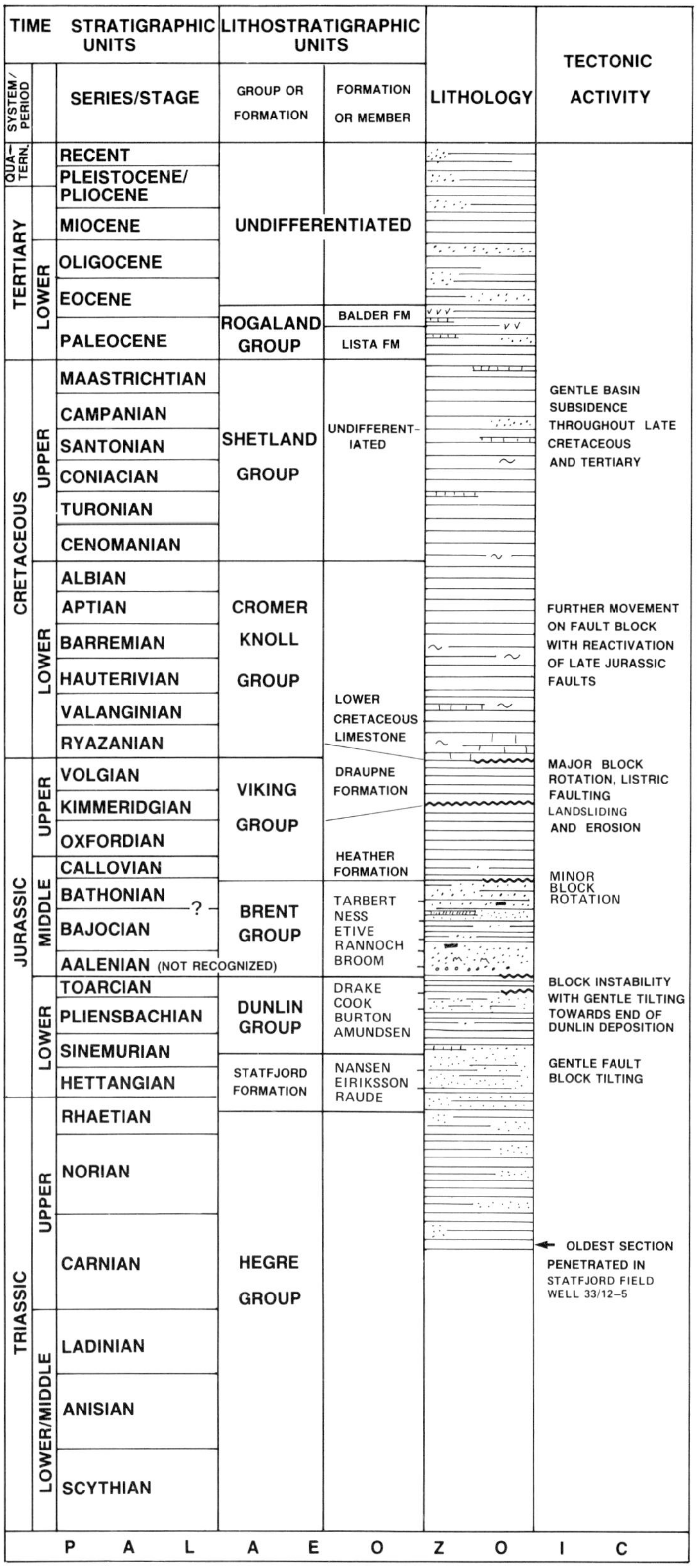

Fig. 5. Simplified lithostratigraphic column in the Statfjord Field.

and is subdivided into four formations, all present in the Statfjord Field. From oldest to youngest, these are the Amundsen, Burton, Cook and Drake.

The Dunlin Group has an average depositional thickness of around 250 m in the Statfjord Field area. However, post-depositional faulting has thinned the unit due to tectonic stretching over a wide area on the eastern part of the field.

Marine conditions, initiated during the deposition of the upper part of the Statfjord

Formation, prevailed throughout the deposition of the Dunlin Group. The Amundsen Formation is predominantly calcareous shale with local arenaceous content, especially in the south-western portion of the field, where a calcareous sand unit ranges up to 15 m in thickness. A Late Pliensbachian unconformity occurs close to the base of the Cook Formation, indicating a rapid fall in sea level. Rejuvenation of source-areas resulted from this eustatic sea-level change and, coupled with local tectonism, led to the influx of Cook Formation sands as shallow-marine, distal, deltaic sequences and tidal sand-bars. Winnowing of fines with a resultant improvement of sand quality is apparent over the structurally high areas of the Statfjord Field. The Cook Formation sands are hydrocarbon bearing in the Statfjord Field but are considered only as of secondary importance to the major Brent Group and Statfjord Formation hydrocarbon accumulations.

Unconformities and hiatuses, evidenced by missing section and phosphatic oolite beds, also occur within the Cook and Drake Formations due to relative changes of sea level. An eustatic sea-level rise, combined with continued subsidence of the Viking Graben, lead to a reduction in clastic input and the accumulation of the marine shales of the Drake Formation. The upper part of the Drake Formation represents prodelta clays, and this sedimentary regime continued with the deposition of the overlying Brent Group deltaic sediments.

Middle/Upper Jurassic

During Middle Jurassic times regional doming occurred. It was a precursor of the incipient break-up of the area into the Viking and Central grabens and the inner Moray Firth basin, and was centred in the outer Moray Firth area of the North Sea. This marked the major phase of Middle Jurassic regression, an episode which began in Late Toarcian times. Erosion accompanied the doming, and shed sediment northwards into the Viking Graben where sedimentation took place in a fluvio-deltaic setting. Middle and Upper Jurassic sedimentation in the Statfjord Field is represented by the Brent and Viking Groups, respectively.

Brent Group

The Brent Group comprises an areally extensive complex of deltaic strata representing a regressive–transgressive sequence. It ranges in age from Late Toarcian to Mid-Bathonian. The group was deposited over most of the northern Viking Graben area and ranges from a normal pre-erosional average thickness of 190 m in the Statfjord Field area to over 300 m in the central parts of the Graben. The Brent Group in the Statfjord Field is discussed in detail on pp. 332–335.

Viking Group

The Viking Group overlies the Brent Group and ranges in age from Mid-Bathonian to Early Ryazanian and comprises primarily marine shales laid down in the slowly subsiding Viking Graben. The group is subdivided into the Heather and Draupne formations.

Extensional tectonic activity continued within the East Shetland Basin throughout the deposition of the Viking Group. Uplift, rotation and faulting occurred in conjunction with differential subsidence, and resulted in considerable thickness variations within the Viking Group strata. In addition, erosion of the pre-Viking Group strata is apparent on the crestal portion of the major Statfjord fault-block. Continued rotation, uplift and erosion of the fault-blocks is evidenced by a westward migration of the structural palaeocrest away from the major boundary faults during Late Jurassic times. All these features are evidenced by missing section, stratigraphic relationships and repeat sections.

Heather Formation

The lowermost Heather Formation conformably overlies the Brent Group in the downflank areas of the field while in structurally high areas the Callovian Heather Formation unconformably overlies the Bajocian Ness Formation. Regionally, sedimentation of the Heather Formation continued until Middle to Late Oxfordian times. The uppermost Heather is absent over the crestal areas of the Statfjord Field and is relatively thin in the western-most water injector wells.

Draupne Formation

Deposition of the Draupne Formation commenced during Late Oxfordian and continued until Early Ryazanian times. The high organic content and predominantly planktonic microfaunal assemblages are indicative of deposition in a restricted marine basin, with anaerobic conditions at the sediment–water interface.

Differential subsidence continued in the

Viking Graben throughout the deposition of the Draupne Formation. This subsidence influenced sedimentation and caused significant lateral thickness variations apparent in the Statfjord Field. The Draupne Formation is generally present, but is very thin, across the crest of the field where it onlaps and unconformably overlies older strata.

The post-Brent, Late Jurassic–Early Cretaceous period was the most tectonically active period in the structural evolution of the Statfjord Field as shown by numerous discontinuities, unconformities and episodes of major faulting. The initial response to tilting and uplift of the fault-block was down-to-the-east faulting, deep rotational faulting, landsliding and erosion during Callovian–Oxfordian times, primarily affecting the Brent and Dunlin sequences. Further movements on the boundary fault saw the initiation of additional rotational faults and further erosion of the scarp face, cutting deeper into the succession. Figure 6 is a generalized structure cross-section to illustrate the style of faulting observed in Statfjord Field. This period of major faulting had its greatest effect in the central portion of the field.

The net result was to create, by Late Kimmeridgian times, a central collapse zone with relatively unfaulted, high-standing areas to the north and south (Fig. 7). Following tilting and uplift of the fault-block, deposition of Draupne shales provided a seal for the Brent, Statfjord and Dunlin reservoirs.

Post-Jurassic

Following the deposition of early Lower Cretaceous sediments there was a period of renewed tectonic activity which was probably due to a reactivation of existing faults. This final phase of activity, followed by onlap of Upper Cretaceous strata, resulted in the final field configuration as mapped at the present day.

The remainder of Cretaceous and Tertiary times saw essentially uninterrupted subsidence and sedimentation.

GEOPHYSICS

Basic seismic data acquisition over the Statfjord Field was conducted during the period 1973–77. The data were acquired in six surveys totalling 3600 km of reflection profiles. These data were supplemented by the acquisition of a 3300-km, three-dimensional survey in 1979–80 over the Statfjord Unit area. The basic seismic grid over the licence is 1 km × 1 km with a much denser grid (75 m × 25 m) over the actual field.

The number of seismic horizons mapped has varied with each successive vintage of survey. The early attempts mapped up to seven events from the Near Top Paleocene to the 'X' horizon within the Triassic. The most recent interpretation has identified the Base Cretaceous unconformity and the Top Statfjord as the primary reflection events in terms of delineation of the reservoir. A number of lesser reflectors have also been mapped including the Top Brent, Near Top Brent Zone 4 (Intra-Brent) and Top Dunlin–Base Brent (Fig. 8).

Data quality for all horizons is moderate up to the structural crest of the field. To the east of this point, interference and diffractions (due to the increased structural complexity and to decreasing thicknesses) lead to greatly reduced data quality.

TRAP

The Statfjord Field trap has structural and stratigraphic components. The field comprises Triassic–Jurassic strata in a WNW-dipping (6–8°) fault-block which has been cut by several episodes of faulting and landsliding, and has been truncated by erosional events.

A major down-to-the-east, block-boundary fault system limits the field to the east-south-east. The total displacement across this fault system is estimated to be 1500 m at Top Statfjord level. Upper Jurassic and Cretaceous shales rest unconformably on the subcropping Brent and Statfjord formations and provide a seal. Closure on the structure to the south is provided by a combination of a structural saddle and down-to-the-south normal faulting. Closure to the west and north is defined by structural dip.

The field has an overall north-easterly trend, is 24 km in length and has an average width of 4 km.

The oil–water contact for the Brent Group has been established at 2585.8 mss and at 2805.7 mss for the Statfjord Formation. The secondary hydrocarbon accumulation in the Dunlin DII (Cook Formation sandstone of Vollset and Doré 1984) has had a provisional oil–water contact of 2585.8 mss assigned to it. The reservoirs may not be filled to the spill-point. The total productive area for the field is over 80 km^2.

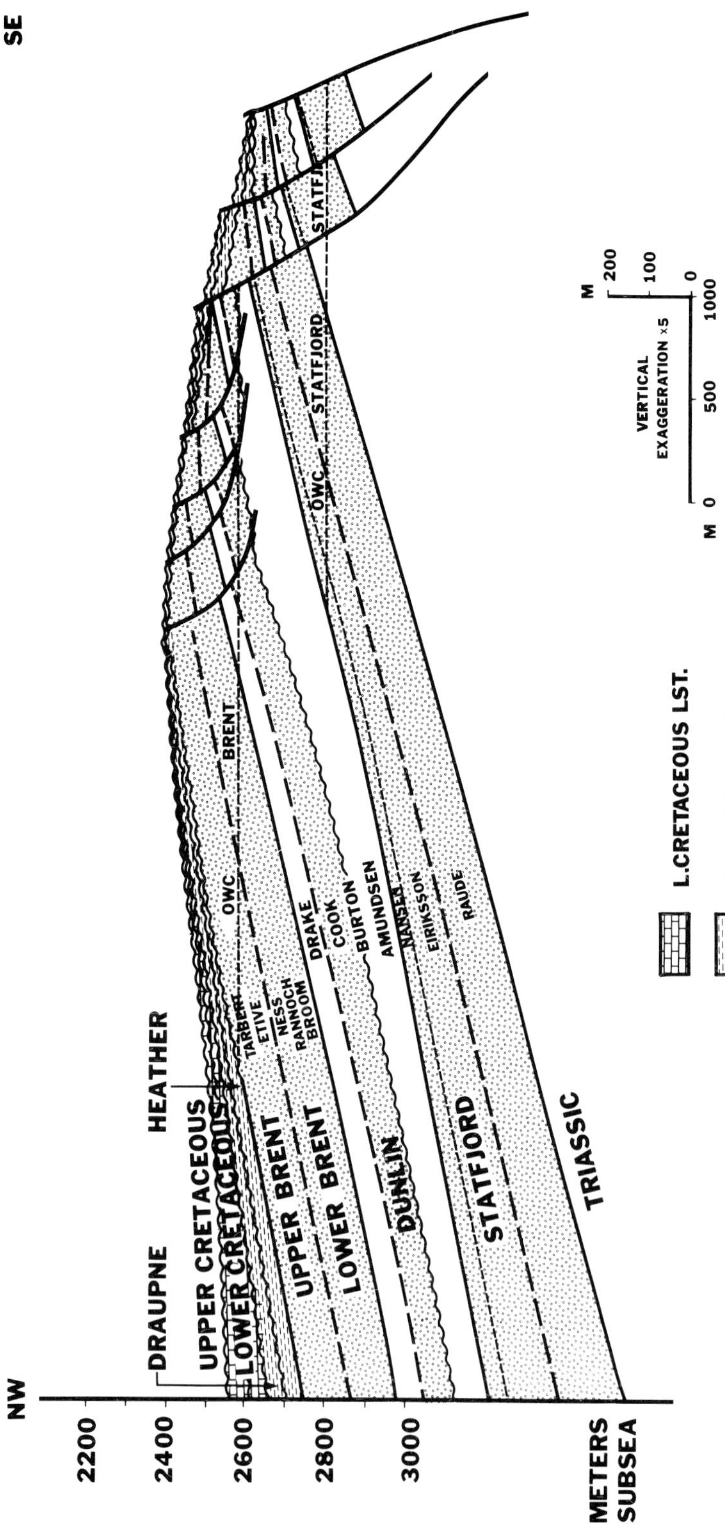

Fig 6. Generalized cross-section of the Statfjord Field.

RESERVOIRS

The Statfjord Field has two principal reservoirs, sandstones of the Statfjord Formation and of the Brent Group (Fig. 9). A subsidiary sandstone reservoir development occurs in the Cook Formation of the Dunlin Group.

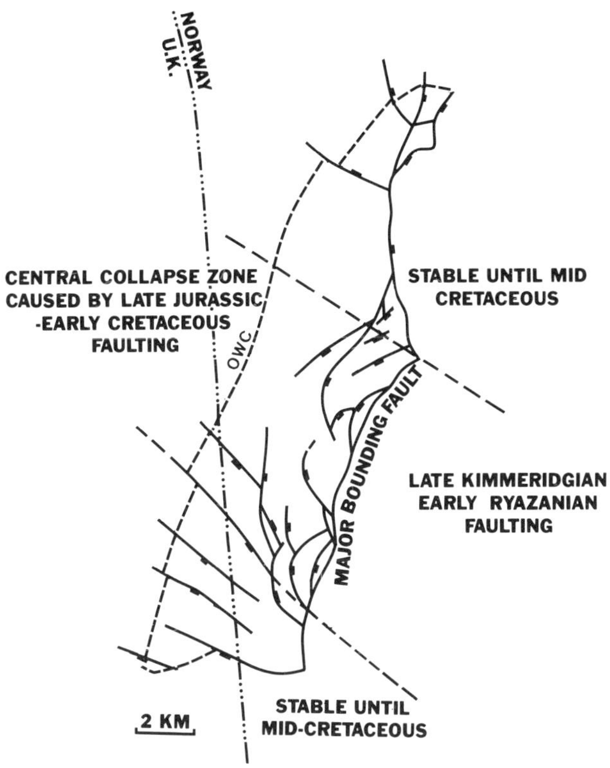

Fig. 7. Field outline showing major tectonic elements.

Statfjord Formation

The Statfjord Formation consists of an overall coarsening-upwards sequence of interbedded sandstones, siltstones and shales (Fig. 10). The formation is subdivided into three members, according to the terminology of Deegan and Scull (1977), later endorsed by Vollset and Doré (1984). From oldest to youngest, these are the Raude, Eiriksson and Nansen members. The Statfjord Formation in the Statfjord Field area has a gross average thickness of 205 metres. Wells 211/24-1 and 33/12-2 are still regarded as the type sections for the Nansen and Raude/Eiriksson members, respectively. For reservoir purposes, the three Statfjord Formation members, Raude, Eiriksson and Nansen, are referred to as Statfjord S3, S2 and S1 respectively.

The Raude Member consists of red and green shales and siltstones interbedded with generally thin, fine- to medium-grained sandstones, which coarsen upwards into the thicker, more massive sandstones of the Eiriksson Member. These latter sandstones are white to grey, fine to very coarse and conglomeratic, feldspathic to arkosic and contain both a kaolinitic matrix and calcite cement. They are interbedded with varicoloured shales and siltstones and, typically,

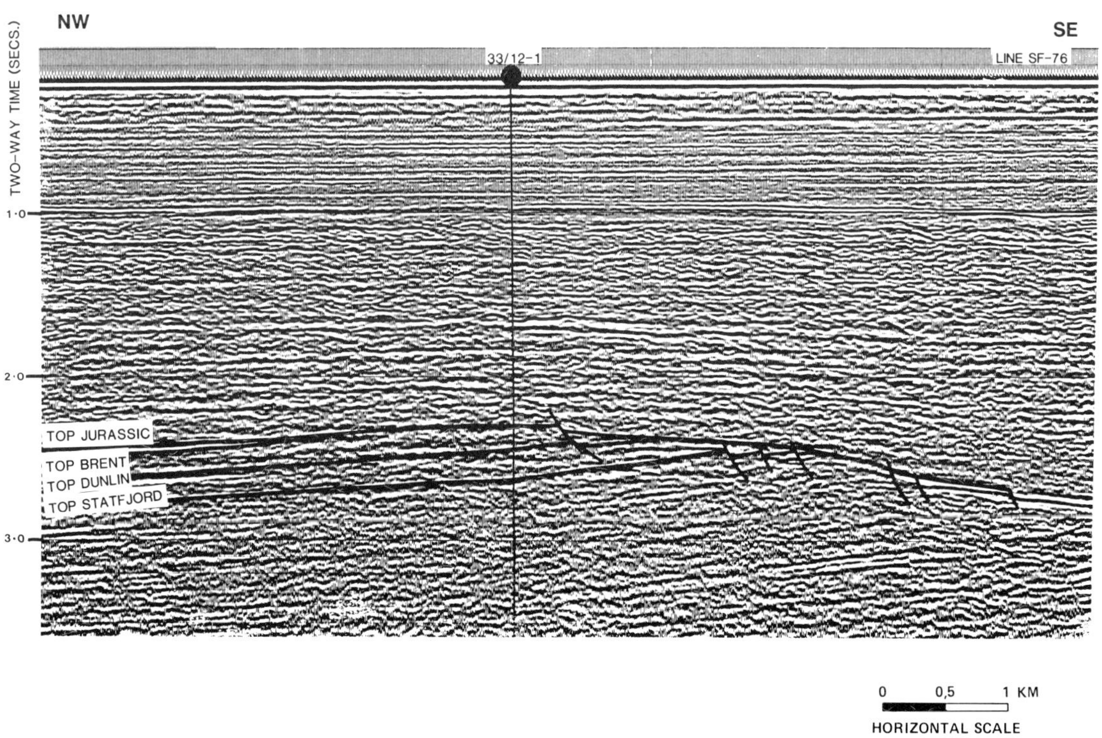

Fig. 8. Seismic line SF–76 through discovery Well 33/12-1. See Fig. 13 for location.

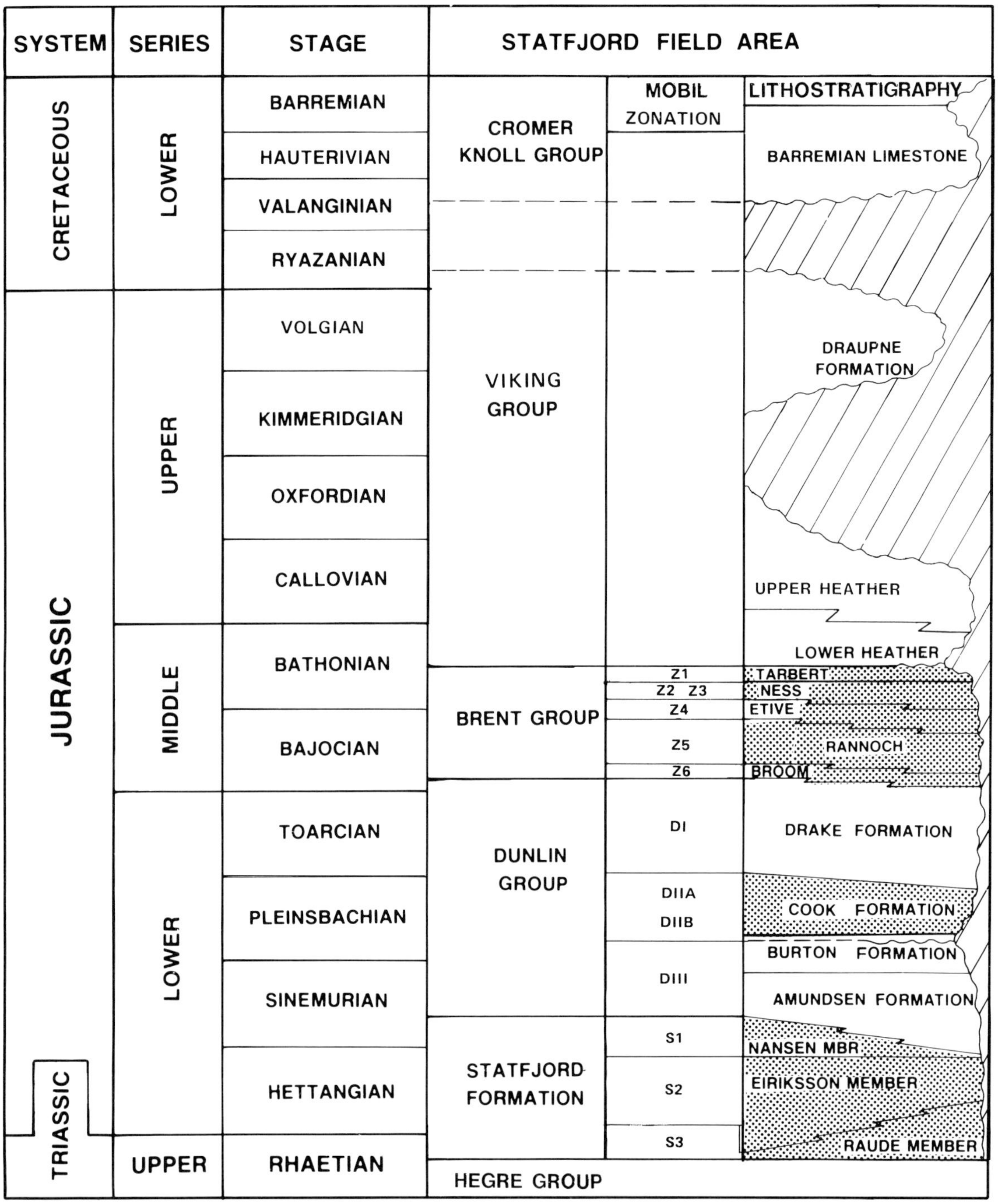

Fig. 9. Jurassic lithostratigraphic nomenclature. Modified after Deegan and Scull (1977) and Vollset and Doré (1984).

dark grey, carbonaceous shales and coals. The Nansen Member of the Statfjord Formation consists of a relatively homogeneous, calcite-cemented sandstone with occasional glauconite and marine fossils. In the extreme south-west part of the Statfjord Field, extending south-west through to the Brent Field, sand-grade clastic deposition continued into the Amundsen Formation of the Dunlin Group. Northwards and eastwards within the Statfjord Field these sandstones pass laterally into the more normal calcareous shales and siltstones of the basal Amundsen.

Reservoir properties generally improve upwards in the Statfjord section.

Depositional sequence analysis suggests that the Raude and Eiriksson members in the area of the Statfjord Field were laid down during two major, progradational, phases. Fluvial deposition prograded into a broad, sandy, alluvial flood basin (mud-plain followed by braid-plain). A major climatic

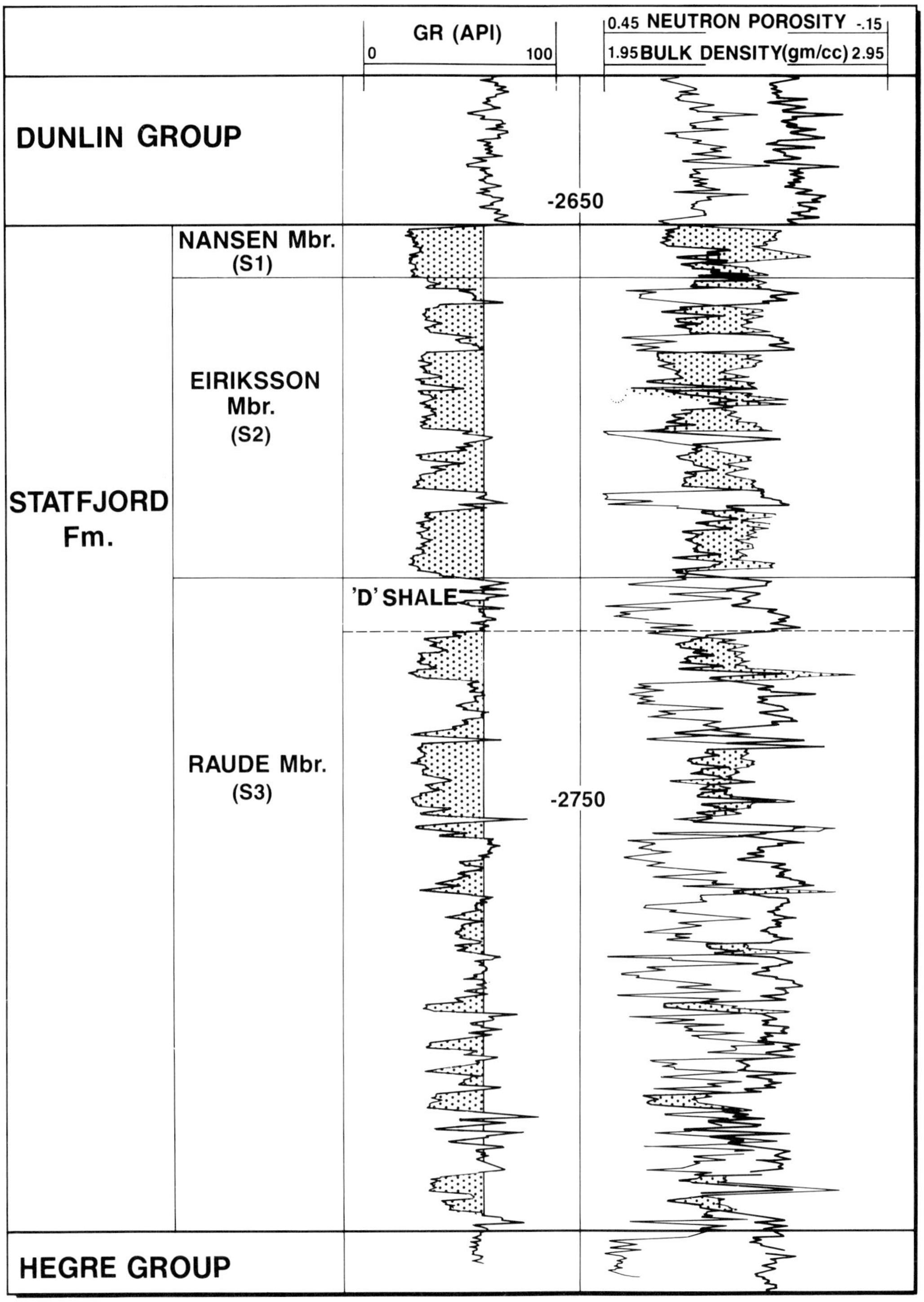

Fig. 10. Statfjord Formation log section.

change from semi-arid conditions in Raude times to humid conditions in Eiriksson times is interpreted (Chauvin and Valachi, 1980). The boundary between the Raude and Eiriksson members is marked by the development of caliché/soil profiles followed by a field-wide lacustrine shale bed. This acts as a pressure communication barrier over much of the field.

Other Statfjord Formation studies have proposed a fan-delta depositional environment, with intermittent marine inundation for the Raude and Eiriksson members (Røe and Steel, 1985). Mobil is not in agreement with this particular interpretation.

The Nansen Member was laid down as a transgressive-marine sheet-sand over the alluvial flood basin. The direction of transgression was probably from north to south and was caused by a combination of local tectonics and a major sea-level rise. Regional data indicate the Nansen Member to be present over a much wider area than the Raude and Eiriksson members, having been deposited outside the confines of the major tectonic flood basin into which these two members were deposited. This suggests a temporary cessation of major fault activity around the basin margins.

Brent Group

The Brent Group is subdivided into five formations, from oldest to youngest: Broom, Rannoch, Etive, Ness, Tarbert (Fig. 11). For reservoir management purposes in Statfjord Field, the Brent Group has been divided into two major units designated Upper and Lower

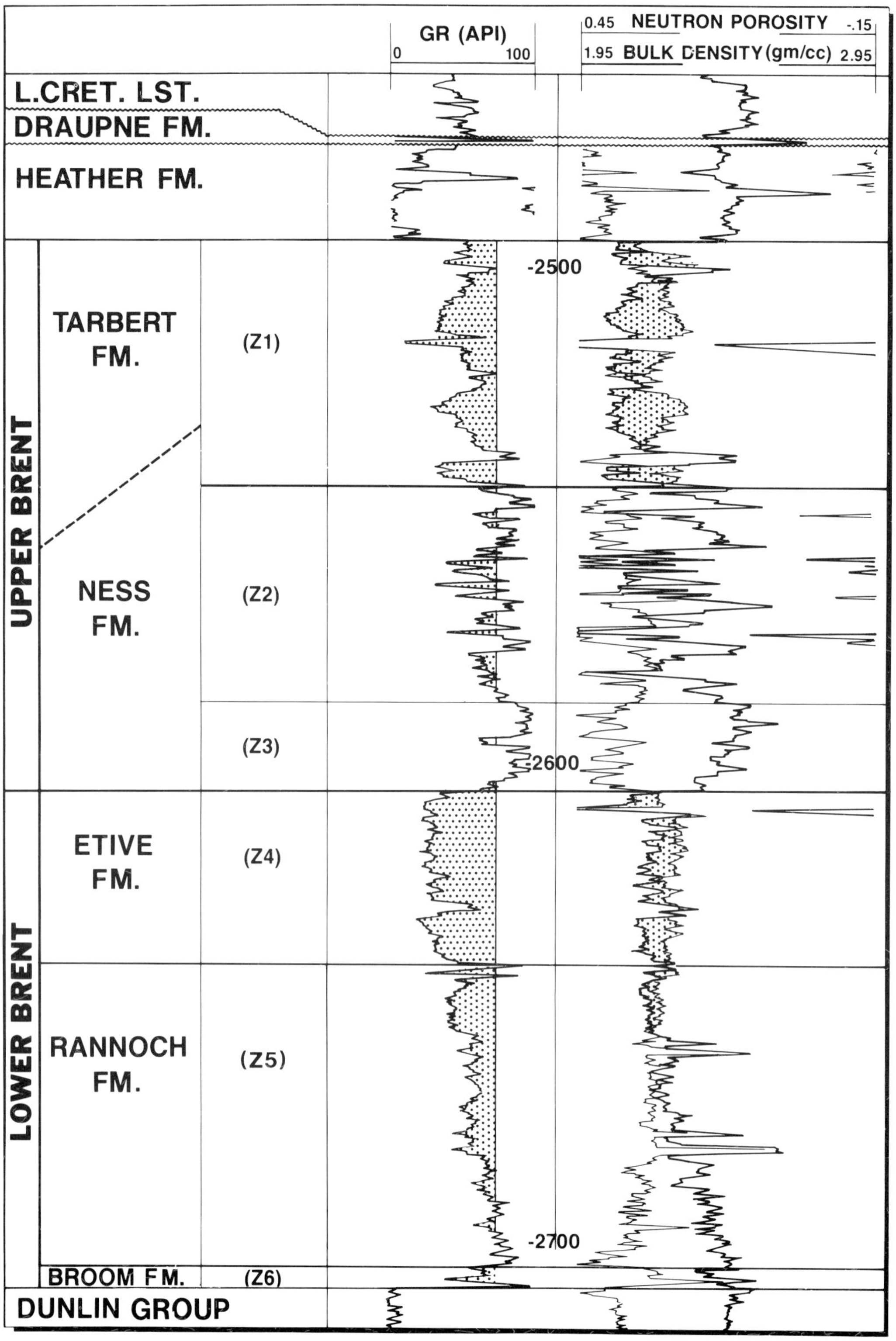

Fig. 11. Brent Group log section.

Brent which are separated by a major Lower Ness (Zone 3) shale barrier. The Mobil subdivision of these major units into six reservoir zones, together with the equivalent subdivisions of Deegan and Scull and a brief summary of facies and reservoir characteristics of each are outlined below.

Broom Formation (Zone 6)

This consists of a coarse-grained, poorly sorted conglomeratic sandstone which is generally 1–10 m thick and is not present at all locations in the field. It represents shallow-marine, storm-derived deposits which mark the onset of the regressive sequence that follows. Reservoir properties are generally poor.

Rannoch Formation (Zone 5)

This consists of fine, micaceous, kaolinitic and feldspathic sandstones and siltstones grading upwards into cleaner, fine-grained micaceous sandstones. Environmentally, it represents prodelta sediments grading upwards to delta-front and lower-shoreface sandstones. Reservoir characteristics improve from poor to moderate at the base to good at the top. Rock properties are generally better than indicated by logs, as evidenced by core and well-test data.

Etive Formation (Zone 4)

This consists of white, fine- to medium-, occasionally coarse-grained sandstones with low mica content. It was deposited as a beach barrier and tidal channel complex, with wash-over sands at the top of the sequence and with occasional coal beds. Reservoir properties are generally excellent.

Lower Ness Formation (Zone 3)

This is a ubiquitous dark grey shale with occasional thin sandstones and coals at the base. It was deposited in a saline bay–lagoonal environment. This shale acts as a pressure barrier between Upper and Lower Brent reservoirs. The shale becomes less well defined in the northern area of the field and is interbedded with thick sandstone bodies.

Upper Ness Formation (Zone 2)

This consists of an interbedded sequence of sandstones, shales and coals. The sandstones are fine to medium grained, frequently carbonaceous and show both coarsening-upwards and fining-upwards trends. The sequence was deposited in a delta-top environment with its associations of swamp, lagoon, distributary channel, levée and wash-over fans. The overall sand net-to-gross ratio is approximately 50% with generally good reservoir properties within each sand-body. As a reservoir unit, it suffers from the limited communication between many of the sand-bodies.

Tarbert Formation (Zone 1)

This unit generally consists of a massive, white, fine- to medium-grained sandstone with subordinate siltstones, shales, coals and occasional calcareous bands. The formation was deposited in a marginal-marine environment with delta-top facies predominating to the south and interfingering with marine sediments to the north. Communication between sand-bodies is much better than in the underlying Ness Formation and reservoir properties, especially in the northern area, are very good.

The Brent Group represents the sediments associated with the northerly progradation through the East Shetland Basin of a wave–tide interaction delta system initiated on the flanks of the Middle Jurassic, outer Moray Firth basin volcanic updoming. The delta system in Statfjord Field area is characterized by a single progradational phase followed by a retrogradational, destructive phase of sedimentation. The open-marine environment lay to the north, outside of the Statfjord Field area, and is delineated by both the delta-front and delta-plain regional thickness and facies associations.

The Brent Group is thus divided into five facies assemblages characteristic of its five recognized formations. Deltaic progradation commenced over the shallow-marine, distal bars of the Broom Formation, resulting in the shoreface sands of the Rannoch Formation. The Rannoch silts and sands coarsened upwards as sedimentation from the fluvial source exceeded the capacity for sediment dispersal by marine agents. Proximal mouth-bar deposits were not preserved in the deltaic cycle present in the Statfjord Field.

The Etive Formation consists of extensive tidal-channel sequences with a laterally equivalent shoreline barrier system. These form a complete coastal barrier and inlet complex. The Ness Formation delta-plain deposits include a large, coalescing, distributary channel system, together with two, less extensive, younger fluvial channel deposits. The axial orientation of the distributary channel sands suggests topographic influence. The clean, texturally mature sands of the Tarbert Formation cap the Ness delta-plain assemblage and represent delta abandonment.

Dunlin DII Formation

The DII Formation is the field terminology for the sandstone unit within the Dunlin Group and is equivalent to the Cook Formation of Vollset and Doré (1984). For reservoir evaluation purposes it has been subdivided into the DIIA and DIIB (Fig. 12) Each cycle is represented by a finely laminated sequence of sandstones, siltstones and shales, exhibiting

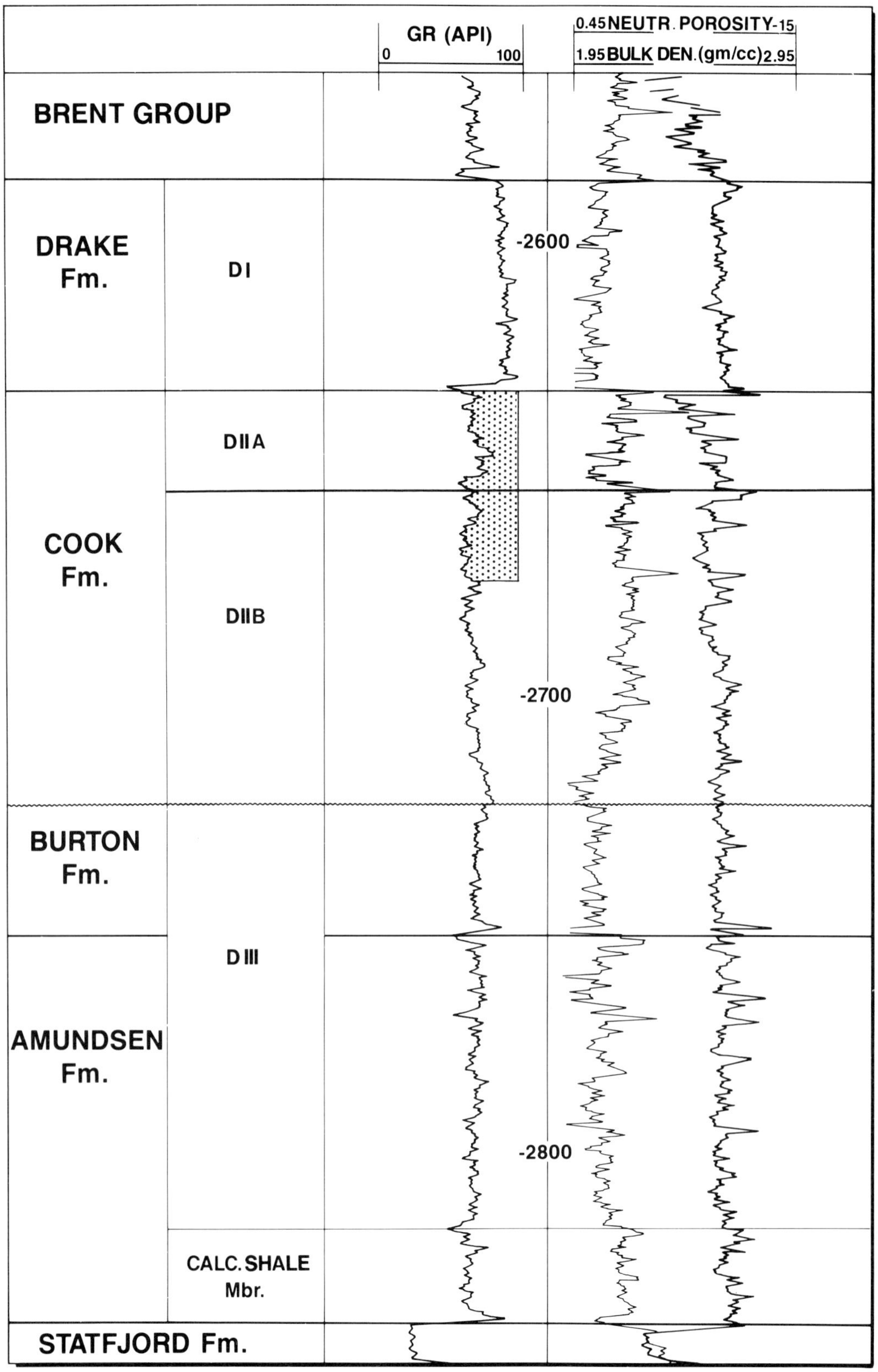

Fig. 12. Dunlin Group log section.

an overall coarsening-upwards trend. The sandstones are generally fine to medium grained. They represent prograding tidal sand-ridges deposited in a marine, midshelf, sublittoral position.

Reservoir quality varies considerably but the best-quality sandstones are developed in the northern part of the field in the 'C' platform area. Porosities of over 30% with horizontal permeabilities in excess of 1000 mD have been recorded from the better sandstones.

SOURCE

The Draupne Formation (Vollset and Doré, 1984), previously termed the Kimmeridge Clay Formation (Deegan and Scull, 1977), is believed to be the principal source-rock for the Statfjord Field reservoirs and most of the related fields within the East Shetland Basin. It ranges in age from Oxfordian (Upper Jurassic) to Ryazanian (Lower Cretaceous). The formation varies in thickness from several metres over the crest of the field to a few hundred metres to the east of the large bounding fault system. On the west flank of the field, sections up to 40 m thick have been encountered in water injection wells and it is expected that the formation will continue to thicken down-flank. Within the Statfjord unit area the Draupne Formation is considered to be immature, and hence the field is thought to have been sourced from further down-flank along the graben.

The Draupne shales are unlikely to have entered the oil generation window before Late Cretaceous–Early Tertiary times. Very little is known of migration routes, but the bounding fault system and the Base Upper Jurassic and Base Cretaceous unconformity surfaces would seem to offer potential pathways.

HYDROCARBONS

There are three separate oil accumulations in the Statfjord Field. These are in the Statfjord Formation, Dunlin II and the Brent Group. Tables 2, 3, and 4 provide the best estimates of pertinent petrophysical and engineering data available at present.

The Brent Group accumulation extends from a crestal high of 2360 mss to an oil–water contact at 2585.8 mss. The oil–water contact in the Brent is very sharp, with only a very thin transition zone. The areal extent of the hydrocarbon accumulation in the Brent Group is shown best by Figs 13 and 14. As noted previously, the Brent Group contains two separate reservoirs (Upper and Lower) based on correlation and pressure differences across a Mid-Brent shale. However, the oil–water contacts, oil characteristics and initial pressure gradients in the two reservoirs are the same. Thus the Brent Group accumulation is considered as a single accumulation within the context of geological time.

The Dunlin II accumulation is considered to be separate from the Brent Group accumulation. Initial pressures and gradients in the two accumulations are the same. However, based on a limited data set, there are differences in the crude gravity (albeit minor), GOR, bubble-point and oil–water contacts of the two accumulations. Initially it was considered, through lack of firm evidence, that the oil–water contacts in the Brent and Dunlin were the same. Recent data, however, suggest that the Dunlin accumulation has a lower oil–water contact than the Brent.

The Statfjord Formation accumulation (Fig. 15) is completely separate from the overlying Dunlin and Brent accumulations. Initial reservoir pressures, GOR, bubble-point and oil–water contacts are all markedly different.

Hydrocarbon type is remarkably similar in all three accumulations, being of low sulphur content, with no H_2S and with a gravity of around 0.82–0.86 g/cm^3. There is no natural gas-cap in any of the accumulations though a secondary gas-cap has most likely been generated in the Statfjord Formation as a result of gas injection. Formation water salinities and resistivities are very similar, with around 14 000 ppm total chlorides.

Initial well production capacity for the Statfjord accumulation was around 4000 Sm^3/day, although one well has produced 9380 Sm^3/day from the topmost sand in the Statfjord Formation (Nansen Member). Initial well production capacity for the Brent Group was around 4800 Sm^3/day though, again, significantly higher rates have been recorded from production wells.

Test data for the Dunlin are confined to the three Wells 33/12-2, 33/9-9 and 33/9-C38. Well C38 flowed 1730 Sm^3/day through a 3/4-inch choke giving a productivity index of 65 Sm^3/day/bar.

RESERVES

Original reservoir pressures indicated that both Brent and Statfjord reservoirs would

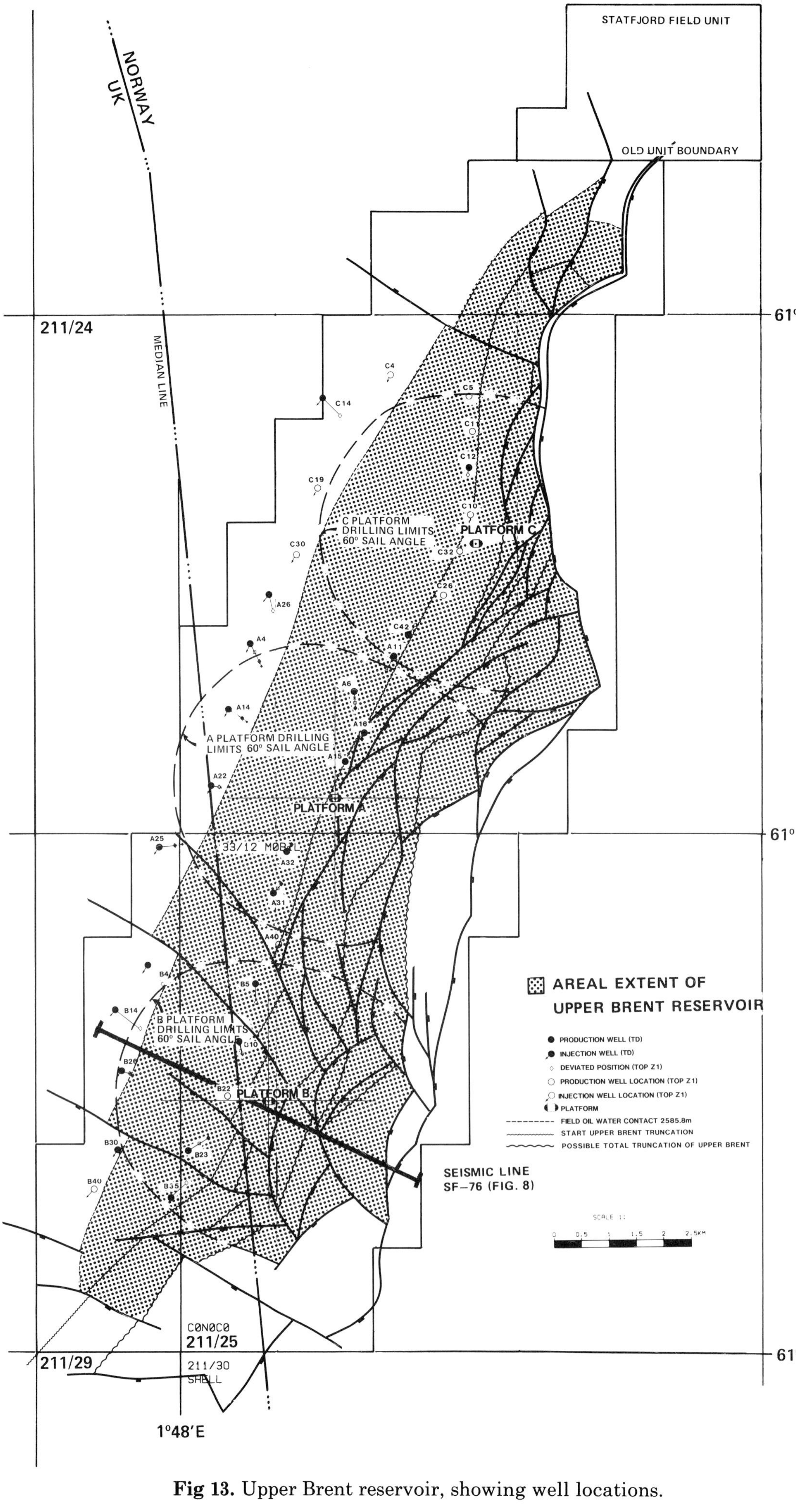

Fig 13. Upper Brent reservoir, showing well locations.

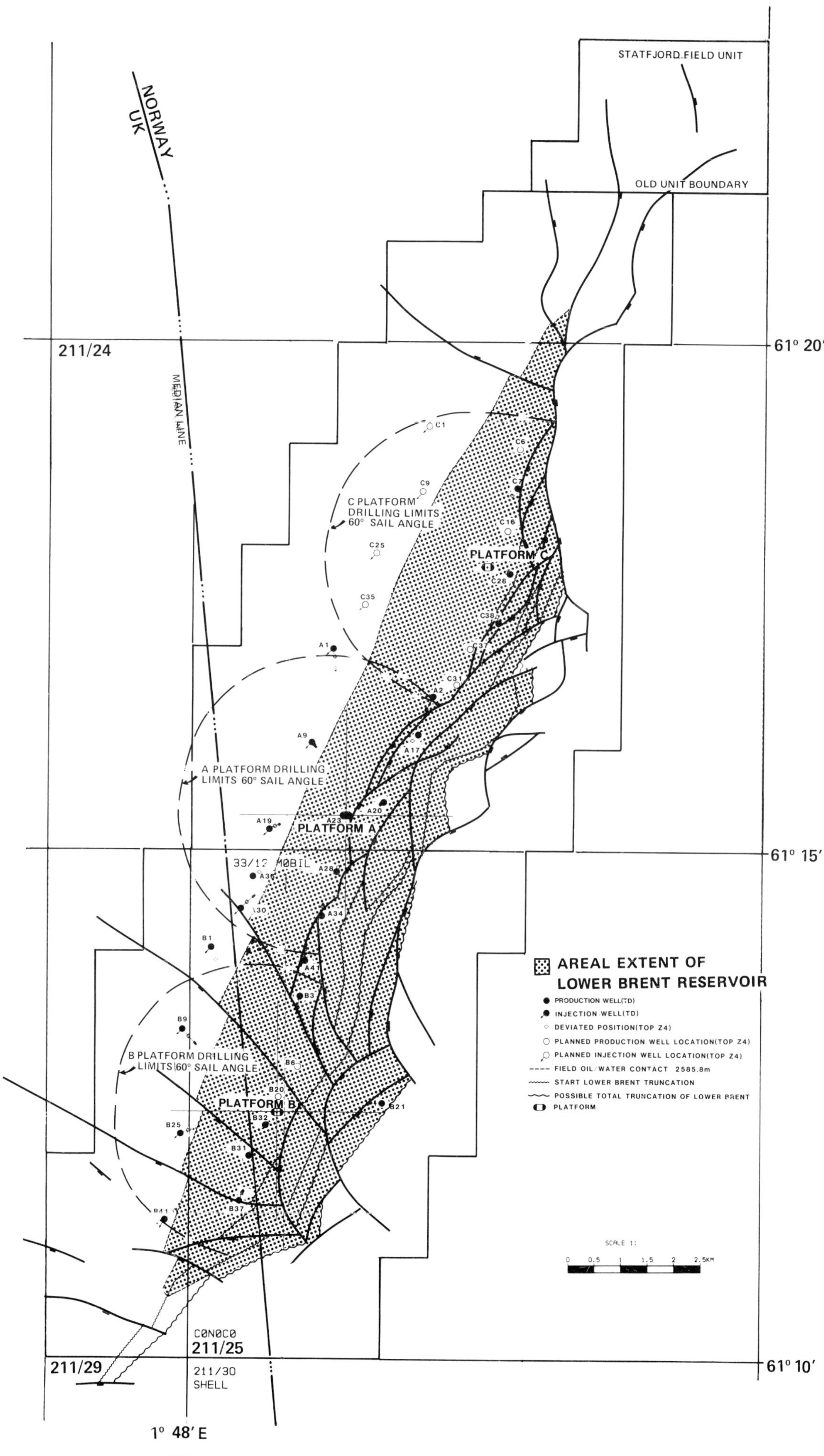

Fig. 14 Lower Brent reservoir, showing well locations.

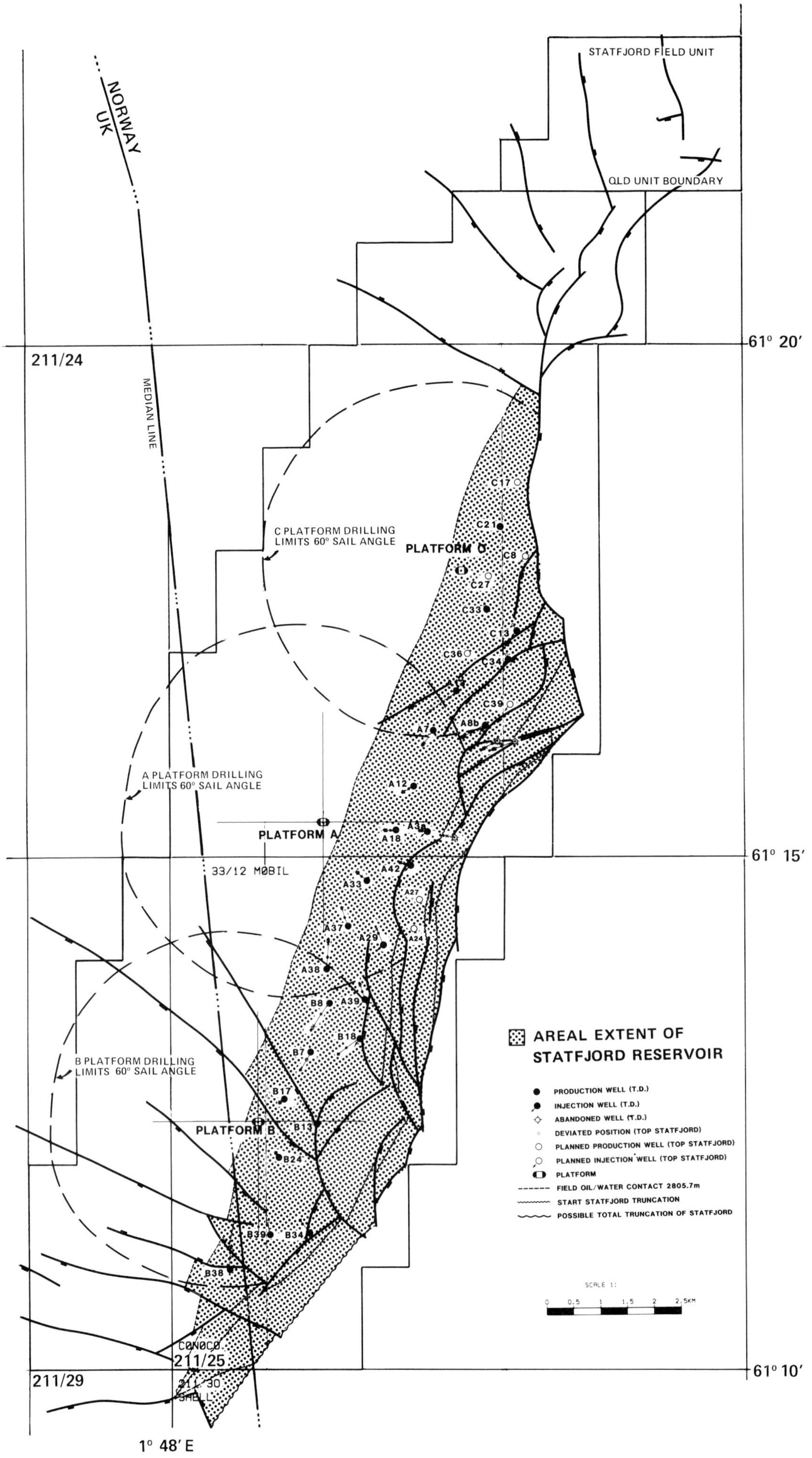

Fig. 15. Statfjord reservoir, showing well locations.

have limited aquifer support and that injection programmes would be required to enhance hydrocarbon recovery. As a result of a 1977 full-field study it was concluded that optimal results would be achieved on all three platforms by injecting water into the Brent reservoir, and injecting produced gas into the Statfjord reservoir. Figures 13, 14, and 15 show the configuration of injection and production wells for the field's reservoirs. The Dunlin reservoir is not yet developed. The development philosophy for each of the field's reservoirs was as described below.

Statfjord reservoir

Laboratory analyses had indicated the suitability of the Statfjord reservoir to high-pressure, miscible gas flood as a means of improving ultimate recovery. The combined gas production from Brent and Statfjord reservoirs provided ample gas for a complete gas flood of the Statfjord reservoir.

Early high production rates from the Statfjord reservoir were necessary to accommodate all produced gas prior to the commencement of gas sales. For this reason, initial development emphasis was on the Statfjord reservoir withdrawal/injection strategy in order to maintain the 352-bar reservoir pressure to ensure injection-gas miscibility without overpressuring the reservoir and driving oil into the aquifer. Statfjord production philosophy was to minimize premature gas breakthrough and to maximize gas-sweep efficiency.

To achieve this, production wells were completed in a stratigraphically high position and as close to the oil–water contact as possible without incurring early water breakthrough or coning. Gas injectors were located structurally high but as stratigraphically low as possible to allow the gas to migrate to the top of the structure and, in time, form a secondary gas cap which would push the oil down towards the oil producers. This strategy has produced excellent results to date.

The lacustrine shale at the top of the Raude member provides a field-wide pressure communication barrier, and for future development plans the reservoir will be subdivided into Upper and Lower Statfjord reservoirs at this horizon. The Lower Statfjord thus comprises the Raude Member (S3) while the Upper Statfjord comprises the Eiriksson (S2) and Nansen (S1) members. For detailed reservoir management purposes the reservoir has been further subdivided into a total of 10 layers.

Brent reservoir

The Brent Formation is subdivided into two reservoirs designated Upper and Lower Brent. The pressure barrier between Upper and Lower Brent is provided by the laterally extensive shales of Zone 3. Production wells are completed structurally high and stratigraphically low. Wells are recompleted, where possible, in stratigraphically higher positions when the water-flood front reaches the well. Water injectors are generally completed in lower-permeability intervals (Zone 5 in the Lower Brent and Zone 2 in the Upper Brent) in an attempt to prevent premature water breakthrough. Initial data suggest that the waterfront is acting very much as anticipated and only limited water breakthrough problems have occurred, due to fingering or coning.

Reservoir management has been assisted by the extensive application of reservoir simulation. Successful history matching has provided confidence in the ability of reservoir simulation to predict water and gas breakthrough and reservoir pressure trends. Many of the production strategies now incorporate the predictions of the simulation models.

Figure 16 shows the oil production history for the Brent and Statfjord reservoirs.

Conclusion

Following its discovery in 1974 the Statfjord Field has been efficiently and quickly brought to current production levels of around 119 240 Sm^3 of oil per day. An extensive data acquisition programme has enabled the formulation of a sound geological model and the application of sophisticated reservoir simulation and engineering practices. This prudent

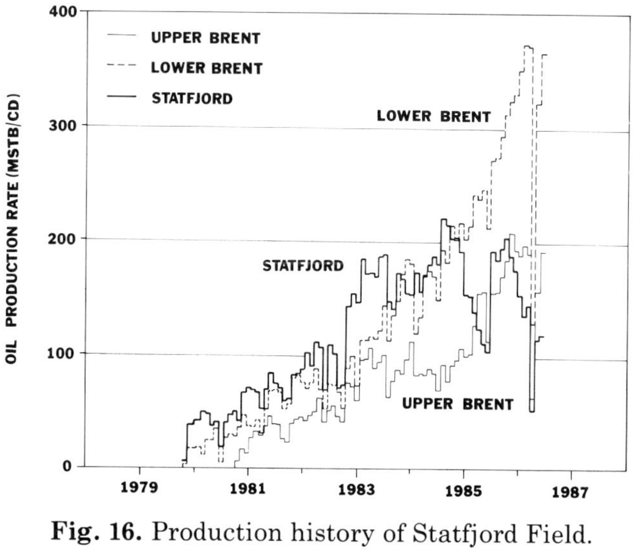

Fig. 16. Production history of Statfjord Field.

approach to reservoir management has enabled the safe and effective development of Western Europe's largest and most prolific oil field.

Acknowledgements

The authors wish to thank the Statfjord Group of companies for their co-operation and permission to publish this report. Particular thanks are offered to the Statfjord Division Development Engineering Group, especially L. A. Høyland and J. E. Inderhaug, and to J. W. Buza of Mobil Exploration Norway.

Every effort possible has been made to ensure the accuracy of the statistical information provided in this report. The views and opinions expressed in this report are those of the authors and do not necessarily reflect the views and opinions of all members of the Statfjord Group of companies.

REFERENCES

Bjørlykke, K. and Brendsdal, A. 1986. Diagenesis of the Brent sandstone in the Statfjord field, North Sea, In D.L. Gautier (Ed.), *Roles of Organic Matter in Sediment Diagenesis,* SEPM Spec. Publication no. 38, pp. 157–167, Tulsa.

Buza, J.W. and Unneberg, A. 1986. Statfjord geology, reservoir performance, *Oil Gas Journal* **84** (29), 68–72.

Buza, J.W. and Unneberg, A. 1986. Statfjord field development strategies, *Oil Gas Journal* **84** (30), 110–114.

Chauvin, A.L. and Valachi, L.Z. 1980. Sedimentology of the Brent and Statfjord Formations of Statfjord Field, In *The Sedimentation of the North Sea Reservoir Rocks,* Norsk Petroleumsforening, Article XVI.

Deegan, C.E. and Scull, B.J. 1977. A standard lithostratigraphic nomenclature for the Central and Northern North Sea, *Report Inst. Geol. Sci. 77/25; Bulletin, Norwegian Petroleum Directorate 1.*

Hamar, G.P. and Hjelle, K. 1984. Tectonic framework of the Møre Basin and the Northern North Sea, In Spencer A.M. *et al.* (Eds), *Petroleum Geology of the North European Margin,* Graham and Trotman, London, pp. 349–358.

Hancock, W.P. and Hagen, D.A. 1986. Fine tuning increases Statfjord B output 39%, *World Oil* **202**(1), 79–84.

Jones, E.L., Raveling, H.P. and Taylor, H.R. 1975. Statfjord Field, In *Jurassic Northern North Sea Symposium,* Norsk Petroleumsforening, Article 20.

Kirk, R.H. 1980. Statfjord Field—a north sea giant, In Halbouty, M.T. (Ed.) *Giant Oil and Gas Fields of the Decade 1969–78, AAPG Memoir 30,* pp. 95–116.

Malm, O.A., Furnes, H. and Bjørlykke, K. 1979. Volcaniclastics of Middle Jurassic age in the Statfjord oil-field of the North Sea, *N.Jb. Geol. Paleont. Mh.* **10,** 607–618.

Røe, S.L. and Steel, R. 1985. Sedimentation, sealevel rise and tectonics at the Triassic–Jurassic boundary (Statfjord Formation), Tampen Spur, Northern North Sea, *Journal Petroleum Geology* **8** (2), 163–186.

Rønnevik, H.C., van den Bosch, W. and Bandlien, E.H. 1975. A proposed nomenclature for the main structural features in the Norwegian North Sea, In *Jurassic Northern North Sea Symposium,* Norsk Petroleumsforening, Article 18.

Vollset, J. and Doré, A.G. 1984. A revised Triassic and Jurassic Lithostratigraphic nomenclature for the Norwegian North Sea, *Norwegian Petroleum Directorate bulletin No. 3.*

29

Statfjord Nord

S. J. Gradijan and M. Wiik

Mobil Exploration Norway Incorporated, PO Box 510, N-4001 Stavanger, Norway

The Block 33/9 Beta* Discovery, a 'satellite' structure to the Statfjord Field, was discovered to be hydrocarbon bearing in 1977. The 33/9-8 discovery well location was based on interpretation of seismic reflection surveys and an extrapolation of the productive regional trend. Two principal sandstone reservoirs, a previously unnamed Upper Jurassic sandstone of the Draupne Formation for which the name Munin sand is proposed, and the Middle Jurassic sandstones of the Brent Group, may contain recoverable resources in the order of 36–66 × 10^6 Sm^3 oil. The productive area is approximately 13 km^2. Reservoir properties are excellent.

Beta is a typical East Shetland Basin tilted Jurassic fault-block–stratigraphic trap, formed by the westward tilting of a minor fault-block. Brent deltaic sandstones are truncated by Middle to Late Kimmerian unconformities on the south-east of the structure. The Upper Jurassic Munin sand was deposited on the western part of the structure. The seal is provided by overlying and onlapping Jurassic and Cretaceous shales. Upper Jurassic shales are thought to be the oil source. Additional appraisal is planned and the structure could be declared commercial in the late 1980s. Development may utilize a subsea atmospheric system tied back to Statfjord Field for processing.

INTRODUCTION

The Beta discovery lies in 300 m of water seven kilometres north of Statfjord Field and five kilometres north-east of Murchison Field (see Fig. 1). The discovery is expected to be a medium-sized oil field by North Sea standards, and recoverable resource estimates are 36–66 × 10^6 Sm^3 oil. Two reservoirs are present. One is a Volgian-age sandstone, difficult to map, here termed the Munin sand, that is part of the Draupne Formation. It has excellent reservoir characteristics. The second is a Brent Group sequence of sandstones. The field is known by its exploration designation as the 33/9 Beta* prospect. Unofficially the structure had been known as Munin, after the second of the pair of ravens of Norwegian mythology said to perch on Odin's shoulder.

HISTORY

Licence 037, which includes Block 33/9, was awarded to a group of nine companies in 1973. Mobil Exploration Norway Incorporated is the operator for Statoil (50%), Mobil (15%), Conoco (10%), Shell (10%), Esso (10%), Saga (45/24%), Amoco (25/24%), Amerada (25/24%) and Texas Eastern (25/24%). Currently, all proven hydrocarbons on the Beta structure occur in this licence, but hydrocarbons may extend into Block 34/7. This block was awarded to a different group of seven companies in 1984 as Licence 089. Saga is the

* Editor's note. Throughout this article the discovery is referred to as 33/9 Beta, but in January 1987 the name 'Statfjord Nord' was officially approved.

Geology of the Norwegian Oil and Gas Fields © A. M. Spencer *et al.*, eds (Graham & Trotman, 1987) pp. 341–350.

Statfjord Nord
Field data summary

Trap		
Type	stratigraphical, tilted fault-block	
Depth to crest	2600 mss	
Lowest closing contour	2800 mss	
Hydrocarbon contacts	Munin ODT 2680 mss; Brent OWC 2716 mss	
Gas column	unknown	
Oil column	47 m known (Munin + Brent)	
Pay zone		
Formation	Munin sand	Brent Group
Age	Late Jurassic	Mid-Jurassic
Gross thickness (average)	46 m	140 m
Sand/shale ratio	0.80	0.75
Porosity (average)	23%	22%
Net/gross ratio (average)	0.80	0.75
Hydrocarbon saturation	81%	81%
Permeability (average)	2900 mD	1500 mD
Hydrocarbons		
Oil gravity	0.835 g/cm^3	0.838 g/cm^3
Oil type	black oil	black oil
Gas gravity (relative to air)	0.757	0.743
Bubble-point	137.9 bar	
Gas/oil ratio	61 Sm^3/Sm^3	61 Sm^3/Sm^3
Formation volume factor	1.39	1.23
Formation water		
Salinity	—	22 000 ppm NaCl equivalents
Resistivity	—	0.110 ohm m at 77 °C
Reservoir conditions		
Temperature	96 °C	96 °C
Pressure	393.3 bar	393.3 bar
Field size		
Area	13.2 km^2	
Gross rock volume	1.9×10^9 m^3	
Recovery factor—oil	40%	40%
Drive mechanism	water injection	water injection
Recoverable hydrocarbons		
oil	36–66 $\times 10^6$ Sm^3	
gas	3.4×10^9 Sm^3	
Production		
Start-up date (estimated)	1994	
Development scheme	subsea atmospheric system	
Production rate (estimated)	21 400 Sm^3/day	

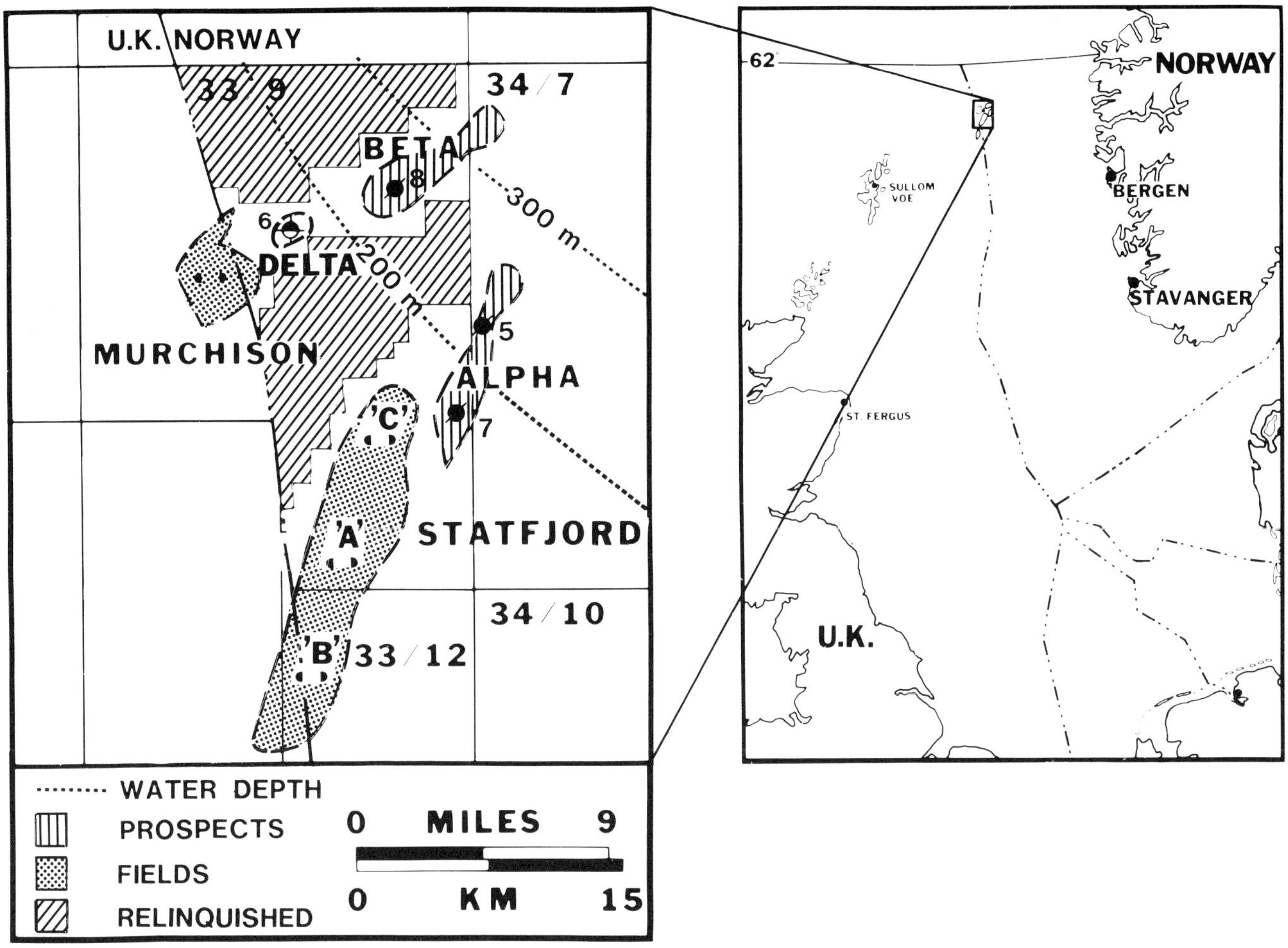

Fig. 1. Location of PL037 Block 33/9 Beta discovery.

operator for this block on behalf of Statoil (50%), Norsk, Hydro (12%), Saga (10%), Elf (8%), Esso (15%), Deminex (4%) and DNO (1%).

Licence 037 expires in August 2009. All relinquishments have been made. The Licence 089 primary term expires in 1990, at which time 50% must be relinquished. Established commerciality should provide the basis to extend the licence term up to the end of 2020.

The Mobil 33/9-6 Well tested the Delta structure adjacent to the Beta structure in 1976. This well encountered hydrocarbons in the Middle Jurassic Brent sandstones lying on the same hydrocarbon trend as Murchison Field. Well 33/9-8 was drilled in January 1977 by Mobil with an objective to test the Block 33/9 Beta Prospect and was located in a higher structural position than the 33/9-6 Well. The 33/9-8 Well established reserves in an Upper Jurassic sandstone which had not been previously encountered in this area, and in the Brent Group. No hydrocarbons were found in the Statfjord Formation. It is the only well drilled on the Beta structure.

Log analysis indicated a gross pay zone of 46.8 m with 35.4 m net pay. Three cores were cut in the Jurassic sandstones to evaluate hydrocarbon shows. The well was tested and flowed oil at a rate of 1997 Sm^3/day with a WHP of 2000 psi through a 3/4-inch choke during a 4 hour 12 minute test. Perforations were at 2683–2686 m in the Upper Jurassic sandstone. The well also tested oil from two thin sandstone intervals at the top of the Brent Group. A third test, lower in the Brent Group, tested water. The well was plugged and abandoned on 21 February 1977 at a total depth of 3058 m in the Lower Jurassic Statfjord Formation.

It is anticipated that development of the Beta discovery will be tied to the Statfjord Field production facilities once the Statfjord Field goes into decline. Further appraisal and development of the Beta discovery will be timed to provide first production from Beta coincident with the availability of excess production capacity at Statfjord Field. Production start is now expected in 1994. A three-dimensional seismic programme was conducted in 1986 and is to be followed by appraisal drilling in 1987.

In 1984, Statoil exercized its option to assume the operatorship of Licence 037, effective from 1 January 1987. This includes the operatorship of Beta.

GEOLOGICAL SETTING

The Beta discovery lies in the northern part of the Viking Graben. The Beta structure is part of the complex of north-west-tilted fault-blocks on the north-west flank of the Viking Graben. Here, a series of tilted blocks, which include the Beta and Delta discoveries, the Murchison Field and the Dunlin Field, follow a major, generally south-west-dipping fault complex (Fig. 2). These structures lie about ten kilometres north-west from the parallel complex of major faults bounding the huge Statfjord and Brent fields.

Several authors have described the tectonic history of this area of the North Sea, specifically Ziegler (1975) and, in more detail, Kirk (1980). Jurassic sediments were deposited by complexes of deltas and streams and segregated into sandy and shaly deposits by fluvial and marine processes. The deposits were uplifted on several occasions, eroded and subsequently submerged. Sediments were then redeposited or other sediments deposited upon them.

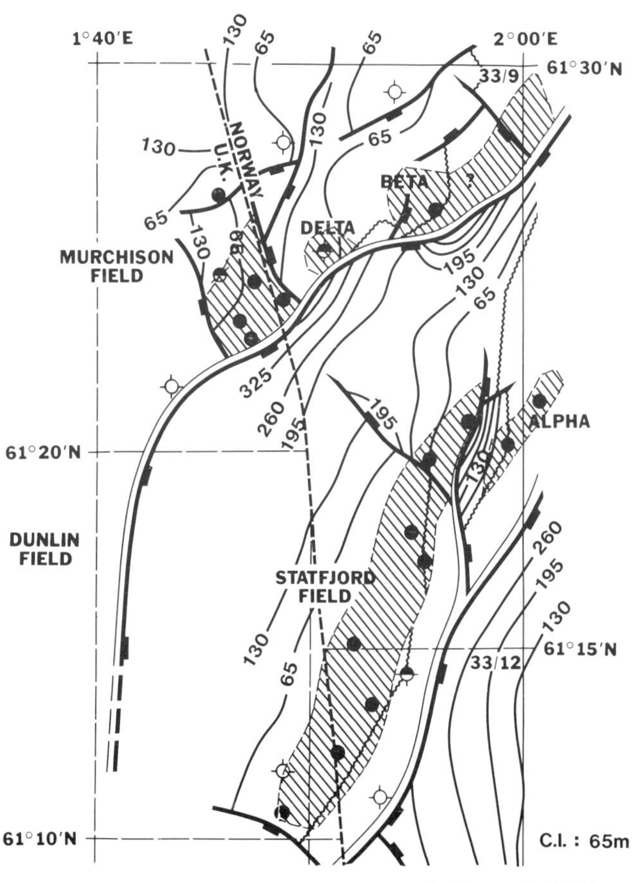

Fig. 2. Isopach map in metres of Upper/Middle Jurassic shales (Viking Group) showing major Jurassic fault-block trends and truncation areas (modified after Kirk, 1980).

Some of the shaly sections were rich in kerogen and provided the primary source for hydrocarbons in the area. The sandy sequences, where structurally high, have become reservoirs.

The area was faulted during the Late Kimmerian creating numerous tilted, rotated fault-blocks responsible for the major characteristics of the present-day Beta structure. The Beta structure remained a positive feature until it was buried by Late Cretaceous sediments.

The generalized stratigraphic column for the Beta area shown in Fig. 3 uses the lithostratigraphic nomenclature proposed by Deegan and Scull (1977) and contains the revisions proposed in Vollset and Doré (1984). Figure 3 also shows the stratigraphy of the Jurassic sequence in the 33/9-8 Well. Figure 4 describes the regional stratigraphy of the Mesozoic sequence.

The oldest sediments the 33/9-8 Well penetrated were Early Jurassic Statfjord Formation sandstones of continental origin. These sandstones are medium grained, kaolinitic and calcareous with calcilutite stringers. They are overlain by shales and siltstones of the Dunlin Group. Sandstones of reservoir quality known from the Statfjord Field Dunlin section are not present here.

The Middle Jurassic Brent Group contains the main reservoir units at Beta. This interval is 140 m thick of which about 75% consists of sandstones deposited in a marginal-marine shelf environment.

Overlying the Brent Group is the Heather Formation, consisting of 35 m of shales which may effectively isolate the main reservoir from the overlying secondary reservoir, which is a Volgian sandstone. This sandstone is 46 m thick and contains about 80% sand. It is interpreted as a possible offshore-storm or storm-beach deposit.

Reservoir sandstones of Volgian age are uncommon in the North Sea region. Rawson and Riley (1982) have documented their occurrence in widely separate areas of the North Sea. The closest occurrence is the major Upper Jurassic deep-water sandstones which form the reservoir in the Magnus Field, north

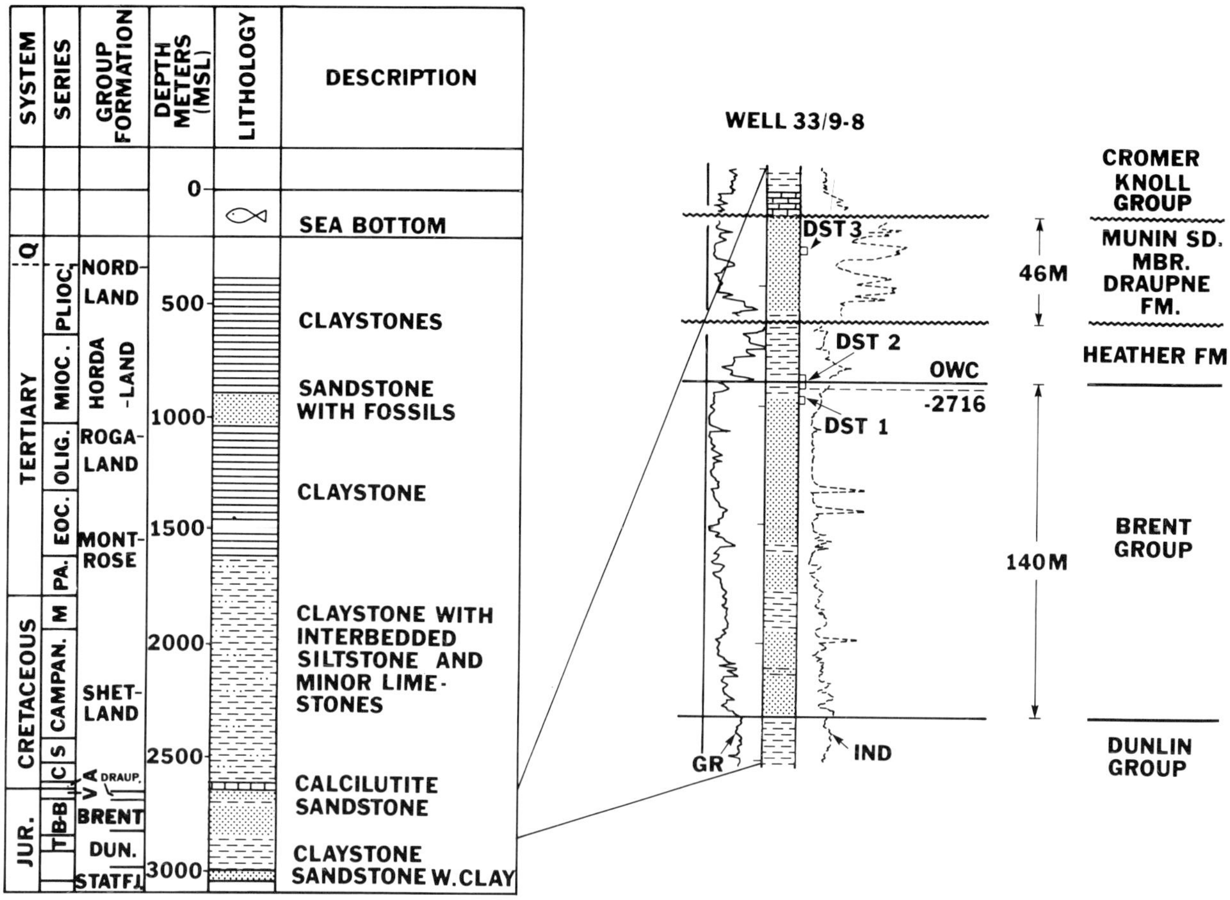

Fig. 3. Stratigraphic section present in 33/9-8 Beta discovery well.

and west of Beta. These sands have a limited distribution, according to Ziegler (1979). Proximal fan deposits (turbidites) are also recognized in the southern Viking Graben. The presence of rip-up clasts in the upper part of the Volgian sand in Beta indicates that parts of it may be an offshore-storm deposit or a storm-beach deposit and not a turbidite. Based on the sand's stratigraphic position, it could be considered part of the Magnus, Piper or Draupne formations. It is suggested, after looking at all the data, that the sand be known as the Munin sand member of the Draupne Formation, Fig. 3 Munin is one of Odin's ravens and was one of the names proposed for this field in which the sand is first described.

The Jurassic rocks are overlain by a monotonous section of clays and silts of Cretaceous to end Oligocene age belonging to the Cromer Knoll, Shetland, Montrose and Rogaland groups. These are in turn overlain by interbedded sands and shales of the Nordland Group.

GEOPHYSICS

The 33/9-8 Well, which is the only well on the Beta structure, was drilled in 1977 and was located on the basis of maps made from the 1973/74 MNG/MNG.5 survey which covered the whole Licence 037 area. The Beta prospect was completely reinterpreted in 1978 incorporating about 100 km of 1976/77 MNG.25 and MNG.75 data and about 15 km of Statoil data purchased from the adjoining Block 34/7. The total seismic database now formed a triangular grid with the lines orientated N–S, NW–SE and NE–SW with line-spacing approximately $1 \times 1 \times 1/2$ km. The seismic quality is poor to fair at the level of the objective Jurassic horizons. Figure 5 shows a dip-line through Well 33/9-8.

No additional seismic has been shot over the Beta structure since 1977, except for parts of the two lines SG8146-407 (1981) and SG8231-412 (1982), both from the Tampen Spur Group seismic, which both go through the 33/9-8 Well. Also, no new major interpretation has

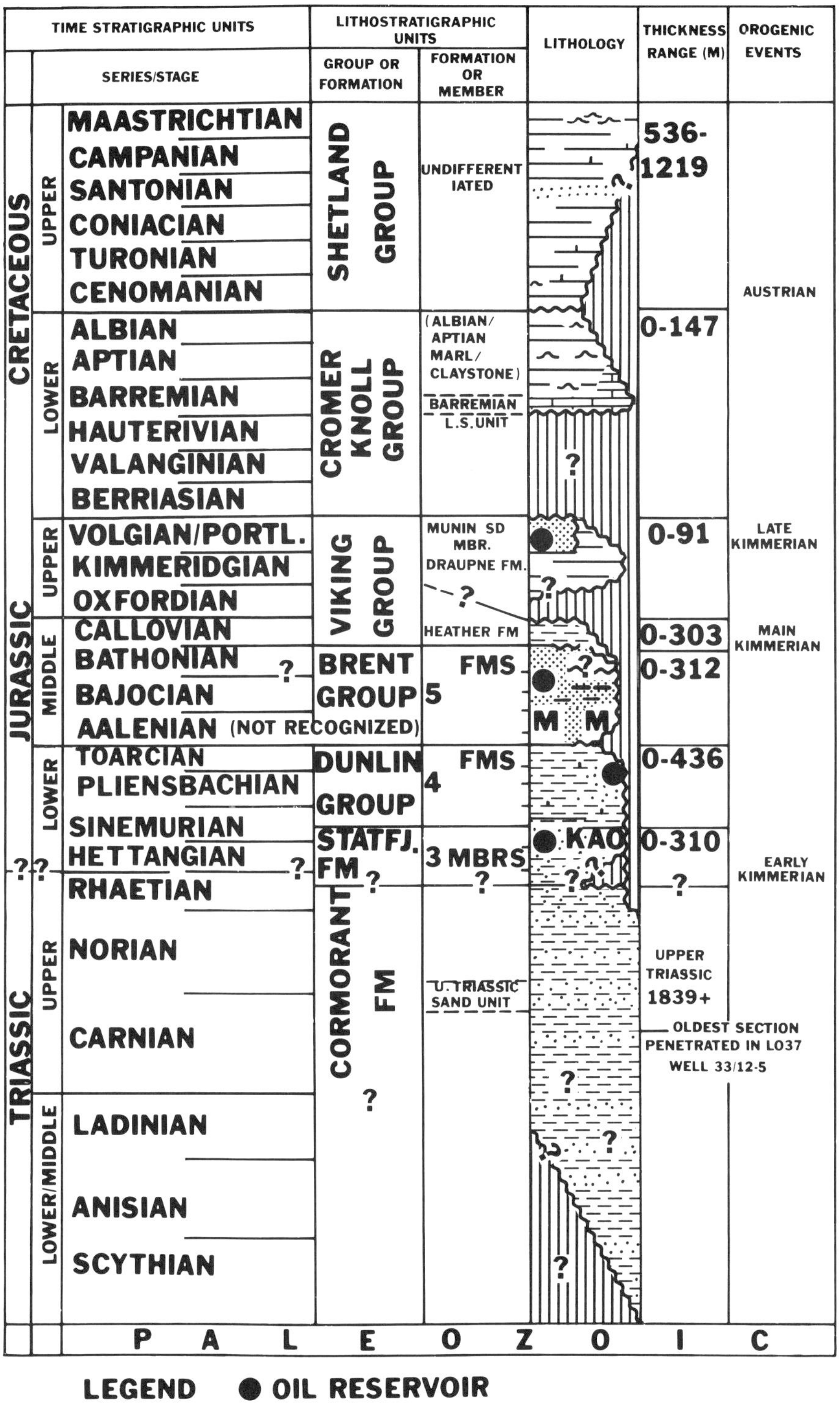

Fig. 4. Lithostratigraphic column for Licence 037 area (modified after Kirk, 1980).

been done since 1978. The next mapping of the Beta prospect will be based on the three-dimensional survey shot in 1986.

TRAP

The 33/9 Beta structure at the Brent Group and Statfjord Formation levels is a NW-tilted fault-block bounded on the south-east by the NE-trending Murchison fault (Fig. 6). It is a complex structural–stratigraphic trap.

The Upper Jurassic Munin sand reservoir appears to be a localized sand build-up and may be an isolated sand body. Stratigraphically equivalent shales are present down-dip in the nearby 33/9-6 Well. The areal extent of this sandstone is unknown and it may be

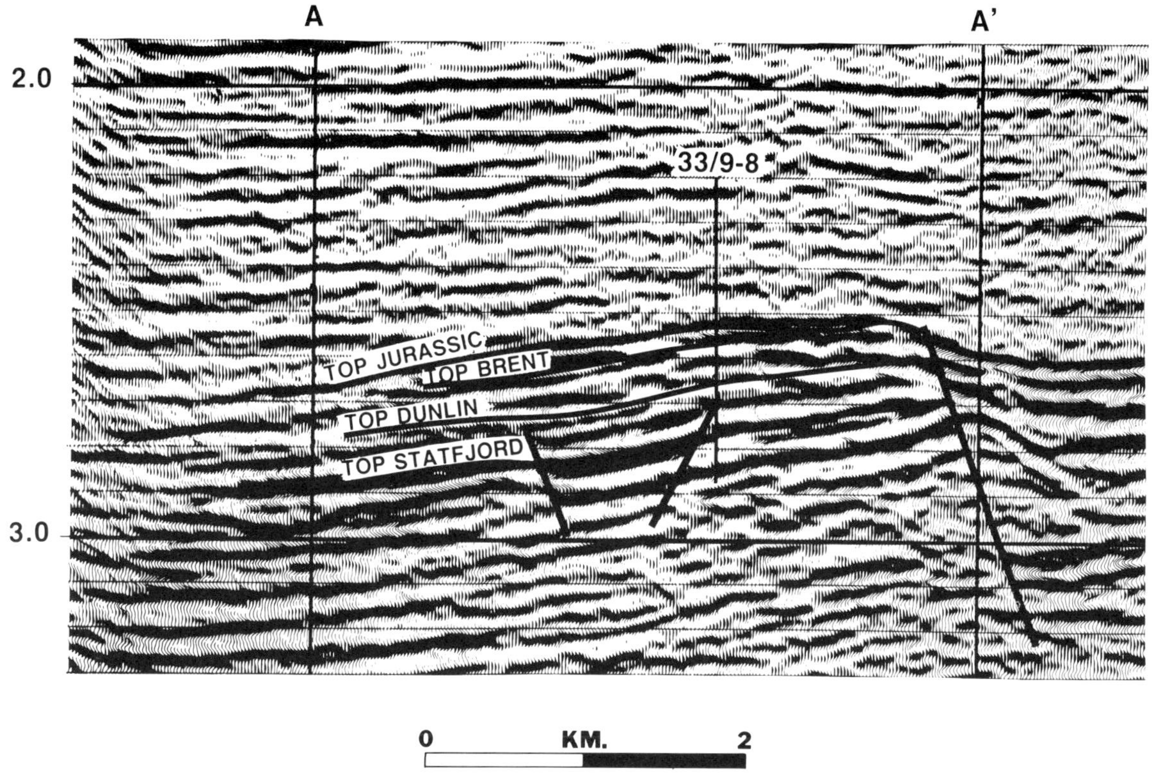

Fig. 5. Seismic profile along A–A' shown in Fig. 6.

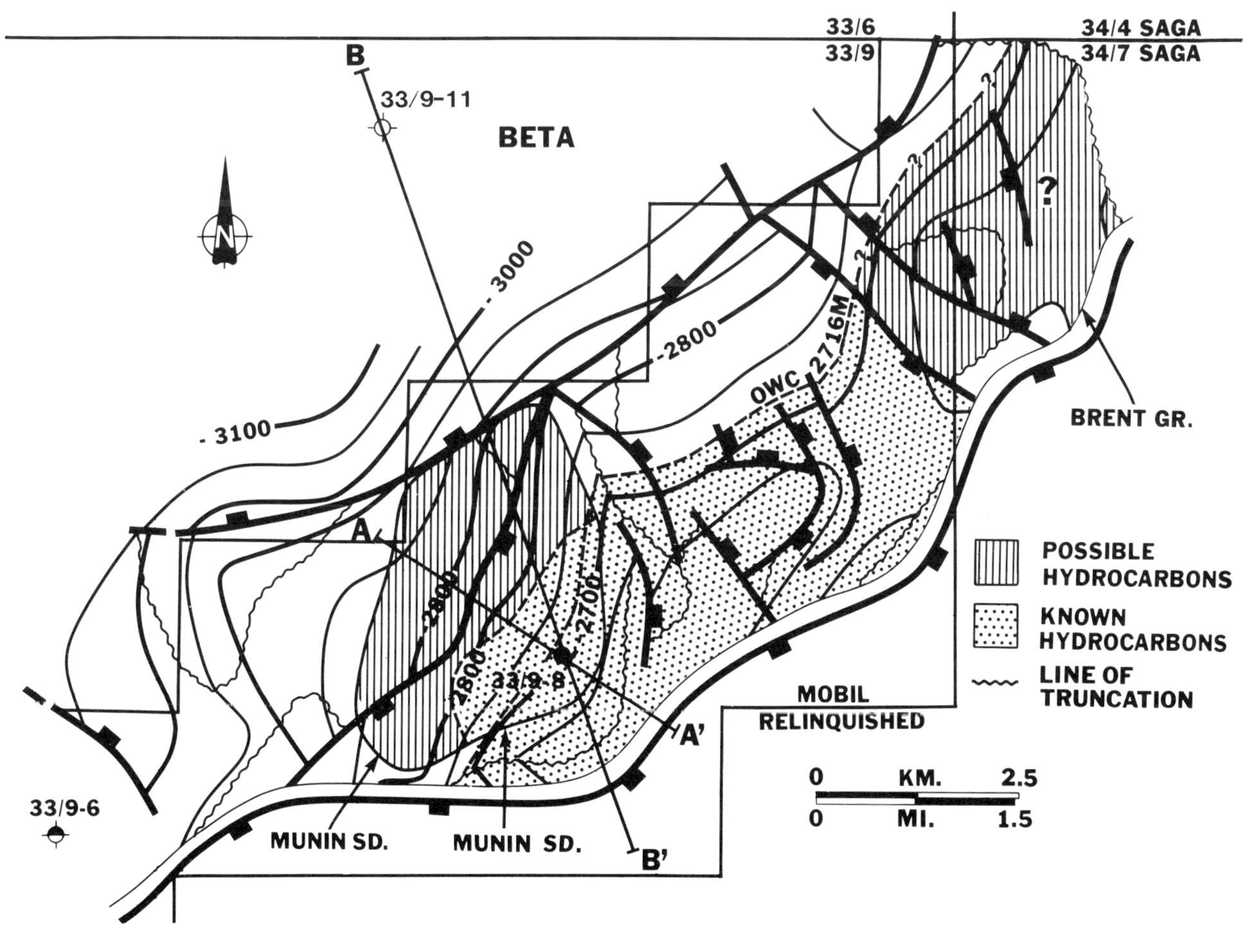

Fig. 6. Beta discovery structure at top of Brent Group level. Depth in metres. Contour interval 50 m. Munin sand member is superimposed.

restricted to the south-western part of the Beta high. It has not been drilled elsewhere within the licence area, although a possible coeval 'beach deposit' has been identified in the 33/9-9 Well in the Statfjord Field. Drill-stem tests indicate the probability that the sandstone is areally extensive but of indeterminate size based upon the well test data available. The sand-body geometry is unknown, although it may be indicated by isopaching between the top Brent and 'J' unconformity seismic horizons.

It is believed that the Brent Group is truncated south and east of the 33/9-8 Well, in which a complete Brent sequence is present (Figs 6 and 7). Seismic data indicate extensive erosion of the Brent Group along the southern-bounding fault in areas not yet drilled. However, prior to drilling the south flank of Murchison Field it was believed that the Brent was eroded there also. Since the same major fault trend which borders Murchison Field to the south also borders Beta, it is possible that erosion is minimal here too.

The structure is intensely faulted with major down-to-the-south-east faults paralleling the Beta structural high. An extensive 'secondary' system of faults is perpendicular to this trend. A horst-block formed by the NW–SE-trending fault system is prominent on the north-east part of the structure west of the boundary between Blocks 33/9 and 34/7. The horst's location and probable sealing character may preclude the possibility of hydrocarbons extending into Block 34/7. Harding (1984) described similar geometries from adjacent North Sea oil fields.

The oil–water contact in the Brent Group is at 2716.5 m subsea based on drill-stem tests. The lowest known oil in the Upper Jurassic Munin sand is at 2680 m subsea. However, the oil–water contact in this reservoir may be identical to that of the Brent Group. Although the Statfjord Formation was wet in Well 33/9-8, hydrocarbons may occur in it trapped against the down-faulted Dunlin Group in the south-eastern corner of the structure above 3030 m subsea.

RESERVOIR

The primary reservoir sandstones at Beta belong to the Middle Jurassic Brent Group. The secondary reservoir sandstone is the Munin sand member of the Draupne Formation, found immediately above the Heather Formation. Communication between the two

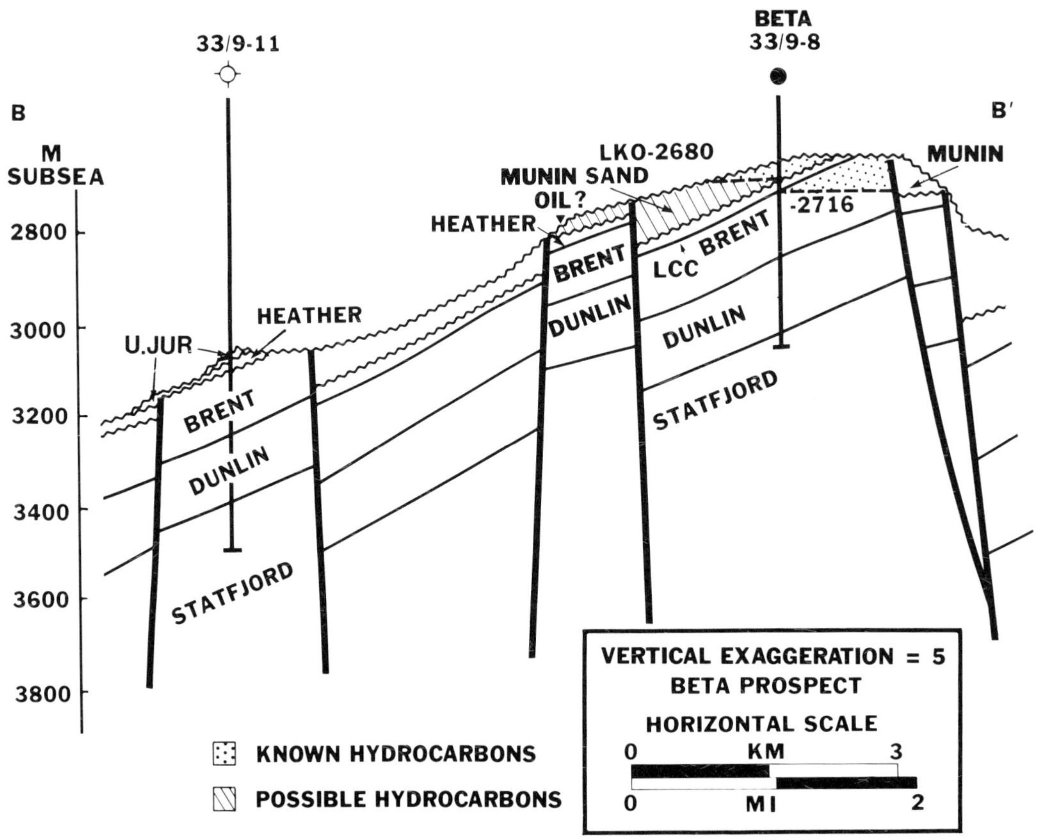

Fig. 7. Structural cross-section through the Beta prospect. Location of the section is shown in Fig. 6.

PERIOD/SERIES GROUP/FM	LITHOLOGY	THICKNESS RANGE(M)	1 TOC	2 EOM	EOM/ TOC	3 CPI	4 RO
PALEOCENE ROGALAND GP		167-306	0.43 (64)	141 (3)	3.3 (3)	1.45 (3)	0.41 (3)
UPPER CRET. SHETLAND GP		536-1219	0.78 (201)	172 (40)	2.64 (40)	1.71 (20)	0.43 (12)
LOWER CRET. CROMER KNOLL GROUP		0-147	1.04 (4)	78 (2)	1.41 (2)	—	0.71 (2)
U. JURASSIC DRAUPNE FM (HOT SHALE)	●	0-91	4.58 (4)	2160 (4)	4.00 (4)	—	—
M. JURASSIC HEATHER FM		0-303	2.24 (16)	350 (9)	1.71 (9)	1.68 (9)	0.40 (2)
M. JURASSIC BRENT GROUP	●	0-312	—	—	—	—	0.44 (5)
L. JURASSIC DUNLIN GP	●	0-436	1.47 (86)	234 (38)	1.95 (38)	1.68 (36)	0.59 (6)
L. JURASSIC /TRIASSIC STATFJORD FM	●	0-310	0.44 (19)	—	—	—	0.68 (4)
TRIASSIC CORMORANT FM		>1839	0.35 (56)	50 (6)	1.71 (6)	1.37 (3)	0.73 (8)

LEGEND

() NUMBER OF VALUES

● OIL RESERVOIR

1. TOTAL ORGANIC CARBON (% ROCK WEIGHT)
2. EXTRACTABLE ORGANIC MATERIAL (PPM)
3. CARBON PREFERENCE INDEX
4. VITRINITE REFLECTANCE

Fig. 8. Licence 037 source evaluation data averaged by stratigraphic unit.

reservoir sand-bodies has not yet been demonstrated. Both the Upper Jurassic Munin sandstones and the Brent sandstones have excellent reservoir properties.

The Munin sand member's gross thickness is 46 m and it has a very high net-to-gross ratio. Rip-up clasts are present within the sandstone matrix in the upper part of the unit. They represent deposits emplaced by short-lived, storm-generated marine currents. These are intercalated with massive to laminated, occasionally glauconitic sands, interpreted as offshore bars and shoals. The lowest part of the sequence consists of intercalated shelf-muds and thin, ripple-laminated sands. Porosities within the reservoir range up to 30% and average 23%. Permeability is as high as 5300 mD and averages 2900 mD. The sand tested at over 1900 Sm^3/day oil.

The Brent sandstone's gross thickness is 140 m and the net-to-gross ratio is high. The lowermost Brent sandstones are fine grained and micaceous. These are overlain by coarser and cleaner sandstones, and at the top of the section are found poorly sorted argillaceous sands. Porosities range up to 27% and average 22%. Permeability averages 1500 mD.

SOURCE

The organically rich shales of the Upper Jurassic (Kimmeridgian) Draupne Formation are interpreted to be the major source of oil for the 33/9 Beta discovery. Other rock units in the area appear to have marginal source potential, as pointed out by Kirk (1980). The Draupne Formation contains predominantly oil-prone kerogen in the Licence 037 area (Fig. 8).

The stratigraphic/structural geometry of the area ensured clear migration paths to both the Munin and Brent sandstones. Draupne Formation oil shales are represent down-dip of the Munin sand member of the Draupne Formation which may be in fluid communication with the Brent Group. In places, the Munin sand may be completely interbedded with the Kimmeridgian source-rocks.

HYDROCARBONS

Drill-stem tests in the 33/9-8 Well established an oil–water contact at 2716 m subsea. Flow rates of up to 1004 Sm^3/day were established in the Brent Group and up to 1997 Sm^3/day in the Upper Jurassic Munin sand member. Oil gravity in the Upper Jurassic Munin sand is 0.835 g/cm^3 and GOR 60 Sm^3/Sm^3. Oil gravity in the Brent is 0.838 g/cm^3 and GOR is 60 Sm^3/Sm^3.

RESOURCES

Recoverable resources in the Beta discovery are estimated at 36–66 × 10^6 Sm^3 oil. The wide range reflects the uncertainties of estimating total resources based solely upon a single well, the complexity of reservoir and structure geometry, and the quality of seismic data acquired to date.

ACKNOWLEDGEMENTS

This article is based extensively on a number of restricted company reports not listed in the references. The writers also gratefully acknowledge the help and assistance of Mobil Exploration Norway Incorporated staff, in particular I. T. Nyberg, for his invaluable assistance and criticisms. The writers are grateful to Mobil, Statoil and all of the Licence 037 partners for permission to publish this paper.

REFERENCES

Deegan, C.E. and Scull, B.J. (Compilers), 1977. A proposed standard lithostratigraphic nomenclature for the Central and Northern North Sea, *Bull. Norwegian Petroleum Directorate No. 1; Inst. Geol. Sci. Rept.* 77/25.

Harding, T.P. 1984. Graben hydrocarbon occurrences and structural style, *American Association Petroleum Geologists Bulletin* **68,** 333–362.

Kirk, R.H. 1980. Statfjord Field—a North Sea giant, *American Association of Petroleum Geologists Memoir 30,* 95–116.

Rawson, P.F. and Riley, L.A. 1982. Latest Jurassic–Early Cretaceous events and the Late Cimmerian unconformity in the North Sea area, *American Association of Petroleum Geologists Bulletin* **66,** 2628–2648.

Vollset, J. and Doré, A.G. (Eds). 1984. A revised Triassic and Jurassic lithostratigraphic nomenclature for the Norwegian North Sea, *Norwegian Petroleum Directorate Bull. 3.*

Ziegler, P.A. 1979. Factors controlling North Sea hydrocarbon accumulations, *World Oil,* November 1979, 111–124.

Ziegler, W.H. 1975. Outline of the geological history of the North Sea, In Woodland, A.W. (Ed.), *Petroleum and the Continental Shelf of North-west Europe* Vol. 1, Applied Science, Barking, pp. 165–190.

30

Statfjord Øst

I. T. Nyberg

Mobil Exploration Norway Incorporated, PO Box 510, N-4001 Stavanger, Norway

The 33/9 Alpha* discovery, a satellite structure to Statfjord Field, is located on the western margin of the Viking Graben. The hydrocarbon accumulation straddles the boundary between Blocks 33/9 and 34/7. The potential field was discovered in late 1976 by the 33/9-7 Well which was drilled to a total depth of 3101 mss into Triassic sediment and proved oil in the Middle Jurassic Brent Group. A complete Brent sequence with 60 m of gross pay was encountered. Hydrocarbons are of black oil type and with a gravity of 0.84 g/cm^3 at 15.5 °C. The well also penetrated the Dunlin Group and the Statfjord Formation. The Alpha structure is a NE–SW-orientated fault-block tilted to the west. Resources are estimated to be in the range of 14.5–40 × 10^6 Sm^3 oil, depending both on whether the Statfjord Formation is hydrocarbon bearing and on the horizontal extent of the Brent Group. The trap is of a structural/stratigraphic nature, as the reservoir is both bounded by faults and truncated by the Upper Jurassic unconformity. At the present time two wells have been drilled on the Alpha structure, 33/9-7 and 34/7-5. The development of the Alpha structure is tied to the start of the Statfjord Field decline and may be done by a conventional subsea system.

INTRODUCTION

The 33/9 Alpha structure is located close to the Norwegian–UK boundary, approximately 200 km west of Bergen on the Norwegian mainland and 500 km north-east of St Fergus on the Scottish coast. Its closest neighbour is the giant Statfjord oil field, three kilometres to the west. The Alpha structure straddles the boundary between Blocks 33/9 and 34/7. Water depth in the area ranges from 200 to 300 m (Fig. 1).

So far, two wells, 33/9-7 and 34/7-5, have been drilled in the structure. The structure is an elongated, west-tilted fault-block bounded to the east by faulting and truncated by the Upper Jurassic Unconformity.

Resources are estimated at approximately 14.5 × 10^6–40 × 10^6 m^3 of oil, depending upon whether the Statfjord Formation is oil bearing in addition to the Brent Group, and on the possibilities of horizontal communication between the Brent Formation in the different areas.

The field has not yet been declared commercial but, depending upon the results of further appraisal drilling in 1987, plans are to put it on stream in early 1990. The present development plan of the Alpha structure is tied to the start of the Statfjord Field decline in the beginning of the 1990s. The production may be done by a conventional subsea system and connected to the Statfjord 'C' platform.

* Editor's note: Throughout this article the discovery is referred to as 33/9 Alpha, but in January 1987 the name 'Statfjord Øst' was officially approved.

Geology of the Norwegian Oil and Gas Fields © A. M. Spencer *et al.*, eds (Graham & Trotman, 1987) pp. 351–362.

Statfjord Øst
Field data summary — 33/9-7 Well

Trap	
Type	Structural/stratigraphical
Depth to crest	2430 mss
Lowest closing contour	2450 m Top Jurassic Level
Oil–water contact	2507 mss
Gas column	Not present
Oil column	*c.* 60 m
Pay zone	
Formation	Brent Formation
Age	Middle Jurassic
Gross thickness (average)	223 m (33/9-7 Well)
Sand/shale ratio	0.78
Net/gross ratio (average)	0.78
Cut off for N/G	0%
Porosity (average)	26.3%
Hydrocarbon saturation	82.1%
Permeability (average; range)	1120; 460–1700 mD
Productivity index	74 Sm^3/day/bar
Hydrocarbons	
Oil gravity	0.84 g/cm^3 at 15.5 °C
Oil type	black oil
Gas gravity (relative to air)	0.69
Bubble-point	245.2 bar
Gas/oil ratio	(160 Sm^3/Sm^3)
Formation volume factor	1.5
Formation water	
Salinity	22 000 ppm NaCl equivalent
Resistivity	0.110 ohm m at 63 °C
Reservoir conditions	
Temperature	76 °C
Pressure	337 bar
Field size	
Area	13.2 km^2
Gross rock volume	1.6×10^9 m^3
Recovery factor—oil	40%
Recoverable hydrocarbons	
oil	14.5–40 $\times 10^6$ m^3
gas	2.5–5.4 $\times 10^9$ m^3
Production	
Start-up date	Early in 1990s
Development scheme	8–11 subsea wells
Production rate	16 500 Sm^3/day

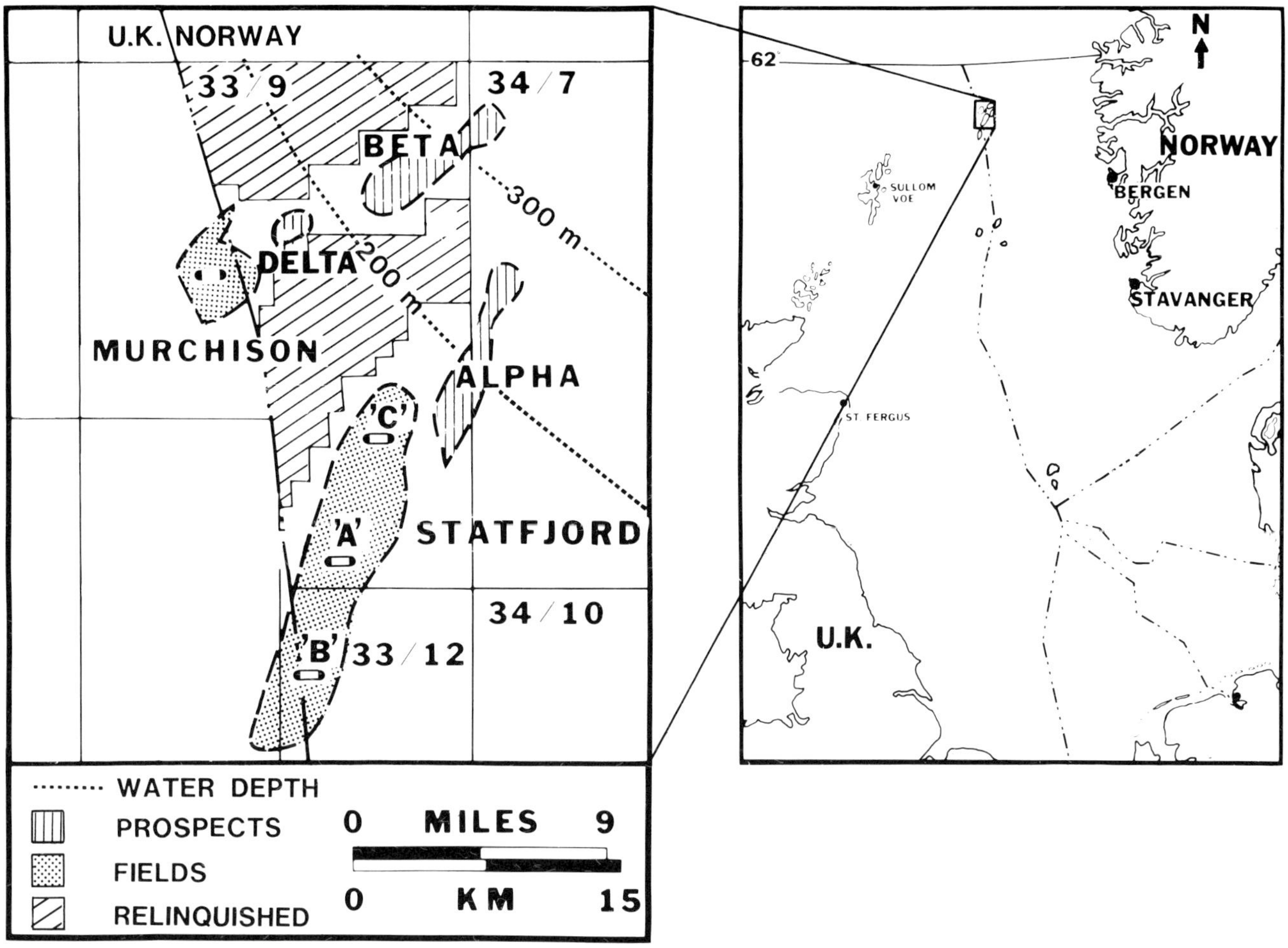

Fig. 1. Location map.

HISTORY

Licence 037, which includes Blocks 33/9 and 33/12, was awarded in August 1973. Mobil Exploration Norway Incorporated is the operator for Statoil (50%), Mobil (15%), Conoco (10%), Shell (10%), Esso (10%), Saga (45/24%), Amoco (25/24%), Amerada (25/24%) and Texas Eastern (25/24%). The Alpha structure extends into Block 34/7, which was awarded to a different group of seven companies in 1984 as Licence 089. Saga is the operator for this block on behalf of Statoil (50%), Norsk Hydro (12%), Saga (10%), Elf (8%), Esso (15%), Deminex (4%) and DNO (1%). Licence 037 expires in August 2009. All obligatory relinquishments were completed by August 1979. The primary term of Licence 089 expires in 1990, at which time 50% must be relinquished. Established production should extend the licence term to the end of 2020.

During the time-period 1973–75, several seismic surveys were carried out. Based on these seismic surveys, the Alpha structure discovery well, 33/9-7, was drilled in late 1976 by Mobil as the last obligation well in Licence 037. The well reached TD in Triassic sediments at 3101 mss, and proved hydrocarbons in the Middle Jurassic, Brent Group.

Additional seismic was shot in 1979. The reinterpretation of these data better defined the Alpha prospect outline and structure. In 1982 and 1983 more seismic was shot in the area and provided evidence of internal faulting within the structure.

In early 1985 Saga drilled the 34/7-5 Well north-east of 33/9-7. The 34/7-5 Well penetrated oil-bearing sandstone in the same Middle Jurassic Brent Group as in 33/9-7. The well increased the understanding of the Alpha prospect and confirmed the overall evaluation.

Development/appraisal plans have been ongoing during 1984 and 1985. Effort is being made to time the production start-up to coincide with the Statfjord Field decline and to unitize the structure between licences 037 and 089.

GEOLOGICAL SETTING

The Alpha structure is located in the northern part of the Viking Graben system within the East Shetland Basin close to the Statfjord Field (Fig. 2).

The 33/9-7 Well penetrated a Triassic to present-day sedimentary sequence (Figs 3 & 4). Sedimentation in the East Shetland area probably began during Permo-Triassic time and continued through the Mesozoic and Tertiary. Interruptions of sedimentation occurred during the Jurassic and Early Cretaceous (Ziegler, 1975).

During Permian time, establishment of an intracratonic basin took place. The following rifting during Permo-Triassic time established the East Shetland Basin and began an extensional stress regime which influenced North Sea tectonics and sedimentation throughout Mesozoic time. Reactivation of older NE-trending fault systems was probably the prime control of Early Jurassic–Late Triassic sedimentation. This taphrogenic tectonic activity caused block faulting, tilting and erosion during the entire Jurassic period. This activity was most probably of major influence on Late Jurassic and Early Cretaceous sedimentation. It has been interpreted that the main period of rotational block-faulting and erosion took place during the deposition of the Middle to Upper Jurassic Heather Formation.

In the Licence 037 area, continuing syndepositional fault-block rotation and gradual basin subsidence, together with transgression, were the main factors controlling the distribution of the organically rich Upper Jurassic shale/claystone sediments. Periods of uplift and erosion occurred at the end of the Jurassic period and in the Early Cretaceous, creating the post-rift geometry shown in Fig. 2. During the Late Cretaceous, continuous basin subsidence became the main factor controlling sedimentation in the area.

In the Alpha prospect area the depositional environment for the Statfjord Formation represents a transition from continental-environment, alluvial-plain and braided-stream deposits, to coast-environment, transgressive deposits (Fig. 3). The resulting sediment consisted of interbedded sand, silt and clay. The Dunlin Group was deposited in a marine environment. Water depth changed several times during the deposition, and the sediments were mainly clay and shale with minor silt and sand. The Brent Group sands reflect a wave-tide interactive delta system. Facies of thick sand-beds as well as interbedded sands, silts and clays exist.

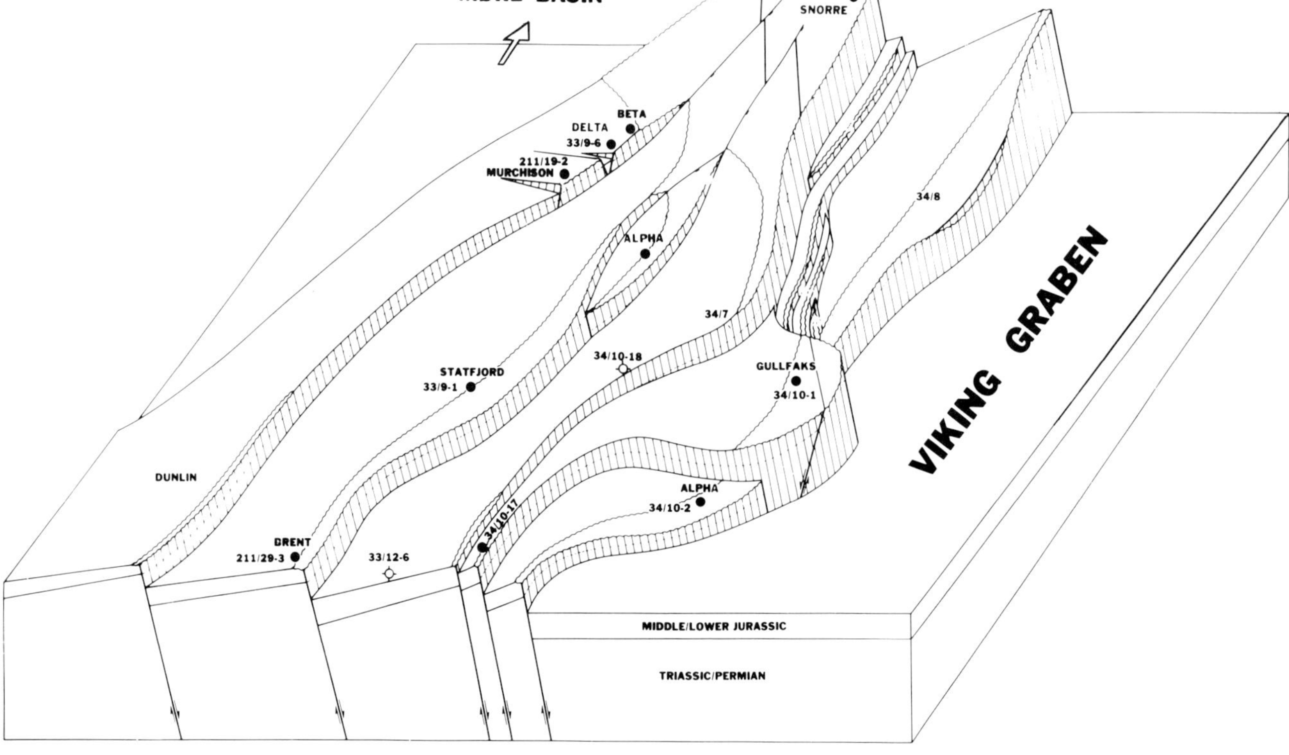

Fig. 2. Structural setting of 33/9-Alpha: block diagram showing post-rift geometry and erosion.

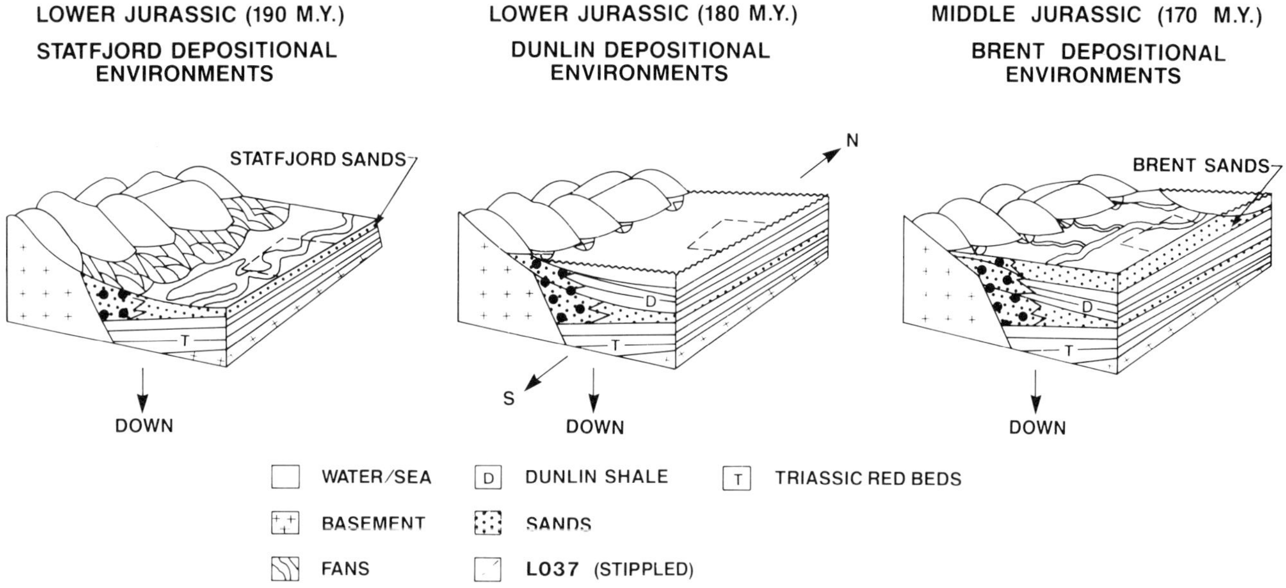

Fig. 3. Simplified Lower and Middle Jurassic depositional models.

GEOPHYSICS

Data from the 1973 MNG and the 1974 MNG.5 seismic surveys, which covered the whole Licence 037 area in a 1 × 1 km grid, were used for the first mapping of Alpha structure. The quality of the data was poor to fair. Well 33/9-7 was located from maps based on these surveys.

The remapping done in 1978 included some MNG.25 and MNG.75 seismic lines shot in 1976 and some SF lines shot in 1977. About 50 km of the MNG.25 and .75 surveys cover the northern part of the Alpha structure, while about 100 km of the SF survey, which was shot over the Statfjord Field, cover the southern part of the Alpha structure. Over the Alpha structure both these surveys consist of parallel lines with 1/2-km line-spacing and WNW–ESE orientation. The quality is poor to good at the objective Jurassic horizons.

The latest (1983) seismic mapping of the Alpha structure was based on SG8231 data from the 'Tampen Spur Group Seismic' shot in 1982. The lines from this survey form a 1 × 1 km grid with line orientation E–W and NW–SE over the whole Alpha structure. Qualitatively, the 1982 data are the best data acquired so far (Fig. 5). A three-dimensional survey was shot over the Alpha structure in 1986.

TRAP

The trap in the Alpha structure area is a NE-trending fault-block and is of a stratigraphical/structural type (Figs 6–9). The hydrocarbon-bearing formations have been tilted to the west and truncated by Upper Jurassic unconformities. The Brent and Dunlin Group have also been faulted on several occasions and show internal faulting. Most of the tilting occurred during Jurassic times. The truncation of the tilted structure occurred in the late Middle Jurassic to Late Jurassic interval. The reservoir sandstones of the Brent Group are sealed by the overlying Viking Group shales and clays. Faulting was active during the whole period of Brent and Dunlin Group development.

Closure of the structure is provided by the eastern truncation line and by the eastern and northern faults. The last closing contour, at Top Jurassic level, lies at 2450 mss (Fig. 9).

RESERVOIR

The proven hydrocarbon-bearing reservoir in the Alpha structure are deltaic sandstones of the Middle Jurassic Brent Group. Up-dip, possible hydrocarbon-bearing Statfjord Formation remains yet to be proven.

The lithostratigraphic nomenclature described by Deegan and Scull (1977) for the Brent Group is valid in the Alpha structure area.

The Brent Group is up to 220 m thick and can be subdivided into:

- lower Brent unit consisting of very fine- to coarse-grained sandstone, showing a coarsening-upwards sequence, containing mica,

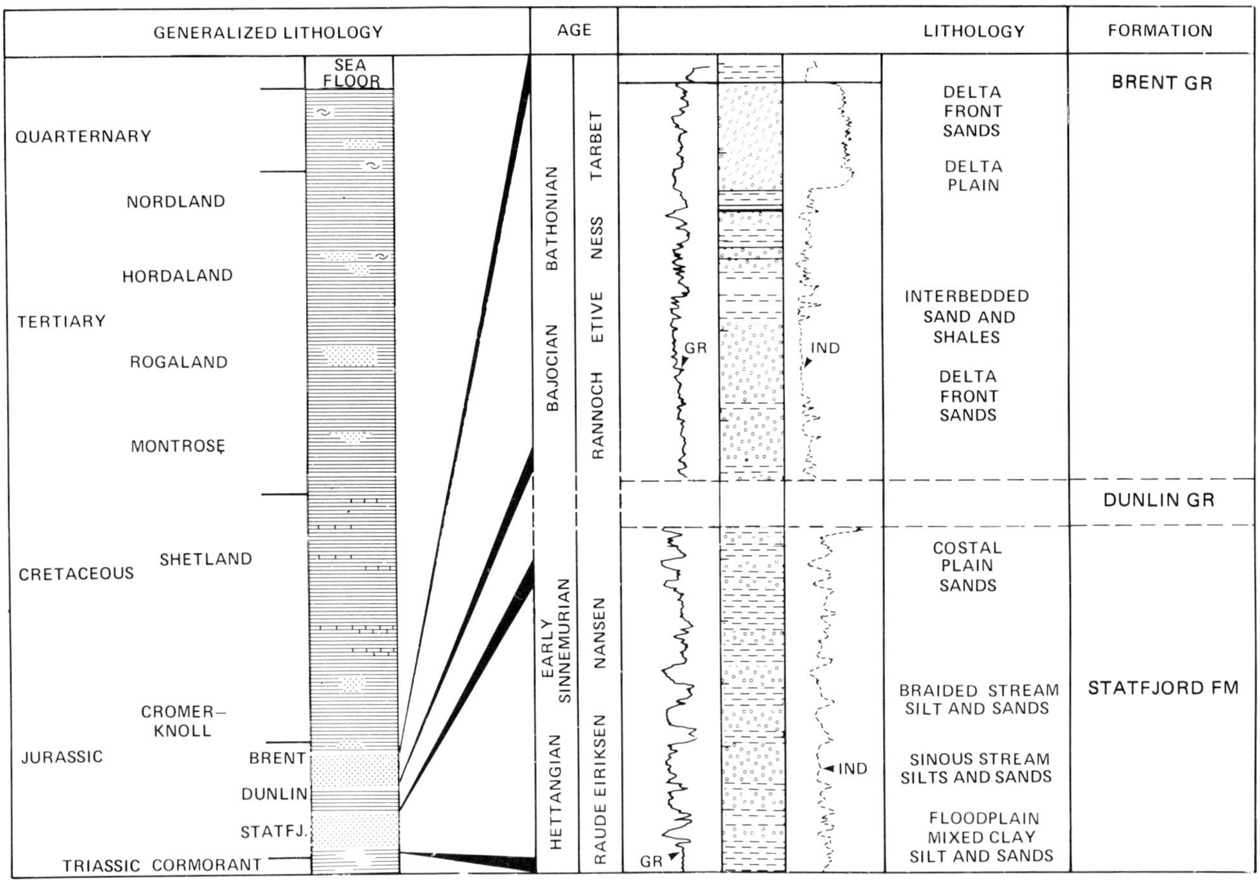

Fig. 4. Generalized stratigraphic column for the 33/9-7 Well.

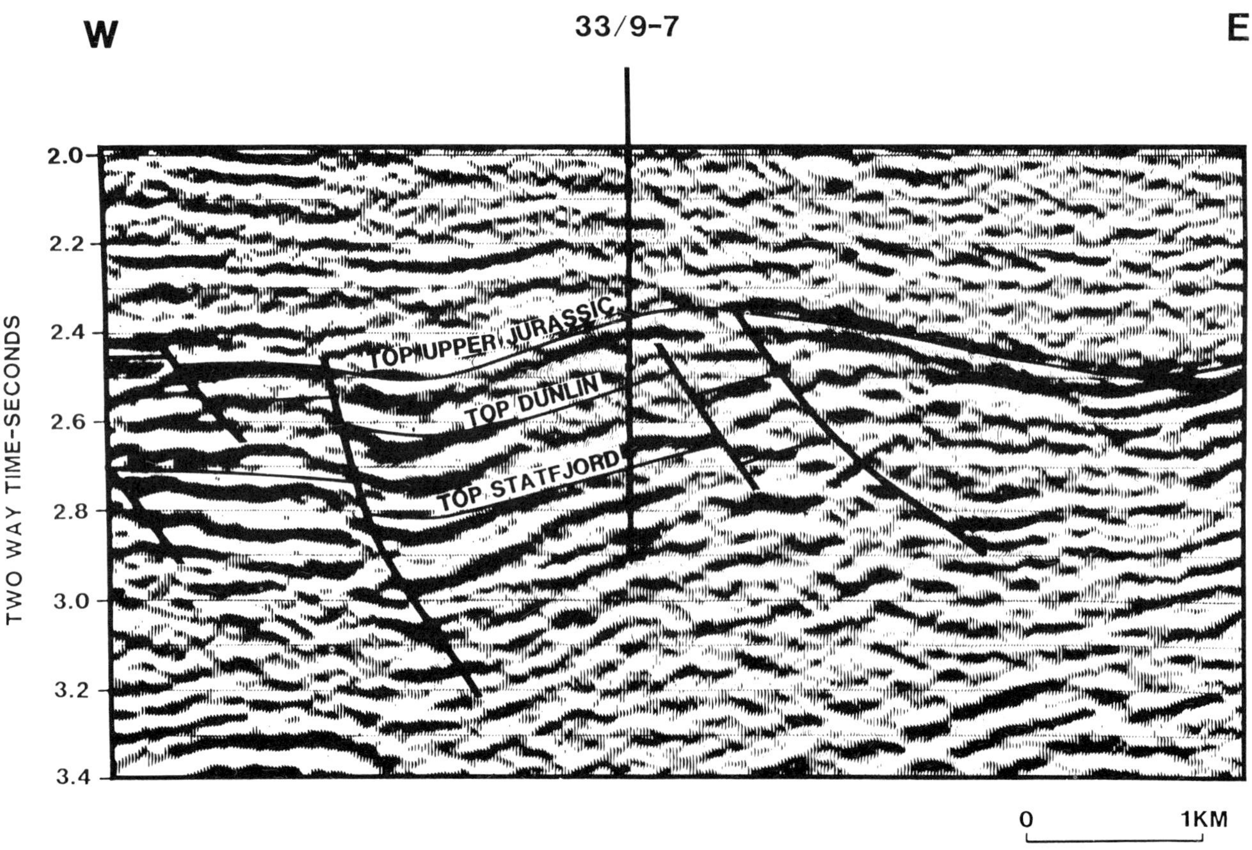

Fig. 5. 1982 Seismic line through Well 33/9-7. See Fig. 9 for location.

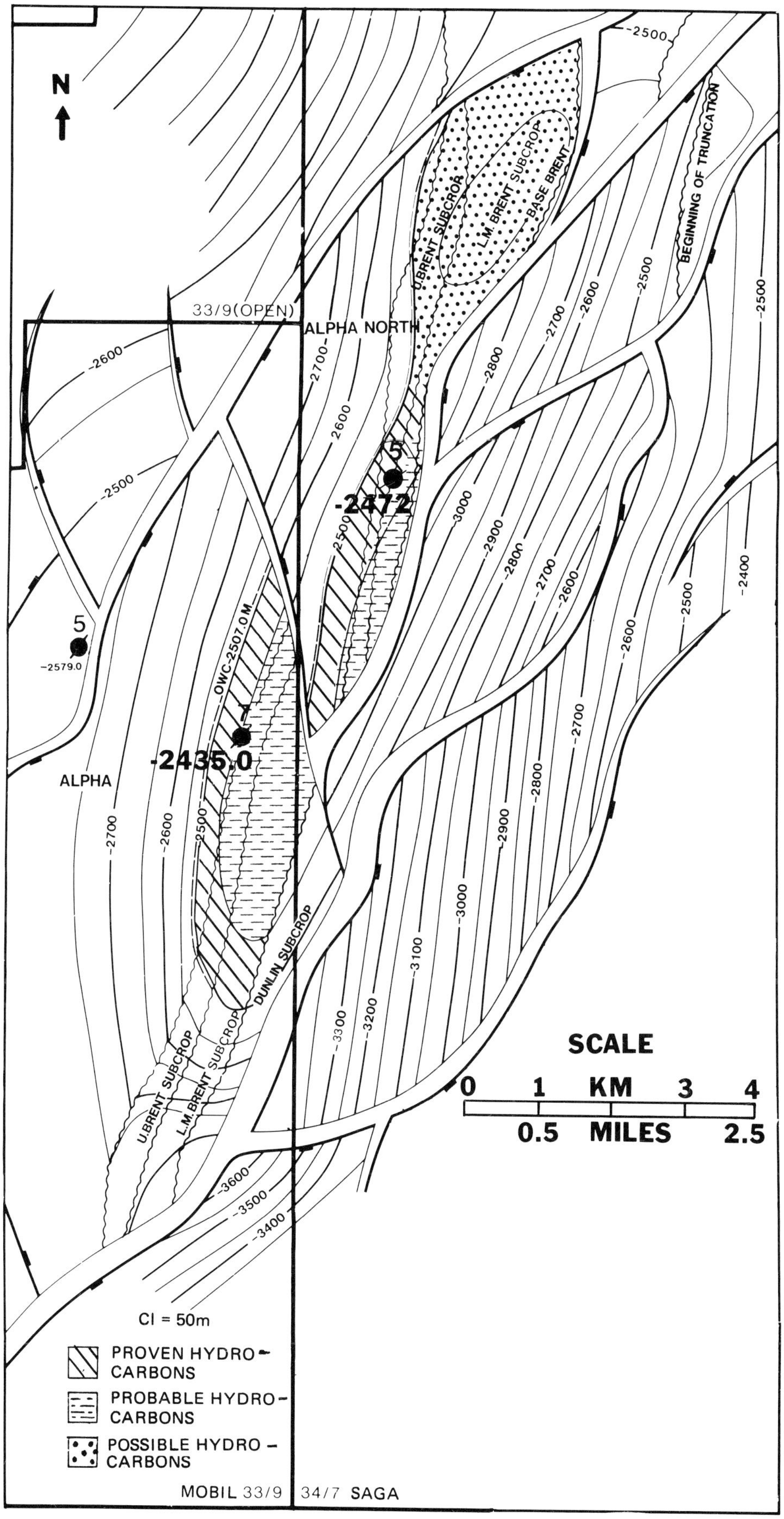

Fig. 6. Structure map of Top Brent Group.

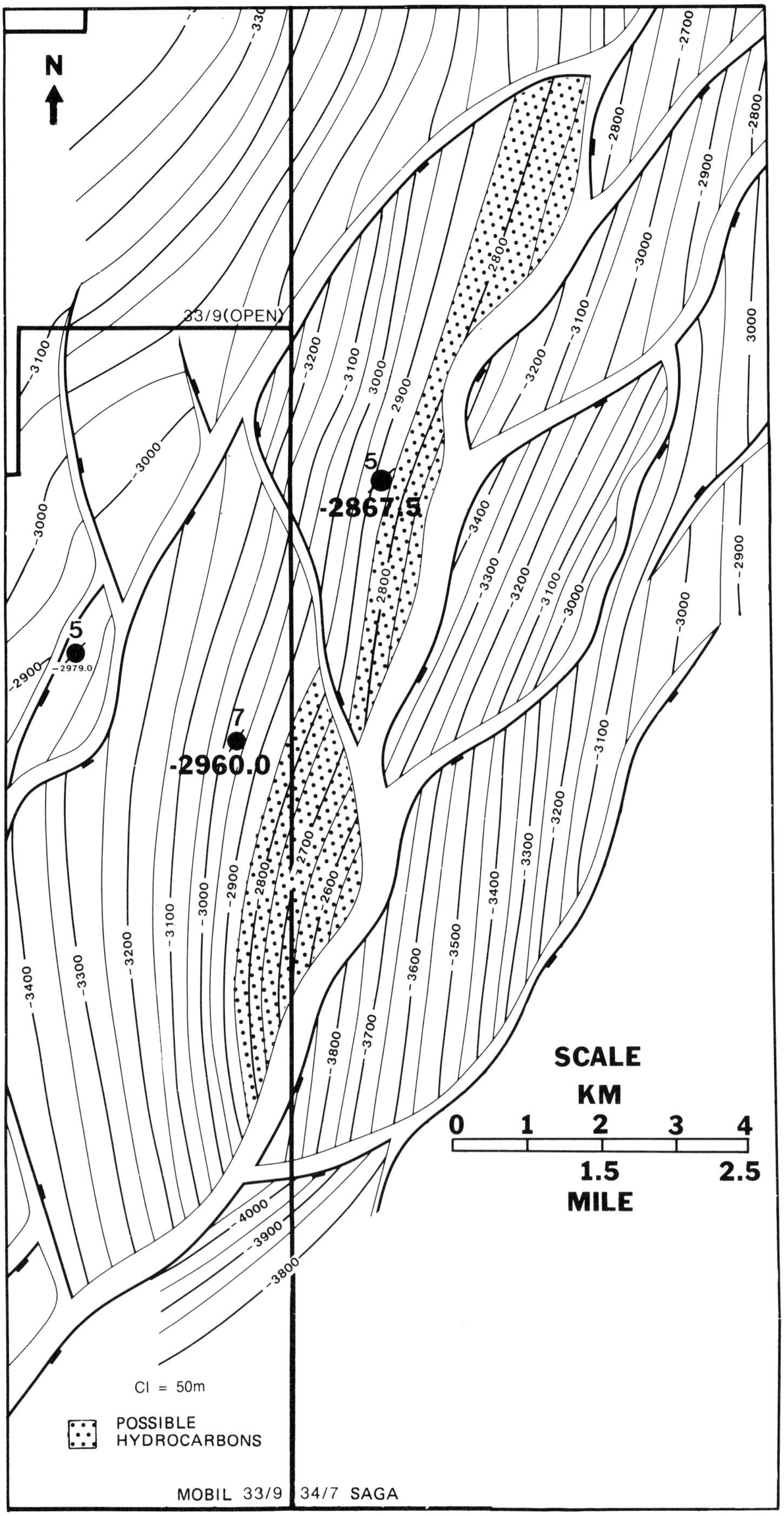

Fig. 7. Structure map of Top Statfjord Formation.

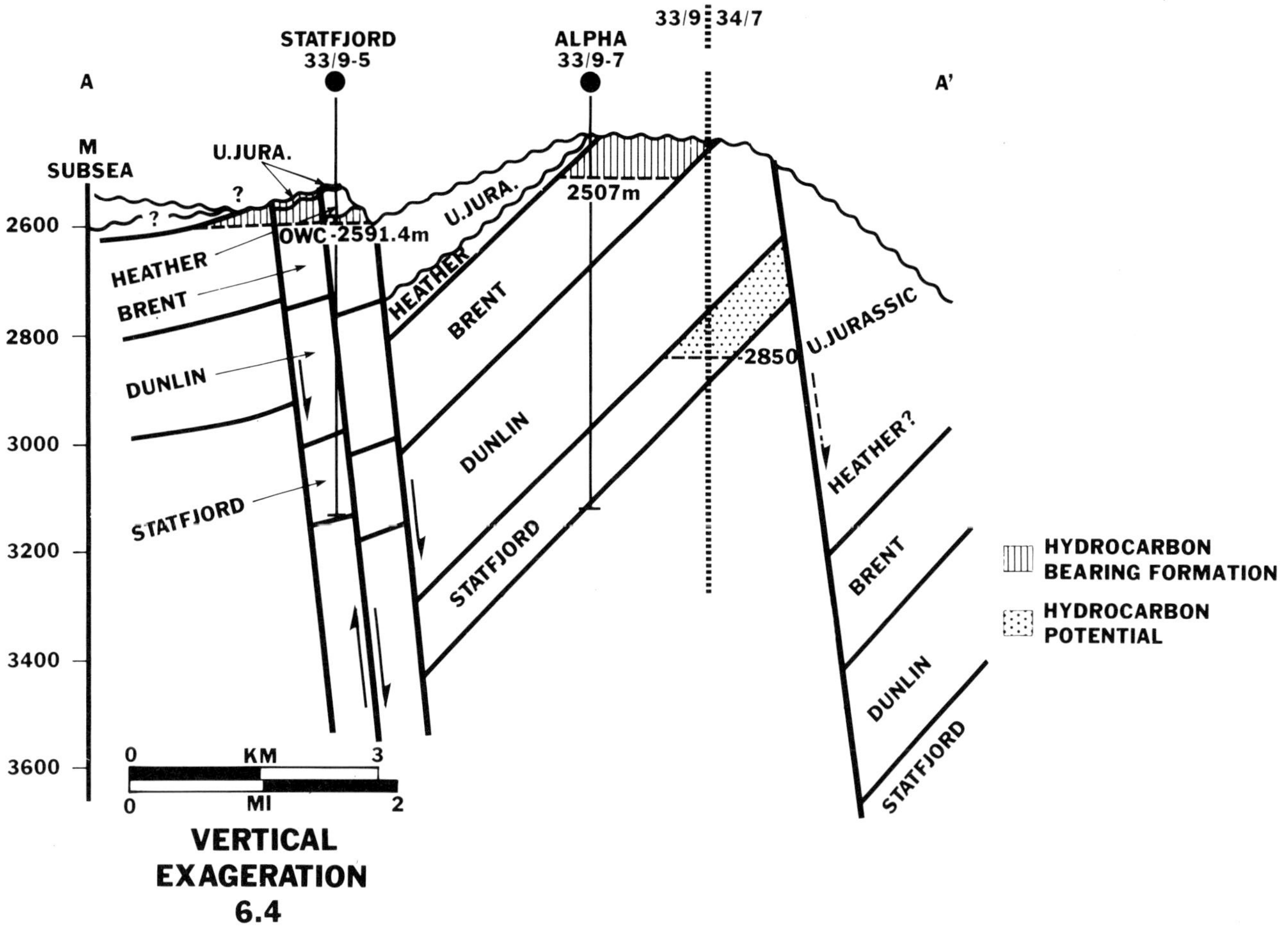

Fig. 8. Structural profile. See Fig. 9 for location.

pyrite and, locally, calcite cemented sediment. This unit represents the build-out for a progradational bay.

- middle Brent unit with interbedded fine- to medium-grained sandstone, siltstone, claystone and thin coal beds. It represents delta-top deposition behind a coastal barrier sand.
- upper Brent unit sandstone representing delta-top and various shoreline environments.

A further subdivision into five reservoir units is also made and was described by Deegan and Scull (1977), where the lower Brent consists of the Broom, Rannoch and Etive members, the middle Brent is the Ness Member, and the upper Brent is the Tarbert Member. In 1984 Vollset and Doré upgraded the Brent Group members to formations. The lithostratigraphic description will then be, from oldest to youngest:

1. The Broom Formation is a coarse, unsorted basal sandstone. In the Alpha prospect it is not present, probably due to nondeposition.

2. The Rannoch Formation is a basal micaceous sandstone showing a coarsening upward sequence. The top of this unit is in vertical communication with the overlying massive sandstone. The porosity of this zone averages above 20%. A study by Robertson Research (1980) postulated that the mica content may cause minor permeability problems in the basal parts or where the mica has been locally concentrated. A further permeability problem in the water zone might be caused by diagenetic illite. Illite has been found in several cores where new core-analysis techniques have been used to determine it. When the illite is present near the oil–water contact, restriction in the water-drive mechanism may occur.

3. The Etive Formation is a massive sandstone reflecting a coastal-barrier environment. This sand is vertically very homogeneous and has an average porosity greater than 25%.

4. The Ness Formation consists of interbedded sandstone, siltstone and clay with minor coal. This unit contains several reservoir sandstone bodies of varying geometry and

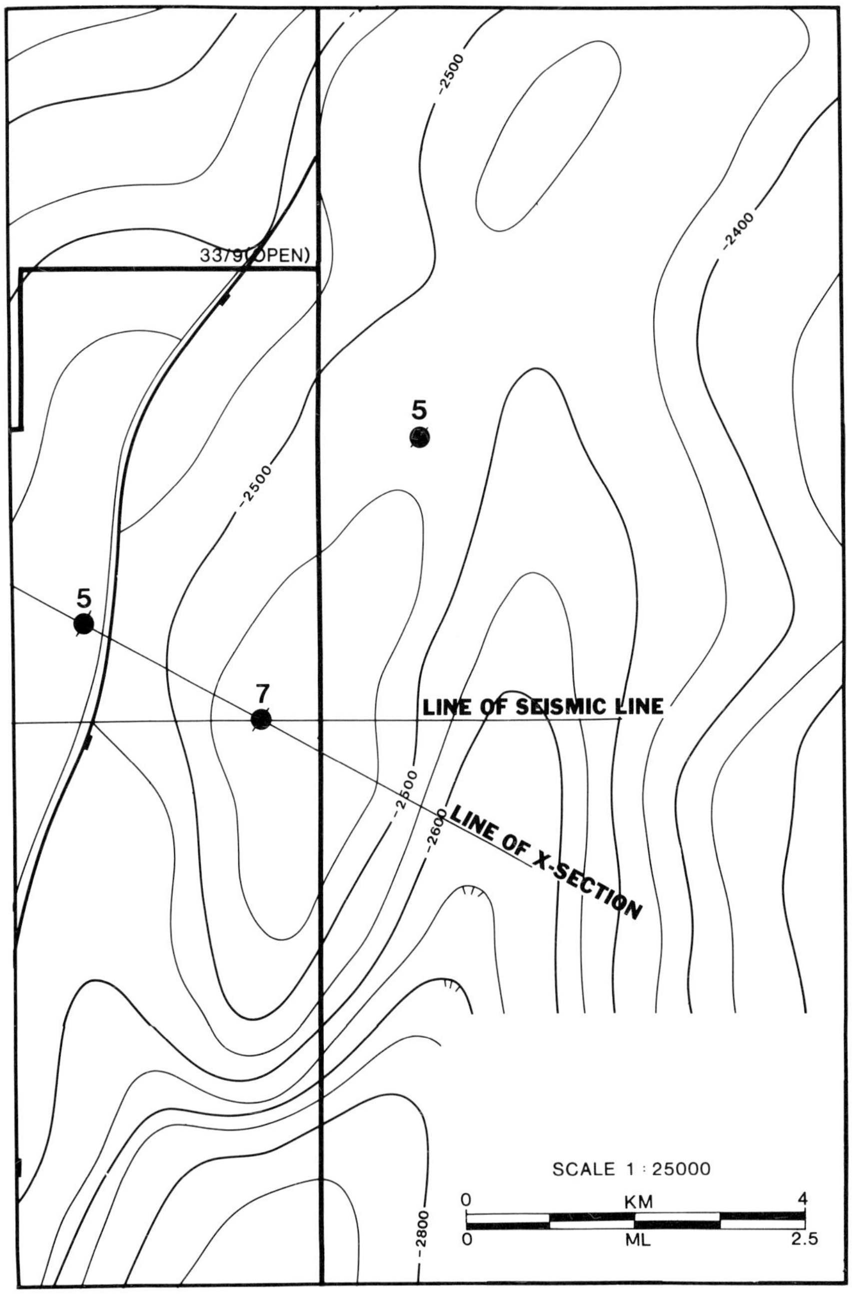

Fig. 9. Structure map of Top Jurassic.

continuity. These sandstone bodies are either channels or distributary mouth-bar deposits. Porosity is generally high. Quartz cementation may reduce the porosity locally as its overgrowths reduce the pore-throat size. Minor calcite cementation may also occur. Average porosity is greater than 20%.

5. The Tarbert Formation is a transgressive-marine sandstone. Within this unit several sorting-energy states may exist. However, no vertical barrier is believed to be present. The porosity is high and averages above 25%.

A second reservoir in the Licence 037 area is the Lower Jurassic Statfjord Formation. No hydrocarbons have yet been proven in this reservoir in the Alpha structure, but a large up-dip hydrocarbon potential remains to be tested. The Statfjord Formation was subdivided into three members by Deegan and Scull (1977). Oldest to youngest, in the Alpha structure area the Statfjord Formation section is subdivided as follows:

1. The Raude Member consists of interbedded sandstone, siltstone and clay, reflecting a

major progradation of sandy deposits to an alluvial flood basin. However, good horizontal permeability is believed still to exist. Both permeability and porosity vary with changes in facies type.

2. The Eiriksen Member consists of interbedded sandstone, siltstone and clay, reflecting a braided alluvial plain similar to the Raude Member. A climatic change is believed to be the main environmental difference compared with the Raude. The porosity of the Eiriksen Member in the neighbouring Statfjord Field is slightly higher than in the Eiriksen Member of the Alpha structure.

3. The Nansen Member was deposited as a transgressive-marine sheet-sand. As this unit contains coarser sand with little or no clay, both permeability and porosity are higher than in the underlying Raude and Eiriksen members.

A total of three different producing reservoir bodies may occur within the Alpha prospect, to judge from conditions in the Statfjord Field. Hence, the Brent Group can be divided into one upper and one lower producible reservoir, and the Statfjord Formation treated as one single producible reservoir.

SOURCE

Based on results from geochemistry, PVT and pressure analysis, the Draupne Formation is shown to be the main source of hydrocarbons in the Brent province, Brent Group (Robertson Research, 1980). Isotope profiles from several crude oils, from the Statfjord Formation, the Dunlin Group and the Brent Group of the Statfjord Field and in the Licence 037 area, show strong evidence for a genetic relationship among these oils (see Kirk 1980, Table 1 and Statfjord Nord this volume, Fig. 8, p. 349).

During continuous basin subsidence, beginning in Late Cretaceous times, conditions for the generation and migration of hydrocarbons were created. Migration probably occurred in several episodes in the Cretaceous and Tertiary periods. The migration paths would have been up-dip from the thicker Draupne sequences which occupy structurally low positions. The major migration routes were probably provided both by high permeable sediments such as the Statfjord Formation Nansen Member and the Etive Formation in the Brent Group, and by fault systems and unconformities developed in the Upper Jurassic to Upper Cretaceous sections.

HYDROCARBONS

The Alpha prospect crude is 'black oil' with a gravity of 0.84 g/cm^3 at 15.5 °C. A strong marine algal component is the dominant kerogen type in the Licence 037 area. The 33/9-7 Well tested, on a 19.05 mm (3/4-inch) choke, 1100 Sm^3/day during a 322-minute test.

RESOURCES

Vertically, three different potentially productive reservoirs may be present based upon experience from the neighbouring Statfjord Field. In the Alpha prospect the Statfjord Formation is potentially the lowest reservoir and the Lower Brent and Upper Brent are shallower. The Statfjord Formation and the Brent Group are separated by the interlying Dunlin Group, while the Upper and Lower Brent Group reservoirs are separated by the shaly Middle Brent unit.

In addition, three different structural areas are recognized within the main structure: Alpha south, Alpha north and Alpha east (see Fig. 10). Communication between these areas may be restricted due to faulting and erosion.

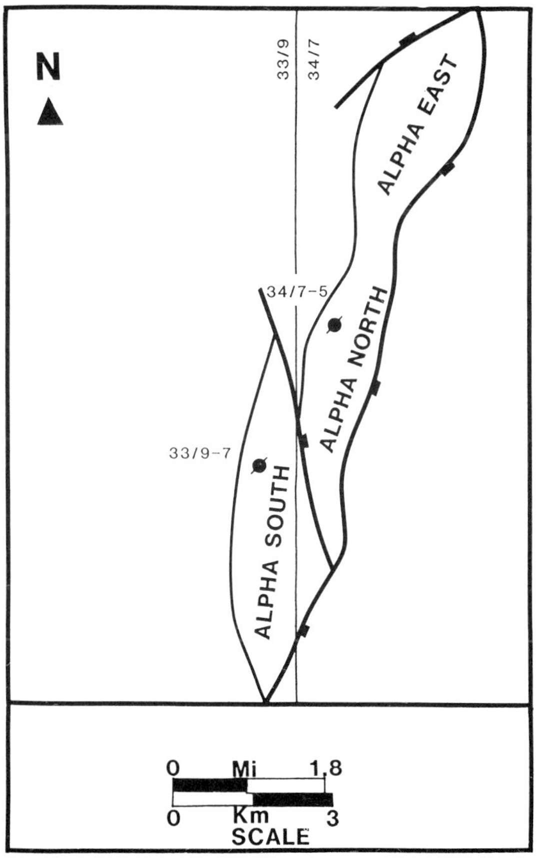

Fig. 10. Alpha prospect outline map.

Uncertainties about up-dip hydrocarbon accumulations in the Statfjord Formation make reserve estimates for the Alpha structure difficult. However, the close relationship with the Brent Field in the UK and with the Statfjord Field within the same licence, in both of which hydrocarbons are present in the Statfjord Formation, makes the possibilities of finding oil in the Alpha Statfjord Formation good.

In view of the above, resources in the Alpha structure have been estimated to be in the range of 14.5×10^6 to 40×10^6 Sm^3 oil. The wide range reflects the problems of defining the horizontal extension of the field at Brent level and of estimating the possible resources in the Statfjord Formation up-dip from the wells.

Each of the three reservoirs will require separate water injectors and oil producers. Depending on the reserves eventually proven, one or two subsea drilling templates and associated wells and equipment will be needed to produce the Alpha structure oilfield. Each subsea template will be tied to a central manifold and connected to the Statfjord 'C' platform.

ACKNOWLEDGEMENTS

This chapter is based extensively on a number of restricted company reports not listed in the references. The writer gratefully acknowledges the help and assistance of Mobil Exploration Norway staff, for their invaluable assistance and criticisms. The writer is grateful to Mobil, Statoil and all of the Licence 037 partners for permission to publish this paper.

REFERENCES

Deegan, C.E. and Scull, B.J. (Compilers). 1977. A proposed standard lithostratigraphic nomenclature for the Central and Northern North Sea, *Bull. Norwegian Petroleum Directorate* No. 1; *Inst. Geol. Sci. Rept.* 77/25.

Kirk, R.H. 1980. Statfjord Field — a North Sea giant, *Memoir of the American Association of Petroleum Geologists* **30,** 95–116.

Rawson, P.F. and Riley, L.A. 1982. Latest Jurassic–Early Cretaceous events and the late Cimmerian unconformity in the North Sea area, *American Association of Petroleum Geologists Bulletin* **66,** 2628–2648.

Robertson Research, 1980. *The Brent Sand in The North Viking Graben, North Sea, a Sedimentological and Reservoir Engineering Study*. Robertson Research International Limited, London, 5 volumes.

Vollset, J. and Doré A.G., 1984. A revised Triassic and Jurassic Lithostratigraphic Nomenclature for the Norwegian North Sea, *Norwegian Petroleum Directorate Bulletin 3.*

Vail, P.R. and Todd, R.G. 1977. Eustatic model for the North Sea during the Mesozoic, In *Mesozoic Northern North Sea Symposium*, Norsk Petroleumsforening, Article 12.

Ziegler, P.A. 1979. Factors controlling North Sea hydrocarbon accumulations, *World Oil,* November 1979, 111–124.

Ziegler, P.A. 1981. Evolution of sedimentary basins in North-west Europe, In Illing, L.V. & Hobson, G.D. (Eds), *Petroleum Geology Of Continental Shelf Of North-west Europe*, Institute of Petroleum, Heyden, London, pp. 3–39.

Ziegler, W.H. 1975. Outline of the geological history of the North Sea, In Woodland, A.W. (Ed.), *Petroleum and the Continental Shelf of North-west Europe*, Vol. 1, Applied Science, Barking, pp. 165–190.

Part VI—North Viking Graben-East

31

Agat

Arve Gulbrandsen

Saga Petroleum a.s., Box 9, N-1322 Høvik, Norway

The Agat Field is the first find within the Lower Cretaceous sequence offshore Norway. The field is located in the north-eastern part of the North Sea to the east of the Sogn Graben. The reservoir interval consists of submarine-fan deposits of Late Albian age. The gas is probably generated from Upper Jurassic shales in the Sogn Graben and is trapped in stratigraphic closures. The lateral extents of the finds are very uncertain. Four wells have been drilled from 1976 to 1982. Two of these have tested gas from two different reservoirs. The total gas-in-place estimate is 65×10^9 Sm^3. The Operator is Saga Petroleum and Statoil and BP are partners.

INTRODUCTION

Block 35/3 is located approximately 50 km west of the Norwegian coastline, immediately south of 62°N and 160 km north-north-west of Bergen (Fig. 1). The gas field has been named Agat. The find is the first within the Lower Cretaceous offshore Norway. The water depth varies from 250 to 300 m.

The reservoir interval consists of gravity-flow deposits, such as turbidites, debris-flow deposits and grain-flow deposits. The reservoir has a combination of stratigraphic and structural closure. The interval is not affected by faulting. The most likely possible resources are estimated to be 65×10^9 Sm^3 gas-in-place, but maximum possible resources are much higher. The main uncertainty is in the estimates of the lateral extension of the reservoir sandstones. The well test results, however, are good.

HISTORY

Block 35/3 was allocated in the first batch of the third Norwegian licensing round in June, 1975 as Licence PL 041. Saga became operator with 15% ownership and Statoil and BP were partners with 50% and 35% respectively. This was one of the 'Måløy Blocks', which were the first to be awarded in this part of the North Sea. The main prospective horizons were thought to be the Mid–Upper Jurassic sandstones arranged in rotated fault-blocks. According to the Licence Agreement, 50% of the area was subject to relinquishment in 1981.

The first well, 36/1-2 drilled in 1975, was a joint well with the neighbouring licence to the east. It was located just east of the block boundary aiming for a Jurassic prospect (Fig. 2). Oil shows were encountered in very thin, fairly tight sandstone stringers in the Cenomanian–Turonian; poor shows were also encountered in a 40-m-thick Jurassic sandstone sequence, which was tight due to cementation.

Well 35/3-1 was drilled in 1976 and had also a rotated Jurassic fault block as the main target. It was located in the western part of the block. The well did not test the Jurassic properly but encountered about 400 m of Lower Cretaceous strata with well-developed

Geology of the Norwegian Oil and Gas Fields © A. M. Spencer *et al.*, eds (Graham & Trotman, 1987) pp. 363–370.

Agat Field data summary

Trap	
Type	stratigraphic trap
Depth to crest	2800 m
Gas–water contacts	3455 mss and 3561 mss
Gas column	maximum *c.* 700 m
Pay zones	
Formation	Lower Cretaceous sandstone
Age	Late Albian
Gross thickness	*c.* 200 m
Net/gross ratio	0.1–0.3
Cut-off for N/G	Ø = 12%, $V_{sh=0.4}$
Porosity	up to 20%
Hydrocarbon saturation	up to 54%
Permeability	200 mD (Well 35/3-2) and 400 mD (Well 35/3-4)
Hydrocarbons	
Condensate gravity	*c.* 0.78 g/cm^3
Gas gravity (relative to air)	0.62
Dew point	431 bar (Well 35/3-2) and 341 bar (Well 35/3-4)
Gas expansion factor	286
Formation water	
Salinity	15 700 mg/l
Resistivity	0.521 ohm m at 23°C
Reservoir conditions	
Temperature	118°C
Pressures	*c.* 455 bar (Well 35/3-2) and 352 bar (Well 35/3-4) at 3550 m (35/3-2) and 3450 m (35/3-4)
Field size	
Area	*c.* 150 km^2
Gross rock volume	$(10 + 5.5) \times 10^{9}$ m^3
Drive mechanism	depletion/waterdrive
Gas-in-place	*c.* 65×10^9 Sm^3
Production	
—	

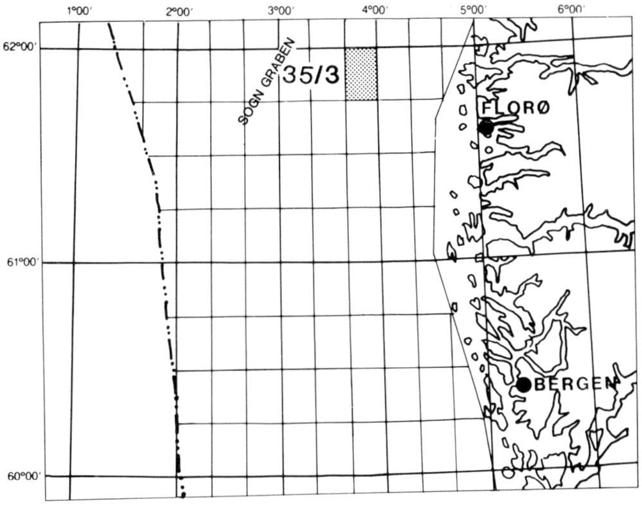

Fig. 1. Location map.

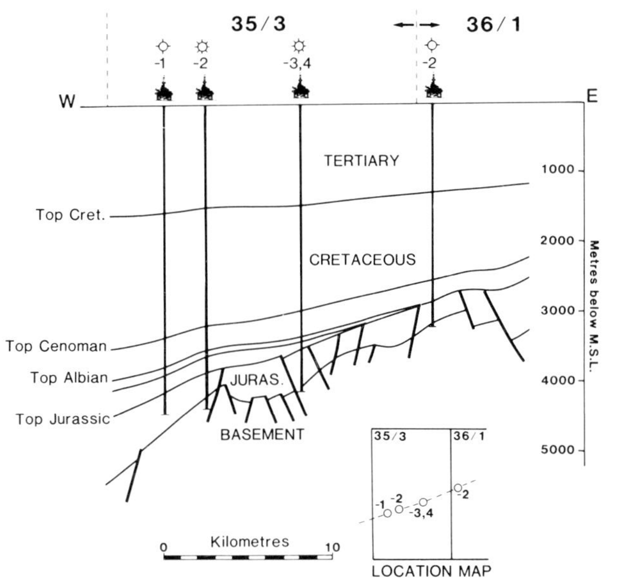

Fig. 2. Generalized correlation diagram showing the eastward on-lap of the Lower Cretaceous interval.

sandstone bodies. The hydrocarbon shows were poor both in the Jurassic and Cretaceous sequences.

Well 35/3-2 was located up-dip, 2.5 km to the east of Well 35/3-1, still aiming for a Jurassic target. It was drilled in 1980 and became the first well to test gas in the Lower Cretaceous in this area. The Jurassic sandstones had shows, but were not tested.

Well 35/3-3 was located 5.5 km farther up-dip to the east in 1980. This was the first well aiming for the Lower Cretaceous prospect. It was respudded as 35/3-4 due to technical problems, and tested hydrocarbons from the same Lower Cretaceous sequence as Well 35/3-2. However, there was no pressure communication with the reservoir in Well 35/3-2.

The last well so far, 35/3-5 drilled in 1982, was located 8 km south-east of Well 35/3-4, but slightly higher up-dip. The well was drilled to test the lateral extension of the field. It proved to be dry, but the reservoir quality within the prospective interval was improved.

GEOLOGICAL SETTING

The gas field in Block 35/3 lies on the up-faulted eastern side of the Sogn Graben. The Jurassic fault-blocks have a general dip towards the west. The unconformably overlying Cretaceous interval is very little influenced by faulting and has, in general, a relatively steep dip towards the west. The Cretaceous strata onlap the unconformity inside and partly to the east of Block 35/3 (Fig. 2).

The stratigraphy of the Lower Cretaceous interval is dominated by Aptian and Albian strata (Fig. 3). Only a few metres of older Cretaceous strata have been observed. The total thickness of Lower Cretaceous sediments is 3–500 m, of which the Late Albian interval constitutes approximately 50%. The lithology varies from shales to sandstone beds. In addition, both grain- and matrix-supported conglomerates are identified. The main sandstones have grain sizes ranging from fine to medium and are moderately to well sorted. Occasionally, calcite cementation is present.

The Jurassic rocks and the shallow basement to the east, towards the coastline of Norway, are probably the sources from which the onlapping sediments were eroded. The reservoir interval in the Lower Cretaceous comprises marine strata with little continental influence. The sediments were probably transported down-slope from the east through submarine canyons and deposited as the slope flattened out (Fig. 4).

Several submarine-fan systems in the Lower Cretaceous seem to have filled up the area inside Block 35/3. There are no variations in the reservoir interval thickness that can be related directly to fans. Based on the available data, the number of fan systems is hard to predict.

The sedimentological model proposed is mainly based on core studies; it shows various types of gravity-flow deposits such as turbidites, debris-flow deposits and grain-flow deposits. In addition, the seismic interpretation clearly shows the onlapping to the east. Seismic stratigraphy studies of the Creta-

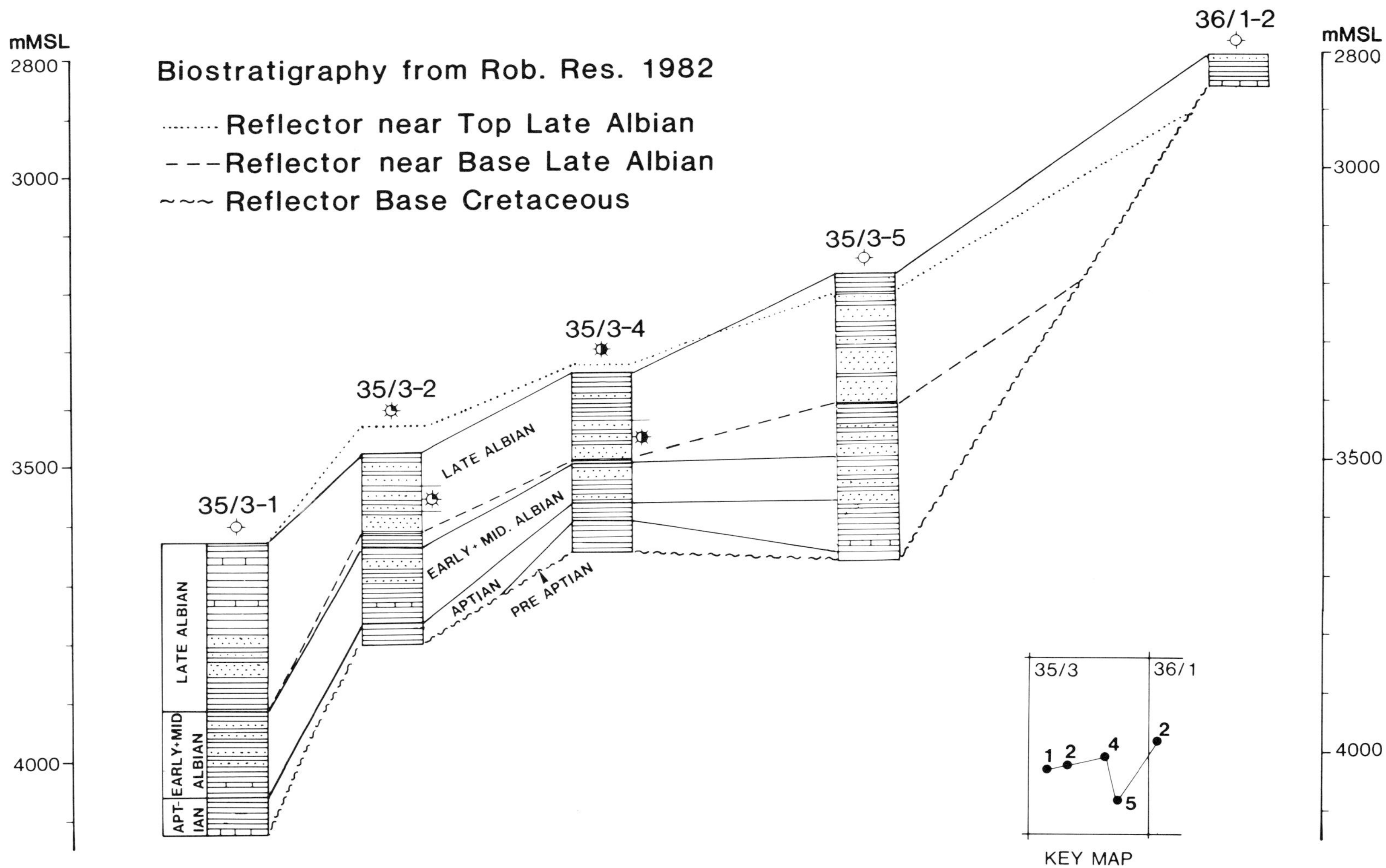

Fig. 3. Correlation of the Lower Cretaceous section, based on biostratigraphic results of the Robertson Research Study in 1982.

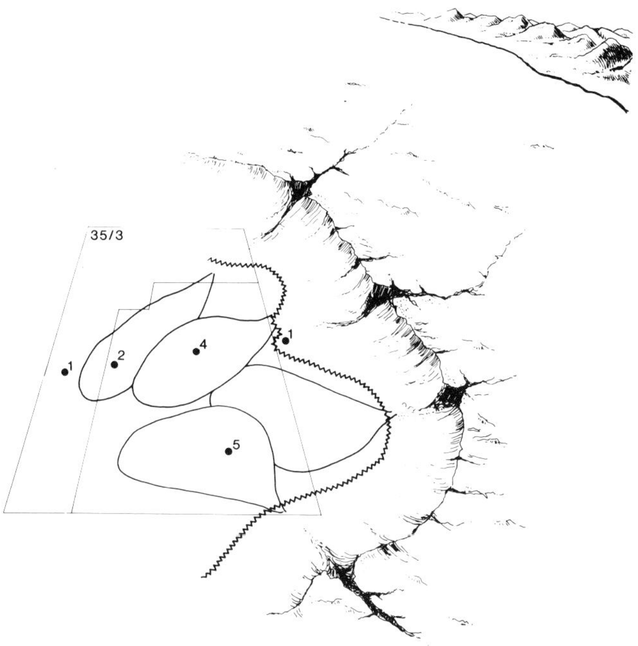

Fig. 4. Hypothetical Early Cretaceous sedimentary environment.

ceous strata have been done and support this model in general. However, it has not proved possible to map fan-bodies directly on the seismic.

GEOPHYSICS

Seismic surveys were undertaken by Chevron in 1972 and Saga in 1975 and 1981. The last Saga survey, SG 8130, shows the best data quality and has a 1 × 1.5 km regular grid in approximately north–south and east–west directions. In addition, one line has been shot between wells 35/3-1, -2, -4 and 36/1-2.

The latest interpretation and mapping were completed in 1986 and the top and base of the Late Albian reservoir interval were interpreted (Fig. 5). Figure 6 shows the depth map of the top of the Late Albian interval.

Based on the palaeogeography, seismic interpretation and well data, the lateral extensions of the two tested reservoirs, A and B, have been interpreted (Fig. 6).

At present high-resolution seismic data and a reprocessing of the 1981 Saga survey are being evaluated.

TRAP

The Lower Cretaceous strata have a general dip to the west and no interesting structural closures. The structurally highest area coincides with the termination of the reservoir interval due to seismic onlap, giving a stratigraphic closure to the east and north. Within the reservoir interval, the dip to the west gives a relief of approximately 700 m.

In view of the lack of pressure communication between the two wells which tested gas, a stratigraphic boundary must exist between these two wells. There must also be a stratigraphic boundary between Well 35/3-4 which tested gas and the structurally higher dry Well 35/3-5; this inference is supported by the pressure data from the two wells. The western reservoir which was tested by Well 35/3-2 is called the 'A discovery', and the eastern reservoir which was tested by Well 35/3-4 is called the 'B discovery' (Fig. 6). The two 'discoveries' have different gas–water contacts and different pressures.

Similar 'prospects' are interpreted both within the Late Albian interval and in the underlying Early to Middle Albian interval up-dip Well 35/3-5. This prospective area is indicated with S on the map (Fig. 6). The total volume calculated for these prospects is very speculative but relatively high.

RESERVOIR

The reservoir interval in Block 35/3 is in the Lower Cretaceous. The proven reserves are all within rocks of Late Albian age. Up-dip, however, there is potential for further discoveries also within the pre-Late Albian sequence. All these are dependent on stratigraphic closures. The sedimentological model suggests that the tested reservoirs may improve in reservoir quality up-dip, i.e. towards the sediment source-area.

The interpreted submarine-fans have accumulated in a depocentre in the southern, central part of Block 35/3. There are relatively small variations in the reservoir interval thickness, except in the eastern area where there is a gradual thinning.

The Late Albian interval shows great variations in reservoir quality. The net-to-gross ratio varies from 0.1 to 0.3, porosity is up to 20%, and water saturation is approximately 50%. The test results, however, have been very good, with no water production.

SOURCE

Geochemical data indicate that all of the hydrocarbons found within the Cretaceous in the wells have a source-rock of Upper Juras-

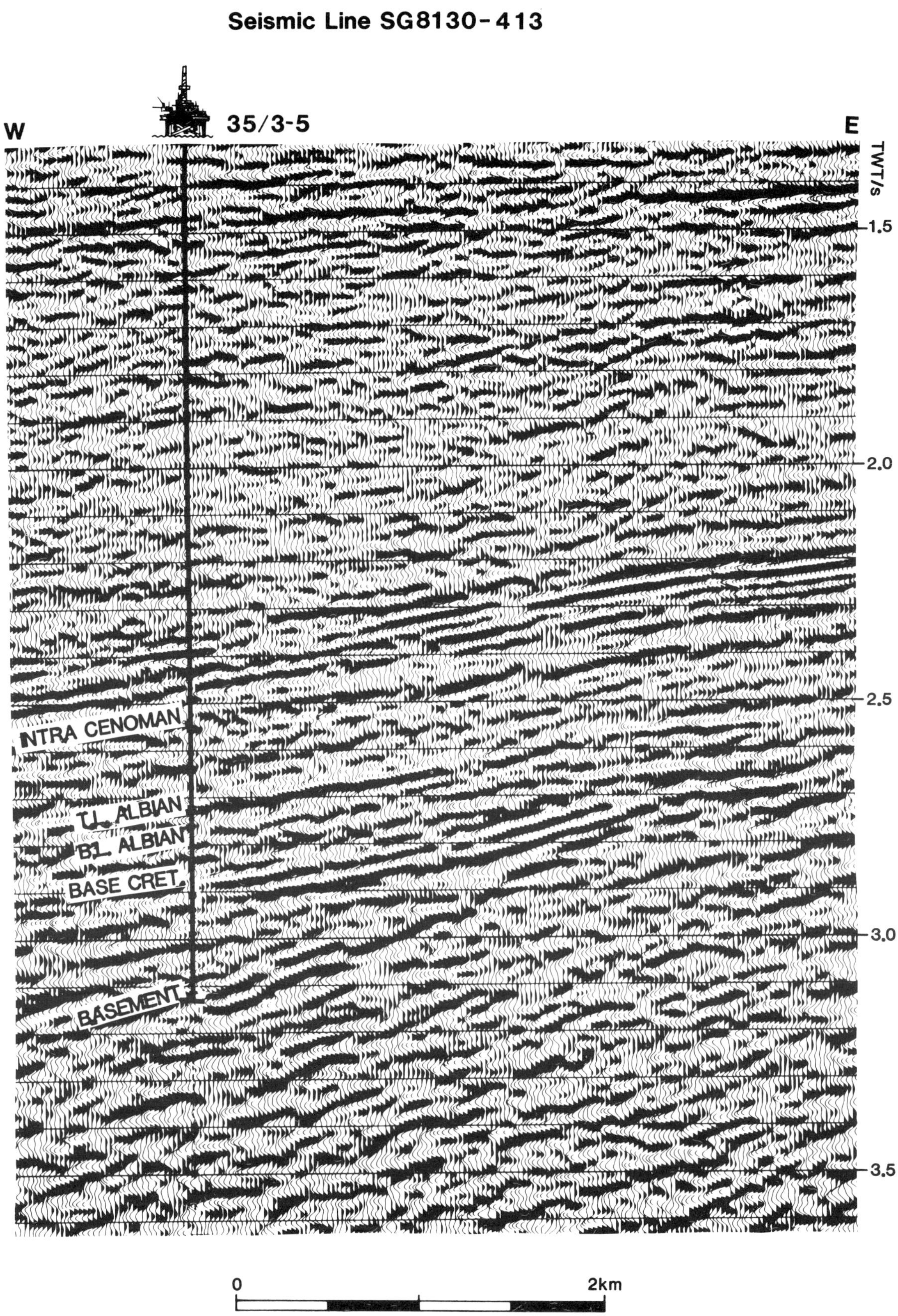

Fig. 5. Seismic line SG 8130-413 with horizon indications (See Fig. 6 for location).

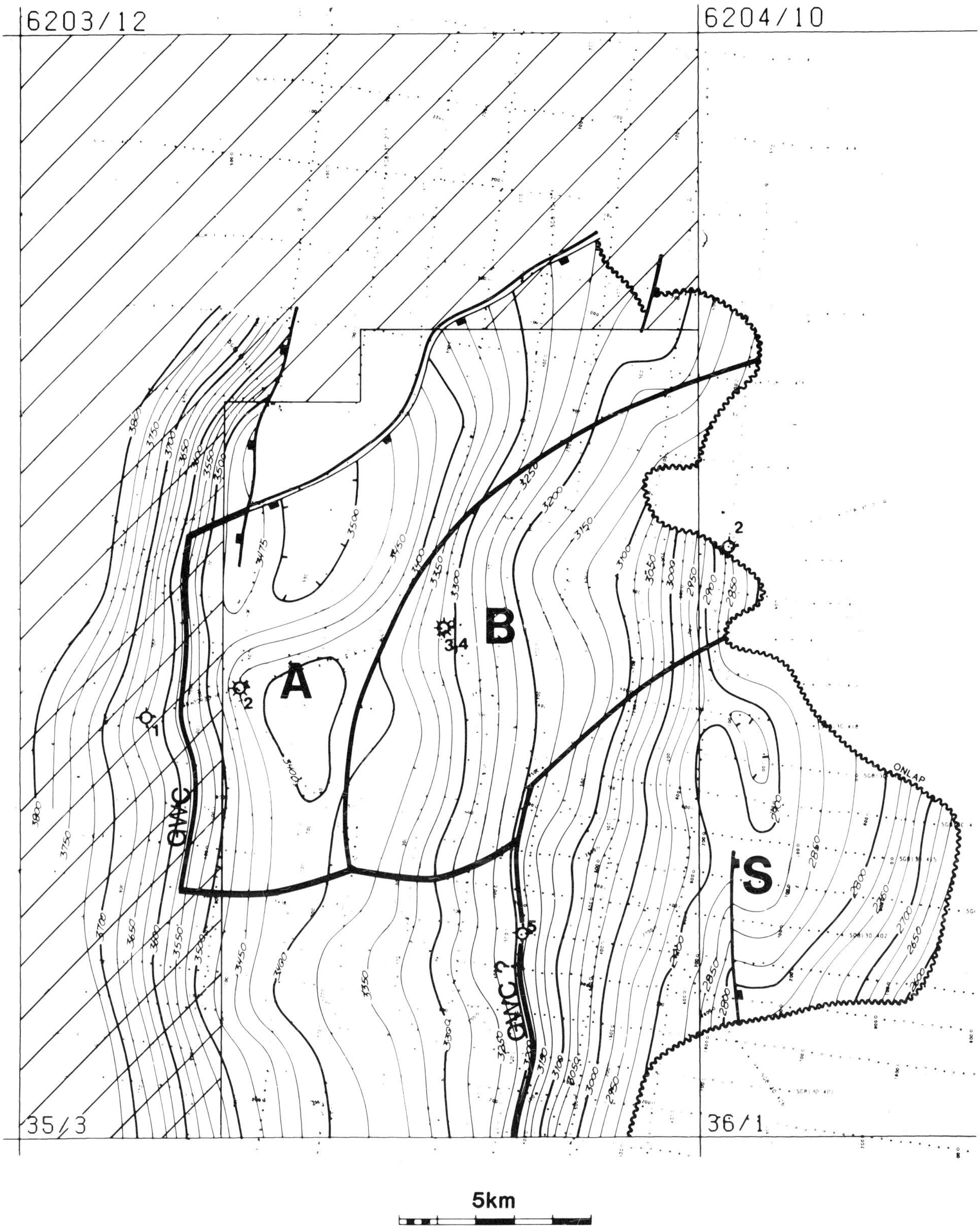

Fig. 6. Depth map in metres subsea of the Top Late Albian reflector. Areas A, B and S are the hypothetical 'prospects' referred to in the text and equivalent to the fans suggested in Fig. 4.

sic, Draupne Formation, type. The structural setting of this area makes it very likely that hydrocarbons have been generated in the basin to the west, the Sogn Graben, and have migrated up and into the Lower Cretaceous.

HYDROCARBONS

Several tests have been conducted from the Late Albian interval. The gas-flow rate of Well 35/3-2 gave 1×10^6 Sm^3/day and a condensate flow-rate of 53 Sm^3/day with a choke of 15 mm. The gas flow-rate of Well 35/3-4 gave 0.73×10^6 Sm^3/day and a condensate flow-rate of 83 Sm^3/day with a choke of 14.3 mm.

The gas gravity was 0.62 (air = 1) for both wells. The condensate gravity was 0.81 g/cm^3 for Well 35/3-2 and 0.78 g/cm^3 for Well 35/3-4. The reservoir pressure in wells 35/3-2 and 35/3-4 was approximately 455 bar and 352 bar respectively. The reservoir temperature is 118°C and the depth of the reservoirs is approximately 3500 m.

RESOURCES

The drive mechanism assumed for the depletion of these gas reservoirs is volumetric expansion and water drive. The gas-in-place estimate is 65×10^9 Sm^3 with a speculative upside potential of 175×10^9 Sm^3.

ACKNOWLEDGEMENTS

I would like to thank Statoil and BP for permission to publish this paper. In Saga a large number of employees have been involved in the work with this licence since 1975 and they are gratefully acknowledged.

REFERENCE

Myrland, R., Messell, K. and Ræstad, N. 1981. Gas Discovery in 35/3 — an exploration history, In *Norwegian Symposium on Exploration,* Norsk Petroleumsforening, Article 17.

32

Brage

A. Hage, K. Bomstad and J. E. Strand

Norsk Hydro a.s., PO Box 490, 1301 Sandvika, Norway

The Brage Field is located in Block 31/4 on the western margin of the Horda Platform, west of the Troll Field and east of the Oseberg Field. Hydrocarbons are trapped in three different horizons. In the main field the Upper Jurassic Sognefjord Formation contains oil with a gas-cap, while the lower Fensfjord Formation contains oil with associated gas. They are separated by a calcareous cemented siltstone. A narrow horst structure in the western part of the field contains oil in the Lower Jurassic Statfjord Formation. The Upper Jurassic reservoir rocks are sandstones with good porosity but low permeability due to their poorly sorted, argillaceous fine-grained nature. These sandstones were deposited in shallow-marine conditions between the offshore inner shelf and the lower shoreface. The Statfjord sandstones are of fluvial/delta-plain origin. The porosities average 24–25% and the permeability varies from 1 to 300 mD in the Sognefjord Formation, 30–200 mD in the Fensfjord Formation and 0.5–2 D in the Statfjord Formation.

So far, four wells have been drilled within the Brage Field. The in-place hydrocarbons are estimated to be in the range of 106–215 × 10^6 Sm^3 oil and 13–21 × 10^9 Sm^3 gas.

INTRODUCTION

The Brage Field is located 125 km west of Bergen in the Norwegian sector of the North Sea. The field covers an area of approximately 55 km^2 and is located in the south-western part of Block 31/4. Hydrocarbons are trapped in three different horizons: in the Upper Jurassic Sognefjord and Fensfjord Formations, and in the Lower Jurassic Statfjord Formation (Figs 1 and 2). Water depth varies from 130 to 170 m. The field consists of a flat domal main part and a horst block in the western part. The Sognefjord and Fensfjord sandstones make up the reservoir in the main part of the field. In the horst area the Upper Jurassic is partly eroded and the oil-bearing reservoir is found in the Statfjord sandstone. All three reservoirs have different oil/water contacts. In the Upper Jurassic reservoirs the contacts are uncertain due to low porosity causing a transitional contact.

The field is named after the god Brage who was the god for eloquence and scaldic art in Norse mythology.

HISTORY

Production Licence 055, comprising Block 31/4, was awarded in April 1979 to a group consisting of Norsk Hydro 15% (operator), Statoil 50%, Esso 20%, Arco 10% and BP 5%. Part of the licence was relinquished in April 1985.

The original mapping of the block showed big structural closures at both Brent and Statfjord levels. These were the objective for Well 31/4-2 (Well 31/4-1 at the same location had to be abandoned at a depth of 1000 m due to technical problems). An intra-Brent Group sandstone 5 m thick was found to be oil bearing. In addition an oil column 4 m thick was encountered in an intra-Dunlin sand-

Geology of the Norwegian Oil and Gas Fields © A. M. Spencer *et al.*, eds (Graham & Trotman, 1987) pp. 371–378.

Brage Field data summary

	Reservoirs		
	Sognefjord	*Fensfjord*	*Statfjord*
Trap			
Type	faulted anticline	faulted anticline	horst block
Depth to crest	1990 mss	2070 mss	2300 mss
Lowest closing contour	2030 mss	2150 mss	
Hydrocarbon contacts	2004/2029 mss	2147 mss	2381 mss
Gas column	14 m	—	—
Oil column	25 m	77 m	81 m
Pay zone			
Age	Upper Jurassic	Upper Jurassic	Lower Jurassic
Gross thickness	0–70 m	70–110 m	150–200 m
Net/gross ratio (average range)	0.70	0.70; 0.6–0.9	0.6; 0.4–0.8
Cut-off for N/G	18%, $V_{sh} > 0.5$	Ø < 18%, $V_{sh} > 0.5$	Ø < 12%, $V_{sh} > 0.4$
Porosity	25%	27%	24%
Hydrocarbon saturation	60%	60%	65%
Permeability	1–300 mD	30–200 mD	500–2000 mD
Productivity index	—	20 Sm^3/day/bar	500 Sm^3/day/bar
Hydrocarbons			
Oil gravity	0.83 g/cm^3	0.85 g/cm^3	0.83 g/cm^3
Gas gravity (relative to air)	0.70	0.9	0.8
Bubble-point	180 bar	167 bar	90 bar
Gas/oil ratio	145 Sm^3/Sm^3	93 Sm^3/Sm^3	60 Sm^3/Sm^3
Formation volume factor	1.45	1.29	1.22
Formation water			
Salinity		45000 ppm NaCl equivalent	
Resistivity		0.0557 ohm m at 87 °C	
Reservoir conditions			
Temperature	85 °C	87 °C	98 °C
Pressure	205 bar	215 bar	245 bar
Pressure gradient in reservoir	0.07 bar/m	0.072 bar/m	0.068 bar/m
Field size			
Area	25 km^2	47 km^2	7.7 km^2
Oil in place	$16–50 \times 10^6$ Sm^3	$70–110 \times 10^6$ Sm^3	$15–50 \times 10^6$ Sm^3
Gas in place	$7–10 \times 10^9$ Sm^3	$6–10 \times 10^9$ Sm^3	$1–2 \times 10^9$ Sm^3
Recovery factor—oil		expected to be 25–35% for the whole field	
Drive mechanism	water inj./depl.	water injection	water injection
Recoverable oil	$4–13 \times 10^6$ Sm^3	$15–35 \times 10^6$ Sm^3	$4–20 \times 10^6$ Sm^3
Production			
—			

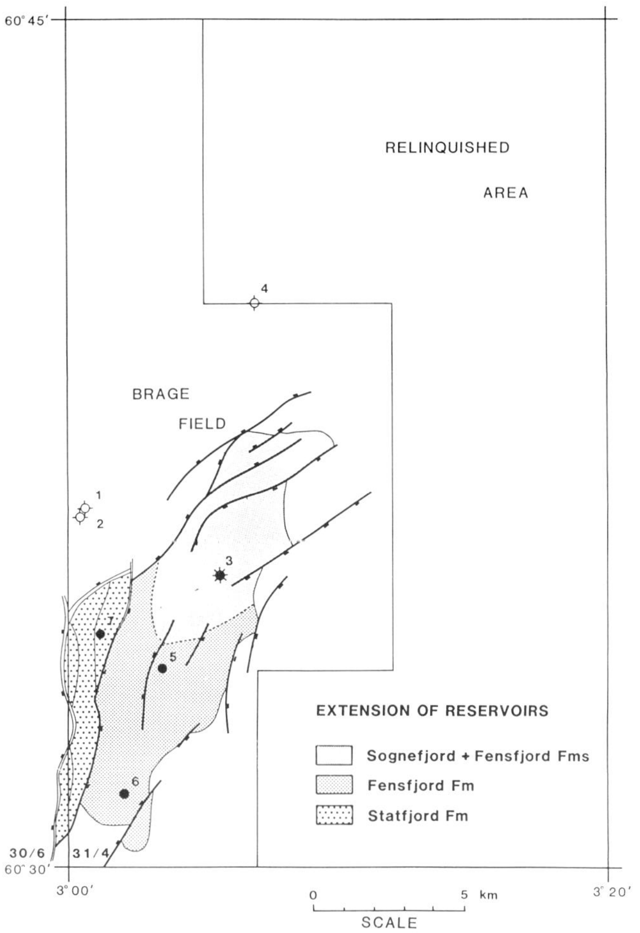

Fig. 1. Generalized map of the Brage Field.

stone with an observed oil–water contact. The Statfjord Formation was found to be water bearing.

Well 31/4-3, drilled in 1980, was the discovery well for the Brage Field. A 25-m oil column with a gas-cap of 11 m was encountered in the Sognefjord Formation, while in the Fensfjord Formation a 31-m oil column was found. The fluid contacts were poorly defined due to argillaceous fine-grained sandstone in the contact zones.

Wells 31/4-5 and -6 confirmed the discovery, the Fensfjord Formation having oil columns of approximately 37 and 22 m respectively, whilst the Sognefjord Formation was penetrated below the oil–water contact. In these two wells the Fensfjord oil–water contact again coincided with low-porosity reservoir horizons.

Well 31/4-7 was drilled in 1984 in a separate horst structure on the western flank of the Brage Field. The Upper Jurassic was partly eroded and the sandstones were not of commercial reservoir quality. The Statfjord Formation sandstones were, however, of good reservoir quality and an oil column of 22 m was encountered.

The seismic data on the block were collected from 1976 to the end of 1983, with the latest processing in 1983. This seismic grid is on average denser than 1 × 1 km. Recently, a new seismic survey with a 200-m line-spacing has been shot over the field to try to define the reservoirs more accurately.

Further appraisal drilling will be necessary before the field can be declared commercial.

GEOLOGICAL SETTING

The Brage Field is situated on the western part of the Horda Platform. Two rift phases have affected the Horda Platform. The effects of the early phase, which occurred in Early Triassic or Late Permian times is poorly known in the Brage Field.

The later rift phase occurred in early Bathonian to Berriasian times and can be divided into early (Early Bathonian–Kimmeridgian) and late (Portlandian–Berriasian) episodes. The early rifting episode is recognized by listric movements on the major faults, e.g. the western fault defining the Brage Horst. The later rifting began after the deposition of the Brent Group, which is onlapped by the synrift Lower Viking Group. The fault movements followed the orientation from the early episode. During the late episode the rifting waned.

The Early Cretaceous post-rift movements were less pronounced and showed less faulting activity than the rift phase and only planar normal fault movements are recognized. The Draupne Formation was eroded and removed in some places, such as on the Brage Horst. Fault activity along the Brage Horst decreased gradually and stopped in Paleocene time.

The oldest strata encountered by drilling in the Brage Field are of disputed Permian–Lower Triassic age, and consist of arkosic sandstones of terrestrial origin. Triassic strata vary from claystones to sandstones and occasionally limestones, and are of terrestrial fluvial origin. The upper part of the Rhaetian reflects more marginal terrestrial conditions with mainly alluvial-plain deposits, but some marine–brackish elements occur in Well 31/4-2. The Jurassic sequence starts with the fluvio-deltaic deposits of the Statfjord Formation. The Dunlin Group has claystones and sandstones of mainly shallow-marine origin. The Brent Group is eroded on the Brage Horst and is thin in the rest of the Brage Field. The

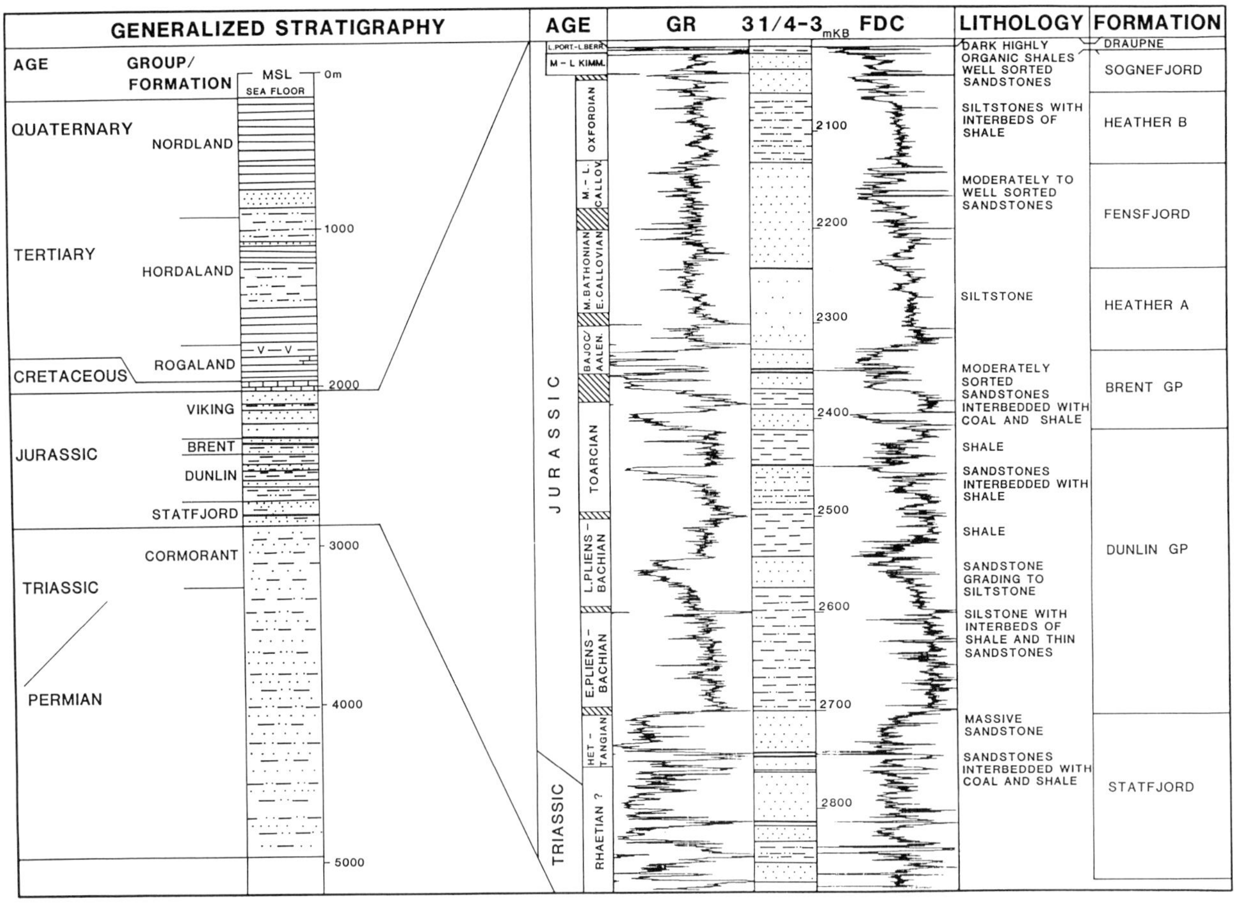

Fig. 2. Stratigraphy of Well 31/4-3.

deposits are of marginal-marine to mouth-bar and delta-plain origin. The Viking Group contains two shallow-marine sandstones between fine-grained clastic deposits. The Draupne Formation is very thin and even absent on the Brage Horst. The Upper and Lower Cretaceous sections are also relatively thin in the field area.

GEOPHYSICS

The block is covered by 2430 km of seismic data shot during the years 1976–83. The Brage Field itself has been covered by an additional 1700 km of seismic data shot during 1985 with a line spacing of 200 m and three-dimensional reconnaissance processing.

Seismic data quality varies both vertically and horizontally. Below the Cretaceous sequence the seismic response has been rather poor due to reduced energy penetration and low impedance contrasts in the zones of interest in the Jurassic sequence. The data from the 1985 survey have strongly improved the quality in the pre-Cretaceous section, but these data are not yet interpreted in the main field area.

The Top Cretaceous, the Base Cretaceous and the Top Brent horizons give the best seismic reflectors. From the maps of these horizons and combining well data, seismic information and sedimentological interpretations, Top Sognefjord and Top Fensfjord depth maps have been constructed. It is hoped that the new 1985 data will allow direct mapping of intra-reservoir reflectors (Fig. 3).

In the western horst area seismic reflectors below the Base Cretaceous Unconformity could hardly be defined on the previous data. The horst is steep and narrow and gave sideswipe and interference effects which destroyed the seismic signal. On the 1985 data, however, the top reservoir can consistently be interpreted in the whole horst area.

TRAP

The Brage Field trap is an elongated flat dome combined with a horst-block in the western part (Figs 4, 5). The doming started in late

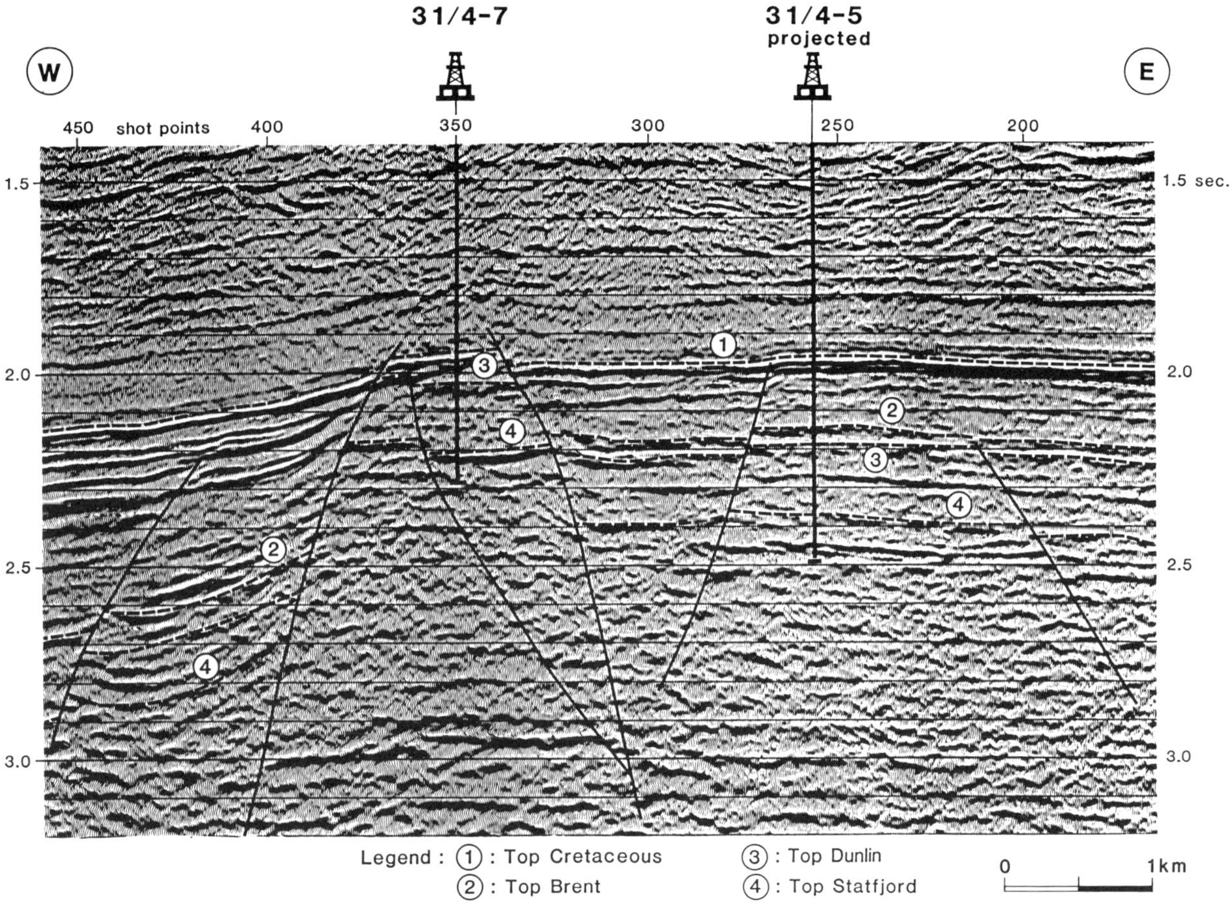

Fig. 3. Seismic line NH 8501-181 (1985 survey). For location, see Fig. 4.

Middle Jurassic times and lasted into the Early Cretaceous. The doming was a result of the Bathonian to Berriasian rift phase, which also formed the horst at the edge of the Horda Platform.

In the main field area the reservoirs are truncated towards the south. The Sognefjord reservoir is capped by the Draupne shales, and the Fensfjord reservoir is capped partly by the intra-Heather siltstones and partly by the Draupne shales.

Both the Sognefjord and Fensfjord reservoirs are assumed to be filled to spill-point. The structure is interpreted to have structural dip-closed spill-points for both reservoirs to the north-east. The Sognefjord sandstone has a gas–oil contact at 2004 mss and an oil–water contact at 2029 mss. The Fensfjord sandstone has an oil–water contact at 2146 mss. The contact zones are found in low-porosity sandstones, and the transition zones range up to 10 m.

The Statfjord reservoir in the horst is capped by the Dunlin shales. The structural closure is fault related and no spill-point has been defined. The oil–water contact is found at 2381 mss in high-porosity reservoir.

The total Brage Field covers an area of approximately 55 km^2.

RESERVOIR

The Upper Jurassic sandstones

The lower sandstone, the Fensfjord Formation, consists of offshore siltstones and fine-grained, fairly well-sorted, clean sandstones of lower shoreface environment. A coarse sandstone at the top in Well 31/4-5 has been interpreted as a possible distal part of a distributary mouth-bar (Fig. 2).

The Fensfjord reservoir has been divided into seven sedimentary units based on log and core analysis. Most of the units can be further divided into two subunits, with the upper one usually cleaner and/or coarser than the lower one. Normally both subunits are bioturbated. The upper one is interpreted as representing a lower shoreface or transitional offshore inner shelf – lower shoreface environment, while the lower part represents an offshore inner shelf environment with occasional storm-wave/current influence. The Fensfjord Formation is Late Bathonian to Callovian in age.

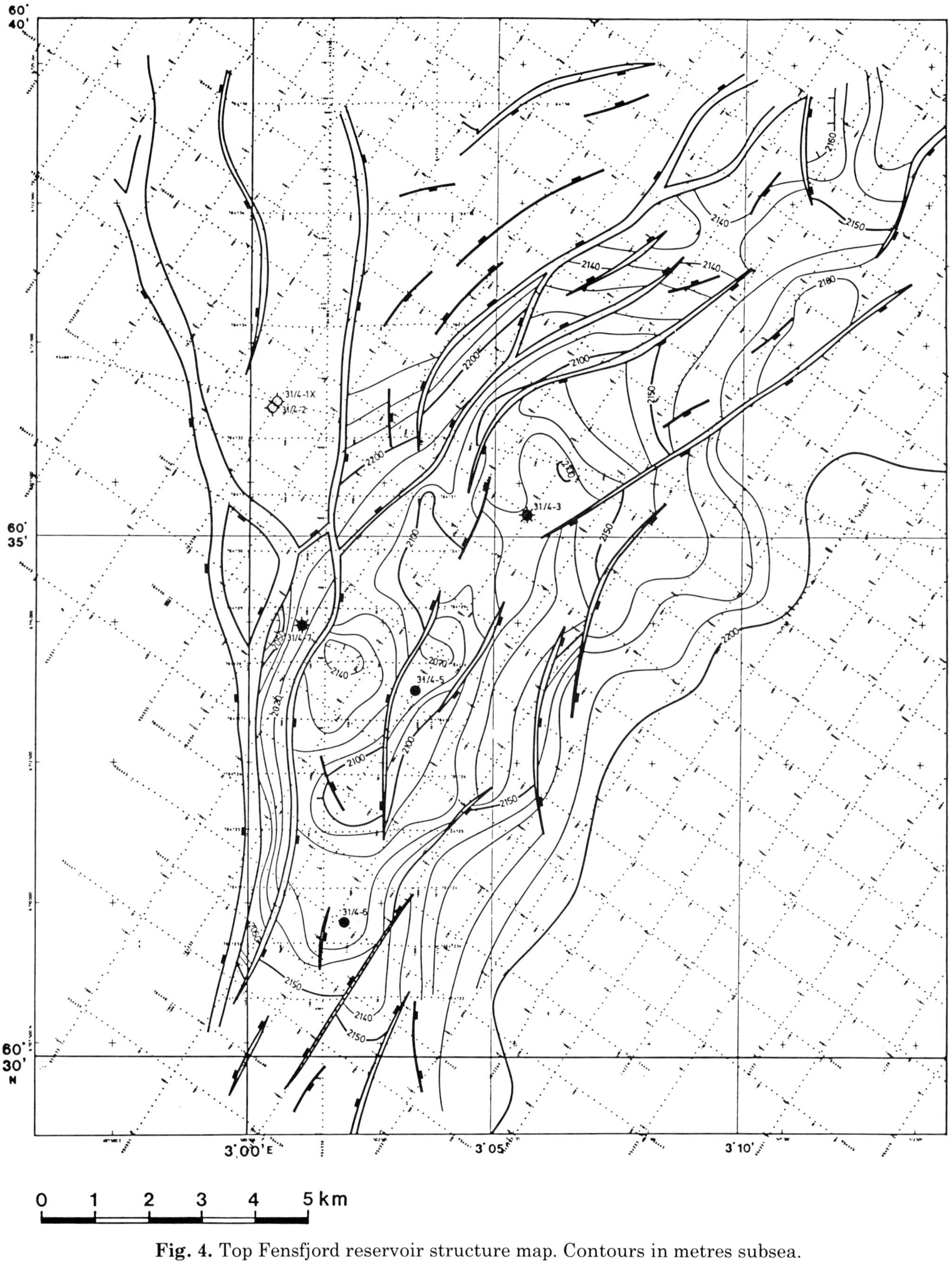

Fig. 4. Top Fensfjord reservoir structure map. Contours in metres subsea.

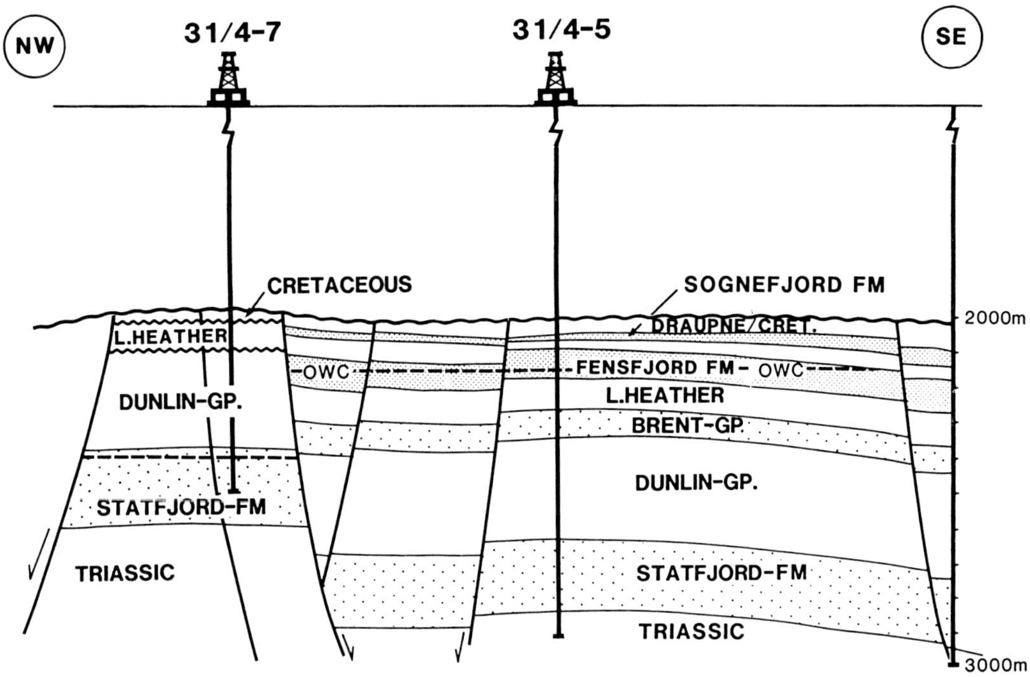

Fig. 5. Structural cross-section SE–NW. For location, see Fig. 4.

The upper sandstone, the Sognefjord Formation, is probably truncated in the southern part of the field. It contains offshore siltstones and fine-grained, fairly clean sandstones of lower-shoreface environment.

The Sognefjord Formation is of Oxfordian to Kimmeridgian age. It has been divided into five sedimentary units, based upon log correlation and core analysis.

The overall depositional environment of the Upper Jurassic sandstones implies that the facies have large lateral extents showing only gradual transitions. The depositional pattern was cyclic, reflecting oscillations of a nearby shoreline.

The Statfjord Sandstones

Well 31/4-7, on the western-most horst feature, detected hydrocarbons in the Statfjord Formation. The sandstones range in age from Hettangian (possibly older) to Early Pliensbachian at the top. The sandstones are yellow–brown to light grey and are predominantly very fine to medium, but in part coarse grained. The grain shape is subangular to subrounded. They are of fluvial to deltaic origin.

SOURCE

The source rocks for the hydrocarbons are mainly supposed to be the Upper Jurassic shales of the Draupne Formation. The hydrocarbons are thought to have been generated in the main Viking Graben and in the deeper parts of the surrounding sub-basins and to have migrated from there into the Brage Field, which is located on the north-western edge of the Horda Platform.

HYDROCARBONS

The uppermost reservoir, the Sognefjord Formation, contains an oil-leg with an overlying gas-cap. The oil gravity is approximately 0.83 g/cm^3, which is slightly lighter than in the Fensfjord reservoir. Gas gravity is around 0.70 (air = 1).

The Fensfjord reservoir contains an undersaturated oil with a gas/oil ratio of 93 Sm^3/Sm^3 and a specific gravity of 0.85 g/cm^3. Reservoir temperature is around 87 °C with a reservoir pressure around 215 bar.

The oil in the Statfjord reservoir has a specific gravity of 0.83 g/cm^3 and a reservoir temperature around 100 °C. The gas/oil ratio is 60 Sm^3/Sm^3.

RESOURCES

Estimating the recoverable oil and gas in the Brage Field is very difficult due to uncertainties both in the distribution of the reservoir properties and in the seismic mapping of the reservoirs. The structure is very flat, and a minor change in seismic velocities will have a large impact on the estimated hydrocarbon distribution. The in-place hydrocarbons in the Brage Field are estimated to be:

Oil in-place	
Sognefjord Fm	$16–50 \times 10^6$ Sm^3
Fensfjord Fm	$70–110 \times 10^6$ Sm^3
Statfjord Fm	$15–50 \times 10^6$ Sm^3
Gas in-place	
Sognefjord Fm	$7–10 \times 10^9$ Sm^3
Fensfjord Fm	$6–10 \times 10^9$ Sm^3
Statfjord Fm	$1–2 \times 10^9$ Sm^3

The recovery factor, with water injection, for the oil in the total Brage Field is expected to be in the region of 25–35%. The range reflects the uncertainty in productivity and reservoir continuity.

ACKNOWLEDGEMENT

This chapter is based on internal reports made by the Exploration Department, Research Centre and Production Geology Department in Norsk Hydro. The authors thank Norsk Hydro and the partners in PL 055 (ARCO, BP, Esso and Statoil) for giving permission to publish this article.

REFERENCES

Deegan, C.E. and Scull, B.J. (Compilers). 1977. A standard lithostratigraphic nomenclature for the Central and Northern North Sea, *Institute of Geological Sciences, Report* 77/25.

Vollset, J. and Doré, A.G. (Eds). 1984. A revised Triassic and Jurassic lithostratigraphic nomenclature for the Norwegian North Sea, *Norwegian Petroleum Directorate. Bull. 3.*

33

Oseberg

Olav Nipen

Oseberg Production, Norsk Hydro, PO Box 646,
N-5001 Bergen, Norway

The Oseberg Field is a large oil and gas accumulation located in Norwegian Blocks 30/6 and 30/9 around 140 km west-north-west of Bergen. The field is located on the Bergen High, in the transition zone between the tectonically active Viking Trough to the west and the stable Horda Platform to the east. The field consists of three major eastwardly rotated fault-blocks, Alpha, Gamma and Alpha North. The hydrocarbons occur in the Brent Group, and the main reservoir interval is the Etive Formation. The presence of a gas cap has been proven by wells on Alpha and Gamma. The gas column on Alpha is up to 380 m and the oil column is around 210 m, thus giving a total hydrocarbon column of about 600 m. The Etive reservoir has a gross thickness of 30–45 m on Alpha and 50–70 m on Gamma. The total Brent thickness is 46–190 m. The structural dip is normally 6–10° ENE. The projected horizontal distance from the GOC to OWC is 2 km.

By December 1986, 11 exploration and appraisal wells and seven predrilled development wells had been drilled within the Oseberg Unit (which is the unitized part of Blocks 30/6 and 30/9). Many satellite structures exist around the field, and discoveries have been made in the Brent Group, the Cook Formation and in the Statfjord Formation. The production mechanism initially described in the Field Development Plan (FDP) is pressure maintenance by water injection in the Alpha and Alpha North structures, and associated gas injection in the Gamma structure. Subsequently, in spring 1986, this was changed to external gas injection on the Alpha and Gamma structures.

Recoverable reserves were initially estimated to be 159×10^6 Sm^3 based upon the FDP. Later, in spring 1986, this estimate was increased to 203×10^6 Sm^3 due to the changed production mechanism and new geological/geophysical data.

INTRODUCTION

The Oseberg Field was discovered in 1979. The field includes now the Alpha, Alpha North and Gamma structures within Norwegian Blocks 30/6 and 30/9 around 140 km west-north-west of Bergen. The water depth is around 105 m.

The reservoir rocks are found within the Brent Group, where the Etive Formation is the main reservoir.

Recoverable oil reserves are presently estimated to be 203×10^6 Sm^3.

The Oseberg Field was named after the Oseberg Longboat, a 9th century Viking ship discovered in 1904. It is widely considered to be the greatest archaeological find ever made in Norway.

HISTORY

Production Licence 053 (Block 30/6) was awarded in April 1979 to the following companies:

Den norske stats oljeselskap a.s. (Statoil)	50 %

Geology of the Norwegian Oil and Gas Fields © A. M. Spencer *et al.*, eds (Graham & Trotman, 1987) pp. 379–387.

Oseberg Field data summary

Trap	
Type	Truncated fault-block
Depth to crest	2120 mss
Gas–oil contact	2497 mss
Oil–water contact	2700–2719 mss
Gas column	*c.* 380 m
Oil column	203–222 m
Pay zone	
Group	Brent
Age	Middle–Early Jurassic
Thickness	46–190 m
Net/gross	
Tarbert	> 0.5
Ness	0.4–1.0
Etive	0.8–1.0
Porosity	20–26%
Hydrocarbon saturation	60–89%
Permeability (average; range)	Around 2 D; 1 mD–3.5 D
Productivity index (ideal)	
oil	415–1950 Sm^3/day/bar
gas	360–780 × 10^3 Sm^3/day/bar
Hydrocarbons	
Stock tank oil density	0.857 g/cm^3
Gas/oil ratio of gas-cap gas	3200 Sm^3/Sm^3
Bubble-point	280.2 bar at gas–oil contact
Formation volume factor	1.34–1.47
Solution gas–oil ratio	130–160 Sm^3/Sm^3
Formation water	
—	
Reservoir conditions	
Temperature	99.8°C at 2497 mss
Pressure	280.7 bar at 2497 mss
Field size	
Recovery factor—oil	0.52 in Etive
Recovery factor—gas	0.52
Recoverable oil	203 × 10^6 Sm^3
Imported gas	25.5 × 10^9 Sm^3
Reproduced imported gas	19 × 10^9 Sm^3
Recoverable gas (total)	89 × 10^9 Sm^3
Drive mechanism	Gas injection by external gas and associated gas
Production	
Start-up date	1 April 1989
Production rate	204000 BOPD

Elf Aquitaine Norge A/S	$13\frac{1}{3}$%
Total Marine Norsk A/S	$6\frac{2}{3}$%
Norsk Hydro Produksjon a.s.	12.5%
Mobil Exploration Norge Incorporated	10%
Saga Petroleum a.s.	7.5%

Statoil was appointed operator with Elf as technical assistant. In April 1982 the operatorship was transferred from Statoil to Norsk Hydro.

The first well, 30/6-1, was drilled in 1979 in a structurally high position on the Alpha Structure (Figs 1, 2 and 3) to evaluate the potential of the Jurassic and Paleocene strata. Gas-bearing sandstones were found in a thin Brent Group. The Statfjord Formation was water bearing, whilst the Paleocene had no significant sandstone development.

Two additional wells in the gas cap were drilled in 1979. Then, in 1981, Well 30/6-4 was drilled in a down-dip position and a full oil column was found in the Brent Group. In addition, the reservoir quality of the Etive Formation was found to be significantly better and the thickness greater than in the previous wells. In 1982 Well 30/6-6 established an OWC at about 2710 mss.

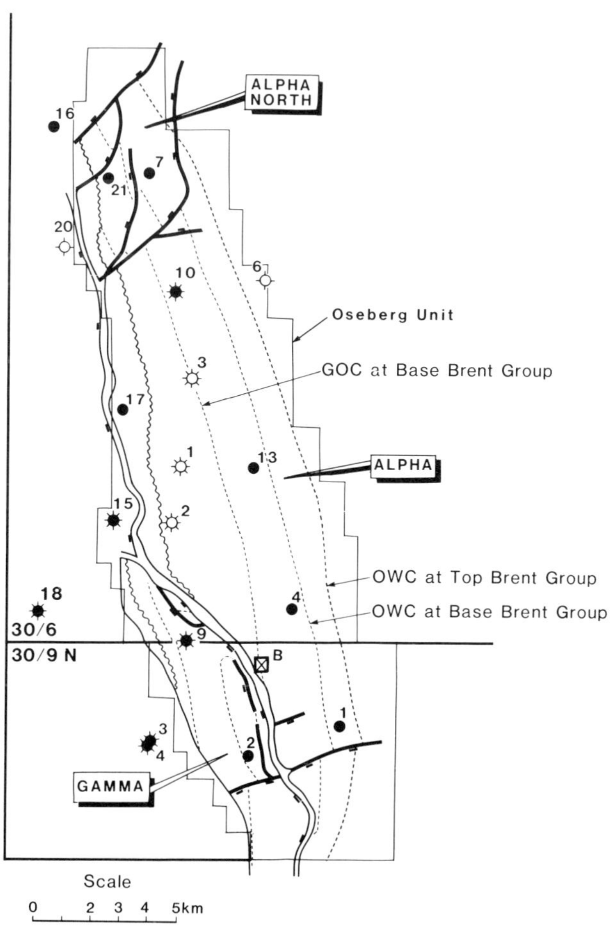

Fig. 1. Oseberg Field summary map.

Well 30/6-7 was located on the Alpha North structure and was the first well drilled by Norsk Hydro as the new operator on this licence. The well proved oil in this structure. The next well, 30/6-9, proved very promising results on the Gamma structure where the reservoir thickness had increased significantly from that at Alpha, and excellent reservoir properties were found.

It became apparent that the Oseberg discovery extended into Block 30/9 and that this would require a unified approach across the two blocks. In August 1982, PL 079 was awarded covering the north-west corner of Block 30/9 (about 20% of the block). The licences were:

Den norske stats oljeselskap a.s. (Statoil)	73.5%
Norsk Hydro Produksjon a.s.	16%
Saga Petroleum a.s.	10.5%

and Norsk Hydro was appointed the operator.

The last four appraisal wells on the field (before the FDP was submitted) had the following objectives:

30/9-1 (Alpha):	to determine the OWC and reservoir properties in the southern part of the Alpha Structure;
30/6-10 (Alpha):	to determine the GOC in the Etive Formation;
30/6-13 (Alpha):	to determine reservoir properties and improve reserve estimation in the middle part of the field;
30/9-2 (Gamma):	to determine the OWC and the reservoir properties in the south of Gamma.

The operator presented a Commerciality Report to the partners, and the Oseberg Field was declared commercial on 11 August 1983. The Field Development Plan (FDP) was completed in late 1983 and was approved by the Storting in the spring session of 1984.

The Field Centre is located in the southern part of the field. It is designed as a split concept consisting of a production and accommodation platform with a concrete gravity base substructure, and a drilling platform with a steel jacket substructure installed above a predrilled template. The oil transportation system will be a pipeline from Oseberg to Sture at the western coast of Norway. This will be the first oil pipeline to the Norwegian coast.

Predrilling of development wells started in

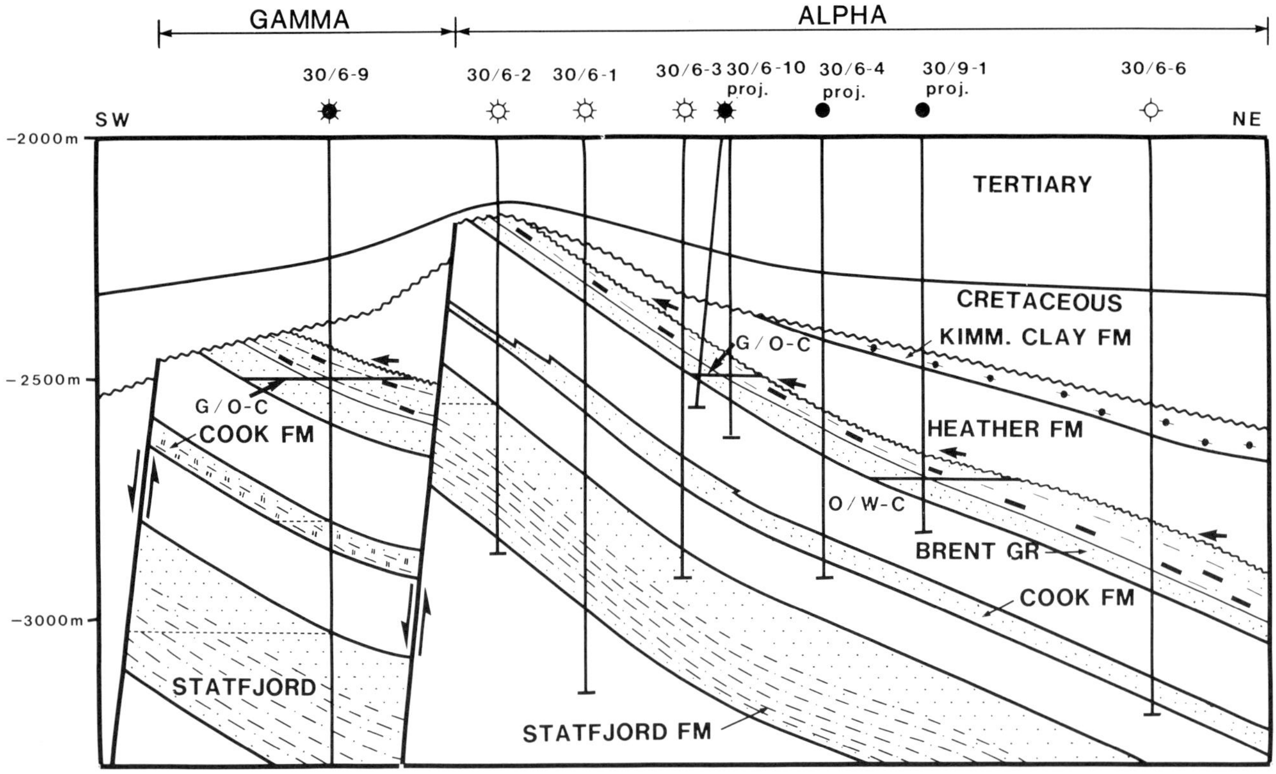

Fig. 2. Schematic SW–NE cross-section.

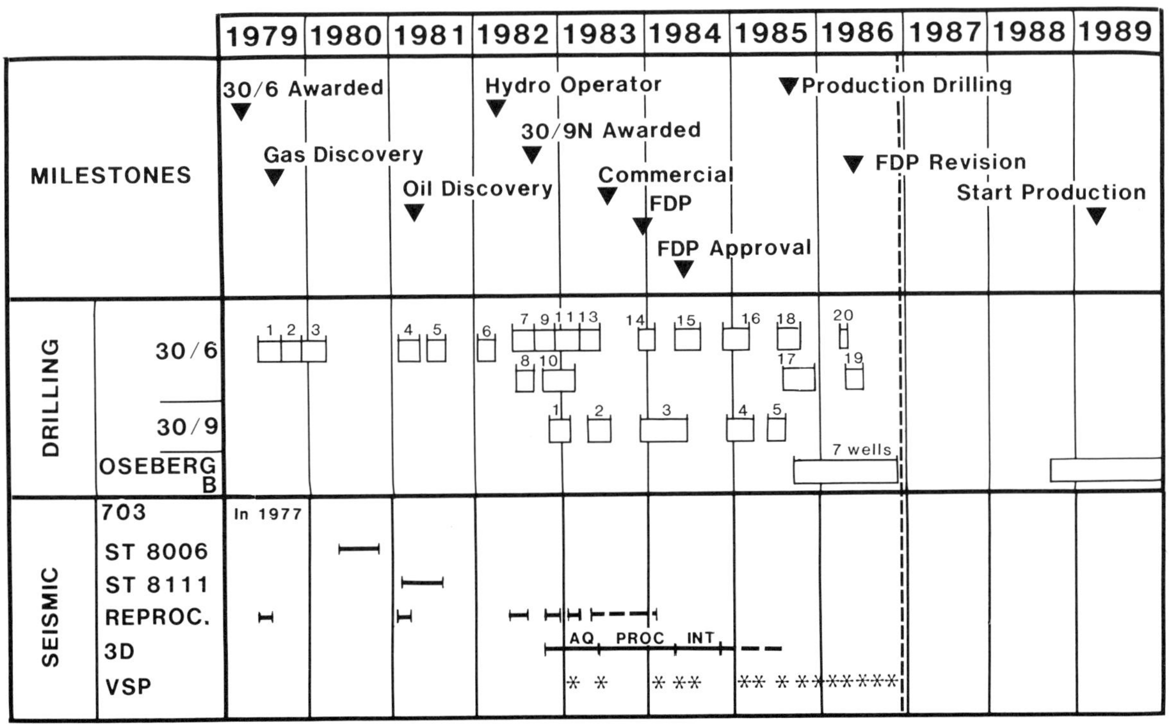

Fig. 3. Oseberg time schedule overview.

the autumn of 1985 and to date (December 1986) seven wells from the template and one subsea well have been drilled. Start of production is planned for 1 April 1989.

An Oseberg unit agreement has been established. Assuming 60% of the hydrocarbons lie within Block 30/6 and applying the sliding scale in Block 30/6, the partners' share in the Oseberg Field are:

Den norske stats oljeselskap a.s. (Statoil)	63.24%
Norsk Hydro Produksjon a.s.	13.75%
Saga Petroleum a.s.	8.61%

Elf Aquitaine Norge A/S	6.40%
Total Marine Norsk A/S	3.20%
Mobil Exploration Norge Incorporated	4.80%

In spring 1986 the planned production mechanism was changed to external gas injection on the Alpha and Gamma structures.

Blocks 30/6 and 30/9 contain several interesting discoveries and prospects. Discoveries include wells 30/6-5 (Beta South), 30/6-15 (Gamma North), 30/6-17 (Alpha; Cook Formation), 30/6-18 (Kappa), 30/6-19 (Beta Saddle) and 30/9-3 (Omega). Oseberg Phase II studies have been initiated; an integrated development approach for the northern part of the Oseberg Field and surrounding discoveries is being evaluated.

GEOLOGICAL SETTING

The Oseberg Field is located on the Bergen High in the transition zone between the tectonically active Viking Trough to the west and the more stable Horda Platform to the east. A general structural description of the Oseberg area and the fault scarp to the west has been given by Badley *et al.* (1984).

The major NNW–SSE-trending faults have a generally long ranging activity and also show differential subsidence between Alpha and Gamma during Brent deposition. The major NE–SW faults have a much more punctuated and generally later activity and are also influenced by non-extensional forces.

A generalized stratigraphic column is shown in Fig. 4. The Triassic and Lower Jurassic strata are, in a broad sense, typical for this area. Discoveries on satellite structures have been made in the Cook Formation and at two levels in the Statfjord Formation. The Brent Group has so far been subdivided into the Etive, Ness and Tarbert formations. They seem to conform to a deltaic model in which the Etive and Ness represent a major constructional phase, which upon abandonment was transgressed by the Tarbert Formation.

The Upper Jurassic section is composed predominantly of fine-grained rocks, including the organic-rich deposits of the Draupne Formation which are present in a down-flank position. The Cretaceous section consists of fine-grained argillaceous material and some carbonates. Hydrocarbon shows have been found in limestone stringers in many wells. The Tertiary and Quaternary section is generally clay and claystone, but important sand development occurs in the Utsira Formation, and geometrically complex sands are present in the Oligocene section.

GEOPHYSICS

The main seismic exercises are illustrated in Fig. 3. By 1980 a 1 × 1 km seismic grid had been established. During the FDP studies in 1982–83 several reprocessing exercises were carried out in areas of poor data quality. These efforts achieved significant improvement in data quality and led to improvements in interpretation and mapping, and the two-dimensional data were considered sufficient for the FDP.

A large three-dimensional survey was shot and processed by Geco from October 1982 to April 1984. The survey consisted of 389 E–W lines 75 m apart and amounts to 7500 km of migrated data.

A typical seismic row from the three-dimensional survey is shown as Fig. 5, and Fig. 6 illustrates the major sequence boundaries around the reservoir level. All these boundaries can be mapped either from the three-dimensional processed data or from specially reprocessed inverted data.

The present three-dimensional interpretation shows a general similarity to the two-dimensional interpretation, but important changes are found at the outer boundaries of the field (in the south and along the crest) and at the major fault-zone between Alpha and Gamma.

TRAP

The Oseberg Field is a fault trap with about 600 m of structural relief at top reservoir level. The structural spill-point is found at the southern end of the field in an area of poor data quality. The present interpretation is that the mapped spill-point is at the same level as, or very close to, the OWC. Seal is provided by the Heather Formation and the Cretaceous limestones and shales. The crest is found south-west of Well 30/6-2 at a level of about 2120 mss.

RESERVOIR

The Brent Group, which is the reservoir for the Oseberg Field, may be partly Lower Jurassic in age, spanning the Late Toarcian to Bathonian stages. In addition the Brent Group's lithostratigraphy is atypical at Oseberg where the usual five-fold subdivision,

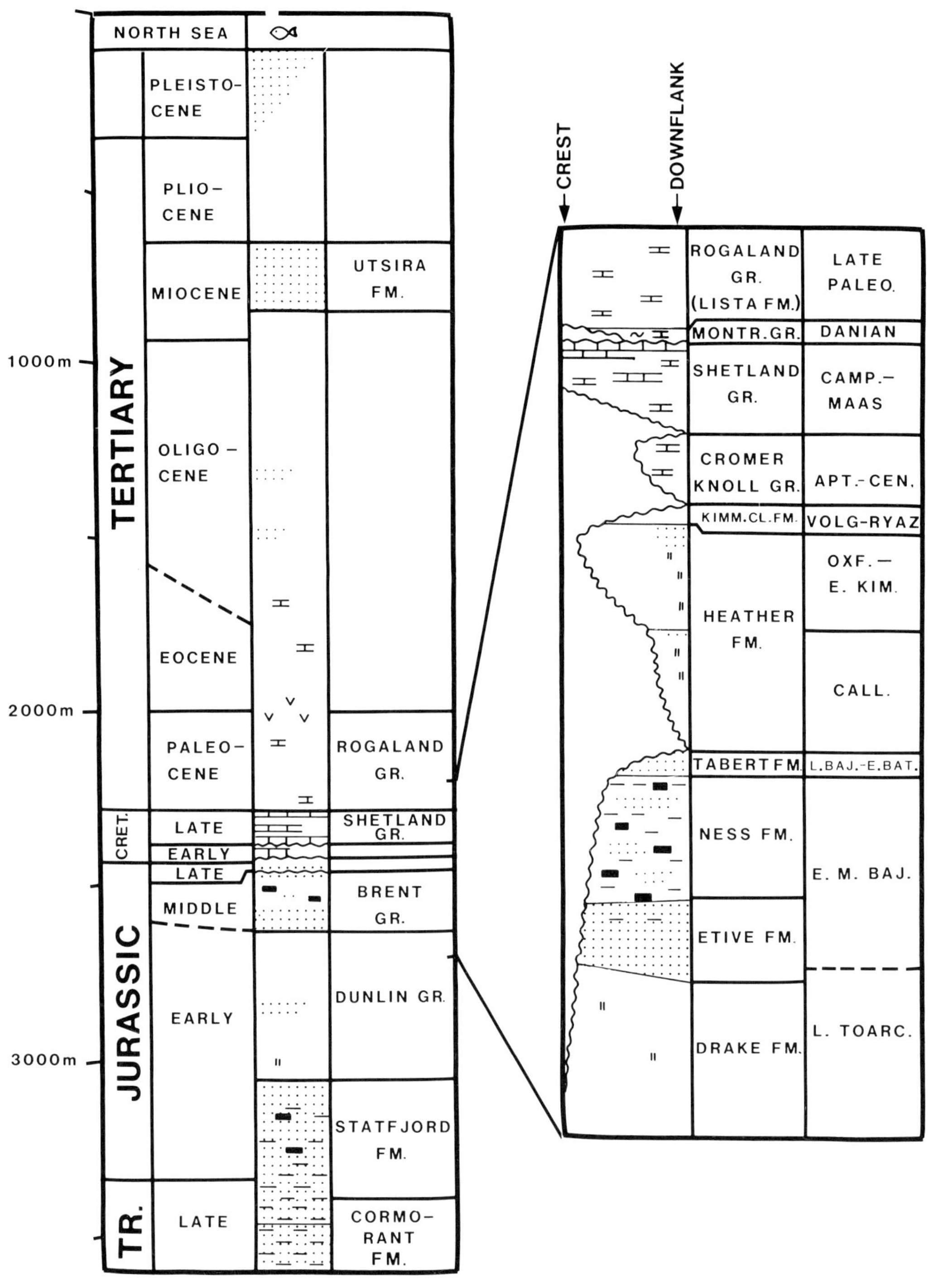

Fig. 4. General stratigraphy.

into (in ascending order) Broom, Rannoch, Etive, Ness and Tarbert formations, is not possible. The following Brent Group terminology has been adopted by the Operator for the Oseberg Field:

- *Tarbert Formation*
  Generally micaceous sandstones overlain by coarser, locally pebbly sandstone. The formation is more thickly developed down-flank and is thought to be a transgressive sandstone.
- *Ness Formation*
  A complex interbedded sequence of sandstones, siltstones, shales and coals representing a delta-plain palaeoenvironment. The sandstones are variable in origin and character, representing fluvial channels, crevasse splays, levees and bay fills. The finer-grained deposits are usually attri-

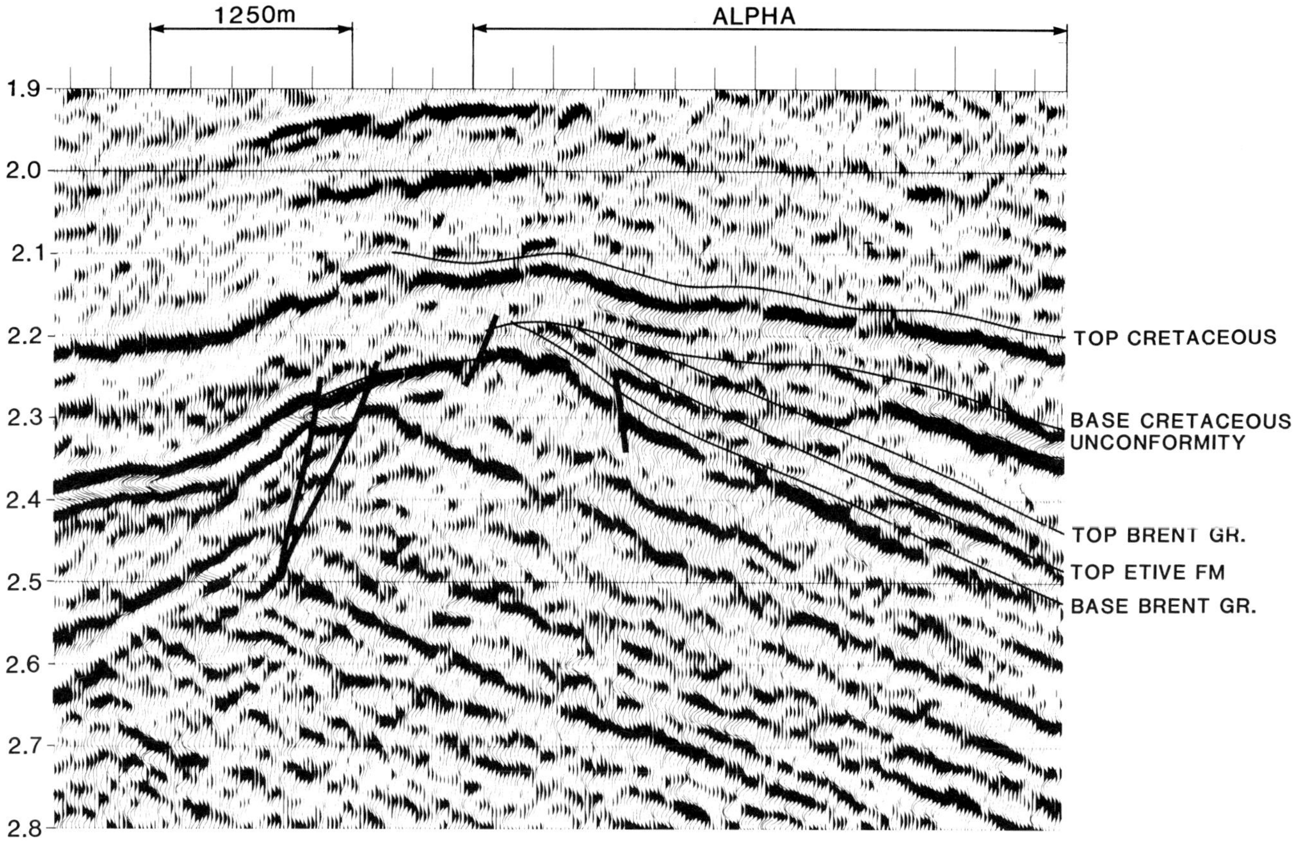

Fig. 5. Seismic section to show data quality in central part of Oseberg three-dimensional survey.

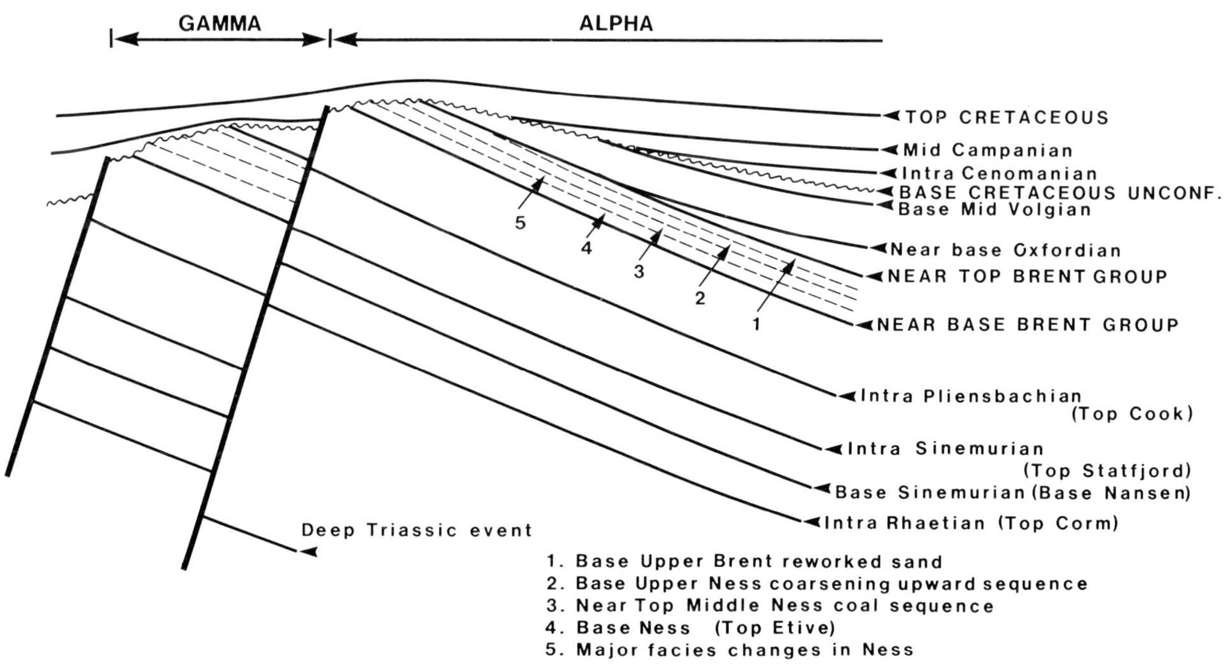

Fig. 6. Main sequence boundaries.

buted to overbank/floodplain deposition, coal swamps and lagoons/bays of variable salinities.

- *Etive Formation*

Typically comprising poorly sorted, pebbly, coarse-grained, massive sandstones representing delta-front/lower-delta-plain deposition. This unit is composed of three coarsening-upwards cycles in wells 30/6-1 and -2. In each cycle, wave-influenced micaceous, fine-grained sandstones pass up into coarser-grained, less

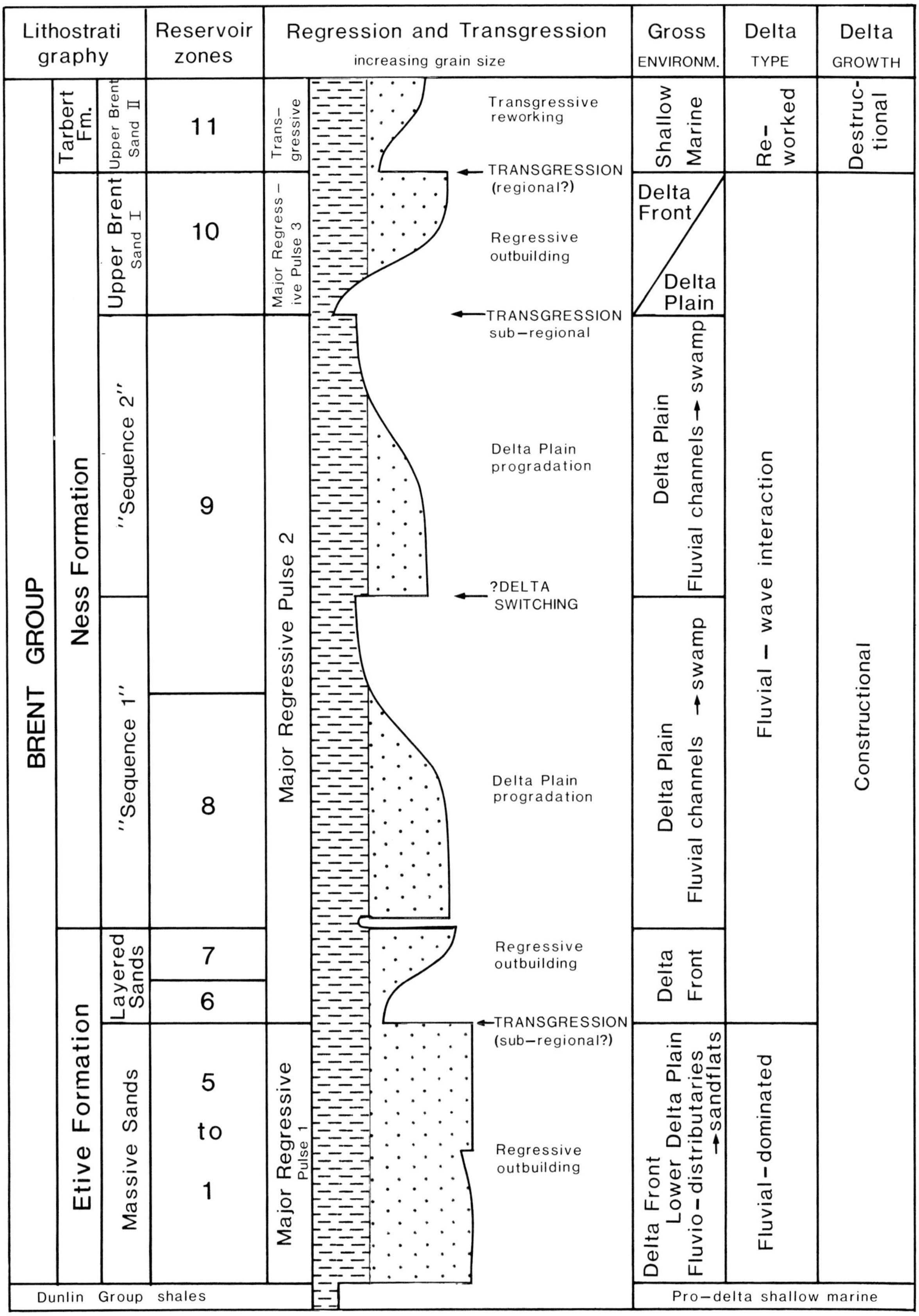

Fig. 7. Brent Group evolution.

micaceous sandstones that are not unlike the more typical massive sands.

The Brent Group would seem to conform to a deltaic model in which the Etive and Ness represent a major regressive, constructional phase which, upon abandonment, was transgressed by the Tarbert Formation (Fig. 7).

A new model and nomenclature as proposed by Steel *et al.* (1986) will be implemented on the Oseberg Field.

Across Oseberg the Brent Group thickness is structurally controlled. The easterly tilting and post-Middle Jurassic erosion of the Oseberg fault-blocks (Alpha, Alpha North and Gamma) have resulted in a thickening downflank of the Brent Group (with progressive incoming of younger lithostratigraphic units down-dip). Thus the near-crestal Well 30/6-2 penetrated only Etive and a thin interval of Ness (total Brent = 46.0 m) whilst the downflank Well 30/6-6 penetrated Tarbert as well, and a total Brent thickness of nearly 150 m. Similar thickening occurs between Alpha and the downfaulted structures, Gamma and Alpha North.

The Etive Formation is the main reservoir, but hydrocarbons are also found in the sandstones of the Ness and Tarbert formations. For the most part the laterally continuous, coarse-grained sandstones of the Etive Formation, which is approximately 40 m thick, have excellent reservoir properties. They are found in Alpha, Alpha North and Gamma. For reservoir description purposes, the Etive sandstone body has been subdivided into three zones. The Ness Formation sandstones, which have fair to good reservoir properties in some cases, are interbedded with shales and have uncertain lateral and vertical interconnectivity. The Tarbert Formation is found in the north-eastern part of the field, where it consists of continuous sand-sheets.

SOURCE

The main source-rock is regarded as the Draupne Formation. Surrounding the field there is, and has been, access to large volumes of this potential source-rock in various stages of maturity, including maturity levels sufficient to generate oil and gas.

HYDROCARBONS

The gas in the gas-cap of the Oseberg Field is a moderately wet gas with a gas–oil ratio of about 3200 Sm^3/Sm^3. A slight trend of decreasing methane content with depth is observed. The properties of the oil show a variation with depth identified by an increase in heavy components, a corresponding decrease in methane content, and consequently also a decrease of saturation pressure with depth.

Extremely high productivity has been observed in the hydrocarbon-bearing zones of the Etive Formation. Ideal productivity indices of 415–1950 Sm^3/day/bar for the oil zone, and 360 000–780 000 Sm^3/day/bar for the gas zone, have been derived from well-test data interpretations.

RESERVES

In the Field Development Plan the recoverable oil reserves for the Oseberg Field were estimated to be 153×10^6 Sm^3, about 43% of original oil-in-place. In addition, approximately 7×10^6 Sm^3 of condensate will be recovered from the gas production. Sales gas volumes have been estimated to be $50\text{–}70 \times 10^9$ Sm^3.

In the spring of 1986 the operator's estimate of recoverable oil reserves was increased to 203×10^6 Sm^3. This was due to a change in the proposed production mechanism, to three-dimensional seismic interpretation, to new geological information, and to new studies of recovery efficiency.

REFERENCES

Badley, M.E., Egeberg, T. and Nipen, O. 1984. Development of rift basins illustrated by the structural evolution of the Oseberg feature, Block 30/6, offshore Norway, *Journal of the Geological Society of London* **141** (4), 639–649.

Larsen, V., Aasheim, S.M. and Masset, J.M. 1981. 30/6-Alpha structure, a field case-study in the Silver Block, In *Norwegian Symposium on Exploration*, Norsk Petroleumsforening Oslo, Article 14, 36 pp.

Steel, R., Graue, E., Helland-Hansen, W., Johnsen, J.R., Lømo, L., Nøttvedt, A., Rønning, K. and Ryseth, A. 1987. Advance and retreat of Brent Delta System, Norwegian North Sea, In (J. Brooks and K. Glennie, eds) *3rd Conference on Petroleum of North-west Europe,* London Oct 27–29, 1986, Graham & Trotman.

Torvund, T. 1986. Oseberg—late stage developments, In *Offshore Northern Seas—Reservoir Engineering Conference*, Norsk Petroleumsforening, Article R2, 21 pp.

Torvund, T. and Nipen, O. 1987. The Oseberg Field: potential for increased oil recovery by gas injection in a high permeability environment, In Kleppe, J. *et al.* (Eds), *North Sea Oil and Gas Reservoirs*, Graham and Trotman, London.

34

Troll

D. I. Gray

A/S Norske Shell, PO Box 40, N-4056, Tananger, Norway

The Troll oil and gas field is located on the north-western edge of the Horda Platform, at the eastern margin of the Viking Graben in 300–350 m of water. It lies in Norwegian Blocks 31/2, 31/3, 31/5 and 31/6 covering an area of 770 km^2. The accumulation is contained within a series of north-easterly and south-westerly tilted fault-blocks and is sealed by claystones and marls of Late Jurassic to Tertiary age. The field contains a major gas cap with a maximum column of some 230 m, overlying an oil rim of variable thickness. Three main communicating hydrocarbon-bearing structures are identified: Troll West, which is divided into (a) a south-western oil province and (b) an areally larger gas province; and (c) the major gas accumulation of Troll East. The reservoir interval lies within the Middle to Upper Jurassic Viking Group sands of the Fensfjord, Middle Heather, Sognefjord and Upper Heather formations. This shallow-marine sequence is cyclic, with alternations of transgressive phase sands and silts and progradational shoreface facies, themselves often reworked by subsequent transgression. Eastwards the cyclic nature changes, and westwards the sands pinch out diachronously into their deeper marine silt equivalents. The reservoir rocks have excellent properties, with porosities ranging up to 34% and permeabilities up to tens of Darcies. To date 26 exploration and appraisal wells have been drilled in the Troll Blocks, delineating in-place volumes of 1788×10^9 Sm^3 gas and 970×10^6 m^3 oil. Some 165×10^6 m^3 oil lies in the oil column, 22–28 m thick, of the Troll West oil province. Field development planning is currently in hand.

INTRODUCTION

The Troll Field is a major gas accumulation underlain by a thin oil rim. It is situated some 80 km north-west of Bergen and 90 km south-east of the Statfjord Field. The field is located in water depths of 300–350 m, and extends to four licence Blocks: 31/2, 31/3, 31/5 and 31/6 (Fig. 1). Troll West covers some 290 km^2 and Troll East 480 km^2.

The Troll reservoir interval comprises a thick, poorly consolidated shallow-marine sandstone sequence of Middle to Late Jurassic age, assigned to the normally shaly Viking Group (Vollset and Doré, 1984), and sealed by Upper Jurassic to Tertiary claystones and marls. Hydrocarbons are trapped in a series of north-easterly and south-westerly tilted fault-blocks. These divide the accumulation into three communicating hydrocarbon-bearing structures (Fig. 2):

- Troll West oil province containing a small gas-cap and an oil column 22–28 m thick;
- Troll West gas province with some 200 m (maximum) of gas cap and an oil rim of 12 m;
- Troll East, with some 230 m (maximum) of gas-cap and an oil rim of up to 4 m thick.

The name 'Troll' is derived from Norse mythology and is used to emphasize the size of the field and, with some imagination, its shape.

HISTORY

Production Licence 054 was awarded in 1979 for Block 31/2 to the following:

Geology of the Norwegian Oil and Gas Fields © A. M. Spencer *et al.*, eds (Graham & Trotman, 1987) pp. 389–401.

Troll Field data summary

Trap	
Type	structural; series of tilted fault-blocks
Depth to crest	
Troll West	1340 mss
Troll East	1315 mss
Lowest closing contour	1551 mss
Hydrocarbon contacts	
— Troll West oil province	GOC 1543.3 mss OWC 1570.3 mss
— Troll West gas province	GOC 1547.0 mss OWC 1559.0 mss
— Troll East	GOC/GLC 1547.0 mss OWC 1551.0 mss
Gas column (Troll East)	maximum 230 m
Oil column	
— Troll West oil province	22–28 m
— Troll West gas province	12 m
— Troll East	0–4 m
Pay zone	
Formations	Upper Heather, Sognefjord Middle Heather, Fensfjord
Age	Callovian to Portlandian
Gross thickness	maximum 230 m
Porosity range	19–34%
Permeability range	1 mD to >10 D
Hydrocarbons	
Oil gravity	0.882–0.910 g/cm^3
Oil type	saturated, slightly waxy
Gas type	dry, lean and sweet gas
Gas gravity (relative to air)	0.61
Condensate gravity	0.802–0.825 g/cm^3
Bubble-point/dew-point	158 bar at 68°C
Gas/oil ratio	64 Sm^3/Sm^3
Condensate yield	35–40 Sm^3/10^6 Sm^3 gas at 56°C, 69 bar
Formation volume factor	
— gas	152.1
— oil	1.18
Formation water	
Salinity	47000 ppm NaCl equivalents
Resistivity	0.16 ohm m at 20°C
Reservoir conditions	
Temperature	68°C
Pressure	158 bar at 1547 mss
Pressure gradient in reservoir	
— gas	0.0113 bar/m
— oil	0.075–0.079 bar/m, gravity dependent
— water	0.102 bar/m
Field size	
Area	770 km^2
Gross rock volume	53×10^9 m^3
Hydrocarbon pore volume	12.5×10^9 m^3
Drive mechanism	gas-cap expansion and aquifer influx
Recovery factor—oil 0.2	
Recovery factor—gas 0.7	
Recoverable hydrocarbons	
— oil	33×10^6 Sm^3
— gas	1252×10^9 Sm^3

Production

A plan for Phase I Development of the gas in Troll East has been submitted to the Norwegian Authorities. Phase I Development calls for first gas production in 1996.

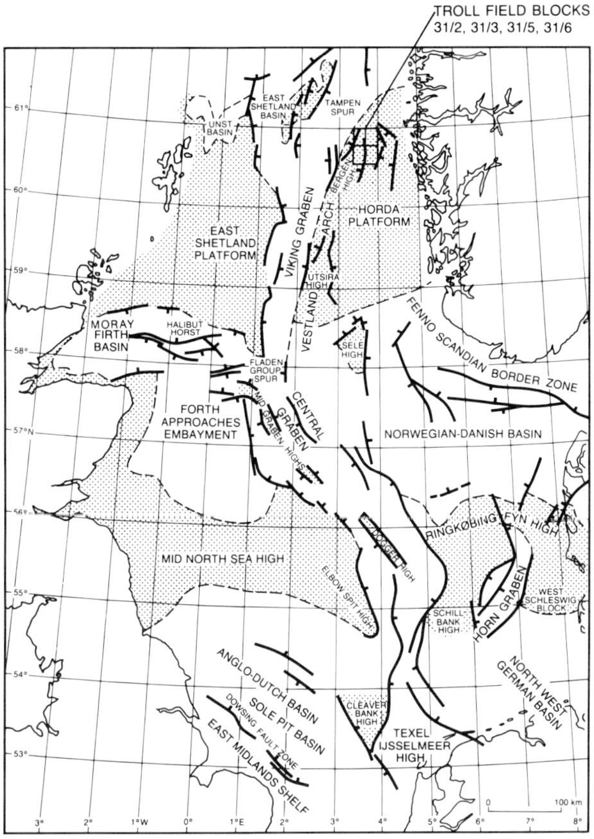

Fig. 1. Location map showing regional structural elements.

Statoil (50%)
A/S Norske Shell (35%; operator)
Norsk Hydro Produksjon A/S (5%)
Norske Conoco A/S (5%)
Superior Oil (Norge) A/S (5%)

The Troll Field was discovered in the same year with the first exploration well in Block 31/2 (Brekke *et al.*, 1981). Prior to this, seismic data across the area had revealed a flat seismic event that cut across the tilted fault-blocks. This flatspot was mapped and interpreted as a hydrocarbon–water contact (Birtles, 1986).

Well 31/2-1 was located in a tilted fault-block containing the most pronounced flatspot (Fig. 3). This well encountered 130 m of Middle to Upper Jurassic gas sand with traces of oil and reached total depth in the Triassic. It confirmed that the flatspot corresponded to the level of the hydrocarbon–water contact.

A further four exploration wells were drilled on the structure in Block 31/2 (Troll West), followed by nine appraisal wells up to the end of 1984. The main aim of these appraisal wells was to determine the thickness variation of the reservoir and producibility of the oil rim. A total of 17 seismic surveys have been conducted over Troll between 1971 and 1982.

In 1983 Production Licence 085 was awarded for the adjacent Blocks 31/3, 31/5 and 31/6 to Statoil (85%), Norske Hydro Produksjon A/S (9%), and Saga Petroleum a.s. (6%). Operator responsibility for Licence 085 is shared among the three companies. Since 1983, a total of nine wells have been drilled on the structure within PL 085: two wells in 31/5, two in 31/3 and five in 31/6 (Fig. 2). An additional three exploration wells were drilled off the Troll structure. A Statement of Commerciality was issued for Troll West on the 15 November 1983. Since this date, efforts have been directed towards the formulation of a field-development plan.

GEOLOGICAL SETTING

The Troll Field is located at the north-western edge of the Horda Platform (Fig. 1) on the Bergen High which lies close to the eastern margin of the Viking Graben. The field consists of north-easterly and south-westerly tilted fault-blocks sealed by Upper Jurassic to Lower Tertiary claystones and marls. The hydrocarbons are entrapped within shallow-marine sands of the Fensfjord, Middle Heather, Sognefjord and Upper Heather formations of the Middle to Upper Jurassic Viking Group (Fig. 4).

Triassic inception of the Viking Graben rifting imposed a structural grain of easterly dipping megablocks over this area. The progressive northward dextral offset of the graben axis produced a series of arcuate NE–SW faults during the Middle to Upper Jurassic. These influenced Viking Group deposition.

Early Cretaceous/Late Kimmerian rift-related movements occurred up to and including Aptian times, resulting in a series of unconformities and imposing a conjugate set of N–S- and NW–SE-trending faults. Earlier faults were at this time reactivated and there was a degree of dextral rotation. The ensuing regional graben subsidence which commenced during the Upper Cretaceous led to further reactivation of the major N–S faults and caused renewed megablock tilt and rotation. This resulted in an absence of much of the Late Cretaceous over the Troll area. Cretaceous unconformities condense westwards, where the succession is eroded down to the uppermost part of the Viking Group (Fig. 5).

During Tertiary times, stress releases

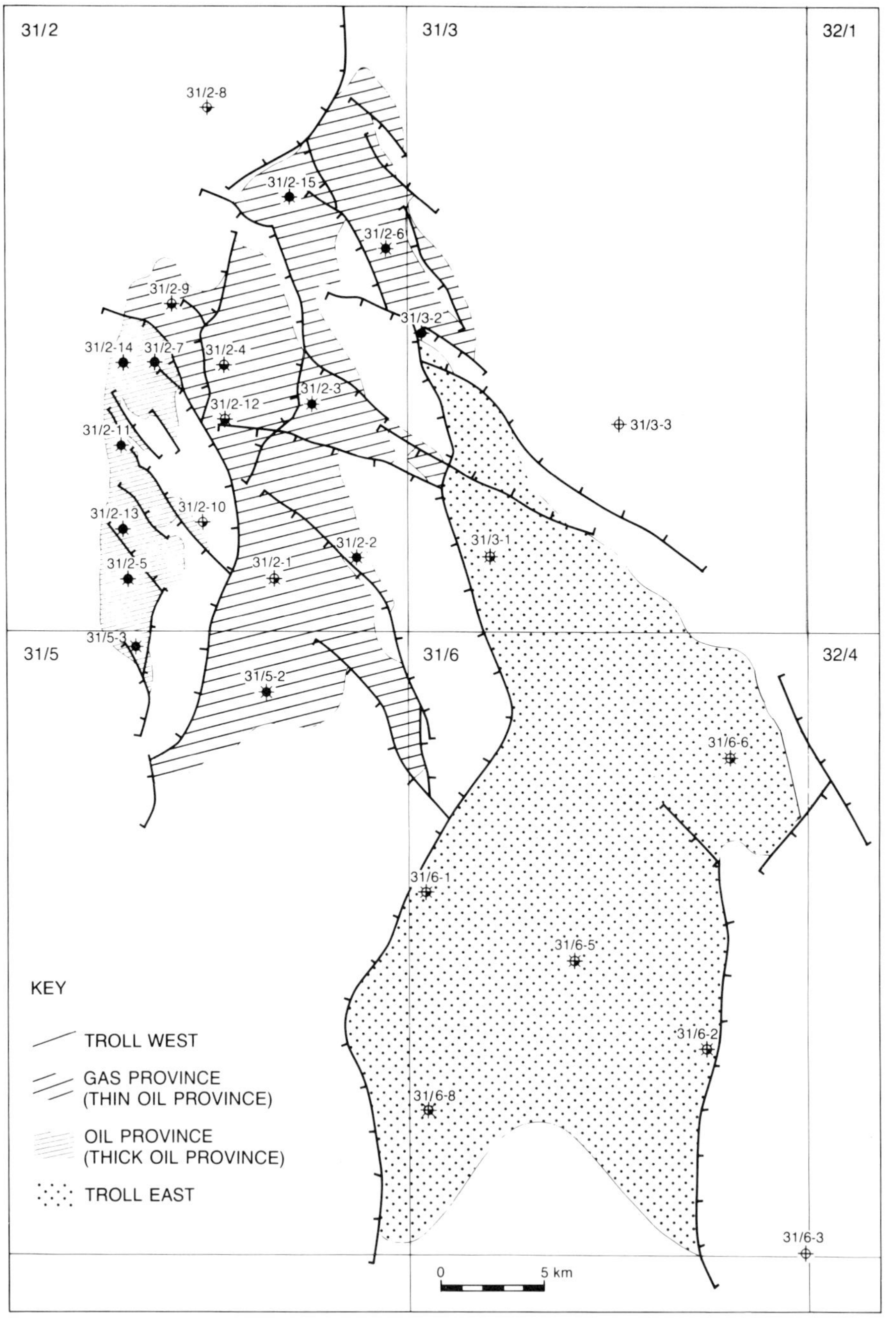

Fig. 2. Troll Field geography and well locations.

between the down-warping graben and the Horda Platform imposed a NW–SE fault component and developed a hinge zone through Block 31/2.

The Viking Group and its stratigraphic relationships are shown in Fig. 4. The deepest exploration wells on the structure penetrated Triassic rocks of the Hegre Group.

The Hegre Group comprises a continental series of claystones and sandstones. These are transitional to and overlain by continental fluvio-deltaic and shallow-marine sandstones, siltstones, claystones and coals of the early to Mid-Jurassic Statfjord Formation and Dunlin and Brent groups.

The Viking Group represents a period of incremental transgression over the earlier Jurassic, gradually changing from paralic and marginal-marine to higher-energy conditions in Late Jurassic times. Sediment supply during this period was from coastal distributaries situated some way east of Troll. The Lower Heather Formation comprises fine micaceous siltstones and is overlain by coarse sandstones of the Krossfjord Formation. The Fensfjord Formation marked an onset of more

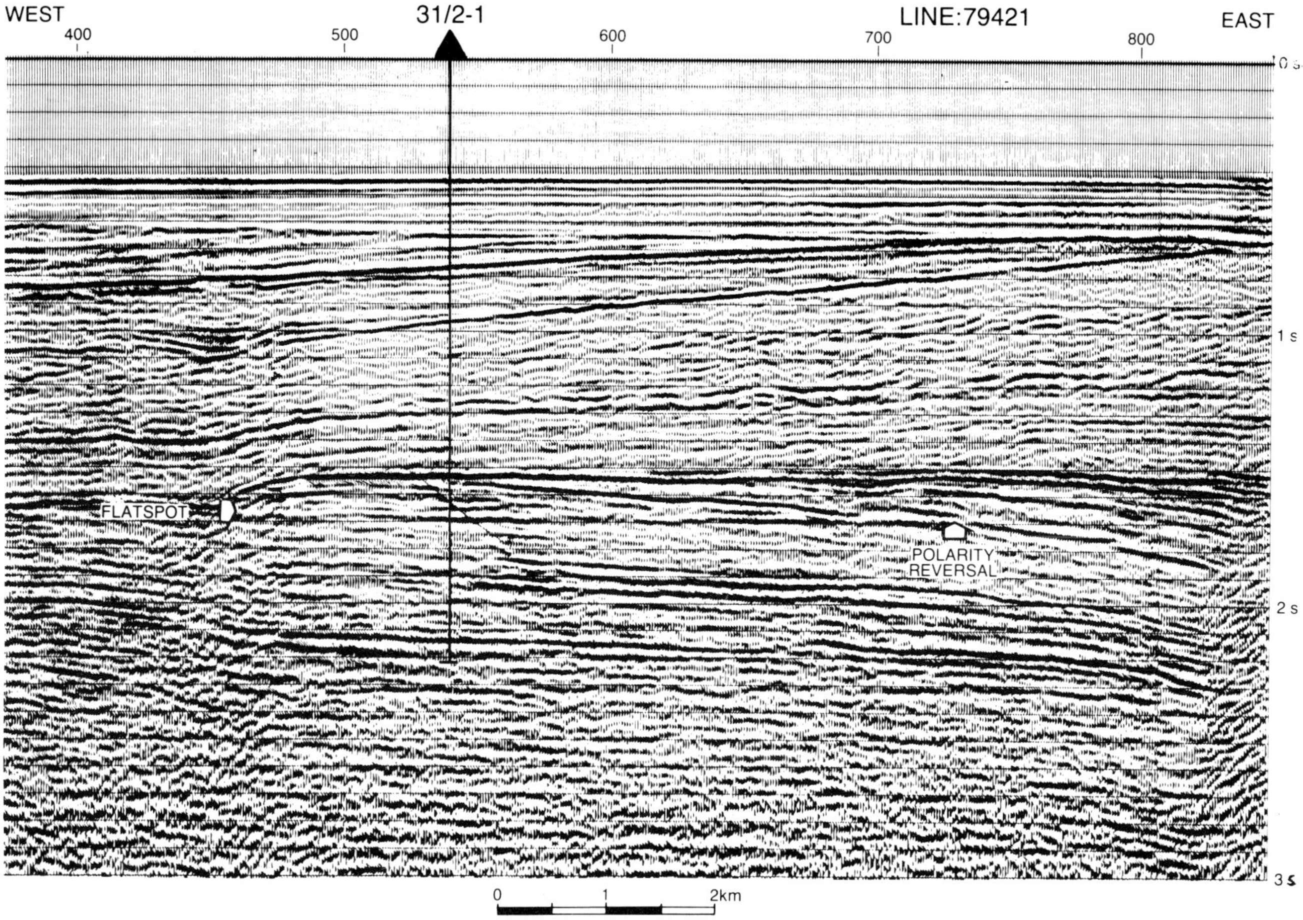

Fig. 3. Troll Field prospect identification: Line 79421.

variable coarsening-upwards sequences through shoreface progradation. During latest Callovian and earliest Oxfordian times a marked increase in transgressive activity led to deposition of the micaceous sands and silts of the Middle Heather Formation, themselves overlain by southwards- and westwards-advancing coarse, clean, transgressive sands successively off-lapped by westward-prograding shoreface sand facies of the Sognefjord Formation, the principal hydrocarbon-bearing interval of Troll (see type logs in Figs 6 and 7). In the Troll East area the Upper Heather siltstones are present and pinch-out westwards towards the Troll West area.

The Draupne (Kimmeridge Clay) Formation was deposited over the area but subsequent Kimmerian-related erosion during Early Cretaceous times has removed it from much of the western part of the structure, where Cretaceous to Tertiary shales and silts rest uncomformably on the Upper Jurassic. The Cromer Knoll Group (Barremian to Cenomanian) is present as dark claystones and marls, condensing westwards as a result of intra-Aptian erosion. The Upper Cretaceous Shetland Group is represented by silts and marls that overstep earlier Cretaceous and Jurassic rocks. They are themselves overstepped by the Paleocene Maureen Formation.

GEOPHYSICS

Between 1971 and 1982, 17 seismic surveys were undertaken across the Troll Field. The present coverage is refined to a 1×1 km or denser grid across Troll. The current seismic interpretation is based primarily on the 1980 and 1981 surveys. A number of seismic lines have been processed to band limited acoustic impedance.

Seismic coverage shot in the 1970s revealed a 'flatspot' which was interpreted as a direct hydrocarbon indicator (Fig. 5). This flatspot coincided with the interpreted spill-point of the structure and suggested a closure of some 770 km^2. Polarity reversals at the top reservoir reflection were also observed (Fig. 3)

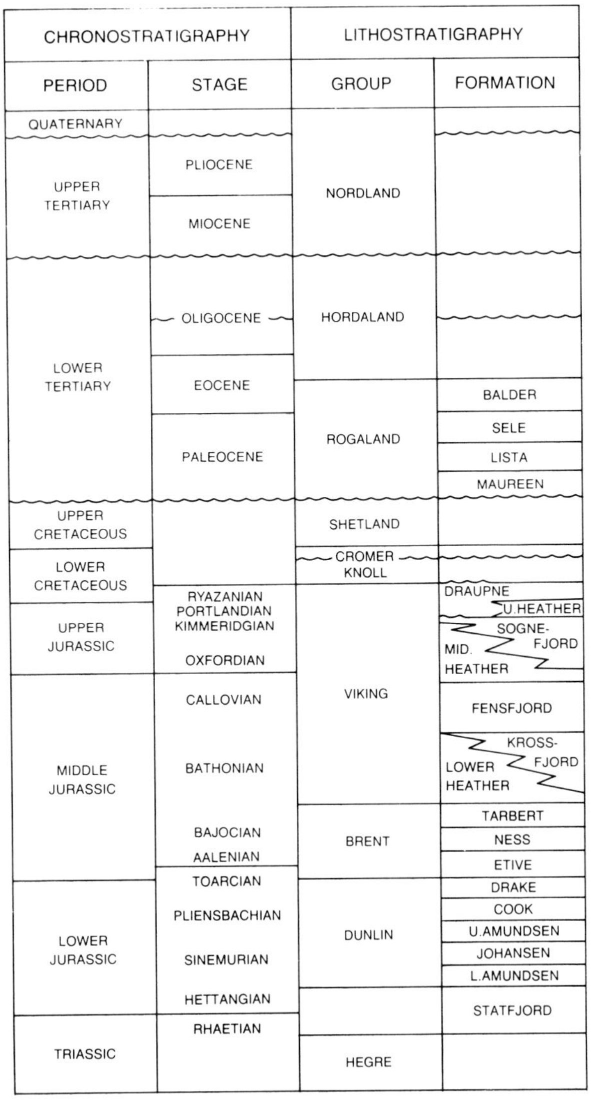

Fig. 4. Stratigraphic summary of the Troll Field area.

along the accumulation edges (Birtles, 1986).

The seismic data available are of a high quality. The top reservoir reflector is clearly defined over the structure where the gas column is thicker than 10 m, by a strong negative impedance contrast. It is less clear over the oil province.

TRAP

The Troll Field is a structural trap produced by multiphase faulting associated with oblique slip and extensional tectonics, located along the margins of the positive Horda Platform, adjacent to the Viking Graben. The accumulation, some 50 × 30 km, is divided into a number of fault-blocks dipping north-east and south-west: one megablock in Troll East and a number of smaller blocks in Troll West. The structural crest in Troll West is at 1340 mss and is at 1315 mss in Troll East. Owing to the late Kimmerian movements the cap-rock seal systematically becomes younger westwards from Late Jurassic claystones in the east to Paleocene claystones in the west.

Hydrocarbon communication between Troll East and Troll West is restricted to two relatively narrow corridors (Fig. 2), although effective pressure communication exists via the underlying aquifer. The accumulation is believed to be filled to a spill-point in the south-east corner of Block 31/6. The observed variations in oil-column thickness are believed to be the result of a two-stage migration history in which the oil, displaced by

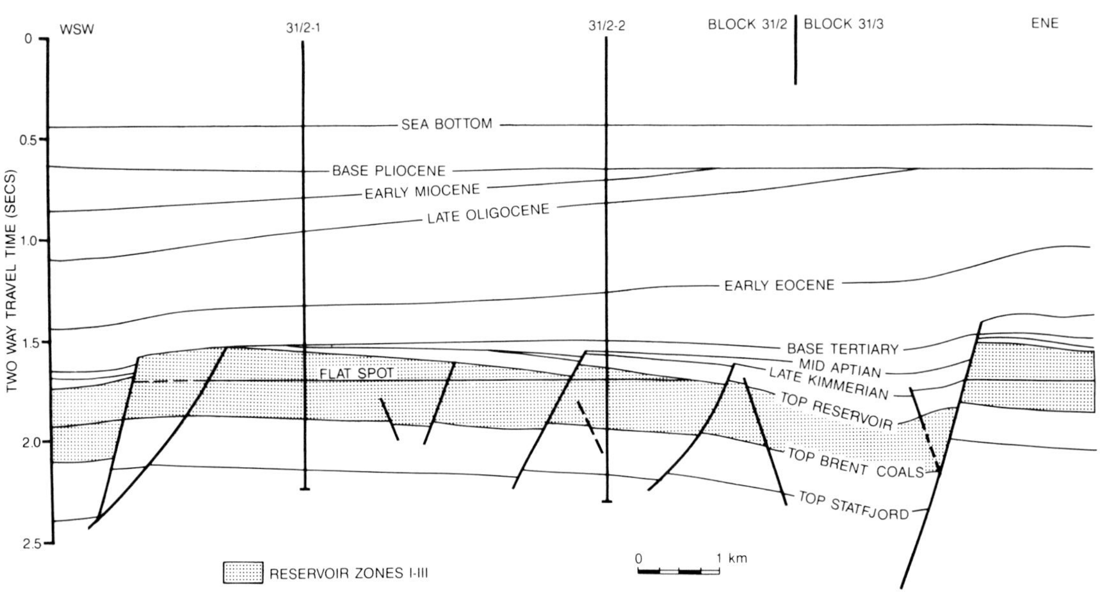

Fig. 5. Geological interpretation of WSW–ENE of Line 8107–207 showing major structural units and sequence boundaries.

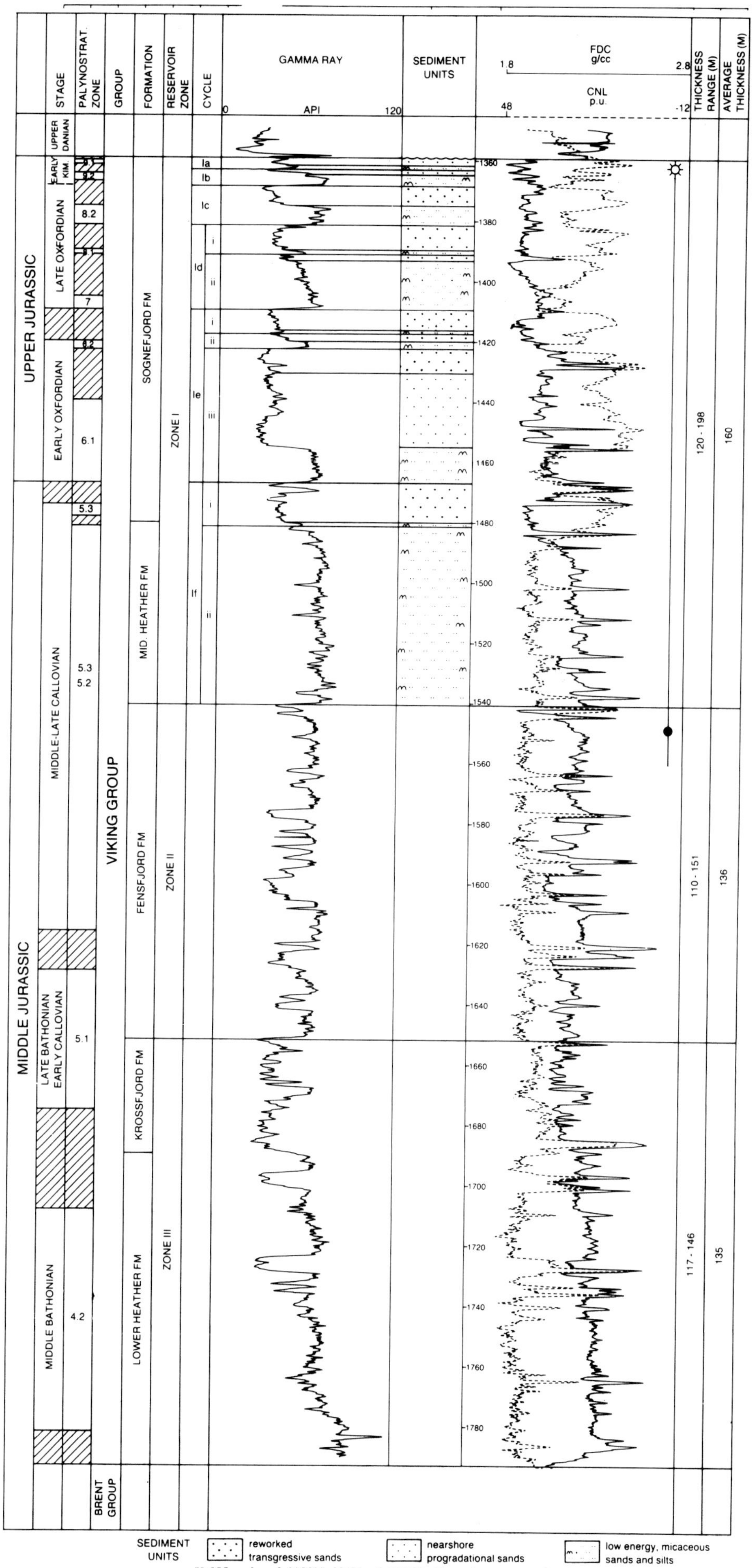

Fig. 6. Type log Troll West (Block 31/2).

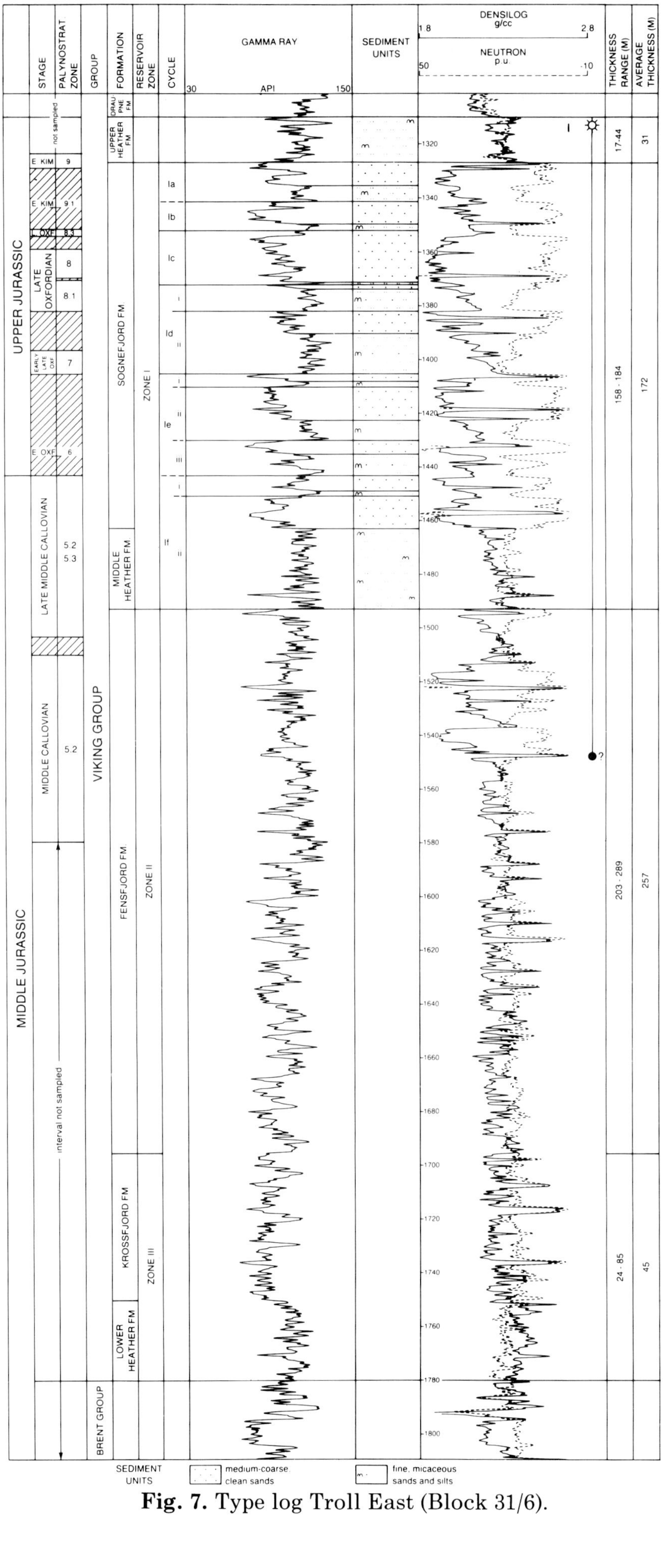

Fig. 7. Type log Troll East (Block 31/6).

gas, was prevented from achieving equilibrium by entry pressure-barriers across low-permeability lithologies and faults. The fault planes are considered to be non-sealing.

Hydrocarbon contacts are:

Troll West	OWC 1570.3 mss
Oil Province	GOC 1543.3–1548.3 mss
Gas Province	OWC 1559.0 mss
	GOC 1547.0 mss
Troll East	OWC 1551.0–1547.0 mss
	GOC/GLC 1547.0 mss

RESERVOIR

Some 3400 m of core have been cut over the reservoir interval. Because of the unconsolidated nature of the reservoir, fibre-glass sleeve coring techniques were used in all but the earliest exploration wells.

The reservoir sands belong to the Middle to Upper Jurassic Viking Group. The hydrocarbons are entrapped within the Fensfjord, Middle Heather, Sognefjord, and Upper Heather formations, of Callovian to Portlandian age. The Sognefjord Formation contains more than 90% of the hydrocarbons-in-place.

The reservoir has been subdivided into three zones in which the genetic and chronostratigraphic links are highlighted:

- Zone I combines the Sognefjord and Middle Heather formations. It contains more than 90% of the hydrocarbons-in-place.
- Zone II is equivalent to the Fensfjord Formation.
- Zone III combines the Krossfjord and Lower Heather formations.

This subdivision is most characteristic of the western part of the Troll Field (Blocks 31/2, 31/5 and part of 31/3—Fig. 6). In the east, formation boundaries are more variable and, in addition, the Upper Heather Formation is present (Fig. 7).

Fensfjord Formation

The Fensfjord Formation has a porosity range of 23–30%, a gross thickness of 110–290 m and comprises a stacked series of small, coarsening-upwards units, passing from micaceous siltstones to fine sandstones and medium to coarse sands. Palynofacies analyses based on the abundance and character of acid insoluble organics of both terrestrial and marine origins (Whitaker, 1984) show a fresh terrestrial source with little reworking and degradation. The sequence is interpreted as progradational shoreface facies.

Middle Heather Formation and Sognefjord Formation

The Middle Heather Formation has a porosity range of 19–26%, average permeability of 10 mD and a gross thickness of 15–110 m. The Sognfjord Formation has a porosity range of 25–34%, permeability in the Darcy range and a gross thickness of 24–160 m. The Middle Heather and Sognefjord formations comprise six depositional cycles, each marked by a rapid rise in sea level. A depositional cycle comprises both a progradational and a transgressive component (Fig. 8). In each cycle, low-energy, distal, fine micaceous sediments coarsen upwards into shoreface medium to coarse sand sequences representing shoreface progradation of a complex sand-body across the shelf. The Heather Formation represents just one major transgressive phase of the late Callovian. This sequence is often overlain or partly replaced by coarse, well-sorted clean sands, storm-reworked as offshore bars from more proximal settings during subsequent transgressive phases.

Palynofacies analyses recognize alternations between well-preserved assemblages of a low-energy environment dominated by coastal progradation, and poorly preserved high-energy assemblages interpreted to have an 'oceanic' character, which correlate to transgressive intervals. Palynostratigraphy reveals that the Middle Heather/Sognefjord Formation boundary is strongly diachronous (Fig. 8). Clean sands pinch out and, towards the west, interfinger with micaceous sands and silts. Whilst this is characteristic of Troll West, in Troll East the Sognefjord Formation sands become finer grained and thinner bedded, with repetitions of small, coarsening-upwards sequences possibly reflecting the eastwards transition to an upper shoreface environment. The Middle Heather also thins considerably eastwards causing formation boundaries to become less distinct.

Upper Heather Formation

The Upper Heather Formation has an average porosity of 19–23% and a gross thickness of 0–44 m. This formation is present only in Troll East and comprises an open-marine siltstone facies.

Tightly cemented, calcite-rich zones are recognized throughout the reservoir sequence. These are generally of a nodular

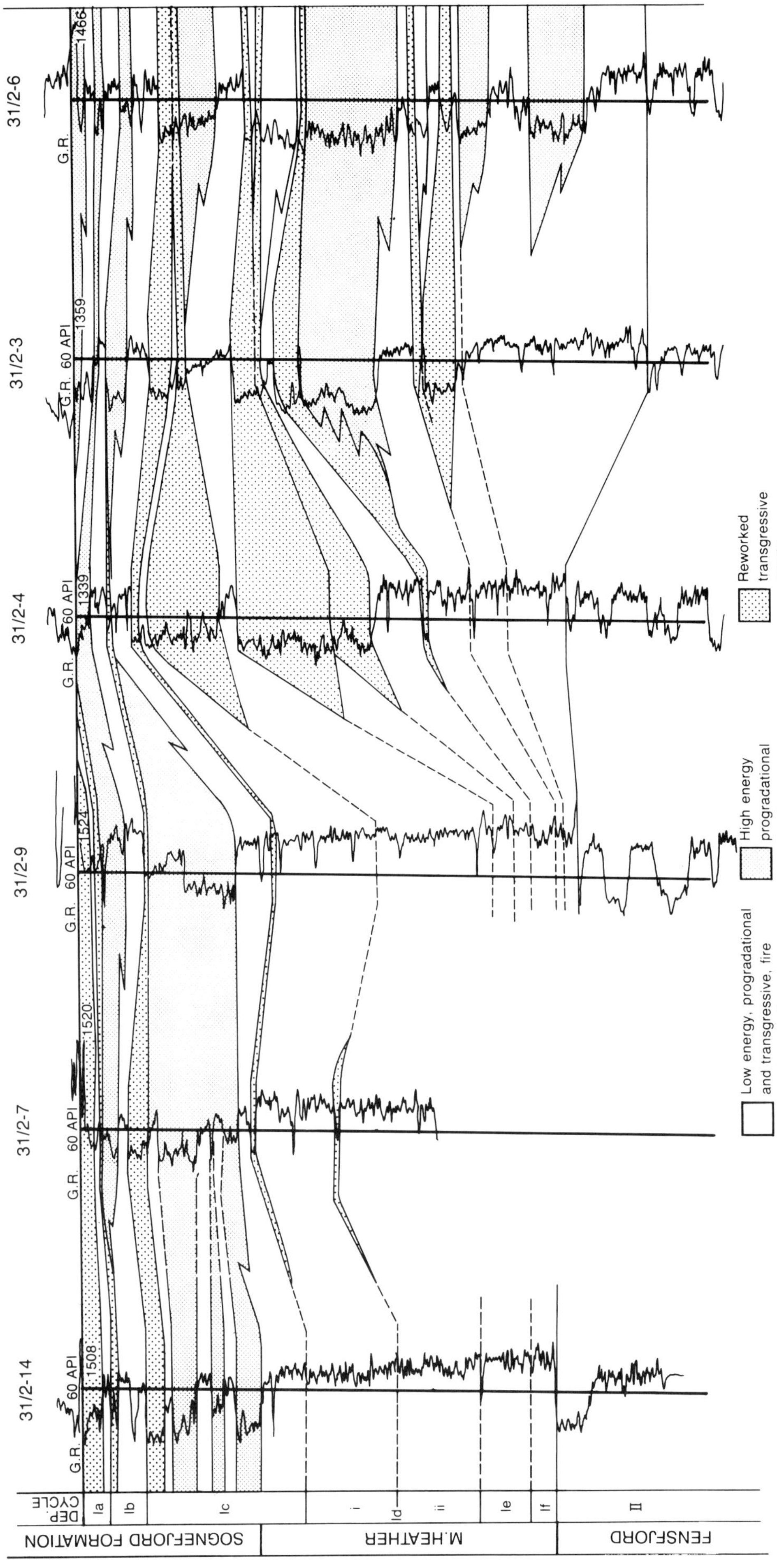

Fig. 8. Troll West stratigraphic correlation.

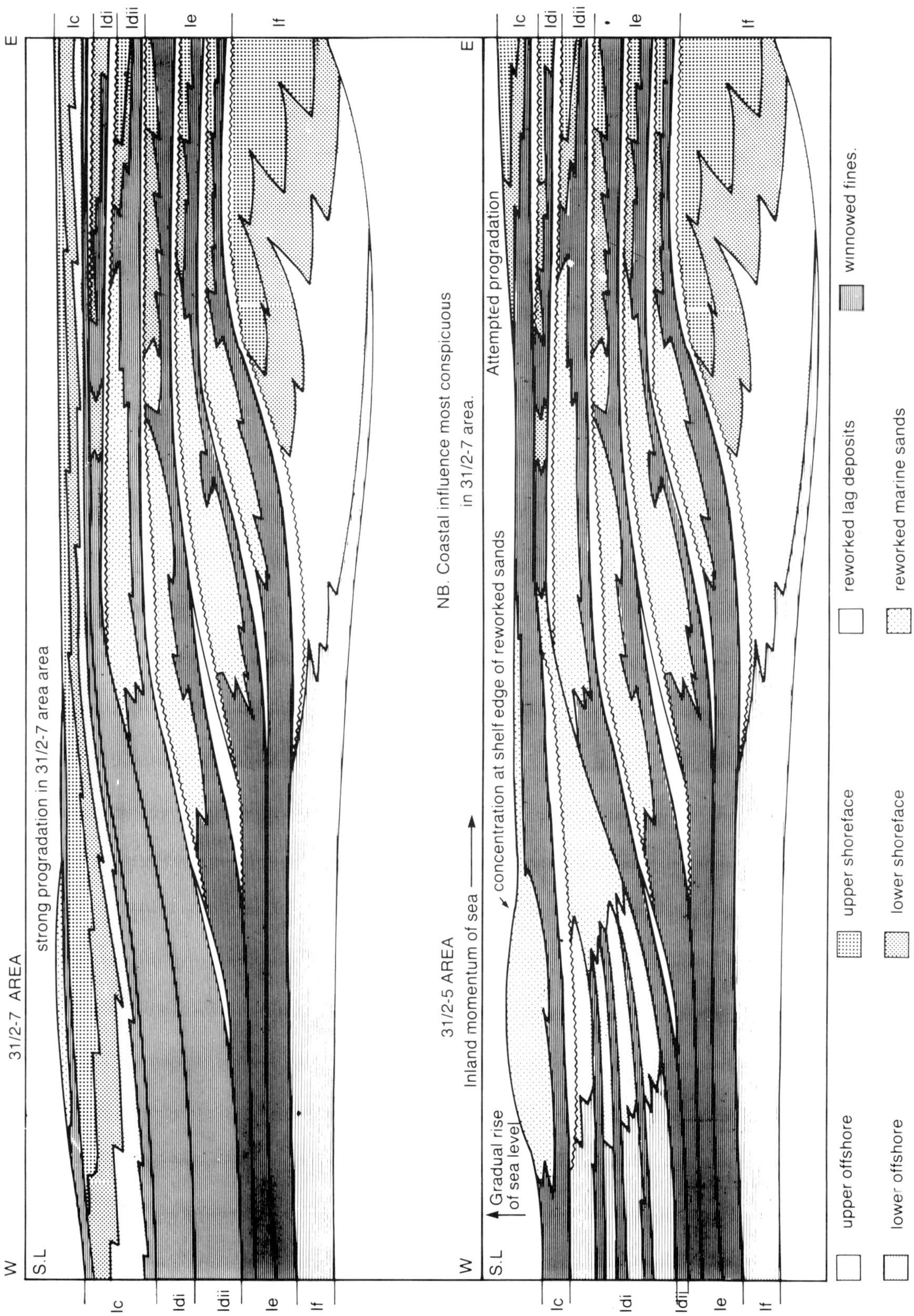

Fig. 9. Troll Field depositional model.

nature and are believed to have a limited lateral extent. With the exception of these calcite zones, the full gross hydrocarbon column is treated as net productive rock.

SOURCE

The dominant source-rock is the organic-rich Draupne (Kimmeridge Clay) Formation. A large kitchen is present to the west and northwest of the field. Oil migration accompanied by minor gas expulsion began in Late Cretaceous/Early Tertiary times, concomitant with structural tilting towards the east. The major phase of gas migration was initiated in Late Tertiary times and led to an eastwards displacement of the oil by gas.

HYDROCARBONS

The Troll crude has a gravity range of 0.882–0.910 g/cm^3. The oil is saturated, and slightly waxy with a GOR of 64 Sm^3/m^3 and a viscosity at reservoir conditions of 1.4–2.0 cp. Reservoir temperature is 68°C, with an initial pressure at 1547 mss of 158 bara. The system is in equilibrium with bubble- and dew-points at 158 bara. The gas is dry, lean and sweet and contains some 93% methane. Water salinity is 51000 ppm NaCl equivalents.

The gas zone in Troll West has been tested with four wells, up to 3.4×10^6 Sm^3/day. The oil zone in West Troll has been tested with nine wells, up to 1250 m^3/day. A recent gas test on a Troll East well achieved 3.7×10^6 Sm^3/day.

RESOURCES

Gas-cap expansion and aquifer influx are the primary gas recovery mechanisms. Although the formations are loosely consolidated and fairly compressible, compaction drive will not contribute directly to hydrocarbon recovery, although compaction in the aquifer will contribute to water influx. In the absence of any sealing faults or severe lateral discontinuities, the gas accumulation will act to provide expansion energy with very low pressure gradients throughout the system. Since the gas is dry and lean, retrograde condensation will not be of concern, and the gas can be treated as a single-phase expansion source.

Gas ultimate recovery, assuming a reasonable 70% recovery factor, is as follows:

Troll West 440×10^9 Sm^3
Troll East 810×10^9 Sm^3

Oil recovery production mechanisms will be a combination of water drive and gas-cap expansion drive. Ultimate oil recovery from the Oil Province, assuming development and a reasonable 20% recovery factor, could be 33×10^6 m^3.

Production from the thin oil rims (12–0 m) of Troll Field is not at present considered viable.

The shallow reservoir depth restricts the reach of wells drilled from a platform. Interference effects, however, are considered negligible for gas producers due to the high permeabilities. Gas development is being considered from fixed-platform clusters. Oil development is being considered via subsea wells tied into a central production unit.

The reservoir sands are mostly unconsolidated and gravel pack completions will be required for sand control.

ACKNOWLEDGEMENTS

This paper has been compiled from numerous sources within A/S Norske Shell. The ideas expressed reflect the results of studies carried out by A/S Norske Shell in its role of the Operator in Licence PL 054 (Block 31/2). In certain areas, especially with respect to the Troll East, these views depart slightly from those held by the Operators in Licence PL 085 (Blocks 31/3, 31/5 and 31/6). The consent to publish given by Statoil, Norsk Hydro Produksjon A/S, Saga Petroleum AS, Norske Conoco A/S and Mobil Development Norway A/S is gratefully acknowledged.

REFERENCES

Brekke, T., Pegrum, R.M. and Watts, P.B. 1981. First exploration results in Block 31/2, In *Norwegian Symposium on Exploration*, Norsk Petroleumsforening, Article 16.

Birtles, R. 1986. The seismic flatspot and the discovery and delineation of the Troll Field, In Spencer, A.M. *et al.* (Eds), *Habitat of Hydrocarbons on the Norwegian Continental Shelf*, Graham and Trotman, London, pp. 207–215.

Hellem, T., Kjemperud, A. and Øvrebø, O.K. 1986. The Troll Field: a geological/geophysical model established by the PL 085 Group, In Spencer, A.M. *et al.* (Eds), *Habitat of Hydrocarbons on the Norwegian Continental Shelf*, Graham and Trotman, London, pp. 217–238.

Reinholdtsen, B. 1984. Improved recovery from thin oil sands. Part 2—Troll Field, In *Offshore Northern Seas—Improved Offshore Recovery*, Norsk Petroleumsforening, Article R2.

Rønnevik, H. and Johnsen, S. 1984. Geology of the Greater Troll Field area, *Oil Gas Journal* **82** (4), 100–106.

Vollset, J. and Doré, A.G. (Eds). 1984. A revised Triassic and Jurassic Lithostratigraphic nomenclature for the Norwegian North Sea, *Norwegian Petroleum Directorate Bulletin No. 3.*

Whitaker, M.F. 1984. The usage of palynostratigraphy and palynofacies in definition of Troll Field geology, In *Offshore Northern Seas—Reduction of Uncertainties by Innovative Reservoir Geomodelling*, Norsk Petroleumsforening, Article G6.

Part VII—Halten Terrace

35

Midgard

Ole F. Ekern

Saga Petroleum AS, Box 9, N–1322 Høvik, Norway

The Midgard Field is located on the Halten Terrace, primarily in the Norwegian Blocks 6507/11 and 6407/2. The main part of the field is found within a N–S-trending horst-block. The reservoir is represented by the Middle Jurassic Tomma Formation with additional reserves in the Lower Jurassic Aldra Formation to the north where the Middle Jurassic strata have been eroded.

The sandstones in both units have very good reservoir characteristics with porosities and permeabilities of 24–28% and 1–6D respectively. The main hydrocarbon phase is gas–condensate. A gas-in-place value of 116×10^9 Sm^3 has been estimated. The potentially recoverable gas resources are 87×10^9 Sm^3.

INTRODUCTION

The Midgard Field is located 230 km northwest of Trondheim (Fig. 1). The shortest distance to the coastline is 160 km. The water depth in the area is about 250–300 m.

The field consists of four fault segments, which have been named Alpha, Beta, Gamma and Delta from north to south, and covers an area of 47 km^2 (Fig. 2). The field is mainly a gas–condensate accumulation, and in-place resources of 116×10^9 Sm^3 gas and 27×10^6 Sm^3 condensate represent the present estimates. In addition, an oil-leg has been encountered in the Beta segment. Provided that the oil layer is also present in the Gamma and Delta segments, an in-place value of 38×10^6 Sm^3 is possible. The hydrocarbons are found in marine sandstones of Early and Middle Jurassic age.

'Midgard' is the name of the human world in Norse mythology.

HISTORY

The Midgard Field is located in Blocks 6507/11 and 6407/2 which were awarded as PL 062 and PL 074 respectively. Licence PL 062 was awarded in 1981 following the second part of the fifth concession round and PL 074 one year later following the third part of the fifth round.

The partners in the two licences are:

	Company	*Interest* (%)
PL 062:	Saga Petroleum AS (operator)	10
	Statoil	50
	A/S Norske Shell	25
	Arco Norge A/S	10
	Norsk Hydro AS	5
PL 074:	Saga Petroleum AS (operator)	10
	Statoil	50
	Norsk Agip A/S	15
	Arco Norge A/S	15
	Deminex (Norge) A/S	10

The first well in the field, 6507/11-1, was drilled in 1981 as the third well in the Haltenbank area. The well proved the presence of gas–condensate in the Alpha segment. This was the first hydrocarbon discovery offshore Mid-Norway.

Two additional wells have since been

Geology of the Norwegian Oil and Gas Fields © A. M. Spencer *et al.*, eds (Graham & Trotman, 1987) pp. 403–410.

Midgard Field Data Summary

Trap	
Type	Fault-bounded horst-block
Depth to crest	2270 mss in Alpha
Lowest closing contour	2500 mss; 2490 mss in Alpha
Hydrocarbon contacts	GOC 2488 mss; GWC 2493 mss in Alpha; OWC 2499.5 mss
Gas column	225 m in Alpha
Oil column	11.5 m
Pay zone	
Formations	Tomma and Aldra formations
Ages	Middle and Early Jurassic
Gross thicknesses	Tomma 124–142 m; Aldra 183–202 m
Net/gross ratios	Tomma 0.75–0.94; Aldra 0.71
Cut off for N/G	Ø = 12%; V_{sh} = 40%
Porosity (average)	Tomma 28%; Aldra 24%
Hydrocarbon saturation	Tomma S_g = 90–95%, S_o = 76%; Aldra S_g = 78%
Permeability (average)	Tomma 6D; Aldra 1D
Hydrocarbons (*data from single-stage flash to 15°C, 1 bar)	
Oil gravity*	0.849 g/cm^3
Condensate gravity*	0.809 g/cm^3
Gas gravity* (relative to air)	0.776
Bubble point*	255 bar
Dew point*	256 bar
Gas/oil ratio	172.5 Sm^3/Sm^3
Condensate yield	231 Sm^3/10^6 Sm^3 gas
Formation volume factor, oil	1.520
Gas expansion factor	210
Formation water	
Salinity	75 000 ppm NaCl equivalents
Resistivity	0.1 ohm m at 24°C
Reservoir conditions	
Temperature	91°C at 2450 mss
Pressure	251 bar at 2450 mss
Field size	
Area	47 km^2
Gross rock volume	2.5 × 10^9 m^3
Recovery factor—condensate	0.63
—gas	0.75
Drive mechanism	Depletion/weak waterdrive
Recoverable hydrocarbons	
—oil	?
—NGL/condensate	17 × 10^6/Sm^3
—gas	87 × 10^9/Sm^3
Production	
—	

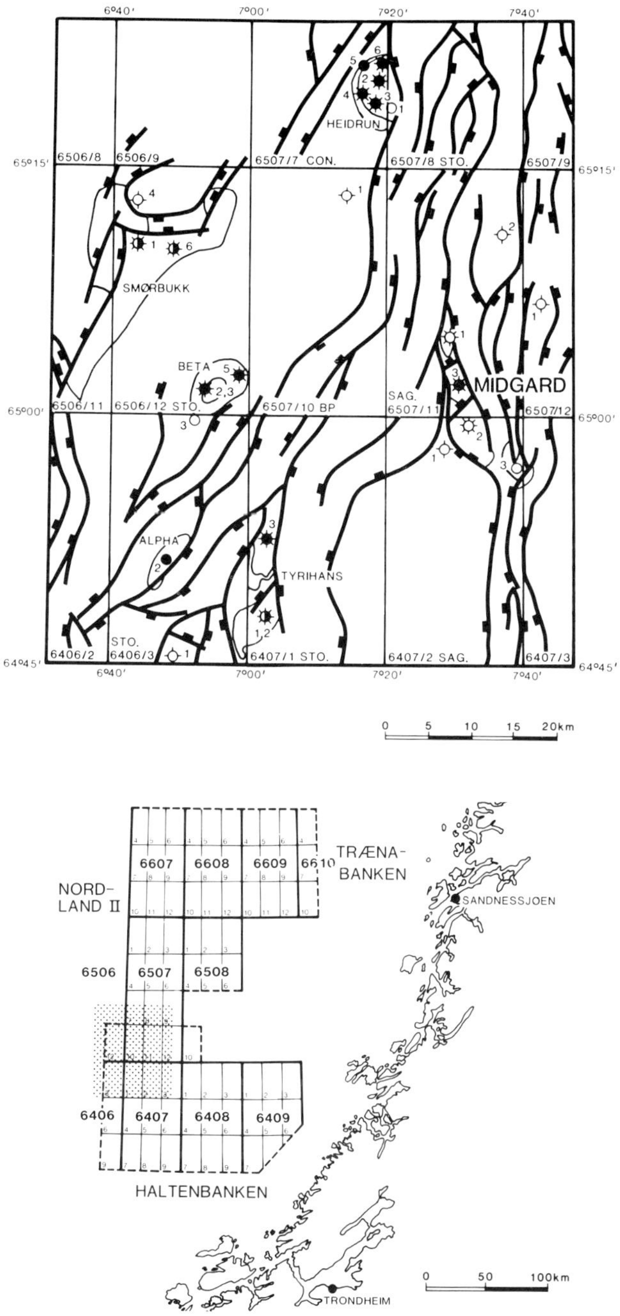

Fig. 1. Location map.

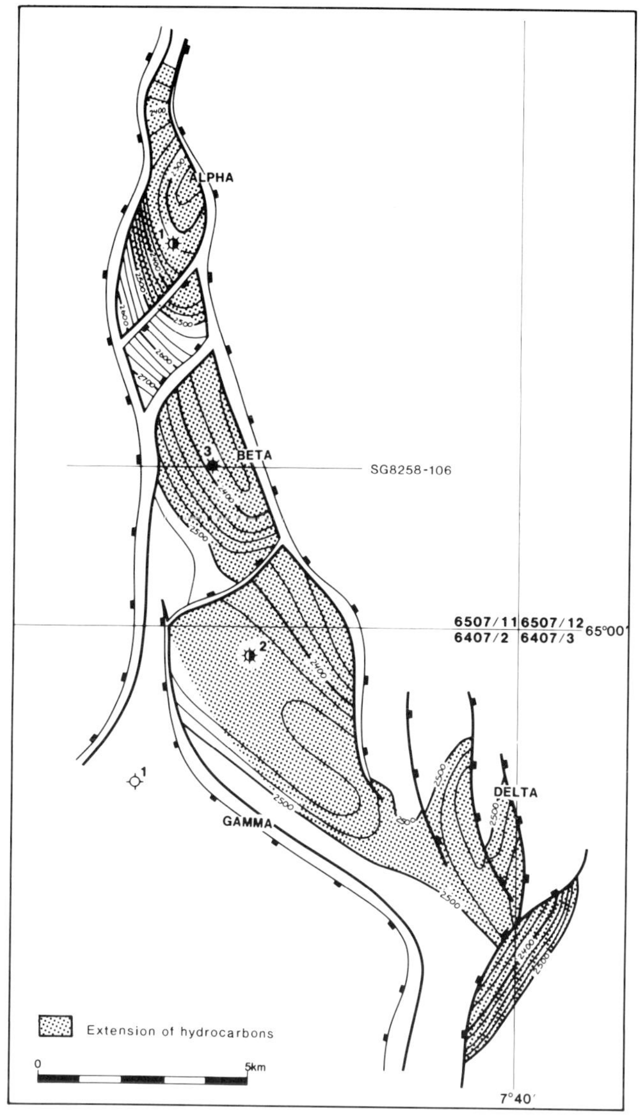

Fig. 2. Midgard Field. Top Middle Jurassic Unconformity (Top Tomma Formation)—depth map. Contours in metres subsea.

drilled. Well 6407/2-2 was drilled in 1983, proving gas–condensate in the Gamma segment. In 1985 Well 6507/11-3 was drilled, proving gas–condensate with a basal oil-leg in the Beta segment. The Delta segment will probably be tested in 1986.

GEOLOGICAL SETTING

The Midgard Field is located on the Halten Terrace. This area represents a complex basin-and-range province in front of the southern part of the Trøndelag Platform (Hagevang and Rønnevik, 1986). The Halten Terrace was a part of a regional basin until Middle–Late Jurassic time. The main tectonic activity took place during the Middle to Late Jurassic Kimmerian tectonic phases. Cretaceous deposits are mainly found in basin areas which were structurally enhanced by differential subsidence. In post-Late Cretaceous times rapid subsidence and sedimentation took place, only broken by a few unconformities.

A generalized stratigraphic column is shown in Fig. 3. The nomenclature is according to a proposal from Statoil and has not yet been formally approved.

Only a few metres of rock identified as 'Triassic Grey Beds' have been penetrated in

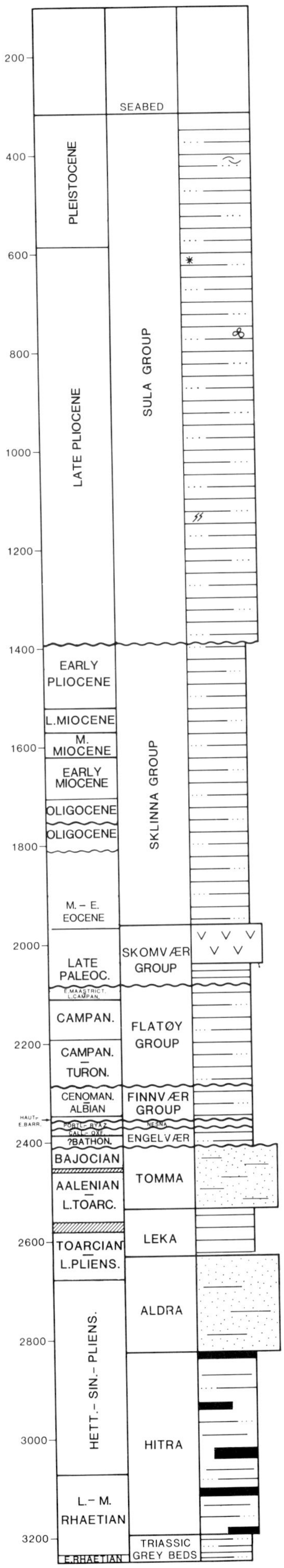

the Midgard Field. In other locations in the Halten Area however, more than 2.5 km of Triassic rocks have been drilled (e.g. Well 6507/12-2). The Triassic Grey Bed sequence is overlain by the Lower Jurassic Hitra Formation which consists of interbedded sandstones, shales and coals deposited in a fluvial–coastal plain environment.

A marine transgression started during deposition of the upper part of the Hitra Formation. The depositional environment changed to more marginal- to shallow-marine conditions. The sands deposited during this period have been defined as the Aldra Formation.

The transgression continued during Early–Late Toarcian times and more marine conditions developed. Predominantly claystones and siltstones known as the Leka Formation were deposited. This unit is overlain by a massive regressive sandstone complex termed the Tomma Formation which was deposited in a fluctuating shallow marine environment.

The boundary between the Tomma Formation and the Upper Jurassic Engelvær and Nesna formations is represented by a major unconformity. In the northern part of the field the unconformity cuts down into the Lower Jurassic.

Marine conditions prevailed during the Cretaceous and the Tertiary. The Cretaceous sequence is dominated by argillaceous limestones and marls in the lower part, changing into more uniform claystones in the upper part. The Tertiary sequence is also dominated by argillaceous deposits.

GEOPHYSICS

The main interpretation of Blocks 6507/11 and 6407/2 was carried out in 1979 prior to the fifth round, 1st part and in 1981 prior to the fifth round, third part. The interpretation was mainly based on a Statoil survey from 1977 (prefix 713- and 715-) which has a nearly block-diagonal orientation. Additional E–W-orientated data were acquired in 1982 (SG 8258 and SG 8271) (see Fig. 4). Together these surveys constitute the main database for the Midgard Field, totalling approximately 240 km.

In 1985 a three-dimensional survey was shot covering the Gamma and Delta segments in Block 6407/2. The data, which total 4240 km, will be interpreted in 1986 and will serve as

Fig. 3. Generalized stratigraphic column, Midgard Field

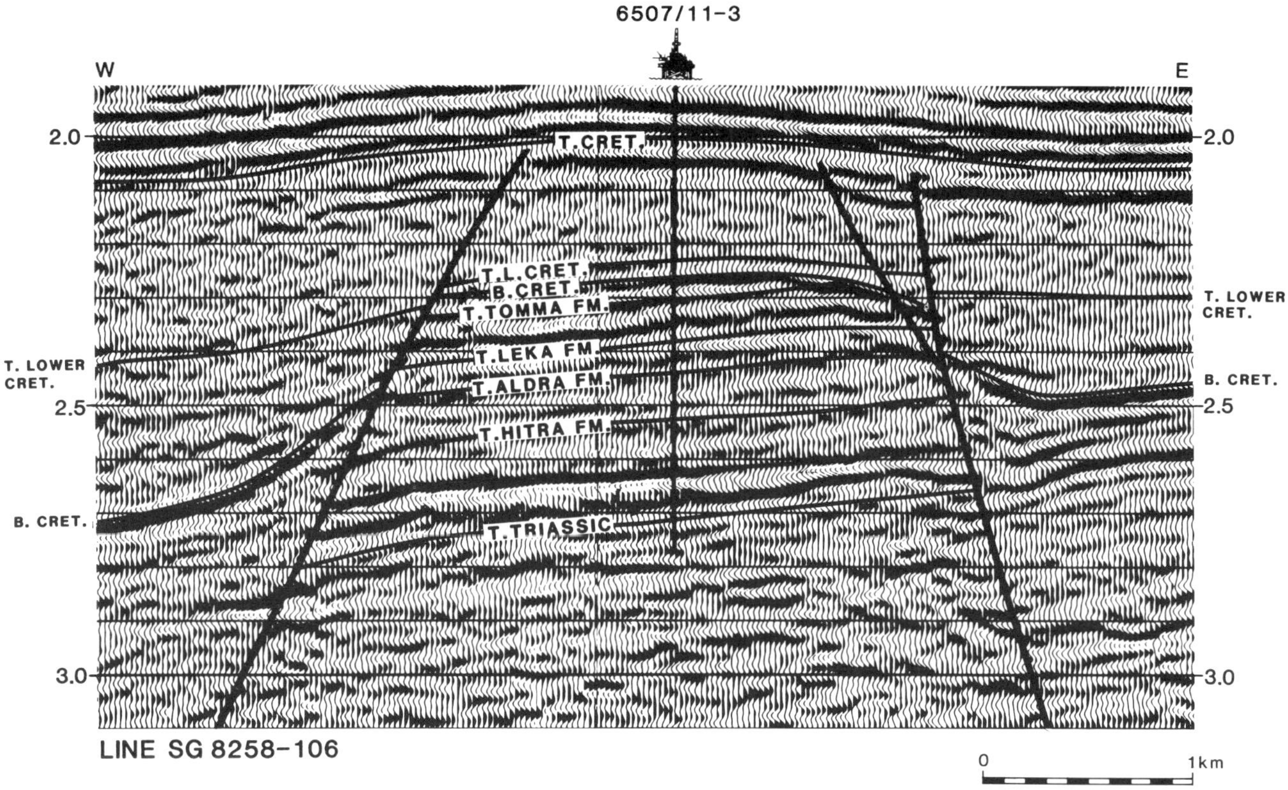

Fig. 4. Seismic line, SG 8258–106. The position of the line is indicated on Fig. 2.

a basis for decisions on a test of the Delta prospect.

TRAP

The Midgard Field must be classified as a structural trap. It is formed by a NNW-trending horst which is cut by cross-faults in a NNE–SSW direction. These faults are responsible for the delineation of three compartments, termed the Alpha, Beta and Gamma segments (Fig. 5). An additional prospect is located to the south-south-east on the same trend. It is closely associated with the 'Midgard horst', has been regarded as a part of the field and named the Delta segment.

The development of the Midgard horst was initiated by the Middle–Late Jurassic tectonic events. The structure was further developed during Cretaceous times through rapid differential subsidence. A Cretaceous argillaceous sequence also forms the cap-rock for the Midgard Field.

The model for hydrocarbon distribution, contacts and spill-points is still being evaluated. Thus major elements in this model have so far not been proved by well results. How-

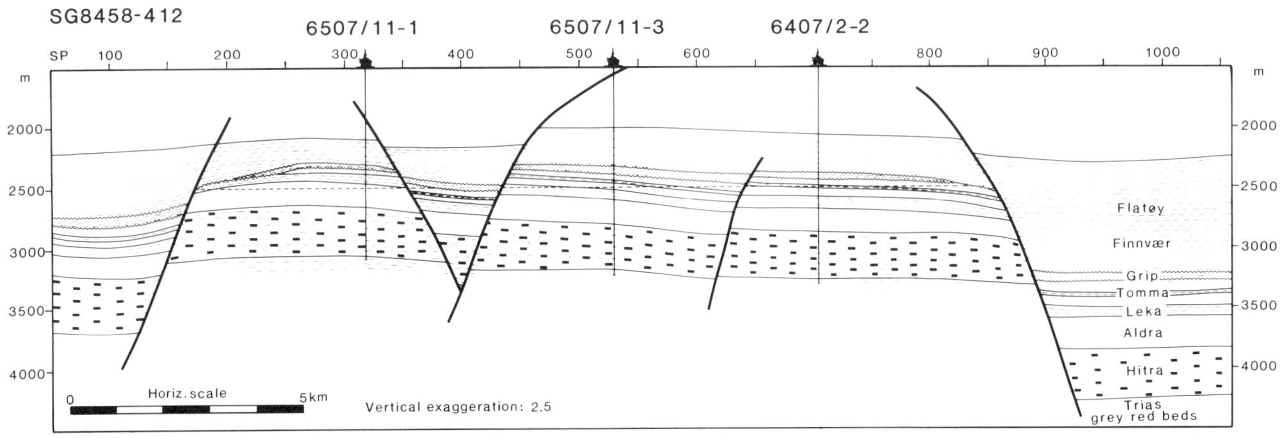

Fig. 5. Geological cross-section; Alpha, Beta and Gamma segments.

ever, present mapping indicates that the Alpha and Beta segments are separate. In Well 6507/11-1 on the Alpha segment, a gas–water contact was identified at 2493 mss. This depth corresponds with the spill-point to the north.

In Well 6507/11-3 on the Beta segment a gas–oil contact was found at 2488 mss, followed by a 11.5 m oil zone, resulting in an oil–water contact at 2499.5 mss.

In Well 6407/2-2, located on the Gamma segment, only a gas-down-to level was identified. The same contacts are likely to apply for the Gamma segment as for the Beta segment, however.

Between the Gamma and Delta segments a saddle has been identified. This saddle is apparently shallower than the oil–water contact. This increases the possibility of having communication between the three southernmost compartments. Possibly the spill-point for this part of the Midgard trap is in the northern flank of the Delta segment.

RESERVOIR

The reservoir in the Midgard Field is represented by sandstone sequences of Early and Middle Jurassic age. The distribution of hydrocarbons in these two units is a function of the structural development and subsequent erosion of the Upper Jurassic strata. The Lower Jurassic sequence, the Aldra Formation, contains the main hydrocarbon reserves in the Alpha segment while the Middle Jurassic Tomma Formation constitutes the reservoir in the rest of the Midgard Field.

The Jurassic units show only small thickness variations along the strike of the Midgard Field (Fig. 6). The Aldra Formation has been found to be 202 m, 191 m and 183 m in Wells 6507/11-1, 6507/11-3 and 6407/2-2 respectively. The Tomma Formation has thicknesses of 124 m and 142 m in 6507/11-3 and 6407/2-2. In Well 6507/11-1 an eroded sequence of 54 m was found.

Both reservoir formations are products of transgressive and regressive cycles. The Lower Jurassic transgression introduced marginal to shallow marine conditions. The Aldra Formation is thus represented by coastal lagoonal deposits in the lower part, changing into bar-type deposits upwards in the sequence.

During Late Toarcian times a regressive phase developed, changing the environment into a tidally influenced coast-line. This system is represented by the lower part of the Tomma Formation. During deposition of the upper part of the Tomma Formation, the marine influence increased, producing upper–lower shoreface deposits. These fluctuations during Middle Jurassic times allow a subdivision of the Tomma Formation into three units. These units can be identified over large parts of the Haltenbank area. A further subdivision of unit I of the Tomma Formation is possible and is currently being evaluated.

The reservoir units have very good reservoir properties. Porosity values for the Tomma Formation are in the range 26–29%. Permeabilities range up to several darcies. The properties for the Aldra Formation are slightly poorer, with porosities ranging from 22 to 25% and permeabilities in the range 10–1500 mD.

SOURCE

There are two marine source-rocks which have reached various degrees of maturity within the drainage area of the Midgard Field: the Hitra and Nesna formations (Elvsborg *et al.*, 1985).

The Lower Jurassic Hitra Formation is characterized by a high content of carbonaceous shales and laminae of coal. The sequence is dominated by terrestrially derived, humic organic matter with a rich potential for gas generation and also for generation of heavier hydrocarbons. This unit is thus the obvious source for the gas–condensate of the Midgard Field.

The oil zone proved in well 6507/11-3 may however, have its origin in the Upper Jurassic Nesna Formation. This unit consists of a dark bituminous shale very rich in organic matter. It is characterized as a rich type-II kerogen with potential for oil generation.

The Hitra Formation shows lateral variations from immature to late mature within the Midgard Field drainage area. This indicates the existence of an active gas-generation kitchen. The Nesna Formation is predominantly immature to early mature.

HYDROCARBONS

A total of five drill-stem tests have been conducted in the wells drilled on the Midgard Field. The maximum production of gas measured was 1.5×10^6 Sm^3/day through a 32-mm choke from the Tomma Formation in Well 6507/11-3. The tests from the gas zone have demonstrated variations in the produ-

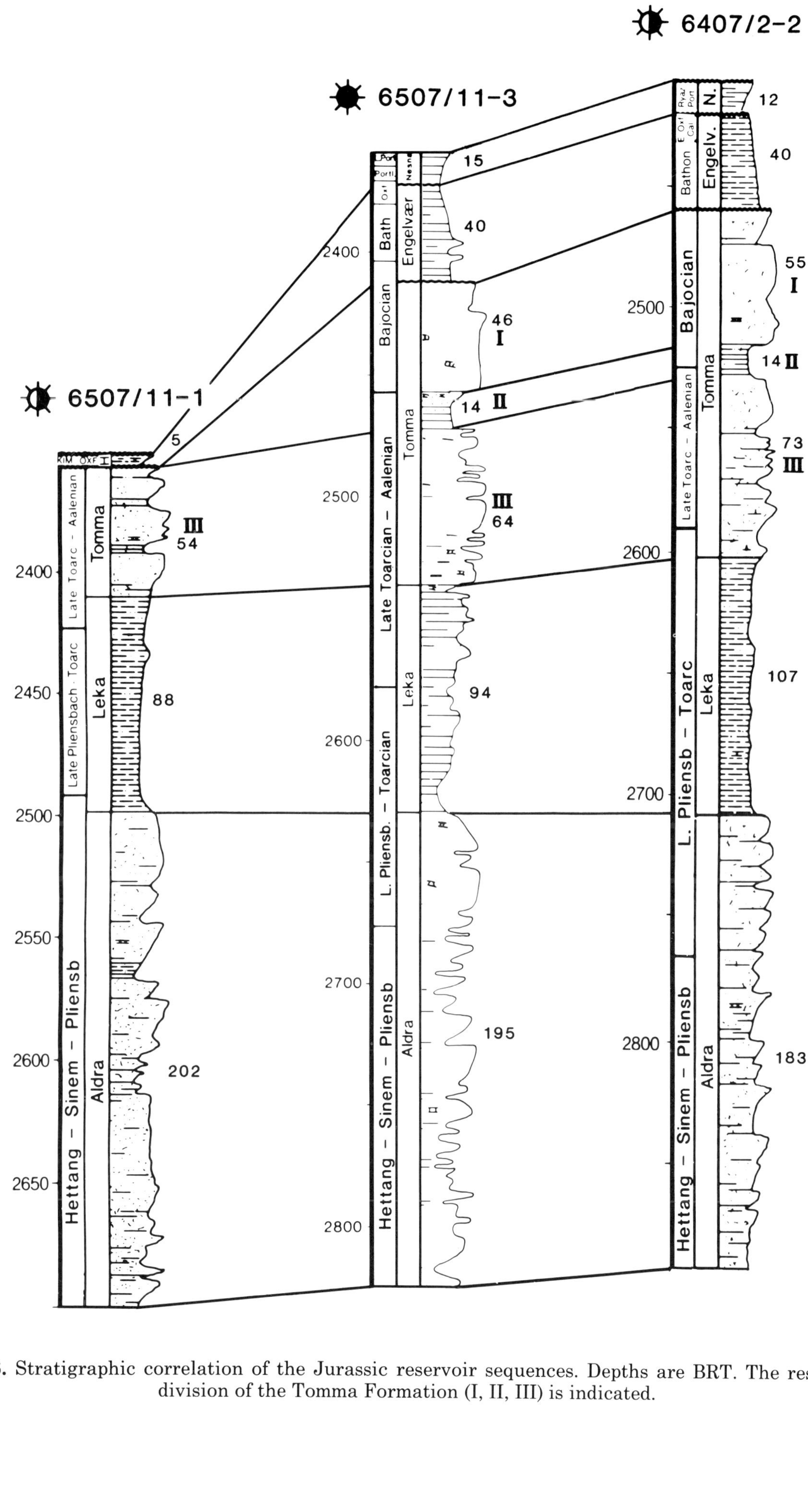

Fig. 6. Stratigraphic correlation of the Jurassic reservoir sequences. Depths are BRT. The reservoir division of the Tomma Formation (I, II, III) is indicated.

ced liquid content. Most of the gas–condensate ratios which have been calculated from the test results are in the range 6000–9000 Sm^3/Sm^3.

The oil zone was tested in Well 6507/11-3. A maximum rate of 1500 Sm^3/day was produced. At this rate however, gas-coning was experienced. Rates of the order of 200 Sm^3/day cannot be exceeded if gas breakthrough after a few days is to be avoided. A GOR of 126.5 Sm^3/Sm^3 was experienced during the testing phase. The oil gravity is 0.85 g/cm^3.

RESOURCES

Two drive mechanisms are possible in the Midgard reservoir: volumetric expansion and water drive. Water drive, however, is likely to be restricted by the major faults surrounding the structure.

Assuming a weak water drive, a recovery factor of 0.75 has been determined for the gas, and 0.63 for the condensate. The recoverable gas resources for the Midgard Field have been estimated at 87×10^9 Sm^3, with a figure for associated condensate of 17×10^6 Sm^3, using this assumption.

There are obvious uncertainties as to whether the thin oil zone constitutes a production target.

ACKNOWLEDGEMENTS

I would like to thank Arco Norge, Norsk Agip, Norsk Hydro, Norske Shell, Deminex and Statoil for permission to publish this paper. Several present and former Saga employees have contributed to the results on which this article is based; in particular should be mentioned T. Hagevang and Å. Karlsen.

REFERENCES

Elvsborg, A., Hagevang, T., and Throndsen, T. 1985. Origin of the gas-condensate of the Midgard Field at Haltenbanken, In Thomas B.M. *et al.* (Eds), *Petroleum Geochemistry in the Exploration of the Norwegian Shelf*, Graham and Trotman, London, pp. 213–219.

Hagevang, T. and Rønnevik, H.C. 1986. Basin development and hydrocarbon occurrence offshore Mid-Norway, In Halbouty, M.T. (Ed.), *Future Petroleum Provinces of the World, Memoir of the American Association of Petroleum Geologists,* 40.

36

Tyrihans

Vidar Larsen,* Per Olav Mørkeseth† and Sjur M. Aasheim*

*Statoil, Box 1212, 5001 Bergen, Norway
†Statoil, Box 300, 4000 Stavanger, Norway

The Tyrihans Field is located in the Haltenbanken area, which lies north-east of the Møre Basin, south of the Nordland Ridge and west of the Trøndelag Platform (Fig. 1). The first Well 6407/1-2, tested gas and condensate in 1983 in the Tyrihans South structure finding a gas–water contact at 3687 mss. The second Well, 6407/1-3, tested oil and gas from the Tyrihans North structure in January 1984; oil was proven down to 3680 mss and the oil–gas contact was at 3658 mss.

Tyrihans South is an anticlinal structure within a south-easterly tilted fault-block. The structure is separated by a saddle from Tyrihans North, which is a horst structure. The reservoir is a sandstone in the upper part of the Middle Jurassic Tomma Formation, which was deposited in a delta environment. It has an average thickness of 100 m and an average porosity of 15%. The range of permeabilities is 58–720 mD.

The recoverable resources are estimated to be 8×10^6 Sm^3 oil, 18×10^6 Sm^6 condensate and 45×10^9 Sm^3 gas.

INTRODUCTION

The Tyrihans Field is located between 130 and 150 km west of the coast of Norway. The water depth varies from 290 to 300 m. The size of the entire field is 47 km^2 and it is in a structural trap. The reservoir is a sandstone in the upper part of the Tomma Formation, which is interpreted to be a fan-delta deposit.

The name Tyrihans is taken from a Norwegian fairy tale. Tyrihans is described as a poor young man who wins the princess and half of the kingdom with the help of luck and magic.

HISTORY

Block 6407/1 was awarded on 23 April 1986, under Petroleum Licence 073, to a group made up of the following companies:

Den norske stats oljeselskap a.s. (operator)	50%
Norske Conoco A/S	20%
Amoco Norway A/S	20%
Norsk Hydro Produksjon a.s.	10%

The work obligation required a seismic survey on a 1 × 1 km grid, and two wells drilled into Triassic rocks or to 5000 m. These two wells have now been drilled. The first well, 6407/1-1, was spudded on Tyrihans South on 19 October 1982 by drilling rig *Dyvi Delta*. The well had to be plugged at approximately 900 m due to operational problems. It was respudded as 6407/1-2 a few days later, and a total depth of 4560 m was reached in April 1983. This was the first well to be drilled through the rough winter season, north of 62 °N. The well encountered a gas–condensate column in the Upper Tomma Formation with a gas–water contact at 3687 mss.

Geology of the Norwegian Oil and Gas Fields © A. M. Spencer *et al.*, eds (Graham & Trotman, 1987) pp. 411–418.

Tyrihans Field data summary

	North	South
Trap		
Type	horst	faulted dome
Depth to crest (top Tomma Fm)	3490 m	3540 m
Gas–oil contact	3658 m	
Oil–water contact	3680 m (minimum)	3687 m
Gas column	168 m	
Oil column	22 m (minimum)	147 m
Pay Zone		
Formation	Upper Tomma Formation	
Age	Middle Jurassic	
Thickness	80–120 m	
Net/gross ratio	0.85–1.00	
Porosity	13–20%	
Hydrocarbon saturation	69%	75%
Permeability (average)	450 mD	150 mD
Hydrocarbons		
Oil gravity	0.883 g/cm^3	
Gas gravity (relative to air = 1.0)	0.79	1.08
Condensate gravity		0.80 g/cm^3
Gas/oil ratio	3889 Sm^3/Sm^3	871 Sm^3/Sm^3
Dew point	377 bar	377 bar
Gas expansion factor	250	217
Formulation water		
—		
Reservoir conditions		
Temperature	132 °C	136 °C
Pressure	378 bar	378 bar
Pressure gradient, gas	0.022 bar/m	
Pressure gradient, oil	0.069 bar/m	0.04 bar/m
Field size		
Area	47 km^2	
Gross rock volume	1.4 × 10^{9}m^3	1·5 × 10^{9}m^3
Hydrocarbon pore volume	205 × 10^{6}m^3	130 × 10^{6}m^3
Recovery factor–oil	18%	
Recovery factor–gas	65%	65%
Drive mechanism	aquifer support—water/gas injection	
Recoverable oil	8 × 10^6 Sm^3	
Recoverable gas	24 × 10^9 Sm^3	21 × 10^9 Sm^3
Recoverable condensate		18 × 10^6 Sm^3
Production		
—		

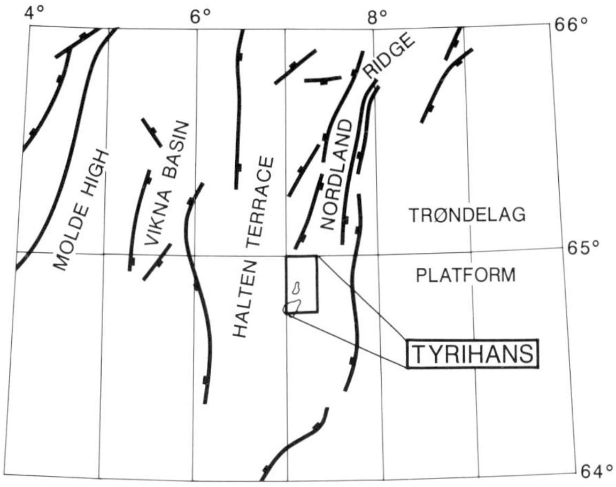

Fig. 1. Location of the Tyrihans Field and the main structural elements in the area.

In January 1984 Well 6407/1-3 reached its total depth in Triassic rocks and tested oil with a gas-cap on the Tyrihans North structure. The oil was proven down to 3680 mss and the gas–oil contact was determined at 3658.5 mss (Fig. 2).

GEOLOGICAL SETTING

Block 6407/1 is situated in a complex zone of faults where several structural elements intersect. Tectonic movements, which culminated at the Jurassic–Cretaceous transition, gave rise to rotated fault-blocks in the southern part of Block 6407/1. In the eastwardly tilted fault-block, an anticlinal structure is superimposed and forms a horst structure northwards. Well 6407/1-2 was drilled on this anticlinal structure, whereas Well 6407/1-3 was located on the northern horst-like structure.

The thickness variations of individual stratigraphic units indicate a synsedimentary fault activity. Truncation of the crests of the rotated fault blocks was caused by erosion in the late and main Kimmerian periods. A

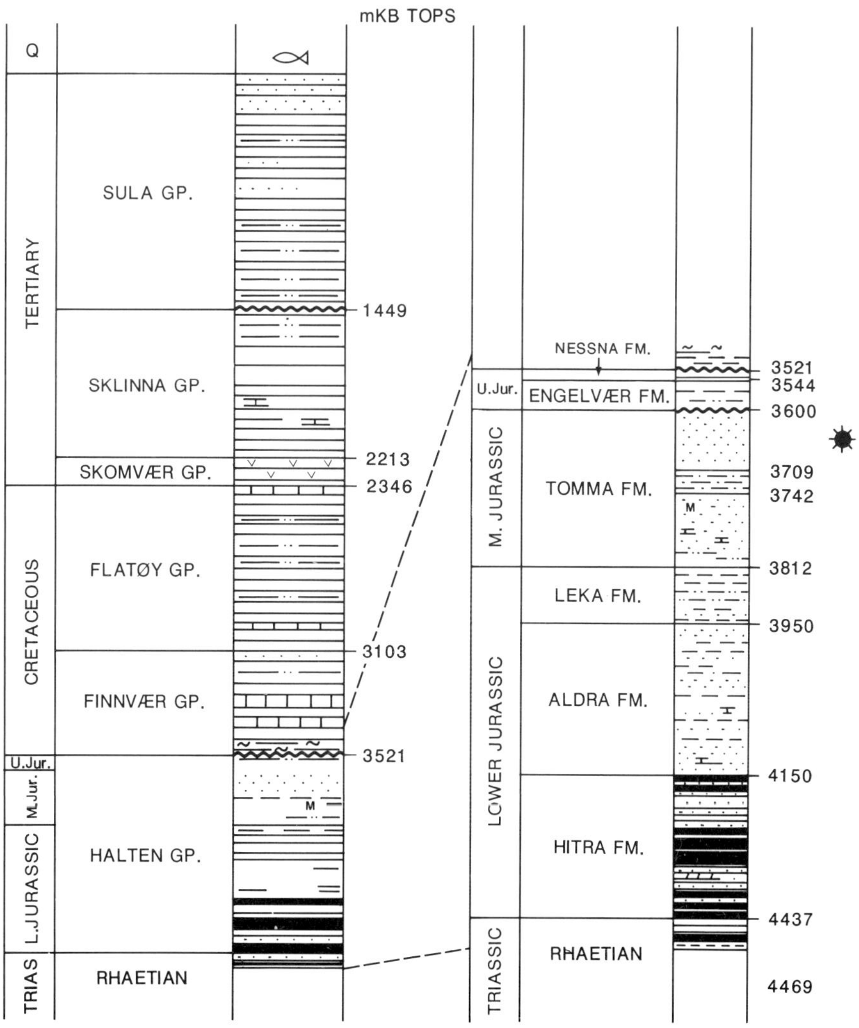

Fig. 2. Generalized stratigraphy of the Well 6407/1-3.

graben-like trough can be traced from Block 6406/3, diagonally through Blocks 6407/1–6507/10 and into Block 6507/11.

The stratigraphy of the Upper Palaeozoic is little known. Total sediment thickness in the area is about 7000 m, of which about 3000 m is attributed to pre-Middle Jurassic strata. A N–S tectonic stress regime is thought to have developed after the deposition of Lower Permian clastic sediments. Faulting initiated a transgression which deposited what are inferred to be evaporites and salt in the Møre Basin. The Permian evaporites were restricted to the west of the Nordland High. Above lies a Triassic sequence containing two salt sections more than 300 m thick, separated by more than 500 m of mudstones and siltstones. The salt sections of both Triassic and Permian age are thought to have been mobilized during Jurassic–Cretaceous times.

The salt section is overlain by 850 m of Upper Triassic red claystones and siltstones interbedded with some stringers of continental sandstone. In the upper 100 m of this sequence, there is a change of colour from red to grey (grey beds). Well 6407/1-2 stopped above these grey beds within the Triassic part of an overlying coal unit, a sequence of coal, mudstones and sandstones. A more humid climate in a fluvio-deltaic environment is thought to be responsible for the change from Triassic red beds to the continental post-Rhaetian coal units. The Lower Jurassic coal unit is 400–500 m thick. The associated sandstones are normally fine to medium grained, and the unit sometimes contains calcareous horizons. Dating gives a Rhaetian–Pliensbachian age, thus implying a more prolonged continental environment in this area compared with the North Sea. A 200 m thick Lower Jurassic sand sequence overlying the coal unit is often referred to as the Cook equivalent. However, we think that these fluvio-deltaic, fine to coarse sandstones, with marine incursions towards the top, represent deposits similar to the Statfjord and Johansen Formations.

On top of this sandstone lies a Lower Jurassic marine mudstone unit 150 m thick with sand stringers. The sand content in this unit increases towards the top, a fact which can possibly be related to the regressive nature of the overlying Middle Jurassic sand unit. The mudstone sequence appears to decrease in thickness towards the east.

Well 6407/1-2 found a Middle Jurassic sand sequence 250 m thick. This is more than twice the thickness previously encountered for this sequence in the area. Thickening of individual units, as well as preservation of younger sandstones, is due to the relative subsidence of this area compared to the areas drilled previously. The penecontemporaneous subsidence and the relatively immature texture of the sandstone could suggest local drainage patterns and source-areas, which again would be important for the facies evolution through the sequence.

Regionally, the Middle Jurassic sand sequence is part of a major regressive episode found in all the wells on Haltenbanken. Datings give a Bajocian–Bathonian age which is thus comparable to the North Sea Brent Group.

The top of the sequence is defined by an apparently regional unconformity which is transgressed by Oxfordian marine shales belonging to the Heather Formation. They are overlain by a dark brown to black organic shale of Portlandian age, which is comparable to the 'hot shale' of the North Sea, thus making it an excellent hydrocarbon source-rock for the area where it is sufficiently buried for maturation.

The Lower Cretaceous is generally thin on the Trøndelag Platform. A period of non-deposition seems to have occurred between Portlandian and Barremian times. However, thicker sequences may be expected in down-faulted areas. The sediments consist of mudstones, marl and limestones with occasional sand stringers, and were deposited in a sublittoral environment. A regional westerly tilt during Mid- and Late Cretaceous times resulted in an overall onlap situation, with more than 1700 m of clay west of the Trøndelag Platform but non-deposition in the east. The tilting was possibly related to Cretaceous faulting and gave rise to several unconformities in the Cretaceous sequence.

In Well 6407/1-2 an Early Paleocene mudstone, 45 m thick, overlies the Maastrichtian sequence, which is in turn overlain by about 70 m of tuffaceous sediments in the Skomvær Group, correlatable with the North Sea Balder Formation.

A second group of Tertiary sediments, the Sklinna Group (Early Eocene—Early Pliocene), some 770 m thick, consists of mudstones with some silt and limestone streaks. Some interfingering fine-grained sandstones are occasionally observed. An unconformity is present some 300 m below the top of the succession. This group is found in all wells on Haltenbanken and can be correlated with the Hordaland Group in the northern North Sea. Another unconformity defines the top of the group.

The Sklinna Group is overlain by light grey claystones and poorly sorted sandstones of the Sula Group deposited during Pliocene times. These rocks are equivalent to the Nordland Group in the North Sea. The Pliocene sequence is some 900 m thick and the Quaternary about 200 m.

GEOPHYSICS

The following seismic surveys are available:

727	SL/SSL	1977
991	GECO/SSL	1979
ST8110	GECO/SSL	1981
ST8306	GECO/SSL	1983

At the time of the first evaluation the seismic grid consisted of approximately 2 × 2 km overage. When the licence was awarded in 1982, the cover was increased to 1 × 1 km and a new interpretation showed a horst structure with a saddle-point. The interpretation prior to Well 6507/1-2 (1982–83) and the reinterpretation prior to Well 6507/1-3 (1983–84) were carried out on conventional, migrated two-dimensional sections.

Seven horizons have been interpreted in Block 6407/1: top Palaeocene; top Cretaceous; a Lower Cretaceous reflector; late Kimmerian unconformity; Middle Jurassic unconformity; top Dunlin equivalent; near top Hettangian (Fig. 3).

The main uncertainty is connected with the picking of the Middle Jurassic unconformity in the south-western part of the structure. The interpretation presented here can be regarded as a maximum case for the Tyrihans South structure.

TRAP

The Tyrihans South trap is an anticlinal structure which, towards the north, transforms into a horst structure across a saddle area. Tyrihans North comprises this northern horst. Both the anticline and the horst are superimposed on a larger scale eastward tilted fault-block which is part of the Halten

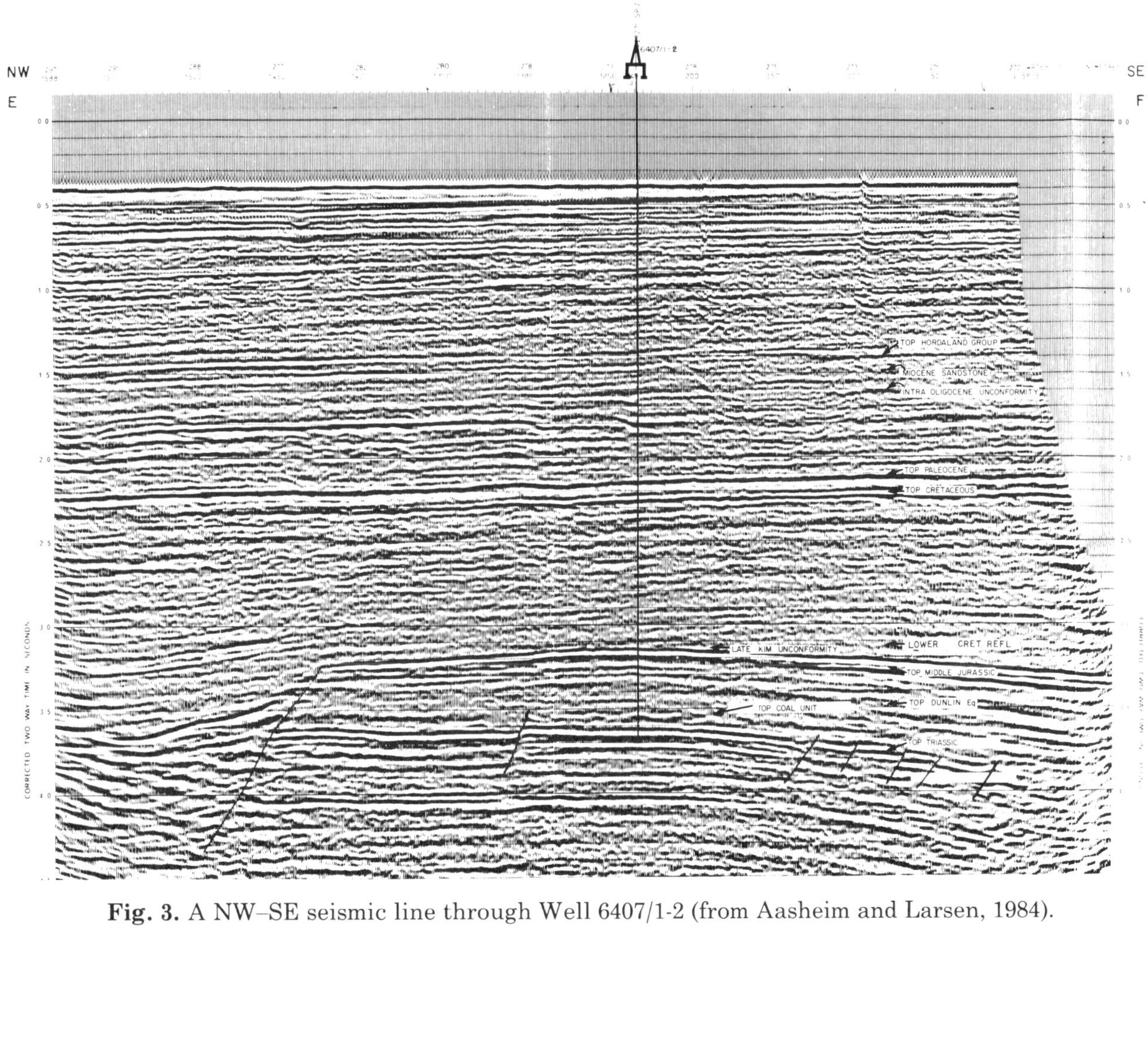

Fig. 3. A NW–SE seismic line through Well 6407/1-2 (from Aasheim and Larsen, 1984).

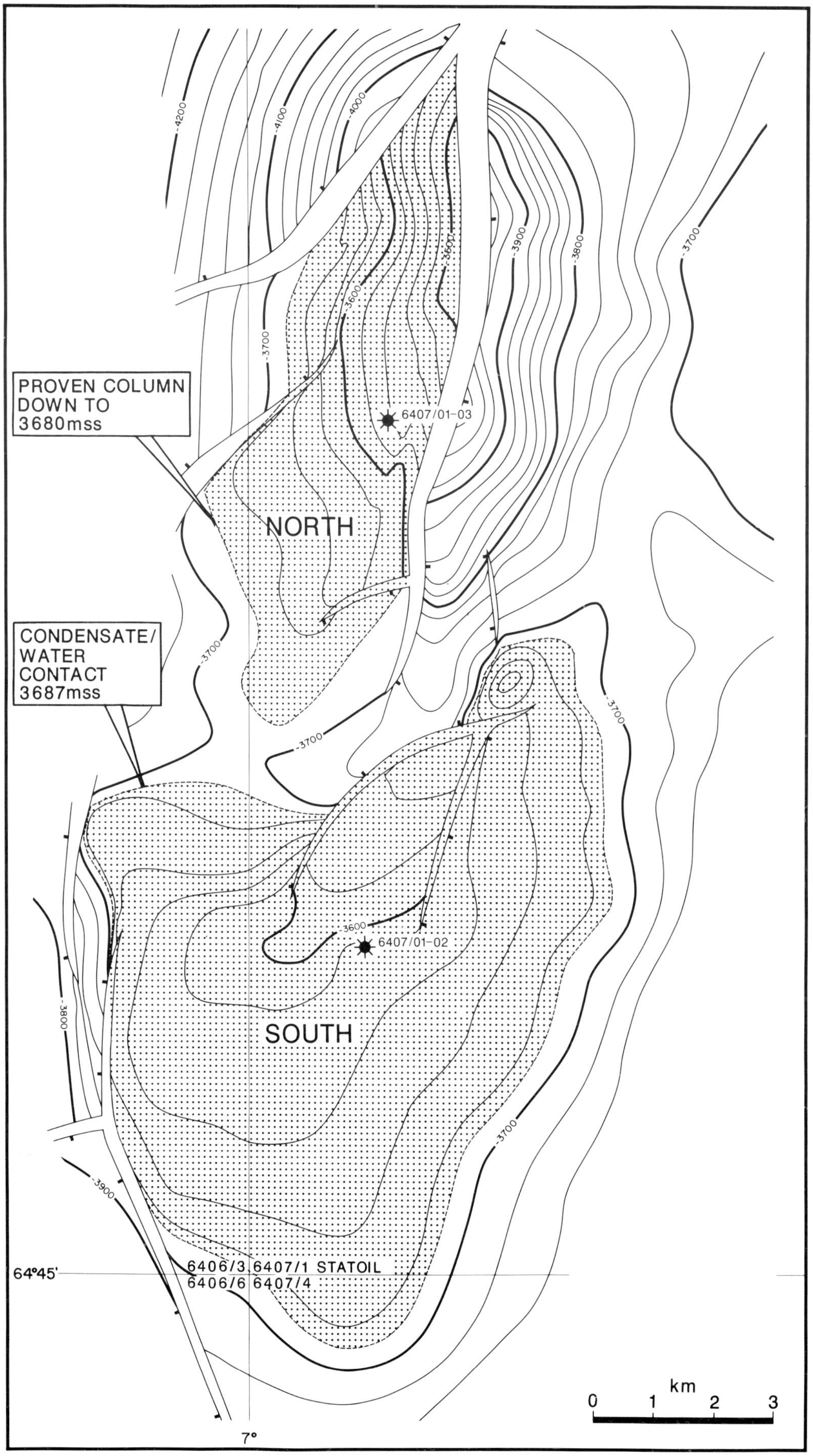

Fig. 4. Tyrihans Field—Top Tomma Formation depth map. C.I. 25 m.

Terrace. The cap-rocks of both structures are Upper Jurassic and Lower Cretaceous mudstones (Fig. 4).

Well 6407/1-3 proved hydrocarbons in Middle Jurassic sandstones on the Tyrihans North structure. The hydrocarbons consist of gas overlying an oil leg. The gas–oil contact is defined at 3658.5 mss and the oil–water contact at 3680 mss. The hydrocarbon column in the well is 109 m thick. The oil–water contact apparently coincides with the transition from the upper part of Tomma Formation into the underlying silt/mudstone unit. This makes it difficult to define the total hydrocarbon column in the Tyrihans North structure (Fig. 5).

Well 6407/1-2 at Tyrihans South found gas–condensate, with a gas–water contact at 3687 m. The hydrocarbon column in the well is 57.5 m. The total column in Tyrihans south is expected to be approximately 100 m.

The depth map for the Tyrihans South structure indicates that it is not filled to spill-point. The critical spill-point to the Tyrihans North is along a NE–SW-trending fault in the northern part of the structure.

RESERVOIR

The 250-m thick Middle Jurassic sandstone of the Tomma Formation, which is the reservoir rock in the wells, has been subdivided into three units.

The basal unit consists of subrounded–subangular, fine–medium, occasionally coarse, sandstone, interbedded with scattered stringers of limestone and claystone. The unit has a low net/gross ratio of 15%, and the average porosity is 14%. The unit is probably of Bajocian age and represents a regressive shallow-marine sequence, possibly tidally influenced.

An apparently regional transgression defines the base of the middle unit, which is a 43 m thick sequence of mudstones, interbedded with some calcareous sand layers and streaks of dolomitic limestones. Preliminary pressure data indicate that this sequence may act as a seal.

The upper reservoir unit consists of an angular–subangular, fine–medium sandstone with frequent layers of coarse–pebbly sandstone. The unit gives a strikingly uniform log pattern reflecting a very high net/gross ratio. The average porosity is 15%.

The cored section of the Middle Jurassic sandstone contains subarkosic arenites with 5–10% mica and K-feldspar as the dominant feldspar mineral. Quartz overgrowth and clay minerals are the major diagenetic porosity reducing agents. Secondary porosity is observed, resulting from dissolution of the feldspar. Investigation of the cored section reveals sedimentary structures indicating deposition in standing shallow water. Massive reworked sandstone, occasionally arranged into coarsening-upwards sets less

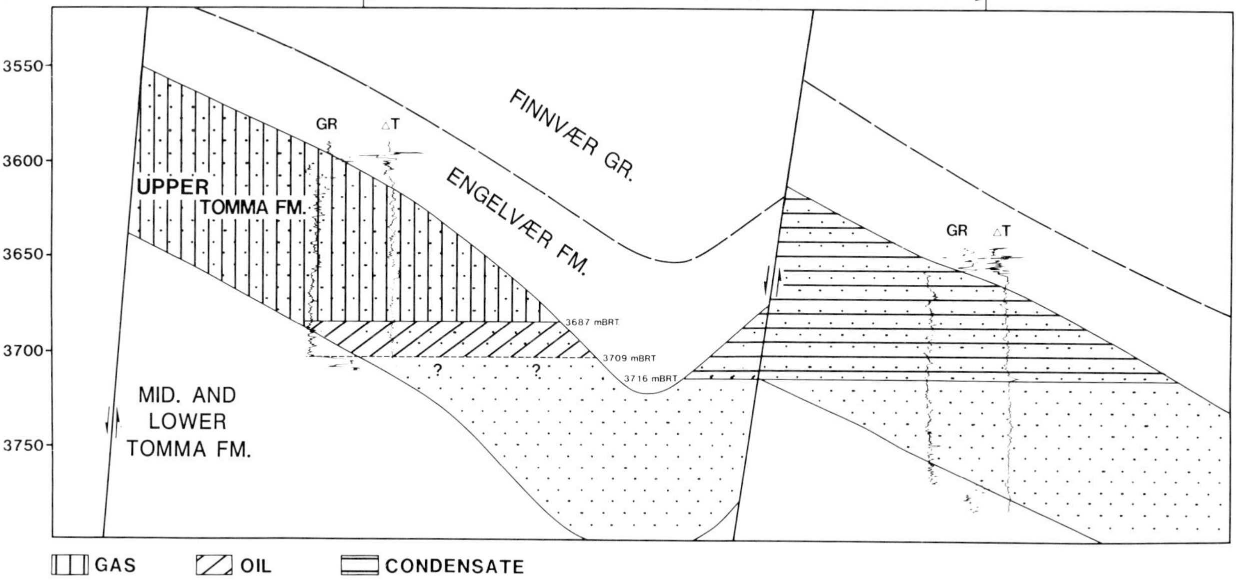

Fig. 5. Schematic cross-section of the Tyrihans Field.

than one metre thick and with very few wave-influenced structures, characterizes the lower part of the cored section, suggesting a rather constant supply of clastics into the subsiding basin. The upper half of the cored section shows frequent layers of pebbly sandstones, arranged both in a fining-up (erosive base) and coarsening-up pattern; the sandstones are also frequently cross-bedded.

As a preliminary depositional model for the upper reservoir unit, we suggest a regressive fan-delta building out into fairly shallow waters, with decreasing wave influence and increasing fluvial processes towards the top. The immature texture and the poor rounding of quartz grains, together with dip-meter indications as well as regional data, could support a westerly source for the Middle Jurassic sandstones.

SOURCE-ROCK

The source-rock of the field is thought to be the anoxic shale named the Nesna Formation, which is equivalent to the Draupne and the Kimmeridge Clay Formations of the North Sea. This has reached a sufficient degree of maturation to generate hydrocarbons within a structural low area west of the structure.

However, the Lower Jurassic Hitra Formation coal unit may also have contributed to the generation of the hydrocarbons in the Tyrihans Field.

HYDROCARBONS

The fluid system in Well 6407/1-2 (Tyrihans South) is defined as a gas–condensate system. PVT analysis of the bottom-hole samples from test No. 1 in well 6407/1-3 indicates a gas-saturated oil system at reservoir conditions, with a reservoir liquid density of 0.65 g/cm^3 and a gas/oil ratio of 189 Sm^3/Sm^3. Drill-stem test No. 2 in this well represents the gas-cap.

A summary of the test results are presented in Table 1.

Table 1. Summary of drill-stem test results

Well	6407/1-2	6407/1-3	6407/1-3
Test No.	1	1	2
Interval (m RKB)	3659–3669	3697.5–3702.5	3665–3670
Choke (mm)	19	9.5	19
Liquid rate (Sm^3/day)	454	396.1	191.8
Liquid density (g/cm^3)	0.79	0.87	0.78
Gas rate (Sm^3/day)	394 000	58 430	725 400
Relative gas density (air = 1)	0.735	0.665	0.657
GOR (Sm^3/Sm^3)	869	147.8	3782

ACKNOWLEDGEMENTS

The authors wish to thank Conoco, Amoco, Norsk Hydro and Statoil for permission to use the data presented.

REFERENCES

Aasheim, S.M. and Larsen, V. 1984. The Tyrihans discovery—preliminary results from well 6407/1-2, In Spencer, A.M. *et al.* (Ed), *Petroleum Geology of the North European Margin*, Graham and Trotman, London, pp. 285–291.

Part VIII—Hammerfest Basin

37

Askeladd

R. Grung Olsen and O. K. Hanssen

Statoil, PO Box 40, 9401 Harstad, Norway

The Askeladd gas field was the first significant gas discovery offshore northern Norway. The field is situated in Block 7120/8 on Tromsøflaket in the western part of the Hammerfest Basin. The gas is trapped in Lower to Middle Jurassic sandstones which were deposited as an overall transgressive marine, clastic, coastal and inner-shelf sequence. Three wells have been drilled in the block and two of the wells penetrated different gas–water contacts. With the third well the eastern extension of Askeladd North was delineated. The reservoir thickness is of the order of 100 m. Reservoir properties are fair with porosities around 15% and permeabilities of up to 250 mD. Gas in-place is estimated to be 73×10^9 Sm^3. Currently there are no firm development plans.

INTRODUCTION

The Askeladd gas field is located on Tromsøflaket, approximately 100 km off the Norwegian coast and 200 km north-east of Tromsø (Fig. 1). Water depth in the area is 240–280 m. The field covers a large part of the block, approximately 105 km^2.

The field has in-place resources of approximately 73×10^9 Sm^3. The field is named after the Norwegian fairy-tale hero Askeladd. He seems not to amount to much, but he turns out to be the smartest because his heart is in the right place. Askeladd therefore conquers every obstacle and always succeeds in the end.

HISTORY

Production licence 064, comprising Block 7120/8, was awarded in the fifth licensing round in 1981 to a group of companies with Statoil (50%) as operator. The other companies are Esso (25%), Norsk Hydro (15%), Elf Aquitaine (5%) and Phillips (5%).

By the end of 1983 the obligatory work commitment had been completed. The initial licence period expires in 1987. To date, three wells have been drilled within the block. The objective of Well 7120/8-1, which was drilled in 1981, was to test for possible hydrocarbon accumulations in Middle Jurassic sandstones in a seismically defined closure east of a major fault. The well turned out to be the first significant gas discovery on Tromsøflaket with 88 m of gas-bearing sandstones of Jurassic age.

Well 7120/8-2 drilled in 1982, confirmed the southern extension of the Askeladd field, but the gas–water contact was found to be 19 m shallower than in the first well.

The purpose of the 7120/8-3 well, which was drilled in 1983, was to test the north-easterly extension of the Askeladd North structure. The reservoir interval was below the gas–water contact of Well 7120/8-1 and was thus water bearing. By the end of 1981 the block was covered by a seismic grid of 1 × 1 km. In 1983 a three-dimensional seismic survey was shot by Geco. Field development studies on

Geology of the Norwegian Oil and Gas Fields © A. M. Spencer *et al.*, eds (Graham & Trotman, 1987) pp. 419–428.

Askeladd Field data summary

Trap	
Type	tilted fault-blocks
Depth to crest (Top Stø Fm)	2010 mss
Lowest closing contour—Askeladd N	2155 mss
—Askeladd S	2165 mss
Gas–water contacts—Askeladd N	2155 mss
—Askeladd S	2136 mss
Gas column—Askeladd N	145 m
—Askeladd S	106 m
Pay zone	
Formation	Stø Formation
Age	Early/Middle Jurassic
Gross thickness	87–95 m
Net/gross ratio (average; range)	0.82; 0.67–1.0
Porosity (average; range)	15%; 13–17%
Gas saturation	80%
Permeability (average; range)	100 mD; 5–250 mD
Hydrocarbons	
Gas gravity (relative to air)	0.666
Gas type	dry
Dew point	210 bar
Condensate yield	50 $Sm^3/10^6$ Sm^3 gas
Gas expansion factor	210
Formation water	
Salinity	80 000 ppm NaCl equivalents
Resistivity	0.04 ohm m
Reservoir conditions	
Temperature	80 °C
Pressure—Askeladd N	237.7 bar
—Askeladd S	235.7 bar
Pressure gradient in reservoir	0.021 bar/m
Field size	
Area	105 km^2
Gross rock volume	$3.6 \times 10^9 m^3$
Hydrocarbon pore volume	$0.35 \times 10^9 m^3$
Recovery factor—gas	0.70
Drive mechanism	gas expansion
Recoverable gas	$50 \times 10^9 Sm^3$
Production	
—	

Askeladd have been carried out, but there are currently no firm plans for development.

GEOLOGICAL SETTING

The Askeladd structures are located in the very western part of the Hammerfest Basin close to the transition zone to the TromsøBasin (Fig. 2). The Hammerfest Basin is an E–W-trending basin between the Loppa High to the north and the Troms–Finnmark Fault Complex to the south. The basin widens and deepens westwards towards the Ringvassøy–Loppa Fault Complex, which is a network of N–S-trending listric faults dipping towards the very deep Tromsø Basin (Sund *et al.*, 1986). The major N–S fault which delineates Askeladd in the west reflects the eastern-most progradation of this major fault system. The development of the Ringvassøy–Loppa Fault complex is related to an episode of extreme subsidence of the Tromsø Basin in late Early Cretaceous times.

From Middle Permian to Middle Jurassic times the Hammerfest Basin was part of a regional intracratonic basin with deposition of marginal carbonates and thick clastic sequences. The Hammerfest Basin as a structural element is a result of the Late Kimmerian tectonic phase. The main subsidence occurred during Early Cretaceous times along the northern and southern margins of the basin. Mostly argillaceous sediments were deposited during this time.

From Eocene to Oligocene times the open-

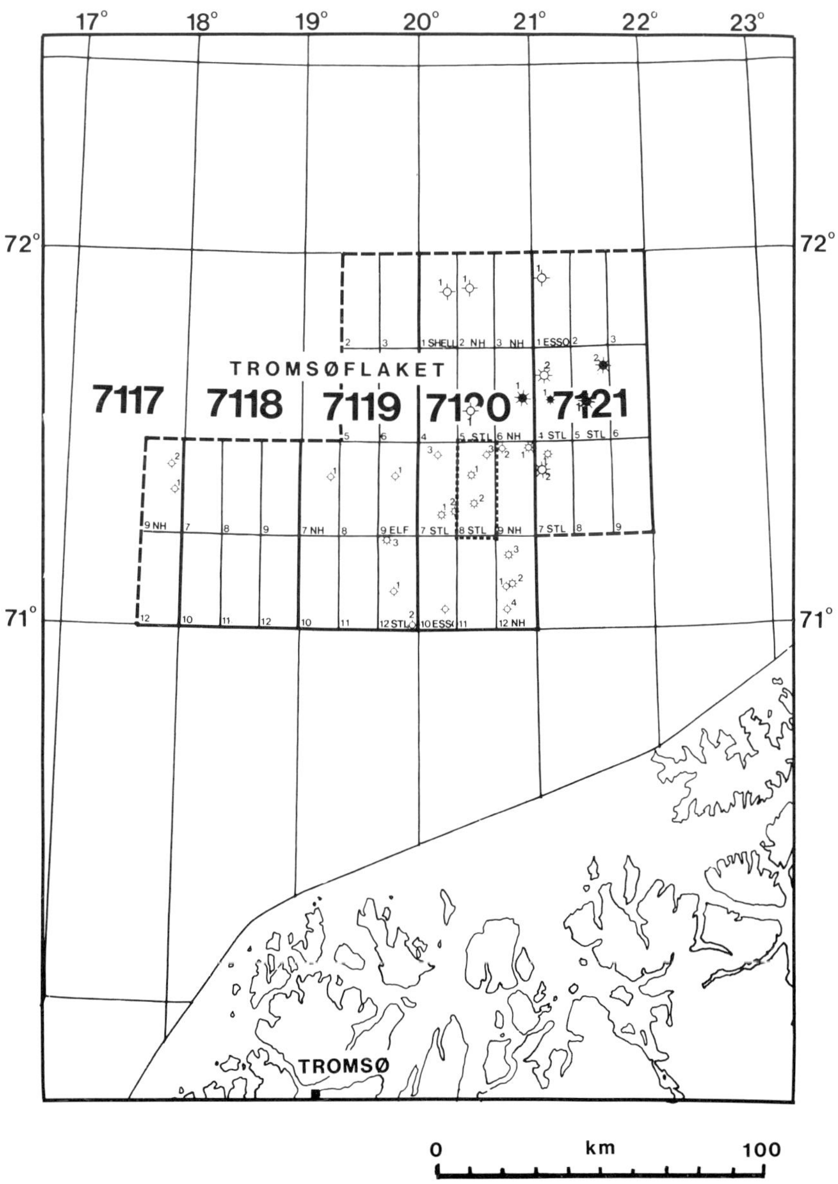

Fig. 1. Location map, Troms I.

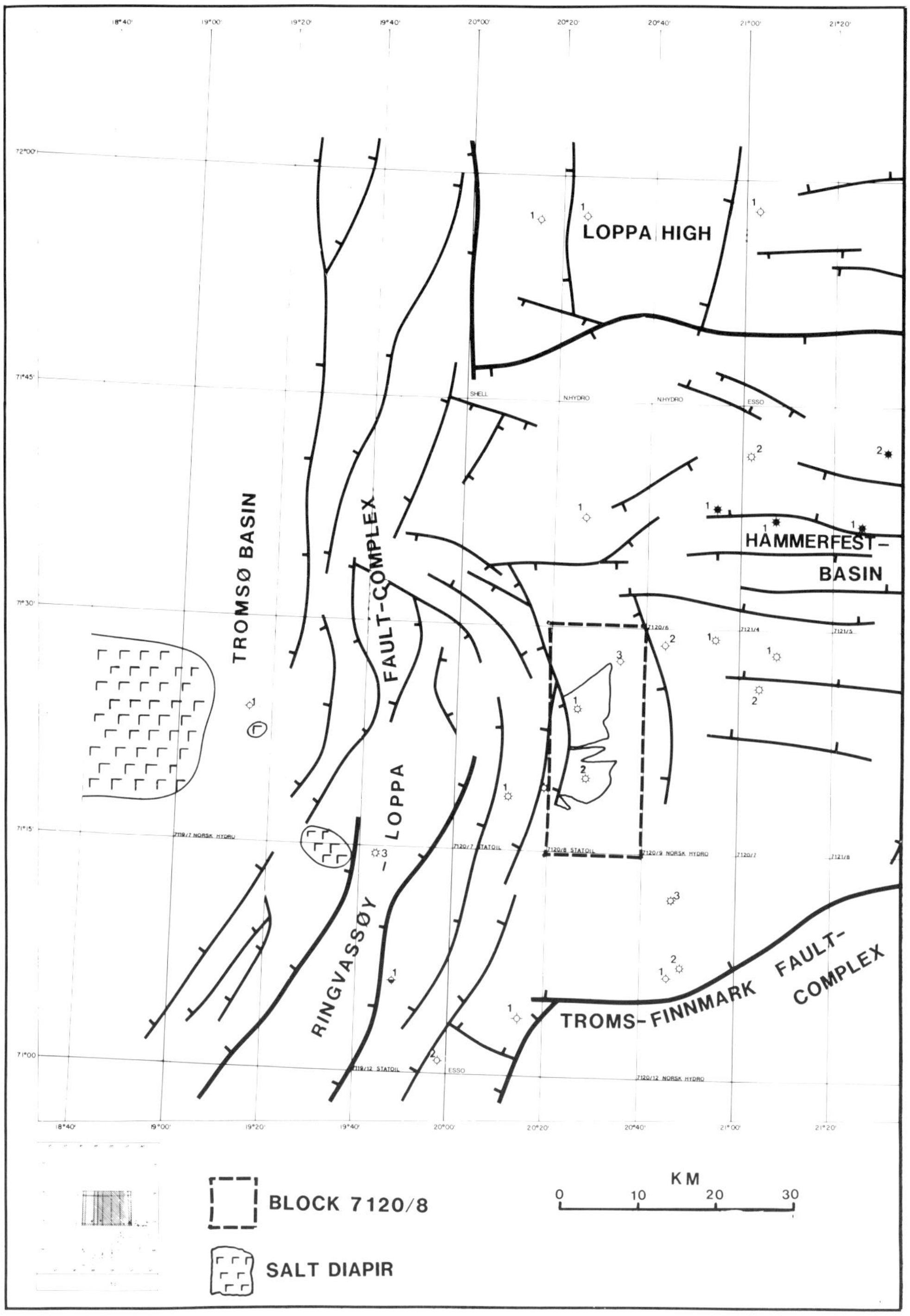

Fig. 2. Location map showing the major structural elements.

ing of the Norwegian–Greenland Sea resulted in further subsidence of the Tromsø Basin relative to the Hammerfest Basin. The sea withdrew westwards resulting in strong erosion of the Paleogene sediments. A sequence of Pliocene and Quaternary sediments is preserved in the area. A stratigraphic log correlation is shown in Fig. 3.

GEOPHYSICS

The evaluation of the Askeladd Gas Field is based on data from several seismic surveys shot between 1973 and 1983. In 1975, 875 km of seismic lines were shot by GSI and processed by Statex. These data were reprocessed by Geco in 1981. A new survey consisting of

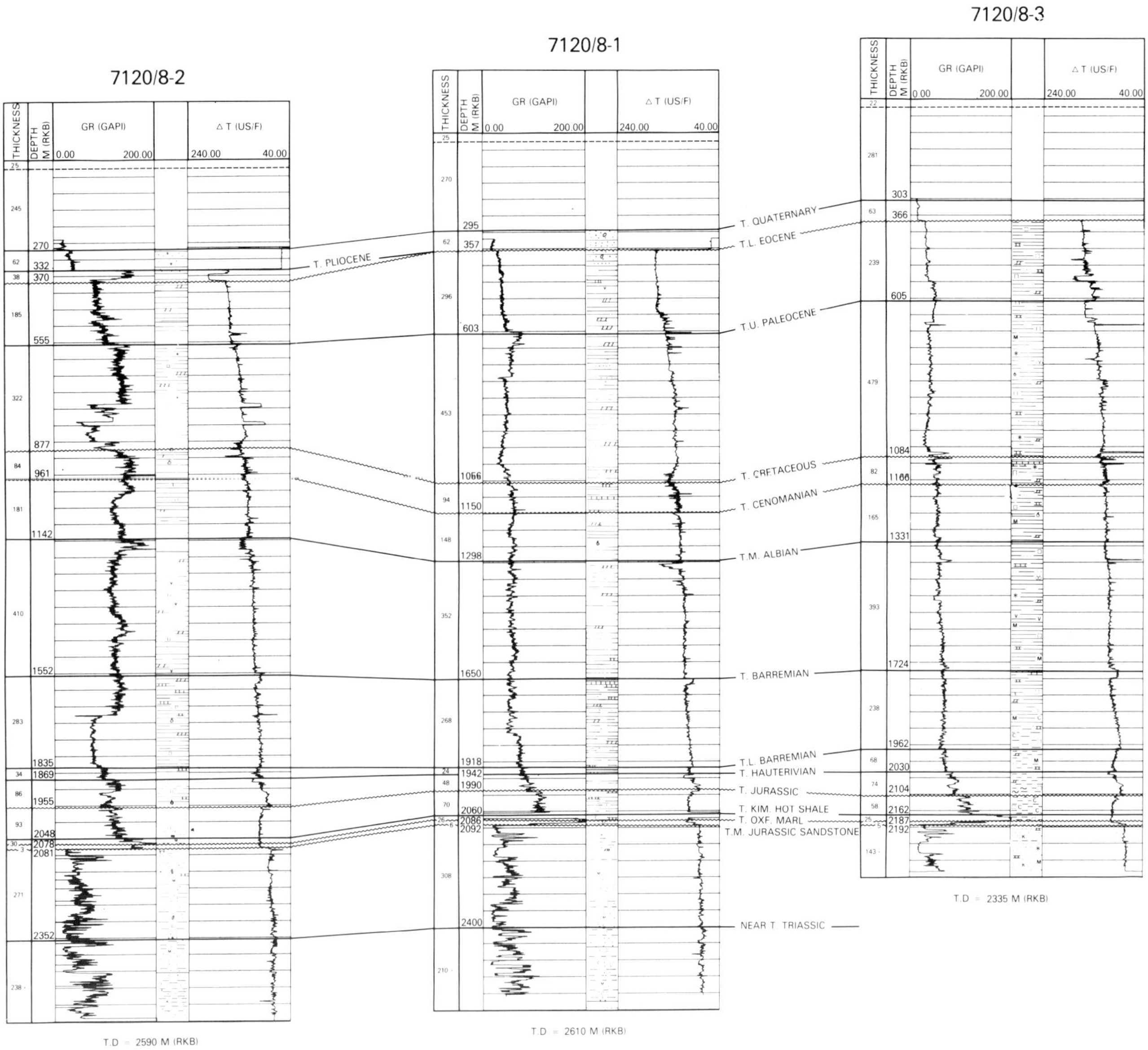

Fig. 3. Stratigraphic log correlation, Block 7120/8.

644 km of seismic lines was shot and processed by Geco in 1981. The reprocessed 1975 data and the 1981 survey constitute a 1 × 1 km seismic grid. These surveys were the basis for seismic interpretation until 1983 when a three-dimensional seismic survey was shot and processed by Geco. This survey consists of 201 lines oriented N–S, 75 m apart, and they amount to 6029 km of three-dimensional data. Figure 4 shows a N–S seismic line through Well 7120/8-2.

The quality of the data is good to excellent, and the interpretation has been carried out on migrated sections. It is not possible to identify the top Middle Jurassic sand on the seismic data. The seismic reflector used to map the top reservoir is judged to originate from the top of the Oxfordian shale. As this unit is a maximum of 6 m thick, the depth map of the underlying reservoir has been constructed by adding the shale isopach to the depth-converted top Oxfordian shale time-map (Fig. 5).

TRAP

The Askeladd Field consists of two structural traps situated in tilted fault-blocks. The area is intensely tectonized with two fault patterns —one approximately N–S and the other E–W. The tilted fault-blocks are a result of tensional movements in the area during the subsidence of the Tromsø Basin in late Early Cretaceous times. The traps are sealed westwards against Upper Jurassic shales along a N–S listric fault-plane. Structural dip is 1–2°

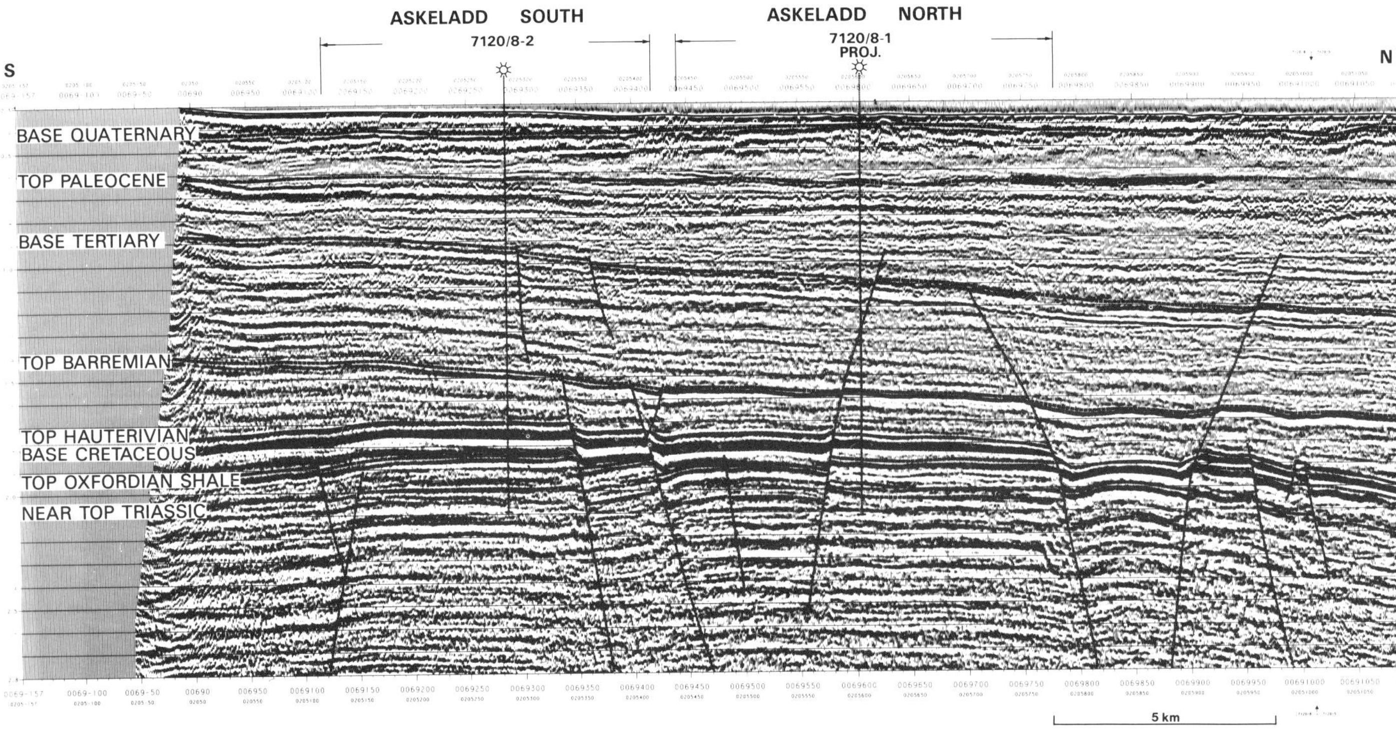

Fig. 4. Seismic line ST 8320–069 through Well 7120/8-2.

towards directions varying from north–east to south.

The field is structurally separated by an E–W fault dipping northwards between Wells 7120/8-1 and 7120/8-2. The Askeladd structures are not filled to spill-point. Askeladd North has a gas–water contact at 2155 mss (2180 m RKB) whereas Askeladd South has a gas–water contact at 2136 mss (2161 m RKB). The structures cover an area of about 105 km^2. A structural cross-section across the reservoir section in Askeladd North and South is presented in Fig. 6.

RESERVOIR

The reservoir sands in the Askeladd field are of Early to Middle Jurassic age and belong to the Stø Formation. Based on core and well-log data, three main lithofacies are defined: clean sandstones (quartz arenites); argillaceous sandstones; and dark mudstones.

The clean sandstones are medium to fine grained, occasionally with coarse-grained lags. The sedimentary structures that have been distinguished range from low-energy wave-ripples to storm-energy, hummocky cross-stratification. Low-angle cross-stratification and parallel lamination are abundant. Biogenic structures occur as vertical burrows and shell lags are observed occasionally.

The argillaceous sandstones are very fine to fine grained and strongly burrowed with few primary structures preserved. Minor normal-graded, non-burrowed beds, often with shell lags, suggest high-energy deposition induced by storm episodes.

The dark mudstones are well laminated but are frequently mixed with intervals of lenticular, bedded, silty mudstones. Cores indicate a good relationship between lithology and gamma-ray log response. The highest readings correspond to the mudstone facies, the lowest readings correlate with the clean sandstones, whilst the medium readings correspond to the argillaceous sandstones.

Based on this relationship three units have been defined as mudstone facies. These can be correlated between all the Askeladd wells. The sandstones are classified into five correlatable units based on characteristic gamma-

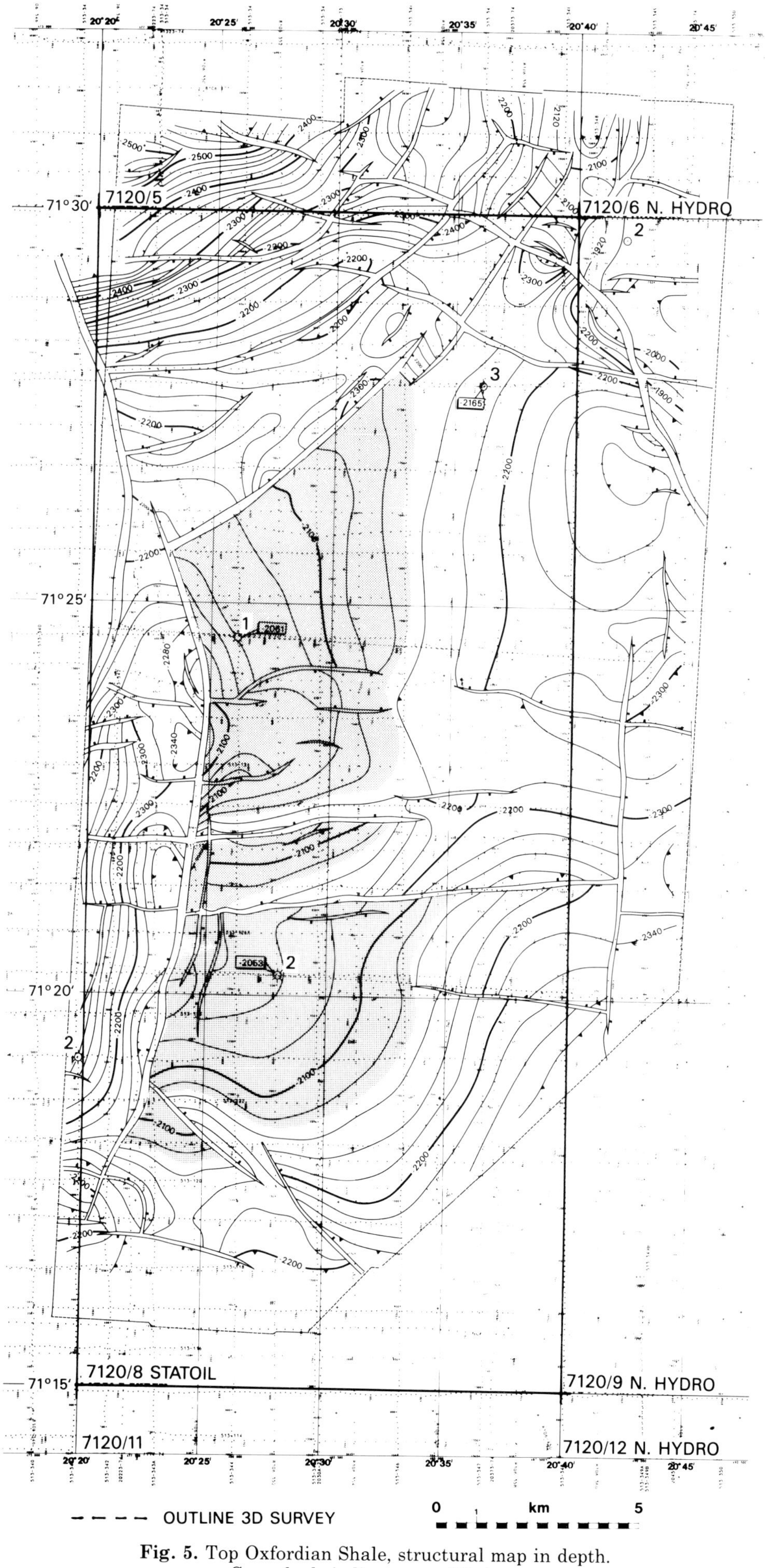

Fig. 5. Top Oxfordian Shale, structural map in depth. Grey shade indicates gas-filled area.

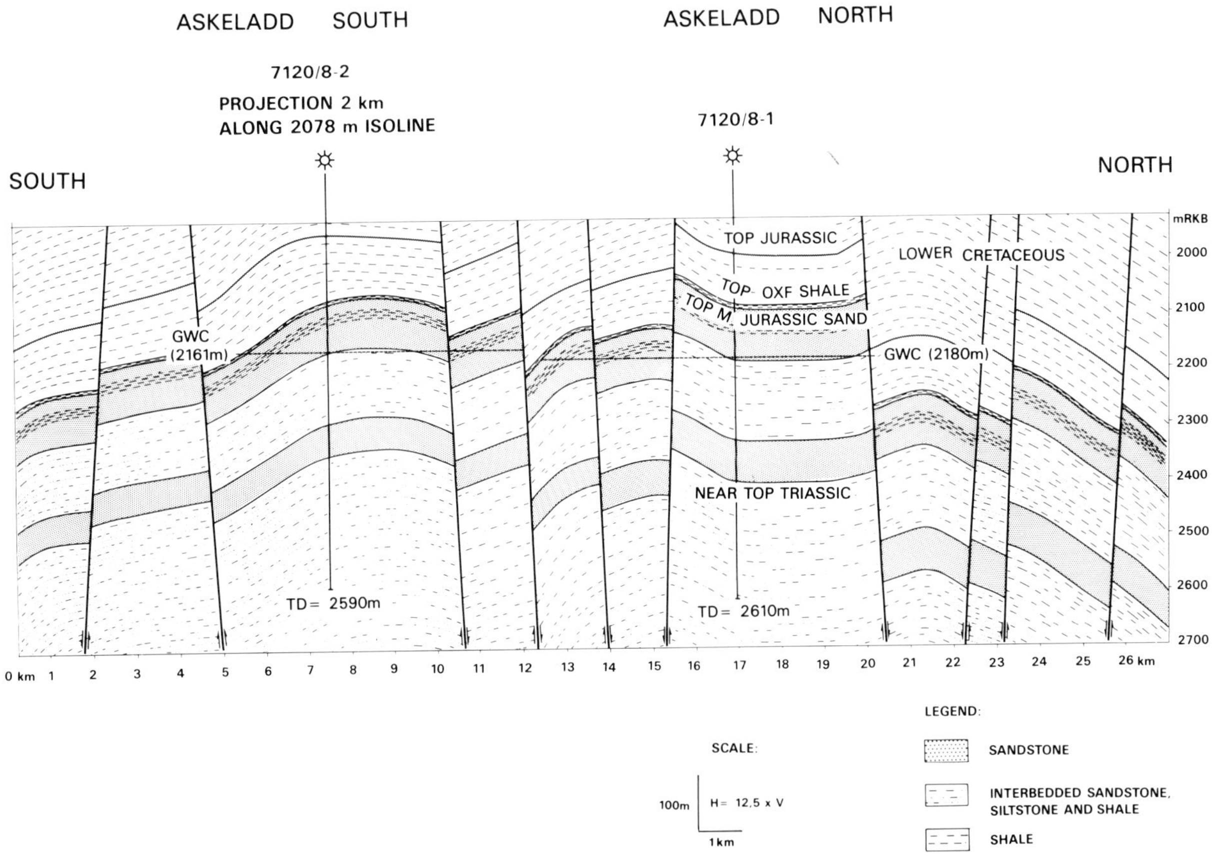

Fig. 6. Structural cross-section across the reservoir.

ray log responses. The gamma-ray log has a characteristic cylinder shape in the sand intervals with medium and low readings. The three mudstone units split the reservoir in the upper part (Fig. 7).

The deposition of the reservoir sandstone seems to have taken place in a near-shore to inner-shelf marine environment. Indicators of a marine environment are phosphate and authigenic minerals like pyrite and glauconite. The strong bioturbation and the shell lags clearly demonstrate a shallow-marine origin for the reservoir. The unusual thickness of the sandstone sequences and their regularity on the gamma-ray log can be explained by tectonic stability in the area. A delicate balance between sediment supply and basin subsidence gave a stationary coastline. Sedimentation took place in a shoreface environment and formed laterally extensive sand-bodies that can be correlated throughout the Askeladd area. The relatively thin mudstone units indicate rapid transgressive episodes due to low sediment input. Olaussen *et al.* (1983) suggested regional metamorphic rocks and low-grade or unmetamorphosed sedimentary rocks as the major provenance for the sandstones. They suggested that the transport direction was from the east and north-east. Subsequent information from other fields in the Tromsøflaket area indicates a wide, undulating coastline and a more northerly source area for the central part of the sedimentary basin. The marine redistribution of the sediments followed an E–W trend with transgressions entering the area from the south-west and north-west.

Net/gross ratio for the individual sand units averages 0.82. The excellent primary porosity in the mature sandstones has been variably reduced by compaction and silica and carbonate cementation (Olaussen *et al.*, 1983). Thus the porosity in the reservoir averages 15%.

SOURCE

Based on its maturity level, the gas is most likely to have been derived from Lower Jurassic/Upper Triassic source-rocks. These are carbonaceous mudstones and coals interbed-

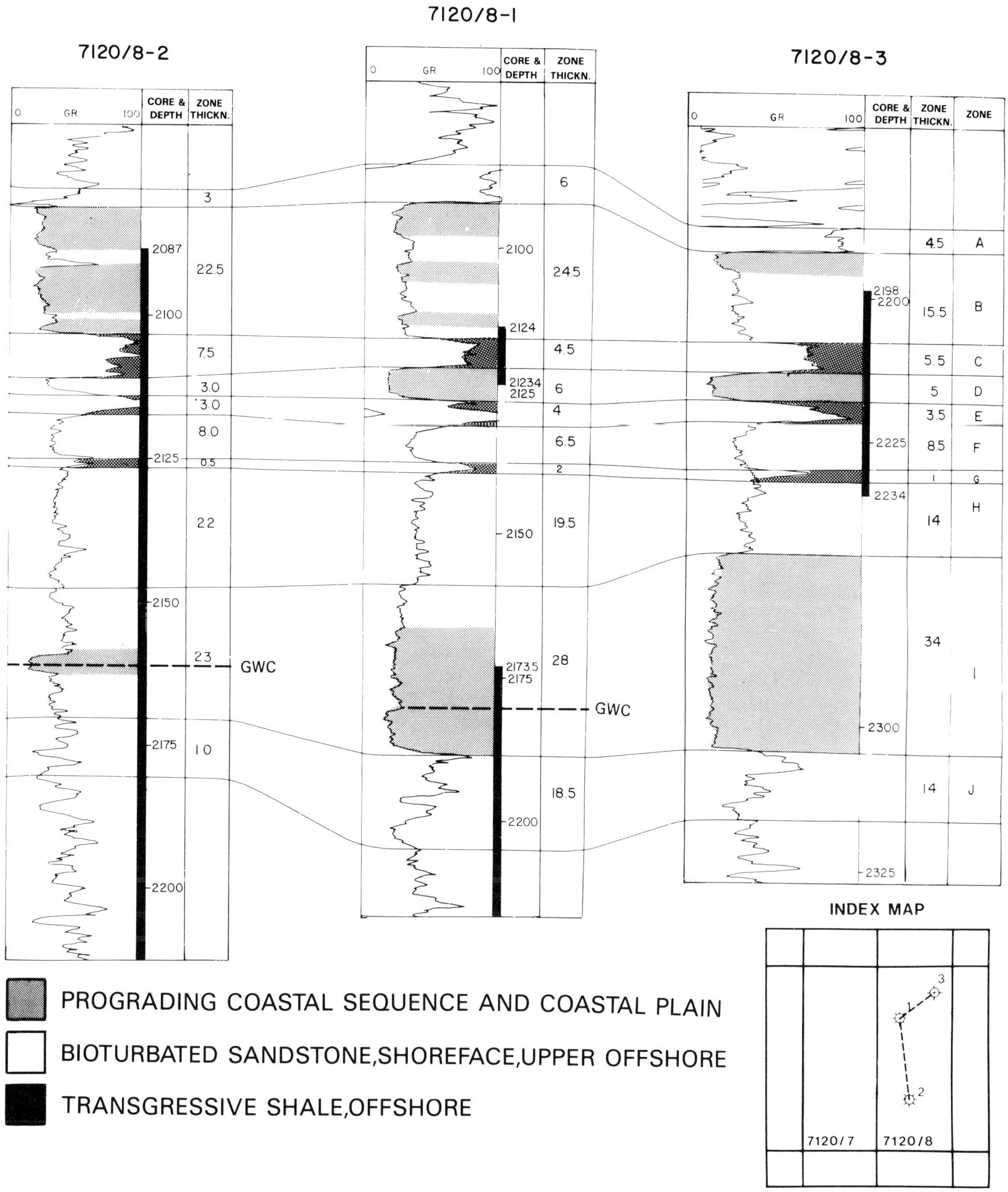

Fig. 7. Reservoir log correlation and palaeogeographic interpretation.

ded with sandstones. The possibility cannot be excluded, however, that the source is Upper Jurassic shales if the migration occurred along a drainage path from west to east. It is most likely that the gas started to migrate into the reservoir in Early Tertiary times.

HYDROCARBONS

The Askeladd Field consists of dry gas with a pressure gradient of 0.021 bar/m. It contains approximately 5% CO_2 and no H_2S. Reservoir temperature is 80°C and the reservoir has an initial pressure of 237 bar. Flow-rates obtained during drill-stem tests reached 953 100 Sm^3/day on 25-mm choke.

RESOURCES

The recoverable resources in the Askeladd Field are estimated to be in the order of 50×10^9 Sm^3 with an overall recovery factor of about 70%. Gas expansion should be the dominant production mechanism in the reservoir. Simulation studies show that the recovery factor will be independent of perforation strategy. Ultimately the recovery factor will be dependent on the shut-in criteria for the wells, well-head pressure and/or rate and also on the development strategy.

ACKNOWLEDGEMENTS

This article is based extensively on previously published papers about the Askeladd Field and the Tromsøflaket area, especially by S. Westre (1983). The authors thank the licensees in PL 064 for giving permission to publish this article.

REFERENCES

Olaussen, S., Dalland, A., Gloppen, T.G., and Johannesen, E. 1983. Depositional environment and diagenesis of Jurassic reservoir sandstones in the eastern part of Troms I area, In Spencer, A.M. *et al.* (Eds), *Petroleum Geology of the North European Margin*, Graham and Trotman, London, pp. 61–79.

Sund, T., Skarpnes, O., Jensen, L.N. and Larsen, R.M. 1986. Tectonic development and hydrocarbon potential offshore Troms, Northern Norway, *Memoir of the American Association of Petroleum Geologists* **40,** pp. 615–627.

Westre, S. 1983. The Askeladd gas field—Troms I, In Spencer, A.M. *et al.* (Eds), *Petroleum Geology of the North European Margin*, Graham and Trotman, London, pp. 33–39.

Part IX—Bibliography

38
Bibliography of petroleum geology for the Norwegian North Sea (56°–62°N)

B. Jacobsen* and A. M. Spencer†

*Conoco Norway Inc, Postboks 488, 4001 Stavanger, Norway
†Statoil, Forus (Geo), Postboks 300, 4001 Stavanger, Norway

INTRODUCTION

This bibliography aims to list the published articles relevant to the petroleum geology of the Norwegian sector of the North Sea, i.e. from 56° to 62°N. The references are restricted to those articles describing geology, geophysics and petroleum occurrences which contain significant information on the Norwegian sector itself. We have excluded references dealing solely with Recent and Quaternary strata and have included only fully published articles and not abstracts and unpublished technical reports. The present list is complementary to our earlier bibliography of the Norwegian continental shelf to the north of 62°N (Jacobsen and Spencer, 1984).

The list is ordered according to the Norwegian alphabet which comprises 29 characters: all those of the English alphabet plus æ, ø, å/aa at the end.

SYMPOSIA

There have been eight main symposia on Norwegian petroleum geology, all of which—except for the first—have been organized by the Norwegian Petroleum Society (Norsk Petroleumsforening—NPF). The resulting publications are these:

1. Whiteman, A.J., Roberts, D. and Sellevoll, M.A. (Eds). 1975. Petroleum geology and geology of the North Sea and North-east Atlantic continental margin, *Norges geol. Unders.* **316,** 376 pp.
2. *Jurassic Northern North Sea Symposium.* 1975. 22 separate articles. Norsk Petroleumsforening.
3. *Mesozoic Northern North Sea Symposium.* 1977. 16 separate articles. Norsk Petroleumsforening.
4. *Sedimentation of the North Sea Reservoir Rocks.* 1980. 21 separate articles. Norsk Petroleumsforening.
5. *Norwegian Symposium on Exploration.* 1981. 18 separate articles. Norsk Petroleumsforening.
6. Spencer, A.M. *et al.* (Eds). 1984. *Petroleum Geology of the North European Margin*, Graham and Trotman, London. 436 pp.
7. Thomas, B.M. *et al.* (Eds). 1985. *Petroleum Geochemistry in Exploration of the Norwegian Shelf*, Graham and Trotman, London. 337 pp.
8. Spencer, A.M. *et al.* (Eds). 1986. *Habitat of Hydrocarbons on the Norwegian Continental Shelf.* Graham and Trotman, London. 354 pp.

Publications 2 to 5 were issued as bound photocopies and can be obtained from: Norsk Petroleumsforening, Postboks 1897, Vika, 0124 Oslo 1, Norway.

Geology of the Norwegian Oil and Gas Fields © A. M. Spencer *et al.*, eds (Graham & Trotman, 1987) pp. 429–443.

ONS

At the biennial technology exhibitions held in Stavanger since 1974, there have been technical conferences on many subjects. The main conference proceedings on petroleum geological topics relevant to this bibliography are those listed below. The conference proceedings have been issued in bound photocopies by Norsk Petroleumsforening.

1. Offshore North Sea 1974. Exploration geology and geophysics. Articles G-IV/1 to G-IV/8.
2. Offshore North Sea 1976. Exploration—geology and geophysics. Articles G-IV/1 to G-IV/6.
3. Offshore North Sea 1980. Exploration—Geological and geophysical highlights. Articles E-III/1 to E-III/10.
4. Offshore Northern Seas 1982. The geological framework and hydrocarbon potential of basins in Northern Seas. Articles E/1 to E/8.
5. Offshore Northern Seas 1984.
   - Reduction of uncertainties by innovative reservoir geo-modelling. Articles G1 to G8.
   - Improved offshore recovery. Articles R1 to R4.
6. Offshore Northern Seas 1986.
   - Reservoir Engineering Conference. Articles R1 to R4.
   - Exploration conference. Articles G1 to G7.

In addition to the above conferences, the ONS label has been used for 'advanced projects' conferences, and those in 1983 and 1985 contained papers on petroleum geological topics. The conference proceedings have been issued as bound photocopies and are obtainable from Norsk Petroleumsforening.

- Offshore Northern Seas. Advanced Projects Conference. 1983. *The Troll Field.* Articles T1 to T16.
- Offshore Northern Seas. Advanced Projects Conference. 1985. *Technology Applications—Today and Tomorrow.* Articles A1 to A15, B1 to B13.

BIBLIOGRAPHY

Alhilali, K.A. and Damuth J.E. 1987. Slide block (?) of Jurassic sandstone and submarine channels in the basal Upper Cretaceous of the Viking Graben: Norwegian North Sea, *Marine and Petroleum Geology* **4** (1), 35–48.

Al-Kasim, F., Rønnevik, H.C. and Ulleberg, K. 1975. Review of the Jurassic offshore Norway, In *Jurassic Northern North Sea Symposium*, Norsk Petroleumsforening, Article 3, 18 pp.

Andersen, O.B. 1978. Bouguer anomaly map of Denmark and surrounding waters, *Geodætisk Instituts Skr. 3. Række*, Bind 44.

Badley, M.E., Egeberg, T. and Nipen, O. 1984. Development of rift basins illustrated by the structural evolution of the Oseberg feature, block 30/6, offshore Norway, *Journal of the Geological Society of London* **141** (4), 639–649. Discussion by D. Soutter and J.S. Ritchie **142** (5), 933–934.

Bailey, C.C., Price, I. and Spencer, A.M. 1981. The Ula Oil Field, block 7/12, Norway, In *Norwegian Symposium on Exploration*, Norsk Petroleumsforening, Article 18, 26 pp.

Baird, R.A. 1986. Maturation and source-rock evaluation of Kimmeridge Clay, Norwegian North Sea, *Bulletin of the American Association of Petroleum Geologists* **70,** (1), 1-11.

Balling, N.P. 1976. Geothermal models of the Crust and uppermost Mantle of the Fennoscandian Shield in south Norway and the Danish embayment, *Journal of Geophysics* **42,** 237–256.

Barnard, P.C. and Cooper, B.S. 1981. Oils and source rocks of the North Sea area, In Illing, L.V. and Hobson, G.D. (Eds), *Petroleum Geology of the Continental Shelf of North-west Europe*, Heyden, London, pp. 169–175.

Barril, R. 1985. Reservoir management considerations in the Frigg Field, In *Offshore Northern Seas—Advanced Projects Conference—Technology Applications—Today and Tomorrow*, Article A4, 16 pp.

Barton, P. 1985. Comparison of deep reflection and refraction structures in the North Sea, In Barazangi, M. (Ed.), *Reflection Seismology: A Global Perspective, American Geophysical Union Geodynamics Series* 13, 297–300.

Barton, P.J. and Matthews, D.H. 1984. Deep structure and geology of the North Sea region interpreted from a seismic refraction profile, *Annales Geophysicae* 1984, **2** (6), 663–668.

Barton, P., Matthews, D., Hall, J. and Warner, M. 1984. Moho beneath the North Sea compared on normal incidence and wide-angle seismic records, *Nature, London* **308,** 55–56.

Barton, P.J. and Wood, R. 1984. Tectonic evolution of the North Sea basin: crustal stretching and subsidence, *Geophysical Journal of the Royal Astronomical Society* **79,** 987–1022.

Bath, P.G.H. 1987. Enhanced oil recovery in the North Sea, In Kleppe, J. *et al.* (Eds). *North Sea Oil and Gas Reservoirs*, Graham and Trotman, London.

Beach, A. 1985. Some comments on sedimentary basin development in the Northern North Sea, *Scottish Journal of Geology* **21** (4), 493–512.

Beach, A. 1986. A deep seismic reflection profile across the northern North Sea, *Nature, London*

323 (6083), 53–55.

Beaumont, C. 1978. The evolution of sedimentary basins on a viscoelastic lithosphere: theory and examples, *Geophysical Journal of the Royal Astronomical Society* **55,** 471–497.

Behrens, K., Hansen, J., Flüh, E.R., Goldflam, S. and Hirschleber, H. 1986. Seismic investigations in the Skagerrak and Kattegat, *Tectonophysics* **128,** 209–228.

Bergsager, E. 1980. Status of exploration south of 62°, In *Offshore North Sea—Exploration— Geological and Geophysical Highlights*, Norsk Petroleumsforening, Article E-III/10, 23 pp.

Bergsager, E. 1983. Exploration results, Production and Development Plans on the Norwegian Continental Shelf, *NPD contribution* No. 13.

Bergsager, E. 1984. The Norwegian Continental Shelf—discoveries and related research challenges, In *Offshore Northern Seas—Reduction of uncertainties by innovative reservoir geo-modelling*, Norsk Petroleumsforening, Article G8, 31 pp. (Also available as *NPD contribution* no. 21, 34 pp.)

Bergsager, E. 1985. Character of the North Sea, In Bang-Andersen, A., Greenhill, B. and Grude, E.H. (Eds), *The North Sea. A Highway of Economic and Cultural Exchange, Character, History*, Norwegian University Press, Oslo, pp. 9–26.

Berstad, S. and Dypvik, H. 1982. Sedimentological evolution and natural radioactivity of Tertiary sediments from the central North Sea, *Journal of Petroleum Geology* **5** (1), 77–88.

Birtles, R. 1986. The seismic flatspot and the discovery and delineation of the Troll Field, In Spencer, A.M. *et al.* (Eds), *Habitat of Hydrocarbons on the Norwegian Continental Shelf*, Graham and Trotman, London, pp. 207–215.

Bjørlykke, K. 1980. Clastic diagenesis and basin evolution, *Revista del Instituto de Investigaciones Geológicas. Diputación Provincial. Universidad de Barcelona*, **34,** 21–44.

Bjørlykke, K. 1983. Diagenetic reactions in sandstones, In Parker, A. and Sellwood, B.W. (Eds), *Sediment Diagenesis*, Reidel, Dordrecht, pp. 169–213.

Bjørlykke, K. 1984. Formation of secondary porosity: how important is it? In McDonald, D.A. and Surdam, R.C. (Eds), *Clastic Diagenesis, Memoir of the American Association of Petroleum Geologists* **37,** 277–286.

Bjørlykke, K. and Brendsdal, A. 1986. Diagenesis of the Brent sandstone in the Statfjord field, North Sea. In Gautier, D.L. (Ed.), *Roles of Organic Matter in Sediment Diagenesis, Special Publication Society Economic Paleontologists Mineralogists* **38,** pp. 157–167.

Bjørlykke, K., Dypvik, H. and Finstad, K.G. 1975. The Kimmeridgian Shale, its composition and radioactivity. In *Jurassic Northern North Sea Symposium*, Norsk Petroleumsforening, Article 12, 20 pp.

Bjørlykke, K., Elverhøi, A. and Malm, A.O. 1979. Diagenesis in Mesozoic sandstones from Spitsbergen and the North Sea—a comparison, *Geologische Rundschau* **68,** 1152–1171.

Blair, D.G. 1975. Structural styles in North Sea oil and gas fields, In Woodland, A.W. (Ed.), *Petroleum and the Continental Shelf of North-west Europe, Vol. 1, Geology*, Applied Science, Barking, pp. 327–338.

Blundell, D.J. 1981 The Nature of the continental crust beneath Britain, In Illing, L.V. and Hobson, G.D. (Eds), *Petroleum Geology of the Continental Shelf of North-West Europe*, Heyden, London, pp. 58–64.

Brekke, T., Pegrum, R.M. and Watts, P.B. 1981. First exploration results in block 31/2, In *Norwegian Symposium on Exploration*, Norsk Petroleumsforening, Article 16, 34 pp.

Brennand, T.P. 1984. Petroleum geology in North Sea exploration 1964–1983, In Glennie, K.W. (Ed.), *Introduction to the Petroleum Geology of the North Sea*, Blackwell, Oxford, pp. 1–15.

Brewster, J. and Dangerfield, J.A. 1984. Chalk fields along the Lindesnes Ridge, Eldfisk, *Marine and Petroleum Geology* **1** (3), 239–278.

Brewster, J., Dangerfield, J. and Farrell, H. 1986. The geology and geophysics of the Ekofisk Field waterflood, *Marine and Petroleum Geology* **3** (2), 139–169.

Brewster, J. and Jeangeot, G. 1987. The production geology of the Frigg Field, In Kleppe, J. *et al.* (Eds). *North Sea Oil and Gas Reservoirs*, Graham and Trotman, London.

Brooks, J.R.V. and Thomas, P. 1974. A selected bibliography of North Sea geology (up to 1974), *Proceedings of the Geologists Association* **85** (1), 103–111.

Brooks, J.R.V. and Thomas, P. 1977. A classified bibliography of North-west European continental shelf geology, *Scottish Journal of Geology* **13** (1), 1–10.

Brown, S. 1984. Jurassic, In Glennie, K.W. (Ed.), *Introduction to the Petroleum Geology of the North Sea*, Blackwell, Oxford, pp. 103–131.

Bungum, H. and Swearingen, P.H. 1986. Earthquake hazard assessment in the North Sea, *Physics of the Earth and Planetary Interiors* **44,** 201–210.

Buza, J.W. and Unneberg, A. 1986. Statfjord Field 1—Statfjord geology, reservoir performance. *Oil and Gas Journal* **84** (29), 68–72.

Buza, J.W. and Unneberg, A. 1986. Statfjord Field 2—Statfjord field development strategies. *Oil Gas Journal* **84** (30), 110–114.

Buza, J.W. and Unneberg, A. 1987. Geological and reservoir engineering aspects of the Statfjord Field, In Kleppe, J. *et al.* (Eds). *North Sea Oil and Gas Reservoirs*, Graham and Trotman, London.

Byrd, W.D. 1975. Geology of the Ekofisk Field, offshore Norway, In Woodland, A.W. (Ed.) *Petroleum and the Continental Shelf of North-west Europe*, Vol. 1. Geology, Applied Science, Barking, pp. 439–446.

Campbell, C.J. and Ormaasen, E. 1987. The discovery of oil and gas in Norway: an historical synopsis, In Spencer, A.M. *et al.* (Eds), *Geology of the Norwegian Oil and Gas Fields*, Graham and Trotman, London, pp. 1–37.

Carstens, H. and Dypvik, H. 1981. Abnormal formation pressure and shale porosity, *Bulletin of the American Association of Petroleum Geologists* **65** (2), 344–350.

Carstens, H. and Finstad, K.G. 1981. Geothermal gradients of the northern North Sea Basin, 59–62°N, In Illing, L.V. and Hobson, G.D. (Eds), *Petroleum Geology of the Continental Shelf of North-west Europe*, Heyden, London, pp. 152–161.

Cassell, B.R., Mykkeltveit, S., Kanestrøm, R. and Husebye, E.S. 1983. A North Sea–southern Norway seismic crustal profile, *Geophysical Journal of the Royal Astronomical Society* **72,** 733–753.

Chauvin, A.L. and Valachi, L.Z. 1980. Sedimentology of the Brent and Statfjord formations of Statfjord Field, In *The Sedimentation of the North Sea Reservoir Rocks*, Norsk Petroleumsforening, Article XVI, 17 pp.

Chew, K.J. and Stephenson, H. 1986. Exploration success offshore Norway compared with the remainder of the North Sea Graben System and with other hydrocarbon provinces, In Spencer, A.M. *et al.* (Eds), *Habitat of Hydrocarbons on the Norwegian Continental Shelf*, Graham and Trotman, London, pp. 61–74.

Chiarelli, A and Duffaud, F. 1980. Pressure origin and distribution in Jurassic of Viking Basin (United Kingdom–Norway), *Bulletin of the American Association of Petroleum Geologists* **64** (8), 1245–1250.

Christiansen, J.B. and Strand, J.E. 1987. 15/5-1, In Spencer, A.M. *et al.* (Eds), *Geology of the Norwegian Oil and Gas Fields*, Graham and Trotman, London, pp. 265–272.

Clemmensen, L., Jacobsen, V. and Steel, R. 1980. Some aspects of Triassic sedimentation and basin development, East Greenland, North Sea, In *The Sedimentation of the North Sea Reservoir Rocks*, Norsk Petroleumsforening, Article XVII, 21 pp.

Conort, A. 1986. Habitat of Tertiary hydrocarbons, South Viking Graben, In Spencer, A.M. *et al.* (Eds), *Habitat of Hydrocarbons on the Norwegian Continental Shelf,* Graham and Trotman, London, pp. 159–170.

Cooper, B.S. and Barnard, P.C. 1984. Source rocks and oils of the central and northern North Sea, In Demaison, G. and Murris, R.J. (Eds), *Petroleum Geochemistry and Basin Evaluation,* ***Memoir of the American Association of Petroleum Geologists*** **35,** **303–314.**

Cornelius, C.D. 1975. Geothermal aspects of hydrocarbon exploration in the North Sea area, In Whiteman, A.J., Roberts, D. and Sellevoll, M.A. (Eds), *Petroleum Geology and Geology of the North Sea and North-east Atlantic Continental Margin, Norges geoliske Undersøkelse* **36,** 29–67.

Cornford, C. 1984. Source rocks and hydrocarbons of the North Sea. In Glennie, K.W. (Ed.), *Introduction to the Petroleum Geology of the North Sea*, Blackwell, Oxford, pp. 171–204.

Cornford, C., Needham, C.E.J. and de Walque, L. 1986. Geochemical habitat of North Sea oils and gases, In Spencer, A.M. *et al.* (Eds), *Habitat of Hydrocarbons on the Norwegian Continental Shelf*, Graham and Trotman, London, pp. 39–54.

Curtin, D.P. and Ballestad, S. 1986. South Viking Graben: habitat of Mesozoic hydrocarbons, In Spencer, A.M. *et al.* (Eds), *Habitat of Hydrocarbons on the Norwegian Continental Shelf*, Graham and Trotman, London, pp. 153–157.

Dahl, B. and Speers, G.C. 1985. Organic geochemistry of the Oseberg Field (I), In Thomas, B.M. *et al.* (Eds), *Petroleum Geochemistry in Exploration of the Norwegian Shelf*, Graham and Trotman, London, pp. 185–195.

Dangerfield, J. and Brown, D.A. 1987. The Ekofisk Field, In Kleppe, J. *et al.* (Eds), *North Sea Oil and Gas Reservoirs*, Graham and Trotman, London.

Darley, J., Røssland, L., van Golf-Racht, T. and Krakstad, O. 1983. Reservoir simulation of the Troll Field, In *Offshore Northern Seas Advanced Projects Conference—The Troll Field*, Article T4, 65 pp.

Davies, E.J. and Watts, T.R. 1977. The Murchison oil-field, In *Mesozoic Northern North Sea Symposium*, Norsk Petroleumsforening, Article 15, 24 pp.

Day, G.A., Cooper, B.A., Andersen, C., Burgers, W.F.J., Rønnevik, H.C. and Schöneich, H. 1981. Regional seismic structure maps of the North Sea, In Illing, L.V. and Hobson, G.D. (Eds), *Petroleum Geology of the Continental Shelf of North-west Europe*, Heyden, London, pp. 76–84.

Deegan, C.E. and Scull, B.J. (Compilers). 1977. A standard lithostratigraphic nomenclature for the central and northern North Sea, *Report Institute Geological Science London* 77/25 (also designated *Norwegian Petroleum Directorate Bull.* 1).

Derumaux, F. 1980. Le Permien évaporitique de Mer du Nord relations entre tectonique et sédimentologie, *Bulletin de Centre Recherches Elf Aquitaine* **4** (1), 495–510.

D'Heur, M. 1980. Chalk reservoir of the West Ekofisk Field, In *The Sedimentation of the North Sea Reservoir Rocks*, Norsk Petroleumsforening, Article X, 20 pp.

D'Heur, M. 1984. Porosity and hydrocarbon distribution in the North Sea chalk reservoirs, *Marine and Petroleum Geology* **1** (3), 211–237.

D'Heur, M. 1986. The Norwegian chalk fields, In Spencer, A.M. *et al.* (Eds), *Habitat of Hydrocarbons on the Norwegian Continental Shelf*, Graham and Trotman, London, pp. 77–89.

D'Heur, M. 1987a. Albuskjell, In Spencer, A.M. *et al.* (Eds), *Geology of the Norwegian Oil and Gas* ***Fields*****. Graham and Trotman, London, pp. 39–50.**

D'Heur, M. 1987b. Cod, In Spencer, A.M. *et al.* (Eds), *Geology of the Norwegian Oil and Gas Fields*, Graham and Trotman, London, pp. 51–62.

D'Heur, M. 1987c. Tor, In Spencer, A.M. *et al.*

(Eds), *Geology of the Norwegian Oil and Gas Fields*, Graham and Trotman, London, pp. 129–142.

D'Heur, M. 1987d. West Ekofisk, In Spencer, A.M. *et al.* (Eds), *Geology of the Norwegian Oil and Gas Fields*, Graham and Trotman, London, pp. 165–175.

D'Heur, M. and Michaud F. 1987. Edda, In Spencer, A.M. *et al.* (Eds), *Geology of the Norwegian Oil and Gas Fields*, Graham and Trotman, London, pp. 63–72.

D'Heur, M. and Pekot, L.J. 1987. Tommeliten, In Spencer, A.M. *et al.* (Eds), *Geology of the Norwegian Oil and Gas Fields*, Graham and Trotman, London, pp. 117–128.

D'Heur, M. and de Walque, L. 1987. Bream–Brisling, In Spencer, A.M. *et al.* (Eds), *Geology of the Norwegian Oil and Gas Fields*, Graham and Trotman, London, pp. 185–192.

D'Heur, M., de Walque, L. and Michaud, F. 1985. Geology of the Edda Field reservoir, *Marine and Petroleum Geology* **2** (4), 327–340.

Dixon, J.E., Fitton, J.G. and Frost, R.T.C. 1981. The tectonic significance of post-Carboniferous igneous activity in the North Sea Basin, In Illing, L.V. and Hobson, G.D. (Eds), *Petroleum Geology of the Continental Shelf of North-west Europe*, Heyden, London, pp. 121–137.

Donato, J.A. and Tully, M.C. 1981. A regional interpretation of North Sea gravity data, In Illing, L.V. and Hobson, G.D. (Eds), *Petroleum Geology of the Continental Shelf of North-west Europe*, Heyden, London, pp. 65–75.

Doré, A.G., Vollset, J. and Hamar, G.P. 1985. Correlation of the offshore sequences referred to the Kimmeridge Clay Formation—relevance to the Norwegian sector, In Thomas, B.M. *et al.* (Eds), *Petroleum Geochemistry in Exploration of the Norwegian Shelf*, Graham and Trotman, London, pp. 27–37.

Dunn, W.W. 1975. North Sea basinal area, Europe —an important oil and gas province, In Whiteman, A.J., Roberts, D. and Sellevoll, M.A. (Eds), *Petroleum Geology and Geology of the North Sea and North-east Atlantic Continental Margin, Norges geoliske undersøkelse* **36,** 69–97.

Dypvik, H. 1983. Clay mineral transformations in Tertiary and Mesozoic sediments from North Sea, *Bulletin of the American Association of Petroleum Geology* **67** (1), 160–165.

Dypvik, H. and Brunfelt, A.O. 1979. Distribution of rare earth elements in some North Atlantic Kimmeridgian black shales, *Nature, London* **278,** 339–341.

Dypvik, H. and Vollset, J. 1979. Petrology and diagenesis of Jurassic sandstones from Norwegian Danish Basin, North Sea, *Bulletin of the American Association of Petroleum Geology* **63** (2), 182–193.

Dypvik, H. and Vollset, J. 1980. Deltaic sedimentation during the Jurassic in the Norwegian–Danish basin (North Sea), *Geologie en Mijnbouw* **59** (1), 25–32.

Ekern, O.F. 1986. Late Oligocene gas accumulations, Block 2/2, Norway, In Spencer, A.M. *et al.* (Eds), *Habitat of Hydrocarbons on the Norwegian Continental Shelf*, Graham and Trotman, London, pp. 143–149.

Ekern, O.F. 1987. Midgard, In Spencer, A.M. *et al.* (Eds), *Geology of the Norwegian Oil and Gas Fields*, Graham and Trotman, London, pp. 403–440.

Engelstad, N. 1987. Murchison, In Spencer, A.M. *et al.* (Eds), *Geology of the Norwegian Oil and Gas Fields*, Graham and Trotman, London, pp. 295–305.

Enjolras, J.M., Gouadain, J., Mutti, E. and Pizon, J. 1986, New turbiditic model for the Lower Tertiary sands in the South Viking Graben, In Spencer, A.M. *et al.* (Eds), *Habitat of Hydrocarbons on the Norwegian Continental Shelf*, Graham and Trotman, London, pp. 171–178.

Erichsen, T., Helle, M., Henden, J. and Rognebakke, A. 1987. Gullfaks, In Spencer, A.M. *et al*, (Eds), *Geology of the Norwegian Oil and Gas Fields*. Graham and Trotman, London, pp. 273–286.

Eynon, G. 1980. Basin development and sedimentation in the Middle Jurassic in the northern North Sea, In *The Sedimentation of the North Sea Reservoir Rocks*, Norsk Petroleumsforening, Article IV, 19 pp.

Eynon, G. 1981. Basin development and sedimentation in the Middle Jurassic of the northern North Sea, In Illing, L.V. and Hobson, G.D. (Eds), *Petroleum Geology of the Continental Shelf of North-west Europe*, Heyden, London, pp. 196–204.

Faber, E. and Stahl, W. 1984. Geochemical surface exploration for hydrocarbons in North Sea, *Bulletin of the American Association of Petroleum Geologists* **68** (3), 363–386.

Fagerland, N. 1983. Tectonic analysis of a Viking Gorben border fault, *Bulletin of the American Association of Petroleum Geologists* **67** (11), 2125–2136.

Fagerland, N. 1984. A rift-related stratigraphic trap, *Journal Petroleum Geol.* **7** (2), 227–236.

Feazel, C.T., Keany, J. and Peterson, R.M. 1985. Cretaceous and Tertiary chalk of the Ekofisk field area, central North Sea, In Roehl, P.O. and Choquette, P.W. (Eds), *Carbonate Petroleum Reservoirs*, Springer-Verlag, New York. pp. 495–507.

Field, J.D. 1985. Organic geochemistry in exploration of the northern North Sea, In Thomas, B.M. *et al.* (Eds), *Petroleum Geochemistry in Exploration of the Norwegian Shelf*, Graham & Trotman, London, pp. 39–57.

Fisher, M.J. 1984. Triassic, In Glennie, K.W. (Ed.), *Introduction to the Petroleum Geology of the North Sea*, Blackwell, Oxford, pp. 85–101.

Floden, T. and Sellevoll, M.A. 1972. Two seismic profiles across the Norwegian Channel west of Bergen, *Stockholm Contributions in Geology* **24** (2), 25–32.

Frost, R.T.C., Fitch, F.J. and Miller, J.A. 1981. The age and nature of the crystalline basement of the North Sea Basin, In Illing, L.V. and Hobson, G.D. (Eds), *Petroleum Geology of the Continental Shelf of North-West Europe*, Heyden, London, pp. 43–57.

Fuller, J.G.C.M. 1975. Jurassic source-rock potential and hydrocarbon correlation, North Sea, In *Jurassic Northern North Sea Symposium*, Norsk Petroleumsforening, Article 11, 18 pp.

Furnes, H., Elvsborg, A. and Malm, O.A. 1982. Lower and Middle Jurassic alkaline magmatism in the Egersund Sub-basin, North Sea, *Marine Geology* **46,** 53–69.

Færseth, R.B., Oppebøen, K.A. and Sæbøe, A. 1986. Trapping styles and associated hydrocarbon potential in the Norwegian North Sea, In Halbouty, M.T. (Ed). *Future Petroleum Provinces of the World, Memoir of the American Association of Petroleum Geologists* **40,** 585–597.

Gabrielsen, R.H. 1986. Structural elements in graben systems and their influence on hydrocarbon trap types, In Spencer, A.M. *et al.* (Eds), *Habitat of hydrocarbons on the Norwegian Continental Shelf*, Graham and Trotman, London, pp. 55–60.

Gabrielsen, R.H., Ekern, O.F. and Edvardsen, A. 1986. Structural development of hydrocarbon traps, Block 2/2, Norway. In Spencer, A.M. *et al.* (Eds), *Habitat of hydrocarbons on the Norwegian Continental Shelf*, Graham and Trotman, London pp. 129–141.

Gabrielsen, R.H., Ulvøen, S., Elvsborg, A. and Fredrik, O. 1985. The geological history and geochemical evaluation of Block 2/2, offshore Norway, In Thomas, B.M. *et al.* (Eds), *Petroleum Geochemistry in Exploration of the Norwegian shelf*, Graham and Trotman, London, pp. 165–178.

Gage, M.S. and Doré, A.G. 1986. A regional geological perspective of the Norwegian offshore exploration provinces, In Spencer, A.M. *et al.* (Eds), *Habitat of Hydrocarbons on the Norwegian Continental Shelf*, Graham and Trotman, London, pp. 21–38.

Gevirtz, J.L., Carey, B.D. Jr and Blanco, S.R. 1985. Regional geochemical analysis of the southern portion of the Norwegian sector of the North Sea, In Thomas, B.M. *et al.* (Eds), *Petroleum Geochemistry in Exploration of the Norwegian Shelf*, Graham and Trotman, London, pp. 247–261.

Glennie, K.W. 1983. Lower Permian Rotliegend desert sedimentation in the North Sea area, In Brookfield, M.E. and Ahlbrandt, T.S. (Eds), *Eolian Sediments and Processes*, Developments in Sedimentology No. 38. Elsevier, pp. 521–541.

Glennie, K.W. (Ed.). 1984. *Introduction to the Petroleum Geology of the North Sea*, Blackwell, Oxford.

Glennie, K.W. 1984. The structural framework and the pre-Permian history, In Glennie, K.W. (Ed.), *Introduction to the Petroleum Geology of the North Sea*, Blackwell, Oxford, pp. 17–39.

Glennie, K.W. 1984. Early Permian–Rotliegend, In Glennie, K.W. (Ed.), *Introduction to the Petroleum Geology of the North Sea*, Blackwell, Oxford, pp. 41–60.

Goff, J.C. 1983. Hydrocarbon generation and migration from Jurassic source rocks in the E. Shetland Basin and Viking Graben of the northern North Sea, *Journal of the Geological Society of London* **140** (3), 445–474. (Reprinted in *Memoir of the American Association of Petroleum Geologists* **35,** 273–303.)

Gowers, M.B. and Sæbøe, A. 1985. On the structural evolution of the Central Trough in the Norwegian and Danish sectors of the North Sea, *Marine and Petroleum Geology* **2,** 298–318.

Gradijan, S.J. and Wiik, M. 1987. Statfjord Nord, In Spencer, A.M. *et al.* (Eds), *Geology of the Norwegian Oil and Gas Fields*, Graham and Trotman, London, pp. 341–350.

Gray, D.I. 1987. Troll, In Spencer, A.M. *et al.* (Eds), *Geology of the Norwegian Oil and Gas Fields*, Graham and Trotman, London, pp. 389–401.

Gray, I. and Handley, R.L. 1980. The integration of geological and seismic data in the Murchison oil field by the use of synthetic seismograms and seismic modelling, In *Offshore North Sea—Exploration—Geological and Geophysical Highlights*, Norsk Petroleumsforening, Article E-III/6, 13 pp.

Griggs, J.F. and Wiborg, R. 1984. Improving recoveries from offshore chalk reservoirs, In *Offshore Northern Seas — Improved offshore Recovery*, Norsk Petroleumsforening, Article R1, 13 pp.

Grung Olsen, R. and Hanssen, O.K. 1987. Askeladden, In Spencer, A.M. *et al.* (Eds), *Geology of the Norwegian Oil and Gas Fields*, Graham and Trotman, London, pp. 420–428.

Gulbrandsen, A. 1987. Agat, In Spencer, A.M. *et al.* (Eds), *Geology of the Norwegian Oil and Gas Fields*, Graham and Trotman, London, pp. 363–370.

Hage, A., Bomstad, K. and Strand, J.E. 1987. Brage, In Spencer, A.M. *et al.* (Eds), *Geology of the Norwegian Oil and Gas Fields*, Graham & Trotman, London, pp. 371–378.

Hall, P.B., Schou, L. and Bjorøy, M. 1985. Aromatic hydrocarbon variations in North Sea wells, In Thomas, B.M. *et al.* (Eds), *Petroleum Geochemistry in Exploration of the Norwegian Shelf*, Graham and Trotman, London, pp. 293–301.

Halstead, P.H. 1975. Northern North Sea faulting, In *Jurassic Northern North Sea Symposium*, Norsk Petroleumsforening, Article 10, 38 pp.

Hamar, G.P. 1975. A Jurassic structure complex in northern North Sea, In *Jurassic Northern North Sea Symposium*, Norsk Petroleumsforening, Article 17, 18 pp.

Hamar, G.P. 1982. Tektonisk historie av Nordsjøen, *Dansk geologisk Forening Årsskrift for 1981*, pp. 161–164.

Hamar, G.P., Fjæran, T. and Hesjedal, A. 1983. Jurassic stratigraphy and tectonics of the south-south-eastern Norwegian offshore, In Kaasschieter, J.P.H. and Reijers, T.J.A. (Eds), *Petro-*

leum Geology of the South-eastern North Sea and the Adjacent Onshore Areas, Geol. Mijnbouw **62,** (1), 103–114.

Hamar, G.P. and Hjelle, K. 1984. Tectonic framework of the Møre Basin and the northern North Sea, In Spencer, A.M. *et al.* (Eds), *Petroleum Geology of the North European Margin*, Graham and Trotman, London, pp. 349–358.

Hamar, G.P., Jakobsson, K.H. Ormaasen, D.E. and Skarpnes, O, 1980. Tectonic development of the North Sea north of Central Highs, In *The Sedimentation of the North Sea Reservoir Rocks*, Norsk Petroleumsforening, Article III, 11 pp.

Hamre, S. 1984. Improved recovery from thin oil sands. Part 1—Frigg Field, In *Offshore Northern Seas—Improved Offshore Recovery*, Norsk Petroleumsforening, Article R2, 12 pp.

Hancock, J. 1984. Cretaceous, In Glennie, K.W. (Ed.), *Introduction to the Petroleum Geology of the North Sea*, Blackwell, Oxford, pp. 133–150.

Hancock, J. 1985. Oil in Chalk, *Geology Today* **1** (5), 146–150.

Hancock, J.M. and Scholle, P.A. 1975. Chalk of the North Sea, In Woodland, A.W. (Ed.), *Petroleum and the Continental Shelf of North-West Europe*, Vol. 1, *Geology*, Applied Science, Barking, pp. 413–427.

Handley, R.L. and Gray, I. 1980. The integration of geological and seismic data in the Murchison oilfield by the use of synthetic seismograms and seismic modelling, In *Offshore North Sea—Exploration—Geological and Geophysical Highlights*, Norsk Petroleumsforening, Article E-III/6, 13 pp.

Hanslien, S. 1987. Balder, In Spencer, A.M. *et al.* (Eds), *Geology of the Norwegian Oil and Gas Fields*, Graham and Trotman, London, pp. 193–201.

Harding, T.P. 1984. Graben hydrocarbon occurrences and structural style, *Bulletin of the American Association of Petroleum Geologists* **68** (3), 333–362.

Hardman, R.F.P. 1982. Chalk reservoirs of the North Sea, *Bulletin of the Geological Society of Denmark* **30,** 119–137.

Hardman, R.F.P. and Eynon, G. 1977. Valhall Field—a structural/stratigraphic trap, In *Mesozoic Northern North Sea Symposium*, Norsk Petroleumsforening, Article 14, 33 pp.

Hardman, R.F.P. and Kennedy, W.J. 1980. Chalk reservoirs of the Hod Field, Norway, In *The Sedimentation of the North Sea Reservoir Rocks*, Norsk Petroleumsforening, Article XI, 31 pp.

Harper, M.L. and Shaw, B.E. 1974. Cretaceous–Tertiary carbonate reservoirs in the North Sea, In *Offshore North Sea: Exploration Geology and Geophysics*, Article G IV/4, 20 pp.

Hatton, I.R. 1986. Geometry of allochthonous Chalk Group members, Central Trough, North Sea, *Marine and Petroleum Geology* **3** (2), 79–98.

Havskov, J. 1987. Recent North Sea seismicity studies, *Physics of the Earth and Planetary Interiors* **45** (1), 37–44.

Hay, J.T.C. 1978. Structural development in the northern North Sea, *Journal of Petroleum Geology* **1** (1), 65–77.

Hazeu, G.J.A. 1981. 34/10 Delta structure geological evaluation and appraisal, In *Norwegian Symposium on Exploration*, Norsk Petroleumsforening, Article 13, 36 pp.

Heiberg, S. 1980. The reservoirs on the Norwegian continental shelf and their importance to oil and gas production, In *The Sedimentation of the North Sea Reservoir Rocks*, Norsk Petroleumsforening, Article XXI, 12 pp.

Helgøy, T. 1984. Prediction techniques from logging tools: are the results confirmed by testing? In *Offshore Northern Seas—Reduction of uncertainties by innovative reservoir geo-modelling*, Norsk Petroleumsforening, Article G4, 24 pp.

Hellem, T., Kjemperud, A. and Øvrebø, O.K. 1986. The Troll Field: a geological/geophysical model established by the PL085 Group, In Spencer, A.M. *et al.* (Eds), *Habitat of Hydrocarbons on the Norwegian Continental Shelf*, Graham and Trotman, London, pp. 217–238.

Héritier, F.E., Lossel, P. and Wathne, E. 1979. Frigg Field—large submarine-fan trap in Lower Eocene rocks of the Viking Graben, North Sea, *Bulletin of the American Association of Petroleum Geology* **63** (11), 1999–2020. (Reprinted in *Memoir of the American Association of Petroleum Geologists* **30,** 59–79.)

Héritier, F.E., Lossel, P. and Wathne, E. 1980. The Frigg gas field, In Illing, L.V. and Hobson, G.D. (Eds), *Petroleum Geology of the Continental Shelf of North-west Europe*, Heyden, London, pp. 380–391.

Hesjedal, A. and Hamar, G.P. 1983. Lower Cretaceous stratigraphy and tectonics of the south-south-eastern Norwegian offshore. In Kaasschieter, J.P.H. and Reijers, T.J.A. (Eds), *Petroleum Geology of the south-eastern North Sea and the Adjacent Onshore Areas, Geologie en Mijnbouw* **62** (1), 135–144.

Hollander N.B. 1987. Snorre, In Spencer, A.M. *et al.* (Eds), *Geology of the Norwegian Oil and Gas Fields*, Graham and Trotman, London, pp. 307–318.

Holtedahl, O. 1950. Supposed marginal fault lines in the shelf area off some high northern lands, *Bulletin of the Geological Society of America* **61,** 493–500.

Holtedahl, H. and Sellevoll, M.A. 1970. Geology of the continental margin of the eastern Norwegian Sea and of the Skagerrak, *Institute Geological Science Report* 70/14, pp. 34–52.

Home, P.C. 1987, Ula, In Spencer, A.M. *et al.* (Eds), *Geology of the Norwegian Oil and Gas Fields*, Graham and Trotman, London, pp. 143–151.

Hopkinson, J.P. 1974. North Sea petroleum geology, In *Offshore North Sea—Exploration Geology and Geophysics*, Article G-IV/1, 31 pp.

Hospers, J. 1985. Sideswipe reflections and other external and internal reflections from salt plugs in the Norwegian-Danish Basin. *Geophysical Prospecting* **33,** 52–71.

Hospers, J. and Finnstrøm, E.G. 1984. The gravity

field of the Norwegian Sector of the North Sea, *Norges geoliske Undersøkelse* **396,** 25–34.

Hospers, J., Finnstrøm, E.G. and Rathore, J.S. 1985. A regional gravity study of the northern North Sea, *Geophysical Prospecting* **33** (4), 543–566.

Hospers, J. and Holthe, J. 1980. Salt tectonics in block 8/8 of the Norwegian Sector of the North Sea, *Tectonophysics* **68,** 257–282.

Hospers, J. and Holthe, J. 1984. Growth of salt-induced structures in the Norwegian–Danish Basin. *Tectonophysics* **107,** 81–104.

Hospers, J. and Rathore, J.S. 1984. Interpretation of aeromagnetic data from the Norwegian sector of the North Sea, *Geophysical Prospecting* **32,** 929–942.

Hospers, J., Rathore, J.S., Feng, J. and Finnstrøm, E.G. 1986. Thickness of pre-Zechstein salt Palaeozoic sediments in the southern part of the Norwegian sector of the North Sea, *Norsk Geolisk Tidsskrift* **66** (4), 295–304.

Hovland, M. and Sommerville, J.H. 1985. Characteristics of two natural gas seepages in the North Sea, *Marine and Petroleum Geology* **2** (4), 319–326.

Hovland, M., Talbot, M., Olaussen, S. and Aasberg, L. 1985. Recent formed methane-derived carbonates from the North Sea floor, In Thomas, B.M. *et al.* (Eds), *Petroleum Geochemistry in Exploration of the Norwegian Shelf*, Graham and Trotman, London, pp. 263–266.

Huc, A.Y., Irwin, H. and Schoell, M. 1985. Organic matter quality changes in an Upper Jurassic shale sequence from the Viking Graben, In Thomas, B.M. *et al.* (Eds), *Petroleum Geochemistry in Exploration of the Norwegian Shelf*, Graham and Trotman, London, pp. 179–183.

Hughes, W.B., Holba, A.G., Miller, D.E. and Richardson, J.S. 1985. Geochemistry of Greater Ekofisk crude oils, In Thomas, B.M. *et al.* (Eds), *Petroleum geochemistry in exploration of the Norwegian Shelf*, Graham and Trotman, London, pp. 75–92.

Hurst, A. 1987. Problems of reservoir characterization in some North Sea sandstone reservoirs solved by the application of microscale geological data, In Kleppe, J. *et al.* (Eds), *North Sea Oil and Gas Reservoirs*, Graham and Trotman, London.

Hurst, A. and Buller, A.T. 1984. Dish structures in some Paleocene deep-sea sandstones (Norwegian sector, North Sea): origin of the dish forming clays and their effect on reservoir quality, *Journal of Sedimentary Petrology* **54** (4), 1206–1211.

Illing, L.V. and Hobson, G.D. (Eds). 1981. *Petroleum Geology of the Continental Shelf of North-west Europe*. Heyden, London.

Jacobsen, B. and Spencer, A.M. 1984. Bibliography, In Spencer, A.M. *et al.* (Eds), *Petroleum Geology of the North European Margin*, Graham and Trotman, London, pp. 425–436.

Jacqué, M. and Thouvenin, J. 1975. Lower Tertiary tuffs and volcanic activity in the North Sea, In Woodland, A.W. (Ed.), *Petroleum and the Continental Shelf of North-west Europe*, Vol. 1, Geology. Applied Science, Barking, pp. 455–465.

Jakobsson, K.H., Hamar, G.P., Ormaasen, D.E. and Skarpnes, O. 1980. Triassic facies in the North Sea north of the Central Highs, In *The Sedimentation of the North Sea Reservoir Rocks*, Norsk Petroleumsforening, Article XVIII, 10 pp.

Johnson, H.D. and Stewart, D.J. 1985. Role of clastic sedimentology in the exploration and production of oil and gas in the North Sea, In Brenchley, P.J. and Williams, B.P.J. (Eds), *Sedimentology, Recent Developments and Applied Aspects*, Blackwell, Oxford, pp. 249–310.

Johnson, R.J. and Dingwall, R.G. 1981. The Caledonides; their influence on the stratigraphy of the North-west European continental shelf, In Illing, L.V. and Hobson, G.D. (Eds), *Petroleum Geology of the Continental Shelf of North-west Europe,* Heyden, London, pp. 85–97.

Jones, E.L., Raveling, H.P. and Taylor, H.R. 1975. Statfjord Field, In *Jurassic Northern North Sea Symposium*, Norsk Petroleumsforening, Article 20, 19 pp.

Karlsson T. and Gausland I. 1986. Application of new 3D technology to recent field development in Norway, In *Offshore Northern Seas—Exploration Conference*, Norsk Petroleumsforening, Article G4, 16 pp.

Karlsson, W. 1986. The Snorre, Statfjord and Gullfaks oilfields and the habitat of hydrocarbons on the Tampen Spur, offshore Norway, In Spencer, A.M. *et al.* (Eds), *Habitat of Hydrocarbons on the Norwegian Continental Shelf*, Graham and Trotman, London, pp. 181–197.

Karlsson, W., Vollset, J., Bjørlykke, K. and Jørgensen, P. 1979. Changes in mineralogical composition of Tertiary sediments from North Sea wells, In Mortland, M.M. and Farmer, V.C. (Eds), *International Clay Conference 1978*, Elsevier, pp. 281–289.

Kemp, A.C. and Grindstad, K.C. 1981. Computer modelling of a faulted reservoir, *51st Annual Meeting, Society of Exploration Geophysicists*, Vol. 5, pp. 2935–2951.

Kennedy, W.J. 1980. Aspects of chalk sedimentation in the southern Norwegian offshore, In *The Sedimentation of the North Sea Reservoir Rocks*, Norsk Petroleumsforening, Article IX, 29.

Kent, P.E. 1975. The tectonic development of Great Britain and the surrounding seas, In Woodland, A.W. (Ed.), *Petroleum and the Continental Shelf of North-west Europe,* Vol. 1,*Geology*, Applied Science, Barking, pp. 3–28.

Kent, P.E. 1980. The North Sea—evolution of a major oil and gas play, In Miall, A.D. (Ed.), *Facts and Principles of World Petroleum Occurrence, Memoir of the Canadian Society of Petroleum Geologists* **6,** 633–652.

Kessler, L.G., Zang, R.D., Englehorn, J.A. and Eger, J.D. 1980. Stratigraphy and sedimentology of a Paleocene submarine fan complex, Cod Field, Norwegian North Sea, In *The Sedimen-*

tation of the North Sea Reservoir Rocks, Norsk Petroleumsforening, Article VIII, 19 pp.

Kirk, R.H. 1980. Statfjord Field—a North Sea giant, In Halbouty, M.T. (Ed.), *Giant Oil and Gas Fields of the Decade 1968–1978, Memoir of the American Association of Petroleum Geologists* **30,** 95–116.

Knox, R.W.O'B., Morton, A.C. and Harland, R. 1981. Stratigraphical relationships of Palaeocene sands in the UK sector of the central North Sea, In Illing, L.V. and Hobson, G.D. (Eds), *Petroleum Geology of the Continental Shelf of North-west Europe*, Heyden, London, pp. 267–281.

Larese, R.E., Haskell, N.L., Prezbindowski, D.R. and Beju, D. 1984. Porosity development in selected Jurassic sandstones from the Norwegian and North Seas, Norway—an overview, In Spencer, A.M. *et al.* (Eds), *Petroleum Geology of the North European Margin*, Graham and Trotman, London, pp. 81–95.

Larsen, R.M. and Jaarvik, L.J. 1981. The geology of the Sleipner Field complex, In *Norwegian Symposium on Exploration*, Norsk Petroleumsforening, Article 15, 31 pp.

Larsen, V., Aasheim, S.M. and Masset, J.M. 1981. 30/6—Alpha structure, a field case study in the Silver block, In *Norwegian Symposium on Exploration*, Norsk Petroleumsforening, Article 14, 36 pp.

Larsen, V., Mørkeseth, P.O. and Aasheim, S.M. 1987. Tyrihans, In Spencer, A.M. *et al.* (Eds), *Geology of the Norwegian Oil and Gas Fields*, Graham and Trotman, London, pp. 411–418.

Larter, S.R. 1985. Integrated kerogen typing in the recognition and quantitative assessment of petroleum source rocks, In Thomas, B.M. *et al.* (Eds), *Petroleum Geochemistry in Exploration of the Norwegian Shelf*, Graham and Trotman, London, pp. 269–286.

Leadholm, R.H., Ho, T.T.Y. and Sahai, S.K. 1985. Heat flow, geothermal gradients and maturation modelling on the Norwegian continental shelf using computer methods, In Thomas, B.M. *et al.* (Eds), *Petroleum Geochemistry in Exploration of the Norwegian Shelf*, Graham and Trotman, London, pp. 131–143.

Leckie, G.G. 1982. Lithology and subsidence in the North Sea, In Kent, P.E., Bott, M.H.P., Mckenzie, D.P. and Williams, C.A. (Eds), *The Evolution of Sedimentary Basins, Philosophical Transactions of the Royal Society, London,* **A, 305,** 85–99.

Leebeeck, A.D. 1987. The Frigg Field reservoir: characteristics and performance, In Kleppe, J. *et al.* (Eds), *North Sea Oil and Gas Reservoirs*, Graham and Trotman, London.

Leeder, M.R. 1983. Lithospheric stretching and North Sea Jurassic clastic sourcelands, *Nature, London* **305,** 510–513.

Leonard, R.C. and Munns, J.W. 1987. Valhall, In Spencer, A.M. *et al.* (Eds), *Geology of the Norwegian Oil and Gas Fields*, Graham and Trotman, London, pp. 153–163.

Lilleng, T. 1980. Lower Tertiary (Danian–Lower Eocene) reservoir sand developments between 59° 30′N and 61° 30′ of the northern North Sea, In *The Sedimentation of the North Sea Reservoir Rocks*, Norsk Petroleumsforening, Article VI, 22 pp.

Lindberg, P., Riise, R. and Fertl, W.H. 1980. Occurrence and distribution of over pressures in the northern North Sea area, *Society of Petroleum Engineers, 55th Annual Conference*, SPE 9339.

Lovell, J.P.B. 1984. Cenozoic, In Glennie, K.W. (Ed.), *Introduction to the Petroleum Geology of the North Sea*, Blackwell, Oxford, pp. 151–169.

Malm, O.A., Christensen, O.B. Østby, K.L., Furnes, H., Løvlie, R. and Rueslåtten, H. 1984. The Lower Tertiary Balder Formation: an organogenic and tuffaceous deposit in the North Sea region, In Spencer, A.M. *et al.* (Eds), *Petroleum Geology of the North European Margin*, Graham and Trotman, London, pp. 149–170.

Malm, O.A., Furnes, H. and Bjørlykke, K. 1979. Volcaniclastics of Middle Jurassic age in the Statfjord oil field of the North Sea, *Neues Jahrbuch für Geologie und Paleontologie Monatschefte* **10,** 607–618.

Mapstone, N.B. 1975. Diagenetic history of a North Sea chalk, *Sedimentology* **22,** 601–614.

Maragna, B., Zaro, G. and Pessina, P. 1985. Block 33/6 geochemical evaluation, In Thomas, B.M. *et al.* (Eds), *Petroleum Geochemistry in Exploration of the Norwegian shelf*, Graham and Trotman, London, pp. 197–204.

Massie, I., Beardall, T.J., Hemmens, P.D. and Fox, M.J. 1985. Murchison: a review of reservoir performance during the first five years, *Paper at 60th Annual Meeting, Society of Petroleum Engineers*, SPE 14343, 5 pp.

McGovney, J.E. and Radovich, B.J. 1985. Seismic stratigraphy and facies of the Frigg Fan complex, In Berg, O.R. and Woolverton, D.G. (Eds), *Seismic stratigraphy II, Memoir of the American Association of Petroleum Geology* **39,** 139–156.

Michaud, F. 1987. Eldfisk, In Spencer, A.M. *et al.* (Eds), *Geology of the Norwegian Oil and Gas Fields*, Graham and Trotman, London, pp. 89–105.

Michelsen, O. 1977. Jurassic deposits of the Norwegian–Danish basin, In *Mesozoic Northern North Sea Symposium*, Norsk Petroleumsforening, Article 6, 26 pp.

Michelsen, O. 1978. Stratigraphy and distribution of Jurassic deposits of the Norwegian–Danish Basin, *Danmarks Geologiske Undersøgelse*, Series B, 2, 28 pp.

Minturn, L.W. 1982. Ekofisk: first of the giant oil fields in western Europe: discussion. *Bulletin of the American Association of Petroleum Geologists* **66** (9), 1408–1411.

Morton, A.C. 1982. Lower Tertiary sand development in Viking Graben, North Sea, *Bulletin of the American Association of Petroleum Geologists* **66** (10) 1542–1559. Discussion, by J. Williams **67** (11), 2150–2151.

Morton, A.C. and Humphreys, B. 1983. The petrology of the Middle Jurassic sandstones from the Murchison Field, North Sea, *Journal of Petroleum Geology* **5** (3), 245–260.

Mudge, D.C. and Bliss, G.M. 1983. Stratigraphy and sedimentation of the Paleocene sands in the Northern North Sea, In Brooks, J. (Ed), *Petroleum Geochemistry and Exploration of Europe*, Special Publication Geological Society London **12,** 95–111.

Munns, J.W. 1984. The Valhall field—a new geological description, In *Offshore Northern Seas—reduction of Uncertainties by Innovative Reservoir Geo-modelling*, Norsk Petroleumsforening, Article G7, 54.

Munns, J.W. 1985. The Valhall field: a geological overview, *Marine and Petroleum Geology* **2** (1), 23–43.

Mure, E. 1987a. Frigg, In Spencer, A.M. *et al.* (Eds), *Geology of the Norwegian Oil and Gas Fields*, Graham and Trotman, London, pp. 203–213.

Mure, E. 1987b, Øst Frigg Alpha and Beta, In Spencer, A.M. *et al.* (Eds), *Geology of the Norwegian Oil and Gas Fields*, Graham and Trotman, London, pp. 215–222.

Mure, E. 1987c. Nord-Øst Frigg, In Spencer, A.M. *et al.* (Eds), *Geology of the Norwegian Oil and Gas Fields*, Graham and Trotman, London, pp. 223–228.

Mure, E. 1987d. Heimdal, In Spencer, A.M. *et al.* (Eds), *Geology of the Norwegian Oil and Gas Fields*, Graham and Trotman, London, pp. 229–234.

Myrland, R., Messel, K. and Ræstad, N. 1981. Gas discovery in 35/3—an exploration history, In *Norwegian Symposium on Exploration*, Norsk Petroleumsforening, Article 17, 17 pp.

Nagy, J. 1985. Jurassic foraminiferal facies in the Statfjord area, Northern North Sea.—I. *Journal of Petroleum Geology* **8** (3), 273–295.

Nagy, J. 1985. Jurassic foraminiferal facies in the Statfjord area, Northern North Sea—II. *Journal of Petroleum Geology* **8** (4), 389–404.

Nagy, J., Dypvik, H. and Bjærke, T. 1984. Sedimentological and paleontological analyses of Jurassic North Sea deposits from deltaic environments, *Journal of Petroleum Geology* **7** (2), 169–188.

Nielsen, O.B., Sørensen, S., Thiede, J. and Skarbø, O. 1986. Cenozoic differential subsidence of North Sea, *Bulletin of the American Association of Petroleum Geologists* **70** (3), 276–298.

Nipen, O. 1987. Oseberg, In Spencer, A.M. *et al.* (Eds), *Geology of the Norwegian Oil and Gas Fields*, Graham and Trotman, London, pp. 379–387.

Norbury, I. 1987. Hod, In Spencer, A.M. *et al.* (Eds), *Geology of the Norwegian Oil and Gas Fields*, Graham and Trotman, London, pp. 107–116.

Nordberg, H.E. 1981. Seismic Hydrocarbon Indicators in the North Sea, In *Norwegian Symposium on Exploration*, Norsk Petroleumsforening, Article 8, 40 pp.

Nordgård Bolås, H.M. 1987. Odin, In Spencer, A.M. *et al.* (Eds), *Geology of theNorwegian Oil and Gas Fields*, Graham and Trotman, London, pp. 235–242.

Northam, M.A. 1985. Correlation of northern North Sea oils: the different facies of their Jurassic source, In Thomas, B.M. *et al.* (Eds), *Petroleum Geochemistry in Exploration of the Norwegian Shelf*, Graham and Trotman, London, pp. 93–99.

Norwegian Petroleum Directorate Annual reports 1973, 1974, 1975, 1976, 1977, 1978, 1979, 1980, 1981, 1982, 1983, 1984, 1985.

Norwegian Petroleum Directorate Papers.

1. Lithology, Well 8/3-1. 1975.
2. Lithology, Well 25/11-1. 1976.
3. Lithology, Well 16/2-1. 1976.
4. Lithology, Well 16/11-1. 1977.
5. Lithology, Well 9/8-1. 1977.
6. Lithology, Well 16/7-1. 1977.
7. Lithology, Wells 2/11-1 and 2/8-1. 1977.
8. Lithology, Well 16/1-1. 1977.
9. Lithology, Well 16/6-1. 1977.
10. Lithology, Wells 7/11-1, 7/11-2, 7/11-3 and 7/11-4. 1977.
11. Lithology, Well 16/9-1. 1977.
12. Lithology. Well 17/11-1. 1977. Compiler, Randi C. Olsen.
13. Lithology, Well 2/8-2. 1977. Compiler, Roald Riise.
14. Lithology, Well 17/4-1. 1978. Compiler, Randi C. Olsen.
15. Lithology, Wells 1/3-1 and 1/3-2. 1978. Compiler, Svein Frodesen.
16. Lithology, Well 7/12-1. 1978. Compiler, Roald Riise.
17. Lithology, Wells 2/3-1, 2/3-2 and 2/3-3. 1978. Compiler, Roald Riise.
18. Lithology, Well 7/8-1. 1978. Compiler Inger Flesland Strass.
19. Lithology, Well 2/6-1. 1979. Compiler, Svein Frodesen.
20. Lithology, Well 7/3-1. 1979. Compiler, Inger Flesland Strass.
21. Lithology, Well 17/10-1. 1979. Compiler, Randi C. Olsen.
22. Lithology, Well 8/10-1. 1979. Compiler, Svein-Erik Sjulsen.
23. Lithology, Well 11/10-1. 1979. Compiler, Inger Flesland Strass.
24. Lithology, Wells 9/4-1, 9/4-2 and 9/4-3. 1979. Compiler, Svein Frodesen.
25. Lithology, Wells 2/4-1, 2/4-2, 2/4-3, 2/4-4 and 2/4-5. 1980. K. Ofstad.
26. Lithology, Well 10/8-1. 1980. I.F. Strass.
27. Lithology, Well 9/12-1. 1980. R.C. Olsen.
28. The Balder Area. 1981. S. Frodesen, Aa. Moe, K. Ofstad, E. Ormaasen, S.-E. Sjulsen, K. Ulleberg.
29. The Heimdal Area. Wells 25/4-1, 25/4-2, 25/4-3 and 25/4-4. 1981. S. Frodesen.
30. The Eldfisk Area. Wells 2/7-1, 2/7-2, 2/7-3, 2/7-6, 2/7-7, 2/7-8 and 2/7-9. 1981. K. Ofstad.

31. The Norwegian–Danish Basin. Wells 7/11-1, 7/8-2, 7/9-1, 8/1-1, 8/4-1, 8/9-1, 8/11-1, 8/12-1, 9/4-4, 9/10-1, 9/11-1, 10/5-1, 11/9-1, 16/8-1, 16/11-2, 17/9-1, 17/11-2, 17/12-1, 17/12-2, 18/11-1. 1982. R.C. Olsen, I.F. Strass.

32. Geology of the southernmost part of the Norwegian section of the Central Trough. Wells 2/7-11, 2/8-3, -4, -6, -7, -8, -9, -10, -11, 2/9-1, 2/10-1, 2/11-2 and 3/7-1. 1983. K. Ofstad.

Nyberg, I.T. 1987. Statfjord Øst, In Spencer, A.M. *et al.* (Eds), *Geology of the Norwegian Oil and Gas Fields*. Graham and Trotman, London, pp. 351–362.

Oljedirektoratet: Well Data Summary Sheets.
Volume 1: Wells completed prior to September 1970. Issued January 1976.
Volume 2: Wells completed between October 1970 and December 1971. Issued December 1976.
Volume 3: Wells completed between November 1971 and December 1972. Issued December 1977.
Volume 4: Wells completed 1973. Issued December 1978.
Volume 5: Wells completed 1974. Issued December 1979.
Volume 6: Wells completed 1975. Issued December 1980.
Volume 7: Wells completed 1976. Issued December 1981.
Volume 8: Wells completed 1977. Issued December 1982.
Volume 9: Wells completed 1978. Issued December 1984.
Volume 10: Wells completed 1979. 2 volumes 1985.
Volume 11: Wells completed 1980. 2 volumes 1986.

Olje- og energidepartementet (The Royal Ministry of Petroleum and Energy, Norway) 1986. Faktaheftet (Fact sheet, the Norwegian Continental Shelf). (Now published annually.)

Ormaasen, D.E., Hamar, G.P., Jakobsson, K.H. and Skarpnes, O. 1980. Permo-Triassic correlations in the North Sea area north of the Central Highs, In *The Sedimentation of the North Sea Reservoir Rocks*, Norsk Petroleumsforening, Article XIX, 15 pp.

Osborne, P. and Evans, S. 1987. The Troll Field, In Kleppe, J. *et al.* (Eds), *North Sea Oil and Gas Reservoirs*, Graham and Trotman, London, pp.

Parker, J.R. 1975. Lower Tertiary sand development in the central North Sea, In Woodland, A.W. (Ed.), *Petroleum and the Continental Shelf of North-west Europe*, Vol. 1, *Geology*, Applied Science, Barking, pp. 447–453.

Parsley, A.J. 1984. North Sea hydrocarbon plays, In Glennie, K.W. (Ed.), *Introduction to the Petroleum Geology of the North Sea*, Blackwell, Oxford, pp. 205–230.

Pegrum, R.M. 1984. The extension of the Tornquist Zone in the Norwegian North Sea, *Norsk Geolisk Tidsskrift* **64** (1), 39–68.

Pegrum, R.M. 1984. Structural development of the south-western margin of the Russian–Fennoscandian Platform, In Spencer, A.M. *et al.* (Eds), *Petroleum geology of the North European Margin*, Graham and Trotman, London, pp. 359–369.

Pegrum, R.M. and Ljones, T.E. 1984. 15/9 Gamma gas field offshore Norway, new trap type for North Sea basin with regional structural implications, *Bulletin of the American Association of Petroleum Geologists* **68** (7), 874–902.

Pekot, L.J. and Gersib, G.A. 1987. Ekofisk, In Spencer, A.M. *et al.* (Eds), *Geology of the Norwegian Oil and Gas Fields*, Graham and Trotman, London, pp. 73–87.

Ramberg, I.B. and Smithson, S.B. 1975. Geophysical interpretation of crustal structure along the south-eastern coast of Norway and Skagerrak, *Bulletin of the Geological Society of America* **86,** 769–774.

Ranaweera, H.K.A. 1987. Sleipner Vest, In Spencer, A.M. *et al.* (Eds), *Geology of the Norwegian Oil and Gas Fields*, Graham and Trotman, London, pp. 253–264.

Rathore, J.S. and Hospers, J. 1986. A lineaments study of southern Norway and adjacent offshore areas, *Tectonophysics* **131**, 257–285.

Rawson, P.F. and Riley, L.A. 1982. Latest Jurassic–Early Cretaceous events and the 'Late Cimmerian unconformity' in North Sea area, *Bulletin of the American Association of Petroleum Geologists* **66** (12), 2628–2648.

Reeder, M.L. and Scotchman, I.C. 1985. Hydrocarbon generation—central and northern North Sea, *Oil and Gas Journal* **83** (12), 137–144.

Reinholdtsen, B. 1984. Improved recovery from, thin oil sands Part 2—Troll field, In *Offshore Northern Seas—Improved offshore Recovery,* Norsk Petroleumsforening, Article R2, 18 pp.

Revoy, M. 1984. Frigg field: Production history and seismic response, In *Offshore Northern Seas —Reduction of uncertainties by innovative reservoir geo-modelling*, Norsk Petroleumsforening, Article G5, 44 pp.

Rickards, L.M. 1974. The Ekofisk area, discovery to development, In *Offshore North Sea—Exploration Geology and Geophysics*, Article G-IV/3, 16 pp.

Roberts, J.D., Mathieson, A.S. and Hampson, J.M. 1987. Statfjord, In Spencer, A.M. *et al.* (Eds), *Geology of the Norwegian Oil and Gas Fields*, Graham and Trotman, London, pp. 319–340.

Rochow, K.A. 1981. Seismic stratigraphy of the North Sea 'Palaeocene' deposits, In Illing, L.V. and Hobson, G.D. (Eds), *Petroleum Geology of the Continental Shelf of North-west Europe*, Heyden, London, pp. 255–266.

Rokoengen, K. and Ronningsland,T.M. 1983. Shallow bedrock geology and Quarternary thickness in the Norwegian sector of the North Sea between 60° 30′N and 62° N. *Norsk Geolisk Tidsskrift*, **63** (2–3), 83–102.

Rounsaville, R.R. 1986. Preliminary reservoir development evaluation—Snorre Field, In *Offshore Northern Seas—Reservoir Engineering Conference*, Norsk Petroleumsforening, Article

R1, 30 pp.
Røe, S.L. and Steel, R. 1985. Sedimentation, sea-level rise and tectonics at the Triassic–Jurassic boundary (Statfjord Formation), Tampen Spur, northern North Sea, *Journal of Petroleum Geology* **8** (2), 163–186.
Rønnevik, H., Bergsager, E.I., Moe, A., Øvrebø, O., Navrestad, T. and Stangenes, J. 1975. The geology of the Norwegian Continental Shelf, In Woodland, A.W. (Ed.), *Petroleum and the Continental Shelf of North-west Europe,* Vol. 1, *Geology,* Applied Science, Barking, pp. 117–130.
Rønnevik, H., Eggen, S. and Vollset, J. 1983. Exploration of the Norwegian Shelf, In Brooks, J. (Ed.), *Petroleum Geochemistry and Exploration of Europe*, Special Publication Geological Society London **12,** 71–93. (Also available as *NPD Contribution* no. 5, 43 pp).
Rønnevik, H. and Johnsen, S. 1984. Geology of the Greater Troll Field area, *Oil and Gas Journal* **82** (4), 100–106.
Rønnevik, H.C., Johnsen, S. and Al-Kasim, F. 1983. Geology of the Greater Troll area, In *Offshore Northern Seas advanced projects conference—The Troll Field,* Article T3, 25 pp.
Rønnevik, H.C., van den Bosch, W. and Bandlien, E.H. 1975. A proposed nomenclature for the main structural features in the Norwegian North Sea, In *Jurassic Northern North Sea Symposium*, Norsk Petroleumsforening, Article 18, 16 pp.
Rønning, K., Johnston, C.D., Johnstad, S.E. and Songstad, P. 1986. Geology of the Hild Field, In Spencer, A.M. *et al.* (Eds), *Habitat of Hydrocarbons on the Norwegian Continental Shelf,* Graham and Trotman, London, pp. 199–206.
Rønning, K., Johnston, C.D., Johnstad, S.E. and Songstad, P. 1987. Hild, In Spencer, A.M. *et al.* (Eds), *Geology of the Norwegian Oil and Gas Fields*, Graham and Trotman, London, pp. 287–294.
Rønning, K. and Steel, R. 1987. Depositional sequences within a 'transgressive' reservoir sandstone unit: the Middle Jurassic Tarbert Formation, Hild area, northern North Sea, In Kleppe, J. *et al.* (Eds), *North Sea Oil and Gas Reservoirs*, Graham and Trotman, London.
Saeland, G.T. and Simpson, G.S. 1981. 3-D seismic data help to delineate subunconformity trap in Norwegian N. Sea, *Oil and Gas Journal* **79** (45), 381–398.
Saeland, G.T. and Simpson, G.S. 1982. Interpretation of 3-D Data in delineating a subunconformity trap in block 34/10, Norwegian North Sea, In Halbouty, M.T. (Ed.), *The deliberate search for the subtle trap. Memoir of the American Association of Petroleum Geologists* **32,** 217–236.
Sarg, J.F. and Skjold, L.J. 1982. Stratigraphic traps in Paleocene sands in the Balder Area, North Sea, In Halbouty, M.T. (Ed.), *The deliberate search for the subtle trap, Memoir American Association Petroleum Geologists* **32,** 197–206.
Schatzinger, R.A., Feazel, C.T. and Henry, W.E. 1985. Evidence of resedimentation in chalk from the Central Graben, North Sea, *Proceedings SEPM Core Workshop No. 6, Deep Water Carbonates*, pp. 342–385.
Schiener, E.J., Stober, G. and Faber, E. 1985. Surface geochemical exploration for hydrocarbons in offshore areas—principles, methods and results, In Thomas, B.M. *et al.* (Eds), *Petroleum geochemistry in Exploration of the Norwegian Shelf*, Graham and Trotman, London, pp. 223–238.
Scholle, P.A. 1977. Chalk diagenesis and its relation to petroleum exploration: a modern miracle? *Bulletin of the American Association of Petroleum Geologists* **6** (7), 982–1009.
Schou, L., Eggen, S. and Schoell, M. 1985. Oil–oil and oil–source-rock correlation, northern North Sea, In Thomas, B.M. *et al.* (Eds), *Petroleum Geochemistry in Exploration of the Norwegian Shelf*, Graham and Trotman, London, pp. 101–117.
Sclater, J.G. and Christie, P.A.F. 1980. Continental stretching: an explanation of the post-Mid-Cretaceous subsidence of the Central North Sea Basin, *Journal of Geophysical Research* **85,** 3711–3739.
Scull, B.J. 1975. Geochemical surveys for oil and gas, Norway, *IX International Congress of Sedimentology, Proceedings*, theme 3, pp. 79–83.
Sellevoll, M.A. 1973. Mohorovičić discontinuity beneath Fennoscandia and adjacent parts of the Norwegian Sea and the North Sea, *Tectonophysics* **20,** 359–366.
Sellevoll, M.A. and Aalstad, I. 1971. Magnetic measurements and seismic profiling in the Skagerrak. *Marine Geophysical Researches* **1,** 284–302.
Sellevoll, M.A. and Sundvor, E. 1974. The origin of the Norwegian Channel—a discussion based on seismic measurements, *Canadian Journal of Earth Science* **11,** 224–231.
Selley, R.C. 1978. Porosity gradients in North Sea oil-bearing sandstones, *Journal of the Geological Society, London* **135,** 119–132.
Sharma, P.V. 1970. Geophysical evidence for a buried volcanic mount in the Skagerrak, *Bulletin of the Geological Society of Denmark* **19,** 368–377.
Shephard, F.P. 1931. Glacial troughs of the continental shelves, *Journal of Geology* **39,** 345–360.
Simpson, R.D.H. and Whitley, P.K.J. 1981. Geological input to reservoir simulation of the Brent Formation, In Illing, L.V. and Hobson, G.D. (Eds), *Petroleum Geology of the Continental Shelf of North-west Europe*, Heyden, London, pp. 310–314.
Skaar, F.E. 1973. Recent scientific investigations and literature. The Norwegian Shelf and adjacent areas, *NTNFK, Oslo, Publication* no. 33, 86 pp.
Skarpnes, O., Briseid, E. and Milton, D.I. 1982. The 34/10 Delta prospect of the Norwegian North Sea: Exploration study of an unconformity trap, In Halbouty, M.T. (Ed.), *The deliberate search for the subtle trap, Memoir of the American*

Association of Petroleum Geologists **32,** 207–216.

Skarpnes, O., Hamar, G.P., Jakobsson, K.H. and Ormassen, D.E. 1980. Regional Jurassic setting of the North Sea north of the Central Highs, In *The Sedimentation of the North Sea Reservoir Rocks*, Norsk Petroleumsforening, Article XIII, 8 pp.

Skaug, M. and Gundesø, R. 1986. Geological modelling of the Frigg field with special emphasis on shale mapping, *Society of Petroleum Engineers, European Petroleum Conference Proceedings* pp. 111–121 (SPE 15859).

Skjerven, J., Rijs, F. and Kalheim, J.E. 1983. Late Palaeozoic to Early Cenozoic structural development of the south-south-eastern Norwegian North Sea, In Kaasschieter, J.P.H. and Reijers, T.J.A. (Eds), *Petroleum Geology of the South-eastern North Sea and the Adjacent Onshore Areas, Geologie en Mijnbouw* **62** (1), 35–45.

Skjold, L.J. 1980. Paleocene sands of the Balder Field, In *The Sedimentation of the North Sea Reservoir Rocks*, Norsk Petroleumsforening, Article VII, 7 pp.

Skovbro, B. 1983. Depositional conditions during chalk sedimentation in the Ekofisk area Norwegian North Sea, In Kaasschieter, J.P.H. and Reijers, T.J.A. (Eds), *Petroleum Geology of the south-eastern North Sea and the Adjacent Onshore Areas, Geol. Mijnbouw* **62** (1), 169–175.

Sorgenfrei, T. 1969. Geological perspectives in the North Sea area, *Bulletin of the Geological Society of Denmark* **19,** 160–196.

Spencer, A.M., Home, P.C. and Wiik, V. 1986. Habitat of hydrocarbons in the Jurassic Ula Trend, Central Graben, Norway, In Spencer, A.M. *et al.* (Eds), *Habitat of Hydrocarbons on the Norwegian Continental Shelf*, Graham and Trotman, London, pp. 111–127.

Strand, J.E. and Slatsveen, P.R. 1987. 7/11-5, In Spencer, A.M. *et al.* (Eds), *Geology of the Norwegian Oil and Gas Fields*, Graham and Trotman, London, pp. 177–183.

Sundvor, E. and Nysæther, E. 1975. Geological outline of the Norwegian continental margin between 60° and 68° north, *Memoir of the Canadian Society of Petroleum Geologists* **4,** 267–281.

Sørensen, S., Jones,M., Hardman, R.F.P, Leutz, W.K. and Schwarz, P.H. 1986. Reservoir characteristics of high- and low-productivity chalks from the Central North Sea, In Spencer, A.M. *et al.* (Eds), *Habitat of Hydrocarbons on the Norwegian Continental Shelf*, Graham and Trotman, London, pp. 91–110.

Tansø, A., King, G.E. and Holman, G.B. 1983. Production from high porosity chalk, In *Offshore Northern Seas—Reservoir Engineering Conference*, Norsk Petroleumsforening, Article R3, 34 pp.

Taylor, J.C.M. 1981. Zechstein facies and petroleum prospects in the central and northern North Sea, In Illing, L.V. and Hobson, G.D. (Eds), *Petroleum Geology of the Continental Shelf of North-west Europe*, Heyden, London, pp. 176–185.

Taylor, J.C.M. 1984. Late Permian–Zechstein, In Glennie, K.W. (Ed.), *Introduction to the Petroleum Geology of the North Sea*, Blackwell, Oxford, pp. 61–83.

Thomas, B.M., Møller-Pedersen, P., Whitaker, M.F. and Shaw, N.D. 1985. Organic Facies and hydrocarbon distributions in the Norwegian North Sea, In Thomas, B.M. *et al.* (Eds), *Petroleum Geochemistry in Exploration of the Norwegian Shelf*, Graham and Trotman, London, pp. 3–26.

Thouand, A., Brewster, J. and Mure, E. 1986. Frigg Field mid-life appraisal programme, In *Offshore Northern Seas—Reservoir Engineering Conference*, Norsk Petroleumsforening, Article R4, 34 pp.

Torvund, T. 1986. Oseberg—late stage developments, In *Offshore Northern Seas—Reservoir Engineering Conference*, Norsk Petroleumsforening, Article R2, 21 pp.

Torvund, T. and Nipen, O. 1987. The Oseberg Field: potential for increased oil recovery by gas injection in a high-permeability environment, In Kleppe, J. *et al.* (Eds), *North Sea Oil and Gas Reservoirs*, Graham and Trotman, London.

Ungerer, P., Chiarelli, A. and Oudin, J.L. 1985. Modelling of petroleum genesis and migration with a bidimensional computer model in the Frigg sector, Viking Graben, In Thomas, B.M. *et al.* (Eds), *Petroleum Geochemistry in Exploration of the Norwegian Shelf* Graham and Trotman, London, pp. 121–129.

Vail, P.R., Mitchum, R.M. and Todd, R.G. 1977. An eustatic model for the North Sea during the Mesozoic, In *Mesozoic Northern North Sea Symposium*, Norsk Petroleumsforening, Article 12, 35 pp.

Vail, P.R. and Todd, R.G. 1981. Northern North Sea Jurassic unconformities, chronostratigraphy and sea-level changes from seismic stratigraphy, In Illing, L.V. and Hobson, G.D. (Eds), *Petroleum Geology of the Continental Shelf of North-west Europe*, Heyden, London, pp. 216–235.

Van den Bark, E. and Thomas, O.D. 1980. Ekofisk: first of the giant oil fields in western Europe, *Memoir of the American Association of Petroleum Geologists* **30,** 195–224; reprinted in *Bulletin of the American Association of Petroleum Geologists* **65** (11), 2341–2363.

Vollset, J. and Doré, A.G. (Eds). 1984. A revised Triassic and Jurassic Lithostratigraphic Nomenclature for the Norwegian North Sea, *Norwegian Petroleum Directorate Bulletin* **3.**

Warner, M.R. 1987. Seismic reflections from the Moho—the effect of isostasy, *Geophysical Journal of the Royal Astronomical Society* **88,** 425–435.

Watts, N.L. 1983. Microfractures in chalks of Albuskjell field, Norwegian Sector, North Sea: possible origin and distribution, *Bulletin of the American Association of Petroleum Geologists* **67** (2), 201–234.

Watts, N.L., Lapré, J.F., Van Schijndel-Goester, F.S. and Ford, A. 1980. Upper Cretaceous and Lower Tertiary chalks of the Albuskjell area, North Sea: deposition in a slope and a base-of-slope environment, *Geology* **8** (5), 217–221.

Weigel, W., Hjelme, J. and Sellevoll, M. 1970. A refraction profile through the Skagerrak from northern Jutland to southern Norway, *Meddelelser Geodaetisk Institott København* **45**.

Whitaker, M.F. 1984. The usage of palynostratigraphy and palynofacies in definition of Troll field geology, In *Offshore Northern Seas—Reduction of uncertainties by innovative reservoir geo-modelling*, Norsk Petroleumsforening, Article G6, 28 pp.

Whiteman, A.J., Rees, G., Naylor, D. and Pegrum, R.M. 1975. North Sea troughs and plate tectonics, In Whiteman, A.J., Roberts, D. and Sellevoll, M.A. (Eds), *Petroleum Geology and Geology of the North Sea and North-east Atlantic Continental Margin, Norges geol. Unders.* **316,** 137–161.

Wiborg, R. and Jewhurst, M. 1985. Ekofisk—reservoir management in a compacting environment, In *Offshore Northern Seas—Advanced Projects conference—Technology Applications—Today and Tomorrow*. Article A5, 24 pp.

Williams, J.J. 1977. Review of the structures and stratigraphy of oil occurrences in the northern North Sea, In *Mesozoic Northern North Sea Symposium*, Norsk Petroleumsforening, Article 13, 25 pp.

Wood, R. and Barton, P. 1983. Crustal thinning and subsidence in the North Sea, *Nature, London* **302,** 134–136.

Woodland, A.W. (Ed.). 1975. *Petroleum and the Continental Shelf of North-west Europe,* Vol. 1, *Geology*, Applied Science, Barking.

Younge, B., Myrland, R., Taylor, H. and Ruble, C.L. 1975. A preliminary report from the Jurassic stratigraphic nomenclature subcommittee in Norway, In *Jurassic Northern North Sea Symposium*, Norsk Petroleumsforening, Article 19, 14 pp.

Zagar, J.J. 1984. Improved recovery from thin oil sands. Part 3—Balder field, In *Offshore Northern Seas—Improved Offshore Recovery*, Norsk Petroleumsforening, Article R2, 8 pp.

Ziegler, P.A. 1975. North Sea basin history in the tectonic framework of north-western Europe, In Woodland, A.W. (Ed.), *Petroleum and the Continental Shelf of North-west Europe*, Vol. 1, *Geology*, Applied Science, Barking, pp. 131–149.

Ziegler, P.A. 1975. Geologic evolution of the North Sea and its tectonic framework, *Bulletin of the American Association of Petroleum Geology* **59,** 1073–1097.

Ziegler, P.A. 1975. The geological evolution of the North Sea area in the tectonic framework of north-western Europe, In Whiteman, A.J., Roberts, D. and Sellevoll, M.A. (Eds), *Petroleum Geology and Geology of the North Sea and North-east Atlantic Continental Margin, Norges Geol. Unders.* **36,** 1–27.

Ziegler, P.A. 1977. Geology and hydrocarbon provinces of the North Sea, *GeoJournal* **1,** 7–32.

Ziegler, P.A. 1978. North Sea rift and basin development, In Ramberg, I.B. and Neumann, E-R. (Eds), *Tectonics and Geophysics of Continental Rifts*, NATO Advanced Study Institute, Series C, Mathematical Physical Science, Reidel, Dordrecht, pp. 249–277.

Ziegler, P.A. 1978. North-western Europe: tectonics and basin development, *Geologie en Mijnbouw* **57,** 589–626.

Ziegler, P.A. 1979. Factors controlling North Sea hydrocarbon accumulations, *World Oil* **189** (6), 111–124.

Ziegler, P.A. 1980. North-west European basin: geology and hydrocarbon provinces, In Miall, A.D. (Ed.), *Facts and Principles of World Petroleum Occurrence*, Canadian Society of Petroleum Geologists, Special Memoir **6,** pp. 653–706.

Ziegler, P.A. 1980. North-Western Europe: subsidence patterns of post-Variscan basins, In Cogne, J. and Slansky, M. (Eds), *Géologie de l'Europe du Pré-cambrien aux bassins sédimentaires post-Hercyniens*, Mémoire Bureau Recherche Géologique minières, No. 108, 249–280.

Ziegler, P.A. 1981. Evolution of sedimentary basins in north-west Europe, In Illing, L.V. and Hobson, G.D. (Eds), *Petroleum Geology of the Continental Shelf of North-west Europe*, Heyden, London, pp. 3–39.

Ziegler, P.A. 1982. *Geological Atlas of Western and Central Europe*, Shell Internationale Petroleum Maatschappij B.V., The Hague.

Ziegler, P.A. 1982. Faulting and graben formation in western and central Europe, In Kent, P.E., Bott, M.H.P., McKenzie, D.P. and Williams, C.A. (Eds), *The Evolution of Sedimentary Basins, Philosophical Transactions of the Royal Society, London Series A* **305,** 113–143.

Ziegler, P.A. 1985. Late Caledonian framework of western and Central Europe, In Gee, D.G. and Sturt, B.A. (Eds), *The Caledonide Orogen — Scandinavia and Related Areas*. Wiley Interscience, Chichester. Part I, pp. 3–18.

Ziegler, P.A. and Louwerens, C.J. 1979. Tectonics of the North Sea, In Oele, E., Schuttenhelm, R.T.E. and Wiggers, A.J. (Eds), *The Quaternary History of the North Sea. Acta Univ. Ups. Symp. Univ. Ups. Annum Quingentesimum Celebrantis*: 2. Uppsala, 7–22.

Ziegler, W.H. 1975. Outline of the geological history of the North Sea, In Woodland, A.W. (Ed.), *Petroleum and the Continental Shelf of North-west Europe*, Vol. 1, *Geology*. Applied Science, Barking, pp. 165–190.

Ziegler, W.H., Doery, R. and Scott, J. 1986. Tectonic habitat of Norwegian oil and gas, In Spencer, A.M. *et al.* (Eds), *Habitat of Hydrocarbons on the Norwegian Continental Shelf*, Graham and Trotman, London, pp. 3–19.

Zimmerman, J. 1980. Seismic trace inversion—examples from the North Sea, In *Offshore North Sea—Exploration—Geological and Geophysical*

Highlights, Norsk Petroleumsforening, Article E-III/2, 5 pp.
Østvedt, O.J. 1987. Sleipner Øst, In Spencer, A.M. *et al.* (Eds), *Geology of the Norwegian Oil and Gas Fields*, Graham and Trotman, London, pp. 243–252.
Åm, K. 1973. Geophysical indications of Permian and Tertiary igneous activity in the Skagerrak, *Norges geolisk Undersøkelse* **287,** 1–25.

ACKNOWLEDGEMENTS

We wish to thank the following colleagues who have helped by supplying or checking references: K. Abelsnes (Librarian, Statoil); K. Bjørlykke, J. D. Collinson, T. Fjæran, R. Lilletveit and the members of the editorial subcommittee.

We would like to thank specially Kirsti Rolfsen for perfectly typing the Bibliography.

Part X
Compilation Tables

TABLE I. Tertiary oil and gas fields

	ODIN	FRIGG	NORD-ØST FRIGG	ØST FRIGG ALPHA	ØST-FRIGG BETA	HEIMDAL	BALDER	SLEIPNER ØST	COD
TRAP									
(1) Type	Structural/ Stratigraphic	Structural/ Stratigraphic	Structural/ Stratigraphic	Structural/ Stratigraphic	Structural/ Stratigraphic	Dome	Structural/ Stratigraphic	Structural/ Stratigraphic	Dome over salt swell
(2) Depth to crest (mss)	1964	1785	1923	1891	1900	2010	1650	2260	2870
(3) Lowest closing contour (mss)	2025	1955	1990	1955	1950	2150	—	2417	3070
(4) GOC or GWC (mss)	2021	—	1980	1948	1937	2145	—	2417	3070 +
(5) OWC (mss)	2025	—	1983.5	1956	1947	2150	1760	—	—
(6) Gas column (m)	< 57	160	57	56	37	135	none	157	170 +
(7) Oil column (m)	4	9	3.5	8	11	5	*c.* 80	none	none
PAY ZONE									
(11) Lithology	Sandstone	Sandstone	Sandstone	Sandstone	Sandstone	Sandstone	Sandstone	Sandstone	Sandstone
(12) Formation	Frigg	Frigg	Frigg	Frigg	Frigg	Heimdal	Balder + Heimdal	Heimdal + Maureen	Forties
(13) Age	E. Eoc.	E. Eoc.	E. Eoc.	E. Eoc.	E. Eoc	Paleoc.	Eoc. + Paleoc.	Paleoc.	L. Paleoc.
(14) Gross thickness (m)	< 57	55–170	—	40	25	—	20–30, < 56	0–120	< 100
(15) Net/gross ratio (av., range)	0.93, 0.81–1.0	0.95	0.9	0.99	1.0	0.79	0.75–1.0	0.95	—
(16) Porosity (av., range) (%)	28–31	27–30	28	29	27–30	25	32–35	26	17, 15–24
(17) Hydrocarbon saturation (av., range) (%)	81, 76–86	91	85	90–94	86–93	89	85	67	57
(18) Permeability (av., range) (mD)	> 1000	1500, 900–3000	1000, 750–1500	1000, 700–1500	1200, 400–2000	1000	1000–3000	400	15–100
(19) Productivity index (Sm³/d/bar)	—	—	—	—	—	—	—	—	—
HYDROCARBONS									
(21) Oil gravity (g/cm³)	0.83	0.835	0.896	0.83	0.83	—	0.91	—	—
(22) Gas gravity (air = 1)	0.60	0.581	0.585	0.586	0.578	0.7	—	0.93	0.69
(23) Condensate gravity (g/cm³)	—	—	—	—	—	0.72	—	—	0.76
(24) Bubble point (bar)	—	—	—	—	—	—	148–165	—	—
(25) Dew point (bar)	83–103	—	—	—	—	210	—	—	—
(26) Gas/oil ratio (Sm³/Sm³)	—	—	—	—	—	—	53	—	—
(27) Condensate yield (Sm³/10⁶ Sm³ gas)	9.54	4.3	7.3	5.1	5.1	156	—	560	425
(28) Formation volume factor	—	—	—	—	—	—	1.15	—	—
(29) Gas expansion factor	201	197	195	196	196	208	—	208	—
FORMATION WATER									
(31) Salinity (NaCl equivalent) (ppm)	50 000	63 000	65 000	55 000	55 000	70 000	—	—	
(32) Resistivity (Ohm.m)	0.075	—	—	—	—	—	—	—	—
RESERVOIR CONDITIONS									
(41) Temperature (°C)	65	61	61.5	63	63	76	77	93	132
(42) Pressure (bar)	206	198	201	197	197	217	178	244	366
(43) Pressure gradient (bar/m)	0.029	0.013	0.013	0.013	0.013	0.020	—	0.023	—
FIELD SIZE									
(51) Area (km²)	40	100	19	13	7	33	44	59	13
(52) Gross rock volume (10⁶m³)	1026	5454	420–450	295	132	1372	—	—	—
(53) Hydrocarbon pore volume (10⁶m³)	230	1366	90–96	77	34	—	—	371	46
(54) Recovery factor—oil	0	0	0	0	0	0	0.30	—	—
(55) Recovery factor—gas	0.7–0.8	0.72	0.6	0.59	0.61	0.71	—	—	—
(56) Drive mechanism	Depletion + water drive	Depletion + water drive	Depletion + water drive	Depletion + water drive	Depletion + water drive	Depletion + water drive	Water drive and water injection	Depletion + water drive	—
(57) Recoverable—oil (10⁶Sm³)	0	0	0	0	0	—	11–27	—	—
(58) Recoverable—NGL/cond. (10⁶Sm³)	—	—	—	—	—	4.7	—	—	2
(59) Recoverable—gas (10⁹Sm³)	32–36	191	11	8.9	3.7	33.9	—	53	8.3
PRODUCTION									
(61) Start-up date	1984	1977	1983	1988	1988	1986	—	—	1977
(62) Development scheme	1 platform	5 platforms	6 subsea wells	subsea	Subsea	1 platform	—	—	1 platform
(63) Cumulative production (end 1986)									
—oil (10⁶Sm³)	—	—	—	—	—	—	—	—	—
—gas (10⁹Sm³)	5.8 (end 85)	135	5.7	0	0	2.2	—	—	6.1

TABLE II. Cretaceous oil and gas fields

	ALBUSKJELL	EDDA	EKOFISK	ELDFISK North	ELDFISK South	HOD West	HOD East	TOMMELITEN Alpha	TOMMELITEN Gamma	TOR	VALHALL	WEST EKOFISK	AGAT
(1)	Anticline over salt dome	Dome over salt swell	Dome over salt swell	Dome over salt swell	Dome over salt swell	Dome	Dome over salt swell	Dome over salt swell	Dome over salt swell	Dome over salt swell	Dome	Dome over salt swell	Stratigraphic
(2)	3070	3063	2888	2800	2700	2575	2650	3000	3025	2872	2400	3065	2800
(3)	3260	3231	3155	2896	2896	2750	2750	3140	3260	3155	c. 2700	3275	—
(4)	3400 +	—	—	—	—	—	—	3180	—	—	—	—	3455, 4561
(5)	—	variable	≃ 3288	variable		?	?	—	—	c.3292	variable	3320	—
(6)	330 +	none	none	none		none	none	180	> 295	none	none	none	< 700?
(7)	none	128	400	380		78	166	none	none	c.320	c.200	210	—
(11)	Chalk	Chalk	Chalk	Chalk		Chalk		Chalk		Chalk	Chalk	Chalk	Sandstone
(12)	Ekofisk + Tor	Ekofisk + Tor	Ekofisk + Tor	Ekofisk, Tor, Hod		Ekofisk, Tor, Hod		Ekofisk, Tor, Hod		Ekofisk, Tor	Tor, Hod	Ekofisk, Tor	
(13)	Paleoc.–L.Cret.	Paleoc.–L.Cret.	Paleoc.–L.Cret.	Paleoc.–L.Cret		Paleoc.–L.Cret.		Paleoc.–L.Cret		Paleoc.–L.Cret	L.Cret	Paleoc.–L.Cret	E.Cret
(14)	—	170–253	250	200			—	125	> 165	213–275	—	164–268	c.200
(15)	—	—	0.7, 0.3–1.0	0.8, 0.2–1.0			0.9–1.0			0.62, 0.1–1.0	1.0	0.86	0.1–0.3
(16)	21	25, 10–27	15–48	8–42			33, 24–38	27, < 42	24, < 43	21, 13–36	< 45	31, 18–34	< 20
(17)	65	64	90	85			80 (40 in Hod)			63	98	78	< 54
(18)	0.01–1	< 24	0.06–8.0	0.01–35, < 140			1.4, 0.5–6.5	0.02–4.0	0.01–4 0	0.04–5.0, < 200	0.1–10		200, 400
(19)	—	—		—	—		935	—	—	—	—	—	—
(21)	—	0.83–0.84	0.84	0.84			0.68	0.78–0.85	0.78	0.816	0.845	0.80–0.81	—
(22)	0.70	—	0.68	0.70			—	0.71	0.68	0.73	—	0.70	0.62
(23)	0.79	—	—	—			—	—	—	—	—	—	c. 0.78
(24)	—	348	383	414	352		270	—	—	305	324	478	—
(25)	—	—	—	—	—		—	—	—	—	—	444	431, 341
(26)	890, 1600	307	276	377	264		156	—	—	260	231	731	—
(27)	—	—	—	—	—		—	—	—	—	—	—	—
(28)	—	1.89	1.72	2.08	1.74		1.405	—	—	1.85	1.65	1.37	—
(29)	—	—	—	—	—		—	—	—	—	—	—	286
(31)	—	52 000	61 000	52 000			60 000	70 000	78 000	117 000		52 000	15 700
(32)	—	0.042	0.038	0.042			0.046	0.040	0.039	0.037, 0.016	0.045	0.042	0.521
(41)	138	132	131	119			100	124	130	136		131	118
(42)	500	490	485	470			178	485	487	492		498	c.455, 352
(43)		0.15–0.16	0.155	0.156–0.163			0.079	—	—	0.057		0.022–0.026	—
(51)	25	10	49	18		7	7	7	4.5	22	—	12	?150
(52)	—	—	—	1963		242	—	—	—	—	—	615	10 000 + 5500
(53)	—	34.4	1828	475		—	60.9	—	—	165	259	148	—
(54)	—	0.22	0.21	0.18	0.22		0.16	—	—	0.21	0.12–0.15	—	—
(55)	—	—	—	—	—		—	—	—	—	—	0.75	—
(56)	—	—	Solution gas drive, gas injection, water flood	Solution gas drive			Water drive	—	—	Solution gas drive	—	—	—
(57)	—	4	237	48			6	} 9		18	—	—	—
(58)	8	—	13.5	—			—			—	—	10	—
(59)	18	2.2	126	39			—	21		12–15	—	28	c. 65 in-place
(61)	1979	1979	1971	1979		—	—	1987		1978	1982	1977	—
(62)	2 platforms	1 platform	3 production platforms	2 production platforms		—	—	Subsea, linked to Edda		1 platform	—	1 platform	—
(63)	—	3 (Jan 86)	120 (Feb 86)	26 (Oct 85)		—	—	—		14.6 (Oct 85)	6.3 (Dec 85)	8.6	—
	13.5	1.7 (Jan 86)	63 (Feb 86)	10 (Oct 85)		—	—	—		8.8 (Oct 85)	1.0 (Dec 85)	21.7	—

TABLE III. Jurassic/Triassic oil and gas fields

	GULLFAKS			HILD	TROLL West	TROLL East	BRAGE	OSEBERG
TRAP								
(1) Type (mss)	Tilted fault blocks which are deeply eroded			Tilted fault blocks	Tilted fault blocks	Tilted fault block	Tilted fault block	Tilted, eroded fault blocks
(2) Depth to crest (mss)	1740	1785	1865	3600	1340	1315	1990, 2070	2120
(3) Lowest closing contour (mss)	—	—	—	?	1551		2030,2150	—
(4) GOC or GWC (mss)	—	—	—	Several	1543, 1547	1547	2004	2497
(5) OWC (mss)	1947	2090	2043	—	1570, 1559	1551	2029,2147	2700–2719
(6) Gas column (m)	none	none	none	< 245	203, 207	< 230	14, 0	c.380
(7) Oil column (m)	207	305	178	none	22–28, 12	0–4	·25, 77	203–222
PAY ZONE								
(11) Lithology	Sandstone	Sandstone	Sandstone	Sandstone	Sandstone		Sandstone	Sandstone
(12) Formation	Brent	Cook	Statfjord	Tarbert	W. Heather, Sognefjord, M. Heather Fensfjord		Sognefjord, Fensfjord	Brent
(13) Age	M. Jur.	E. Jur.	E. Jur.	M. Jur.	L. Jur.		L. Jur.	M. Jur.
(14) Gross thickness (m)	—	—	—	100	< 230		0–70,70–110	46–190
(15) Net/gross ratio (av., range)	—	—	—	0.4–0.7	—		0.7, 0.6–0.9	0.4–1.0
(16) Porosity (av. range) (%)	20–34	24–36	19–30	13–20	19–34		25, 27	20–26
(17) Hydrocarbon saturation (av., range) (%)	85	75	85	54–87	—		60, 60	60–89
(18) Permeability (av., range) (mD)	20–5000	5–6000	20–3000	0.01–2000	1–10 000		1–300, 30–200	2000, 1000–3500
(19) Productivity index (Sm3/d/bar)	100–200	100–200	100–200	—	—		20	gas 36 000–780 000 oil 415–1950
HYDROCARBONS								
(21) Oil gravity (g/cm3)	0.88	0.85	0.83	0.802–0.819	0.882–0.910		0.83,0.85	0.857
(22) Gas gravity (air = 1)	0.70	0.82	0.83	0.657–1.067	0.61		0.70–0.90	—
(23) Condensate gravity (g/cm3)	—	—	—	—	0.802–0.825		—	—
(24) Bubble point (bar)	244	236	234	—	158		180,167	280
(25) Dew point (bar)	—	—	—	630–737	158		—	—
(26) Gas/oil ratio (Sm3/Sm3	100	155	162	1264–4660	64		145,93	130–160
(27) Condensate yield Sm3/106 Sm3 gas	—	—	—	—	35–40		—	—
(28) Formation volume factor	1.26	1.498	1.507	—	1.18		1.45,1.29	1.34–1.47
(29) Gas expansion factor	—	—	—	—	152		—	—
FORMATION WATER								
(31) Salinity (NaCl equivalent) (ppm)	25 000	25 000	25 000	—	47 000		45 000	—
(32) Resistivity (Ohm.m)	0.168	0.168	0.168	—	0.16		0.056	—
RESERVOIR CONDITIONS								
(41) Temperature (°C)	72	76	80	126–145	68		85, 87	100
(42) Pressure (bar)	310	317	325	730–775	158		205, 215	281
(43) Pressure gradient (bar/m)	0.075	0.079	0.058	—	0.011 (gas),	0.075–0.079 (oil)	0.07, 0.072	—
FIELD SIZE								
(51) Area (km2)	35	15	10	< 120	770		25,47	—
(52) Gross rock volume (106m3)	—	—	—	—	53 000		—	—
(53) Hydrocarbon pore volume (106m3)	503	64	68	—	12 500		—	—
(54) Recovery factor—oil	0.445	0.26	0.42	—	0.2		0.25–0.35	0.52 (Etive)
(55) Recovery factor—gas	—	—	—	—	0.7			0.70
(56) Drive mechanism	Water injection	Water injection	Water injection	—	Water drive + gas expansion		Water injection, depletion	gas injection
(57) Recoverable—oil (106Sm3)	179	12	19	—	33		4–13 + 15–35	203
(58) Recoverable—NGL/cond. (106Sm3)	—	—	—	—		—	—	—
(59) Recoverable—gas (109Sm3)	—	—	—	30–40 in place	1252		—	89
PRODUCTION								
(61) Start-up date	1986	1990	1990	—	Phase I: 1996		—	1989
(62) Development scheme	Three platforms and subsea wells			—	—		—	—
(63) Cumulative production—oil (10^6Sm^3)	—	—	—	—		—	—	—

TABLE III. *(continued)*

	ASKELADD	TYRIHANS North	TYRIHANS South		SNORRE		MURCHISON	STATFORD NORD	STATFJORD ØST	STATFJORD		
(1)	Tilted fault blocks	Horst	Faulted dome	Faulted horst	Tilted fault blocks, which are deeply eroded		Tilted fault block	Tilted, eroded fault block	Tiltec, eroded fault block	Titled, eroded fault block		
(2)	2010	3490	3540	2270	2300	2375	2904, av.2956	2600	2430	2360	2475	2575
(3)	2155(N), 2165(S)	—	—	2490, 2500	2400	2400	3200	2800	2450	—	—	—
(4)	2155(N), 2136(S)	3658	—	2493, 2488	—	—	—	—	—	—	—	—
(5)	—	> 3680	3687	2499.5	2595	2561–2595	3080	2716	2507	2586	c.2600	2806
(6)	145(N),106(S)	168	—	225	none	none	none	?	none	none	none	none
(7)	none	> 22	147	11.5	295	220	176	47 +	c.60	226	111	231
(11)	Sandstone	Sandstone		Sandstone	Sandstone	Sandstone	Sandstone	Sandstone	Sandstone	Sandstone	Sandstone	Sandstone
(12)	Stø	Upper Tomma		Tomma, Aldra	Statfjord	Lunde	Brent	Munin, Brent	Brent	Brent	Dunlin	Statfjord
(13)	E.–M.Jur.	M.Jur		E.–M.Jur.	E.Jor.	L.Trias	M.Jur.	L., M.Jur.	M.Jur.	M.Jur.	E.Jur.	E.Jur.–L.Trias.
(14)	87–95	80–120		124–142, 183–202	70, 24–73	570, 400–600	130, 75–220	46, 140	223	156	70	123
(15)	0.82, 0.67–1.0	0.85–1.0		0.74–0.94, 0.71	0.53, 0.40–0.63	0.49, 0.35–0.68	0.7, 0.45–0.8	0.80, 0.75	0.78	0.75	0.40	0.58
(16)	15, 13–17	13–20		28, 24	25, 19–29	24, 19–29	22, 16–26	23, 22	26	27	23	22
(17)	80	69	75	90–95, 78	89, 79–94	68, 57–86	70, 65–82	81, 81	82	84	40	64
(18)	100, 5–250	450	150	6000, 1000	1300–2000	320–535	500–1000 (< 10 000)	2900, 1500	1120, 460–1700	2500	150, 10–300	470
(19)	—	—	—	—	70–175	30–90	160		74	231	2–65	208
(21)	—	0.883	—	0.849	0.711–0.764	0.674–0.701	0.83	0.835, 0.838	0.84	0.82	0.86	0.83
(22)	0.666	0.79	1.08	0.776	—	—	—	0.757, 0.743	0.69	0.76	0.75	0.78
(23)	—	—	—	0.809	—	—	—	—	—	—	—	—
(24)	—	—	—	255	90–126	179	131	138	245	275	248	196
(25)	210	377	377	256	—	—	—	—	—	—	—	—
(26)	—	3889	871	173	58–85	132, 91–137	96	61, 61	160	190	140	155
(27)	50	—	—	231	—	—	—	—	—	—	—	—
(28)	—	—	—	1.52	1.24, 1.19–1.27	1.40, 1.32–1.52	1.313	1.39, 1.23	1.5	1.58	1.44	1.54
(29)	210	250	217	210	—	—	—	—	—	—	—	—
(31)	80 000	—	—	75 000	34 000	42 050	14 000	22 000	22 000	14 000	—	13 600
(32)	0.04	—	—	0.1	0.196	0.175	0.312	0.110	0.110	0.11	—	0.128
(41)	80	132	136	91	90	93	110	96	76	89	92	100
(42)	237.7(N), 235.7(N)	378	378	251	383	383	444	393	337	383	383	404
(43)	0.021	0.069	0.04	—	0.070–0.075	0.068	0.070	—	—	0.064	0.064	0.065
(51)	105	47		47	30	50	17	13.2	13.2	80	26	35
(52)	3600	1400	1500	2500	—	—	c.1400	1900	1600	6650	1400	4200
(53)	350	205	130	—	—	—	96	—	—	1130	33	343
(54)	—	0.18	—	cond. 0.63	0.38, 0.32–0.42	0.35, 0.21–0.41	0.41	0.4	0.4	—	—	—
(55)	0.70	0.65	0.65	0.75	—	—	—	—	—	—	—	—
(56)	Gas expansion	Water drive + water/gas injection		Weak water drive, depletion	Water injection	Water injection	Water drive + water injection	Water injection	—	water drive + water injection	Solution gas + water injection	Water drive + gas injection, later water injection
(57)	—	8	—	?	39 (35–46)	77 (63–102)	51	36–66	14–40	482		
(58)	—	—	18	17	—	—	2×10^6 tonnes	—	—	—		
(59)	50	24	21	87	—	—	39	34	2.5–5.4	88		
(61)	—	—	—	—	1992	1993	1980	1994	1990s	1979	1993	1979
(62)	—	—	—	—	Tension leg platform	TLP or Subsea	1 platform	Subsea atmospheric	Subsea	3 platforms + tanker loading	Recompletion of existing wells	3 platforms + tanker loading
(63)	—	—	—	—	—	—	26.5	—	—	84 $19^{(7/86)}$	—	46 $11^{(7/86)}$

TABLE III *(continued)*

	15/5–1	SLEIPNER ØST	SLEIPNER VEST	BREAM	BRISLING	7/11–5	ULA
1	Faulted dome	Faulted dome	Faulted domes over salt swells	Dome over salt swell	Faulted anticline	Dome over salt swell	Dome over salt swell
2	3380	2645	3375, 3450	2260	2120	4040	3345
3	3670	*c.* 2800	3648	2400	2250	4260	*c.* 3700
4	?	*c.* 2800	3538–3637				
5				2315	2138	?	3508
6	58	155	250	—	—		—
7		—	—	55	18	46 +	163
11	Sandstone	Sandstone	Sandstone	Sandstone	Sandstone	Sandstone	Sandstone
12	Hugin	Hugin	Hugin	Sandnes	Sandnes	Ula	Ula
13	L. Jur	L. Jur	M. Jur	M. Jur	M. Jur	L. Jur	L. Jur
14	58	35	54–203	51	8	46	84–166
15	0.75	0.78	0.80	0.3	0.9	0.8	0.93, 0.85–1.0
16	14	22	18–23	22	10	17.5	12–22
17	86	78	79			74	75–97
18	28, 1–180	100–400	10–430			11, 1–320	0.2–2800
19	25 000					3.6	120–350
21				0.855–0.882	0.89	0.83	0.83
22	0.73	0.94	0.76			0.88	
23			0.79				
24						319	159
25	471						
26						267	89
27	544	490	3754				
28						1.75	1.33
29	256	234	285				
31			70 000			190 000	175 000
32	0.038		0.031			0.013	0.014
41	127	104	123			161	143
42	474	300	440			590	483
43	0.038	0.024	0.032			0.051	0.067
51	8.3	22	80	20	5	26	11
52	376		6960			350–900	631
53	30	85	688			38–97	
54	cond. 0.22					0.20–0.35	
55	0.76		0.75			0.3–0.5	0.40
56	depletion	gas expansion				solution gas drive, water drive	water drive
57				⩽1	⩽1	2–10	25
58	0.9		45×10^6 Sm^3				
59	6.7	7	126			1–4	
61							1986
62							3 steel platforms
63							(PDQ)

Part XI
Wells Drilled

A list of exploration and appraisal wells,
from the Norwegian Petroleum Directorate

BOREHOLE LIST EXPLORATION (with utm-coord.) Date: 23.Januar 1987 PB/ABS

R=re-entry, X=did not reach prospect, S=deviated hole, A=side-tracked new hole

BOREHOLE	NPD NO LIC NO	POSITION	SPUD COMPL.	OPERATOR RIG	SPUD CLASS. COMPLETION CLASS.	WD KB	TD (RKB) FORM. AT T
1/03-01	15 011	56 51 21.00 02 51 05.00 6301488 utm-n/s 490936 utm-e/w	68.07.06 68.11.11	SHELL ORION	WILDCAT P&A. GAS DISCOVERY	71 26	4898 PERMIAN
1/03-02	26 011	56 56 10.00 02 45 00.00 6310443 utm-n/s 484786 utm-e/w	69.05.14 69.07.27	SHELL SEDNETH I	WILDCAT P&A. DRY HOLE	73 26	4298 CRETACEOUS
1/03-03	343 065	56 57 08.56 02 58 54.05 6312226 utm-n/s 498885 utm-e/w	82.08.22 83.03.24	ELF BORGSTEN DOLPHIN	WILDCAT P&A. OIL/GAS DISC.	68 25	4875 PERMIAN
1/03-04	361 065	56 56 39.21 02 43 00.23 6311353 utm-n/s 482765 utm-e/w	83.02.15 83.05.08	ELF DYVI ALPHA	WILDCAT P&A. DRY HOLE.	72 25	3198 PERMIAN
1/03-05	435 011	56 46 16.02 02 53 38.85 6292053 utm-n/s 493528 utm-e/w	84.10.01 85.02.11	SHELL NEDDRILL TRIGON	WILDCAT P&A. DRY HOLE.	71 35	4710 PERMIAN.
1/05-01 X	99 018	56 34 40.50 02 38 28.00 6270599 utm-n/s 477951 utm-e/w	73.10.12 73.10.19	PHILLIPS OCEAN VIKING	WILDCAT P&A. JUNKED	70 27	501 TERTIARY
1/05-02	100 018	56 34 40.50 02 38 28.00 6270599 utm-n/s 477951 utm-e/w	73.10.19 74.04.15	PHILLIPS OCEAN VIKING	WILDCAT P&A. OIL/GAS DISC.	70 27	4285 PERMIAN
1/06-01	75 011	56 38 02.40 02 59 50.30 6276785 utm-n/s 499834 utm-e/w	72.07.10 72.10.26	SHELL ZAPATA NORDIC	WILDCAT P&A. GAS/COND DISCOV	69 35	4822 PERMIAN
1/06-02	78 011	56 33 50.60 02 51 53.20 6269007 utm-n/s 491689 utm-e/w	72.11.28 73.01.12	SHELL ZAPATA NORDIC	WILDCAT P&A. DRY HOLE	69 36	3383 L. CRET.
1/06-03	105 011	56 38 36.40 02 55 49.70 6277839 utm-n/s 495735 utm-e/w	74.04.12 74.09.11	SHELL ZAPATA NORDIC	APPRAISAL P&A. GAS/CONDENSATE	69 34	3343 L.CRET
1/06-04	144 011	56 44 52.06 02 42 23.64 6289491 utm-n/s 482053 utm-e/w	75.12.29 76.04.09	SHELL CHRIS CHENERY	WILDCAT P&A. DRY HOLE	72 33	2810 L.CRET.
1/09-01	167 044	56 24 05.07 02 54 06.49 6250900 utm-n/s 493939 utm-e/w	76.10.14 77.02.17	STATOIL ROSS RIG	WILDCAT SUSP. OIL/GAS DISCOV	78 25	3706 L.CRET.
1/09-02	175 044	56 23 52.77 02 55 34.63 6250517 utm-n/s 495449 utm-e/w	77.06.01 77.08.12	STATOIL ROSS RIG	APPRAISAL P&A. DRY HOLE	75 25	3459 L.CRET.
1/09-03	181 044	56 24 56.20 02 54 15.15 6252481 utm-n/s 494089 utm-e/w	77.08.13 77.11.27	STATOIL DYVI GAMMA	WILDCAT SUSPENDED.	75 36	2782 TERTIARY
1/09-03 R	181 044	56 24 56.20 02 54 15.15 6252481 utm-n/s 494089 utm-e/w	78.05.27 78.09.30	STATOIL DYVI BETA	WILDCAT P&A. GAS/COND. DISC.	75 36	4570 L.JURASSI
1/09-04	182 044	56 29 03.76 02 56 00.29 6260133 utm-n/s 495899 utm-e/w	77.08.13 78.01.12	STATOIL ROSS RIG	WILDCAT SUSP. GAS/COND. DISC	75 25	3710 PERMIAN
1/09-05	203 044	56 29 27.69 02 57 55.27 6260870 utm-n/s 497866 utm-e/w	78.10.03 78.12.11	STATOIL DYVI BETA	APPRAISAL P&A. DRY HOLE	75 35	3450 L. CRET.
1/09-06 S	318 044	56 29 03.85 02 56 00.14 6260135 utm-n/s 495896 utm-e/w	82.03.21 82.12.01	STATOIL SEDCO 707	APPRAISAL SUSP. GAS/COND. DISC	75 25	3529 L.CRET.

BOREHOLE LIST EXPLORATION (with utm-coord.) Date: 23.Januar 1987 PB/ABS

R=re-entry, X=did not reach prospect, S=deviated hole, A=side-tracked new hole

BOREHOLE	NPD NO LIC NO	POSITION	SPUD COMPL.	OPERATOR RIG	SPUD CLASS. COMPLETION CLASS.	WD KB	TD (RKB) FORM. AT TD
2/01-01	73 019	56 45 26.60 03 12 18.80 6290539 utm-n/s	72.08.28 72.11.19 512548 utm-e/w	CONOCO OCEAN TIDE	WILDCAT P&A. GAS DISCOVERY	66 34	4178 JURASSIC
2/01-02	187 019	56 57 30.76 03 12 32.07 6312931 utm-n/s	77.12.14 78.02.26 512705 utm-e/w	BP NORSKALD	WILDCAT P&A. DRY HOLE	70 24	3550 PERMIAN
2/01-03	229 019	56 54 41.39 03 06 30.39 6307681 utm-n/s	79.11.03 80.03.29 506603 utm-e/w	BP SEDCO H	WILDCAT P&A. OIL DISCOVERY	66 30	4297 PERMIAN
2/01-04	322 019	56 54 39.82 03 04 02.25 6307628 utm-n/s	82.04.12 82.08.03 504097 utm-e/w	BP ALADDIN	APPRAISAL P&A. OIL DISCOVERY.	66 25	4525 PERMIAN
2/01-05	353 019	56 47 54.06 03 12 21.39 6295098 utm-n/s	82.11.13 83.04.05 512578 utm-e/w	BP SEDCO 707	WILDCAT P&A. OIL DISCOVERY	66 25	4454 TRIASSIC
2/01-06	413 019	56 55 14.54 03 01 37.84 6308701 utm-n/s	84.04.30 84.08.12 501654 utm-e/w	BP DYVI ALPHA	APPRAISAL P&A. OIL DISCOVERY.	66 25	4588 TRIASSIC
2/01-07	431 019	56 51 49.85 03 05 41.88 6302374 utm-n/s	84.09.06 85.03.06 505790 utm-e/w	BP GLOMAR MORAY F.I	WILDCAT P&A. DRY HOLE.	69 38	5464 PERMIAN.
2/01-08	473 019	56 53 40.63 03 06 55.95 6305803 utm-n/s	85.07.28 85.11.24 507039 utm-e/w	BP NEDDRILL TRIGON	APPRAISAL P&A. OIL/GAS DISC.	66 38	4158 TRIASSIC
2/02-01	320 066	56 47 15.69 03 38 15.62 6294074 utm-n/s	82.04.09 82.07.03 538959 utm-e/w	SAGA DYVI ALPHA	WILDCAT P&A. OIL/GAS DISC.	59 25	4003 PERMIAN
2/02-02	336 066	56 56 40.61 03 22 46.81 6311426 utm-n/s	82.07.04 82.08.27 523099 utm-e/w	SAGA DYVI ALPHA	WILDCAT P&A. GAS DISCOVERY	66 25	3127 PERMIAN
2/02-03	362 066	56 47 36.42 03 25 38.14 6294616 utm-n/s	83.02.04 83.05.11 526100 utm-e/w	SAGA TREASURE SAGA	WILDCAT P&A. DRY HOLE	65 27	4100 TRIASSIC
2/03-01	23 022	56 53 09.46 03 51 38.08 6305163 utm-n/s	69.02.10 69.04.03 552439 utm-e/w	MURPHY OCEAN TRAVELER	WILDCAT P&A. GAS DISCOVERY	59 27	2934 PERMIAN
2/03-02	30 022	56 54 53.41 03 49 02.28 6308344 utm-n/s	69.07.26 69.08.13 549764 utm-e/w	MURPHY OCEAN TRAVELER	APPRAISAL P&A. DRY HOLE	58 27	2298 TERIARY
2/03-03	61 022	56 48 18.90 03 58 11.50 6296267 utm-n/s	71.10.08 71.11.20 559226 utm-e/w	MURPHY OCEAN TIDE	WILDCAT P&A. DRY HOLE	56 31	2973 PERMIAN
2/03-04	418 022	56 46 45.61 03 48 27.89 6293254 utm-n/s	84.05.28 84.07.24 549360 utm-e/w	GULF GLOMAR MORAY F.I	WILDCAT P&A. DRY HOLE	56 38	3386 PERMIAN
2/04-01 X	33 018	56 32 40.23 03 11 55.22 6266841 utm-n/s	69.08.25 69.09.16 512216 utm-e/w	PHILLIPS OCEAN VIKING	WILDCAT P&A. JUNKED	71 27	1662 TERTIARY
2/04-02	34 018	56 32 08.65 03 14 54.57 6265875 utm-n/s	69.09.18 69.12.24 515283 utm-e/w	PHILLIPS OCEAN VIKING	WILDCAT P&A. OIL/GAS DISC.	70 27	3306 CRETACEOUS
2/04-03	35 018	56 32 42.12 03 14 55.49 6266909 utm-n/s	70.01.27 70.05.31 515295 utm-e/w	PHILLIPS OCEAN VIKING	WILDCAT P&A. OIL DISCOVERY	71 27	3432 CRETACEOUS

BOREHOLE LIST EXPLORATION (with utm-coord.) Date: 23.Januar 1987 PB/ABS

R=re-entry, X=did not reach prospect, S=deviated hole, A=side-tracked new hole

BOREHOLE	NPD NO / LIC NO	POSITION	SPUD / COMPL.	OPERATOR / RIG	SPUD CLASS. / COMPLETION CLASS.	WD / KB	TD (RKB) / FORM. AT
2/04-04	37	56 30 48.20	70.06.02	PHILLIPS	APPRAISAL	71	3425
	018	03 12 43.50	70.08.01	OCEAN VIKING	P&A. OIL DISCOVERY	27	CRETACEOU
		6263380 utm-n/s		513051 utm-e/w			
2/04-05	40	56 34 29.77	70.06.21	PHILLIPS	APPRAISAL	68	3320
	018	03 12 13.03	70.08.27	NEPTUNE 7	P&A. OIL DISCOVERY	24	CRETACEOU
		6270229 utm-n/s		512510 utm-e/w			
2/04-06	44	56 34 00.40	70.08.28	PHILLIPS	WILDCAT	70	3412
	018	03 05 08.20	70.12.10	NEPTUNE 7	P&A. GAS/COND. DISC.	9	L.CRET.
		6269305 utm-n/s		505261 utm-e/w			
2/04-07	59	56 38 05.40	71.07.24	PHILLIPS	APPRAISAL	68	3494
	018	03 16 44.90	71.10.08	MAERSK EXPLORER	P&A. OIL DISCOVERY	34	L.CRET.
		6276912 utm-n/s		517123 utm-e/w			
2/04-08	63	56 38 16.43	71.11.25	PHILLIPS	APPRAISAL	66	4075
	018	03 18 58.67	72.03.29	MAERSK EXPLORER	P&A. OIL/GAS DISC.	37	PERMIAN
		6277263 utm-n/s		519401 utm-e/w			
2/04-09	91	56 36 26.60	73.08.13	PHILLIPS	WILDCAT	69	3752
	018	03 05 08.50	73.10.09	ZAPATA EXPLORER	P&A. OIL/GAS DISC.	35	L.CRET.
		6273826 utm-n/s		505260 utm-e/w			
2/04-10	94	56 40 41.60	73.10.13	PHILLIPS	WILDCAT	67	3418
	018	03 13 21.70	73.12.20	ZAPATA EXPLORER	P&A. OIL DISCOVERY	35	L.CRET.
		6281730 utm-n/s		513645 utm-e/w			
2/04-11	103	56 44 45.90	73.12.25	PHILLIPS	WILDCAT	66	2481
	018	03 13 21.80	74.04.09	ZAPATA EXPLORER	P&A. OIL/GAS DISC.	35	PERMIAN
		6289283 utm-n/s		513622 utm-e/w			
2/04-12	133	56 31 26.60	75.07.28	PHILLIPS	WILDCAT	70	3383
	018	03 08 36.90	75.09.08	OCEAN VIKING	P&A. DRY HOLE	27	L.TRIASS
		6264556 utm-n/s		508833 utm-e/w			
2/04-19 B	149	56 33 54.85	75.12.16	PHILLIPS	APPRAISAL	69	4637
	018	03 12 13.20	76.03.15	EKOFISK B	P&A. OIL DISC.	35	JURASSIC
		6269149 utm-n/s		512516 utm-e/w			
2/05-01	41	56 38 19.95	70.08.01	AMOCO	WILDCAT	68	3972
	006	03 20 07.94	70.11.22	ORION	P&A. OIL/GAS DISC.	27	JURASSIC
		6277378 utm-n/s		520581 utm-e/w			
2/05-02	51	56 38 41.60	71.01.15	AMOCO	APPRAISAL	66	3597
	006	03 22 20.60	71.04.20	MAERSK EXPLORER	P&A. OIL DISCOVERY	35	L.CRET.
		6278059 utm-n/s		522837 utm-e/w			
2/05-03	67	56 35 50.80	72.04.02	AMOCO	WILDCAT	67	3731
	006	03 26 58.80	72.06.15	ZAPATA EXPLORER	P&A. OIL/GAS DISC.	34	PERMIAN
		6272806 utm-n/s		527611 utm-e/w			
2/05-04	76	56 33 54.40	72.09.03	AMOCO	WILDCAT	68	3490
	006	03 36 56.40	72.10.31	ZAPATA EXPLORER	P&A. OIL DISC.	35	L.CRET.
		6269286 utm-n/s		537837 utm-e/w			
2/05-05	81	56 34 54.40	72.12.09	AMOCO	APPRAISAL	65	3456
	006	03 25 49.10	73.02.09	ZAPATA EXPLORER	P&A OIL DISC.	36	L.CRET.
		6271055 utm-n/s		526433 utm-e/w			
2/05-06	196	56 34 13.24	78.05.14	AMOCO	WILDCAT	68	4132
	006	03 37 16.57	78.08.16	NORSKALD	P&A. DRY HOLE	25	TRIASSIC
		6269872 utm-n/s		538176 utm-e/w			
2/05-07	384	56 43 16.69	83.08.11	SHELL	WILDCAT	66	4531
	067	03 22 01.07	84.02.24	NEDDRILL TRIGON	P&A. OIL DISCOVERY	34	TRIASSIC
		6286563 utm-n/s		522459 utm-e/w			
2/06-01	21	56 35 52.00	69.04.21	ELF	WILDCAT	69	3336
	008	03 54 55.00	69.05.30	OCEAN VIKING	P&A. DRY HOLE	29	PERMIAN
		6273127 utm-n/s		556201 utm-e/w			

REHOLE LIST   EXPLORATION (with utm-coord.)   Date: 23.Januar 1987   PB/ABS

OREHOLE	NPD NO LIC NO	POSITION	SPUD COMPL.	OPERATOR RIG	SPUD CLASS. COMPLETION CLASS.	WD KB	TD (RKB) FORM. AT TD
R=re-entry, X=did not reach prospect, S=deviated hole, A=side-tracked new hole							
2/06-02	238 008	56 30 48.90 03 42 39.66 6263608 utm-n/s 543756 utm-e/w	80.01.03 80.05.25	ELF DYVI ALPHA	WILDCAT P&A. OIL DISCOVERY	70 25	4760 PERMIAN.
2/06-03	352 008	56 31 49.32 03 49 25.38 6265553 utm-n/s 550669 utm-e/w	82.11.09 83.02.25	ELF BYFORD DOLPHIN	WILDCAT P&A. DRY HOLE	71 25	4060 PERMIAN
2/07-01	42 018	56 25 44.68 03 12 14.21 6253994 utm-n/s 512579 utm-e/w	70.08.03 70.12.11	PHILLIPS OCEAN VIKING	WILDCAT P&A. OIL/GAS DISC.	71 27	4573 JURASSIC
2/07-02	50 018	56 16 00.85 03 09 09.04 6235933 utm-n/s 509446 utm-e/w	70.12.13 71.03.02	PHILLIPS OCEAN VIKING	WILDCAT P&A. DRY HOLE	71 27	3965 JURASSIC
2/07-03	68 018	56 23 02.90 03 14 45.90 6248999 utm-n/s 515195 utm-e/w	72.04.14 72.10.11	PHILLIPS ORION	WILDCAT P&A. OIL/GAS DISC.	69 30	4359 PERMIAN
2/07-04	71 018	56 27 52.03 03 06 11.53 6257917 utm-n/s 506359 utm-e/w	72.06.23 72.09.03	PHILLIPS OCEAN VIKING	WILDCAT P&A. OIL/GAS DISC.	72 27	3356 L.CRET.
2/07-05	77 018	56 28 43.00 03 09 28.00 6259499 utm-n/s 509718 utm-e/w	72.09.03 73.01.02	PHILLIPS OCEAN VIKING	APPRAISAL P&A. DRY HOLE	72 27	3353 L. CRET.
2/07-06	80 018	56 25 00.95 03 14 35.82 6252650 utm-n/s 515010 utm-e/w	73.03.31 73.06.08	PHILLIPS OCEAN VIKING	APPRAISAL P&A. OIL DISC.	70 27	3381 E. CRET.
2/07-07	82 018	56 22 06.75 03 15 27.86 6247266 utm-n/s 515922 utm-e/w	73.01.02 73.03.29	PHILLIPS OCEAN VIKING	APPRAISAL P&A. OIL/GAS DISC.	70 27	3083 E CRET.
2/07-08	85 018	56 23 08.60 03 19 06.60 6249195 utm-n/s 519666 utm-e/w	73.06.08 73.08.12	PHILLIPS ZAPATA EXPLORER	WILDCAT P&A. OIL/GAS DISC.	70 34	3318 L.CRET.
2/07-09	87 018	56 19 50.60 03 14 29.30 6243054 utm-n/s 514931 utm-e/w	73.12.25 74.04.10	PHILLIPS ZAPATA NORDIC	APPRAISAL P&A OIL DISC.	71 37	4448 JURASSIC
2/07-10	96 018	56 28 33.70 03 05 03.30 6259205 utm-n/s 505190 utm-e/w	73.10.06 73.12.24	PHILLIPS ZAPATA NORDIC	APPRAISAL P&A OIL DISC.	72 37	3370 L.CRET.
2/07-11	97 018	56 19 49.38 03 18 29.13 6243032 utm-n/s 519051 utm-e/w	75.02.24 75.03.30	PHILLIPS OCEAN VIKING	WILDCAT P&A DRY HOLE	69 27	3377 E.CRET.
2/07-12	208 018	56 29 43.76 03 00 36.37 6261368 utm-n/s 500622 utm-e/w	78.12.17 79.01.30	PHILLIPS DYVI BETA	WILDCAT P&A. DRY HOLE	75 35	1828 L.PERMIAN
2/07-13	210 018	56 29 34.74 03 01 58.38 6261089 utm-n/s 502024 utm-e/w	79.02.01 79.04.21	PHILLIPS DYVI BETA	WILDCAT P&A. DRY HOLE	73 35	3388 E.CRET.
2/07-14	221 018	56 29 20.30 03 14 02.65 6260666 utm-n/s 514414 utm-e/w	79.08.09 80.01.20	PHILLIPS HAAKON MAGNUS	APPRAISAL SUSPENDED	71 25	3690 L.CRET.
2/07-15	239 018	56 23 46.82 03 18 54.63 6250375 utm-n/s 519456 utm-e/w	80.01.29 80.06.02	PHILLIPS HAAKON MAGNUS	WILDCAT P&A. DRY HOLE.	69 25	4423 JURASSIC
2/07-16	245 018	56 25 21.68 03 05 41.87 6253268 utm-n/s 505858 utm-e/w	80.03.20 80.07.12	PHILLIPS NORTRYM	WILDCAT P&A. DRY HOLE.	71 25	4818 E.CRET.

BOREHOLE LIST EXPLORATION (with utm-coord.) Date: 23.Januar 1987 PB/ABS

BOREHOLE	NPD NO LIC NO	POSITION	SPUD COMPL.	OPERATOR RIG	SPUD CLASS. COMPLETION CLASS.	WD KB	TD (RKB) FORM. AT

R=re-entry, X=did not reach prospect, S=deviated hole, A=side-tracked new hole

BOREHOLE	NPD NO / LIC NO	POSITION	SPUD / COMPL.	OPERATOR / RIG	SPUD CLASS. / COMPLETION CLASS.	WD / KB	TD (RKB) / FORM. AT
2/07-17 X	253	56 20 13.00	80.06.24	PHILLIPS	WILDCAT	73	1600
	018	03 06 04.00	80.07.12	BORGSTEN DOLPHIN	P&A. JUNKED	25	MIOCENE
		6243724 utm-n/s	506251 utm-e/w				
2/07-18 X	259	56 20 15.43	80.08.20	PHILLIPS	WILDCAT	73	601
	018	03 06 12.77	80.08.31	BORGSTEN DOLPHIN	P&A. JUNKED	25	E.CRET.
		6243800 utm-n/s	506401 utm-e/w				
2/07-19	262	56 20 18.36	80.09.02	PHILLIPS	WILDCAT	73	4876
	018	03 06 13.61	81.02.02	BORGSTEN DOLPHIN	SUSP. OIL DISCOVERY.	25	E.CRET.
		6243891 utm-n/s	506416 utm-e/w				
2/08-01	7	56 17 47.20	67.11.28	AMOCO	WILDCAT	70	2596
	006	03 26 59.70	68.06.02	DRILLSHIP	P&A. DRY HOLE	17	TERTIARY
		6239302 utm-n/s	527846 utm-e/w				
2/08-02	16	56 29 52.85	70.06.28	AMOCO	WILDCAT	69	3246
	006	03 28 33.60	70.07.30	ORION	P&A. DRY HOLE	26	CRETACEOU
		6261749 utm-n/s	529305 utm-e/w				
2/08-03	72	56 18 31.00	72.06.16	AMOCO	APPRAISAL	69	4115
	006	03 26 54.10	72.09.03	ZAPATA EXPLORER	P&A. OIL/GAS DISC.	34	JURASSIC
		6240656 utm-n/s	527741 utm-e/w				
2/08-04	83	56 17 53.80	73.04.28	AMOCO	APPRAISAL	70	2852
	006	03 20 38.06	73.06.07	ZAPATA EXPLORER	P&A OIL DISC.	34	E.CRET.
		6239468 utm-n/s	521284 utm-e/w				
2/08-05	107	56 29 58.70	74.04.11	AMOCO	WILDCAT	69	3304
	006	03 27 19.70	74.05.23	ZAPATA EXPLORER	P&A. DRY HOLE	34	E.CRET.
		6261922 utm-n/s	528040 utm-e/w				
2/08-06	124	56 15 42.91	75.04.07	AMOCO	WILDCAT	68	2669
	006	03 23 40.81	75.06.30	WAAGE DRILL I	P&A. OIL/GAS DISC.	27	E.CRET.
		6235439 utm-n/s	524449 utm-e/w				
2/08-07	134	56 26 50.90	75.07.23	AMOCO	WILDCAT	68	2868
	006	03 36 48.50	75.08.24	ZAPATA EXPLORER	P&A. DRY HOLE	33	L.PERMIAN
		6256192 utm-n/s	537819 utm-e/w				
2/08-08	143	56 16 50.28	75.11.10	AMOCO	APPRAISAL	70	2667
	006	03 24 15.93	76.03.15	DEEPSEA SAGA	P&A. OIL DISC.	25	E.CRET.
		6237525 utm-n/s	525041 utm-e/w				
2/08-09	155	56 17 48.50	76.04.09	AMOCO	APPRAISAL	69	2703
	006	03 22 04.33	76.06.27	SEDCO 135 G	P&A. OIL/GAS DISC.	33	L.CRET.
		6239312 utm-n/s	522768 utm-e/w				
2/08-10	157	56 15 53.34	76.06.30	AMOCO	APPRAISAL	69	2683
	006	03 23 04.33	76.08.29	SEDCO 135 G	P&A. OIL/GAS DISC.	33	L.CRET.
		6235758 utm-n/s	523819 utm-e/w				
2/08-11	164	56 16 56.15	76.08.10	AMOCO	APPRAISAL	70	2655
	006	03 22 15.68	76.10.11	ROSS RIG	P&A. OIL/GAS DISC.	25	L.CRET.
		6237696 utm-n/s	522972 utm-e/w				
2/09-01	79	56 22 11.88	72.11.03	AMOCO	WILDCAT	67	3551
	032	03 39 59.77	72.12.07	ZAPATA EXPLORER	P&A. DRY HOLE	34	L. CRET.
		6247595 utm-n/s	541178 utm-e/w				
2/09-02	220	56 20 56.66	79.07.07	AMOCO	WILDCAT	68	4367
	032	03 56 01.70	79.09.04	DYVI ALPHA	P&A. DRY HOLE	25	TRIASSIC
		6245461 utm-n/s	557715 utm-e/w				
2/10-01	141	56 07 32.15	75.09.16	PHILLIPS	WILDCAT	66	4609
	031	03 18 06.68	76.04.23	OCEAN VIKING	P&A. DRY HOLE	27	CARBON
		6220236 utm-n/s	518765 utm-e/w				
2/11-01	29	56 14 16.98	69.07.14	AMOCO	WILDCAT	68	4693
	033	03 27 07.05	69.10.03	ORION	P&A. OIL/GAS DISC.	27	JURASSIC
		6232803 utm-n/s	528015 utm-e/w				

REHOLE LIST EXPLORATION (with utm-coord.) Date: 23.Januar 1987 PB/ABS

OREHOLE	NPD NO LIC NO	POSITION	SPUD COMPL.	OPERATOR RIG	SPUD CLASS. COMPLETION CLASS.	WD KB	TD (RKB) FORM. AT TD

R=re-entry, X=did not reach prospect, S=deviated hole, A=side-tracked new hole

2/11-02	120 033	56 10 51.20 03 25 25.90 6226430 utm-n/s	74.11.06 74.12.29 526312 utm-e/w	AMOCO ZAPATA EXPLORER	WILDCAT P&A. OIL/GAS DISC.	66 36	2806 E.CRET.
2/11-03	184 033	56 10 53.74 03 27 51.78 6226525 utm-n/s	77.10.10 77.12.05 528827 utm-e/w	AMOCO DYVI BETA	APPRAISAL P&A. DRY HOLE	72 36	3052 L.CRET.
2/11-03 A	192 033	56 10 53.74 03 27 51.78 6226525 utm-n/s	77.12.05 78.03.14 528827 utm-e/w	AMOCO DYVI BETA	APPRAISAL P&A. OIL DISCOVERY	72 36	3332 L.CRET.
2/11-04	188 033	56 14 50.37 03 23 18.92 6233812 utm-n/s	78.03.20 78.05.18 524081 utm-e/w	AMOCO DYVI BETA	APPRAISAL P&A. OIL DISC.	69 34	2858 E.CRET.
2/11-05	214 033	56 09 57.70 03 26 06.42 6224780 utm-n/s	79.05.18 79.07.05 527022 utm-e/w	AMOCO DYVI ALPHA	APPRAISAL P&A. DRY HOLE	69 25	2945 E.CRET.
2/11-06	305 033	56 10 35.50 03 27 36.72 6225959 utm-n/s	81.09.03 82.02.28 528572 utm-e/w	AMOCO SEDCO 703	APPRAISAL SUSP. OIL DISCOVERY	72 26	4076 E.CRET.
2/11-07	513 068	56 14 06.09 03 37 38.84 6232552 utm-n/s	86.04.16 86.09.06 538896 utm-e/w	HYDRO TREASURE SCOUT	WILDCAT P&A. DRY HOLE	73 23	5042 L.JURASSIC
2/12-01	530 113	56 14 04.07 03 42 27.50 6232538 utm-n/s	86.10.14 00.00.00 543868 utm-e/w	HYDRO TREASURE SCOUT	WILDCAT	70 23	
3/05-01	195 022	56 37 25.78 04 25 11.97 6276555 utm-n/s	78.05.03 78.06.28 587130 utm-e/w	GULF ODIN DRILL	WILDCAT P&A. DRY HOLE	67 25	3425 PERMIAN
3/05-02	199 022	56 32 34.46 04 23 11.10 6267505 utm-n/s	78.06.29 78.08.20 585252 utm-e/w	GULF ODIN DRILL	WILDCAT P&A. DRY HOLE	67 25	3825 TRIASSIC
3/07-01	89 023	56 27 43.50 04 00 07.80 6258099 utm-n/s	73.08.01 73.09.13 561756 utm-e/w	ELF OCEAN TIDE	WILDCAT P&A. DRY HOLE	63 31	3227 BASEMENT
3/07-02	273 023	56 29 58.37 04 04 08.36 6262330 utm-n/s	81.03.30 81.06.20 565809 utm-e/w	ELF DYVI ALPHA	WILDCAT P&A. DRY HOLE.	62 25	4330 E.PERMIAN
3/07-03	293 023	56 24 54.77 04 10 54.33 6253058 utm-n/s	81.06.21 81.08.31 572913 utm-e/w	ELF DYVI ALPHA	WILDCAT P&A. DRY HOLE	67 25	3542 PERMIAN
6/03-01	441 086	57 58 10.20 01 55 30.38 6425966 utm-n/s	84.11.02 85.02.01 436408 utm-e/w	STATOIL DEEPSEA BERGEN	WILDCAT P&A. OIL/GAS DISC.	86 23	3560 TRIASSIC
6/03-02	492 086	57 54 25.99 01 59 14.19 6418976 utm-n/s	85.11.21 86.03.10 439982 utm-e/w	STATOIL ROSS ISLE	WILDCAT P&A. DRY HOLE	89 22	4091 PERMIAN
7/01-01	55 004	57 48 10.30 02 10 08.80 6407211 utm-n/s	71.07.03 71.08.05 450615 utm-e/w	AMOCO MAERSK EXPLORER	WILDCAT P&A. DRY HOLE	82 31	2808 TRIASSIC
7/03-01	25 005	57 50 35.25 02 44 55.61 6411417 utm-n/s	69.04.07 69.06.10 485084 utm-e/w	AMOCO ORION	WILDCAT P&A. DRY HOLE	64 31	4700 DEVONIAN
7/08-01	19 016	57 19 20.30 02 28 27.50 6353529 utm-n/s	68.08.31 69.02.05 468340 utm-e/w	PHILLIPS OCEAN VIKING	WILDCAT P&A. DRY HOLE	82 27	3335 TRIASSIC

BOREHOLE LIST EXPLORATION (with utm-coord.) Date: 23.Januar 1987 PB/ABS

R=re-entry, X=did not reach prospect, S=deviated hole, A=side-tracked new hole

BOREHOLE	NPD NO LIC NO	POSITION	SPUD COMPL.	OPERATOR RIG	SPUD CLASS. COMPLETION CLASS.	WD KB	TD (RKB) FORM. AT
7/08-02	88 016	57 23 39.35 02 35 30.21 6361492 utm-n/s 475459 utm-e/w	73.07.24 73.08.29	CONOCO OCEAN VIKING	WILDCAT P&A. DRY HOLE	81 27	3006 PERMIAN
7/08-03	387 069	57 15 30.58 02 32 45.78 6346394 utm-n/s 472613 utm-e/w	83.09.12 83.12.12	CONOCO BORGNY DOLPHIN	WILDCAT P&A. OIL/GAS DISC.	81 25	4320 PERMIAN
7/08-04	446 069	57 15 28.03 02 25 46.23 6346369 utm-n/s 465582 utm-e/w	84.12.11 85.02.20	CONOCO NORTRYM	WILDCAT P&A. DRY HOLE.	82 25	4400 TRIASSIC.
7/09-01	54 020	57 20 37.10 02 51 21.40 6355791 utm-n/s 491329 utm-e/w	71.04.22 71.05.29	CONOCO MAERSK EXPLORER	WILDCAT P&A. DRY HOLE	70 31	2931 PERMIAN
7/11-01	10 018	57 04 15.60 02 26 24.40 6325570 utm-n/s 466051 utm-e/w	68.02.28 68.06.15	PHILLIPS OCEAN VIKING	WILDCAT P&A. GAS/COND. DISC.	78 27	3974 PERMIAN
7/11-02	17 018	57 04 15.20 02 24 26.50 6325574 utm-n/s 464065 utm-e/w	68.07.21 68.10.14	PHILLIPS OCEAN TRAVELER	APPRAISAL P&A. GAS/COND. DISC.	82 27	3428 CRETACEOU
7/11-03	20 018	57 02 58.60 02 28 21.50 6323174 utm-n/s 468005 utm-e/w	68.10.17 69.02.07	PHILLIPS OCEAN TRAVELER	APPRAISAL P&A. OIL/GAS DISC.	79 27	3351 TERTIARY
7/11-04	28 018	57 07 32.70 02 29 26.00 6331641 utm-n/s 469155 utm-e/w	69.07.02 69.07.31	PHILLIPS OCEAN VIKING	WILDCAT P&A. DRY HOLE	79 27	3322 CRETACEOU
7/11-05	316 070	57 08 06.39 02 29 53.53 6332679 utm-n/s 469626 utm-e/w	82.02.09 82.06.10	HYDRO TREASURE SEEKER	WILDCAT P&A. OIL/GAS DISC.	80 25	4478 TRIASSIC
7/11-06	342 070	57 06 59.45 02 36 55.16 6330563 utm-n/s 476703 utm-e/w	82.08.09 82.10.20	HYDRO TREASURE SEEKER	WILDCAT P&A. DRY HOLE.	71 25	4500 TRIASSIC
7/11-07	359 018	57 04 10.80 02 26 05.04 6325424 utm-n/s 465724 utm-e/w	82.12.29 83.12.25	PHILLIPS COD PLATFORM	WILDCAT SUSPENDED. OIL DISC.	78 41	4927 PERMIAN
7/11-07 R	359 018	57 04 10.80 02 26 05.04 6325424 utm-n/s 465724 utm-e/w	84.09.29 84.10.08	PHILLIPS COD PLATFORM	WILDCAT SUSP. OIL DISCOVERY	78 42	4927 PERMIAN
7/11-08	389 070	57 12 49.94 02 31 46.50 6341433 utm-n/s 471586 utm-e/w	83.09.22 83.12.12	HYDRO TREASURE SCOUT	WILDCAT DRY HOLE.	79 23	4750 TRIASSIC
7/11-09	491 070	57 12 11.79 02 24 50.18 6340308 utm-n/s 464590 utm-e/w	85.11.26 86.03.09	HYDRO BYFORD DOLPHIN	WILDCAT P&A. DRY HOLE	91 25	4272 TRIASSIC
7/12-01 S	18 019	57 06 02.90 02 51 06.10 6328759 utm-n/s 491014 utm-e/w	68.08.31 68.10.18	GULF GULFTIDE	WILDCAT P&A. DRY HOLE	71 28	3312 JURASSIC
7/12-02	161 019	57 06 41.34 02 50 50.73 6329948 utm-n/s 490758 utm-e/w	76.07.04 76.09.24	CONOCO NORSKALD	WILDCAT SUSP. OIL/GAS DISC.	71 25	3676 TRIASSIC
7/12-02 R	161 019	57 06 41.34 02 50 50.73 6329948 utm-n/s 490758 utm-e/w	84.03.24 84.04.29	BP DYVI ALPHA	WILDCAT P&A. OIL DISCOVERY	71 25	3676 TRIASSIC
7/12-03	174 019	57 06 24.54 02 48 41.56 6329433 utm-n/s 488584 utm-e/w	77.04.05 77.06.03	BP NORSKALD	APPRAISAL P&A. DRY HOLE	71 24	3686 JURASSIC

BOREHOLE LIST EXPLORATION (with utm-coord.) Date: 23.Januar 1987 PB/ABS

R=re-entry, X=did not reach prospect, S=deviated hole, A=side-tracked new hole

BOREHOLE	NPD NO LIC NO	POSITION	SPUD COMPL.	OPERATOR RIG	SPUD CLASS. COMPLETION CLASS.	WD KB	TD (RKB) FORM. AT TD
7/12-03 A	191 019	57 06 24.54 02 48 41.56 6329433 utm-n/s 488584 utm-e/w	77.06.04 77.09.06	BP NORSKALD	APPRAISAL P&A. DRY HOLE	71 25	4191 PERMIAN
7/12-04	183 019	57 05 36.66 02 51 36.96 6327946 utm-n/s 491532 utm-e/w	77.09.11 77.12.12	BP NORSKALD	APPRAISAL SUSPENDED	71 25	3621 TRIASSIC
7/12-04 R	183 019	57 05 36.66 02 51 36.96 6327946 utm-n/s 491532 utm-e/w	84.03.16 84.03.20	BP DYVI ALPHA	APPRAISAL P&A. OIL DISCOVERY	71 25	3621 TRIASSIC
7/12-05	276 019	57 09 49.01 02 46 42.35 6335763 utm-n/s 486598 utm-e/w	81.02.06 81.06.07	BP BORGSTEN DOLPHIN	WILDCAT P&A. OIL DISCOVERY	73 25	4440 PERMIAN
7/12-06	282 019	57 07 16.85 02 50 04.14 6331047 utm-n/s 489977 utm-e/w	81.04.10 81.07.24	BP SEDCO 707	APPRAISAL P&A. OIL DISC.	69 25	3700 TRIASSIC
8/01-01	66 016	57 51 43.53 03 12 27.64 6413520 utm-n/s 512323 utm-e/w	71.12.02 72.02.07	PHILLIPS OCEAN TIDE	WILDCAT P&A. DRY HOLE	67 31	2971 PERMIAN
8/03-01	1 003	57 59 13.16 03 40 13.29 6427604 utm-n/s 539640 utm-e/w	66.07.19 66.10.10	ESSO OCEAN TRAVELER	WILDCAT P&A. DRY HOLE	94 26	3015 PRE-DEVON
8/03-02	345 071	57 54 06.61 03 47 45.41 6418205 utm-n/s 547178 utm-e/w	82.10.04 82.12.04	STATOIL WEST VANGUARD	WILDCAT P&A. DRY HOLE	78 22	2657 TRIASSIC
8/04-01	177 005	57 44 50.24 03 00 03.86 6400720 utm-n/s 500063 utm-e/w	77.06.21 77.07.25	UNION NORJARL	WILDCAT P&A. DRY HOLE	63 24	2632 L.JURASSIC
8/09-01	145 013	57 26 27.28 03 51 03.48 6366931 utm-n/s 551083 utm-e/w	75.12.22 76.02.10	CONOCO OCEAN VICTORY	WILDCAT P&A. DRY HOLE	69 24	2375 L.PERMIAN
8/10-01	27 017	57 08 01.04 03 03 55.09 6332404 utm-n/s 503952 utm-e/w	69.05.30 69.07.01	PHILLIPS OCEAN VIKING	WILDCAT P&A. DRY HOLE	66 27	3099 PERMIAN
8/10-02	240 017	57 08 06.90 03 18 09.50 6332623 utm-n/s 518318 utm-e/w	80.02.05 80.03.17	PHILLIPS NORTRYM	WILDCAT P&A. DRY HOLE.	66 25	2997 PERMIAN
8/11-01	126 017	57 06 30.50 03 38 26.00 6329785 utm-n/s 538800 utm-e/w	75.04.24 75.06.29	PHILLIPS OCEAN VIKING	WILDCAT P&A. DRY HOLE	64 27	3810 L.TRIASSIC
8/12-01	56 014	57 13 18.60 03 46 45.13 6342490 utm-n/s 547054 utm-e/w	71.05.31 71.07.23	CONOCO MAERSK EXPLORER	WILDCAT P&A. DRY HOLE	63 34	2875 TRIASSIC
9/03-01	525 115	57 49 45.19 04 45 58.38 6411211 utm-n/s 604894 utm-e/w	86.07.29 86.09.04	SHELL BORGNY DOLPHIN	WILDCAT P&A. DRY HOLE	124 25	1971 TRIASSIC
9/04-U1	11 013	57 35 02.00 04 01 13.00 6382987 utm-n/s 561007 utm-e/w	68.03.31 68.05.19	CONOCO ENDEAVOUR	WILDCAT P&A. DRY HOLE	68 29	2963 PERMIAN
9/04-02	39 013	57 41 11.05 04 02 34.85 6394420 utm-n/s 562191 utm-e/w	70.07.19 70.08.29	TEXACO GULFTIDE	WILDCAT P&A. DRY HOLE	70 29	3026 TRIASSIC
9/04-03	74 013	57 36 54.50 04 18 57.70 6386771 utm-n/s 578623 utm-e/w	72.07.14 72.08.19	CONOCO OCEAN TIDE	WILDCAT P&A. DRY HOLE	72 27	2681 TRIASSIC

BOREHOLE LIST   EXPLORATION (with utm-coord.)   Date: 23.Januar 1987   PB/ABS

BOREHOLE	NPD NO LIC NO	POSITION	SPUD COMPL.	OPERATOR RIG	SPUD CLASS. COMPLETION CLASS.	WD KB	TD (RKB) FORM. AT

R=re-entry, X=did not reach prospect, S=deviated hole, A=side-tracked new hole

9/04-04	178	57 42 01.48	77.07.19	SAGA	WILDCAT	78	2902
	013	04 13 20.79	77.08.20	DEEPSEA SAGA	P&A. DRY HOLE	25	TRIASSIC
		6396158 utm-n/s	572861 utm-e/w				
9/08-01	5	57 20 33.97	68.05.23	ESSO	WILDCAT	68	2176
	003	04 20 12.98	68.06.29	ENDEAVOUR	P&A. DRY HOLE	39	PERMIAN
		6356476 utm-n/s	580468 utm-e/w				
9/10-01	46	57 00 04.02	70.08.30	MURPHY	WILDCAT	59	2205
	022	04 00 33.52	70.09.18	GULFTIDE	P&A. DRY HOLE	29	L.CRET.
		6318105 utm-n/s	561313 utm-e/w				
9/11-01	57	57 00 41.40	71.07.03	CONOCO	WILDCAT	65	2197
	014	04 31 40.60	71.08.19	TRANSWORLD RIG61	P&A. DRY HOLE	25	TRIASSIC
		6319845 utm-n/s	592790 utm-e/w				
9/12-01	24	57 11 40.00	69.03.28	SHELL	WILDCAT	61	2698
	012	04 57 21.00	69.05.06	SEDNETH I	P&A. DRY HOLE	26	TRIASSIC
		6340869 utm-n/s	618188 utm-e/w				
10/05-01	156	57 34 51.90	76.05.31	ELF	WILDCAT	98	1843
	015	05 35 25.00	76.06.26	NORJARL	P&A. DRY HOLE	24	L.PERMIAN
		6385172 utm-n/s	654878 utm-e/w				
10/08-01	38	57 25 00.01	70.12.12	ELF	WILDCAT	81	2861
	009	05 34 21.71	71.01.17	NEPTUNE 7	P&A. DRY HOLE	24	PERMIAN
		6366836 utm-n/s	654520 utm-e/w				
11/09-01	147	57 16 33.20	76.01.16	ELF	WILDCAT	73	1922
	009	06 44 52.60	76.02.28	DEEPSEA DRILLER	P&A. DRY HOLE	25	TRIASSIC
		6350484 utm-n/s	364214 utm-e/w				
11/10-01	31	57 00 46.00	69.08.02	SYRACUSE	WILDCAT	68	2430
	021	06 10 04.70	69.08.19	OCEAN VIKING	P&A. DRY HOLE	27	TRIASSIC
		6322515 utm-n/s	328039 utm-e/w				
15/02-01	310	58 45 19.75	81.11.26	HYDRO	WILDCAT	109	4600
	048	01 35 40.54	82.02.24	NORTRYM	P&A. DRY HOLE	25	PERMIAN
		6513827 utm-n/s	418684 utm-e/w				
15/03-01 S	121	58 50 57.00	74.11.27	ELF	WILDCAT	109	5132
	025	01 43 13.25	75.07.06	DEEPSEA DRILLER	P&A. GAS/COND. DISC.	25	M.JURASSI
		6524112 utm-n/s	426158 utm-e/w				
15/03-02	168	58 59 00.50	76.10.29	ELF	WILDCAT	116	4262
	025	01 47 12.60	77.01.24	POLYGLOMAR DRILL	SUSPENDED	25	L.JURASSI
		6538995 utm-n/s	430266 utm-e/w				
15/03-02 R	168	58 59 00.50	77.07.26	ELF	WILDCAT	116	4989
	025	01 47 12.60	77.11.27	PENTAGONE 84	P&A. DRY HOLE	24	M.JURASSI
		6538995 utm-n/s	430266 utm-e/w				
15/03-03	209	58 52 31.25	79.01.05	ELF	APPRAISAL	25	5116
	025	01 46 46.24	79.08.09	PENTAGONE 84	P&A. GAS/COND. DISC.	25	TRIASSIC
		6526964 utm-n/s	429625 utm-e/w				
15/03-04	304	58 49 05.28	81.10.03	ELF	WILDCAT	107	4259
	025	01 52 59.82	82.03.30	BORGSTEN DOLPHIN	P&A. OIL/GAS DISC.	25	TRIASSSIC
		6520488 utm-n/s	435502 utm-e/w				
15/03-05	397	58 47 13.04	83.12.28	ELF	WILDCAT	103	4130
	025	01 50 34.11	84.05.13	BYFORD DOLPHIN	P&A. OIL DISCOVERY.	25	JURASSIC
		6517057 utm-n/s	433105 utm-e/w				
15/05-01	186	58 35 04.39	77.11.26	HYDRO	WILDCAT	119	3800
	048	01 39 08.35	78.04.07	TREASURE SEEKER	P&A. GAS DISCOVERY	25	JURASSIC
		6494728 utm-n/s	421641 utm-e/w				
15/05-02	200	58 38 56.70	78.08.16	HYDRO	WILDCAT	120	4322
	048	01 36 16.50	78.12.10	TREASURE SEEKER	P&A. OIL/GAS DISC.	25	TRIASSIC
		6501969 utm-n/s	419015 utm-e/w				

BOREHOLE LIST EXPLORATION (with utm-coord.) Date: 23.Januar 1987 PB/ABS

BOREHOLE	NPD NO LIC NO	POSITION	SPUD COMPL.	OPERATOR RIG	SPUD CLASS. COMPLETION CLASS.	WD KB	TD (RKB) FORM. AT TD
R=re-entry, X=did not reach prospect, S=deviated hole, A=side-tracked new hole							
15/05-03	260 048	58 43 47.93 01 38 12.05 6510937 utm-n/s 421061 utm-e/w	80.08.21 80.12.06	HYDRO NORTRYM	WILDCAT P&A. DRY HOLE.	110 25	5042 PERMIAN.
15/06-01 X	60 029	58 32 48.26 01 41 32.67 6490471 utm-n/s 423890 utm-e/w	71.08.07 71.09.08	ESSO GLOMAR GRAND ISL	WILDCAT P&A. JUNKED	116 9	1616 TERTIARY
15/06-02	62 029	58 32 47.58 01 41 54.09 6490444 utm-n/s 424236 utm-e/w	71.09.09 71.10.26	ESSO GLOMAR GRAND ISL	WILDCAT SUSPENDED	115 9	3331 E.CRET.
15/06-02 R	62 029	58 32 47.58 01 41 54.09 6490444 utm-n/s 424236 utm-e/w	74.05.09 74.08.01	ESSO DRILLMASTER	WILDCAT P&A. GAS/COND. DISC.	115 24	4779 PERMIAN
15/06-03	114 029	58 30 15.84 01 42 39.49 6485738 utm-n/s 424879 utm-e/w	74.09.05 74.12.19	ESSO DRILLMASTER	WILDCAT P&A. GAS/COND. DISC.	113 24	3795 TRIAS
15/06-04	159 029	58 37 30.97 01 48 19.18 6499092 utm-n/s 430617 utm-e/w	76.06.28 76.08.16	ESSO NORJARL	WILDCAT P&A. DRY HOLE	111 25	3530 TRIASSIC
15/06-05	185 029	58 30 29.67 01 45 50.40 6486106 utm-n/s 427978 utm-e/w	77.10.10 77.11.29	ESSO DRILLMASTER	APPRAISAL P&A. DRY HOLE	108 24	3824 TRIASSIC
15/06-06	308 029	58 31 14.57 01 44 53.78 6487513 utm-n/s 427087 utm-e/w	82.04.10 82.06.09	ESSO GLOMAR BISCAY II	APPRAISAL P&A. GAS/COND. DISC.	105 25	3760 TRIASSIC
15/08-01	298 046	58 21 48.09 01 34 22.31 6470197 utm-n/s 416499 utm-e/w	81.07.18 82.01.07	STATOIL GLOMAR BISCAY II	WILDCAT P&A. GAS/COND. DISC.	112 25	4300 TRIASSIC
15/09-01	171 046	58 28 19.00 01 45 11.70 6482077 utm-n/s 427276 utm-e/w	77.02.24 77.05.30	STATOIL ROSS RIG	WILDCAT P&A. OIL/GAS DISC.	106 25	3734 L.TRIASSIC
15/09-02	194 046	58 25 34.06 01 42 28.20 6477026 utm-n/s 424529 utm-e/w	78.04.11 78.06.17	STATOIL ROSS RIG	WILDCAT P&A. GAS/COND. DISC.	108 25	3764 TRIASSIC
15/09-03	206 046	58 29 10.04 01 41 38.46 6483722 utm-n/s 423852 utm-e/w	78.12.16 79.04.03	STATOIL ROSS RIG	APPRAISAL P&A. HYDROCARBONES	115 25	3796 TRIASSIC
15/09-04	213 046	58 24 00.26 01 47 06.93 6474041 utm-n/s 428999 utm-e/w	79.04.05 79.06.14	STATOIL ROSS RIG	WILDCAT P&A. GAS/COND. DISC.	102 25	3716 TRIASSIC
15/09-05	232 046	58 24 12.47 01 42 29.20 6474503 utm-n/s 424497 utm-e/w	79.11.20 80.04.09	STATOIL NORSKALD	APPRAISAL P&A. GAS/COND. DISC.	108 25	3946 TRIAS
15/09-06	244 046	58 27 13.52 01 41 31.60 6480121 utm-n/s 423671 utm-e/w	80.05.06 80.09.07	STATOIL NORSKALD	APPRAISAL P&A. DRY HOLE	112 25	3146 TRIASSIC.
15/09-07	271 046	58 22 56.07 01 42 45.99 6472135 utm-n/s 424724 utm-e/w	80.12.26 81.04.29	STATOIL NORDRAUG	APPRAISAL P&A.GAS/COND. DISC.	107 25	3776 TRIASSIC
15/09-08	279 046	58 22 55.81 01 46 28.32 6472060 utm-n/s 428335 utm-e/w	81.03.04 81.05.25	STATOIL NORTRYM	APPRAISAL P&A. GAS/COND. DISC.	106 25	3730 TRIASSIC
15/09-09	285 046	58 20 47.81 01 53 24.98 6467983 utm-n/s 435038 utm-e/w	81.05.04 81.07.14	STATOIL NORDRAUG	WILDCAT P&A. GAS/COND. DISC.	83 25	3044 PERMIAN

BOREHOLE LIST EXPLORATION (with utm-coord.) Date: 23.Januar 1987 PB/ABS

R=re-entry, X=did not reach prospect, S=deviated hole, A=side-tracked new hole

BOREHOLE	NPD NO LIC NO	POSITION	SPUD COMPL.	OPERATOR RIG	SPUD CLASS. COMPLETION CLASS.	WD KB	TD (RKB) FORM. AT T
15/09-10	302 046	58 19 42.92 01 46 20.61 6466096 utm-n/s 428101 utm-e/w	81.09.15 81.11.07	STATOIL NORDRAUG	WILDCAT P&A. DRY HOLE.	98 25	3289 TRIASSIC
15/09-11	303 046	58 24 02.53 01 53 41.79 6474001 utm-n/s 435410 utm-e/w	81.09.18 81.12.23	STATOIL ROSS RIG	APPRAISAL P&A. GAS/COND. DISC.	88 25	2950 TRIASSIC
15/09-12	309 046	58 27 22.71 01 43 04.84 6480376 utm-n/s 425188 utm-e/w	81.11.26 82.02.23	STATOIL NORDRAUG	APPRAISAL SUSPENDED	107 25	3740 TRIASSIC
15/09-12 R	309 046	58 27 22.71 01 43 04.84 6480376 utm-n/s 425188 utm-e/w	82.03.26 82.04.27	STATOIL DEEPSEA SAGA	APPRAISAL P&A. GAS/COND. DISC.	107 25	3740 TRIASSIC
15/09-13	319 046	58 22 25.96 01 56 02.86 6470978 utm-n/s 437653 utm-e/w	82.03.21 82.05.27	STATOIL ROSS RIG	APPRAISAL P&A. GAS/COND. DISC.	81 25	3282 PERMIAN
15/09-14	327 046	58 17 23.19 01 41 28.06 6461866 utm-n/s 423258 utm-e/w	82.05.01 82.06.27	STATOIL DEEP SEA SAGA	WILDCAT P&A. DRY HOLE.	101 25	3563 TRIASSIC
15/09-15	330 046	58 18 07.35 01 55 19.28 6462991 utm-n/s 436817 utm-e/w	82.05.28 82.08.01	STATOIL ROSS RIG	WILDCAT P&A. GAS/COND. DISC.	83 25	3200 TRIASSIC
15/09-16	335 046	58 22 09.97 01 52 43.16 6470535 utm-n/s 434400 utm-e/w	82.06.28 82.08.24	STATOIL DEEP SEA SAGA	APPRAISAL P&A. GAS/COND. DISC.	85 25	3120 PERMIAN
15/09-17	356 046	58 26 44.19 01 56 53.58 6478950 utm-n/s 438602 utm-e/w	82.12.09 83.03.30	STATOIL WEST VANGUARD	WILDCAT SUSP. GAS/COND. DISC	86 22	3120 TRIASSIC
15/09-18	396 046	58 21 01.95 01 48 12.27 6468508 utm-n/s 429961 utm-e/w	83.12.16 84.03.02	STATOIL DEEPSEA BERGEN	WILDCAT P&A. DRY HOLE	97 23	3622 TRIASSIC
15/12-01	132 038	58 10 32.60 01 44 23.60 6449113 utm-n/s 425880 utm-e/w	75.07.07 75.09.06	STATOIL ROSS RIG	WILDCAT P&A. DRY HOLE	94 25	3469 L.TRIASSI
15/12-02	146 038	58 08 31.10 01 55 47.47 6445163 utm-n/s 436993 utm-e/w	76.01.07 76.02.27	STATOIL ROSS RIG	WILDCAT P&A. DRY HOLE	86 25	2924 TRIASSIC
15/12-03	252 038	58 14 36.59 01 52 45.67 6456514 utm-n/s 434208 utm-e/w	80.06.21 80.12.22	STATOIL NORDRAUG	WILDCAT P&A. DRY HOLE.	86 25	4450 PERMIAN
15/12-04	433 038	58 03 09.16 01 54 11.61 6435232 utm-n/s 435264 utm-e/w	84.09.13 84.10.31	STATOIL DEEPSEA BERGEN	WILDCAT P&A. OIL DISCOVERY.	87 23	3179 TRIASSIC
15/12-05	507 038	58 04 53.36 01 54 54.24 6438443 utm-n/s 436014 utm-e/w	86.03.12 86.05.04	STATOIL ROSS ISLE	WILDCAT P&A. OIL DISCOVERY	84 22	3105 TRIASSIC
16/01-01	8 001	58 59 18.38 02 02 02.97 6539316 utm-n/s 444490 utm-e/w	67.09.26 67.12.10	ESSO OCEAN TRAVELER	WILDCAT P&A. DRY HOLE	115 26	3214 CRETACEOU
16/01-02	160 001	58 56 16.98 02 13 29.56 6533562 utm-n/s 455385 utm-e/w	76.07.05 76.08.07	ESSO ROSS RIG	WILDCAT P&A. DRY HOLE	138 25	2919 BASEMENT
16/01-03	340 001	58 48 48.90 02 02 56.45 6519834 utm-n/s 445067 utm-e/w	82.07.29 82.09.27	ESSO GLOMAR BISCAY II	WILDCAT P&A. DRY HOLE.	108 25	3498 PALEOZOI(

OREHOLE LIST EXPLORATION (with utm-coord.) Date: 23.Januar 1987 PB/ABS

R=re-entry, X=did not reach prospect, S=deviated hole, A=side-tracked new hole

OREHOLE	NPD NO LIC NO	POSITION	SPUD COMPL.	OPERATOR RIG	SPUD CLASS. COMPLETION CLASS.	WD KB	TD (RKB) FORM. AT TD
16/02-01	3 001	58 53 35.19 02 21 25.77 6528477 utm-n/s	67.07.11 67.08.09 462951 utm-e/w	ESSO OCEAN TRAVELER	WILDCAT P&A. DRY HOLE	113 26	1906 PRE-PERM
16/03-01 X	148 007	58 47 12.88 02 47 32.25 6516492 utm-n/s	76.01.31 76.02.10 487992 utm-e/w	ELF POLYGLOMAR DRILL	WILDCAT P&A. JUNKED	118 24	453 TERTIARY
16/03-02	151 007	58 47 12.80 02 47 34.70 6516491 utm-n/s	76.02.11 76.03.08 488031 utm-e/w	ELF POLYGLOMAR DRILL	WILDCAT P&A. DRY HOLE	118 24	2021 L.JURASSIC
16/04-01	432 087	58 38 18.33 02 08 17.03 6500262 utm-n/s	84.09.08 84.11.18 449960 utm-e/w	HYDRO TREASURE SEEKER	WILDCAT P&A. DRY HOLE.	96 25	2909 BASEMENT.
16/05-01	52 007	58 38 53.66 02 29 39.69 6501144 utm-n/s	71.01.17 71.02.08 470652 utm-e/w	ELF NEPTUNE 7	WILDCAT P&A. DRY HOLE	103 24	1944 BASEMENT
16/06-01	9 007	58 42 06.00 02 54 44.00 6506986 utm-n/s	67.11.07 68.01.19 494913 utm-e/w	ELF OCEAN VIKING	WILDCAT P&A. DRY HOLE	117 28	2061 PRE-PERM
16/07-01	6 001	58 20 22.38 02 18 49.96 6466867 utm-n/s	67.08.11 67.09.24 459826 utm-e/w	ESSO OCEAN TRAVELER	WILDCAT P&A. DRY HOLE	80 26	2781 PERMIAN
16/07-02	314 072	58 28 26.50 02 02 03.12 6482040 utm-n/s	82.01.11 82.03.30 443667 utm-e/w	ESSO GLOMAR BISCAY II	WILDCAT P&A. GAS DISCOVERY	83 25	3146 PERMIAN
16/07-03	331 072	58 24 21.96 02 08 08.06 6474397 utm-n/s	82.06.11 82.07.27 449482 utm-e/w	ESSO GLOMAR BISCAY II	WILDCAT P&A. DRY HOLE.	85 25	3141 PERMIAN
16/07-04	347 072	58 17 09.75 02 01 30.07 6461117 utm-n/s	82.10.15 82.12.06 442828 utm-e/w	ESSO GLOMAR BISCAY II	WILDCAT P&A. GAS/COND. DISC.	78 25	2781 TRIASSIC
16/07-05	423 072	58 21 06.30 02 04 56.16 6468385 utm-n/s	84.06.27 84.08.03 446284 utm-e/w	ESSO ZAPATA UGLAND	WILDCAT P&A. DRY HOLE.	80 25	2900 TRIASSIC ?
16/08-01	166 020	58 27 24.80 02 25 56.80 6479866 utm-n/s	76.09.25 76.10.29 466879 utm-e/w	CONOCO NORSKALD	WILDCAT P&A. DRY HOLE	81 25	2301 TRIASSIC
16/08-02	246 020	58 20 59.81 02 24 59.58 6467968 utm-n/s	80.04.03 80.08.13 465848 utm-e/w	CONOCO SEDCO H	WILDCAT P&A. DRY HOLE.	73 30	3585 PERMIAN
16/09-01	12 002	58 22 41.48 02 48 16.99 6470981 utm-n/s	68.05.08 68.07.12 488578 utm-e/w	ESSO GLOMAR GRAND ISL	WILDCAT P&A. DRY HOLE	89 9	3655 PERM
16/10-01	515 101	58 03 23.68 02 03 14.05 6435546 utm-n/s	86.05.25 86.07.14 444163 utm-e/w	AGIP DYVI STENA	WILDCAT P&A. DRY HOLE	84 25	3151 PERMIAN
16/11-01 S	4 016	58 03 55.74 02 34 00.65 6436229 utm-n/s	67.07.14 67.10.31 474442 utm-e/w	PHILLIPS OCEAN VIKING	WILDCAT P&A. DRY HOLE	67 27	3050 PERMIAN
16/11-02	86 016	58 11 07.90 02 28 13.50 6449634 utm-n/s	73.06.16 73.07.23 468857 utm-e/w	PHILLIPS OCEAN VIKING	WILDCAT P&A. DRY HOLE	70 27	2377 PERMIAN
17/04-01	14 007	58 35 54.00 03 16 05.00 6495508 utm-n/s	68.06.15 68.08.22 515580 utm-e/w	ELF OCEAN VIKING	WILDCAT P&A. DRY HOLE	105 27	3997 PERMIAN

BOREHOLE LIST EXPLORATION (with utm-coord.) Date: 23.Januar 1987 PB/ABS

BOREHOLE	NPD NO LIC NO	POSITION	SPUD COMPL.	OPERATOR RIG	SPUD CLASS. COMPLETION CLASS.	WD KB	TD (RKB) FORM. AT

R=re-entry, X=did not reach prospect, S=deviated hole, A=side-tracked new hole

BOREHOLE	NPD NO / LIC NO	POSITION	SPUD / COMPL.	OPERATOR / RIG	SPUD CLASS. / COMPLETION CLASS.	WD / KB	TD (RKB) / FORM. AT
17/09-01	93	58 28 27.26	73.09.16	ESSO	WILDCAT	159	2818
	002	03 50 16.18	73.10.23	GLOMAR GRAND ISL	SUSPENDED	9	JURASSIC
		6481963 utm-n/s	548868 utm-e/w				
17/09-01 R	93	58 28 27.26	74.05.18	ESSO	WILDCAT	159	3161
	002	03 50 16.18	74.06.11	GLOMAR GRAND ISL	P&A. DRY HOLE	9	TRIASSIC
		6481963 utm-n/s	548868 utm-e/w				
17/10-01	22	58 01 54.00	68.12.30	SHELL	WILDCAT	88	3591
	010	03 09 58.00	69.03.24	SEDNETH I	P&A. DRY HOLE	26	TRIASSIC
		6432394 utm-n/s	509810 utm-e/w				
17/11-01	13	58 12 36.00	68.05.24	SHELL	WILDCAT	77	3270
	010	03 20 34.90	68.06.30	ORION	P&A. DRY HOLE	27	PERMIAN
		6452289 utm-n/s	520158 utm-e/w				
17/11-02	154	58 06 54.91	76.04.12	SHELL	WILDCAT	79	2644
	010	03 22 09.81	76.05.17	CHRIS CHENERY	P&A. DRY HOLE	33	TRIASSIC
		6441748 utm-n/s	521765 utm-e/w				
17/12-01	64	58 11 15.40	71.10.27	PHILLIPS	WILDCAT	62	
	016	03 56 22.20	71.11.01	MAERSK EXPLORER	SUSP. OIL DISCOVERY	27	
		6450129 utm-n/s	555244 utm-e/w				
17/12-01 R	64	58 11 15.40	72.03.14	PHILLIPS	WILDCAT	62	4298
	016	03 56 22.20	72.06.21	OCEAN VIKING	P&A. OIL DISCOVERY	27	PERMIAN
		6450129 utm-n/s	555244 utm-e/w				
17/12-02	92	58 08 19.70	73.08.31	PHILLIPS	WILDCAT	66	2334
	016	03 42 10.90	73.10.09	OCEAN VIKING	P&A. OIL/GAS DISC.	27	DEVONIAN
		6444526 utm-n/s	541396 utm-e/w				
17/12-03	233	58 11 32.84	79.12.14	PHILLIPS	APPRAISAL	112	2730
	016	03 51 44.06	80.02.03	NORTRYM	P&A. DRY HOLE	25	TRIASSIC
		6450608 utm-n/s	550694 utm-e/w				
18/10-01	227	58 04 38.25	79.10.19	ELF	WILDCAT	96	2800
	008	04 07 00.83	80.01.01	DYVI ALPHA	P&A. OIL/GAS DISC.	25	JURASSIC
		6438007 utm-n/s	565878 utm-e/w				
18/11-01	104	58 04 21.30	74.02.17	ELF	WILDCAT	116	2086
	008	04 32 00.10	74.03.31	DEEPSEA DRILLER	P&A. DRY HOLE	25	BASEMENT
		6437964 utm-n/s	590452 utm-e/w				
24/06-01	452	59 36 00.25	85.02.09	TOTAL	WILDCAT	122	4937
	088	01 55 22.16	85.08.25	ZAPATA UGLAND	P&A. OIL/GAS DISC.	25	E.JURASSI
		6607520 utm-n/s	439192 utm-e/w				
24/09-01	152	59 16 09.48	76.02.29	CONOCO	WILDCAT	143	4907
	039	01 47 31.18	76.07.03	ROSS RIG	P&A. DRY HOLE	25	L.JURASSI
		6570814 utm-n/s	431139 utm-e/w				
24/09-02	176	59 16 12.43	77.06.13	CONOCO	WILDCAT	118	2743
	039	01 55 12.26	77.07.22	BORGNY DOLPHIN	P&A. DRY HOLE	25	CRETACEOU
		6570781 utm-n/s	438440 utm-e/w				
24/09-03	275	59 22 25.38	81.01.28	CONOCO	WILDCAT	120	3051
	039	01 46 27.28	81.04.15	SEDCO 704	P&A. OIL DISCOVERY	26	E.CRET.
		6582461 utm-n/s	430341 utm-e/w				
24/12-01	189	59 02 29.80	78.01.16	STATOIL	WILDCAT	113	
	045	01 52 57.93	78.04.09	ROSS RIG	SUSPENDED	25	
		6545372 utm-n/s	435888 utm-e/w				
24/12-01 R	189	59 02 29.80	78.06.07	STATOIL	WILDCAT	113	4825
	045	01 52 57.93	78.08.14	TREASURE SEEKER	P&A. DRY HOLE	25	TRIASSIC
		6545372 utm-n/s	435888 utm-e/w				
24/12-02	294	59 12 00.75	81.06.23	STATOIL	WILDCAT	119	5100
	045	01 52 53.34	82.01.21	DYVI DELTA	P&A. DRY HOLE	32	JURASSIC
		6563033 utm-n/s	436110 utm-e/w				

BOREHOLE LIST EXPLORATION (with utm-coord.) Date: 23.Januar 1987 PB/ABS

BOREHOLE		NPD NO LIC NO	POSITION	SPUD COMPL.	OPERATOR RIG	SPUD CLASS. COMPLETION CLASS.	WD KB	TD (RKB) FORM. AT TD
R=re-entry, X=did not reach prospect, S=deviated hole, A=side-tracked new hole								
25/01-01		53 024	59 53 17.40 02 04 42.70 6639470 utm-n/s	71.03.30 71.07.22 448427 utm-e/w	ELF NEPTUNE 7	WILDCAT P&A. GAS DISCOVERY	101 24	4570 JURASSIC
25/01-02		58 024	59 56 08.00 02 04 54.60 6644744 utm-n/s	71.07.22 71.08.25 448685 utm-e/w	ELF NEPTUNE 7	APPRAISAL P&A. GAS/COND. DISC.	103 24	2772 L.CRET.
25/01-03		65 024	59 54 05.00 02 10 08.50 6640875 utm-n/s	71.11.14 72.01.27 453510 utm-e/w	ELF PENTAGONE 81	APPRAISAL P&A. GAS DISCOVERY	105 24	2872 L.CRET.
25/01-04		108 024	59 59 52.51 02 15 54.80 6651563 utm-n/s	74.04.01 74.05.30 459011 utm-e/w	ELF DEEPSEA DRILLER	WILDCAT P&A. OIL/GAS DISC.	107 25	2795 L.CRET.
25/01-05		136 024	59 52 27.92 02 06 36.99 6637914 utm-n/s	75.07.24 75.09.12 450183 utm-e/w	ELF DEEPSEA DRILLER	APPRAISAL P&A. OIL/GAS DISC.	102 28	2262 TERTIARY
25/01-06		190 024	59 46 57.68 02 05 42.23 6627710 utm-n/s	78.01.26 78.03.18 449192 utm-e/w	ELF PENTAGONE 84	WILDCAT P&A. DRY HOLE	105 24	2895 L.CRET.
25/01-07		455 024	59 55 08.28 02 04 52.33 6642897 utm-n/s	85.03.09 85.05.27 448624 utm-e/w	ELF BYFORD DOLPHIN	APPRAISAL SUSP.GAS DISCOVERY	351 25	2720 CRETACEOUS
25/01-08	S	466 024	59 54 03.28 02 06 09.79 6640870 utm-n/s	85.05.28 85.07.25 449800 utm-e/w	ELF BYFORD DOLPHIN	APPRAISAL SUSP. CONTROL HOLE	102 25	2650 PALEOCENE
25/01-09		528 129	59 50 13.34 02 13 30.24 6633671 utm-n/s	86.09.09 86.10.11 456562 utm-e/w	HYDRO TREASURE SCOUT	WILDCAT P&A. DRY HOLE	112 23	2807 E.CRET.
25/02-01		90 026	59 55 30.20 02 23 02.00 6643380 utm-n/s	73.08.04 73.09.21 465555 utm-e/w	ELF NEPTUNE 7	WILDCAT P&A. GASS DISCOVERY	104 24	2741 L.CRET.
25/02-02		110 026	59 53 34.80 02 20 15.80 6639834 utm-n/s	74.05.31 74.07.11 462939 utm-e/w	ELF DEEPSEA DRILLER	WILDCAT P&A. GAS DISCOVERY	104 25	2740 L.CRET.
25/02-03		115 026	59 49 09.70 02 21 21.40 6631623 utm-n/s	74.09.01 74.10.09 463879 utm-e/w	ELF DEEPSEA DRILLER	WILDCAT P&A. DRY HOLE	113 25	2795 L.CRET.
25/02-04		129 026	59 58 44.09 02 22 58.88 6649378 utm-n/s	75.05.14 75.10.20 465563 utm-e/w	ELF NEPTUNE 7	WILDCAT P&A. GAS/COND. DISC.	113 24	4384 L.TRIASSIC
25/02-05		153 026	59 48 01.40 02 28 18.30 6629454 utm-n/s	76.03.08 76.03.08 470356 utm-e/w	ELF POLYGLOMAR DRILL	WILDCAT P&A. OIL/GAS DISC.	117 24	4002 TRIASSIC
25/02-06		180 026	59 45 33.55 02 33 05.96 6624847 utm-n/s	77.08.01 77.11.15 474809 utm-e/w	ELF POLYGLOMAR DRILL	WILDCAT P&A. OIL DISC.	117 24	3750 TRIASSIC
25/02-07		321 026	59 54 20.41 02 38 25.87 6641116 utm-n/s	82.04.01 82.07.12 479891 utm-e/w	 BORGSTEN DOLPHIN	WILDCAT P&A. DRY HOLE	117 25	4110 TRIASSIC
25/02-08		420 026	59 54 45.14 02 20 27.62 6642009 utm-n/s	84.06.18 84.08.01 463144 utm-e/w	ELF LE PELERIN	APPRAISAL P&A. OIL/GAS DISC.	106 12	2380 E.TERTIARY
25/02-09		470 026	59 53 10.16 02 22 15.09 6639055 utm-n/s	85.06.14 85.07.24 464786 utm-e/w	ELF NORTRYM	APPRAISAL SUSP. GAS DISCOVERY	112 25	2297 PALEOCENE

BOREHOLE LIST EXPLORATION (with utm-coord.) Date: 23.Januar 1987 PB/ABS

R=re-entry, X=did not reach prospect, S=deviated hole, A=side-tracked new hole

BOREHOLE		NPD NO LIC NO	POSITION	SPUD COMPL.	OPERATOR RIG	SPUD CLASS. COMPLETION CLASS.	WD KB	TD (RKB) FORM. AT T
25/02-10	S	494 112	59 53 11.80 02 30 08.33 6639042 utm-n/s 472144 utm-e/w	85.12.02 86.03.19	ELF HENRY GOODRICH	WILDCAT SUSPENDED	120 21	2967 TERTIARY
25/04-01		69 036	59 34 27.30 02 13 22.60 6604408 utm-n/s 456100 utm-e/w	72.07.01 72.12.09	ELF NEPTUNE 7	WILDCAT P&A. GAS/COND. DISC.	122 23	4060 TRIASSIC
25/04-02		98 036	59 35 46.51 02 18 51.40 6606801 utm-n/s 461285 utm-e/w	73.10.18 73.12.06	ELF NEPTUNE 7	WILDCAT P&A. OIL/GAS DISC.	121 24	2775 L.CRET.
25/04-03		119 036	59 30 44.20 02 05 28.90 6597601 utm-n/s 448572 utm-e/w	74.10.13 74.11.22	ELF DEEPSEA DRILLER	WILDCAT P&A. GAS/COND. DISC.	122 25	2714 L.CRET.
25/04-04		127 036	59 36 08.63 02 12 26.64 6607553 utm-n/s 455259 utm-e/w	75.05.16 75.07.07	ELF ODIN DRILL	APPRAISAL P&A. OIL/GAS DISC.	116 25	2681 DANIAN
25/04-05		254 036	59 34 04.75 02 11 39.21 6603729 utm-n/s 454469 utm-e/w	80.06.27 81.03.26	ELF DYVI ALPHA	APPRAISAL P&A. OIL DISCOVERY.	121 25	4355 TRIASSIC
25/06-01		493 117	59 31 32.04 02 48 02.07 6598746 utm-n/s 488717 utm-e/w	85.12.18 86.02.03	SAGA TREASURE SAGA	WILDCAT P&A. OIL/GAS DISC.	120 26	2281 BASEMENT
25/07-01	S	512 103	59 18 35.23 02 16 05.37 6574928 utm-n/s 458330 utm-e/w	86.04.14 86.07.19	CONOCO NORTRYM	WILDCAT P&A. DRY HOLE	127 25	3592 BASEMENT
25/08-01		36 027	59 15 02.19 02 29 38.18 6568218 utm-n/s 471135 utm-e/w	70.04.28 70.07.04	ESSO GLOMAR GRAND ISL	WILDCAT P&A. OIL DISCOVERY	129 9	2606 TRIASSIC
25/08-02		139 027	59 23 43.69 02 22 36.82 6584407 utm-n/s 464611 utm-e/w	75.09.06 75.10.01	ESSO ROSS RIG	WILDCAT P&A. DRY HOLE	25 25	2578 L.TRIASSIC
25/08-03		280 027	59 16 22.17 02 31 18.26 6570681 utm-n/s 472739 utm-e/w	81.03.20 81.04.17	ESSO GLOMAR BISCAY II	WILDCAT P&A. OIL DISCOVERY	130 25	1868 L.CRETACEC
25/10-01		32 028	59 11 21.98 02 19 14.37 6561496 utm-n/s 461183 utm-e/w	69.08.04 69.10.16	ESSO GLOMAR GRAND ISL	WILDCAT SUSPENDED	124 10	1747 TERTIARY
25/10-01 R		32 028	59 11 21.98 02 19 14.37 6561496 utm-n/s 461183 utm-e/w	70.07.07 70.08.03	ESSO GLOMAR GRAND ISL	WILDCAT P&A. DRY HOLE	124 10	2092 JURASSIC
25/10-02		43 028	59 09 38.37 02 11 38.18 6558371 utm-n/s 453904 utm-e/w	70.08.05 70.08.25	ESSO GLOMAR GRAND ISL	WILDCAT SUSPENDED	121 9	2191
25/10-02 R		43 028	59 09 38.37 02 11 38.18 6558371 utm-n/s 453904 utm-e/w	72.05.02 72.07.08	ESSO GLOMAR GRAND ISL	WILDCAT P&A. OIL DISCOVERY	121 9	3181 SILURIAN
25/10-03		45 028	59 12 58.96 02 19 41.98 6564491 utm-n/s 461651 utm-e/w	70.08.27 70.09.13	ESSO GLOMAR GRAND ISL	APPRAISAL P&A. OIL DISCOVERY	126 9	1920 TERTIARY
25/10-04		270 028	59 11 26.04 02 19 50.14 6561615 utm-n/s 461752 utm-e/w	80.12.08 81.01.18	ESSO GLOMAR BISCAY II	APPRAISAL SUSPENDED	126 25	2349 PERMIAN
25/10-04 R		270 028	59 11 26.04 02 19 50.14 6561615 utm-n/s 461752 utm-e/w	81.06.01 81.06.13	ESSO GLOMAR BISCAY II	APPRAISAL P&A. OIL DISCOVERY	126 25	2550 PERMIAN

BOREHOLE LIST EXPLORATION (with utm-coord.) Date: 23.Januar 1987 PB/ABS

BOREHOLE	NPD NO LIC NO	POSITION	SPUD COMPL.	OPERATOR RIG	SPUD CLASS. COMPLETION CLASS.	WD KB	TD (RKB) FORM. AT TD
R=re-entry, X=did not reach prospect, S=deviated hole, A=side-tracked new hole							
25/10-05	291 028	59 10 49.65 02 19 27.31 6560494 utm-n/s 461378 utm-e/w	81.06.15 81.07.17	ESSO GLOMAR BISCAY II	APPRAISAL P&A. OIL DISC.	125 25	2011 L.CRET.
25/11-01	2 001	59 10 57.39 02 24 28.18 6560688 utm-n/s 466157 utm-e/w	66.10.18 67.07.09	ESSO OCEAN TRAVELER	WILDCAT P&A. OIL DISCOVERY	127 27	2460 SILURIAN
25/11-02	47 001	59 13 01.59 02 24 35.06 6564528 utm-n/s 466300 utm-e/w	70.09.14 70.09.24	ESSO GLOMAR GRAND ISL	APPRAISAL P&A. OIL DISC.	130 9	1824 TERTIARY
25/11-03	48 001	59 10 36.19 02 26 19.06 6560016 utm-n/s 467911 utm-e/w	70.09.25 70.10.14	ESSO GLOMAR GRAND ISL	APPRAISAL P&A. DRY HOLE	127 9	1856 DANIAN
25/11-04	49 001	59 09 38.77 02 23 25.08 6558265 utm-n/s 465133 utm-e/w	70.10.15 70.11.20	ESSO GLOMAR GRAND ISL	APPRAISAL P&A. DRY HOLE	127 9	1896 DANIAN
25/11-05	106 001	59 12 10.29 02 21 51.87 6562964 utm-n/s 463697 utm-e/w	74.04.03 74.05.08	ESSO DRILLMASTER	WILDCAT P&A. OIL DISCOVERY	124 24	2164 TRIASSIC
25/11-06	201 001	59 10 59.19 02 23 39.37 6560750 utm-n/s 465382 utm-e/w	78.08.20 78.10.01	ESSO NORSKALD	APPRAISAL P&A. OIL DISC.	126 25	1949 E.CRET.
25/11-07	204 001	59 10 33.25 02 21 46.72 6559965 utm-n/s 463587 utm-e/w	78.10.03 78.10.27	ESSO NORSKALD	APPRAISAL P&A. OIL DISC.	124 25	1946 E.CRET.
25/11-08	223 001	59 11 17.45 02 21 40.04 6561332 utm-n/s 463494 utm-e/w	79.09.08 79.10.16	ESSO DYVI ALPHA	APPRAISAL P&A. OIL DISC.	125 25	1950 E.CRET.T
25/11-09	266 001	59 10 14.36 02 22 31.00 6559374 utm-n/s 464284 utm-e/w	80.11.04 80.12.07	ESSO NORSKALD	APPRAISAL P&A. DRY HOLE.	125 25	1910 LATE CRET.
25/11-10	274 001	59 10 08.51 02 20 57.80 6559206 utm-n/s 462802 utm-e/w	81.01.20 81.02.17	ESSO GLOMAR BISCAY II	APPRAISAL P&A. DRY HOLE	125 25	1998 L.CRET.
25/11-11	278 001	59 11 47.75 02 20 47.60 6562278 utm-n/s 462670 utm-e/w	81.02.19 81.03.19	ESSO GLOMAR BISCAY II	APPRAISAL P&A. OIL DISC.	126 25	1960 L.CRET.
25/11-12	281 001	59 12 02.64 02 22 50.85 6562720 utm-n/s 464630 utm-e/w	81.04.18 81.05.07	ESSO GLOMAR BISCAY II	APPRAISAL P&A. DRY HOLE	127 25	1918 L.CRET.
25/11-13	287 001	59 11 26.67 02 23 23.14 6561602 utm-n/s 465133 utm-e/w	81.05.11 81.05.29	ESSO GLOMAR BISCAY II	APPRAISAL P&A. OIL DISC.	127 25	1932 L.CRET.
25/12-01	95 010	59 01 47.00 02 49 17.10 6543526 utm-n/s 489748 utm-e/w	73.10.03 73.12.27	SHELL SEDCO 135 G	WILDCAT P&A. DRY HOLE	135 24	2855 DEVONIAN
29/03-01	514 119	60 57 50.24 01 56 13.25 6759399 utm-n/s 442441 utm-e/w	86.05.20 86.09.15	TOTAL BYFORD DOLPHIN	WILDCAT P&A. OIL/GAS DISC.	131 25	4427 L.JURASSIC
29/06-01	307 043	60 32 17.94 01 59 24.05 6711947 utm-n/s 444581 utm-e/w	81.10.12 82.05.09	BP SEDCO 707	WILDCAT P&A. GAS/COND. DISC.	124 25	4832 E.JURASSIC
29/09-01	390 040	60 28 33.19 01 59 58.95 6704986 utm-n/s 445008 utm-e/w	83.09.23 84.02.24	HYDRO TREASURE SEEKER	WILDCAT P&A. GAS DISCOVERY	104 25	4703 JURASSIC

BOREHOLE LIST EXPLORATION (with utm-coord.) Date: 23.Januar 1987 PB/ABS

BOREHOLE	NPD NO LIC NO	POSITION	SPUD COMPL.	OPERATOR RIG	SPUD CLASS. COMPLETION CLASS.	WD KB	TD (RKB) FORM. AT T
R=re-entry, X=did not reach prospect, S=deviated hole, A=side-tracked new hole							
30/02-01	328	60 52 05.42	82.05.17	STATOIL	WILDCAT	125	4243
	051	02 38 49.16	82.10.12	DYVI DELTA	SUSP. GAS/COND. DISC	29	JURASSIC
		6748314 utm-n/s	480827 utm-e/w				
30/02-02	447	60 49 53.89	84.12.19	STATOIL	WILDCAT	123	4170
	051	02 39 51.65	85.05.04	DYVI DELTA	P&A. GAS/COND. DISC.	29	JURASSIC
		6744240 utm-n/s	481749 utm-e/w				
30/03-01	215	60 54 15.35	79.06.11	STATOIL	WILDCAT	138	3716
	052	02 40 11.71	79.09.07	NORSKALD	SUSPENDED	25	CRETACEOUS
		6752328 utm-n/s	482092 utm-e/w				
30/03-01 R	215	60 54 15.35	82.01.01	STATOIL	WILDCAT	138	4421
	052	02 40 11.71	82.04.26	DYVI DELTA	P&A. HYDROCARBONES	27	TRIASSIC
		6752328 utm-n/s	482092 utm-e/w				
30/03-02	256	60 47 49.23	80.07.05	STATOIL	WILDCAT	186	955
	052	02 55 18.06	80.08.15	DEEPSEA SAGA	SUSPENDED	25	MIOCENE
		6740339 utm-n/s	495736 utm-e/w				
30/03-02 R	256	60 47 49.23	80.09.02	STATOIL	WILDCAT	186	3567
	052	02 55 18.06	81.02.16	DEEPSEA SAGA	P&A. OIL/GAS DISC.	25	TRIASSIC
		6740339 utm-n/s	495736 utm-e/w				
30/03-03	369	60 46 11.44	83.04.01	STATOIL	WILDCAT	208	3347
	052	02 57 05.58	83.05.30	ROSS ISLE	P&A. DRY HOLE.	22	JURASSIC
		6737312 utm-n/s	497360 utm-e/w				
30/03-04	451	60 45 56.85	85.02.05	STATOIL	APPRAISAL	164	3287
	052	02 51 30.20	85.06.12	DEEPSEA BERGEN	SUSP. OIL/GAS DISC.	23	JURASSIC
		6736867 utm-n/s	492284 utm-e/w				
30/04-01	205	60 37 20.87	78.11.01	BP	WILDCAT	120	5455
	043	02 09 34.61	79.05.14	SEDCO 707	P&A. DRY HOLE	25	JURASSIC
		6721189 utm-n/s	454007 utm-e/w				
30/04-02	230	60 31 01.10	79.11.16	BP	WILDCAT	123	4800
	043	02 02 46.22	80.05.15	SEDCO 707	P&A. GAS/COND. DISC.	25	TRIASSIC
		6709524 utm-n/s	447628 utm-e/w				
30/05-01	70	60 31 48.38	72.05.17	SHELL	WILDCAT	106	4124
	034	02 21 59.23	72.07.29	TRANSWORLD RIG61	P&A. DRY HOLE	25	E.CRET.
		6710776 utm-n/s	465227 utm-e/w				
30/06-01	216	60 33 15.10	79.06.19	STATOIL	WILDCAT	106	3175
	053	02 46 38.36	79.09.22	DEEPSEA SAGA	P&A. GAS/COND. DISC.	25	TRIASSIC
		6713312 utm-n/s	487787 utm-e/w				
30/06-02	225	60 32 12.16	79.09.24	STATOIL	APPRAISAL	106	2890
	053	02 46 11.11	79.12.10	DEEPSEA SAGA	P&A. OIL/GAS DISC.	25	TRIASSIC
		6711366 utm-n/s	487365 utm-e/w				
30/06-03	234	60 34 52.98	79.12.16	STATOIL	APPRAISAL	105	2940
	053	02 47 01.41	80.03.07	DEEPSEA SAGA	P&A. GAS DISCOVERY	25	E.JURASSIC
		6716339 utm-n/s	488148 utm-e/w				
30/06-04	277	60 30 35.66	81.02.17	STATOIL	WILDCAT	110	2942
	053	02 50 53.78	81.05.11	DEEPSEA SAGA	P&A. OIL/GAS DISC.	25	JURASSIC
		6708367 utm-n/s	491667 utm-e/w				
30/06-05	290	60 41 20.60	81.06.11	STATOIL	WILDCAT	156	3550
	053	02 57 11.09	81.08.15	DEEPSEA SAGA	P&A. OIL DISCOVERY	25	TRIASSIC
		6728313 utm-n/s	497437 utm-e/w				
30/06-06	311	60 36 43.43	82.01.09	STATOIL	APPRAISAL	113	3220
	053	02 49 54.11	82.02.24	DEEP SEA SAGA	P&A. DRY HOLE	25	JURASSIC
		6719749 utm-n/s	490785 utm-e/w				
30/06-07	329	60 38 39.49	82.05.28	HYDRO	APPRAISAL	114	3236
	053	02 45 21.74	82.08.24	NORTRYM	P&A. OIL/GAS DISC.	25	JURASSIC
		6723353 utm-n/s	486657 utm-e/w				

BOREHOLE LIST EXPLORATION (with utm-coord.) Date: 23.Januar 1987 PB/ABS

R=re-entry, X=did not reach prospect, S=deviated hole, A=side-tracked new hole

BOREHOLE	NPD NO LIC NO	POSITION	SPUD COMPL.	OPERATOR RIG	SPUD CLASS. COMPLETION CLASS.	WD KB	TD (RKB) FORM. AT TD
30/06-08	333 053	60 32 11.92 02 58 20.59 6711336 utm-n/s	82.06.15 82.08.06 498484 utm-e/w	HYDRO TREASURE SEEKER	WILDCAT P&A. DRY HOLE.	121 25	3600 JURASSIC
30/06-09	339 053	60 30 03.46 02 46 52.62 6707381 utm-n/s	82.08.28 82.12.16 487984 utm-e/w	HYDRO NORTRYM	WILDCAT SUSP. OIL/GAS DISC.	107 25	3476 TRIASSIC
30/06-10	348 053	60 36 29.53 02 46 24.74 6719328 utm-n/s	82.10.04 82.12.02 487600 utm-e/w	HYDRO TREASURE SCOUT	APPRAISAL P&A. OIL/GAS DISC.	109 23	2656 JURASSIC
30/06-10 A	360 053	60 36 29.53 02 46 24.74 6719328 utm-n/s	82.12.02 83.03.04 487600 utm-e/w	HYDRO TREASURE SCOUT	APPRAISAL P&A. OIL/GAS DISC.	109 23	2665 JURASSIC
30/06-11	358 053	60 43 29.38 02 44 41.82 6732323 utm-n/s	82.12.20 83.03.30 486085 utm-e/w	HYDRO NORTRYM	WILDCAT P&A. DRY HOLE.	121 25	4001 L.JURASSIC
30/06-12 X	363 053	60 33 15.59 02 49 23.31 6713319 utm-n/s	83.02.20 83.03.09 490300 utm-e/w	HYDRO TREASURE SEEKER	APPRAISAL P&A. JUNKED.	104 25	612
30/06-13	366 053	60 33 14.59 02 49 21.76 6713288 utm-n/s	83.03.11 83.05.14 490276 utm-e/w	HYDRO TREASURE SEEKER	APPRAISAL SUSP. OIL/GAS DISC.	105 25	2775 L.JURASSIC
30/06-13 R	366 053	60 33 14.59 02 49 21.76 6713288 utm-n/s	86.09.08 86.09.11 490276 utm-e/w	HYDRO TREASURE HUNTER	APPRAISAL P&A. OIL/GAS DISC.	105 25	2725 L.JURASSIC
30/06-14	398 053	60 38 07.69 02 59 21.76 6722343 utm-n/s	83.12.17 84.02.08 499418 utm-e/w	HYDRO TREASURE SCOUT	WILDCAT P&A. OIL DISCOVERY	148 23	2900 JURASSIC
30/06-15	414 053	60 32 18.49 02 45 49.64 6711563 utm-n/s	84.05.02 84.09.05 487038 utm-e/w	HYDRO TREASURE SEEKER	WILDCAT P&A. OIL/GAS DISC.	107 25	3972 TRIASSIC
30/06-16	442 053	60 39 31.72 02 41 37.64 6724983 utm-n/s	84.11.09 85.01.21 483260 utm-e/w	HYDRO TREASURE SCOUT	WILDCAT SUSP. OIL & GAS	108 23	3300 TRIASSIC
30/06-17	478 053	60 34 15.77 02 44 59.84 6715193 utm-n/s	85.08.12 85.08.23 486293 utm-e/w	HYDRO VILDKAT	WILDCAT SUSPENDED	111 24	613
30/06-17 A	504 053	60 34 15.77 02 44 59.84 6715193 utm-n/s	86.02.04 86.03.18 486293 utm-e/w	HYDRO TREASURE HUNTER	WILDCAT P&A. GAS DISCOVERY	111 25	3677 E.JURASSIC
30/06-17 R	478 053	60 34 15.77 02 44 59.84 6715193 utm-n/s	85.11.14 86.02.04 486293 utm-e/w	HYDRO TREASURE HUNTER	WILDCAT P&A. OIL DISCOVERY	111 25	2650 JURASSIC
30/06-18	476 053	60 30 30.71 02 41 21.25 6708245 utm-n/s	85.08.01 85.01.23 482932 utm-e/w	HYDRO BYFORD DOLPHIN	WILDCAT P&A. OIL/GAS DISC.	108 25	3690 TRIASSIC
30/06-19	511 053	60 42 58.03 02 54 54.18 6731330 utm-n/s	86.04.09 86.06.21 495364 utm-e/w	HYDRO POLAR PIONEER	WILDCAT SUSP. OIL/GAS DISC.	136 23	3304 E.JURASSIC
30/06-20	506 053	60 37 20.52 02 42 18.52 6720921 utm-n/s	86.03.10 86.04.13 483862 utm-e/w	HYDRO TREASURE SCOUT	WILDCAT P&A. DRY HOLE	111 23	3042 E.JURASSIC
30/07-01 X	131 040	60 29 27.59 02 01 34.49 6706648 utm-n/s	75.06.29 75.08.05 446491 utm-e/w	HYDRO POLYGLOMAR DRILL	WILDCAT P&A. JUNKED	109 24	1008 TERTIARY

BOREHOLE LIST EXPLORATION (with utm-coord.) Date: 23.Januar 1987 PB/ABS

R=re-entry, X=did not reach prospect, S=deviated hole, A=side-tracked new hole

BOREHOLE	NPD NO LIC NO	POSITION	SPUD COMPL.	OPERATOR RIG	SPUD CLASS. COMPLETION CLASS.	WD KB	TD (RKB) FORM. AT TD
30/07-02	137 040	60 29 26.06 02 01 40.85 6706598 utm-n/s	75.08.06 75.11.09 446588 utm-e/w	HYDRO POLYGLOMAR DRILL	WILDCAT P&A. OIL DISCOVERY	111 24	2592 L.CRET.
30/07-03	163 040	60 17 09.24 02 14 54.44 6683645 utm-n/s	76.08.06 76.10.25 458441 utm-e/w	HYDRO POLYGLOMAR DRILL	WILDCAT P&A. DRY HOLE	98 25	4044 E.CRET.
30/07-04 X	170 040	60 29 29.72 02 03 24.32 6706690 utm-n/s	77.01.25 77.02.05 448169 utm-e/w	HYDRO POLYGLOMAR DRILL	WILDCAT P&A. JUNKED	117 25	777 TERTIARY
30/07-05 X	172 040	60 29 30.19 02 03 27.40 6706703 utm-n/s	77.02.05 77.02.14 448216 utm-e/w	HYDRO POLYGLOMAR DRILL	WILDCAT P&A. JUNKED	117 24	805 TERTIARY
30/07-06	173 040	60 29 29.82 02 03 26.14 6706691 utm-n/s	77.02.14 77.07.29 448196 utm-e/w	HYDRO POLYGLOMAR DRILL	WILDCAT SUSPENDED	116 24	
30/07-06 R	173 040	60 29 29.82 02 03 26.14 6706691 utm-n/s	78.04.18 78.06.03 448196 utm-e/w	HYDRO TREASURE SEEKER	WILDCAT P&A. GAS/COND. DISC.	116 25	4116 JURASSIC ?
30/07-07	207 040	60 16 19.30 02 16 07.30 6682087 utm-n/s	78.12.18 79.07.01 459543 utm-e/w	HYDRO TREASURE SEEKER	WILDCAT P&A. GAS DISC.	110 25	5127 JURASSIC
30/07-08	269 040	60 29 43.08 02 00 20.56 6707143 utm-n/s	80.11.19 81.04.04 445370 utm-e/w	HYDRO TREASURE SEEKER	WILDCAT SUSP. FOR RE-ENTRY.	103 25	4287
30/07-08 R	269 040	60 29 43.08 02 00 20.56 6707143 utm-n/s	81.09.23 82.01.27 445370 utm-e/w	HYDRO TREASURE SEEKER	WILDCAT P&A. GAS/COND.DISC.	103 25	4812 TRIASSIC
30/09-01	351 079	60 28 25.09 02 52 45.38 6704325 utm-n/s	82.10.24 83.01.29 493362 utm-e/w	HYDRO TREASURE SEEKER	WILDCAT P&A. OIL/GAS DISC.	105 25	2895 JURASSIC
30/09-02	370 079	60 27 53.00 02 49 13.03 6703339 utm-n/s	83.04.01 83.07.12 490116 utm-e/w	HYDRO NORTRYM	WILDCAT SUSP. OIL/GAS DISC.	104 25	2830 L.JURASSIC
30/09-02 R	370 079	60 27 53.00 02 49 13.03 6703339 utm-n/s	86.06.01 86.07.07 490116 utm-e/w	HYDRO TREASURE HUNTER	WILDCAT SUSP. FOR TEST W/PTS	104 25	2830 L.JURASSIC
30/09-03	400 079	60 28 16.33 02 45 37.55 6704071 utm-n/s	83.12.31 84.03.06 486827 utm-e/w	HYDRO NORTRYM	WILDCAT P&A. OIL DISCOVERY	110 25	3138 E.JURASSIC
30/09-03 A	406 079	60 28 16.33 02 45 37.55 6704071 utm-n/s	84.03.06 84.05.20 486827 utm-e/w	HYDRO NORTRYM	WILDCAT P&A. OIL & GAS DISC.	111 25	4300 TRIASSIC
30/09-04 S	444 079	60 28 12.83 02 45 32.05 6703964 utm-n/s	84.11.22 85.03.30 486742 utm-e/w	HYDRO TREASURE SEEKER	WILDCAT P&A. OIL/GAS/COND.	110 25	4303 JURASSIC
30/09-05	467 079	60 21 31.34 02 53 37.09 6691522 utm-n/s	85.05.27 85.07.19 494131 utm-e/w	HYDRO TREASURE SEEKER	WILDCAT P&A. GAS DISCOVERY	103 25	2980 TRIASSIC
30/10-01 X	84 030	60 00 20.37 02 04 07.17 6652563 utm-n/s	73.05.05 73.07.18 448059 utm-e/w	ESSO SAIPEM II	WILDCAT P&A. JUNKED	115 14	2917 L.CRET.
30/10-02	102 030	60 04 58.87 02 09 42.76 6661108 utm-n/s	73.12.24 74.03.28 453367 utm-e/w	ESSO DRILLMASTER	WILDCAT P&A. OIL/GAS DISC.	104 24	2755 CRETACEOUS

BOREHOLE LIST EXPLORATION (with utm-coord.) Date: 23.Januar 1987 PB/ABS

BOREHOLE	NPD NO LIC NO	POSITION	SPUD COMPL.	OPERATOR RIG	SPUD CLASS. COMPLETION CLASS.	WD KB	TD (RKB) FORM. AT TD

R=re-entry, X=did not reach prospect, S=deviated hole, A=side-tracked new hole

BOREHOLE	NPD NO LIC NO	POSITION	SPUD COMPL.	OPERATOR RIG	SPUD CLASS. COMPLETION CLASS.	WD KB	TD (RKB) FORM. AT TD
30/10-03	112 030	60 03 12.27 02 10 02.06 6657807 utm-n/s	74.08.08 74.08.31 453623 utm-e/w	ESSO DRILLMASTER	APPRAISAL P&A. GAS DISCOVERY	106 24	2255 TERTIARY
30/10-04 X	116 030	60 00 25.07 02 04 06.19 6652707 utm-n/s	74.09.11 74.09.25 448046 utm-e/w	ESSO NEPTUNE 7	APPRAISAL P&A. JUNKED	104 24	758
30/10-05	118 030	60 00 25.88 02 04 07.17 6652733 utm-n/s	74.08.25 75.05.01 448061 utm-e/w	ESSO NEPTUNE 7	WILDCAT P&A. DRY HOLE	106 24	5185 L.TRIASSIC
30/11-01	122 035	60 00 22.43 02 31 16.20 6652357 utm-n/s	75.02.05 75.03.14 473295 utm-e/w	SHELL OCEAN VOYAGER	WILDCAT P&A. DRY HOLE	114 24	2582 L.CRET.
30/11-02	123 035	60 07 33.20 02 39 05.20 6665638 utm-n/s	75.03.18 75.04.16 480631 utm-e/w	SHELL OCEAN VOYAGER	WILDCAT P&A. DRY HOLE	110 24	2590 L.CRET.
30/11-03	354 035	60 02 38.59 02 32 15.47 6656563 utm-n/s	82.11.17 83.03.14 474243 utm-e/w	SHELL BORGNY DOLPHIN	WILDCAT P&A. DRY HOLE	112 25	4662 L.JURASSIC
30/11-04	402 035	60 02 43.96 02 32 39.47 6656727 utm-n/s	84.01.25 84.07.24 474615 utm-e/w	SHELL DYVI DELTA	WILDCAT P&A. OIL DISCOVERY.	112 29	5255 TRIASSIC
31/02-01	219 054	60 46 19.16 03 33 15.87 6737678 utm-n/s	79.07.17 79.11.09 530201 utm-e/w	SHELL BORGNY DOLPHIN	WILDCAT SUSPENDED	324 24	2433 TRIASSIC
31/02-01 R	219 054	60 46 19.16 03 33 15.87 6737678 utm-n/s	81.10.26 81.11.09 530201 utm-e/w	SHELL BORGNY DOLPHIN	WILDCAT P&A. OIL/GAS DISC.	324 25	2433 TRIASSIC
31/02-02	241 054	60 46 47.80 03 37 23.50 6738597 utm-n/s	80.02.27 80.04.23 533940 utm-e/w	SHELL WEST VENTURE	WILDCAT SUSP. GAS DISCOVERY.	323 32	1890
31/02-02 R	241 054	60 46 47.80 03 37 23.50 6738597 utm-n/s	80.05.28 80.09.30 533940 utm-e/w	SHELL WEST VENTURE	WILDCAT P&A. GAS DISCOVERY	323 32	2600 TRIASSIC
31/02-03	247 054	60 50 27.84 03 35 10.82 6745387 utm-n/s	80.03.28 80.07.20 531872 utm-e/w	SHELL BORGNY DOLPHIN	APPRAISAL P&A. OIL & GAS DISC.	334 25	2601 TRIASSIC
31/02-04	261 054	60 51 23.57 03 30 44.33 6747077 utm-n/s	80.09.01 80.09.13 527835 utm-e/w	SHELL BORGNY DOLPHIN	WILDCAT SUSPENDED	335 25	5035 E.JURASSIC
31/02-04 R	261 054	60 51 23.57 03 30 44.33 6747077 utm-n/s	80.10.09 81.04.06 527835 utm-e/w	SHELL BORGNY DOLPHIN	WILDCAT SUSP. OIL/GAS DISC.	335 25	5035 E.JURASSIC
31/02-04 R2	261 054	60 51 23.57 03 30 44.33 6747077 utm-n/s	82.11.04 82.11.12 527835 utm-e/w	SHELL BORGNY DOLPHIN	WILDCAT P&A. OIL/GAS DISC.	335 25	5035 E.JURASSIC
31/02-05	263 054	60 46 16.20 03 25 53.46 6737536 utm-n/s	80.10.26 80.12.21 523507 utm-e/w	SHELL WEST VENTURE	WILDCAT SUSP.OIL & GAS DISC.	333 32	2532 TRIASSIC
31/02-05 R	263 054	60 46 16.20 03 25 53.46 6737536 utm-n/s	81.06.12 81.07.20 523507 utm-e/w	SHELL BORGNY DOLPHIN	WILDCAT SUSP.OIL/GAS DISC.	333 25	2525 TRIASSIC
31/02-05 R2	263 054	60 46 16.20 03 25 53.46 6737536 utm-n/s	84.03.22 84.04.22 523507 utm-e/w	SHELL BORGNY DOLPHIN	WILDCAT P&A. OIL/GAS DISC.	333 25	2525 TRIASSIC

BOREHOLE LIST EXPLORATION (with utm-coord.) Date: 23.Januar 1987 PB/ABS

BOREHOLE		NPD NO LIC NO	POSITION	SPUD COMPL.	OPERATOR RIG	SPUD CLASS. COMPLETION CLASS.	WD KB	TD (RKB) FORM. AT T
R=re-entry, X=did not reach prospect, S=deviated hole, A=side-tracked new hole								
31/02-06		296	60 54 13.57	81.07.21	SHELL	APPRAISAL	343	1760
		054	03 38 49.43	81.10.17	BORGNY DOLPHIN	P&A. OIL/GAS DISC.	25	JURASSIC
			6752401 utm-n/s	535104 utm-e/w				
31/02-07		326	60 51 25.93	82.04.22	SHELL	APPRAISAL	338	1660
		054	03 27 09.21	82.06.14	BORGNY DOLPHIN	P&A. OIL/GAS DISC.	25	JURASSIC
			6747127 utm-n/s	524587 utm-e/w				
31/02-08		334	60 57 33.02	82.06.16	SHELL	WILDCAT	346	3375
		054	03 29 49.84	82.08.18	BORGNY DOLPHIN	P&A. DRY HOLE.	25	TRIASSIC
			6758502 utm-n/s	526926 utm-e/w				
31/02-09		344	60 52 54.34	82.08.29	SHELL	APPRAISAL	339	1770
		054	03 28 03.43	82.10.01	BORGNY DOLPHIN	P&A. OIL/GAS DISC.	25	L.JURASSIC
			6749867 utm-n/s	525386 utm-e/w				
31/02-10		346	60 47 38.94	82.10.02	SHELL	APPRAISAL	357	1833
		054	03 29 39.46	82.10.31	BORGNY DOLPHIN	P&A. DRY HOLE	25	L.JURASSIC
			6740120 utm-n/s	526908 utm-e/w				
31/02-11		365	60 49 27.76	83.03.17	SHELL	APPRAISAL	336	1744
		054	03 25 30.16	83.05.25	BORGNY DOLPHIN	P&A. OIL DISCOVERY.	25	JURASSIC
			6743459 utm-n/s	523116 utm-e/w				
31/02-12		373	60 50 05.40	83.05.26	SHELL	APPRAISAL	334	1615
		054	03 30 48.47	83.09.08	BORGNY DOLPHIN	P&A. GAS DISCOVERY.	25	JURASSIC
			6744659 utm-n/s	527916 utm-e/w				
31/02-13		401	60 47 14.15	84.01.02	SHELL	APPRAISAL	334	1727
		054	03 26 03.43	84.03.15	BORGNY DOLPHIN	P&A. OIL & GAS DISC.	25	JURASSIC
			6739330 utm-n/s	523646 utm-e/w				
31/02-14		411	60 51 26.52	84.04.23	SHELL	APPRAISAL	339	1725
		054	03 25 37.68	84.06.21	BORGNY DOLPHIN	P&A. OIL DISCOVERY.	25	JURASSIC
			6747136 utm-n/s	523206 utm-e/w				
31/02-15		434	60 55 25.45	84.09.10	SHELL	APPRAISAL	343	1677
		054	03 34 02.57	84.11.15	BORGNY DOLPHIN	P&A. OIL/GAS DISC.	25	JURASSIC
			6754585 utm-n/s	530762 utm-e/w				
31/03-01		379	60 46 47.86	83.07.17	STATOIL	WILDCAT	334	2374
		085	03 44 03.40	83.10.13	DEEPSEA BERGEN	P&A. OIL/GAS DISC.	23	TRIASSIC
			6738662 utm-n/s	539989 utm-e/w				
31/03-02		403	60 52 11.41	84.03.04	HYDRO	APPRAISAL	340	2090
		085	03 40 41.79	84.04.30	TREASURE SEEKER	P&A. OIL & GAS DISC.	25	JURASSIC
			6748638 utm-n/s	536836 utm-e/w				
31/03-03		438	60 50 02.29	84.10.13	SAGA	WILDCAT	332	2573
		085	03 50 38.16	84.11.18	TREASURE SAGA	P&A. DRY HOLE.	26	TRIASSIC
			6744748 utm-n/s	545884 utm-e/w				
31/04-01	X	224	60 36 18.20	79.09.09	HYDRO	WILDCAT	170	1000
		055	03 00 32.30	79.09.25	NORSKALD	P&A. JUNKED	25	TERTIARY
			6718955 utm-n/s	500491 utm-e/w				
31/04-02		226	60 36 14.60	79.09.25	HYDRO	WILDCAT	142	2900
		055	03 00 26.70	79.11.15	NORSKALD	P&A. OIL DISC.	25	TRIASSIC
			6718845 utm-n/s	500406 utm-e/w				
31/04-03		236	60 35 12.20	79.12.24	HYDRO	WILDCAT	170	4981
		055	03 05 38.10	80.05.11	TREASURE SEEKER	P&A. OIL/GAS DISC.	25	PERMIAN
			6716917 utm-n/s	505145 utm-e/w				
31/04-04		267	60 40 01.12	80.12.17	HYDRO	WILDCAT	214	3150
		055	03 06 54.12	81.02.17	NORTRYM	P&A. DRY HOLE.	25	E.JURASSIC
			6725858 utm-n/s	506287 utm-e/w				
31/04-05		288	60 33 30.92	81.05.27	HYDRO	WILDCAT	148	2930
		055	03 03 25.17	81.07.29	NORTRYM	P&A. OIL DISC.	25	TRIASSIC
			6713781 utm-n/s	503125 utm-e/w				

BOREHOLE LIST EXPLORATION (with utm-coord.) Date: 23.Januar 1987 PB/ABS

R=re-entry, X=did not reach prospect, S=deviated hole, A=side-tracked new hole

BOREHOLE		NPD NO LIC NO	POSITION	SPUD COMPL.	OPERATOR RIG	SPUD CLASS. COMPLETION CLASS.	WD KB	TD (RKB) FORM. AT TD
31/04-06		317 055	60 31 17.09 03 01 59.84 6709640 utm-n/s 501827 utm-e/w	82.02.28 82.04.20	HYDRO NORTRYM	APPRAISAL P&A. OIL/GAS DISC.	130 25	2447 JURASSIC
31/04-07		425 055	60 34 08.21 03 01 05.82 6714935 utm-n/s 501002 utm-e/w	84.07.26 84.09.11	HYDRO VILDKAT	WILDCAT P&A. OIL/GAS DISC.	139 24	2505 JURASSIC
31/04-08		508 055	60 31 22.82 03 00 09.59 6709817 utm-n/s 500146 utm-e/w	86.03.21 86.05.11	HYDRO TREASURE HUNTER	APPRAISAL P&A. OIL/GAS DISC.	124 25	2570 TRIASSIC
31/05-01	X	386 085	60 43 33.17 03 32 57.01 6732540 utm-n/s 529959 utm-e/w	83.09.05 83.09.30	SAGA TREASURE SAGA	WILDCAT PLUGGED & JUNKED.	316 26	860
31/05-02		391 085	60 43 33.29 03 32 52.74 6732542 utm-n/s 529894 utm-e/w	83.10.05 83.11.11	SAGA TREASURE SAGA	WILDCAT SUSP. OIL/GAS DISC.	316 26	2500 TRIASSIC
31/05-02 R		391 085	60 43 33.29 03 32 52.74 6732542 utm-n/s 529894 utm-e/w	84.06.11 84.08.30	SAGA TREASURE SAGA	WILDCAT P&A. OIL/GAS DISC.	316 26	2500 TRIASSIC
31/05-03		405 085	60 44 37.17 03 26 20.44 6734474 utm-n/s 523936 utm-e/w	84.04.10 84.06.09	STATOIL TREASURE SAGA	APPRAISAL P&A. OIL & GAS DISC.	326 26	2253 E.JURASSIC
31/06-01		381 085	60 38 44.89 03 40 53.28 6723687 utm-n/s 537268 utm-e/w	83.07.15 83.10.28	HYDRO NORTRYM	WILDCAT P&A. GAS/COND. DISC.	302 25	4070 PALEOZOIC
31/06-02		393 085	60 34 58.24 03 54 55.76 6716832 utm-n/s 550164 utm-e/w	83.10.18 83.12.11	STATOIL DEEPSEA BERGEN	APPRAISAL SUSP. GAS DISCOVERY.	304 23	2020
31/06-02 R		393 085	60 34 58.24 03 54 55.76 6716832 utm-n/s 550164 utm-e/w	84.07.31 84.09.08	STATOIL DEEPSEA BERGEN	APPRAISAL P&A. GAS DISCOVERY	304 23	2235 TRIASSIC
31/06-03		394 085	60 30 04.77 03 59 53.57 6707818 utm-n/s 554834 utm-e/w	83.11.01 83.12.26	HYDRO NORTRYM	WILDCAT P&A: DRY HOLE.	301 25	2250 TRIASSIC
31/06-04	X	407 085	60 37 06.05 03 48 20.36 6720706 utm-n/s 544097 utm-e/w	84.03.05 84.03.16	STATOIL DEEPSEA BERGEN	APPRAISAL P&A. JUNKED.	305 23	820
31/06-05		409 085	60 37 06.80 03 48 20.98 6720729 utm-n/s 544107 utm-e/w	84.03.16 84.05.21	STATOIL DEEPSEA BERGEN	APPRAISAL P&A. OIL & GAS DISC.	304 23	2082 E.JURASSIC
31/06-06		416 085	60 41 57.08 03 56 04.61 6729803 utm-n/s 551028 utm-e/w	84.05.22 84.07.29	STATOIL DEEPSEA BERGEN	APPRAISAL P&A. GAS DISCOVERY.	313 23	2290 JURASSIC.
31/06-07	X	459 085	60 33 19.68 03 40 39.04 6713624 utm-n/s 537156 utm-e/w	85.04.01 85.04.13	HYDRO TREASURE SEEKER	APPRAISAL P&A. JUNKED	300 25	694
31/06-08		462 085	60 33 21.36 03 40 39.07 6713675 utm-n/s 537156 utm-e/w	85.04.14 85.05.25	HYDRO TREASURE SEEKER	APPRAISAL SUSP. OIL/GAS DISC.	298 25	2138 JURASSIC
31/06-08 R		462 085	60 33 21.36 03 40 39.07 6713675 utm-n/s 537156 utm-e/w	85.07.22 85.09.22	HYDRO TREASURE SEEKER	APPRAISAL P&A. OIL/GAS DISC.	298 25	2138 JURASSIC
33/05-01		218 047	61 44 46.10 01 34 47.40 6846882 utm-n/s 424997 utm-e/w	79.07.19 79.10.18	HYDRO TREASURE SEEKER	WILDCAT P&A. DRY HOLE	340 25	3829 TRIASSIC

BOREHOLE LIST EXPLORATION (with utm-coord.) Date: 23.Januar 1987 PB/ABS

R=re-entry, X=did not reach prospect, S=deviated hole, A=side-tracked new hole

BOREHOLE	NPD NO LIC NO	POSITION	SPUD COMPL.	OPERATOR RIG	SPUD CLASS. COMPLETION CLASS.	WD KB	TD (RKB) FORM. AT TD
33/05-02	297 047	61 39 47.14 01 37 17.40 6837585 utm-n/s	81.07.31 81.11.18 427002 utm-e/w	HYDRO NORTRYM	WILDCAT P&A. DRY HOLE	308 25	4520 TRIASSIC
33/06-01	212 049	61 32 14.83 01 51 57.55 6823340 utm-n/s	79.04.09 79.07.06 439704 utm-e/w	AGIP FERNSTAR	WILDCAT P&A. DRY HOLE	306 25	3900 TRIASSIC
33/09-01	109 037	61 15 08.10 01 50 23.28 6791599 utm-n/s	74.04.02 74.06.04 437747 utm-e/w	MOBIL NORSKALD	WILDCAT P&A. OIL/GAS DISC.	146 25	3126 L.TRIASSIC
33/09-02 X	113 037	61 17 53.75 01 53 41.41 6796673 utm-n/s	74.08.25 74.09.14 440786 utm-e/w	MOBIL NORSKALD	APPRAISAL P&A. JUNKED	146 25	941 TERTIARY
33/09-03	117 037	61 17 55.13 01 53 39.94 6796715 utm-n/s	74.09.15 74.11.14 440765 utm-e/w	MOBIL NORSKALD	APPRAISAL P&A. OIL DISC.	146 25	2692 L.TRIASSIC
33/09-04	135 037	61 19 46.57 01 54 43.04 6800148 utm-n/s	75.07.30 75.09.17 441761 utm-e/w	MOBIL NORSKALD	APPRAISAL P&A. OIL DISC.	144 25	3076 L.TRIASSIC
33/09-05	142 037	61 20 47.87 01 56 51.70 6802013 utm-n/s	75.10.27 76.01.01 443705 utm-e/w	MOBIL ROSS RIG	APPRAISAL P&A. OIL DISC.	152 25	3157 L.TRIASSIC
33/09-06	158 037	61 25 12.80 01 48 43.09 6810334 utm-n/s	76.06.22 76.08.31 436594 utm-e/w	MOBIL DYVI ALPHA	WILDCAT P&A. OIL/GAS DISC.	152 26	3355 E.JURASSIC
33/09-07	165 037	61 20 12.17 01 59 12.66 6800875 utm-n/s	76.09.06 76.11.07 445782 utm-e/w	MOBIL DYVI ALPHA	WILDCAT P&A. OIL/GAS DISC.	155 26	3127 TRIASSIC ?
33/09-08	169 037	61 26 17.44 01 55 04.87 6812235 utm-n/s	76.11.11 77.02.18 442286 utm-e/w	MOBIL DEEPSEA SAGA	WILDCAT P&A. OIL/GAS DISC.	265 25	3085 E.JURASSIC
33/09-09	179 037	61 17 10.19 01 54 26.05 6795313 utm-n/s	77.07.27 77.11.17 441428 utm-e/w	MOBIL BORGNY DOLPHIN	APPRAISAL P&A. OIL/GAS DISC.	145 25	3100 TRIASSIC
33/09-10	193 037	61 28 00.21 01 46 29.63 6815550 utm-n/s	78.04.08 78.06.12 434713 utm-e/w	MOBIL FERNSTAR	WILDCAT P&A. DRY HOLE	162 26	3716 E.JURASSIC
33/09-11	198 037	61 29 26.13 01 52 46.01 6818109 utm-n/s	78.06.17 78.08.28 440330 utm-e/w	MOBIL FERNSTAR	WILDCAT P&A. DRY HOLE	287 25	3528 E.JURASSIC
33/12-01	101 037	61 12 15.14 01 49 43.33 6786257 utm-n/s	73.12.01 74.04.18 437055 utm-e/w	MOBIL WAAGE DRILL 1	WILDCAT P&A. OIL/GAS DISC.	146 27	3060 E.TRIASSIC
33/12-02	111 037	61 13 31.18 01 51 25.97 6788583 utm-n/s	74.06.06 74.08.23 438628 utm-e/w	MOBIL NORSKALD	WILDCAT P&A. OIL/GAS DISC.	147 25	4354 TRIASSIC
33/12-03 X	125 037	61 14 19.64 01 53 25.80 6790051 utm-n/s	75.04.23 75.05.16 440441 utm-e/w	MOBIL NORSKALD	APPRAISAL P&A. JUNKED	146 25	1735 TERTIARY
33/12-04	130 037	61 14 19.20 01 53 27.50 6790037 utm-n/s	75.05.17 75.07.17 440466 utm-e/w	MOBIL NORSKALD	APPRAISAL P&A. OIL DISC.	146 26	2896 L.TRIASSIC
33/12-05	140 037	61 11 05.53 01 51 53.34 6784070 utm-n/s	75.10.09 76.02.21 438958 utm-e/w	MOBIL NORSKALD	APPRAISAL P&A. DRY HOLE	143 25	4574 L.TRIASSIC

BOREHOLE LIST   EXPLORATION (with utm-coord.)   Date: 23.Januar 1987   PB/ABS

BOREHOLE	NPD NO LIC NO	POSITION	SPUD COMPL.	OPERATOR RIG	SPUD CLASS. COMPLETION CLASS.	WD KB	TD (RKB) FORM. AT TD

R=re-entry, X=did not reach prospect, S=deviated hole, A=side-tracked new hole

BOREHOLE	NPD NO LIC NO	POSITION	SPUD COMPL.	OPERATOR RIG	SPUD CLASS. COMPLETION CLASS.	WD KB	TD (RKB) FORM. AT TD
33/12-06	150	61 04 36.41	76.03.02	MOBIL	WILDCAT	138	4614
	037	01 57 17.76	76.06.20	DYVI ALPHA	P&A. DRY HOLE	26	TRIASSIC
		6771950 utm-n/s	443612 utm-e/w				
34/02-01 X	235	61 46 28.06	79.12.29	AMOCO	WILDCAT	389	807
	056	02 33 06.20	80.02.19	BYFORD DOLPHIN	P&A. JUNKED	25	
		6849300 utm-n/s	476346 utm-e/w				
34/02-02	268	61 46 16.83	80.11.27	AMOCO	WILDCAT	386	513
	056	02 33 11.38	80.12.12	SEDCO 703	P&A. SUSPENDED	25	
		6848953 utm-n/s	476419 utm-e/w				
34/02-02 R	268	61 46 16.83	80.12.12	AMOCO	WILDCAT	386	4074
	056	02 33 11.38	81.05.01	SEDCO 703	P&A. DRY HOLE	25	TRIASSIC
		6848953 utm-n/s	476419 utm-e/w				
34/02-03	286	61 47 20.91	81.05.15	AMOCO	WILDCAT	388	3741
	056	02 30 04.60	81.08.13	SEDCO 703	P&A. DRY HOLE	25	TRIASSIC
		6850955 utm-n/s	473696 utm-e/w				
34/02-04	458	61 45 20.11	85.03.22	AMOCO	WILDCAT	381	4107
	056	02 39 37.64	85.06.11	WEST VENTURE	P&A. DRY HOLE	33	JURASSIC
		6847162 utm-n/s	482072 utm-e/w				
34/04-01	217	61 32 49.23	79.07.11	SAGA	WILDCAT	375	2955
	057	02 16 23.66	79.12.08	BYFORD DOLPHIN	P&A. OIL DISCOVERY	25	TRIASSIC
		6824096 utm-n/s	461368 utm-e/w				
34/04-02	242	61 30 30.93	80.03.06	SAGA	WILDCAT	322	3600
	057	02 04 17.04	80.05.24	BYFORD DOLPHIN	P&A. DRY HOLE.	25	TRIASSIC.
		6819953 utm-n/s	450579 utm-e/w				
34/04-03	306	61 36 32.95	81.10.16	SAGA	WILDCAT	25	4460
	057	02 07 34.42	82.03.30	DYVI ALPHA	P&A. DRY HOLE.	25	TRIASSIC
		6831115 utm-n/s	453647 utm-e/w				
34/04-04	315	61 30 20.85	82.09.11	SAGA	WILDCAT	244	3800
	057	02 14 09.54	83.02.03	DYVI ALPHA	P&A. OIL/GAS DISC.	25	TRIASSIC
		6819527 utm-n/s	459334 utm-e/w				
34/04-05	395	61 41 09.49	83.11.13	SAGA	WILDCAT	378	3917
	057	02 17 52.00	84.04.06	TREASURE SAGA	P&A. OIL DISCOVERY	26	TRIASSIC
		6839562 utm-n/s	462840 utm-e/w				
34/04-06	497	61 34 15.49	85.12.31	SAGA	APPRAISAL	374	3282
	057	02 13 19.52	86.03.26	VINNI	P&A. OIL DISCOVERY	26	TRIASSIC
		6826796 utm-n/s	458681 utm-e/w				
34/07-01	415	61 28 21.92	84.05.09	SAGA	WILDCAT	328	2929
	089	02 13 24.71	84.07.24	VILDKAT	P&A. OIL DISCOVERY.	24	TRIASSIC
		6815856 utm-n/s	458628 utm-e/w				
34/07-02	424	61 17 57.16	84.09.02	SAGA	WILDCAT	246	2475
	089	02 09 40.90	84.10.11	TREASURE SAGA	P&A. DRY HOLE.	26	TRIASSIC
		6796565 utm-n/s	455067 utm-e/w				
34/07-03	436	61 25 54.05	84.09.14	SAGA	WILDCAT	303	3414
	089	02 07 43.95	85.01.02	VILDKAT	P&A. OIL DISCOVERY.	24	TRIASSIC
		6811343 utm-n/s	453523 utm-e/w				
34/07-04	445	61 29 04.44	84.11.19	SAGA	WILDCAT	319	3115
	089	02 08 00.26	85.01.16	TREASURE SAGA	P&A. OIL DISCOVERY	26	TRIASSIC
		6817231 utm-n/s	453843 utm-e/w				
34/07-05	448	61 21 54.95	85.01.17	SAGA	WILDCAT	244	3146
	089	02 01 23.67	85.03.16	TREASURE SAGA	P&A. OIL DISCOVERY.	26	TRIASSIC
		6804025 utm-n/s	447777 utm-e/w				
34/07-06	457	61 27 10.85	85.03.17	SAGA	APPRAISAL	307	3685
	089	02 08 17.26	85.05.30	TREASURE SAGA	P&A. OIL DISCOVERY	26	TRIASSIC
		6813713 utm-n/s	454048 utm-e/w				

BOREHOLE LIST EXPLORATION (with utm-coord.) Date: 23.Januar 1987 PB/ABS

R=re-entry, X=did not reach prospect, S=deviated hole, A=side-tracked new hole

BOREHOLE	NPD NO LIC NO	POSITION	SPUD COMPL.	OPERATOR RIG	SPUD CLASS. COMPLETION CLASS.	WD KB	TD (RKB) FORM. AT TD
34/07-07	483 089	61 26 54.53 02 05 55.69 6813236 utm-n/s 451945 utm-e/w	85.09.17 85.12.16	SAGA TREASURE SAGA	WILDCAT P&A. OIL DISCOVERY	303 26	3526 TRIASSIC
34/07-08	503 089	61 22 25.93 02 08 33.37 6804894 utm-n/s 454171 utm-e/w	86.02.05 86.04.11	SAGA TREASURE SAGA	WILDCAT P&A. OIL DISCOVERY	133 26	2766 TRIASSIC
34/07-09	510 089	61 29 11.99 02 11 43.60 6817423 utm-n/s 457150 utm-e/w	86.04.13 86.06.12	SAGA TREASURE SAGA	APPRAISAL P&A. OIL DISCOVERY	304 26	3214 TRIASSIC
34/07-10	527 089	61 25 02.64 02 07 36.23 6809754 utm-n/s 453388 utm-e/w	86.08.26 86.10.29	SAGA TREASURE SAGA	WILDCAT P&A. OIL DISCOVERY	300 26	3000 TRIASSIC
34/08-01	488 120	61 21 53.50 02 25 57.57 6803722 utm-n/s 469666 utm-e/w	85.11.08 86.03.08	HYDRO TREASURE SCOUT	WILDCAT P&A. OIL/GAS DISC.	324 23	3610 TRIASSIC
34/08-02	522 120	61 22 10.51 02 31 36.09 6804208 utm-n/s 474697 utm-e/w	86.10.04 86.11.17	HYDRO POLAR PIONEER	WILDCAT P&A. DRY HOLE	378 23	3240 TRIASSIC
34/10-01	197 050	61 10 46.84 02 12 43.67 6783216 utm-n/s 457626 utm-e/w	78.06.20 78.09.08	STATOIL ROSS RIG	WILDCAT P&A. OIL/GAS DISC.	138 25	2460 TRIASSIC
34/10-02	202 050	61 06 07.92 02 13 39.96 6774576 utm-n/s 458365 utm-e/w	78.09.09 78.12.08	STATOIL ROSS RIG	WILDCAT P&A. OIL/GAS/COND.	133 25	3730 TRIASSIC
34/10-03	211 050	61 12 49.48 02 11 55.03 6787020 utm-n/s 456946 utm-e/w	79.03.13 79.06.07	STATOIL NORSKALD	WILDCAT SUSP. OIL DISC.	140 25	2802 TRIASSIC
34/10-04	222 050	61 12 15.65 02 13 55.96 6785952 utm-n/s 458739 utm-e/w	79.08.12 79.10.15	STATOIL ROSS RIG	WILDCAT P&A. OIL DISC.	197 25	2600 TRIASSIC
34/10-05	228 050	61 11 25.35 02 10 34.93 6784432 utm-n/s 455718 utm-e/w	79.10.17 80.01.02	STATOIL ROSS RIG	APPRAISAL SUSPENDED	136 25	2780 TRIASSIC ?
34/10-06	231 050	61 14 37.09 02 13 43.68 6790330 utm-n/s 458607 utm-e/w	79.11.14 80.01.22	STATOIL BORGNY DOLPHIN	APPRAISAL P&A. DRY HOLE	223 25	2363 JURASSIC ?
34/10-07	237 050	61 12 13.94 02 16 29.04 6785872 utm-n/s 461023 utm-e/w	80.01.07 80.03.23	STATOIL ROSS RIG	APPRAISAL SUSP. OIL DISCOVERY	204 25	2250 TRIASSIC
34/10-07 R	237 050	61 12 13.94 02 16 29.94 6785872 utm-n/s 461037 utm-e/w	83.05.31 83.07.14	STATOIL ROSS ISLE	APPRAISAL P&A. OIL DISCOVERY	204 22	2247 TRIASSIC
34/10-08	243 050	61 09 59.53 02 12 03.44 6781761 utm-n/s 457008 utm-e/w	80.03.08 80.05.25	STATOIL DEEPSEA SAGA	APPRAISAL P&A. OIL DISCOVERY.	133 25	2216 E.JURASSIC
34/10-09	248 050	61 12 55.34 02 15 00.54 6787169 utm-n/s 459717 utm-e/w	80.03.24 80.05.09	STATOIL ROSS RIG	APPRAISAL SUSP. OIL DISCOVERY.	202 25	2200 JURASSIC.
34/10-09 R	248 050	61 12 55.80 02 15 00.54 6787184 utm-n/s 459717 utm-e/w	80.05.26 80.07.03	STATOIL DEEPSEA SAGA	APPRAISAL P&A. OIL DISCOVERY	154 25	2424 E.JURASSIC
34/10-10 X	257 050	61 10 27.95 02 14 43.12 6782612 utm-n/s 459404 utm-e/w	80.08.15 80.08.21	STATOIL NORSKALD	WILDCAT SUSPENDED	154 25	816 MIOCENE

BOREHOLE LIST EXPLORATION (with utm-coord.) Date: 23.Januar 1987 PB/ABS

BOREHOLE	NPD NO LIC NO	POSITION	SPUD COMPL.	OPERATOR RIG	SPUD CLASS. COMPLETION CLASS.	WD KB	TD (RKB) FORM. AT TD
R=re-entry, X=did not reach prospect, S=deviated hole, A=side-tracked new hole							
34/10-10 RX	257	61 10 27.95	80.09.19	STATOIL	WILDCAT	154	816
	050	02 14 43.12	80.10.04	BORGNY DOLPHIN	P&A. JUNKED	25	TERTIARY
		6782612 utm-n/s	459404 utm-e/w				
34/10-11	264	61 13 20.80	80.10.24	STATOIL	APPRAISAL	229	2155
	050	02 17 32.60	81.03.04	ROSS RIG	P&A. OIL DISCOVERY	25	TRIASSIC
		6787932 utm-n/s	461994 utm-e/w				
34/10-12	295	61 09 26.69	81.07.16	STATOIL	WILDCAT	138	2800
	050	02 06 21.72	81.09.12	NORDRAUG	P&A. DRY HOLE	25	E.JURASSIC
		6780811 utm-n/s	451887 utm-e/w				
34/10-13	300	61 12 01.93	81.08.24	STATOIL	APPRAISAL	214	3392
	050	02 18 03.03	82.01.05	DEEPSEA SAGA	P&A. OIL DISC.	25	TRIASSIC
		6785487 utm-n/s	462422 utm-e/w				
34/10-14	312	61 14 00.50	81.12.24	STATOIL	APPRAISAL	224	2647
	050	02 15 18.87	82.03.19	ROSS RIG	P&A. OIL/GAS DISC.	25	TRIASSIC
		6789181 utm-n/s	460013 utm-e/w				
34/10-15	349	61 10 30.06	82.10.16	STATOIL	WILDCAT	164	2400
	050	02 14 44.57	82.12.12	NORDRAUG	P&A. DRY HOLE	0	TRIASSIC
		6782676 utm-n/s	459427 utm-e/w				
34/10-16	357	61 05 35.62	82.12.14	STATOIL	WILDCAT	138	4042
	050	02 10 46.66	83.04.11	NORDRAUG	SUSP. OIL/GAS/COND.	25	TRIASSIC
		6773610 utm-n/s	455758 utm-e/w				
34/10-16 R	357	61 05 35.62	83.08.31	STATOIL	WILDCAT	138	4042
	050	02 10 46.66	83.09.27	ROSS ISLE	P&A. OIL/GAS/COND.	25	TRIASSIC
		6773610 utm-n/s	455758 utm-e/w				
34/10-17	364	61 03 58.93	83.02.23	STATOIL	WILDCAT	135	3466
	050	02 00 50.78	83.07.08	DEEPSEA BERGEN	P&A. OIL/GAS/COND.	25	JURASSIC
		6770740 utm-n/s	446787 utm-e/w				
34/10-18	382	61 14 22.48	83.07.16	STATOIL	WILDCAT	141	3025
	050	02 03 18.83	83.08.30	ROSS ISLE	P&A. DRY HOLE	22	L.JURASSIC
		6790000 utm-n/s	449285 utm-e/w				
34/10-19	392	61 11 10.86	83.10.05	STATOIL	APPRAISAL	193	2218
	050	02 15 30.55	83.12.06	ROSS ISLE	P&A. DRY HOLE.	22	TRIASSIC
		6783930 utm-n/s	460128 utm-e/w				
34/10-20 X	399	60 01 20.39	83.12.23	STATOIL	WILDCAT	135	3743
	050	02 17 45.74	84.04.23	ROSS ISLE	P&A. JUNKED	22	
		6654263 utm-n/s	460759 utm-e/w				
34/10-21	426	61 07 33.04	84.07.26	STATOIL	WILDCAT	136	4001
	050	02 08 54.53	84.10.22	DYVI DELTA	P&A. GAS DISCOVERY	29	TRIASSIC
		6777264 utm-n/s	454125 utm-e/w				
34/10-22	440	61 10 33.50	84.10.25	STATOIL	APPRAISAL	135	579
	050	02 11 14.65	84.11.05	DYVI DELTA	P&A. GAS DISCOVERY.	29	TERTIARY
		6782821 utm-n/s	456291 utm-e/w				
34/10-23	465	61 01 06.35	85.05.06	STATOIL	WILDCAT	135	4764
	050	02 19 01.57	85.10.13	DYVI DELTA	P&A. GAS DISCOVERY	29	E.JURASSIC
		6765192 utm-n/s	463084 utm-e/w				
34/10-24	471	61 10 32.65	85.06.28	STATOIL	APPRAISAL	134	600
	050	02 11 28.10	85.08.05	DEEPSEA BERGEN	P&A. GAS DISCOVERY	23	TERTIARY
		6782791 utm-n/s	456492 utm-e/w				
34/10-25	477	61 12 11.60	85.08.06	STATOIL	APPRAISAL	143	600
	050	02 12 12.77	85.08.17	DEEPSEA BERGEN	P&A. DRY HOLE	23	TERTIARY
		6785845 utm-n/s	457197 utm-e/w				
34/10-26	489	61 12 59.90	85.10.18	STATOIL	APPRAISAL	219	600
	050	02 16 23.77	85.10.26	DEEPSEA BERGEN	P&A. DRY HOLE	23	TERTIARY
		6787296 utm-n/s	460960 utm-e/w				

BOREHOLE LIST EXPLORATION (with utm-coord.) Date: 23.Januar 1987 PB/ABS

R=re-entry, X=did not reach prospect, S=deviated hole, A=side-tracked new hole

BOREHOLE	NPD NO LIC NO	POSITION	SPUD COMPL.	OPERATOR RIG	SPUD CLASS. COMPLETION CLASS.	WD KB	TD (RKB) FORM. AT TD
34/10-27	495	61 10 31.12	85.12.02	STATOIL	APPRAISAL	135	450
	050	02 11 23.96	85.12.14	WEST VENTURE	SUSP. GAS DISCOVERY	32	
		6782745 utm-n/s	456430 utm-e/w				
34/10-27 R	495	61 10 31.12	86.01.27	STATOIL	APPRAISAL	135	450
	050	02 11 23.96	86.01.31	WEST VENTURE	P&A. GAS DISCOVERY	32	
		6782745 utm-n/s	456430 utm-e/w				
34/10-28 X	498	61 06 31.99	86.01.02	STATOIL	APPRAISAL	133	521
	050	02 15 23.72	86.01.16	DYVI STENA	JUNKED	25	
		6775303 utm-n/s	459928 utm-e/w				
34/10-29	496	61 10 33.06	85.12.27	STATOIL	APPRAISAL	135	405
	050	02 11 28.76	86.01.26	WEST VENTURE	P&A. GAS DISCOVERY	32	
		6782804 utm-n/s	456502 utm-e/w				
34/10-30	501	61 06 31.07	86.01.16	STATOIL	APPRAISAL	133	3785
	050	02 15 23.92	86.05.10	DYVI STENA	P&A. OIL/GAS DISC.	25	TRIASSIC
		6775275 utm-n/s	459930 utm-e/w				
34/10-31	502	61 10 32.40	86.02.01	STATOIL	APPRAISAL	132	420
	050	02 11 15.46	86.04.10	WEST VENTURE	P&A. GAS DISCOVERY	32	
		6782786 utm-n/s	456303 utm-e/w				
35/03-01	162	61 50 41.89	76.07.19	SAGA	WILDCAT	304	4475
	041	03 43 41.36	76.10.26	DEEPSEA SAGA	P&A. DRY HOLE	26	JURASSIC
		6857288 utm-n/s	538333 utm-e/w				
35/03-02	249	61 51 05.98	80.05.19	SAGA	WILDCAT	273	4400
	041	03 46 28.22	80.10.26	SEDCO 707	P&A.GAS & COND. DISC	25	BASEMENT.
		6858063 utm-n/s	540764 utm-e/w				
35/03-03 X	265	61 51 54.60	80.10.30	SAGA	WILDCAT	252	900
	041	03 52 25.80	80.11.28	BYFORD DOLPHIN	P&A. JUNKED	25	
		6859632 utm-n/s	545972 utm-e/w				
35/03-04	272	61 51 54.54	80.11.30	SAGA	WILDCAT	257	4089
	041	03 52 26.99	81.06.06	BYFORD DOLPHIN	P&A.GAS & COND. DISC	25	BASEMENT
		6859632 utm-n/s	545989 utm-e/w				
35/03-05	313	61 47 46.71	81.12.22	SAGA	WILDCAT	262	4114
	041	03 54 44.01	82.03.31	WEST VENTURE	P&A. DRY HOLE.	33	BASEMENT
		6851991 utm-n/s	548099 utm-e/w				
35/08-01	258	61 21 26.37	80.07.27	GULF	WILDCAT	377	4352
	058	03 21 44.09	81.01.24	SEDCO 704	P&A. GAS/COND. DISC.	25	TRIASSIC
		6802804 utm-n/s	519372 utm-e/w				
35/08-02	299	61 16 15.42	81.09.11	GULF	WILDCAT	381	4335
	058	03 21 58.17	82.05.21	SEDCO 704	P&A. GAS/COND. DISC.	25	JURASSIC
		6793184 utm-n/s	519636 utm-e/w				
35/11-01	417	61 10 59.27	84.05.24	MOBIL	WILDCAT	365	3361
	090	03 39 44.95	84.08.06	NORTRYM	P&A. DRY HOLE.	25	TRIASSEC
		6783526 utm-n/s	535626 utm-e/w				
36/01-01	128	61 56 41.10	75.05.09	AMOCO	WILDCAT	165	1596
	042	04 15 45.60	75.06.14	DYVI ALPHA	P&A. DRY HOLE	26	DEVONIAN
		6868834 utm-n/s	566255 utm-e/w				
36/01-02	138	61 53 00.25	75.08.26	AMOCO	WILDCAT	226	3261
	042	04 00 52.19	75.10.27	DEEPSEA SAGA	P&A. DRY HOLE	30	M.JURASSIC
		6861771 utm-n/s	553340 utm-e/w				
6406/03-01	412	64 45 29.77	84.04.27	STATOIL	WILDCAT	256	4902
	091	06 49 21.77	84.08.14	ROSS ISLE	P&A. DRY HOLE.	22	TRIASSIC
		7183473 utm-n/s	396410 utm-e/w				
6406/03-02	519	64 51 45.78	86.06.28	STATOIL	WILDCAT	300	4527
	091	06 49 51.87	86.11.22	WEST VANGUARD	P&A. OIL/GAS DISC.	22	L.TRIASSIC
		7195096 utm-n/s	397206 utm-e/w				

OREHOLE LIST   EXPLORATION (with utm-coord.)   Date: 23.Januar 1987   PB/ABS

BOREHOLE	NPD NO LIC NO	POSITION	SPUD COMPL.	OPERATOR RIG	SPUD CLASS. COMPLETION CLASS.	WD KB	TD (RKB) FORM. AT TD
R=re-entry, X=did not reach prospect, S=deviated hole, A=side-tracked new hole							
406/03-03	526	64 59 49.19	86.08.04	STATOIL	WILDCAT	303	4416
	091	06 53 27.90	86.10.26	DYVI DELTA	P&A. DRY HOLE	29	E.JURASSIC
		7209958 utm-n/s	400548 utm-e/w				
406/06-01	479	64 43 48.23	85.08.15	STATOIL	WILDCAT	244	4715
	105	06 42 14.69	85.12.30	DYVI STENA	P&A. DRY HOLE	25	E.JURASSIC
		7180532 utm-n/s	390653 utm-e/w				
407/01-01 X	350	64 47 51.66	82.10.19	STATOIL	WILDCAT	273	900
	073	07 02 21.62	82.11.13	DYVI DELTA	PLUGGED & JUNKED	29	
		7187528 utm-n/s	406849 utm-e/w				
407/01-02	355	64 47 50.61	82.11.13	STATOIL	WILDCAT	273	4560
	073	07 02 23.76	83.05.15	DYVI DELTA	P&A. OIL/GAS DISC.	29	TRIASSIC
		7187494 utm-n/s	406877 utm-e/w				
407/01-03	388	64 52 25.48	83.09.17	STATOIL	WILDCAT	286	4469
	073	07 02 53.47	84.01.16	DYVI DELTA	P&A. OIL/GAS DISC.	29	TRIASSIC
		7195989 utm-n/s	407531 utm-e/w				
407/02-01	332	64 58 04.97	82.06.03	SAGA	WILDCAT	265	3870
	074	07 28 28.12	82.08.06	WEST VENTURE	P&A. DRY HOLE.	33	TRIASSIC
		7205941 utm-n/s	427977 utm-e/w				
407/02-02	375	64 59 36.64	83.05.17	SAGA	WILDCAT	258	3351
	074	07 31 53.08	83.07.31	TREASURE SAGA	P&A. GAS/COND. DISC.	26	TRIASSIC
		7208715 utm-n/s	430730 utm-e/w				
407/02-03	531	64 56 01.39	86.11.07	SAGA	WILDCAT	250	
	074	07 39 53.04	00.00.00	TREASURE SAGA		26	
		7201913 utm-n/s	436877 utm-e/w				
407/04-01	474	64 35 45.36	85.07.20	STATOIL	WILDCAT	225	4835
	106	07 08 42.17	85.11.15	ROSS ISLE	P&A. COND./GAS DISC.	22	TRIASSIC
		7164899 utm-n/s	411212 utm-e/w				
407/06-01	437	64 38 29.62	84.09.16	STATOIL	WILDCAT	227	2895
	092	07 52 31.53	84.10.26	ZAPATA UGLAND	P&A. DRY HOLE.	27	JURASSIC
		7169163 utm-n/s	446258 utm-e/w				
407/06-02	484	64 42 29.56	85.10.04	STATOIL	WILDCAT	221	523
	092	07 40 32.59	85.10.06	WEST VANGUARD	ABAND. AFTER BLOWOUT	22	
		7176774 utm-n/s	436871 utm-e/w				
407/06-02 R	484	64 42 29.56	86.05.08	STATOIL	WILDCAT	221	523
	092	07 40 32.58	86.05.20	ROSS ISLE	SUSPENDERT	22	
		7176774 utm-n/s	436871 utm-e/w				
407/06-02 R2 X	484	64 42 29.56	86.06.13	STATOIL	WILDCAT	221	523
	092	07 40 32.59	86.06.24	BUCENTAUR	P&A. JUNKED	2	
		7176774 utm-n/s	436871 utm-e/w				
407/06-03	533	64 42 31.77	86.12.15	STATOIL	WILDCAT	222	
	092	07 40 09.84	00.00.00	DYVI DELTA		29	
		7176848 utm-n/s	436571 utm-e/w				
407/07-01 S	486	64 16 31.49	85.10.19	HYDRO	WILDCAT	330	3950
	107	07 12 21.12	86.04.07	POLAR PIONEER	P&A. OIL/GAS DISC.	23	TRIASSIC ?
		7129107 utm-n/s	413114 utm-e/w				
407/07-02	532	00 00 00.00	86.11.20	HYDRO	WILDCAT	338	
	107	00 00 00.00	87.01.21	POLAR PIONEER		23	
		0 utm-n/s	0 utm-e/w				
407/09-01	422	64 21 56.14	84.06.26	SHELL	WILDCAT	248	2500
	093	07 47 20.86	84.09.07	BORGNY DOLPHIN	P&A. OIL DISCOVERY.	25	TRIASSIC
		7138488 utm-n/s	441548 utm-e/w				
407/09-02	443	64 24 01.31	84.11.18	SHELL	APPRAISAL	247	1865
	093	07 48 11.26	85.02.02	BORGNY DOLPHIN	P&A. OIL DISCOVERY.	25	M.JURASSIC
		7142350 utm-n/s	442297 utm-e/w				

BOREHOLE LIST   EXPLORATION (with utm-coord.)   Date: 23.Januar 1987   PB/ABS

BOREHOLE	NPD NO LIC NO	POSITION	SPUD COMPL.	OPERATOR RIG	SPUD CLASS. COMPLETION CLASS.	WD KB	TD (RKB) FORM. AT T
R=re-entry, X=did not reach prospect, S=deviated hole, A=side-tracked new hole							
6407/09-03	463 093	64 19 48.94 07 46 30.91 7134563 utm-n/s 440802 utm-e/w	85.05.03 85.07.28	SHELL BORGNY DOLPHIN	APPRAISAL SUSP. OIL DISCOVERY	279 25	1868 JURASSAIC
6407/09-04	472 093	64 22 14.00 07 43 21.84 7139104 utm-n/s 438354 utm-e/w	85.07.04 85.09.10	SHELL WEST VENTURE	APPRAISAL P&A. OIL DISCOVERY	244 32	1820 E.JURASSIC
6407/09-05	482 093	64 16 42.35 07 44 14.66 7128825 utm-n/s 438859 utm-e/w	85.09.12 85.11.13	SHELL WEST VENTURE	APPRAISAL SUSP. OIL DISC.	318 33	1819 JURASSIC
6407/09-06	499 093	64 19 58.07 07 44 23.70 7134880 utm-n/s 439100 utm-e/w	86.01.02 86.03.13	SHELL BORGNY DOLPHIN	APPRAISAL SUSP. OIL DISCOVERY	297 25	1800 E.JURASSIC
6506/12-01	430 094	65 10 07.58 06 43 44.07 7229360 utm-n/s 393591 utm-e/w	84.05.15 85.02.06	STATOIL ROSS ISLE	WILDCAT P&A. GAS/COND. DISC.	250 22	4925
6506/12-02 X	453 094	65 01 29.85 06 53 29.41 7213073 utm-n/s 400672 utm-e/w	85.02.12 85.02.28	STATOIL ROSS ISLE	WILDCAT P&A. JUNKED.	310 22	955
6506/12-03	456 094	65 01 31.09 06 53 27.35 7213113 utm-n/s 400646 utm-e/w	85.03.02 85.07.17	STATOIL ROSS ISLE	WILDCAT P&A. OIL/GAS DISC.	307 22	4360 E.JURASSI
6506/12-04	460 094	65 12 46.97 06 43 30.37 7234297 utm-n/s 393591 utm-e/w	85.03.24 85.08.13	STATOIL DYVI STENA	WILDCAT P&A. DRY HOLE	256 25	4457 JURASSIC
6506/12-05	485 094	65 02 28.60 06 58 21.93 7214766 utm-n/s 404557 utm-e/w	85.10.17 86.03.27	STATOIL DYVI DELTA	APPRAISAL P&A. OIL DISCOVERY	301 29	4588 E.JURASSI
6506/12-06	509 094	65 09 57.60 06 46 43.19 7228967 utm-n/s 395911 utm-e/w	86.03.31 86.08.02	STATOIL DYVI DELTA	APPRAISAL P&A. GAS/COND. DISC.	276 29	4741 E.JURASSI
6507/02-01	518 122	65 54 02.48 07 26 16.43 7309911 utm-n/s 428828 utm-e/w	86.06.24 86.09.29	HYDRO POLAR PIONEER	WILDCAT P&A. DRY HOLE	390 23	4477 TRIASSIC
6507/06-01	517 123	65 36 37.59 07 43 43.04 7277264 utm-n/s 441417 utm-e/w	86.06.15 86.08.23	SAGA TREASURE SAGA	WILDCAT P&A. DRY HOLE	419 26	4040 TRIASSIC
6507/07-01	427 095	65 27 16.73 07 12 52.77 7260481 utm-n/s 417247 utm-e/w	84.08.10 84.12.01	CONOCO NORTRYM	WILDCAT P&A. GAS DISCOVERY	367 22	4825 JURASSIC
6507/07-02	454 095	65 20 12.37 07 18 34.52 7247225 utm-n/s 421294 utm-e/w	85.02.25 85.06.10	CONOCO NORTRYM	WILDCAT P&A. OIL&GAS DISC.	351 25	3262 TRIASSIC
6507/07-03	475 095	65 19 01.31 07 17 44.79 7245043 utm-n/s 420591 utm-e/w	85.07.29 85.09.18	CONOCO NORTRYM	APPRAISAL P&A. OIL/GAS DISC.	346 25	2850 JURASSIC
6507/07-04	490 095	65 19 11.56 07 15 44.99 7245401 utm-n/s 419050 utm-e/w	85.11.06 86.01.13	CONOCO NORTRYM	APPRAISAL P&A. OIL/GAS DISC.	345 25	2850 JURASSIC
6507/07-05	500 095	65 21 30.27 07 17 35.08 7249657 utm-n/s 420590 utm-e/w	86.01.16 86.03.06	CONOCO NORTRYM	APPRAISAL P&A. OIL/GAS DISC.	332 25	2660 E.JURASSI
6507/07-05 A	505 095	65 21 30.27 07 17 35.08 7249657 utm-n/s 420590 utm-e/w	86.03.06 86.04.05	CONOCO NORTRYM	APPRAISAL P&A. OIL/GAS DISC.	332 25	2673 E.JURASSI

OREHOLE LIST EXPLORATION (with utm-coord.) Date: 23.Januar 1987 PB/ABS

OREHOLE	NPD NO LIC NO	POSITION	SPUD COMPL.	OPERATOR RIG	SPUD CLASS. COMPLETION CLASS.	WD KB	TD (RKB) FORM. AT TD
R=re-entry, X=did not reach prospect, S=deviated hole, A=side-tracked new hole							
607/07-06	524 095	65 21 30.03 07 19 10.35 7249615 utm-n/s	86.07.23 86.09.06 421821 utm-e/w	CONOCO NORTRYM	APPRAISAL P&A. OIL/GAS DISC.	351 25	2525 E.JURASSIC
607/08-01	529 124	65 18 35.20 07 20 52.80 7244170 utm-n/s	86.10.28 86.12.09 423003 utm-e/w	STATOIL DYVI DELTA	WILDCAT P&A. OIL/GAS DISC.	342 29	2600 E.JURASSIC
607/10-01	337 075	65 13 10.75 07 14 00.47 7234273 utm-n/s	82.07.10 82.10.31 417384 utm-e/w	BP SEDCO 707	WILDCAT P&A. DRY HOLE.	297 25	3693 TRIASSIC
607/11-01	301 062	65 04 43.95 07 29 23.31 7218274 utm-n/s	81.09.13 81.12.10 428996 utm-e/w	SAGA WEST VENTURE	WILDCAT P&A. GAS DISCOVERY	298 33	3139 TRIASSIC
607/11-02	325 062	65 10 59.70 07 36 39.84 7229775 utm-n/s	82.04.18 82.05.30 434952 utm-e/w	SAGA WEST VENTURE	WILDCAT P&A. DRY HOLE	33 33	2905 TRIASSIC
607/11-03	469 062	65 01 59.80 07 30 42.34 7213168 utm-n/s	85.06.03 85.08.12 429908 utm-e/w	SAGA TREASURE SAGA	WILDCAT P&A. OIL/GAS DISC.	290 26	3246 TRIASSIC
607/12-01	255 059	65 07 01.77 07 42 42.89 7222309 utm-n/s	80.07.01 80.10.26 439524 utm-e/w	SAGA BYFORD DOLPHIN	WILDCAT P&A. DRY HOLE.	225 25	3720 TRIASSIC
607/12-02	289 059	65 11 48.90 07 51 35.41 7231064 utm-n/s	81.06.09 81.11.24 446629 utm-e/w	SAGA BYFORD DOLPHIN	WILDCAT P&A. DRY HOLE	261 25	5008 PERMIAN
607/12-03	481 059	65 03 35.82 07 49 06.40 7215836 utm-n/s	85.08.16 85.09.13 444406 utm-e/w	SAGA TREASURE SAGA	WILDCAT P&A. DRY HOLE	240 26	2600 JURASSIC
07/12-01	521 127	66 12 54.15 07 44 26.72 7344636 utm-n/s	86.07.19 86.10.01 443329 utm-e/w	ELF HENRY GOODRICH	WILDCAT P&A. DRY HOLE	391 21	3521 L.CRET.
08/11-01	523 128	66 01 04.02 08 23 38.10 7322210 utm-n/s	86.07.19 86.08.13 472510 utm-e/w	STATOIL DYVI STENA	WILDCAT P&A. DRY HOLE.	382 25	1620 TRIASSIC
09/05-01	439 080	66 37 42.73 09 24 52.17 7390220 utm-n/s	84.11.03 85.01.05 518349 utm-e/w	STATOIL WEST VANGUARD	WILDCAT P&A. DRY HOLE	294 22	3600 TRIASSIC ?
09/07-01	378 081	66 24 56.49 09 01 14.91 7366433 utm-n/s	83.06.21 83.08.03 500929 utm-e/w	PHILLIPS NORDRAUG	WILDCAT P&A. DRY HOLE	250 25	1969 PALEOZOIC.
09/10-01	383 082	66 14 42.27 09 14 59.50 7347435 utm-n/s	83.08.02 83.08.29 511231 utm-e/w	SAGA TREASURE SAGA	WILDCAT P&A. DRY HOLE	263 26	2166 TRIASSIC
09/11-01	374 083	66 08 13.90 09 33 47.89 7335503 utm-n/s	83.05.19 83.07.07 525430 utm-e/w	HYDRO TREASURE SEEKER	WILDCAT P&A. DRY HOLE.	238 25	3068 TRIASSIC.
10/07-01	371 084	66 17 32.82 10 16 52.92 7353283 utm-n/s	83.04.18 83.06.19 557489 utm-e/w	STATOIL NORDRAUG	WILDCAT P&A. DRY HOLE.	265 25	3333 TRIASSIC
10/07-02	385 084	66 27 49.61 10 10 10.92 7372281 utm-n/s	83.08.28 84.03.14 552123 utm-e/w	STATOIL WEST VANGUARD	WILDCAT P&A. DRY HOLE.	235 22	4215 PERMIAN
17/09-01	323 063	71 22 51.05 17 56 05.76 7922626 utm-n/s	82.04.24 82.07.16 604554 utm-e/w	HYDRO TREASURE SCOUT	WILDCAT P&A. DRY HOLE	258 23	3200 E.CRET.

BOREHOLE LIST EXPLORATION (with utm-coord.) Date: 23.Januar 1987 PB/ABS

BOREHOLE	NPD NO LIC NO	POSITION	SPUD COMPL.	OPERATOR RIG	SPUD CLASS. COMPLETION CLASS.	WD KB	TD (RKB) FORM. AT T
R=re-entry, X=did not reach prospect, S=deviated hole, A=side-tracked new hole							
7117/09-02	372 063	71 25 44.75 17 53 32.45 7927929 utm-n/s	83.05.07 83.09.09 602780 utm-e/w	HYDRO TREASURE SCOUT	WILDCAT P&A. DRY HOLE.	271 23	5000 L.CRET.
7119/07-01	380 076	71 23 42.40 19 15 18.43 7922575 utm-n/s	83.07.12 83.09.11 437872 utm-e/w	HYDRO TREASURE SEEKER	WILDCAT P&A. DRY HOLE	238 25	3167 PALEOZOIC
7119/09-01	421 096	71 24 53.19 19 49 43.26 7924276 utm-n/s	84.06.28 84.09.25 458333 utm-e/w	ELF BYFORD DOLPHIN	WILDCAT P&A. DRY HOLE.	201 0	3243 TRIASSIC
7119/12-01	251 060	71 06 08.00 19 47 40.29 7889443 utm-n/s	80.06.14 80.10.10 456424 utm-e/w	STATOIL ROSS RIG	WILDCAT P&A. DRY HOLE.	200 25	3088 JURASSIC
7119/12-02	283 060	71 00 51.81 19 58 20.81 7879530 utm-n/s	81.04.16 81.06.26 462689 utm-e/w	STATOIL ROSS RIG	WILDCAT P&A. DRY HOLE	182 25	1902 TRIASSIC
7119/12-03	376 060	71 14 20.18 19 44 37.92 7904727 utm-n/s	83.05.20 83.09.12 454909 utm-e/w	STATOIL DYVI DELTA	WILDCAT P&A. GAS/COND. DISC.	211 29	3310 JURASSIC
7120/01-01	480 108	71 55 00.83 20 18 07.13 7980020 utm-n/s	85.08.16 85.11.24 475816 utm-e/w	SHELL BORGNY DOLPHIN	WILDCAT SUSPENDED	342 25	2594
7120/01-01 R	480 108	71 55 00.83 20 18 07.13 7980020 utm-n/s	85.12.02 85.12.24 475816 utm-e/w	SHELL BORGNY DOLPHIN	WILDCAT SUSPENDED	342 25	2610
7120/01-01 R2	480 108	71 55 00.83 20 18 07.13 7980020 utm-n/s	86.03.18 86.07.21 475816 utm-e/w	SHELL BORGNY DOLPHIN	WILDCAT P&A. DRY HOLE	342 25	4003 PERMIAN
7120/02-01	464 109	71 58 57.94 20 28 35.09 7987306 utm-n/s	85.05.05 85.10.29 481923 utm-e/w	HYDRO TREASURE SCOUT	WILDCAT P&A. OIL DISC.	387 23	3495 BASEMENT
7120/05-01	461 110	71 34 51.86 20 26 12.26 7942513 utm-n/s	85.04.17 85.06.06 480135 utm-e/w	STATOIL WEST VANGUARD	WILDCAT P&A. DRY HOLE.	318 22	2700 TRIASSIC
7120/06-01	450 097	71 37 11.76 20 55 59.72 7946757 utm-n/s	85.02.02 85.05.02 497650 utm-e/w	HYDRO TREASURE SCOUT	WILDCAT P&A. OIL&GAS DISC.	314 23	2820 TRIASSIC
7120/07-01	341 077	71 18 36.29 20 11 22.21 7912389 utm-n/s	82.07.31 82.10.08 471011 utm-e/w	STATOIL NORDRAUG	WILDCAT P&A. GAS DISCOVERY	236 25	2842 TRIASSIC
7120/07-02	377 077	71 19 11.49 20 19 44.31 7913418 utm-n/s	83.05.26 83.08.21 476011 utm-e/w	STATOIL WEST VANGUARD	WILDCAT P&A. GAS/COND. DISC.	242 22	2523 TRIASSIC
7120/07-03	408 077	71 27 43.31 20 10 46.44 7929340 utm-n/s	84.03.18 84.06.09 470886 utm-e/w	STATOIL WEST VANGUARD	WILDCAT P&A. DRY HOLE.	258 22	3062 JURASSIC
7120/08-01	292 064	71 24 34.43 20 26 06.20 7923385 utm-n/s	81.06.28 81.09.10 479897 utm-e/w	STATOIL ROSS RIG	WILDCAT P&A. GAS/COND. DISC.	270 25	2610 TRIASSIC
7120/08-02	324 064	71 20 15.72 20 27 57.61 7915358 utm-n/s	82.04.15 82.07.29 480927 utm-e/w	STATOIL NORDRAUG	WILDCAT P&A. GAS DISCOVERY	245 25	2590 TRIASSIC
7120/08-03	368 064	71 27 48.77 20 36 09.47 7929358 utm-n/s	83.04.07 83.05.24 485899 utm-e/w	STATOIL WEST VANGUARD	APPRAISAL P&A. DRY HOLE.	280 22	2333 L.JURASS

OREHOLE LIST EXPLORATION (with utm-coord.) Date: 23.Januar 1987 PB/ABS

R=re-entry, X=did not reach prospect, S=deviated hole, A=side-tracked new hole

BOREHOLE	NPD NO LIC NO	POSITION	SPUD COMPL.	OPERATOR RIG	SPUD CLASS. COMPLETION CLASS.	WD KB	TD (RKB) FORM. AT TD
120/09-01	338 078	71 29 26.57 20 56 49.47 7932343 utm-n/s 498124 utm-e/w	82.07.25 82.09.26	HYDRO TREASURE SCOUT	WILDCAT P&A. GAS DISCOVERY	320 23	2300 L.TRIASSIC
120/09-02	410 078	71 29 40.81 20 42 05.38 7932810 utm-n/s 489424 utm-e/w	84.04.18 84.10.20	HYDRO TREASURE SCOUT	WILDCAT P&A. GAS DISCOVERY.	294 23	5082 PERMIAN
120/10-01	429 098	71 03 06.25 20 16 29.80 7883536 utm-n/s 473722 utm-e/w	84.08.10 84.09.08	ESSO ZAPATA UGLAND	WILDCAT P&A. DRY HOLE.	185 25	2000 TRIASSIC
120/12-01	250 061	71 06 48.71 20 45 20.13 7890288 utm-n/s 491169 utm-e/w	80.06.01 80.10.12	HYDRO TREASURE SEEKER	WILDCAT P&A. DRY HOLE.	192 25	3573 TRIASSIC
120/12-02	284 061	71 07 30.30 20 48 19.00 7891571 utm-n/s 492968 utm-e/w	81.04.15 81.09.11	HYDRO TREASURE SEEKER	WILDCAT P&A. GAS/COND. DISC.	164 25	4680 BASEMENT
120/12-03	367 061	71 11 43.31 20 46 43.70 7899413 utm-n/s 492041 utm-e/w	83.03.16 83.05.05	HYDRO TREASURE SCOUT	WILDCAT P&A. GAS DISCOVERY.	185 23	2523 TRIASSIC
120/12-04	404 061	71 03 01.13 20 42 32.16 7883246 utm-n/s 489450 utm-e/w	84.02.18 84.04.16	HYDRO TREASURE SCOUT	WILDCAT P&A. DRY HOLE.	155 23	2199 BASEMENT
121/01-01	487 111	71 56 25.74 21 04 36.52 7982512 utm-n/s 502657 utm-e/w	85.10.10 85.11.17	ESSO ZAPATA UGLAND	WILDCAT SUSPENDED	370 25	916
121/01-01 R	487 111	71 56 25.74 21 04 36.52 7982512 utm-n/s 502657 utm-e/w	86.03.19 86.08.23	ESSO ZAPATA UGLAND	WILDCAT P&A. DRY HOLE	370 25	5000 KARBON.?
121/04-01	428 099	71 35 59.68 21 09 22.79 7944528 utm-n/s 505507 utm-e/w	84.08.06 84.10.27	STATOIL WEST VANGUARD	WILDCAT P&A. OIL/GAS/COND.	335 22	2609 TRIASSIC
121/04-02	449 099	71 39 26.09 21 03 45.96 7950918 utm-n/s 502204 utm-e/w	85.01.29 85.04.14	STATOIL WEST VANGUARD	WILDCAT P&A. GAS/COND. DISC.	317 22	2800 TRIASSIC
121/05-01	468 110	71 35 54.88 21 24 21.78 7944422 utm-n/s 514306 utm-e/w	85.06.07 85.09.28	STATOIL WEST VANGUARD	WILDCAT P&A. OIL/GAS DISC.	336 22	3200 TRIASSIC
121/05-02	516 110	71 40 20.79 21 39 24.42 7952738 utm-n/s 523051 utm-e/w	86.05.05 86.07.06	STATOIL ROSS ISLE	WILDCAT P&A. OIL/GAS DISC.	328 22	2543 TRIASSIC
121/07-01	419 100	71 28 20.78 21 05 15.18 7930306 utm-n/s 503105 utm-e/w	84.06.11 84.08.05	STATOIL WEST VANGUARD	WILDCAT P&A. GAS/COND. DISC.	326 22	2160 TRIASSIC.
121/07-02	520 100	71 26 37.91 21 03 21.42 7927116 utm-n/s 501987 utm-e/w	86.07.07 86.08.12	STATOIL ROSS ISLE	WILDCAT P&A. GAS DISCOVERY	343 22	2156 TRIASSIC

Number of records listed : 573

Index

Page numbers in *italics* refer to diagrams; page numbers in **bold** refer to those dealing with a specific subject.